Analysis and Mathematical Physics

Mathematics and Its Applications (*East European Series*)

Managing Editor:

M. HAZEWINKEL

Centre for Mathematics and Computer Science, Amsterdam, The Netherlands

Editorial Board:

A. BIALYNICKI-BIRULA, *Institute of Mathematics, Warsaw University, Poland*
H. KURKE, *Humboldt University, Berlin, G.D.R.*
J. KURZWEIL, *Mathematics Institute, Academy of Sciences, Prague, Czechoslovakia*
L. LEINDLER, *Bolyai Institute, Szeged, Hungary*
L. LOVÁSZ, *Eőtvős Loránd University, Budapest, Hungary*
D. S. MITRINOVIĆ, *University of Belgrade, Yugoslavia*
S. ROLEWICZ, *Polish Academy of Sciences, Warsaw, Poland*
BL. H. SENDOV, *Bulgarian Academy of Sciences, Sofia, Bulgaria*
J. T. TODOROV, *Bulgarian Academy of Sciences, Sofia, Bulgaria*
H. TRIEBEL, *University of Jena, G.D.R.*

Hans Triebel

Universität Jena, DDR

Analysis and Mathematical Physics

D. REIDEL PUBLISHING COMPANY

A MEMBER OF THE KLUWER ACADEMIC PUBLISHERS GROUP

Dordrecht / Boston / Lancaster / Tokyo

Library of Congress Cataloging-in-Publication Data

CIP

Triebel, Hans.
Analysis and mathematical physics.
(Mathematics and its applications. East European series)
Translation of: Analysis und mathematische Physik.
Bibliography: p. 445
Includes index.
1. Mathematical physics. 2. Mathematical analysis.
I. Title. II. Series: Mathematics and its applications
(D. Reidel Publishing Company). East European series.
QC20.T72713 1986 530.1′55 86-470
ISBN 90-277-2077-0

Distributors for Albania, Bulgaria, Cuba, Czechoslovakia, German Democratic Republic, Hungary, Korean People's Democratic Republic, Mongolia, People's Republic of China, Poland, Romania, the U.S.S.R., Vietnam, and Yugoslavia by BSB B. G. Teubner Verlagsgesellschaft, Leipzig.

Distributors for the U.S.A. and Canada
Kluwer Academic Publishers,
101 Philip Drive, Assinippi Park, Norwell, MA 02061, U.S.A.

Distributors for all remaining countries
Kluwer Academic Publishers Group,
P.O. Box 322, 3300 AH Dordrecht, Holland.

Original title: *Analysis und Mathematische Physik.*

English Translation: Bernhard Simon and Hedwig Simon, Berlin

Reader: Dorothea Ziegler.

Published by BSB B. G. Teubner Verlagsgesellschaft, Leipzig,
in co-publication with D. Reidel Publishing Company, Dordrecht, Holland.

Printed in the German Democratic Republic

Series Editor's Preface

Approach your problems from the right end and begin with the answers. Then one day, perhaps you will find the final question.

'The Hermit Clad in Crane Feathers' in R. Van Gulik's *The Chinese Maze Murders*.

It isn't that they can't see the solution.
It is that they can't see the problem.

G. K. Chesterton, *The Scandal of Father Brown* 'The Point of a Pin'.

Growing specialization and diversification have brought a host of monographs and textbooks on increasingly specialized topics. However, the "tree" of knowledge of mathematics and related fields does not grow only by putting forth new branches. It also happens, quite often in fact, that branches which were thought to be completely disparate are suddenly seen to be related.

Further, the kind and level of sophistication of mathematics applied in various sciences has changed drastically in recent years: measure theory is used (non-trivially) in regional and theoretical economics; algebraic geometry interacts with physics; the Minkowski lemma, coding theory and the structure of water meet one another in packing and covering theory; quantum fields, crystal defects and mathematical programming profit from homotopy theory; Lie algebras are relevant to filtering; and prediction and electrical engineering can use Stein spaces. And in addition to this there are such new emerging subdisciplines as "experimental mathematics", "CFD", "completely integrable systems", "chaos, synergetics and large-scale order", which are almost impossible to fit into the existing classification schemes. They draw upon widely different sections of mathematics. This programme, Mathematics and its Applications, is devoted to new emerging (sub)disciplines and to such (new) interrelations as exempla gratia:

- a central concept which plays an important role in several different mathematical and/or scientific specialized areas;
- new applications of the results and ideas from one area of scientific endeavor into another;
- influences which the results, problems and concepts of one field of enquiry have and have had on the development of another.

The Mathematics and Its Applications programme tries to make available a careful selection of books which fit the philosophy outlined above. With such books, which are stimulating rather than definitive, intriguing rather than encyclopaedic, we hope to contribute something towards better communication among the practitioners in diversified fields.

Because of the wealth of scholarly research being undertaken in the Soviet Union, Eastern Europe, and Japan, it was decided to devote special attention to the work emanating from these particular regions.

Thus, it was decided to start three regional series under the umbrella of the main MIA programme.

The present volume in the Eastern Europe series can perhaps be best described as a supereconomical complete course in those parts of mathematics which is required for understanding classical and modern mathematical physics and which is needed in order to work in these fields. Thus starting with real numbers the author takes about 100 pages to deal with the calculus of variations and the principles of classical mechanics. Some 150 pages later, after dealing with function theory and other prerequisites, we are equipped to deal with hydrodynamics; then follow geometry, PDEs, operators and distributions, and on that basis classical field theory can be treated. Special relativity and electrodynamics follow and are interlaced with other bits of necessary mathematics, such as tensors, differential forms and distributions on manifolds. The final 250 pages deal with quantum mechanics, general relativity, including black holes and cosmology, wave equations in a curved space-time, singularities and catastrophes, and their physical applications.

In a period wherein the ages-old "marriage" physics-mathematics has taken on new life, this seems a singularly useful volume. Mathematically trained professionals and students of mathematics will find here an almost royal road to physics and for physicists the book has been so written that it can be used as a reference volume for the mathematics needed.

The unreasonable effectiveness of mathematics in science . . .

Eugene Wigner

Well, if you knows of a better 'ole, go to it.

Bruce Bairnsfather

What is now proved was once only imagined.

William Blake

As long as algebra and geometry proceeded along separate paths, their advance was slow and their applications limited.

But when these sciences joined company, they drew from each other fresh vitality and thenceforward marched on at a rapid pace towards perfection.

Joseph Louis Lagrange

Bussum, April 1986 **Michiel Hazewinkel**

Preface to the German Edition

From 1974 to 1979 I had an opportunity, certainly one out of the ordinary, to read a course of lectures extending continuously over ten terms for students of mathematics at the Friedrich Schiller University, Jena. According to the curriculum these lectures had different names (differential and integral calculus, ordinary differential equations, etc.), but certainly the content as well as the objects are best expressed by "Analysis and Mathematical Physics". This book presents the extended skeleton of this course; a skeleton insofar as proofs have largely been omitted (in contrast to major parts of the lectures). On the other hand, Chaps. 27, 32, and 33 have been added subsequently.

The object of the course becomes clear from having a look at the Table of Contents of this book: on the one hand mathematics has developed grandiose, elegant, consistent theories, which need no further justification. On the other hand, frequently it is just the most beautiful of these theories which also form the foundation on which classical as well as modern theoretical physics are based. It was the object not only to describe these foundations, but also to give an impression of the frames that can be erected on them. True to Hilbert's ideal, mathematical theories are carefully separated here from their physical interpretations.

This book is addressed to mathematicians, physicists, and students of mathematics and physics. Especially the mathematical chapters have been so written that they can also serve as a book of reference. Mathematicians will find the principles of classical and modern theoretical physics presented in a language familiar to them. By the concise description of some mathematical foundations of classical and modern theoretical physics, the book is hoped to be also of use for physicists.

Most recently there has been a reapproach between theoretical physics and mathematics. This book is intended to promote this trend: among mathematicians and physicists, but especially among students of both disciplines.

Having told what this book may possibly accomplish, let us refer to what it is not capable of doing. It is neither a textbook nor a collection of concise monographs. A student of mathematics will not be spared cutting his way, line by line, in suitable textbooks through the proofs of those theorems which are only formulated and commented here. Some passages in this book will serve their purpose if they give an appetite for further study. For further details and proofs the reader is referred to the quoted references.

Let us also point out another peculiarity that is connected with the fact that methodical aspects have to be considered in a 10-terms course. (The lectures were compulsory up to the 5th term, which corresponds to Chap. 19, for all students of the class concerned, while the advanced course was compulsory only for the specialists in analysis. The final part of the course was facultative.) Some of the subjects appear repeatedly on ascending levels of abstraction. By way of example, Chap. 19 presents the classical foundations of the theory of partial differential equations, without any embroidery or modern accessories.

Who wants to know more must embark on a deeper study. After the theory of distributions has been presented, partial differential equations are dealt with again on this basis in Chap. 23. Finally, the wave equation in curved space-times is investigated in Chap. 33. This is based on the preceding Chaps. 29 and 32 on geometry. A look at the Table of Contents shows that there are other chains of this kind, such as the stepwise development of the integral concept, or the chapters on tensors, forms, and differential geometry. It has been a principle of this course to develop, on the one hand, the fields of mathematics presented in an exact and (as far as possible) gapless way, but, on the other hand, to aim just at that degree of generality which is absolutely necessary for the understanding of the subsequent physical chapters. Sometimes intuitively geometric arguments are given preference over formally logical conclusions. This is not only a question of time and space but also a matter of personal taste.

The course covers a relatively wide field, and it is almost a matter of course that at many points the author could not give presentations of his own that might make pretence to originality. Many chapters are an adapted interpretation of corresponding monographs. This is the case already with Chaps. 13 and 14, which were read and written following closely the lines of P. R. Halmos [23]. Other examples are Chaps. 32/33 (F.G. Friedlander, [17]), Chap. 34 (Y.-C. Lu, [40]), and Chap. 28 ([66]).

At the end of the book the reader will find a bibliography and references ordered by chapters. They are intended as acknowledgements, but also as recommendations for deeper studies.

Finally, I wish to express my gratitude to the Teubner Verlagsgesellschaft, and particularly to Mrs. Ziegler, for their harmonious co-operation.

Jena, in the fall of 1979 **Hans Triebel**

Preface to the English Edition

In comparison with the German editions we added few new references. Furthermore we included more quotations of eminent mathematicians and scientists, mostly about physics, either as epigraphs or as remarks.

Jena, Spring 1986

Hans Triebel

Table of Contents

This must ye ken!
From one make ten,
Drop two, and then
Make three square, which
Will make you rich;
Skip o'er the four!
From five and six —
In that the trick's —
Make seven and eight,
And all is straight;
And nine is one,
And ten is none.
This is the witch's One
Time's One!

(Witches' Kitchen, Goethe's Faust, I, II, in the Sir Theodore Martin Translation, London, J. M. Dent & Sons Ltd., 1954)

1. Numbers and Spaces

1.1. Real Numbers

1.1.1. Number Systems

The real numbers are familiar from school. Here are some concepts we will need continually in what follows.

Natural numbers: 1, 2, 3, ...

Integers: 0, 1, −1, 2, −2, 3, −3, ...

Rational numbers: $r=\frac{n}{m}$, where n and m are integers, $m\neq 0$.

Real numbers: $\alpha=\pm n_1 n_2 \ldots n_k {\cdot} n_{k+1} n_{k+2} \ldots$, where n_j can assume the values 0, 1, 2, ..., 9 (decimal representation of real numbers).

The fundamental operations of addition, subtraction, multiplication and division as well as the order relations $<$, $\leqq$, $>$, $\geqq$ are supposed to be known.

Powers: Let α be a real number, then we write

$$\alpha^2=\alpha\cdot\alpha,\quad \alpha^3=\alpha^2\cdot\alpha,\quad \ldots,\quad \alpha^j=\alpha^{j-1}\cdot\alpha,\quad \ldots \quad (j=2, 3, \ldots)\,.$$

If, moreover, $\alpha\neq 0$, we may also form the powers

$$\alpha^{-1}=\frac{1}{\alpha},\quad \alpha^{-2}=\alpha^{-1}\cdot\alpha^{-1},\quad \ldots,\quad \alpha^{-j}=\alpha^{-j+1}\cdot\alpha^{-1},\quad \ldots \quad (j=2, 3, \ldots)\,.$$

It is expedient to set $\alpha^0=1$. Then one obtains $\alpha^j\cdot\alpha^k=\alpha^{j+k}$ (where j and k are integers). If one of the two integers j or k is negative, one must additionally require that $\alpha\neq 0$.

Absolute value: For any real number α we define

$$|\alpha|=\begin{cases} \alpha & \text{für} \quad \alpha\geqq 0,\\ -\alpha & \text{für} \quad \alpha<0\,. \end{cases}$$

Numerical axis: As is known from school, the real numbers α can be identified with the points P of a straight line (of infinite extension to both sides) (Fig. 1.1). This is not, however, as unproblematic as it may appear at a first glance. This will be discussed in greater detail in the next subsection.

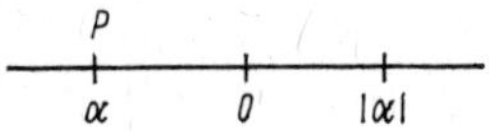

Fig. 1.1

1.1.2. Distance and Axiom of Completeness

Definition 1. *Let α and β be real numbers, then $\varrho(\alpha, \beta) = |\alpha - \beta|$ is the distance between these numbers.*

Lemma. *The distance between real numbers has the following properties:*

1. $\varrho(\alpha, \beta) \geqq 0$, *where* $\varrho(\alpha, \beta) = 0$ *if and only if* $\alpha = \beta$.
2. $\varrho(\alpha, \beta) = \varrho(\beta, \alpha)$. $\quad (\alpha, \beta, \gamma$ *real*).
3. $\varrho(\alpha, \beta) \leqq \varrho(\alpha, \gamma) + \varrho(\gamma, \alpha)$.

Notation: Open interval: $(a, b) = \{\alpha \mid \alpha \text{ real}, a < \alpha < b\}$,
half-open interval: $[a, b) = \{\alpha \mid \alpha \text{ real}, a \leqq \alpha < b\}$,
half-open interval: $(a, b] = \{\alpha \mid \alpha \text{ real}, a < \alpha \leqq b\}$,
closed interval: $[a, b] = \{\alpha \mid \alpha \text{ real}, a \leqq \alpha \leqq b\}$.
$\alpha \in (a, b)$ means $a < \alpha < b$, with analogous interpretations for the other types of intervals; "$\in$" means "is an element of".

Definition 2. *Let M be any set of real numbers. The real number α is called accumulation point of M if every open interval (a, b) with $\alpha \in (a, b)$ contains infinitely many points of M.*

Examples: 1. If $M = \{\alpha_j\}_{j=1}^{\infty}$, with $\alpha_1 = \alpha_3 = \alpha_5 = \ldots = 0$ and $\alpha_2 = \alpha_4 = \alpha_6 = \ldots = 1$, then M has two accumulation points, i.e. 0 and 1.

2. If $M = \{\alpha_j\}_{j=1}^{\infty}$ with $\alpha = \frac{1}{j}$, the 0 is the only accumulation point of M.

3. If $M = (0, 1)$, then the real number α is an accumulation point of M if and only if $\alpha \in [0, 1]$.

Remark 1. At the end of the preceding subsection we had pointed out that the correspondence between real numbers α and points of the numerical axis requires further explanation. Speaking descriptively, the problem is whether the real numbers fill the numerical axis without any gap. We simply forbid the occurrence of gaps by the following

Axiom: *Every bounded infinite set M of real numbers has at least one accumulation point.*

Remark 2. A set M of real numbers is said to be bounded if there exists a non-negative number c such that $|\alpha| \leqq c$ for every $\alpha \in M$.

Remark 3. Our point of view is as follows. The real numbers and the arithmetic operations applicable to them are supposed to be known. The above considerations are not thought to replace this knowledge. On that basis we will introduce complex numbers in a mathematically rigorous manner in the next section. It is possible to begin with the natural numbers instead of the real ones. In this case one has to introduce the integers, the rational and, finally, the real numbers step by step by an axiomatic method. This allows of many comments. We shall not take, however, this time-consuming way.

1.2. Complex Numbers

1.2.1. Definition

Let a and b be real numbers, then the ordered pair $z=\{a, b\}$ is called a complex number. z can be graphically represented in a coordinate system (Fig. 1.2). Addition and multiplication are defined as follows.

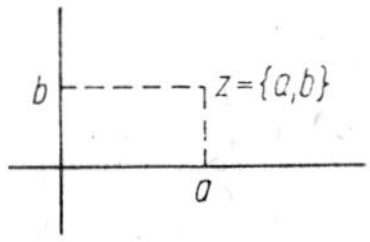

Fig. 1.2

Definition. *Let a_1, a_2, b_1 and b_2 be real numbers.*

$$\{a_1, b_1\}+\{a_2, b_2\}=\{a_1+a_2, b_1+b_2\} \text{ (Addition)};$$

$$\{a_1, b_1\}\cdot\{a_2, b_2\}=\{a_1a_2-b_1b_2, a_1b_2+a_2b_1\} \text{ (Multiplikation)}.$$

Remark. This reduces the addition and multiplication of complex numbers to the addition and multiplication of real numbers.

1.2.2. Properties

Theorem. *Let z_j be complex numbers. Then*

$$z_1+z_2=z_2+z_1 \quad \text{(commutative law of addition)};$$

$$z_1\cdot z_2=z_2\cdot z_1 \quad \text{(commutative law of multiplication)};$$

$$(z_1+z_2)+z_3=z_1+(z_2+z_3) \quad \text{(associative law of addition)};$$

$$(z_1\cdot z_2)\cdot z_3=z_1\cdot(z_2\cdot z_3) \quad \text{(associative law of multiplication)};$$

$$(z_1+z_2)\cdot z_3=z_1\cdot z_3+z_2\cdot z_3 \quad \text{(distributive law)}.$$

Remark 1. The operation put in parentheses must be carried out first.

Notation. From now on we will write z_1z_2 instead of $z_1\cdot z_2$, as well as $z_1+z_2+z_3$ instead of $z_1+(z_2+z_3)$ and $z_1z_2z_3$ instead of $z_1(z_2z_3)$. The theorem shows that it is irrelevant in which order the operations in $z_1+z_2+z_3$ are carried out. Further it is seen that, using the above theorem, it is possible to remove the parentheses from any kind of expression in parentheses, e.g.

$$(z_1+z_2z_3)(z_4+z_5)=z_1(z_4+z_5)+z_2z_3(z_4+z_5)$$
$$=z_1z_4+z_1z_5+z_2z_3z_4+z_2z_3z_5\,.$$

Remark 2. Consequently, with respect to addition and multiplication complex numbers can be treated in the same way as real numbers, including raising to powers,

$$z,\quad z^2=zz,\quad z^3=z^2z,\quad \ldots,\quad z^j=z^{j-1}z,\quad \ldots\quad (j=2, 3, \ldots)\,.$$

1.2.3. Conjugate Elements, Subtraction and Division

Definition. Let $z=\{a, b\}$ *be a complex number. Then* $\bar{z}=\{a, -b\}$ *(conjugate element or conjugate number of* z*) and* $-z=\{-a, -b\}$ (Fig. 1.3).

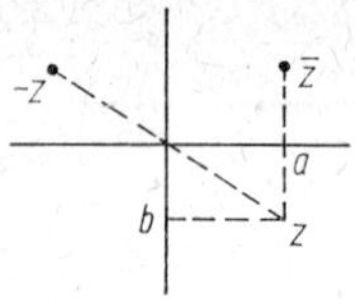

Fig. 1.3

Notation (subtraction). We write

$$z_1 - z_2 = z_1 + (-z_2). \quad \text{One has} \quad z_1 - z_2 = -z_2 + z_1 .$$

Lemma. *Let* $z = \{a, b\}$ *be a complex number and* c *a real number. Then*

$$z + \{0, 0\} = z, \quad z \cdot \{1, 0\} = \{1, 0\} \cdot z = z, \quad z \cdot \{c, 0\} = \{ac, bc\} .$$

Remark 1. $\{0, 0\}$ is the neutral element of addition, and $\{1, 0\}$ is the neutral element of multiplication.

Notation. Let c be real and z complex, then we write cz instead of $\{c, 0\}\, z$. The real numbers c are identified with the special complex numbers $\{c, 0\}$ and will no longer be distinguished from now on. By Def. 1.2.1, this is justified in view of the calculating rules

$$\{c_1, 0\} + \{c_2, 0\} = \{c_1 + c_2, 0\}, \quad \{c_1, 0\}\{c_2, 0\} = \{c_1 c_2, 0\} .$$

Note that all other operations with complex numbers (and hence especially with the special complex numbers $\{c, 0\}$) are introduced step by step, starting from Def. 1.2.1. Thus we have extended the number domain available to us from the real to the complex numbers.

Theorem. (a) *If* z_1 *and* z_2 *are given complex numbers, then the equation* $z_1 + w = z_2$ *has exactly one solution. It is* $w = z_2 - z_1$.

(b) *If* z *is complex, then* $z\bar{z}$ *is real. If* $z \neq \{0, 0\}$, *then* $z\bar{z} > 0$.

(c) *If* $z \neq \{0, 0\}$, *then* $z \cdot w = \{1, 0\}$ *has the unique solution* $w = \frac{1}{z\bar{z}} \cdot \bar{z}$ $\left(\text{where } \frac{1}{z\bar{z}} \text{ is real}\right)$.

(d) *If* $z_1 \neq \{0, 0\}$, *then* $z_1 w = z_2$ *has the unique solution* $w = \frac{1}{z_1 \bar{z}_1} \bar{z}_1 z_2$.

Remark 2. Following the above conventions, from now on we will write 0 instead of $\{0,0\}$, and hence also $z \neq 0$ instead of $z \neq \{0, 0\}$; by analogy we write $zw = 1$ instead of $z \cdot w = \{1, 0\}$.

Remark 3. Using the representation $z = \{a, b\}$, the formulas given in the theorem can be expressed by ordered pairs of real numbers. For example,

$$z\bar{z} = \{a, b\}\, \{a, -b\} = \{a^2 + b^2, 0\} = a^2 + b^2 .$$

Remark 4. The theorem enables complex numbers to be subtracted and divided. $z^{-1} = \frac{1}{z}$ has a well-defined meaning as the solution of $(z^{-1})\, z = 1$ (where $z \neq 0$); by analogy $\frac{z_2}{z_1}$ is the solution of $z_1 w = z_2$ with $z_1 \neq 0$. Thus we can also obtain any powers z^{-1}, $z^{-2} = (z^{-1})^2$, $z^{-3} = z^{-2} z^{-1}$, provided that $z \neq 0$. Then we have extended the four basic arithmetic operations from the real to the complex numbers. As was already pointed out, from now on we will regard the real numbers as special complex numbers.

1.2.4. Normal Representation

Definition. Let $z = \{a, b\}$, *then* $a = \mathrm{Re}\, z$ *is called the real part and* $b = \mathrm{Im}\, z$ *is called the imaginary part of* z (Fig. 1.4). *The complex number* $\mathrm{i} = \{0, 1\}$ *is called the imaginary unit.*

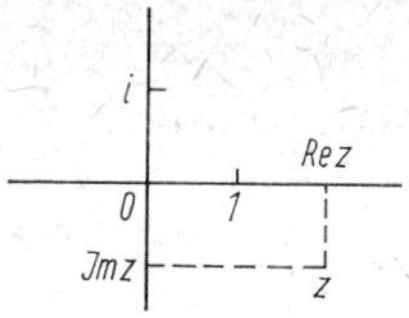

Fig. 1.4

Lemma. *Normal representation:* $z = \text{Re } z + (\text{Im } z)\,\text{i}$.

Example. Let $z = \{a, b\} \neq 0$, then

$$\text{Re } z^{-1} = \frac{a}{a^2+b^2}, \quad \text{Im } z^{-1} = -\frac{b}{a^2+b^2}.$$

The number i. The following relation holds for i:

$$\text{i}^2 = \text{ii} = \{0, 1\}\{0, 1\} = \{-1, 0\} = -1 .$$

Consequently, in this sense i is the square root of -1.

1.3. R_n, C_n, and Metric Spaces

1.3.1. The n-Dimensional Real Space R_n

Definition 1. *Let n be a natural number. Then R_n is the set of ordered n-tuples $(a_1, \ldots, a_n)$ of real numbers a_j. If $\alpha = (a_1, \ldots, a_n)$ and $\beta = (b_1, \ldots, b_n)$ are two elements of R_n, then*

$$\varrho(\alpha, \beta) = \sum_{j=1}^{n} |a_j - b_j|$$

is the distance between these elements.

Remark 1. The elements of R_n are also called points.

Theorem 1. *Let α, β and γ be points of R_n. Then*

1. $\varrho(\alpha, \beta) \geqq 0$, *where* $\varrho(\alpha, \beta) = 0$ *if and only if* $\alpha = \beta$;
2. $\varrho(\alpha, \beta) = \varrho(\beta, \alpha)$;
3. $\varrho(\alpha, \beta) \leqq \varrho(\alpha, \gamma) + \varrho(\gamma, \beta)$.

Remark 2. This theorem is a generalization of Lemma 1.1.2. In particular it is seen that R_1 is the numerical axis described in Subsec. 1.1.1. and 1.1.2.

Definition 2. (a) *Let $\varepsilon > 0$ and $\alpha \in R_n$, then*

$$K = \{\beta \mid \beta \in R_n,\ \varrho(\beta, \alpha) < \varepsilon\}$$

is called an (open) ball (with centre α and radius ε).

(b) *Let M be a set of points in R_n, i.e. $M \subset R_n$. Then $\alpha \in R_n$ is called accumulation point of M if every ball with centre α contains infinitely many points of M.*

(c) *$M \subset R_n$ is said to be bounded if there exist a positive number K and a $\beta \in R_n$ such that $\varrho(\alpha, \beta) \leqq K$ for all $\alpha \in M$.*

Remark 3. If the property (c) holds for a fixed $\beta \in R_n$, then it also holds for every other element of R_n. This follows from property 3 in Theorem 1. Consequently, whether a set $M \subset R_n$ is bounded does not depend on the choice of β.

Theorem 2. *Every bounded infinite set of points of R_n has at least one accumulation point.*

Remark 4. The property of being an infinite point set does not necessarily mean that the set must consist of infinitely many different points. Thus for instance $M=\{\alpha_j\}_{j=1}^{\infty}$, where $\alpha_j = (0, ..., 0)$, is an infinite point set. Its only accumulation point is $(0, ..., 0)$.

Remark 5. In the case $n=1$ Theorem 2 coincides with the axiom of Subsec. 1.1.2.

1.3.2. The n-Dimensional Complex Space C_n

Definition. (a) *The set of complex numbers is denoted by C_1 (complex plane). If $z_1 \in C_1$ and $z_2 \in C_1$, then*

$$\varrho(z_1, z_2) = |\mathrm{Re}\, z_1 - \mathrm{Re}\, z_2| + |\mathrm{Im}\, z_1 - \mathrm{Im}\, z_2|$$

is the distance between the elements z_1 and z_2 (Fig. 1.5).

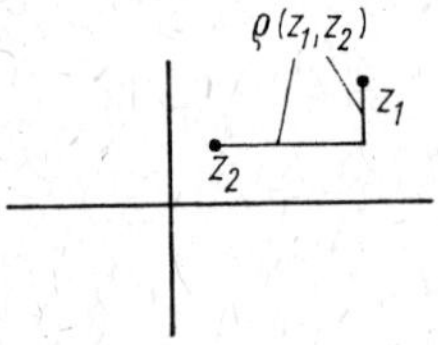

Fig. 1.5

(b) *Let n be a natural number, then C_n is the set of n-tuples $(z_1, ..., z_n)$ of complex numbers z_j. Let $\alpha = (z_1, ..., z_n)$ and $\beta = (w_1, ..., w_n)$ be any two elements of C_n, then*

$$\varrho(\alpha, \beta) = \sum_{j=1}^{n} \varrho(z_j, w_j)$$

is the distance between these elements.

(c) *If in Definition 1.3.1/2 the space R_n is replaced by C_n, one obtains the definition for (open) balls, accumulation points, and bounded sets in C_n.*

Theorem 1. *If one sets* $(z_1, ..., z_n) = (\mathrm{Re}\, z_1, \mathrm{Im}\, z_1, ..., \mathrm{Re}\, z_n, \mathrm{Im}\, z_n)$, *then* $C_n = R_{2n}$.

Remark. Consequently, the n-dimensional complex space C_n may also be regarded as the $2n$-dimensional real space R_{2n}.

Theorem 2. *Theorems 1.3.1/1 and 1.3.1/2 remain valid if C_n is substituted for R_n.*

1.3.3. The Metric Space

Definition 1. *A metric space consists of a set M and a real-valued function which assigns to every pair $a \in M$ and $b \in M$ a non-negative number $\varrho(a, b) \geqq 0$ with the properties*

1. $\varrho(a, b) \geqq 0$, *where* $\varrho(a, b) = 0$ *if and only if* $a = b$,
2. $\varrho(a, b) = \varrho(b, a)$,
3. $\varrho(a, b) \leqq \varrho(a, c) + \varrho(c, b)$ *for all* $c \in M$.

ϱ is called a metric.

Remark 1. R_n and C_n are special metric spaces. This follows from Theorem 1.3.1/1 and Theorem 1.3.2/2. Later on we shall give many other examples of metric spaces, which in part exhibit properties quite different from those of R_n and C_n.

Definition 2. (a) *If in Definition 1.3.1/2 the space R_n is replaced by an arbitrary metric space M, then one obtains the definition of (open) balls, accumulation points, and bounded sets in M.*

(b) *A subset K of the metric space M is said to be closed if K contains all the accumulation points of K.*

Remark 2. If $M = R_1$, then $K = \{0\}$ and $K = [0, 1]$ are closed, whereas $K = (0, 1)$ or $K = (0, 1]$ are not closed.

Lemma. *Every subset K of a metric space M is itself a metric space if the same metric as for M is applied to K.*

2. Convergence and Continuity

2.1. Sequences

2.1.1. Infimum, Supremum, and Limit

Definition 1. *Let $\{a_j\}_{j=1}^{\infty} \subset R_1$ be a sequence of real numbers.*

(a) *$a \in R_1$ is called infimum of the sequence $\{a_j\}$ if a is the greatest real number such that $a \leqq a_j$ for $j = 1, 2, 3, \ldots$*

(b) *$a \in R_1$ is called supremum of the sequence $\{a_j\}$ if a is the smallest real number such that $a \geqq a_j$ for $j = 1, 2, 3, \ldots$*

Lemma 1. (a) *Every bounded below sequence $\{a_j\}_{j=1}^{\infty} \subset R_1$ (i.e., for which there exists a real number M such that $M < a_j$ for $j = 1, 2, \ldots$) has exactly one infimum* (Fig. 2.1 (a)).

(b) *Every bounded above sequence $\{a_j\}_{j=1}^{\infty} \subset R_1$ (i.e., for which there exists a real number M such that $M > a_j$ for $j = 1, 2, \ldots$) has exactly one supremum* (Fig. 2.1 (b)).

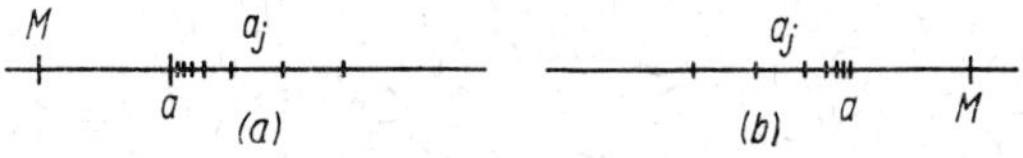

Fig. 2.1

Notation. $\inf\limits_j a_j$ denotes the infimum (if there is any) of a sequence $\{a_j\}$, and $\sup\limits_j a_j$ denotes the supremum (if there is any). As there is at most one infimum and at most one supremum, the notation is meaningful.

Examples. The sequence $1, 0, -1, 1, 0, -1, 1, 0, -1, \ldots$ has the infimum -1 and the supremum 1.

2. For $\{a_j\}_{j=1}^{\infty}$ with $a_j = 1 + \frac{1}{j}$, $\inf_j a_j = 1$ and $\sup_j a_j = 2$. This example shows that the infimum (and, by analogy, the supremum too) need not necessarily be an element of the sequence.

Definition 2. (a) *A sequence $\{a_j\}_{j=1}^{\infty} \subset R_1$ is said to converge if there exists a number $a \in R_1$ with the following property: for any positive number ε there exists a natural number $j_0(\varepsilon)$ such that $|a - a_j| \leqq \varepsilon$ for all $j \geqq j_0(\varepsilon)$* (Fig. 2.2).

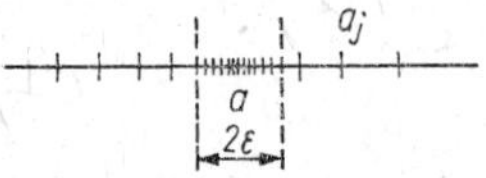

Fig. 2.2

(b) *A sequence $\{z_j\} \subset C_1$ is said to converge if there exists a number $z \in C_1$ with the following property: for any positive number ε there exists a natural number $j_0(\varepsilon)$ such that $\varrho(z_j, z) \leqq \varepsilon$ for all $j = j_0(\varepsilon)$* (Fig. 2.3).

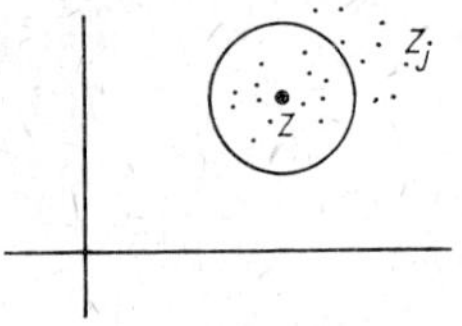

Fig. 2.3

Notation. For convergent sequences we write

$$a = \lim_{j \to \infty} a_j \quad \text{or} \quad a_j \to a$$

$$z = \lim_{j \to \infty} z_j \quad \text{or} \quad z_j \to z\,.$$

This is justified, because the limit points a and z, respectively, are well-determined: a convergent sequence has exactly one limit.

Remark. If $a_j \to a$, then $\{a_j\}_{j=1}^{\infty}$ has the unique accumulation point a. An analogous assertion is true for $z_j \to z$.

Examples.

$$\lim_{j \to \infty} \left(1 + \frac{1}{j}\right) = 1, \quad \lim_{j \to \infty} \left(1 + \frac{(-1)^j}{j}\right) = 1\,.$$

Lemma 2. (a) *A convergent sequence (of elements of R_1 or C_1) is bounded.*

(b) *$\{z_j\}_{j=1}^{\infty} \subset C_1$ is convergent if and only if both $\{\operatorname{Re} z_j\}_{j=1}^{\infty} \subset R_1$ and $\{\operatorname{Im} z_j\}_{j=1}^{\infty} \subset R_1$ converge.*

2.1.2. Properties of Convergent Sequences

In this subsection we shall consider complex sequences. As real sequences are special complex sequences, the results are in particular also true for real sequences. z and w (with or without indices) always denote complex numbers.

Theorem 1. Let $z_j \to z$ and $w_j \to w$.

(a) *If λ and μ are complex numbers, then*

$$\lambda z_j + \mu w_j \to \lambda z + \mu w, \qquad z_j w_j \to z w\,.$$

(b) *If* $z \neq 0$, *then* $z_j \neq 0$ *for* $j \geqq j_0$, *and* $\frac{w_j}{z_j} \to \frac{w}{z}$ $(j \geqq j_0)$.
(c) $\bar{z}_j \to \bar{z}$.

Definition. $\{z_j\}_{j=1}^{\infty} \subset C_1$ *is called a Cauchy sequence* (*or fundamental sequence*) *if for any positive number* ε *there exists a natural number* $j_0(\varepsilon)$ *such that* $\varrho(z_j, z_k) \leqq \varepsilon$ *for all* j *and* k *with* $j \geqq j_0(\varepsilon)$ *and* $k \geqq j_0(\varepsilon)$.

Remark. This also includes the definition of real Cauchy sequences $\{z_j = a_j\}_{j=1}^{\infty}$.

Theorem 2 (*Cauchy's condition*). *A sequence* $\{z_j\}_{j=1}^{\infty} \subset C_1$ *converges if and only if it is a Cauchy sequence.*

2.1.3. Examples

$$\frac{j^2+3j-7}{3j^2+4j+9} = \frac{j^2\left(1+\frac{3}{j}-\frac{7}{j^2}\right)}{3j^2\left(1+\frac{4}{3j}+\frac{3}{j^2}\right)} \to \frac{1}{3} \quad \text{as} \quad j \to \infty .$$

Lemma 1 (*Bernoulli's inequality*). *Let* n *be a natural number and* $a > -1$, *then* $(1+a)^n \geqq 1+na$. *Here the sign of equality is valid if and only if either* $a=0$ *or* $n=1$.

Lemma 2. (a) *Let* k *be an integer and* $0 < a < 1$, *then*

$$n^k a^n \to 0 \quad as \quad n \to \infty .$$

(b) *Let* k *be an integer and* $a > 1$, *then*

$$n^k a^n \to \infty \quad as \quad n \to \infty$$

(*this means that* $n^k a^n$ *is greater than any given positive number if* n *is chosen sufficiently large*).

Lemma 3. $\{a_k\}_{k=1}^{\infty}$ *with* $a_k = \left(1+\frac{1}{k}\right)^k$ *is a monotone increasing bounded sequence;* $\{b_k\}_{k=1}^{\infty}$ *with* $b_k = \left(1+\frac{1}{k}\right)^{k+1}$ *is a monotone decreasing bounded sequence and*

$$\lim_{k\to\infty}\left(1+\frac{1}{k}\right)^k = \lim_{k\to\infty}\left(1+\frac{1}{k}\right)^{k+1} = E .$$

Remark. The lemma says that

$$2 = a_1 < a_2 < \ldots < a_k < a_{k+1} < \ldots < E < \ldots < b_{k+1} < b_k < \ldots < b_2 < b_1 = 4 .$$

Here, for the moment, $E = \sup_k a_k = \inf_k b_k$ is merely an abbreviation, which will, however, be of fundamental interest later on.

2.2. Series

2.2.1. Convergence and Divergence

The series to be discussed here are infinite series of complex numbers, $\sum_{j=1}^{\infty} z_j$, $z_j \in C_1$. If $z_j \in R_1$, one has series of real numbers. The partial sums are denoted by $S_N = \sum_{j=1}^{N} z_j$.

Definition. (a) $\sum_{j=1}^{\infty} z_j$ *is said to converge if the sequence of partial sums,* $\{S_N\}_{N=1}^{\infty}$, *converges. This is written as* $\lim_{N\to\infty} S_N = \sum_{j=1}^{\infty} z_j$. *Otherwise* $\sum_{j=1}^{\infty} z_j$ *is said to diverge.*

(b) $\sum_{j=1}^{\infty} z_j$ *is said to converge absolutely if* $\sum_{j=1}^{\infty} \varrho(z_j, 0)$ *converges.*

Remark 1. If z_j are real numbers, then $\varrho(z_j, 0) = |z_j|$. In this case $\sum_{j=1}^{\infty} z_j$ is absolutely convergent if and only if $\sum_{j=1}^{\infty} |z_j|$ converges.

Theorem. (a) *If* $\sum_{j=1}^{\infty} z_j$ *converges, then* $z_k \to 0$ *as* $k \to \infty$.

(b) *If* $\sum_{j=1}^{\infty} z_j$ *converges absolutely, then* $\sum_{j=1}^{\infty} z_j$ *converges.*

(c) $\sum_{j=1}^{\infty} z_j$ *converges if and only if both* $\sum_{j=1}^{\infty} \operatorname{Re} z_j$ and $\sum_{j=1}^{\infty} \operatorname{Im} z_j$ *converge.*

(d) $\sum_{j=1}^{\infty} z_j$ *converges absolutely if and only if both* $\sum_{j=1}^{\infty} \operatorname{Re} z_j$ *and* $\sum_{j=1}^{\infty} \operatorname{Im} z_j$ *are absolutely convergent.*

Remark 2. A sequence $\{w_j\}_{j=1}^{\infty}$ with $w_j \to 0$ as $j \to \infty$ is called a null sequence. Part (a) of the above theorem means that the terms of a convergent series form a null sequence. The converse of this is not always true, as we shall see in the next subsection.

2.2.2. Examples

Lemma 1. *If* $0 < q < 1$, *then* $\sum_{j=0}^{\infty} q^j$ *(where* $q^0 = 1$*) converges, and* $\sum_{j=0}^{\infty} q^j = \frac{1}{1-q}$ *(geometric series).*

Lemma 2. (a) *The harmonic series* $\sum_{j=1}^{\infty} \frac{1}{j}$ *diverges.*

(b) *For* $k = 2, 3, 4, \ldots,$ $\sum_{j=1}^{\infty} \frac{1}{j^k}$ *converges.*

Lemma 3. *The series* $1 - \frac{1}{2} + \frac{1}{3} - \frac{1}{4} + \frac{1}{5} - \frac{1}{6} + \ldots$ *is convergent, but not absolutely convergent.*

Remark. The series $1 - \frac{1}{2} + \frac{1}{3} - \ldots$ tends oscillatingly towards its limit (Fig. 2.4), whereas $1 + \frac{1}{2} + \frac{1}{3} + \ldots$ increases indefinitely.

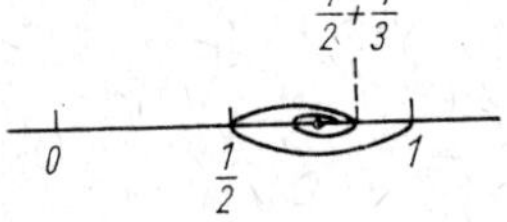

Fig. 2.4

2.2.3. Criteria of Convergence

Theorem 1 (*majorant criterion*). Let $0 \leqq a_j \leqq b_j$. *If* $\sum_{j=1}^{\infty} b_j$ *converges, then* $\sum_{j=1}^{\infty} a_j$ *converges, too.*

Remark 1. $\sum_{j=1}^{\infty} b_j$ is called a majorant of the series $\sum_{j=1}^{\infty} a_j$.

Theorem 2. *Let* $a_j \geqq 0$.

(**a**) (*Ratio test*). *If there exist a number* q *with* $0<q<1$ *and a natural number* j_0, *such that* $a_{j+1} \leqq q a_j$ *for* $j \geqq j_0$, *then* $\sum_{j=1}^{\infty} a_j$ *converges.*

(**b**) (*Root test*). *If there exist a positive number* C, *a number* q *with* $0<q<1$, *and a natural number* j_0, *such that* $a_j \leqq C q^j$ *for* $j \geqq j_0$, *then* $\sum_{j=1}^{\infty} a_j$ *converges.*

Remark 2. Later on we shall explain the symbol $\sqrt[j]{a_j}$ (root). Then (b) can be rewritten as $\sqrt[j]{a_j} \leqq q < 1$ for $j \geqq j_0$. This explains the designation "root test".

Remark 3. As the terms of the series considered here are nonnegative, convergence and absolute convergence coincide.

Example. Let k be an integer and $0<a<1$, then

$$\frac{(n+1)^k a^{n+1}}{n^k a^n} \leqq q < 1$$

for large values of n. Consequently, according to the ratio test, $\sum_{n=1}^{\infty} n^k a^n$ converges (see also Lemma 2.1.3/2).

2.2.4. Rearrangement, Multiplication, and Addition

A permutation of the natural numbers is a mapping $\varphi(n)$ which assigns to every natural number n a natural number $\varphi(n)$ such that for any given natural number k there is exactly one natural number $n(k)$ with $k=\varphi(n(k))$. Hence, in particular, $\varphi(n) \neq \varphi(m)$ for $m \neq n$. Consequently, the numbers 1, 2, 3, ... are only rearranged.

The series presented in this subsection are complex ones, and hence include series of real terms as a special case.

Theorem 1. *Let* $\sum_{i=1}^{\infty} z_j$, *where* $z_j \in C_1$, *be absolutely convergent. Further let* $w_j = z_{\varphi(j)}$, *where* $\varphi(j)$ *is any permutation of the natural numbers. Then* $\sum_{j=1}^{\infty} w_j$ *is absolutely convergent, and* $\sum_{j=1}^{\infty} z_j = \sum_{j=1}^{\infty} w_j$.

Remark 1. The above assertion is false when $\sum_{j=1}^{\infty} z_j$ is only required to be convergent (but not absolutely convergent). This can easily be understood by considering the series $1-\frac{1}{2}+$ $+\frac{1}{3}-\frac{1}{4}+\frac{1}{5}-\cdots$

Theorem 2 (*major rearrangement theorem*). *Let* $\sum_{j=1}^{\infty} z_j$, *where* $z_j \in C_1$, *be absolutely convergent. Let there exist a one-to-one correspondence between the terms* $w_{k,l}$ *and* z_j, *where each of the indices* k, l *and* j *runs through the natural numbers:* $w_{k,l} = z_{j(k,l)}$.

(a) *Then* $\sum_{l=1}^{\infty} w_{k,l}$ *is absolutely convergent for every number* $k=1, 2, 3, \ldots$

(b) *Let* $u_k=\sum_{l=1}^{\infty} w_{k,l}$, *then* $\sum_{k=1}^{\infty} u_k$ *is absolutely convergent, and* $\sum_{l=1}^{\infty} z_l=\sum_{k=1}^{\infty} u_k$.

Remark 2. The sequence $\{z_j\}_{j=1}^{\infty}$ is thus decomposed into infinitely many subsequences $\{w_{k,l}\}_{l=1}^{\infty}$, where $k=1, 2, 3, \ldots$ Then these subsequences are summed up separately (part (a)), and thereafter the sums obtained in this way are added (part (b)). The theorem remains valid when k assumes only finitely many values, i.e., when $\{z_j\}$ is decomposed into a finite number of subsequences.

Theorem 3 (*multiplication theorem*). *Let the complex series* $\sum_{j=1}^{\infty} z_j$ *and* $\sum_{k=1}^{\infty} w_k$ *be absolutely convergent. Further let there exist a one-to-one correspondence between the terms* $z_j w_k$ *and the terms* u_l, *where each of the indices* j, k *and* l *runs through the natural numbers. Then* $\sum_{l=1}^{\infty} u_l$ *is absolutely convergent, and*

$$\left(\sum_{j=1}^{\infty} z_j\right)\left(\sum_{k=1}^{\infty} w_k\right)=\sum_{l=1}^{\infty} u_l .$$

Remark 3. What is meant is the following: find all the products $z_j w_k$ and arrange them in some way or other, e.g., z_1w_1, z_2w_1, z_1w_2, z_3w_1, z_2w_2, z_1w_3, z_4w_1, ..., and then denote these terms by u_1, u_2, u_3, ...

Example. Let $0<q<1$. Then it follows from the above theorem that

$$\left(\sum_{k=0}^{\infty} q^k\right)\left(\sum_{l=0}^{\infty} q^l\right)=\sum_{j=0}^{\infty} (j+1)\, q^j .$$

According to Lemma 2.2.2/1 this gives

$$\sum_{j=0}^{\infty} (j+1)\, q^j=\frac{1}{(1-q)^2}.$$

This example shows that the theorem is also useful in calculating the sums of infinite series.

Theorem 4 (*addition theorem*). *Let the complex series* $\sum_{j=1}^{\infty} z_j$ *and* $\sum_{k=1}^{\infty} w_k$ *be convergent. If* λ *and* μ *are two complex numbers, then* $\sum_{j=1}^{\infty} (\lambda z_j+\mu w_j)$ *converges, and*

$$\sum_{j=1}^{\infty} (\lambda z_j+\mu w_j)=\lambda \sum_{j=1}^{\infty} z_j+\mu \sum_{j=1}^{\infty} w_j .$$

2.3. Real Functions in R_1

2.3.1. Definitions

Let M be a subset of R_1, and $M\neq\emptyset$ (empty set).

Definition 1. *A mapping which assigns to every point* $x\in M$ *a single value* $f(x)\in R_1$ *is called a real function* f *defined on* M (Fig. 2.5). *One writes* $M=D(f)$ (*domain of definition of* f).

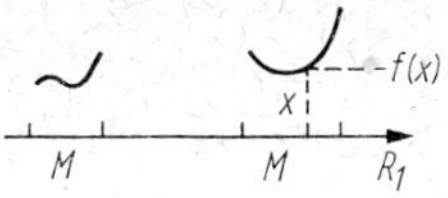

Fig. 2.5

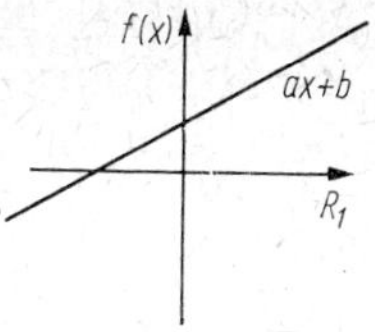

Fig. 2.6

Examples. 1. $D(f)=[0, 1]$ and $f(x)=\begin{cases}1 \text{ for rational } x, \\ 0 \text{ for irrational } x.\end{cases}$ 2. $D(f)=R_1$ and $f(x)=$ $=ax+b$ (Fig. 2.6) (where a and b are real numbers), or $f(x)=|x|$ (Fig. 2.7).

Definition 2. *Let $D(f)=M$ be the domain of definition of the function f.*

(**a**) $\inf_{x\in M} f(x)$ *(infimum of f) denotes the greatest real number α such that $\alpha \leqq f(x)$ for all $x\in M$.* $\sup_{x\in M} f(x)$ *(supremum of f) denotes the smallest real number α such that $f(x)\leqq\alpha$ for all $x\in M$* (Fig. 2.8).

(**b**) *f is said to be bounded if there exists a positive number K such that $|f(x)|\leqq K$ for all $x\in M$.*

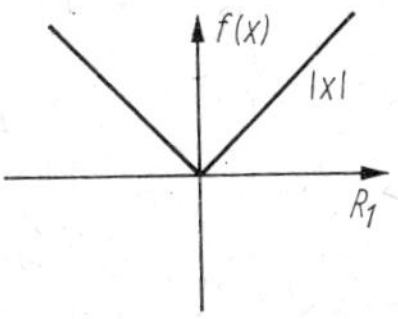

Fig. 2.7

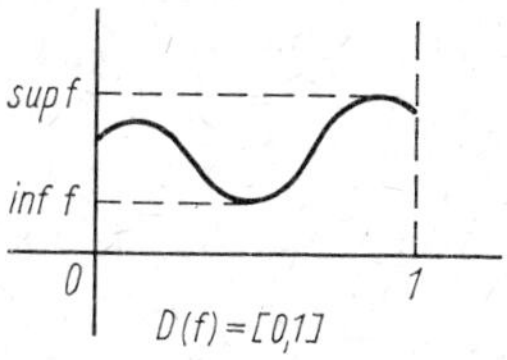

Fig. 2.8

Remark 1. Compare with Def. 2.1.1/1. It is clear that a function f will have at most one infimum and at most one supremum, which justifies the chosen designation.

Lemma 1. *If f is bounded on $D(f)$, then f has a unique infimum and a unique supremum.*

Remark 2. By analogy with Lemma 2.1.1/1, the above Lemma can be extended to functions which are only bounded from either below or above.

Remark 3. If there exists a point $x_0\in M$ such that $f(x_0)=\inf_{x\in M} f(x)$, then f assumes its infimum, which is then the minimum of f, written as

$$f(x_0)=\inf_{x\in M} f(x)=\min_{x\in M} f(x)\,.$$

If there is a point $x_1\in M$ such that $f(x_1)=\sup_{x\in M} f(x)$, then f assumes its supremum, which is then the maximum of f, written as

$$f(x_1)=\sup_{x\in M} f(x)=\max_{x\in M} f(x)\,.$$

Definition 3. *Let f be a function with $D(f)=M$.*

(**a**) *Let $x_0\in M$, then f is said to be continuous at the point x_0 if for any $\varepsilon>0$ there exists a positive number $\delta=\delta(\varepsilon, x_0)$ such that $|f(x)-f(x_0)|\leqq\varepsilon$ for all $x\in M$ with $|x-x_0|\leqq\delta$* (Fig. 2.9).

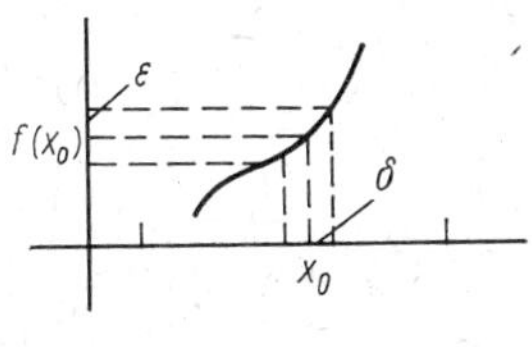

Fig. 2.9

(**b**) *f is called continuous on M if f is continuous at every point $x_0 \in M$.*

(**c**) *f is called uniformly continuous on M if f is continuous on M and if, for any $\varepsilon > 0$, the above number δ can be so chosen that $\delta = \delta(\varepsilon)$ depends on ε alone, being independent of $x_0 \in M$.*

Remark 4. Uniform continuity is understood as follows. f is called uniformly continuous on M if for any $\varepsilon > 0$ there exists a positive number $\delta = \delta(\varepsilon)$ such that $|f(x) - f(y)| \leq \varepsilon$ for all $x \in M$ and $y \in M$ with $|x - y| \leq \delta$.

Examples. The function of the first of the above examples is not continuous at any point $x_0 \in D(f) = [0, 1]$. On the other hand the functions of the second example, where $D(f) = R_1$, are continuous.

Definition 4. *$f(x) \xrightarrow[x \to x_0]{} a$, or $f(x) \to a$ as $x \to x_0$, means that $f(x_j) \to a$ for any sequence $\{x_j\}_{j=1}^{\infty} \subset D(f)$ with $x_j \to x_0$.*

Lemma 2. *f is continuous at the point $x_0 \in D(f)$ if and only if $f(x) \to f(x_0)$ as $x \to x_0$.*

Definition 5. *f is said to be continuous on the right at the point $x_0 \in D(f)$ if there exists a real number a such that $f(x_j) \to a$ for any sequence $\{x_j\}_{j=1}^{\infty} \subset D(f)$ with $x_j \to x_0$ and $x_j > x_0$. f is said to be continuous on the left at the point $x_0 \in D(f)$ if there exists a real number a such that $f(x_j) \to a$ for any sequence $\{x_j\}_{j=1}^{\infty} \subset D(f)$ with $x_j \to x_0$ and $x_j < x_0$.*

Remark 5. Henceforth we shall write $x_j \downarrow x_0$ instead of $x_j \to x_0$ with $x_j > x_0$, and $x_j \uparrow x_0$ instead of $x_j \to x_0$ with $x_j < x_0$. Lemma 2 shows that Definition 5 is a generalization of Def. 3(a). In the case of continuity on the right, the only sequences x_j admitted are those converging to x_0 "from above", whereas in the case of continuity all sequences converging to x_0 are admitted.

Discontinuities. The function $f(x)$ shown in Fig. 2.10 is both left- and right-continuous at the point x_0, but it is not continuous at that point. This is called a jump discontinuity. Apart from jumps there are also other types of discontinuities, e.g., poles (Fig. 2.11) or oscillations of $f(x)$ which increase in frequency as x approaches certain points x_0 (Fig. 2.12).

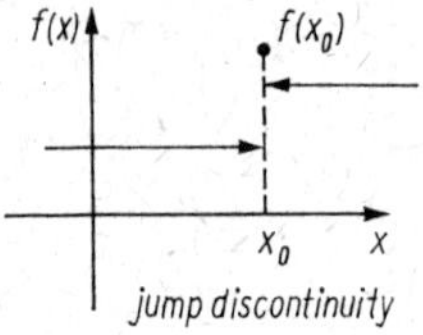

Fig. 2.10

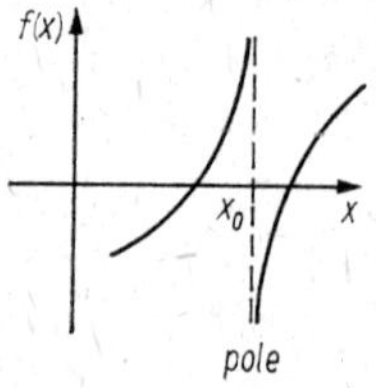

Fig. 2.11

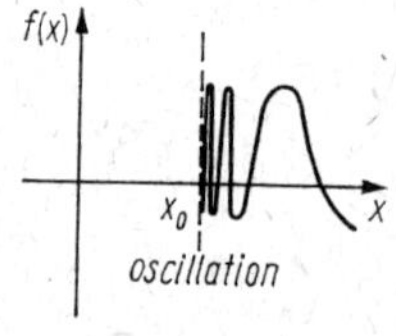

Fig. 2.12

2.3.2. Properties of Continuous Functions

Theorem 1. *Let f and g be functions with $D(f) = D(g)$ in the sense of Def. 2.3.1/1. Let f and g be continuous at the point $x_0 \in D(f)$.*

(**a**) *If λ and μ are real numbers, then $\lambda f(x) + \mu g(x)$ is continuous at x_0.*

(b) *$f(x)\cdot g(x)$ is continuous at x_0.*

(c) *If, moreover, $g(x_0)\neq 0$, then $\dfrac{f(x)}{g(x)}$ is continuous at x_0.*

Theorem 2. *Let f be a continuous function in the sense of Def. 2.3.1/3 (b). Further let $D(f)$ be closed and bounded. Then*

$$\inf_{x\in D(f)} f(x)=\min_{x\in D(f)} f(x),\qquad \sup_{x\in D(f)} f(x)=\max_{x\in D(f)} f(x)\,.$$

Remark 1. If $D(f)$ is regarded as a subset of R_1, then, by Def. 1.3.3/2(b), $D(f)$ is closed if it contains all of its accumulation points in R_1. Examples of closed sets in R_1 are a closed interval or the union of finitely many closed intervals.

Remark 2. Theorem 2 makes two assertions, namely (1) that $f(x)$ is bounded and (2) that it assumes its infimum and supremum (whose existence, by Lemma 2.3.1/1, follows from the boundedness of f).

Theorem 3 (*location principle*). *Let f be continuous (as per Def. 2.3.1/3 (b)), and let $D(f)$ be an interval. If α is a real number such that*

$$\inf f(x)<\alpha<\sup f(x)\,,$$

then there exists at least one point $x_0\in D(f)$ such that $f(x_0)=\alpha$ (Fig. 2.13).

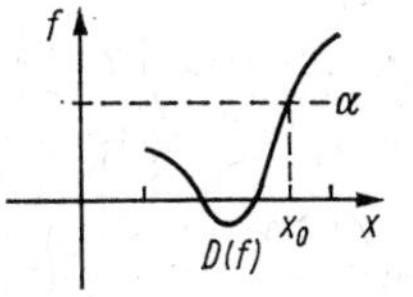

Fig. 2.13

Remark 3. In the last theorem we did not assume that $D(f)$ be closed and bounded. f may be unbounded for open or half-open intervals $D(f)$. Then in the theorem one has to set $\inf f(x)=-\infty$ (if f is unbounded below) or $\sup f(x)=\infty$ (if f is unbounded above) (Fig. 2.14).

Theorem 4. *Let f be continuous (as per Def. 2.3.1/3 (b)), and let $D(f)$ be closed and bounded. Then f is uniformly continuous (Def. 2.3.1/3 (c)).*

Remark 4. The sawtooth-like function shown in Fig. 2.15 is continuous on $D(f)=(0, 1]$, but not uniformly continuous; the curve is made up of triangles of height 1 and base $\left[\dfrac{1}{n+1},\dfrac{1}{n}\right]$, where $n=1, 2, 3, \ldots$ This shows that the assumption made in the last theorem, namely that $D(f)$ be closed, is essential.

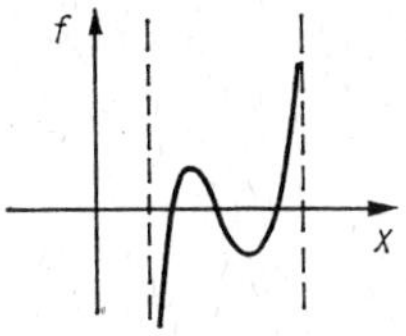

Fig. 2.14

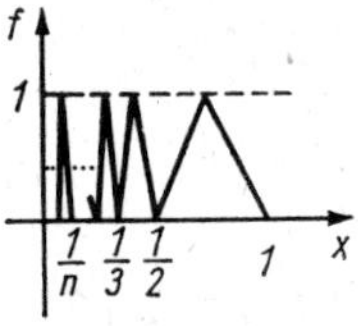

Fig. 2.15

2.4. Continuous Mappings in Metric Spaces

2.4.1. Definition

With $\varrho(x, y)=|x-y|$, R_1 becomes a metric space. If f is a function in the sense of Def. 2.3.1/1, with $D(f)=R_1$, then the continuity of f at the point $x_0\in D(f)$ (Def. 2.3.1/3 (a)) may also be described as follows: for any $\varepsilon>0$ there is some

$\delta > 0$ such that $\varrho(f(x_0), f(x)) \leqq \varepsilon$ for all $x \in R_1$ with $\varrho(x_0, x) \leqq \delta$. This formulation can however be generalized immediately.

Definition. *Let M_1 and M_2 be two metric spaces with the metrics ϱ_1 and ϱ_2, respectively. Further let f be a mapping of M_1 into M_2 (i.e., every $x \in M_1$ has some $f(x) \in M_2$ uniquely assigned to it)* (Fig. 2.16).

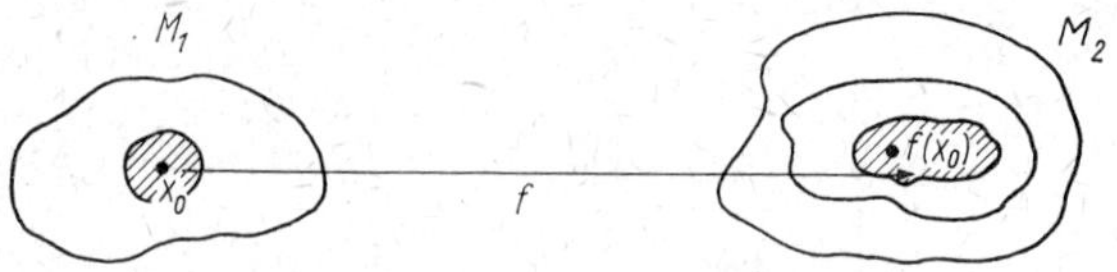

Fig. 2.16

(a) *f is called continuous at the point $x_0 \in M_1$ if for all $\varepsilon > 0$ there exists a positive number $\delta = \delta(\varepsilon, x_0)$ such that $\varrho_2(f(x_0), f(x)) \leqq \varepsilon$ for all $x \in M_1$ with $\varrho_1(x_0, x) \leqq \delta$.*

(b) *f is called continuous on M_1 if f is continuous at every point $x_0 \in M_1$.*

(c) *f is called uniformly continuous on M_1 if f is continuous on M_1 and if, for all $\varepsilon > 0$, the above number δ can be so chosen that $\delta = \delta(\varepsilon)$ depends only on ε, but not on $x_0 \in M_1$.*

Remark 1. This is analogous to Def. 2.3.1/3; see also Remark 2.3.1/4.

Remark 2. A subset N of a metric space M is itself a metric space (with the same metric). If the above mapping f is defined only on a subset $D(f)$ (domain of definition) instead of the whole set M_1, then the above definition can be applied by substituting $D(f)$ for M_1. In this sense the above definition is in fact a generalization of Def. 2.3.1/3.

2.4.2. Examples

As was pointed out at the beginning of Subsec. 2.4.1. (see also Remark 2.4.1/2), Def. 2.4.1 is a generalization of Def. 2.3.1/3. It is of interest to consider other examples, too.

Complex-valued functions. Let M_1 be a subset of R_1, and $M_2 = C_1$. Then $f(x)$ is a complex-valued function and can be decomposed as follows

$$f(x) = \operatorname{Re} f(x) + \mathrm{i} \operatorname{Im} f(x), \qquad x \in M_1. \tag{1}$$

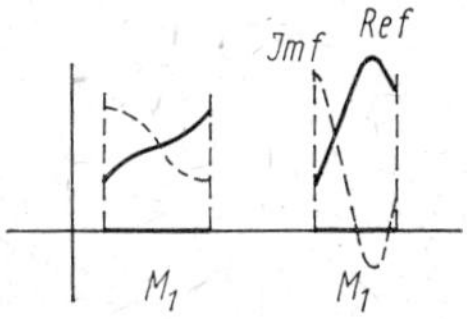

Fig. 2.17

Lemma 1. *The complex-valued function $f(x)$ is continuous if and only if the real functions* Re $f(x)$ *and* Im $f(x)$ *are continuous.*

Real functions in R_n. Let M_1 be a subset of R_n and $M_2 = R_1$. The mapping f can be thought of as a surface represented versus R_n; $n = 2$ has an immediate geometric meaning (Fig. 2.18). It is clear what is meant by continuity in the sense of Def. 2.4.1.

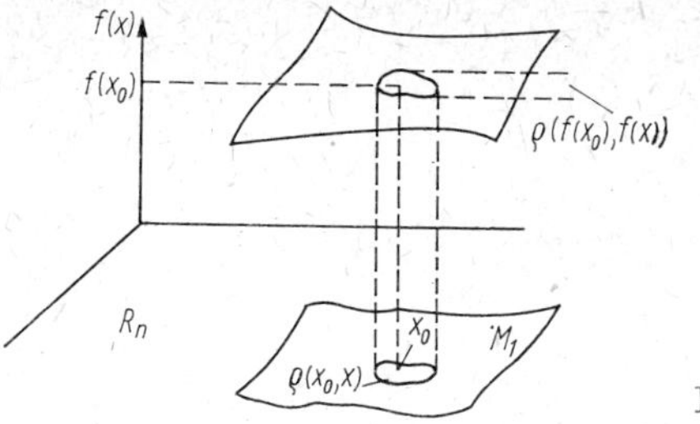

Fig. 2.18

Complex-valued functions in R_n. Let M_1 be a subset of R_n and $M_2=C_1$. The decomposition (1) as well as Lemma 1 are applicable in this case, too.

Mappings in R_n. Let $M_1=M_2=R_n$, and let $f(x)=(f_1(x), ..., f_n(x))$, where $f_j(x)$ are real-valued functions in R_n. $f(x)$ is called a vector function.

Lemma 2. *The vector function $f(x)=(f_1(x), ..., f_n(x))$ is continuous if and only if the real-valued functions $f_j(x)$, $j=1, ..., n$, which are mappings of R_n into R_1, are continuous (componentwise continuity).*

Linear mappings. Let $a_{j,k}$, where $j=1, ..., n$ and $k=1, ..., n$, be real numbers, then the vector function $f(x)=(f_1(x), ..., f_n(x))$, with

$$f_j(x)=\sum_{k=1}^{n} a_{j,k}x_k\,, \quad j=1, ..., n\,,$$

is continuous.

2.4.3. Real-Valued Continuous Functions in R_n

The considerations made in Subsecs. 2.3.1. and 2.3.2. can be transferred from R_1 to R_n. Let f be a function mapping $D(f)\subset R_n$ into R_1. Definition 2.3.1/4 can be adopted word for word.

Lemma. *f is continuous at the point $x_0\in D(f)$ if and only if $f(x)\to f(x_0)$ as $x\to x_0$ (analogue to Lemma 2.3.1/2).*

Theorem 1. *Let f and g be two real-valued continuous functions in R_n, with $D(f)=$ $=D(g)$, and let both functions be continuous at the point $x_0\in D(f)$.*

(a) *Let λ and μ be real numbers. Then $\lambda f(x)+\mu g(x)$ as well as $f(x)\cdot g(x)$ are continuous at the point x_0.*

(b) *If $g(x_0)\neq 0$, then $\dfrac{f(x)}{g(x)}$ is continuous at the point x_0 (analogue to Theorem 2.3.1/1).*

Theorem 2. *Let M be closed and bounded in R_n. Further let f be continuous on $M=$ $=D(f)$ (Def. 2.4.1 (b), where $M=M_1$). Then*

(a) $\inf_{x\in M} f(x)=\min_{x\in M} f(x)$ *and* $\sup_{x\in M} f(x)=\max_{x\in M} f(x)$, *and*

(b) *f is uniformly continuous (Def. 2.4.1 (c)) (analogue to Theorem 2.3.2/2 and Theorem 2.3.2/4).*

2.5. Complete Metric Spaces

2.5.1. Definitions

Definition 1. *Let M be a metric space with the metric ϱ.*

(a) *A sequence $\{x_j\}_{j=1}^{\infty} \subset M$ is said to converge if there exists an element $x_0 \in M$ with the following property: for any positive number ε there is a natural number $j_0(\varepsilon)$ such that $\varrho(x_0, x_j) \leqq \varepsilon$ for all $j \geqq j_0(\varepsilon)$.*

(b) *A sequence $\{x_j\}_{j=1}^{\infty} \subset M$ is called a Cauchy sequence (or fundamental sequence) if for any positive number ε there exists a natural number $j_0(\varepsilon)$ such that $\varrho(x_j, x_k) \leqq \varepsilon$ for all j and k with $j \geqq j_0(\varepsilon)$ and $k \geqq j_0(\varepsilon)$.*

Remark 1. This is a generalization of Def. 2.1.1/2(b) and Def. 2.1.2, which apply to $M = C_1$.

Lemma 1. *Every convergent sequence (in the sense of the above definition) is a Cauchy sequence.*

Remark 2. By Theorem 2.1.2./2, the converse is also true for $M = C_1$. For general metric spaces, however, this is not correct. Consider, for example, $M = (0, 1]$ with the usual distance-type metric, then $x_j = \frac{1}{j}$ is a Cauchy sequence, but does not converge.

Definition 2. *A metric space is called complete if each of its Cauchy sequences is a convergent sequence.*

Remark 3. Remark 2 shows that the definition is meaningful.

Lemma 2. *R_n and C_n are complete metric spaces.*

Remark 4. The lemma follows from Theorems 1.3.1/2 and 1.3.2/2. The space $M = (0, 1]$ mentioned above is an example of a metric space which is not complete.

2.5.2. The Space $C[a, b]$

Definition. *Let $-\infty < a < b < \infty$, then $C[a,b] = \{f \mid f$ real and continuous on $[a, b]\}$, and*

$$\varrho(f, g) = \sup_{x \in [a,b]} | f(x) - g(x) | . \tag{1}$$

Remark 1. Hence $C[a, b]$ is the space consisting of all real-valued continuous functions on the closed interval $[a, b]$. The "points" of this space are functions. If f and g are continuous on $[a, b]$, then $f - g$ is likewise continuous on $[a, b]$. Hence, as $[a, b]$ is closed, $f - g$ is bounded and the sup in Eq. (1) may be replaced by max (Theorem 2.3.2/2).

Theorem. $C[a, b]$ is a complete metric space.

Remark 2. The metric in $C[a, b]$ is given by (1). One has to show that ϱ satisfies the conditions of Def. 1.3.3/1.

Remark 3. Instead of the closed interval $[a, b]$ one may also consider the open interval (a, b). Then $C(a, b) = \{f \mid f$ real, continuous and bounded on $(a, b)\}$. Here it must be required in addition that f be bounded. With the metric (1), $C(a, b)$ is again a complete metric space. It is easy to see that there exist some $f \in C(a, b)$ which cannot be extended to the points a and b in such a way that after the extension they belong to $C[a, b]$ (Fig. 2.19). On the other hand, the restriction of $f \in C[a, b]$ to (a, b) always belongs to $C(a, b)$.

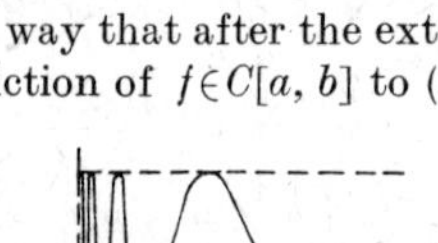

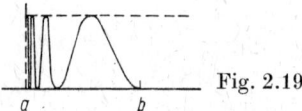

Fig. 2.19

2.5.3. Banach's Contraction Mapping Theorem

Definition. *Let M be a metric space with the metric ϱ. A mapping f of M into M is called contracting if there exists a number α with $0 \leqq \alpha < 1$ such that*

$$\varrho(f(x), f(y)) \leqq \alpha \varrho(x, y) \tag{1}$$

for all $x \in M$ and all $y \in M$ (Fig. 2.20).

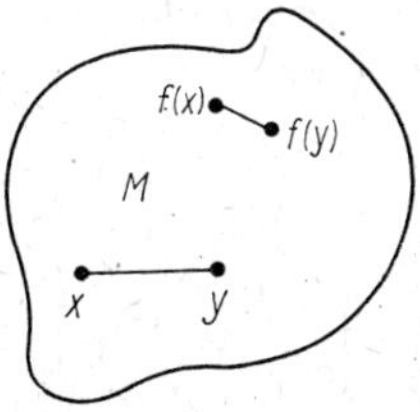

Fig. 2.20

Lemma. *A contracting mapping is continuous.*

Remark. Hence one has to check that Def. 2.4.1 (b) is satisfied for $M_1 = M_2 = M$.

Theorem (*Banach's contraction theorem*). *Let M be a complete metric space, and let f be a contracting mapping in M. Then there exists exactly one point $x \in M$ such that $f(x) = x$ (fixed point).*

Example 1. Let $M = R_1$, and let $f(x)$ be a continuous function on R_1, with $|f(x) - f(y)| \leqq \alpha\, |x - y|$ for all $x \in R_1$ and all $y \in R_1$. Here, $0 \leqq \alpha < 1$. If $f(x)$ is interpreted as a mapping of R_1 into R_1, then the above theorem is applicable. x_0 is the only fixed point with $f(x) = x$ (Fig. 2.21).

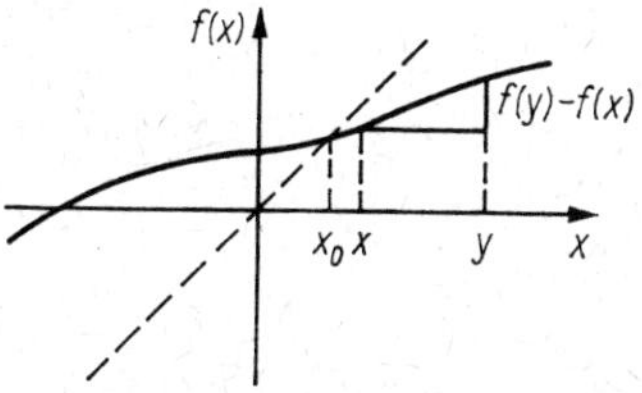

Fig. 2.21

Example 2. The plane R_2, or a subset M of R_2, becomes a metric space if the usual distance between any two points is taken as a metric. If M is a circular ring and f is a rotation about the common centre of the two disks, then (1) holds with $\alpha = 1$ (Fig. 2.22). On the other hand, in general such rotations have no fixed points. This shows that $\alpha < 1$ is essential in the theorem.

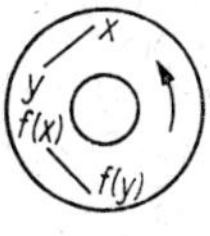

Fig. 2.22

3. Differential and Integral Calculus in R_1 (Basic Concepts)

3.1. Differentiation

3.1.1. Definitions

In this chapter we shall consider (with a few exceptions) only real-valued functions in the sense of Def. 2.3.1/1. If f is not real, this will be expressly mentioned.

Definition. Let *f be a real function, and let* $D(f)=(a, b)$ *be an open interval. Let* $x\in(a, b)$ (Fig. 3.1).

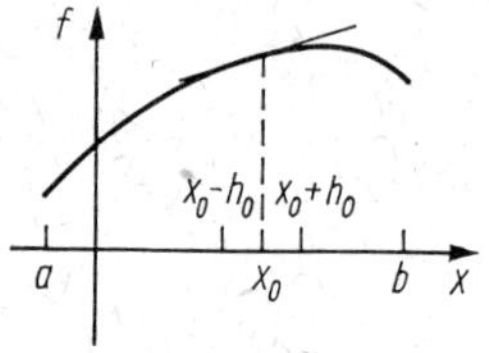

Fig. 3.1

(**a**) $f(x)$ *is called differentiable at the point* x_0 *if*

$$\frac{f(x_0+h)-f(x_0)}{h}\to\alpha \quad \text{as} \quad h\to 0\,,$$

where $|h|\leq h_0$ *and* $h\neq 0$. *One denotes* $\alpha=f'(x_0)=\frac{\mathrm{d}f}{\mathrm{d}x}(x_0)$.

(**b**) $f(x)$ *is called differentiable on* $D(f)=(a, b)$ *if* $f(x)$ *is differentiable at every point* $x_0\in(a, b)$.

(**c**) $f(x)$ *is called right-hand differentiable at the point* x_0 *if*

$$\frac{f(x_0+h)-f(x_0)}{h}\to\beta \quad \text{as} \quad h\downarrow 0\,,$$

where $0<h<h_0$. *One denotes* $\beta=f'_+(x_0)$. $f(x)$ *is called left-hand differentiable at the point* x_0 *if*

$$\frac{f(x_0+h)-f(x_0)}{h}\to\gamma \quad \text{as} \quad h\uparrow 0\,,$$

where $h_0<h<0$. *One denotes* $\gamma=f'_-(x_0)$.

Remark 1. $h\uparrow 0$ and $h\downarrow 0$ were explained in Remark 2.3.1/5. When $D(f)=[a, b)$, one may also find the right-hand derivation at the point $x_0=a$. By analogy, one may find the left-hand derivation at the point b when $D(f)=(a, b]$.

Remark 2 (geometric interpretation). Let $f(x)$ be differentiable at the point x_0, then the straight line

$$g(x)=f(x_0)+f'(x_0)\,(x-x_0) \tag{1}$$

is called tangent (to the curve $f(x)$ at the point x_0) (Fig. 3.2). It is easy to see that $g(x)$ has the approximation property that

$$\left|\frac{f(x)-g(x)}{x-x_0}\right|\to 0 \quad \text{as} \quad x\to x_0\,. \tag{2}$$

Lemma 1. (**a**) *Let $f(x)$ be differentiable at the point x_0, then the line* (1) *is the only line in the plane which has the property* (2).

(**b**) *$f(x)$ is differentiable at the point x_0 if and only if $f(x)$ is both right-hand and left-hand differentiable at x_0 and $f'_+(x_0)=f'_-(x_0)$. In this case, $f'(x_0)=f'_+(x_0)=f'_-(x_0)$.*

Remark 3. The roof-like curve shown in Fig. 3.3 is both right-hand and left-hand differentiable at x_0, but not differentiable at that point.

Lemma 2. *If $f(x)$ is differentiable at the point x_0, it is also continuous at x_0.*

Remark 4. Consequently, differentiability is more than continuity.

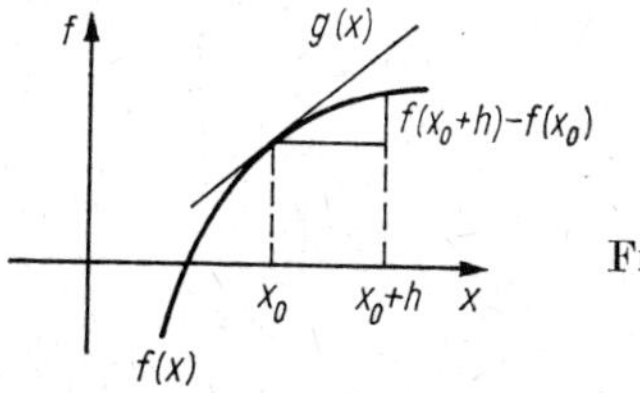

Fig. 3.2

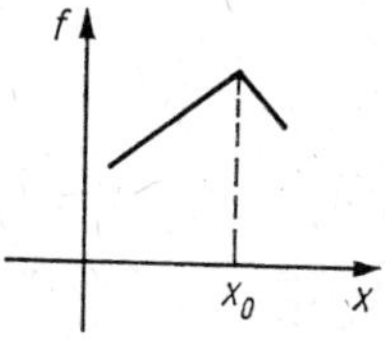

Fig. 3.3

3.1.2. Rules

Theorem 1. *Let f and g be differentiable at the point x_0.*

(**a**) *Let λ and μ be real numbers, then $\lambda f+\mu g$ is differentiable at the point x_0, and $(\lambda f+\mu g)'(x_0)=\lambda f'(x_0)+\mu g'(x_0)$.*

(**b**) *fg is differentiable at the point x_0, and*

$$(fg)'(x_0)=f'(x_0)\,g(x_0)+f(x_0)\,g'(x_0)\,.$$

(**c**) *If $g(x)\neq 0$, then $\frac{f}{g}$ is differentiable at the point x_0, and*

$$\left(\frac{f}{g}\right)'(x_0)=\frac{f'(x_0)\,g(x_0)-f(x_0)\,g'(x_0)}{g^2(x_0)}\,.$$

Theorem 2 (*chain rule*). *If g is differentiable at the point x_0, and if f is differentiable at the point $g(x_0)$, then $h(x)=f(g(x))$ is differentiable at the point x_0, and*

$$h'(x_0)=f'(g(x_0))\,g'(x_0)\,.$$

3.1.3. Examples (Rational Functions)

Let n be a natural number, then

$$(x^n)'=nx^{n-1} \quad \text{for} \quad x\in R_1\,,$$

$$(x^{-n})'=-nx^{-n-1} \quad \text{for} \quad x\in R_1, \quad x\neq 0\,.$$

Hence, for any integer k, the following relation holds:

$$(x^k)'=kx^{k-1} \quad \text{for} \quad x\in R_1, \quad x\neq 0\,.$$

Using this formula together with Theorem 3.1.2/1, it is possible to calculate the derivation of any rational functions $R(x)=\dfrac{\sum_{k=0}^{N} a_k x^k}{\sum_{j=0}^{M} b_j x^j}$, where attention must be paid to the zeros of the polynomial in the denominator.

3.1.4. Inverse Functions

Definition. (a) *The open interval* $(a, b) \subset D(f)$ (Fig. 3.4) *is called monotony interval of the function f if f increases monotonically in* (a, b) (i.e., $f(x_1) \geqq f(x_0)$ *for all* x_0 *and* x_1 *with* $a < x_0 \leqq x_1 < b$) *or decreases monotonically in* (a, b) (*i.e.*, $f(x_1) \leqq f(x_0)$ *for all* x_0 *and* x_1 *with* $a < x_0 \leqq x_1 < b$). *If* $f(x_1) > f(x_0)$ (*or* $f(x_1) < f(x_0)$) *for all* x_0 *and* x_1 *with* $a < x_0 < x_1 < b$, *then f is called strictly monotonic increasing* (*or strictly monotonic decreasing*).

(b) *If* $y = f(x)$ *is strictly monotonic* (*increasing or decreasing*) *in* (a, b), *then the mapping* $y \to x$ *is called inverse function, denoted by* f^{-1} (Fig. 3.5).

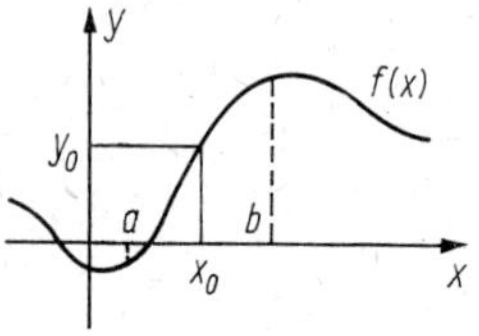

Fig. 3.4

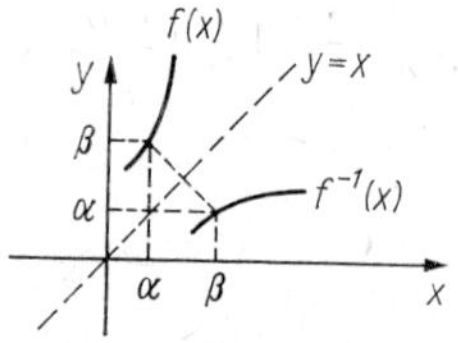

Fig. 3.5

Remark 1. Let f be strictly monotonic increasing in (a, b). Then f is a one-to-one mapping of the points $x_0 \in (a, b)$ onto some set of points on the y-axis. f is determined by the mapping $f: x_0 \to y_0$. The inverse function is determined by the mapping $f^{-1}: y_0 \to x_0$. The roles of the x- and the y-axis are interchanged. Plotting $f^{-1}(x)$ again as a function of x can simply be done by reflection of the curve with respect to the line $y = x$.

Theorem. *Let* $f(x)$ *be differentiable on the interval* $(x_0 - \delta, x_0 + \delta)$, *where* $\delta > 0$. *Let* $f'(x)$ *be continuous at the point* x_0, $f'(x_0) \neq 0$ *and* $y_0 = f(x_0)$. *If* ε *is sufficiently small, then f is strictly monotonic in the interval* $(x_0 - \varepsilon, x_0 + \varepsilon)$. *The inverse function* f^{-1} *is differentiable at the point* y_0, *and* $(f^{-1})'(y_0) = \dfrac{1}{f'(x_0)}$.

Remark 2. It is easy to show that the last assertion follows from the above reflection principle, if one sets $\alpha = x_0$ and $\beta = y_0$, taking into account that the derivative describes the slope of the tangent line.

3.1.5. Mean Value Theorems

The function f is called continuously differentiable on (a, b) if f is differentiable on the open interval (a, b) and the derivative $f'(x)$ is continuous on (a, b).

Theorem 1 (*Rolle's theorem*). *Let f be continuous on* $[a, b]$ *and continuously differentiable on* (a, b). *If* $f(a) = f(b) = 0$, *then there exists at least one point* $x_0 \in (a, b)$ *such that* $f'(x_0) = 0$ (Fig. 3.6).

Remark 1. x_0 may be chosen as a point where $f(x_0) = \sup\limits_{x \in [a,b]} f(x) = \max\limits_{x \in [a,b]} f(x)$.

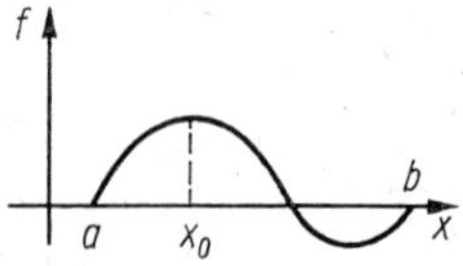

Fig. 3.6

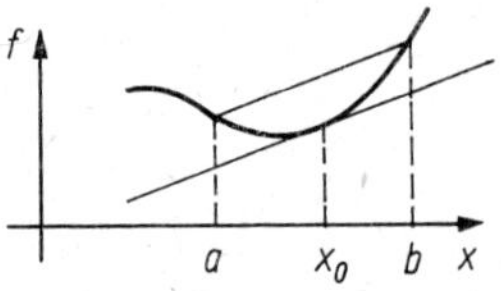

Fig. 3.7

Theorem 2. *Let f and g be continuous on $[a, b]$ and continuously differentiable on (a, b). Further let $g'(x) \neq 0$ on (a, b). Then there exists at least one point $x_0 \in (a, b)$ such that*

$$\frac{f(b)-f(a)}{g(b)-g(a)}=\frac{f'(x_0)}{g'(x_0)}. \tag{1}$$

Remark 2. $g(x)$ is strictly monotonic in $[a, b]$; in particular $g(b) \neq g(a)$.

Remark 3. Let $g(x)=x$. Then it follows from (1) that there exists an $x_0 \in (a, b)$ such that $\frac{f(b)-f(a)}{b-a}=f'(x_0)$. Geometrically, this means that the tangent at x_0 is parallel to the line passing through the points $(a, f(a))$ and $(b, f(b))$ (Fig. 3.7).

Theorem 3 (*l'Hospital's rule*). *Let f and g be continuous on $[a, b]$ and continuously differentiable on (a, b). Further let $f(a)=g(a)=0$, $g(x) \neq 0$ for $x \in (a, b)$, and $g'(x) \neq 0$ for $x \in (a, b)$. If $\lim\limits_{x \downarrow a} \frac{f'(x)}{g'(x)}$ exists, then $\lim\limits_{x \downarrow a} \frac{f(x)}{g(x)}$ exists, too, and*

$$\lim_{x \downarrow a} \frac{f(x)}{g(x)}=\lim_{x \downarrow a} \frac{f'(x)}{g'(x)}.$$

Remark 4. If $f(x) \to \infty$ and $g(x) \to \infty$ as $x \downarrow a$, then the theorem can be applied to $\frac{\frac{1}{g(x)}}{\frac{1}{f(x)}} \left(=\frac{f(x)}{g(x)}\right)$.

Remark 5. It is possible to formulate analogous theorems for $x \uparrow a$ or for $x \to a$.

Examples. Let n be a natural number, then

$$\lim_{x \to 1} \frac{x^n-1}{x-1}=\lim_{x \to 1} \frac{nx^{n-1}}{1}=n,$$

$$\lim_{x \to 1} \frac{(x^n-1)^3}{(x-1)^2}=\lim_{x \to 1} \frac{3n\,(x^n-1)^2\,x^{n-1}}{2\,(x-1)}=\lim_{x \to 1} \frac{(x^n-1)\,[\ldots]}{2}=0.$$

The second example shows that the theorem can also be applied iteratively.

3.1.6. Higher-Order Derivatives; Derivatives of Complex-Valued Functions

Higher-order derivatives. It is clear that Def. 3.1.1 can be used iteratively. If $f(x)$ is differentiable on the interval (a, b), one may ask whether $f'(x)$ is differentiable at the point $x_0 \in (a, b)$. If so, we write $f''(x_0)=(f')'(x_0)$. Iteration: $f^{(n)}(x_0)=(f^{(n-1)})'(x_0)$.

Complex-valued functions. If we consider (by way of exception) complex-valued functions $f(x) \in C_1$ with $D(f)=(a, b)$, then Def. 3.1.1(a) can be adopted word for word, provided that

$$\frac{f(x_0+h)-f(x_0)}{h} \to f'(x_0) \quad \text{as} \quad h \to 0$$

is now understood as convergence in C_1.

Lemma. *The complex-valued function $f(x)$ is differentiable at the point x_0 if and only if the real-valued functions* $\operatorname{Re} f$ *and* $\operatorname{Im} f$ *are differentiable at the point x_0. If this is the case, then*

$$f'(x_0) = (\operatorname{Re} f)'(x_0) + \mathrm{i}(\operatorname{Im} f)'(x_0) .$$

3.2. Integration of Real-Valued Functions

3.2.1. Definition of Riemann's Integral

Let $-\infty < a < b < \infty$. Consider a subdivision Z of the interval $[a, b]$ (Fig. 3.8):

$$a = a_0 < a_1 < \ldots < a_n < a_{n+1} = b . \tag{1}$$

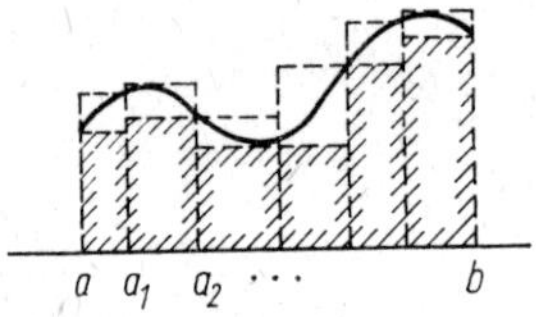

Fig. 3.8

Definition. *Let $f(x)$ be a real-valued bounded function on $D(f) = [a, b]$.*
(**a**) *Then one denotes*

$$\underline{\int} f(x)\,\mathrm{d}x = \sup_Z \Sigma_Z^*, \quad \text{where} \quad \Sigma_Z^* = \sum_{j=1}^{n} (a_{j+1} - a_j) \inf_{y \in [a_j, a_{j+1}]} f(y) ,$$

$$\overline{\int} f(x)\,\mathrm{d}x = \inf_Z \Sigma_Z^{**}, \quad \text{where} \quad \Sigma_Z^{**} = \sum_{j=0}^{n} (a_{j+1} - a_j) \sup_{y \in [a_j, a_{j+1}]} f(y) .$$

(**b**) *$f(x)$ is called integrable if $\underline{\int} f(x)\,\mathrm{d}x = \overline{\int} f(x)\,\mathrm{d}x$.*

Remark 1. $\sup\limits_Z$ is the supremum taken over all subdivisions Z of the form (1), where n is variable (supremum taken over all n and all points of subdivision) ($\inf\limits_Z$ is defined analogously). Since $f(x)$ is bounded, $\inf f(y)$ and $\sup f(y)$ exist for the corresponding subintervals. Σ_Z^* is called lower sum, corresponding to the sum of the shaded rectangles. $\underline{\int}$ is called lower integral. Accordingly, Σ_Z^{**} is called upper sum and $\overline{\int}$ is called upper integral.

Remark 2. If $f(x)$ is integrable, we write

$$\underline{\int} f(x)\,\mathrm{d}x = \overline{\int} f(x)\,\mathrm{d}x = \int f(x)\,\mathrm{d}x = \int_a^b f(x)\,\mathrm{d}x ,$$

where the last notation indicates the integration interval. This is Riemann's integral.

Lemma 1. $\underline{\int} f(x)\,\mathrm{d}x \leqq \overline{\int} f(x)\,\mathrm{d}x$.

Remark 3. A function is thus called integrable if and only if the sign of equality holds in Lemma 1.

Lemma 2. *Let $f(x)$ be integrable over the interval $[a, b]$. Further let y_j be any point of $[a_j, a_{j+1}]$ according to subdivision* (1). *Let*

$$\Sigma_Z = \sum_{j=0}^{n} (a_{j+1} - a_j) f(y_j) .$$

Then there exists a sequence Z_k of subdivisions such that

$$\int f(x)\,dx = \lim_{k\to\infty} \sum\nolimits^*_{Z_k} = \lim_{k\to\infty} \sum\nolimits^{**}_{Z_k} = \lim_{k\to\infty} \sum\nolimits_{Z_k} .$$

Remark 4. One has $\sum^*_{Z_k} \leqq \sum_{Z_k} \leqq \sum^{**}_{Z_k}$, the integral $\int f(x)\,dx$ being approximated from below by $\sum^*_{Z_k}$, and from above by $\sum^{**}_{Z_k}$.

Lemma 3. *Let $f(x)$ be integrable over $[a, b]$, and let $c \in (a, b)$. Let Z^c denote all subdivisions (1) which contain c as a point of subdivision. Then*

(**a**) $\int f(x)\,dx = \sup\limits_{Z^c} \sum^*_{Z^c} = \inf\limits_{Z^c} \sum^{**}_{Z^c}$.

(**b**) *Lemma 2 remains true when the set of possible subdivisions is restricted to those of Z^c.*

3.2.2. Properties

Theorem 1. *Let f and g be integrable over $[a, b]$.*

(**a**) *For any real numbers λ and μ, $\lambda f + \mu g$ is integrable over $[a, b]$, and*

$$\int (\lambda f(x) + \mu g(x))\,dx = \lambda \int f(x)\,dx + \mu \int g(x)\,dx .$$

(**b**) *If, in addition, $f(x) \leqq g(x)$, then*

$$\int f(x)\,dx \leqq \int g(x)\,dx .$$

Theorem 2. *Let f be integrable over $[a, b]$.*

(**a**) *For any $c \in (a, b)$, f is integrable over $[a, c]$ and over $[c, b]$, and*

$$\int_a^b f(x)\,dx = \int_a^c f(x)\,dx + \int_c^b f(x)\,dx .$$

(**b**) *$|f(x)|$ is integrable over $[a, b]$, and*

$$\left| \int_a^b f(x)\,dx \right| \leqq \int_a^b |f(x)|\,dx .$$

Remark. If f is integrable over $[a, b]$ and if $a < c < d < b$, then f is also integrable over $[c, d]$. This follows from (a).

3.2.3. Commutation of Passage to the Limit and Integration

Definition. *Let $-\infty < a < b < \infty$. A sequence $\{f_j(x)\}_{j=1}^{\infty}$ of real- (or complex-)valued functions with $D(f_j) = (a, b)$ is said to converge uniformly towards $f(x)$ with $D(f) = (a, b)$ if for each $\varepsilon > 0$ there exists a natural number $j_0(\varepsilon)$ such that*

$$\sup_{x \in (a,b)} |f(x) - f_j(x)| \leqq \varepsilon$$

for all j with $j \geqq j_0(\varepsilon)$ (Fig. 3.9).

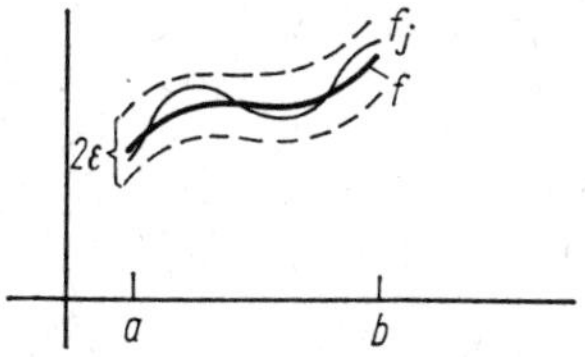

Fig. 3.9

Remark 1. Uniform convergence of f_j towards f is written as $f_j \Rightarrow f$.

Remark 2. If the functions f_j are continuous on $[a, b]$, then uniform convergence coincides with convergence in the space $C[a, b]$ defined in Subsec. 2.5.2. Since $C[a, b]$ is complete (Theorem 2.5.2), it follows that $f \in C[a, b]$.

Theorem. *Let the functions $f_j(x)$, where $j = 1, 2, 3, \ldots$, be integrable over $[a, b]$. If $f_j \Rightarrow f$ on (a, b), then $f(x)$ is also integrable over $[a, b]$, and*

$$\int f(x)\,dx = \lim_{j \to \infty} \int f_j(x)\,dx\,. \tag{1}$$

Remark 3. (1) can also be written as

$$\int \left(\lim_{j \to \infty} f_j(x)\right) dx = \lim_{j \to \infty} \int f_j(x)\,dx$$

(this means that passage to the limit and integration are commutable).

3.2.4. Examples of Integrable Functions, and Counter-Examples

Let $-\infty < a < b < \infty$. A real-valued function $f(x)$ with $D(f) = [a, b]$ is called piecewise continuous if $f(x)$ is continuous on the right at every point $x_0 \in [a, b)$ and continuous on the left at every point $x_0 \in (a, b]$ (Def. 2.3.1/5) and if the number of points of discontinuity is finite (Fig. 3.10). The discontinuities are therefore jump discontinuities as described in Subsec. 2.3.1. Functions of this kind are bounded. Continuous functions are special cases of piecewise continuous functions.

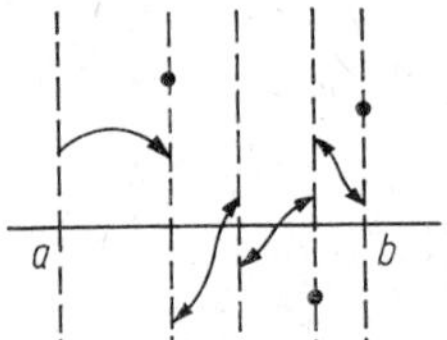

Fig. 3.10

Theorem. *Every function which is piecewise continuous on $[a, b]$ is integrable over $[a, b]$.*

Remark. In particular, $f(x) \equiv c$ is integrable, and

$$\int f(x)\,dx = c\,(b - a)\,.$$

Counter-example. The function $f(x) = \begin{cases} 1 \text{ for } x \text{ rational} \\ -1 \text{ for } x \text{ irrational} \end{cases}$, with $D(f) = [0, 1]$, is not integrable (in the sense of Riemann). For it is immediately clear that all the lower sums are $\sum_Z^* = -1$ and all the upper sums are $\sum_Z^{**} = 1$. On the other hand, $|f(x)| \equiv 1$, and hence is integrable over $[0, 1]$. This shows that the converse of Theorem 3.2.2/2 (b) is not true.

3.2.5. Primitives

Definition. *Let $-\infty < a < b < \infty$. A real-valued function $f(x)$ with $D(f) = [a, b]$ is called Lipschitz-continuous if there exists a positive number M such that $|f(x_1) - f(x_2)| \leq M|x_1 - x_2|$ for all x_1 and all x_2 in $[a, b]$.*

Lemma 1. (**a**) *Every function which is Lipschitz-continuous on* $[a, b]$ *is also uniformly continuous on* $[a, b]$ (*Def.* 2.3.1/3 (c)).

(**b**) *If* f *is continuous on* $[a, b]$ *and differentiable on* (a, b), *with* $\sup_{x\in(a,b)} |f'(x)| < \infty$, *then* f *is Lipschitz-continuous on* $[a, b]$.

Remark. A roof-like function is Lipschitz-continuous on $[a, b]$, but not differentiable at the point x_0 (Fig. 3.11). The inverse function f of $y = x^2$ is continuous on $[0, 1]$, but not Lipschitz-continuous (Fig. 3.12).

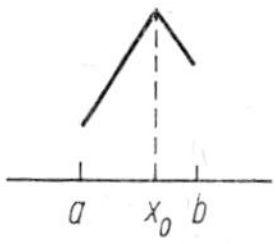

Fig. 3.11

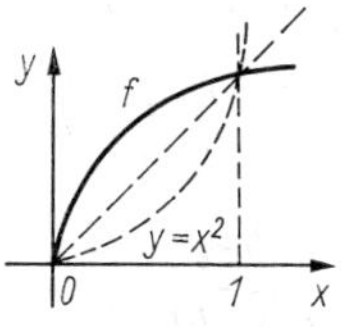

Fig. 3.12

Theorem 1. (**a**) *If* f *is integrable over* $[a, b]$, *then*

$$F(x) = \int_a^x f(y)\,\mathrm{d}y \text{ is Lipschitz-continuous on } [a, b].$$

($F(x)$ *is called the primitive, or Newton's indefinite integral, of* $f(x)$).

(**b**) *If* $f(x)$ *is continuous on* $[a, b]$, *then* $F(x)$ *is differentiable on* (a, b), *and* $F'(x) = f(x)$.

Lemma 2. *If* $f'(x) \equiv 0$ *on* (a, b), *then* $f(x)$ *is constant on* (a, b).

Theorem 2. *Let* f *and* g *be continuous on* $[a, b]$; *let* g *be differentiable on* (a, b) *and* $g'(x) = f(x)$ *for* $x \in (a, b)$. *Then*

$$\int_a^b f(y)\,\mathrm{d}y = g(b) - g(a)\,.$$

Example. For any natural number n, $f(x) = x^n$ and $g(x) = \frac{x^{n+1}}{n+1}$ have the properties required in the theorem. It follows that

$$\int_a^b x^n \mathrm{d}x = \frac{b^{n+1}}{n+1} - \frac{a^{n+1}}{n+1}\,.$$

3.2.6. Integral Operators

Let $-\infty < a < b < \infty$, then to every function $f \in C[a, b]$ (Def. 2.5.2) there is assigned a function Hf by

$$(Hf)(x) = \int_a^x h(y, f(y))\,\mathrm{d}y + c, \quad x \in [a, b]\,. \tag{1}$$

Here c denotes a constant and $h(y, z)$ is a real-valued function with $D(h) = [a, b] \times R_1 = \{(y, z) \mid a \leqq y \leqq b,\ z \in R_1\}$. Naturally, $h(y, z)$ must be subjected to additional conditions in order that (1) be meaningful. What is desired are conditions which make H a continuous mapping of $C[a, b]$ into $C[a, b]$ in the sense of Def. 2.4.1(b). [In the notation used there, $M_1 = M_2 = C[a, b]$, and $\varrho_1 = \varrho_2 = \varrho$ has the meaning defined in Formula (2.5.2/1), while the mapping denoted f in Def. 2.4.1 is now denoted H.]

Theorem 1. *Let $h(y, z)$ be continuous on $D(h)=[a, b]\times R_1$, and let h satisfy the Lipschitz condition*

$$|h(y, z)-h(y, z^*)|\leq M\ |z-z^*|$$

for all $y\in[a, b]$, $z\in R_1$, and $z^*\in R_1$. Here M is a positive number being independent of y, z, and z^*. Then H is a continuous mapping of $C[a, b]$ into $C[a, b]$, and

$$\varrho(Hf, Hg)\leq M\ (b-a)\ \varrho(f, g) \quad \text{for all} \quad f\in C[a, b] \quad \text{and} \quad g\in C[a, b]\,.$$

Moreover, $(Hf)(x)$ is differentiable on (a, b), and

$$(Hf)'(x)=h(x, f(x)) \quad \text{for} \quad x\in(a, b)\,. \tag{2}$$

Remark 1. For $f\in C[a, b]$, $h(x, f(x))$ is continuous on $D(h)$.

Definition. *Let $-\infty<a<b<\infty$, then one denotes*

$$\underbrace{C[a, b]\times\ldots\times C[a, b]}_{n \text{ factors}}=\{f \mid f=\{f_1(x), \ldots, f_n(x)\},\ f_j(x)\in C[a, b]\}\,,$$

$$\varrho_n(f, g)=\sum_{j=1}^{n}\ \sup_{x\in[a,b]}\ |f_j(x)-g_j(x)|\,.$$

Here $g=\{g_1(x), \ldots, g_n(x)\}$.

Lemma. *$C[a, b]\times\ldots\times C[a, b]$ with the metric ϱ_n is a complete metric space.*

Let us consider the vector-valued analogue to (1), namely

$$\begin{aligned}(H_jf)(x)&=H_j\{f_1, \ldots, f_n\}(x)\\ &=\int_a^x h_j(y, f_1(y), \ldots, f_n(y))\,\mathrm{d}y+c_j, \quad x\in[a, b]\,,\end{aligned} \tag{3}$$

where the c_j are again constants, $j=1, \ldots, n$. We set

$$Hf=\{H_1f, \ldots, H_nf\}\,.$$

Here the $h_j(y, z_1, \ldots, z_n)$ are real-valued functions with the domain of definition

$$D(h_j)=[a, b]\times R_n=\{(y, z_1, \ldots, z_n) \mid a\leq y\leq b,\ (z_1, \ldots, z_n)\in R_n\}\,.$$

Theorem 2. *Let $h_j(y, z_1, \ldots, z_n)$ be continuous on $D(h_j)=[a, b]\times R_n$ and satisfy the Lipschitz condition*

$$|h_j(y, z_1, \ldots, z_n)-h_j(y, z_1^*, \ldots, z_n^*)|\leq M\ (|z_1-z_1^*|+\ldots+|z_n-z_n^*|)$$

for all $y\in[a, b]$, $(z_1, \ldots, z_n)\in R_n$, and $(z_1^*, \ldots, z_n^*)\in R_n$. Here M is a positive number being independent of $y, z_1, \ldots, z_n, z_1^*, \ldots, z_n^*$. Then H is a continuous mapping of $C[a, b]\times\ldots\times C[a, b]$ into itself, and

$$\varrho_n(Hf, Hg)\leq nM(b-a)\ \varrho_n(f, g) \tag{4}$$

for all f and g in $C[a, b]\times\ldots\times C[a, b]$. Moreover, $(H_jf)(x)$ is differentiable on (a, b), and

$$(H_jf)'(x)=h_j(x, f_1(x), \ldots, f_n(x))\,.$$

Theorem 3. *If $nM(b-a)<1$, then the mapping H of Theorem 2 has exactly one fixed point in $C[a, b]\times\ldots\times C[a, b]$.*

Remark 2. Theorem 3 follows from (4) and Banach's contraction theorem 2.5.3. Consequently there exists exactly one n-tuple $\{f_1(x), \ldots, f_n(x)\}$ with $f_j(x)\in C[a, b]$, such that

$$f_j(x)=c_j+\int_a^x h_j(x, f_1(x), \ldots, f_n(x))\,\mathrm{d}x, \quad j=1, \ldots, n\,.$$

Then it follows from Theorem 2 that $f_j(x)$ is differentiable on (a, b) and that

$$f_j'(x) = h_j(x, f_1(x), \ldots, f_n(x)), \quad x \in (a, b), \quad j = 1, \ldots, n .$$

Equations of this kind, in which functions $f_1(x), \ldots, f_n(x)$ and derivatives of these functions are connected with one another, are called ordinary differential equations (or, more precisely, systems of ordinary differential equations). The investigations presented in this subsection may also be interpreted in such a way that we have shown the existence of solutions of differential equations, provided that the functions h_j satisfy certain conditions.

4. Ordinary Differential Equations (Existence and Uniqueness Theorems)

4.1. Initial-Value Problems

4.1.1. The Differential Equation $f^{(n)}(x) \equiv 0$

In this chapter we will consider only real functions. Let n be a natural number, then

$$f(x) = \sum_{j=0}^{n-1} a_j x^j , \quad x \in (a, b), \quad -\infty < a < b < \infty , \tag{1}$$

is a polynomial of degree $n-1$, with the real coefficients a_j. One has $f^{(n)}(x) \equiv 0$. If $n = 1$, then Lemma 3.2.5/1 shows that the converse is also true: from $f'(x) \equiv 0$ it follows that $f(x) = a_0$ in (a, b). The following lemma generalizes this assertion.

Lemma 1. *If f is n times differentiable on (a, b) and $f^{(n)}(x) \equiv 0$, then f has the form* (1).

Remark 1. Therefore, polynomials of degree $n-1$ can also be characterized by $f^{(n)}(x) \equiv 0$.

Lemma 2. *Let $x_0 \in R_1$ and $c_j \in R_1$, where $j = 0, \ldots, n-1$. Then there exists a unique solution of $f^{(n)}(x) \equiv 0$ in R_1, where $f^{(j)}(x_0) = c_j$ for $j = 0, \ldots, n-1$ $\quad (f^{(0)} = f)$.*

Remark 2. This solution can be written down immediately:

$$f(x) = \sum_{j=1}^{n-1} \frac{c_j}{j!} (x - x_0)^j \quad (j! = 1 \cdot 2 \cdot \ldots \cdot j, \; 0! = 1).$$

Problem. *The question arises whether the above assertions on the special differential equation $f^{(n)}(x) \equiv 0$ can be extended to more general differential equations of the form*

$$f^{(n)}(x) = h(x, f(x), f'(x), \ldots, f^{(n-1)}(x)) . \tag{2}$$

Does Eq. (2) *have a unique solution with the initial values $f^{(j)}(x_0) = c_j$ for $j = 0, \ldots, n-1$?*

4.1.2. Formulation of the Problems

In Subsec. 3.2.6., $[a, b]\times R_n$ was introduced for $-\infty<a<b<\infty$ (Fig. 4.1). A function was said to be continuously differentiable if it is differentiable and its derivative is continuous.

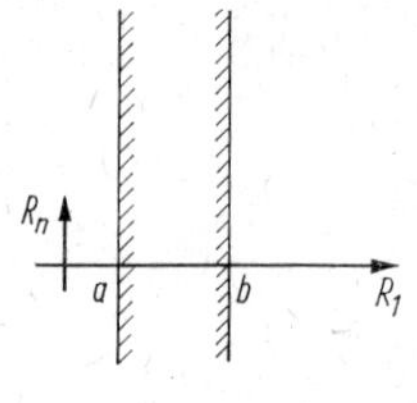

Fig. 4.1

Problem 1 (*Ordinary differential equations of order n*). *Let $h(x_0, x_1, \ldots, x_n)$ be continuous on $[a, b]\times R_n$ and satisfy the Lipschitz condition*

$$|h(x_0, x_1, \ldots, x_n)-h(x_0, x_1^*, \ldots, x_n^*)|\leqq M\,(|x_1-x_1^*|+\ldots+|x_n-x_n^*|) \tag{1}$$

for all $x_0\in[a,b]$, $(x_1, \ldots, x_n)\in R_n$ and $(x_1^*, \ldots, x_n^*)\in R_n$. Here M is a positive number independent of $x_0, x_1, \ldots, x_n, x_1^*, \ldots, x_n^*$. Further let $c_j\in R_1$, $j=0, \ldots, n-1$. Find a function f which is n times continuously differentiable on (a, b) and satisfies the conditions

$$f^{(j)}(x)\in C[a, b] \quad \textit{and} \quad f^{(j)}(a)=c_j \qquad \textit{for} \quad j=0, \ldots, n-1 \tag{2}$$

and

$$f^{(n)}(x)=h(x, f(x), f'(x), \ldots, f^{(n-1)}(x)) \quad \textit{for} \quad x\in(a, b)\,. \tag{3}$$

Remark 1. The space $C[a, b]$ was introduced in Subsec. 2.5.2. For the time being, $f^{(j)}(x)$ can be obtained only for $x\in(a, b)$. $f^{(j)}(x)\in C[a, b]$ means that $f^{(j)}(x)$ has a limit on the right at a and a limit on the left at b and that one sets $f^{(j)}(a)=\lim\limits_{x\downarrow a} f^{(j)}(x)$, $f^{(j)}(b)=\lim\limits_{x\uparrow b} f^{(j)}(x)$ (Fig. 4.2). Problems of the type (2), (3) are called initial-value problems (Cauchy problems) for ordinary differential equations of order n.

Remark 2. If $n=1$, the right-hand side of Eq. (3) is reduced to $f'=h(x, f(x))$. This case allows a simple geometric interpretation. At each point $(x_0, y_0)\in(a, b)\times R_1$, plot a small segment of the straight line defined by $y-y_0=h(x_0, y_0)\cdot(x-x_0)$. This gives a so-called directional field (Fig. 4.3). The problem is to find functions fitting this directional field in such a way that the above straight lines become tangents to the curve $f(x)$.

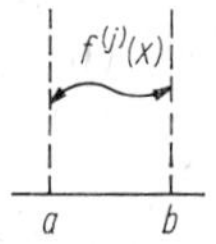

Fig. 4.2

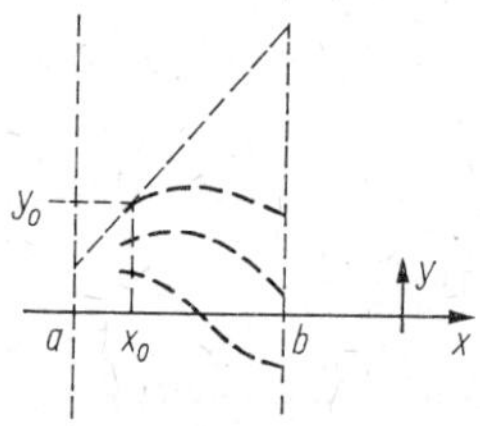

Fig. 4.3

Problem 2 (*Systems of ordinary differential equations of order* 1). *Let $h_j(x_0, x_1, \ldots, x_n)$, where $j=1, \ldots, n$, be continuous on $[a, b]\times R_n$ and satisfy the Lipschitz condition*

$$|h_j(x_0, x_1, \ldots, x_n)-h_j(x_0, x_1^*, \ldots, x_n^*)|\leqq M\,(|x_1-x_1^*|+\ldots+|x_n-x_n^*|)$$

for all $x_0 \in [a, b]$, $(x_1, \ldots, x_n) \in R_n$ *and* $(x_1^*, \ldots, x_n^*) \in R_n$. *Here* M *is a positive number independent of* $x_0, x_1, \ldots, x_n, x_1^*, \ldots, x_n^*$, *and* $j = 1, \ldots, n$. *Further let* $c_j \in R_1$, $j = 1, \ldots, n$. *Find a set of functions* $f_1(x), \ldots, f_n(x)$ *which are continuously differentiable on* (a, b) *and satisfy the conditions*

$$f_j(x) \in C[a, b] \quad \text{and} \quad f_j(a) = c_j \quad \text{for} \quad j = 1, \ldots, n$$

and

$$f_j'(x) = h_j(x, f_1(x), \ldots, f_n(x)) \qquad \text{for} \quad j = 1, \ldots, n \quad \text{and} \quad x \in (a, b)\,.$$

4.2. Existence and Uniqueness Theorems

4.2.1. Systems of First Order

Problem 2 of Subsec. 4.1.2. is solved by reducing it to the problem considered in Subsec. 3.2.6.

Theorem. *Problem 2 of Subsec.* 4.1.2. *has a unique solution.*

4.2.2. Differential Equations of Order n

Theorem. *Problem 1 of Subsec.* 4.1.2. *has a unique solution.*

Remark. Problem 1 is reduced to Problem 2 by setting $f = f_1, f' = f_2, \ldots, f^{(n-1)} = f_n$. Then the differential equation (4.1.2/3) is equivalent to the system of order 1,

$$f_1' = f_2, f_2' = f_3, \ldots, f_{n-1}' = f_n\,,$$
$$f_n' = h(x, f_1(x), \ldots, f_n(x))\,,$$

with the initial conditions $f_j(a) = c_{j-1}$ for $j = 1, \ldots, n$. This means that ordinary differential equations of order n can be regarded as special systems of order 1.

4.2.3. Local Existence and Uniqueness Theorems

For the sake of simplicity we shall restrict ourselves to $f' = h(x, f)$. Till now we have required that h be continuous on $D(h) = [a, b] \times R_1$. Now we only require that $h(x, y)$ be continuous on an open domain Ω of the x,y-plane. The directional field referred to in Remark 4.1.2/2 is then meaningful only on Ω.

Problem. *Let* $h(x, y)$ *be continuous on* $D(h) = \Omega$ *and satisfy the Lipschitz condition*

$$|h(x, y) - h(x, y^*)| \leq M\,|y - y^*|$$

for all points $\{x, y\} \in \Omega$ *and* $\{x, y^*\} \in \Omega$. *Here* M *is a positive number independent of* x, y *and* y^*. *Further let*

$$\sup_{\{x,y\} \in \Omega} |h(x, y)| \leq N\,.$$

For $\{a, c\} \in \Omega$, *find a solution of* $f'(x) = h(x, f(x))$, *with* $f(a) = c$, *in a neighbourhood on the right of* a, (a, b).

Theorem. *Let* $(b - a)\, M < 1$, *and let* Q *be the (closed) rectangle plotted in Fig.* 4.4, *with* $Q \subset \Omega$, *then the problem has a unique solution in* Q.

Remark 1. This is a local existence and uniqueness theorem. The method can however be applied iteratively (sequence of rectangles of the form Q), and in this way the integral curve $f(x)$ can be maximally continued (Fig. 4.5).

Remark 2. A corresponding theorem holds for differential equations of order n and for systems of order 1.

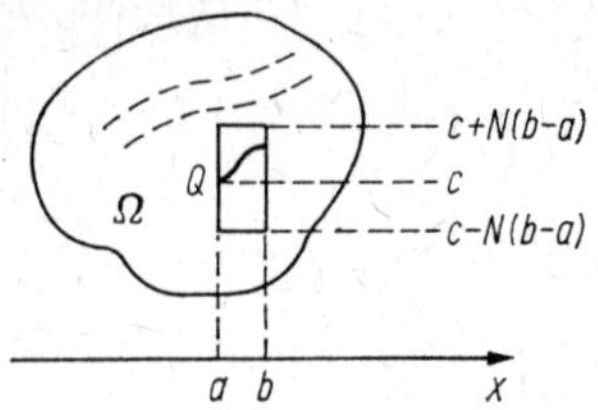

Fig. 4.4

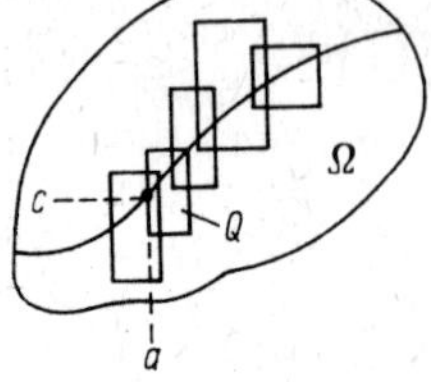

Fig. 4.5

5. Elementary Functions and Power Series

5.1. Exponential Functions and Power Functions (Real-Valued)

5.1.1. The Function e^x

Definition. e(x) *is the solution, being unique on* R_1, *of the differential equation* $f'(x)=f(x)$ *with the initial condition* $f(0)=1$ (Fig. 5.1).

Remark 1. In the sense of Subsec. 4.1.2., $h(x, y)=y$. Hence, according to Theorem 4.2.2, $f'(x)=f(x)$ with $f(0)=1$ has a unique solution in the interval $[-N, N]$. If we let $N\to\infty$, it is observed that this proposition also remains true for R_1 instead of $[-N, N]$.

Theorem. (a) e(x) *is positive, strictly monotone increasing and convex on* R_1. *Further,*

$$\mathrm{e}(x)\to\infty \quad as \quad x\to\infty, \quad \mathrm{e}(x)\to 0 \quad as \quad x\to-\infty .$$

(b) *For all* $x\in R_1$ *and* $y\in R_1$,

$$\mathrm{e}(x+y)=\mathrm{e}(x)\,\mathrm{e}(y) . \tag{1}$$

Remark 2. A real-valued function is called convex if the line segment connecting any two points of the curve lies above the curve. A real-valued function is called concave if the line segment connecting any two points of the curve lies below the curve (Fig. 5.2).

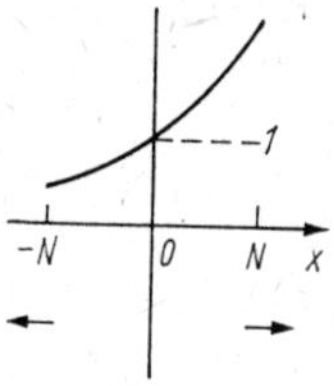

Fig. 5.1

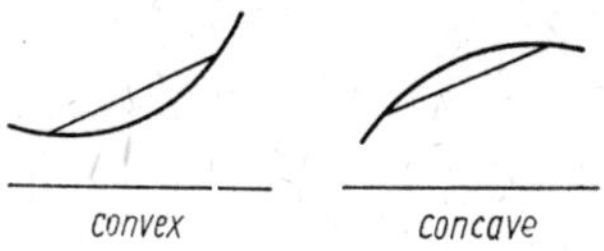

Fig. 5.2

Remark 3. The elegant proof of the theorem uses nothing but the existence and uniqueness theorems stated in Chap. 4 for differential equations: what a pity that proofs are omitted in this book!

Remark 4 (notation). We set $e(1) = e$. Then from (1) it follows that

$$e(n) = e(1)\, e(n-1) = \ldots = (e(1))^n = e^n \quad \text{and} \quad e(-n) = e^{-n},$$

where n is a natural number. This justifies the notation $e(x) = e^x$, because e^x has the previous meaning for integral values of x.

5.1.2. The Function $\log x$

Definition. $\log x$ *is the inverse function of* e^x.

Remark 1. According to Subsec. 3.1.4. and Theorem 5.1.1, the definition is meaningful. The curve of $\log x$ is obtained by reflection of the curve of e^x with respect to the line $y = x$ (Fig. 5.3).

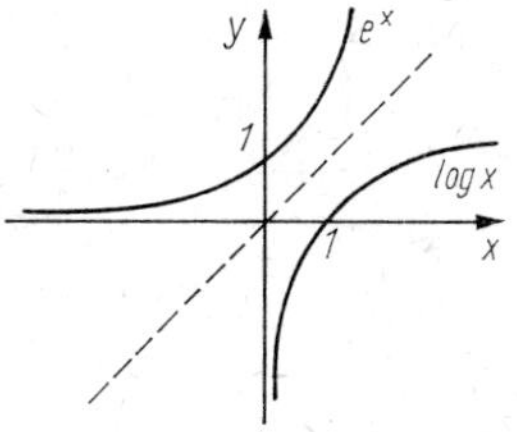

Fig. 5.3

Theorem. (a) *The domain of definition of* $\log x$ *is* $D(\log x) = (0, \infty)$. *The function* $\log x$ *has derivatives up to arbitrary order; it is strictly monotone increasing and concave. Further,* $(\log x)' = \frac{1}{x}$, $\log 1 = 0$, $\log x \to \infty$ *as* $x \to \infty$, *and* $\log x \to -\infty$ *as* $x \downarrow 0$.

(b) *For all* $x > 0$ *and* $y > 0$,

$$\log (xy) = \log x + \log y.$$

Remark 2. In particular, $\log x^k = k \log x$, where k is an integer.

Lemma. $\lim\limits_{x \to 1} \frac{\log x}{x - 1} = 1$.

5.1.3. The Number e

Theorem. $e = \lim\limits_{n \to \infty} \left(1 + \frac{1}{n}\right)^n$.

Remark. $e = e(1)$ has the meaning stated in Subsec. 5.1.1. From Lemma 2.1.3/3 it follows that $e = E$ and $2 < e < 4$. One finds that $e = 2.718\ldots$

5.1.4. The Functions a^x and $\log_a x$

Definition. (a) *For* $a > 0$ *and* $x \in R_1$, *one sets* $a^x = e^{x \log a}$.

(b) *Let* $a > 0$ *and* $a \neq 1$, *then* $\log_a x$ *is the inverse function of* a^x.

Remark 1. For $a = 1$ one has $1^x = a^x = e^0 = 1$. For $a > 0$ and $a \neq 1$, a^x is strictly monotone. Hence there exists an inverse function, which can be obtained by the reflection principle (Fig. 5.4). For $a = e$ one has $\log_e x = \log x$. From now on we will write $\log x = \ln x$.

Remark 2. One has $a^1 = e^{\ln a} = a$. Further, a^k has the usual meaning if k is an integer. This justifies the notation.

Theorem. (**a**) $a^{x+y} = a^x a^y$, $a^x b^x = (ab)^x$, $(a^x)^y = a^{xy}$ *for* $a > 0$, $b > 0$, $x \in R_1$ *and* $y \in R_1$.

(**b**) $(a^x)' = a^x \ln a$ *for* $a > 0$ *and* $x \in R_1$.

(**c**) $\log_a (xy) = \log_a x + \log_a y$ *for* $x > 0$, $y > 0$, $a > 0$, $a \neq 1$.

(**d**) $\log_a x = \dfrac{\ln b}{\ln a} \log_b x$ *for* $x > 0$, $a > 0$, $b > 0$, $a \neq 1$, $b \neq 1$.

Remark 3. The curves a^x, where $a > 0$ and $x \in R_1$, cover the upper half-plane without gaps, except for the set $\{(x, y) \mid x = 0, 0 < y < \infty, y \neq 1\}$. If for instance $x > 0$, then $a^x \to \infty$ as $a \to \infty$ and $a^x \to 0$ as $a \downarrow 0$ (Fig. 5.5).

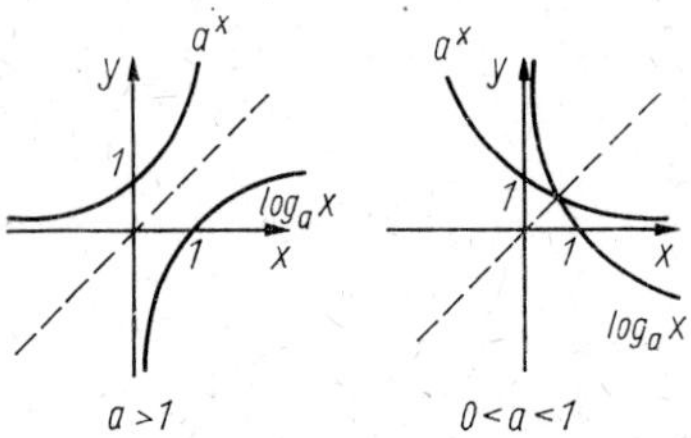

Fig. 5.4

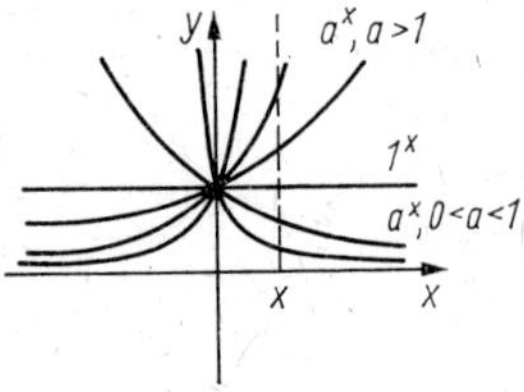

Fig. 5.5

5.1.5. The Function x^α

According to the above considerations, x^y is defined for $x > 0$ and $y \in R_1$. So far we considered $x = a$ fixed and y variable. Let us now fix $y = \alpha \in R_1$ and investigate the function x^α, with the domain of definition $D(x^\alpha) = (0, \infty)$ (Fig. 5.6).

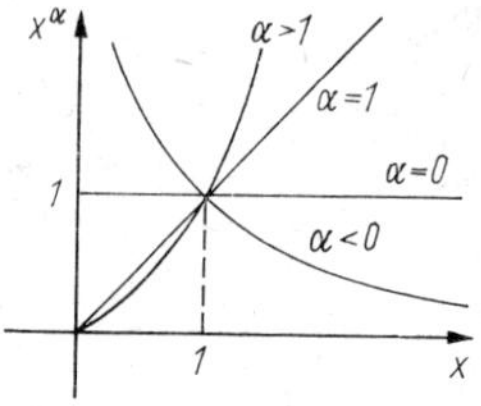

Fig. 5.6

Theorem. (**a**) $x^\alpha > 0$, $x^\alpha y^\alpha = (xy)^\alpha$, $x^\alpha x^\beta = x^{\alpha+\beta}$, $(x^\alpha)^\beta = x^{\alpha\beta}$ *for* $x > 0$, $y > 0$, $\alpha \in R_1$ *and* $\beta \in R_1$.

(**b**) *If* $\alpha > 0$, *then* x^α *is strictly monotone increasing,* $x^\alpha \to \infty$ *as* $x \to \infty$, *and* $x^\alpha \downarrow 0$ *as* $x \downarrow 0$. *If* $\alpha < 0$, *then* x^α *is strictly monotone decreasing,* $x^\alpha \downarrow 0$ *as* $x \to \infty$, *and* $x^\alpha \to \infty$ *as* $x \downarrow 0$. *Further,* $x^0 \equiv 1$.

(**c**) *For* $x > 0$ *and* $\alpha \in R_1$, $(x^\alpha)' = \alpha x^{\alpha - 1}$.

Remark 1. (a) is a mere transcription of Theorem 5.1.4(a).

Remark 2. If $\alpha = \dfrac{p}{q} > 0$ is a rational number (p and q being natural numbers), we may also write $x^\alpha = \sqrt[q]{x^p}$, and further, $x^{\frac{1}{2}} = \sqrt{x}$, $x^{-\frac{1}{2}} = \dfrac{1}{\sqrt{x}}$. Finally, one sets $0^\alpha = 0$ for $\alpha > 0$.

5.2. Trigonometric Functions

5.2.1. The Functions sin x and cos x

Definition. (**a**) sin x *is the solution, being unique on* R_1, *of the differential equation* $f''(x) = -f(x)$ *with the initial conditions* $f(0)=0$, $f'(0)=1$.

(**b**) cos x *is the solution, being unique on* R_1, *of the differential equation* $f''(x) = -f(x)$ *with the initial conditions* $f(0)=1$, $f'(0)=0$.

Remark 1. In the sense of Subsec. 4.1.2., $h(x, y, z) = y$. Hence, according to Theorem 4.2.2, $f''(x) = -f(x)$ with $f(0)=0$ and $f'(0)=1$ has a unique solution on the interval $[-N, N]$. If we let $N \to \infty$, it is observed that this assertion remains valid for R_1 instead of $[-N, N]$.

Remark 2 (geometric interpretation). Later on we will calculate the length of smooth curves L. If, for instance, L is the unit circle (circle of radius 1), then one considers the length of inscribed polygons (Fig. 5.7). The supremum of these lengths is finite, being denoted by 2π. One obtains $\pi = 3.141$ ... The method is also applicable when the length of the arc P_0P_1 shall be determined, which is denoted by x (Fig. 5.8). Consequently, $0 \leq x \leq 2\pi$. Further, $f(x)$ and $g(x)$ have the meaning indicated in Fig. 5.8, where the signs shown in Fig. 5.9 must be taken into account.

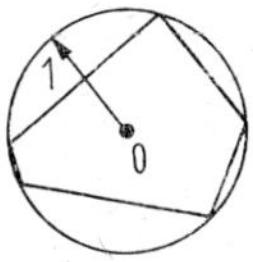

Fig. 5.7

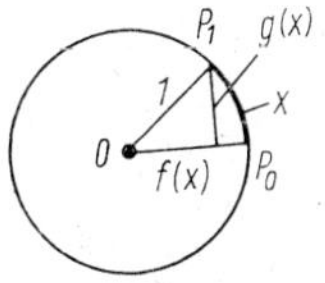

Fig. 5.8

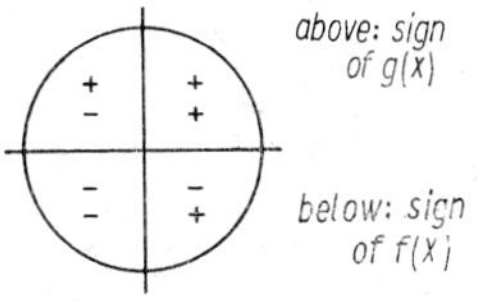

Fig. 5.9

Lemma. *For* $0 \leq x \leq 2\pi$, $g(x) = \sin x$ *and* $f(x) = \cos x$.

Remark 3. To prove this, it is shown that $f(x)$ and $g(x)$ satisfy the above differential equations and initial conditions.

Theorem. *Let* $x \in R_1$ *and* $y \in R_1$.

(**a**) $\sin x = \sin(x + 2\pi)$ (Fig. 5.10), $\cos x = \cos(x + 2\pi)$ (Fig. 5.11) (*periodic functions*).

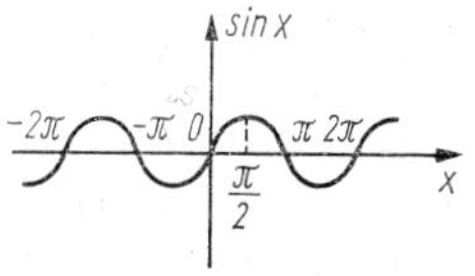

Fig. 5.10

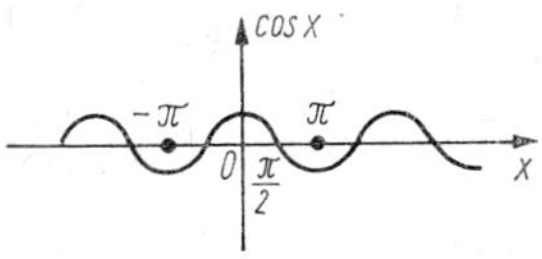

Fig. 5.11

(**b**) $\sin x = 0$ *if and only if* $x = k\pi$ (*where* k *is an integer*); $\cos x = 0$ *if and only if* $x = \frac{\pi}{2} + k\pi$ (*where* k *is an integer*).

(**c**) $\sin x = -\sin(-x)$ (*odd function with respect to* 0), $\cos x = \cos(-x)$ (*even function with respect to* 0).

(d) $\sin^2 x + \cos^2 x \equiv 1$ *for* $x \in R_1$.

(e) $\sin x = \cos\left(x - \frac{\pi}{2}\right)$.

(f) $(\sin x)' = \cos x, \quad (\cos x)' = -\sin x$.

(g) $\sin(x+y) = \sin x \cos y + \sin y \cos x, \quad \cos(x+y) = \cos x \cos y - \sin x \sin y$.

Remark 4. The theorem is proved using the existence and uniqueness theorems of Chap. 4 as well as the above lemma.

Remark 5. In particular, $|\sin x| \leq 1$ and $|\cos x| \leq 1$. Further it is concluded that $\sin(x+\pi) = -\sin x$ and $\cos(x+\pi) = -\cos x$.

5.2.2. The Functions tan *x* and cot *x*

Definition. $\tan x = \frac{\sin x}{\cos x}$, *with the domain of definition* $D(\tan x) = \bigcup_{k=-\infty}^{\infty}\left(-\frac{\pi}{2} + k\pi, \frac{\pi}{2} + k\pi\right)$ (Fig. 5.12); $\cot x = \frac{\cos x}{\sin x}$ *with the domain of definition* $D(\cot x) = \bigcup_{k=-\infty}^{\infty}(k\pi, \pi + k\pi)$ (Fig. 5.13).

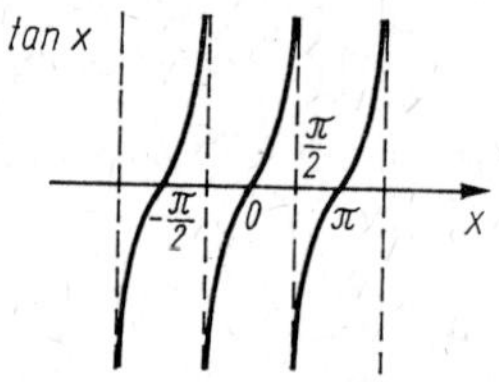

Fig. 5.12

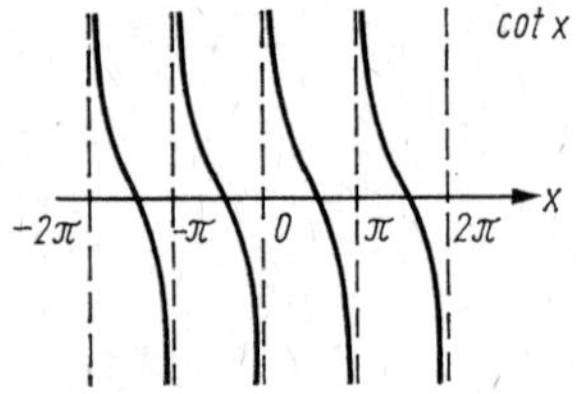

Fig. 5.13

Theorem. $\tan(x+\pi) = \tan x, \quad \cot(x+\pi) = \cot x,$

$$(\tan x)' = 1 + \tan^2 x, \quad (\cot x)' = -1 - \cot^2 x,$$

where $x \in R_1$ *is any point of the respective domain of definition.*

5.2.3. The Functions arcsin *x* and arctan *x*

Definition. $\arcsin x$ *is the inverse function of* $\sin y$, *with* $-\frac{\pi}{2} < y < \frac{\pi}{2}$ (Fig. 5.14). $\arctan x$ *is the inverse function of* $\tan y$, *with* $-\frac{\pi}{2} < y < \frac{\pi}{2}$ (Fig. 5.15).

Remark 1. $\sin y$ and $\tan y$ are strictly monotone on the interval $\left(-\frac{\pi}{2}, \frac{\pi}{2}\right)$. The inverse functions can then be constructed by the reflection principle.

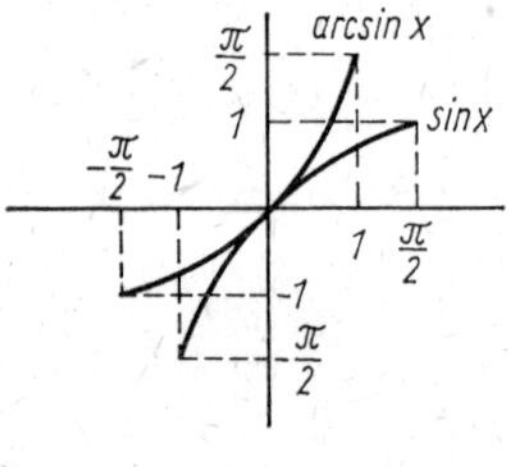

Fig. 5.14

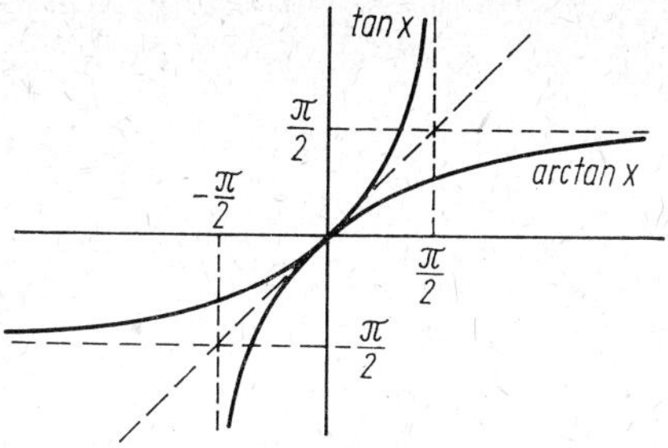

Fig. 5.15

Theorem. *$D(\arcsin x) = (-1, 1)$, and $D(\arctan x) = R_1$. Further,*

$$(\arcsin x)' = \frac{1}{\sqrt{1-x^2}} \quad \textit{and} \quad (\arctan x)' = \frac{1}{1+x^2},$$

where x is any element of the respective domain of definition.

Remark 2. By analogy, $\operatorname{arccos} x$ and $\operatorname{arccot} x$ can be introduced as inverse functions of $\cos y$ and $\cot y$, respectively, on monotony intervals.

5.2.4. The Function $e^{i\varphi}$

The arc length along the unit circle is denoted by φ (cf. Subsec. 5.2.1.), $0 \leq \varphi \leq 2\pi$. Let $z(\varphi) = \cos \varphi + i \sin \varphi$ (Fig. 5.16), where this complex function with the period 2π is extended to the whole R_1: $z(\varphi) = z(\varphi + 2k\pi)$, k being an integer, $\varphi \in R_1$.

Lemma. *$z(\varphi + \psi) = z(\varphi)\, z(\psi)$ for $\varphi \in R_1$ and $\psi \in R_1$.*

Remark 1. The lemma justifies the notation $z(\varphi) = e^{i\varphi}$. One finds $e^0 = e^{2\pi i} = 1$ and $e^{i\pi} = e^{-i\pi} = -1$.

Remark 2 (representation of complex numbers in polar coordinates). Let $z = x + iy$ be a complex number, then the distance to the origin is denoted by $r = |z|$. One has $r^2 = x^2 + y^2$, and hence, using the previously introduced calculating rules, one finds $r = |z| = \sqrt{x^2+y^2}$. The complex number $\frac{z}{r}$ lies on the unit circle (provided that $r \neq 0$; Fig. 5.17) and can thus be represented as $e^{i\varphi}$, i.e. $z = re^{i\varphi}$ (representation of complex numbers in polar coordinates). Here one may substitute $\varphi + 2k\pi$, with any integer k, for φ. Some of the calculating rules for complex numbers are simplified by the introduction of polar co-ordinates. If $z = re^{i\varphi}$, then $\bar{z} = re^{-i\varphi}$. If, additionally, $w = se^{i\psi}$, then

$$z \cdot w = rse^{i(\varphi+\psi)}.$$

Geometrically, the multiplication by w means a dilatation by $s = |w|$, combined with a rotation by the angle ψ (Fig. 5.18).

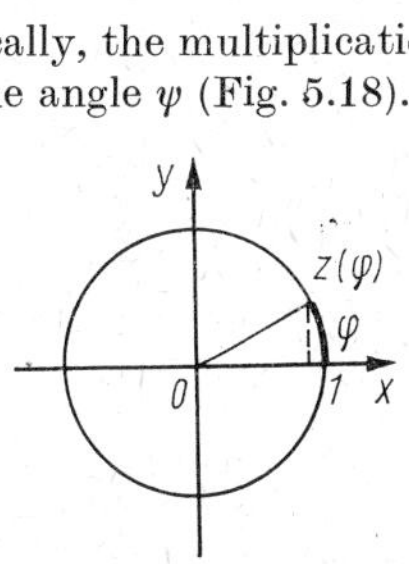

Fig. 5.16

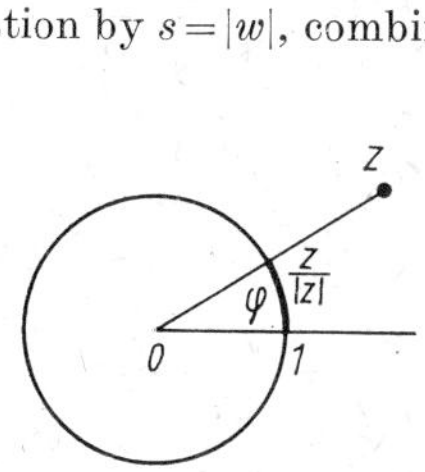

Fig. 5.17

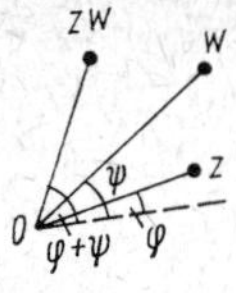

Fig. 5.18

Remark 3 (Moivre's formulas). The representation given above can also be utilized to calculate identities for trigonometric formulas. Let n be a natural number, then

$$\cos nx + \mathrm{i} \sin nx = \mathrm{e}^{\mathrm{i}nx} = (\mathrm{e}^{\mathrm{i}x})^n = (\cos x + \mathrm{i} \sin x)^n .$$

Calculating the right-hand side and comparing the real and the imaginary parts, one finds that

$$\cos nx = \cos^n x - \binom{n}{2} \cos^{n-2} x \sin^2 x + \binom{n}{4} \cos^{n-4} x \sin^4 x - \dots ,$$

$$\sin nx = n \cos^{n-1} x - \binom{n}{3} \cos^{n-3} x \sin^3 x + \binom{n}{5} \cos^{n-5} x \sin^5 x - \dots$$

Here $\binom{n}{k} = \dfrac{n!}{k!\,(n-k)!}$ are binomial coefficients, for $1 \leq k \leq n$.

5.3. Exponential Functions and Power Functions (Complex-Valued)

5.3.1. The Functions e^z and $\ln z$

In this section we will consider complex-valued functions $f(z)$, with the complex plane C_1 or $C_1 - \{0\} = \{w \mid w \in C_1, w \neq 0\}$ as their domains of definition $D(f(z))$. As usual, $z = x + iy = \mathrm{Re}\, z + \mathrm{i}\, \mathrm{Im}\, z = r\mathrm{e}^{\mathrm{i}\varphi}$, where the last expression is the representation in polar co-ordinates as given in Subsec. 5.2.4.

Definition. (a) $\mathrm{e}^z = \mathrm{e}^x \mathrm{e}^{\mathrm{i}y}$, with $D(\mathrm{e}^z) = C_1$.
(b) $\ln z = \ln r + \mathrm{i}\varphi$, $0 \leq \varphi < 2\pi$, with $D(\ln z) = C_1 - \{0\}$.

Remark. e^x, $\mathrm{e}^{\mathrm{i}y}$, $\ln r$ for $r > 0$ were introduced previously.

Riemann surface for $\ln z$. One has

$$z = r\mathrm{e}^{\mathrm{i}\varphi} = r\mathrm{e}^{\mathrm{i}\varphi + 2k\pi\mathrm{i}} ,$$

where k is any integer. The restriction to $0 \leq \varphi < 2\pi$ in the definition of $\ln z$ appears artificial. However, if the definition of $\ln z$ is extended to $\varphi \in R_1$, one obtains a many-valued function, which is undesirable. To avoid this dilemma, one introduces Riemann surfaces. In the case of the function $\ln z$ this is done as follows (Fig. 5.19): an infinite number of complex planes C_1 are laid one on top of another, numbered by $k = 0, 1, -1, 2, -2, \dots$, slit along their positive real half-axes, and the cutting edges of these sheets are cross-connected to one another according to the instruction given in Fig. 5.19. The integers indicated in Fig. 5.19 are the

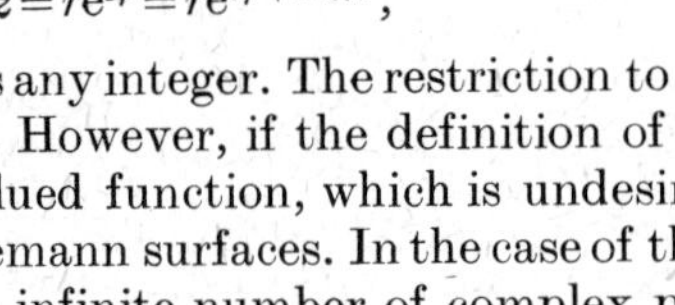

Fig. 5.19

sheet indices, and the pairs of corresponding numbers arranged one above the other show from which sheet to which other sheet one has to move when crossing the slit. Let w be such a path, then this means that, when moving from A to B, one changes from sheet No. 0 to sheet No. 1 or, generally, from sheet No. k to sheet No. $k+1$. Conversely, when moving from B to A, one changes, for example, from sheet No. 17 to sheet No. 16. Now the convention is made that in the k-th sheet z shall be represented by

$$z = re^{i\varphi}, \quad \text{where} \quad 2k\pi \leqq \varphi < 2(k+1)\pi .$$

On this Riemann surface it is now possible to represent

$$\ln z = \ln r + i\varphi, \qquad \varphi \in R_1, \qquad r > 0 ,$$

as a single-valued function (0 does not belong to the surface). If one moves continuously along the Riemann surface, the (single-valued) function varies continuously in z.

Theorem. (a) *For* $z \in C_1$ *and* $w \in C_1$, $e^z e^w = e^{z+w}$.

(b) $\ln z$ *is the "inverse function" of* e^z *in the sense that* $e^{\ln z} = z$ *for all* $z \neq 0$, *independently of the particular sheet selected.*

5.3.2. The Functions z^w, Riemann Surfaces

Till now the function x^y has been defined only for $x > 0$ and $y \in R_1$.

Definition. *Let z and w be any two complex numbers, and let $z \neq 0$. On the Riemann surface for* $\ln z$ *one sets* $z^w = e^{w \ln z}$.

Remark 1. If w is fixed, then the domain of definition of the function $f(z) = z^w$ is the Riemann surface for $\ln z$. If $(z^w)_k$ is the value of the function on the k-th sheet, then using $0 \leqq \varphi < 2\pi$ and $\ln z = \ln r + i\varphi$ one obtains

$$(z^w)_k = e^{w \ln r + wi\varphi + wi2k\pi} = (z^w)_0 \, e^{iw2k\pi} . \tag{1}$$

Remark 2. Let $z = x > 0$ and $w = y \in R_1$, then one obtains $(z^w)_0 = e^{y \ln x} = x^y$. This justifies the above definition.

Lemma. *Let w_1 and w_2 be any two complex numbers, then $z^{w_1} z^{w_2} = z^{w_1 + w_2}$ on the Riemann surface for* $\ln z$.

Remark 3. One finds (Fig. 5.20)

$$i^i = e^{i \ln i} = e^{ii\left(\frac{\pi}{2} + 2k\pi\right)} = e^{-\frac{\pi}{2} - 2k\pi} .$$

Consequently, on the 0-th sheet one obtains $i^i = e^{-\frac{\pi}{2}}$, an interesting formula!

Remark 4. If $w = \alpha$ is real, then $|e^{i\alpha 2\pi k}| = 1$. Then it follows from (1) that the values $(z^\alpha)_k$ lie on a circle of radius $|(z^\alpha)_0|$ (Fig. 5.21).

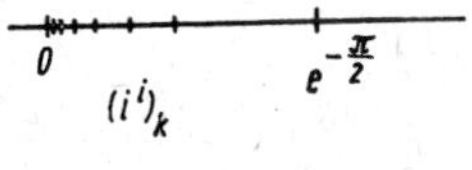

Fig. 5.20

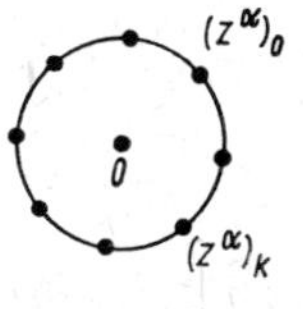

Fig. 5.21

Remark 5. Of special interest is the case where $w=\frac{p}{q}$ is rational, p being an integer and q a natural number, p and q coprime. If l and k are two integers with $l=k+qn$, n integer, then (1) implies that $(z^{\frac{p}{q}})_k=(z^{\frac{p}{q}})_{k+nq}$. Hence $z^{\frac{p}{q}}$ has only q different values, which may be denoted by $(z^{\frac{p}{q}})_0$, $(z^{\frac{p}{q}})_1$, ..., $(z^{\frac{p}{q}})_{q-1}$. Consequently, to represent $z^{\frac{p}{q}}$ as a single-valued function requires not the whole Riemann surface for $\ln z$, but only q sheets of it, e.g., the sheets with the indices 0, 1, ..., $q-1$ (Fig. 5.22). From sheet No. $q-1$ one returns to sheet No. 0, as is shown by the instruction indicated in this figure. The surface of existence (domain of definition) of $\sqrt{z}=z^{\frac{1}{2}}$ then consists of 2 sheets; for example one has $(\sqrt{1})_0=1$ and $(\sqrt{1})_1=e^{i\pi}=-1$ (Fig. 5.23).

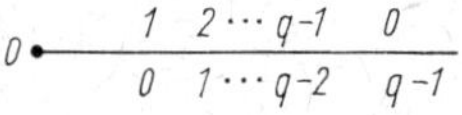

Fig. 5.22

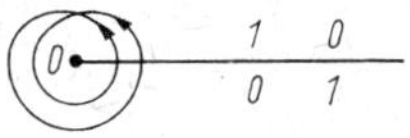

Fig. 5.23

5.3.3. Roots of Unity, Fundamental Theorem of Algebra

Definition. *Let n be a natural number, then the complex numbers $1^{\frac{1}{n}}$ are called the n-th roots of unity.*

Lemma. *There exist n different n-th roots of unity, namely $e^{\frac{2\pi}{n}ki}$, where $k=0, \ldots, n-1$.*

Remark 1. The lemma is an immediate consequence of the considerations given in Subsec. 5.3.2. The roots of unity define the corners of a regular polygon.

Theorem 1. *The polynomial z^n-1 has exactly n different zeros in C_1 (i.e., complex numbers z which satisfy $z^n-1=0$). These zeros are the n-th roots of unity* (Fig. 5.24).

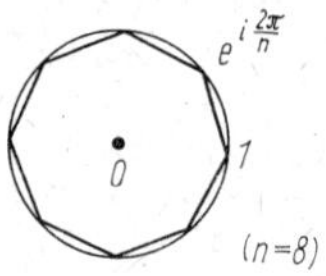

Fig. 5.24

Remark 2. The theorem is a consequence of the above considerations. Now the question arises whether the situation is similar for arbitrary complex polynomials of degree n, $\sum_{j=0}^{n} a_j z^j$, a_j complex. We may suppose that $a_n \neq 0$. Here z_1 is called a zero if $\sum_{j=0}^{n} a_j z_1^j=0$.

Theorem 2 (*fundamental theorem of algebra*). $\sum_{j=0}^{n} a_j z^j$, *a_j complex, $a_n \neq 0$, has exactly n complex zeros. Let $z_1, \ldots, z_n$ be these zeros, then*

$$\sum_{j=0}^{n} a_j z^j = a_n (z-z_1)(z-z_2) \ldots (z-z_n) . \tag{1}$$

Remark 3. It may happen that some (or even all) of the zeros coincide. If, for example, $z_1=z_2=\ldots=z_r$, but $z_k \neq z_1$ for $k=r+1, \ldots, n$, then z_1 is said to be a zero of multiplicity (or order) r. If $\sum_{j=0}^{n} a_j z^j$ has exactly l different zeros $w_1, \ldots, w_l$ with the multiplicities $n_1, \ldots, n_l$, then (1) becomes

$$\sum_{j=0}^{n} a_j z^j = a_n (z-w_1)^{n_1} (z-w_2)^{n_2} \dots (z-w_l)^{n_l}, \quad (a_n \neq 0). \tag{2}$$

Here the n_r's are natural numbers, and $n_1+n_2+\dots+n_l=n$. Hence, taken into account the multiplicities, the polynomial $\sum_{j=0}^{n} a_j z^j$, a_j complex, $a_n \neq 0$, has exactly n complex zeros.

Remark 4. The proof of the theorem is relatively easy if it is supposed that the polynomial has at least one complex zero.The best way to verify this assumption is in terms of complex function theory, cf. Chap. 15. We shall return to this point in Subsec. 15.3.6. and in Remark 15.5.2/2.

Theorem 3. *Let a_j, $j=0, \dots, n$, be real numbers, and let $a_n \neq 0$.*

(**a**) *If w is a zero of multiplicity m of the polynomial $\sum_{j=0}^{n} a_j z^j$, then $\bar{w}$ is likewise a zero of the same multiplicity.*

(**b**) *If $w_1, \dots, w_l$ are the different real zeros with the multiplicities $n_1, \dots, n_l$, and if $v_1, \bar{v}_1, v_2, \bar{v}_2, \dots, v_k, \bar{v}_k$ are the different properly complex zeros* (*i.e.*, $\operatorname{Im} v_r \neq 0$) *with the multiplicities $m_1, \dots, m_k$ for the polynomial $\sum_{j=0}^{n} a_j z^j$, then*

$$\sum_{j=0}^{n} a_j z^j = a_n (z-w_1)^{n_1} \dots (z-w_l)^{n_l} [(z-\operatorname{Re} v_1)^2 + (\operatorname{Im} v_1)^2]^{m_1} \dots \tag{3}$$

$$\dots [(z-\operatorname{Re} v_k)^2 + (\operatorname{Im} v_k)^2]^{m_k}.$$

Remark 5. If $z=x$ is real, then (3) is a purely real decomposition, which is desirable for real polynomials.

5.4. Power Series

5.4.1. Radius of Convergence

Let us consider power series $P(z)=\sum_{j=0}^{\infty} a_j (z-z_0)^j$ defined on the complex plane C_1. Here the coefficients a_j are complex numbers, $z_0 \in C_1$ is fixed and $z \in C_1$ is variable. Trivially, the series converges for $z=z_0$. For what other values of z does the series converge on C_1? Of special interest is absolute convergence, i.e.,

$$\sum_{j=0}^{\infty} |a_j| \, |z-z_0|^j < \infty. \tag{1}$$

Let $\{c_j\}_{j=0}^{\infty}$ be a sequence of real numbes, the $\overline{\lim}\, c_j$ (upper limit) denotes the largest, and $\underline{\lim}\, c_j$ (lower limit) denotes the smallest accumulation point of the numbers c_j. With a_j denoting the coefficients of the above power series, one sets

$$R = (\overline{\lim}\, |a_j|^{\frac{1}{j}})^{-1} \quad \text{(radius of convergence)}, \tag{2}$$

where the convention is made that $R=\infty$ for $\overline{\lim}\, |a_j|^{\frac{1}{j}}=0$ and $R=0$ for $\overline{\lim}\, |a_j|^{\frac{1}{j}}=\infty$.

Theorem. *For* $|z-z_0|<R$, *the series* $\sum_{j=0}^{\infty} a_j (z-z_0)^j$ *converges absolutely. It does not converge absolutely for* $|z-z_0|>R$.

Remark 1. The theorem justifies the term "radius of convergence" for R. Whether or not (1) is true for $|z-z_0|=R$ cannot be stated generally. Either of the two cases may occur.

Remark 2 (examples of power series and associated radii of convergence).

(a) $\sum_{j=0}^{\infty} z^j$ and $R=1$; (b) $\sum_{j=1}^{\infty} j^j z^j$ and $R=0$;

(c) $\sum_{j=1}^{\infty} j^{-j} z^j$ and $R=\infty$.

5.4.2. Addition and Multiplication of Power Series

Let $\sum_{j=0}^{\infty} a_j(z-z_0)^j$ and $\sum_{j=0}^{\infty} b_j(z-z_0)^j$ be two power series with the radii of convergence R_a and R_b respectively.

Theorem. (a) *If* λ *and* μ *are complex numbers, and if* R *is the radius of convergence of* $\sum_{j=0}^{\infty} (\lambda a_j+\mu b_j)(z-z_0)^j$, *then* $R \geqq \min(R_a, R_b)$.

(b) *If* R *is the radius of convergence of the series*

$$\left[\sum_{j=0}^{\infty} a_j (z-z_0)^j\right] \cdot \left[\sum_{j=0}^{\infty} b_j (z-z_0)^j\right] = \sum_{j=0}^{\infty} (a_0 b_j + a_1 b_{j-1} + \ldots + a_j b_0)(z-z_0)^j,$$

then $R \geqq \min(R_a, R_b)$.

5.4.3. Differentiation of Function Sequences and Power Series

In this subsection we consider real functions $f \in C[a, b]$ with $-\infty<a<b<\infty$. As previously, $f' \in C[a, b]$ means that f is continuously differentiable on (a, b) and that f' has a limit on the right at a and a limit on the left at b. The symbol $\Rightarrow$ denotes uniform convergence on $[a, b]$.

Theorem 1. *For* $j=1, 2, 3, \ldots$, *suppose that* $f_j \in C[a, b]$ *and* $f_j' \in C[a, b]$, *and that* $f_j \Rightarrow f$ *and* $f_j' \Rightarrow g$. *Then* $f \in C[a, b]$ *and* $g \in C[a, b]$. *Further,* f *is differentiable on* (a, b), *and* $f'=g$.

Theorem 2. *For* $j=1, 2, 3, \ldots$, *suppose that* $f_j \in C[a, b]$ *and* $f_j' \in C[a, b]$. *Further let* $\sum_{j=1}^{\infty} f_j(x)$ *and* $\sum_{j=1}^{\infty} f_j'(x)$ *converge uniformly. Then* $\sum_{j=1}^{\infty} f_j(x)$ *is differentiable on* (a, b), *and*

$$\left(\sum_{j=1}^{\infty} f_j(x)\right)' = \sum_{j=1}^{\infty} f_j'(x) \quad \textit{(term-by-term differentiation)}.$$

Remark 1. We regard real power series, $\sum_{j=0}^{\infty} a_j (x-x_0)^j$, a_j real, as cut-outs of complex power series. R is again the radius of convergence, which now means absolute convergence on the interval (x_0-R, x_0+R) (Fig. 5.25).

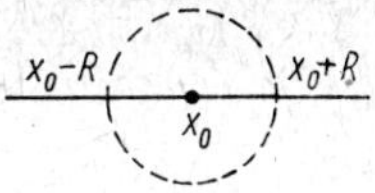

Fig. 5.25

Theorem 3. *Let* $f(x)=\sum_{j=0}^{\infty} a_j (x-x_0)^j$, *$a_j$ real, with $R>0$, then $f(x)$ has derivatives up to arbitrary order on the interval (x_0-R, x_0+R), and*

$$f^{(k)}(x)=\sum_{j=k}^{\infty} j(j-1)\dots(j-k+1)\,a_j\,(x-x_0)^{j-k} \tag{1}$$

(term-by-term differentiation) for $k=1, 2, 3, \dots$, where all of the above series converge absolutely on (x_0-R, x_0+R).

Remark 2. There arises the question whether the converse of the last theorem is also true: consider a function which has derivatives up to arbitrary order on (x_0-R, x_0+R). Can every function with this property be expanded in a power series? In general this is not true.

5.4.4. Taylor Series

The question put at the end of Subsec. 5.4.3. shall now be investigated more closely. Setting $x=x_0$ in Eq. (5.4.3./1), one obtains $f^{(k)}(x_0)=a_k k!$, and hence

$$f(x)=\sum_{k=0}^{\infty}\frac{f^{(k)}(x_0)}{k!}(x-x_0)^k \quad \text{(setting } 0!=1) . \tag{1}$$

This expression is called a Taylor series. The question put at the end of Subsec. 5.4.3. now reads: when does the Taylor series converge, and when does it converge towards $f(x)$?

Theorem 1. *Let $-\infty<x_0<x_1<\infty$. Further let $k=0, 1, 2, \dots$ Suppose that the function f has continuous derivatives up to the order $(k+1)$ on (x_0, x_1), where $f^{(j)}\in C[x_0, x_1]$ for $j=0, \dots, k+1$. Then*

$$f(x)=\sum_{j=0}^{k}\frac{f^{(j)}(x_0)}{j!}(x-x_0)^j+\frac{1}{k!}\int_{x_0}^{x}(x-y)^k f^{(k+1)}(y)\,dy \tag{2}$$

for all x with $x_0<x<x_1$. Further there exists a function $\vartheta=\vartheta(x)$ with $0\leqq\vartheta\leqq 1$, such that

$$f(x)=\sum_{j=0}^{k}\frac{f^{(j)}(x_0)}{j!}(x-x_0)^j+\frac{(x-x_0)^{k+1}}{(k+1)!}f^{(k+1)}(x_0+\vartheta(x-x_0)) \tag{3}$$

for all x with $x_0<x<x_1$.

Remark 1. The last summand in either of the formulas (2) and (3) above is called remainder. Thus the question for the validity of (1) reduces to the question of whether or not these remainders approach zero as $k\to\infty$ (provided that f has derivatives up to arbitrary order).

Remark 2. If f has continuous derivatives up to the order $k+1$ on a neighbourhood $(x_0-\varepsilon, x_0+\varepsilon)$, $\varepsilon>0$, then (2) and (3) hold for $x\in(x_0-\varepsilon, x_0+\varepsilon)$, where in (2) for $x<x_0$ one has to replace $\int_{x_0}^{x}$ by $-\int_{x}^{x_0}$.

Theorem 2. *Let* $-\infty < x_1 < x_2 < \infty$. *Let the function* $f(x)$ *have derivatives up to arbitrary order on* (x_1, x_2), *and let* $x \in (x_1, x_2)$ *and* $x_0 \in (x_1, x_2)$. *If*

$$\frac{|x-x_0|^{k+1}}{(k+1)!} \sup_{y \in I(x_0,x)} |f^{(k+1)}(y)| \to 0 \quad as \quad k \to \infty \tag{4}$$

where $I(x_0, x) = (x_0, x)$ *for* $x_0 < x$ *and* $I(x_0, x) = (x, x_0)$ *for* $x < x_0$, *then* $f(x)$ *can be represented by* (1).

Remark 3. This theorem is an immediate consequence of (3).

5.4.5. Examples of, and Counter-Examples to Taylor Series

Theorem 1. *On* R_1, e^x, $\sin x$ *and* $\cos x$ *can be expanded in absolutely convergent power series. For* $x \in R_1$,

$$e^x = \sum_{j=0}^{\infty} \frac{x^j}{j!}, \quad \sin x = x - \frac{x^3}{3!} + \frac{x^5}{5!} - \frac{x^7}{7!} + \dots, \quad \cos x = 1 - \frac{x^2}{2!} + \frac{x^4}{4!} - \frac{x^6}{6!} + \dots$$

Remark 1. In terms of Theorem 5.4.4/2, one has $x_0 = 0$, and (5.4.4/4) applies.

Theorem 2. *Let* $\alpha > 0$, *then*

$$\lim_{x\to\infty} \frac{e^x}{x^\alpha} = \infty, \quad \lim_{x\to\infty} e^{-x} x^\alpha = 0, \quad \lim_{x\to\infty} \frac{\ln x}{x^\alpha} = 0, \quad and \quad \lim_{x \downarrow 0} x^\alpha |\ln x| = 0 .$$

Remark 2. The first proposition follows from Theorem 1; the other propositions are derived from the first one.

Theorem 3 (*counter-example*). *The function*

$$f(x) = \begin{cases} e^{-\frac{1}{x}} & for \quad x > 0 \\ 0 & for \quad x \leq 0 \end{cases} \qquad \text{Fig. (5.26)}$$

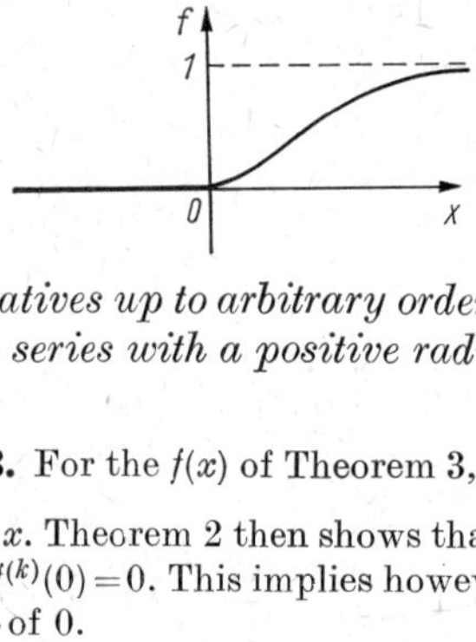

Fig. 5.26

has derivatives up to arbitrary order on R_1, *but cannot, at the point* 0, *be expanded in a Taylor series with a positive radius of convergence.*

Remark 3. For the $f(x)$ of Theorem 3, $f^{(k)}(x) = \dfrac{e^{-\frac{1}{x}}}{x^{k+1}} P_k(x)$ for $x > 0$, where $P_k(x)$ is a polynomial in x. Theorem 2 then shows that f has derivatives up to arbitrary order at the point 0 and that $f^{(k)}(0) = 0$. This implies however that (5.4.4/1) with $x_0 = 0$ cannot be true in a neighbourhood of 0.

5.4.6. Power Series for e^z, Analytic Functions

It turns out that the considerations on real functions, as described in the preceding subsections, can be extended to complex functions. In Chap. 15 we shall deal with the subject systematically. Some statements can however be made already now.

Definition. *Let G be an open domain in the complex plane C_1. A complex-valued function $f(z)$ is called analytic on G if at every point $z_0 \in G$ it is possible to expand $f(z)$ in an absolutely convergent power series $\sum_{j=0}^{\infty} a_j (z-z_0)^j$ with a positive radius of convergence.*

Theorem. e^z *is an analytic function on* $G = C_1$, *and*

$$e^z = \sum_{j=0}^{\infty} \frac{z^j}{j!} \quad \textit{for all} \quad z \in C_1 .$$

Remark. The other complex functions considered previously are analytic, too, provided that the above definition is slightly modified. Thus $\ln z$ and z^w are analytic on the Riemann surface for $\ln z$.

5.4.7. Irrationality of e

A real number which is not rational is called irrational.

Theorem. e *is an irrational number.*

Remark. The proof is based on the representation $e = \sum_{k=0}^{\infty} \frac{1}{k!}$.

6. Banach Spaces

6.1. Definitions and Examples

6.1.1. Definitions

The concept of a vector space is well known from analytical geometry: it is a nonvoid set M in which two operations are defined, namely addition of arbitrary elements of M and multiplication of elements of M by real or complex numbers. The elements of M must have the following properties with respect to addition:

$$x + y = y + x \quad \text{(commutativity)},$$
$$x + o = x \quad \text{(existence of a neutral element } o \in M),$$
$$(x+y)+z = x+(y+z) \quad \text{(associativity)}.$$

Here x, y and z are arbitrary elements of M. Further for every pair of elements $x \in M$ and $y \in M$ there must exist a unique element $z \in M$ such that $x + z = y$. With respect to multiplication by real or complex numbers, the elements of M

must have the following properties:

$$\lambda(x+y)=\lambda x+\lambda y, \quad (\lambda+\mu)x=\lambda x+\mu x ,$$
$$1 \cdot x=x, \quad \lambda(\mu x)=(\lambda\mu)x .$$

Here $x \in M$ and $y \in M$, and λ and μ are real or complex numbers. If only real numbers are admitted for the multiplication, then M is called a real vector space. If complex numbers are admitted for the multiplication, M is called a complex vector space. From now on we shall write 0 instead of o. Further we write $z=-y$ if $y+z=0$.

Definition 1. *A (real or complex) normed space is a (real or complex) vector space M on which a function is defined which assigns to every $x \in M$ a non-negative number $\|x\|$ having the properties*

1. $\|x\| \geqq 0$, *where* $\|x\|=0$ *if and only if* $x=0$,
2. $\|\lambda x\|=|\lambda|\;\|x\|$,
3. $\|y+x\| \leqq \|x\|+\|y\|$ *(triangle inequality).*

Here $x \in M$, $y \in M$, and λ is real or complex. $\|x\|$ is called the norm of x.

Lemma. *With $\varrho(x, y)=\|x-y\|$, a normed space becomes a metric space.*

Remark. The lemma is a consequence of Def. 1.3.3/1. From now on we shall regard normed spaces as special metric spaces.

Definition 2. *A (real or complex) Banach space is a (real or complex) normed space which (regarded as a metric space) is complete.*

6.1.2. Examples

The space R_n. With $\|x\|=\sum_{j=1}^{n}|x_j|$ for $x=(x_1, \ldots, x_n)$, R_n is a Banach space; cf. Lemma 2.5.1/2. Here the addition of $x=(x_1, \ldots, x_n)$ and $y=(y_1, \ldots, y_n)$ as well as the multiplication by real numbers λ are defined in a natural way: $x+y=(x_1+y_1, \ldots, x_n+y_n)$, $\lambda x=(\lambda x_1, \ldots, \lambda x_n)$.

The space C_n. If one considers complex sequences $x=(x_1, \ldots, x_n)$ and multiplication by complex numbers λ, then C_n is a Banach space with respect to the norm $\|x\|$ defined above. (This is a slight modification to Subsec. 1.3.2.)

The space $C[a, b]$. With $\|f\|_{C[a,b]}=\sup_{x \in [a,b]}|f(x)|$, the space $C[a, b]$ defined in Subsec. 2.5.2. is a Banach space; cf. Theorem 2.5.2. Here the addition of functions and the multiplication by real numbers are defined in a natural way:

$$(f+g)(x)=f(x)+g(x), \quad (\lambda f)(x)=\lambda f(x) .$$

6.2. Spaces of Type l_p

6.2.1. Inequalities

Lemma. *For $x>0$, $y>0$, $1<p<\infty$ and $1/p+1/p'=1$, the following relation holds:*
$$xy \leqq x^p/p+y^{p'}/p'.$$

Remark 1. p', with $1/p+1/p'=1$, is also called the conjugate number of p.

Theorem. *Let $x_1, \ldots, x_n$ and $y_1, \ldots, y_n$ be complex numbers, and let $1<p<\infty$ and*

$1/p+1/p'=1$, *then*

$$\left|\sum_{j=1}^{n} x_j y_j\right| \leqq \sum_{j=1}^{n} |x_j|\,|y_j| \leqq \left(\sum_{j=1}^{n} |x_j|^p\right)^{1/p} \left(\sum_{j=1}^{n} |y_j|^{p'}\right)^{1/p'} \tag{1}$$

and

$$\left(\sum_{j=1}^{n} |x_j+y_j|^p\right)^{1/p} \leqq \left(\sum_{j=1}^{n} |x_j|^p\right)^{1/p} + \left(\sum_{j=1}^{n} |y_j|^p\right)^{1/p}. \tag{2}$$

Remark 2. (1) is called Hölder's inequality, (2) is called Minkowski's inequality.

6.2.2. The Spaces $l^n_{p,R}$, $l^n_{p,C}$, $l_{p,R}$, and $l_{p,C}$

Let us consider vector spaces whose elements are finite or infinite sequences $x=(x_1, x_2, \ldots)$ of real or complex numbers. The addition of $x=(x_1, x_2, \ldots)$ and $y=(y_1, y_2, \ldots)$ is defined by $x+y=(x_1+y_1, x_2+y_2, \ldots)$, and the multiplication by real or complex numbers λ is defined by $\lambda x=(\lambda x_1, \lambda x_2, \ldots)$.

Definition 1. (**a**) *Let n be a natural number and let $1 \leqq p < \infty$, then $l^n_{p,R}$ (or $l^n_{p,C}$) is the real (or complex) vector space of the real (or complex) sequences $x=(x_1, \ldots, x_n)$, with* $\|x\|_{l_p^n}=\left(\sum_{j=1}^{n} |x_j|^p\right)^{1/p}$.

(**b**) *Let $1 \leqq p < \infty$, then $l_{p,R}$ (or $l_{p,C}$) is the real (or complex) vector space of the real (or complex) sequences $x=(x_1, x_2, \ldots)$ with* $\|x\|_{l_p}=\left(\sum_{j=1}^{\infty} |x_j|^p\right)^{1/p}<\infty$.

Remark 1. According to Subsecs. 1.3.1. and 6.1.2., $R_n=l^n_{1,R}$ and $C_n=l^n_{1,C}$.

Lemma. *Let n be a natural number, and $1 \leqq p < \infty$, then, for all $x \in l^n_{1,C}$,*

$$\|x\|_{l_1^n} \leqq n\|x\|_{l_p^n} \quad \text{and} \quad \|x\|_{l_p^n} \leqq \|x\|_{l_1^n}.$$

Remark 2. Letting $n \to \infty$ in the second inequality, one obtains $\|x\|_{l_p} \leqq \|x\|_{l_1}$ for all $x \in l_{1,C}$.

Definition 2. *Two norms $\|x\|_1$ and $\|x\|_2$ in a vector space M are said to be equivalent if there exist two positive numbers c_1 and c_2 such that*

$$c_1\|x\|_1 \leqq \|x\|_2 \leqq c_2\|x\|_1$$

for all $x \in M$.

Remark 3. The two normed spaces of Def. 2 are regarded as special metric spaces with the metrics $\varrho_j(x, y)=\|x-y\|_j$, $j=1, 2$. Cauchy sequences with respect to ϱ_1 are also Cauchy sequences with respect to ϱ_2, and vice versa. An analogous statement can be made for convergent sequences. Thus the norms $\|x\|_1$ and $\|x\|_2$ are fully equivalent with respect to convergence behaviour.

Remark 4. The above lemma implies that all norms $\|x\|_{l_p^n}$ with $1 \leqq p < \infty$ are equivalent in $l^n_{1,R}$ or in $l^n_{1,C}$.

Theorem. (**a**) *If n is a natural number and $1 \leqq p < \infty$, then $l^n_{p,R}$ is a real Banach space. If $1 \leqq q < \infty$, then $\|x\|_{l_p^n}$ is an equivalent norm in $l^n_{p,R}$.*

(**b**) *If n is a natural number and $1 \leqq p < \infty$, then $l^n_{p,C}$ is a complex Banach space. If $1 \leqq q < \infty$, then $\|x\|_{l_q^n}$ is an equivalent norm in $l^n_{p,C}$.*

(**c**) *If $1 \leqq p < \infty$, then $l_{p,R}$ is a real, and $l_{p,C}$ is a complex Banach space.*

Convention. R_n can be normed using one of the equivalent norms $\|x\|_{l_p^n}$. From now on we make the convention to set

$$\|x\|_{R_n} = \left(\sum_{j=1}^{n} |x_j|^2\right)^{1/2}, \quad \text{therefore} \quad R_n = l_{2,R}^n .$$

An analogous convention is made for C_n. Further we write l_p^n instead of $l_{p,R}^n$ or $l_{p,C}^n$, because it will be clear from the context whether a real or a complex space is concerned. By analogy we write l_p instead of $l_{p,R}$ or $l_{p,C}$.

7. Integral Calculus in R_1 (continued)

7.1. Classes of Integrable Functions

7.1.1. General Rules (Integration by Parts, Change of the Variable)

Theorem 1. *Let* $-\infty < a < b < \infty$. *If* f, f', g *and* g' *are elements of* $C[a, b]$, *then*

$$\int_a^b fg' \mathrm{d}x = f(b)\, g(b) - f(a)\, g(a) - \int_a^b f' g \mathrm{d}x .$$

Remark 1. This rule is called integration by parts. It can sometimes be used to reduce unknown integrals to known ones. To give an example, let $0 < a < b < \infty$, and let α be real, $\alpha \neq -1$, then

$$\int_a^b x^\alpha \ln x \mathrm{d}x = \int_a^b \left(\frac{x^{\alpha+1}}{\alpha+1}\right)' \ln x \, \mathrm{d}x = \frac{x^{\alpha+1}}{\alpha+1} \ln x|_a^b - \int_a^b \frac{x^{\alpha+1}}{\alpha+1} \frac{1}{x} \mathrm{d}x .$$

Here one sets $h(x)|_a^b = h(b) - h(a)$. The integral on the right-hand side is already known:

$$\int_a^b x^\beta \mathrm{d}x = \begin{cases} \left.\dfrac{x^{\beta+1}}{\beta+1}\right|_a^b & \text{for } \beta \neq -1 \\ \ln x|_a^b & \text{for } \beta = -1 . \end{cases} \tag{1}$$

Change of the Variable. Let $-\infty < a < b < \infty$, and let $\varphi \in C[a, b]$ and $\varphi' \in C[a, b]$ with $\varphi'(x) > 0$ for $x \in [a, b]$. Then φ is strictly increasing (Fig. 7.1). Let $\psi = \varphi^{-1}$ be the inverse function. Then $\psi \in C[\varphi(a), \varphi(b)]$ and $\psi' \in C[\varphi(a), \varphi(b)]$ (cf. Subsec. 3.1.4.).

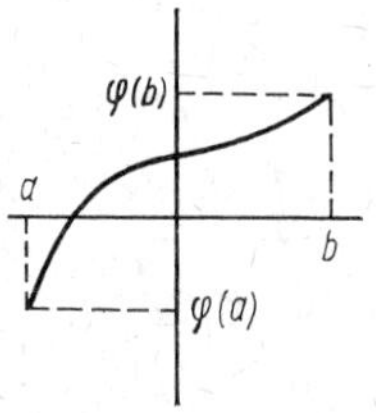

Fig. 7.1

Theorem 2. *Let $f \in C[a, b]$, then $g(x) = f(\psi(x)) \in C[\varphi(a), \varphi(b)]$, and*

$$\int_a^b f(x)\,dx = \int_{\varphi(a)}^{\varphi(b)} g(x)\,\psi'(x)\,dx\,.$$

Remark 2. If $\varphi \in C[a, b]$, $\varphi' \in C[a, b]$ and $\varphi'(x) < 0$ on $[a, b]$, then φ is strictly decreasing. Then it follows from $f \in C[a, b]$ that $g(x) = f(\psi(x)) \in C[\varphi(b), \varphi(a)]$ and

$$\int_a^b f(x)\,dx = -\int_{\varphi(b)}^{\varphi(a)} g(x)\,\psi'(x)\,dx = \int_{\varphi(b)}^{\varphi(a)} g(x)\,|\psi'(x)|\,dx\,.$$

7.1.2. Integration of Rational Functions, Partial Fraction Decomposition

Quotients of polynomials are called rational functions,

$$R(x) = \frac{P(x)}{Q(x)} = \frac{\sum_{k=0}^{n} a_k x^k}{\sum_{l=0}^{m} b_l x^l}, \quad x \in R_1$$

where, for the time being, a_k and b_k are assumed to be arbitrary complex numbers, with $b_m \neq 0$. Further let $n = 0, 1, 2, \ldots$ and $m = 0, 1, 2, \ldots$. Then, according to (5.3.3/2),

$$Q(x) = b_m\,(x-\alpha_1)^{m_1}\,(x-\alpha_2)^{m_2} \ldots (x-\alpha_j)^{m_j}\,,$$

where $\alpha_1, \ldots, \alpha_j$ are the different (complex) zeros of $Q(x)$, with the multiplicities $m_1, \ldots, m_j$.

Theorem (*decomposition into partial fractions*). (**a**) *There exists a unique representation of $R(x)$ as*

$$R(x) = p(x) + \sum_{\mu=1}^{j} \sum_{\nu=1}^{m_\mu} \frac{A_{\mu,\nu}}{(x-\alpha_\mu)^\nu}, \tag{1}$$

where $p(x)$ is a polynomial and the $A_{\mu,\nu}$ are complex numbers.

(**b**) *If all the coefficients a_k and b_l of $R(x)$ are real numbers, and if $\alpha_r = \bar{\alpha}_s$ (conjugate complex zeros of $Q(x)$), then $A_{r,\nu} = \bar{A}_{s,\nu}$ for $\nu = 1, \ldots, m_r$.*

Remark 1. (1) is meaningful for all $x \neq \alpha_\mu$ with $\mu = 1, \ldots, j$. As regards proposition (b), refer to Theorem 5.3.3/3. In particular one has $m_r = m_s$. Further, in this case it follows that $A_{\mu,\nu}$ is real if α_ν is real (keeping the assumptions of (b)).

Remark 2. $p(x)$ and $A_{\mu,\nu}$ can be determined explicitly, although the procedure is a bit tedious when multiple zeros are involved. If we multiply (1) by $(x-\alpha_1)^{m_1}$ and then let $x \to \alpha_1$ we obtain

$$A_{1,m_1} = \lim_{x \to \alpha_1} R(x)\,(x-\alpha_1)^{m_1}\,,$$

where the right-hand side can be calculated by l'Hospital's rule.Thereafter the method is applied to the new (less complicate) rational function $R(x) - \dfrac{A_{1,m_1}}{(x-\alpha_1)^{m_1}}$. In this way $A_{\mu,\nu}$ can be calculated step by step. Finally the expression is reduced to the polynomial $p(x)$.

Integration of $R(x)$. In the integration over intervals $[a, b]$, we have restricted ourselves to real functions, so that now, too, we assume that all the coefficients

a_k and b_l in $R(x)$ are real. Here it is further assumed that $[a, b]$ does not contain any zero of $Q(x)$. Moreover we try to transform (1) into a purely real representation. If α_μ is real, then it follows from Remark 1 that $A_{\mu,\nu}$ is real. If α and A are complex numbers, with $\operatorname{Im} \alpha \neq 0$, then

$$\frac{A}{(x-\alpha)^k} + \frac{\bar{A}}{(x-\bar{\alpha})^k} = q(x) + \sum_{j=1}^{k} \frac{c_j\,(x - \operatorname{Re}\alpha) + d_j}{[(x-\operatorname{Re}\alpha)^2 + (\operatorname{Im}\alpha)^2]^j}, \tag{2}$$

where $q(x)$ is a polynomial and all coefficients on the right-hand side of (2) are real. Part (b) of the theorem now shows that the terms in (1) can be combined in pairs, which gives a purely real representation of $R(x)$. Elementary changes of the variable, of the form $y = ax + b$, show that integration of $R(x)$ can be reduced to integration of x^j, $\frac{1}{(x^2+1)^k}$ and $\frac{x}{(x^2+1)^k}$ over suitable intervals. Here j is an integer and k is a natural number. Integrals of x^j have been dealt with in (7.1.1/1),

$$\int_c^d \frac{x}{x^2+1}\,dx = \int_c^d \frac{1}{2}\,(\ln\,(x^2+1))'\,dx = \frac{1}{2}\ln\,(x^2+1)\,\Big|_c^d,$$

$$\int_c^d \frac{x}{(x^2+1)^k}\,dx = \int_c^d -\frac{1}{2\,(k-1)}\,((x^2+1)^{-k+1})'\,dx$$

$$= -\frac{1}{2\,(k-1)}\,(x^2+1)^{-k+1}\,\Big|_c^d, \quad k = 2, 3, \ldots,$$

$$\int_c^d \frac{dx}{x^2+1} = \arctan x\,\Big|_c^d \quad \text{(Theorem 5.2.3.)}.$$

Finally, the still missing integrals can be calculated by iteration, using the formula

$$\int_c^d \frac{k-1}{(x^2+1)^k}\,dx = \int_c^d \frac{2k-3}{(x^2+1)^{k-1}}\,dx + \frac{x}{(x^2+1)^{k-1}}\Big|_c^d, \quad k = 2, 3, \ldots$$

To sum up, it is seen that the integration of real rational functions can be reduced to that of a few basic types, which in their turn can be integrated by elementary rules.

7.1.3. Integration of $R(\cos x, \sin x)$

Let us consider rational functions of $\cos x$ and $\sin x$,

$$R(\cos x, \sin x) = \frac{\sum\limits_{k,l=0}^{n} a_{k,l} \cos^k x \sin^l x}{\sum\limits_{r,s=0}^{m} b_{r,s} \cos^r x \sin^s x},$$

where $a_{k,l}$ and $b_{r,s}$ are real numbers. As $R(\cos x, \sin x)$ is periodic with the period 2π, the integration can be restricted to intervals $[a, b] \subset (-\pi, \pi)$ which do not contain any zeros of the denominator. Changing the variable by

1st case. For any sufficiently small positive ε, let $f(x)$ be integrable over the interval $[a+\varepsilon, b]$. If $\lim_{x \downarrow a} \int_x^b f(y)\,dy$ exists, then this value is denoted by $\int_a^b f(x)\,dx$, and called improper (Riemann) integral (Fig. 7.7).

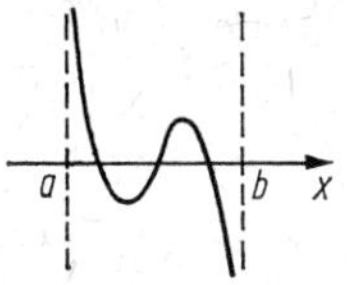

Fig. 7.7

2nd case. For every number N, $N>a$, let $f(x)$ be integrable over the interval $[a, N]$. If $\lim_{N\to\infty} \int_a^N f(y)\,dy$ exists, then this value is denoted by $\int_a^\infty f(x)\,dx$, and also called improper (Riemann) integral.

Remark. Of course these two procedures can be combined or applied repeatedly. In this way it is possible to integrate some kinds of unbounded functions over bounded or unbounded intervals.

Example 1. $\frac{\sin x}{x}$ is continuous on $[0, N]$. Let I_k denote the areas indicated in Fig. 7.8, then one obtains

$$\int_0^x \frac{\sin y}{y}\,dy = I_1 - I_2 + I_3 - I_4 + - \dots I_k + \int_{k\pi}^x \frac{\sin y}{y}\,dy\,,$$

where it has been assumed that $k\pi < x \leqq (k+1)\pi$. As I_k is a monotonically decreasing sequence of positive numbers, it follows that $\int_0^\infty \frac{\sin y}{y}\,dy$ exists. We shall calculate the value of this integral in Subsec. 15.5.1.

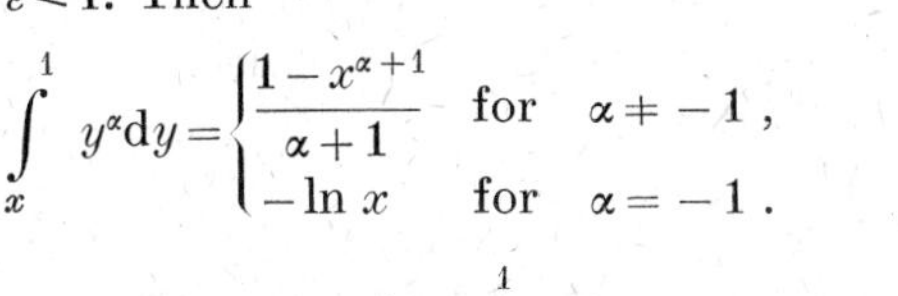

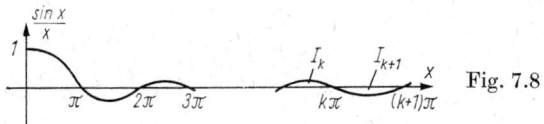

Fig. 7.8

Example 2. If α is a real number, then x^α is continuous on every interval $[\varepsilon, 1]$ with $0<\varepsilon<1$. Then

$$\int_x^1 y^\alpha dy = \begin{cases} \dfrac{1-x^{\alpha+1}}{\alpha+1} & \text{for} \quad \alpha \neq -1\,, \\ -\ln x & \text{for} \quad \alpha = -1\,. \end{cases}$$

From this it follows that $\lim_{x\downarrow 0} \int_x^1 y^\alpha dy$ exists if and only if $\alpha > -1$. For $\alpha \geqq 0$ one has an integral in the proper sense, for $0 > \alpha > -1$ one has an improper integral.

Example 3. Similarly it is seen that $\lim_{x\to\infty} \int_1^x y^\alpha dy$ exists if and only if $\alpha < -1$.

7.2.2. Integral Convergence Test for Series, Euler-Mascheroni Constant

Given the series $\sum_{j=1}^{\infty} a_j$, $a_j > 0$. The question to be answered is whether this series is convergent or divergent. In the way indicated in Fig. 7.9, a step function $h(x)$ with the steps $a_1, a_2, \ldots$ is constructed.

Lemma. (a) *If* $f(x) \geq h(x)$, *with* $\int_0^{\infty} f(x)\,dx < \infty$, *then* $\sum_{j=1}^{\infty} a_j < \infty$.

(b) *If* $0 \leq g(x) \leq h(x)$ *and* $\int_0^{\infty} g(x)\,dx = \infty$, *then* $\sum_{j=1}^{\infty} a_j = \infty$.

Remark 1. The lemma is an implication of $\int_0^{\infty} h(x)\,dx = \sum_{j=1}^{\infty} a_j$. We write "$=\infty$" if the integrals or series concerned do not converge.

Theorem. *There exists a positive number C such that*

$$\lim_{n\to\infty}\left(\sum_{k=1}^{n} \frac{1}{k} - \ln n\right) = C .$$

Remark 2. C is called Euler-Mascheroni constant. The theorem is proved by referring to the above lemma. One sets $g(x) = \frac{1}{x+1}$. Then $\int_0^n g(x)\,dx = \ln(n+1)$, and it can be shown that

$$\lim_{n\to\infty}\left(\sum_{k=1}^{n} \frac{1}{k} - \ln(n+1)\right) = \sum_{k=1}^{n} I_k$$

converges (Fig. 7.10).

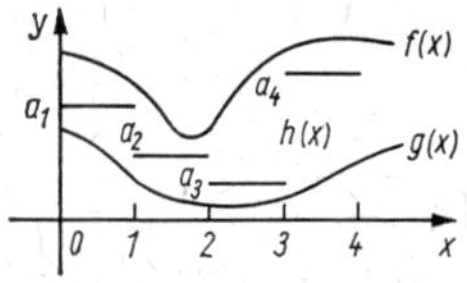

Fig. 7.9

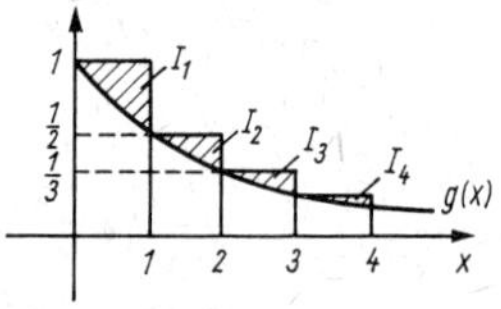

Fig. 7.10

7.2.3. The Γ-Function

Definition. *For* $x > 0$, $\Gamma(x) = \int_0^{\infty} e^{-t} t^{x-1}\,dt$.

Remark 1. $\Gamma(x)$ is a convergent improper integral.

Theorem. *$\Gamma(x)$ is a convex, positive function having derivatives up to arbitrary order on* $(0, \infty)$ (Fig. 7.11).

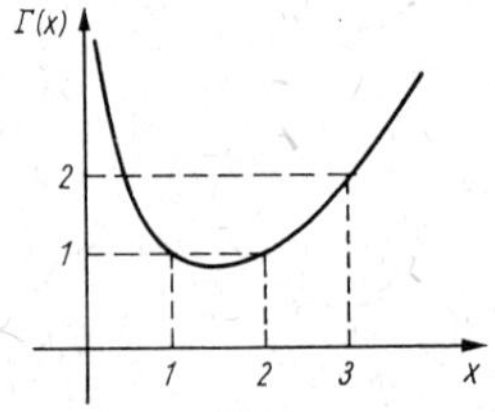

Fig. 7.11

The following relations hold for $\Gamma(x)$.

$$\Gamma(x) \to \infty \quad \text{as} \quad x \to \infty \quad \text{and as} \quad x \downarrow 0\,,$$
$$x\Gamma(x) = \Gamma(x+1) \qquad \text{for} \quad x > 0 \quad \text{and}$$
$$\Gamma(n) = (n-1)! \qquad \text{for} \quad n = 1, 2, 3, \dots$$

Remark 2. The function $\Gamma(x)$ is of great importance. As $\Gamma(n) = (n-1)!$ for natural numbers n, it is also written $\Gamma(x) = (x-1)!$ for $x > 0$.

8. Differential Calculus in R_n

8.1. Partial Derivatives

8.1.1. Definition

As defined in Subsec. 6.2.2., R_n is normed by $|x| = \sqrt{x_1^2 + \dots + x_n^2}$, where $x = (x_1, \dots, x_n) \in R_n$, x_j real. We consider real-valued functions $f(x) = f(x_1, \dots, x_n)$, whose domains of definition, $D(f)$, are open sets in R_n.

Definition 1. $f(x)$ *has a partial derivative with respect to* x_j *at the point* $x^0 = (x_1^0, \dots, x_n^0) \in D(f)$ *if*

$$\varphi(x_j) = f(\underbrace{x_1^0, \dots, x_{j-1}^0}, x_j, \underbrace{x_{j+1}^0, \dots, x_n^0})$$
$$\text{fixed}$$

is differentiable at the point x_j^0; $j = 1, 2, \dots, n$.

Remark 1. This means that $n-1$ co-ordinates are fixed, and $\varphi(x_j)$ is considered real-valued function defined on the x_j-interval $[a, b]$, with $x_j^0 \in (a, b)$ (Fig. 8.1). If $\varphi'(x_j^0)$ exists, it is written $\varphi'(x_j^0) = \dfrac{\partial f}{\partial x_j}(x^0)$. Consequently,

$$\frac{\partial f}{\partial x_j}(x^0) = \lim_{h \to 0} \frac{f(x_1^0, \dots, x_{j-1}^0, x_j^0 + h, x_{j+1}^0, \dots, x_n^0) - f(x^0)}{h}.$$

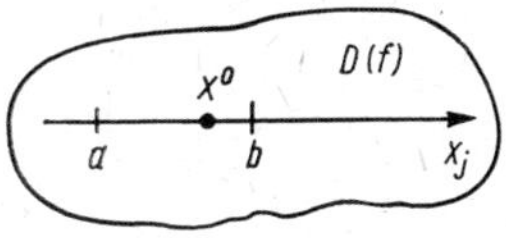

Fig. 8.1

Remark 2. Examples can easily be stated.

$$\frac{\partial}{\partial x_j}\left(\sum_{k=1}^{n} a_k x_k\right) = a_j\,, \quad a_k \text{ real}\,,$$
$$\frac{\partial}{\partial x_j}\left(\sum a_{k_1 \dots k_n} x_1^{k_1} \dots x_n^{k_n}\right) = \sum k_j a_{k_1 \dots k_n} x_1^{k_1} \dots x_j^{k_j - 1} \dots x_n^{k_n}\,,$$

where $\sum$ denotes a finite sum (partial derivatives of polynomials), and the $a_{k_1 \dots k_n}$ are real.

Remark 3. The following example is of fundamental importance.

$$f(x, y)=\begin{cases} \dfrac{2xy}{x^2+y^2} & \text{for} \quad (x, y) \neq (0, 0), \\ 0 & \text{for} \quad x=y=0, \end{cases} \quad x \in R_1, \quad y \in R_1 .$$

$\frac{\partial f}{\partial x}$ and $\frac{\partial f}{\partial y}$ exist for $(x, y) \neq (0, 0)$. From $f(0, y)=f(x, 0)=0$ it follows, however, that $\frac{\partial f}{\partial x}(0, 0)=\frac{\partial f}{\partial y}(0, 0)=0$ exist as well. On the other hand, $f(x, y)$ is not continuous at the point $(0, 0)$. This can be seen if one introduces polar co-ordinates $x=r \cos \varphi$, $y=r \sin \varphi$, which shows that $f(x, y)=2 \sin \varphi \cos \varphi=\sin 2\varphi$ is independent of r.

Lemma. *If $f(x)$ has all its partial derivatives $\frac{\partial f}{\partial x_1}(x), \ldots, \frac{\partial f}{\partial x_n}(x)$ in a neighbourhood U of the point $x^0 \in D(f)$, and if $\sup_{x \in U} \left| \frac{\partial f}{\partial x_j}(x) \right| < \infty$ for $j=1, \ldots, n$, then f is continuous at the point x^0.*

Remark 4. The example of Remark 3 shows that the boundedness of the partial derivatives in a neighbourhood of x^0 is indispensable. In this example, $\frac{\partial f}{\partial x}$ and $\frac{\partial f}{\partial y}$ turn out to be unbounded in a neighbourhood of $(0, 0)$.

Definition 2. *$f(x)$ has the partial derivative of second order $\frac{\partial^2 f}{\partial x_j \partial x_k}$ at the point $x^0 \in D(f)$ if the first partial derivative $\frac{\partial f}{\partial x_k}$ exists in a neighbourhood of x^0 and if the partial derivative $\frac{\partial}{\partial x_j}\left(\frac{\partial f}{\partial x_k}\right)$ exists at the point x^0.*

Remark 5. By iteration, one may then define

$$\frac{\partial^m f}{\partial x_{k_m} \ldots \partial x_{k_1}}=\frac{\partial}{\partial x_{k_m}}\left(\frac{\partial^{m-1} f}{\partial x_{k_{m-1}} \ldots \partial x_{k_1}}\right).$$

8.1.2. Commutativity of Partial Derivatives

Theorem. *If $f(x)$ is continuous in a neighbourhood U of $x^0 \in D(f) \subset R_n$, and if the partial derivatives $\frac{\partial f}{\partial x_j}, \frac{\partial f}{\partial x_k}$ and $\frac{\partial^2 f}{\partial x_j \partial x_k}$ exist and are also continuous in U, then $\frac{\partial^2 f}{\partial x_k \partial x_j}$ exists in U, too, and*

$$\frac{\partial^2 f}{\partial x_j \partial x_k}(x)=\frac{\partial^2 f}{\partial x_k \partial x_j}(x), \quad x \in U .$$

Remark. If $f(x)$ has all its partial derivatives up to the order m in a neighbourhood U, and if all these derivatives are continuous in U, then the above theorem can be applied iteratively. This gives

$$\frac{\partial^m f}{\partial x_{k_m} \ldots \partial x_{k_1}}=\frac{\partial^m f}{\partial x_1^{\alpha_1} \ldots \partial x_n^{\alpha_n}}=D^\alpha f.$$

Here $\alpha=(\alpha_1, \ldots, \alpha_n)$ is a multiple index. α_j indicates how many of the k_r are equal to j; $\alpha_j=0, 1, 2, \ldots$; one has $|\alpha|=\sum_{j=1}^{n} \alpha_j=m$.

8.1.3. Taylor Polynomials

We shall now generalize Theorem 5.4.4/1 to the n-dimensional case. Hence, what is desired are expansions of real-valued functions $f(x)$ at the point $x^0 \in D(f) \subset \subset R_n$. Without loss of generality, we assume that $x^0 = 0$. Let $\alpha = (\alpha_1, \ldots, \alpha_n)$ again be a multiple index with $\alpha_j = 0, 1, 2, \ldots$ and $|\alpha| = \sum_{j=1}^{n} \alpha_j$. Further let $\alpha! = \alpha_1!, \ldots, \alpha_n!$ (with $0! = 1$) and $x^\alpha = x_1^{\alpha_1} \ldots x_n^{\alpha_n}$. Finally let $0(|x|^l)$ denote a function defined in a neighbourhood of 0, which can there be estimated by $|0(|x|^l)| \leqq C|x|^l$, where C is independent of $|x|$.

Theorem. *Let $f(x)$ be $(m+1)$ times continuously differentiable in a neighbourhood U of the origin (i.e., let all the derivatives $(D^\alpha f)(x)$ with $|\alpha| \leqq m+1$ exist and be continuous in a neighbourhood of 0), with $m = 0, 1, 2, \ldots$*

(**a**) *For $x \in U$,*

$$f(x) = \sum_{|\alpha| \leqq m} \frac{1}{\alpha!} (D^\alpha f)(0)\, x^\alpha + O(|x|^{m+1}) \quad (\textit{Taylor polynomial})\,. \tag{1}$$

(**b**) *If*

$$f(x) = \sum_{|\alpha| \leqq m} a_\alpha x^\alpha + O(|x|^{m+1})$$

for $x \in U$, then $a_\alpha = \frac{1}{\alpha!} (D^\alpha f)(0)$.

Remark 1. (b) says that the Taylor polynomial of (a) is the best possible approximation to $f(x)$ at the point 0 by a polynomial of degree m.

Remark 2. By analogy with (5.4.4/3), the remainder $R_m = O(|x|^{m+1})$ in (1) can be represented by

$$R_m(x) = \sum_{|\beta| = m+1} \frac{x^\beta}{\beta!} D^\beta f(\vartheta_j^{(\beta)} x_j) \tag{2}$$

where the $\vartheta_j^{(\beta)}$ are numbers independent of x, with $0 \leqq \vartheta_j^{(\beta)} \leqq 1$.

8.1.4. n-Dimensional Power Series

We shall now generalize Theorem 5.4.4/2: find a representation of $f(x)$ by a Taylor series in R_n in the neighbourhood of the origin:

$$f(x) = \sum_{|\alpha| < \infty} \frac{1}{\alpha!} (D^\alpha f)(0) x^\alpha\,. \tag{1}$$

One possible approach is based on Eqs. (8.1.3/1) and (8.1.3/2). If $R_m(x) \to 0$ as $m \to \infty$, then $f(x)$ can be represented in the form (1). We shall now describe another approach.

Theorem. *Let $a > 0$ and $|x_k| < a$ for $k = 1, \ldots, n$. For every multiple index α and fixed $x_1, \ldots, x_{j-1}, x_{j+1}, \ldots, x_n$, let the (one-dimensional) function $D^\alpha f(x_1, \ldots, x_j, \ldots, x_n)$ be representable as an absolutely convergent power series in $[-a, a]$ (with respect to x_j, and with 0 as the centre of expansion). Then $f(x)$ can be represented as an absolutely convergent Taylor series (in the sense of (1)) in the cube $Q = \{x \mid |x_j| < a\}$.*

8.1.5. Curves and Surfaces in R_n, Chain Rule

Theorem 1 (*chain rule*). *Suppose that $F(x_1, \ldots, x_n)$ has continuous partial derivatives of first order in an open domain U in R_n. Further let $\varphi_j(t)$, $j=1, \ldots, n$, be functions which are continuously differentiable on the interval (a, b). Let $(\varphi_1(t), \ldots, \varphi_n(t)) \in U$ for $t \in (a, b)$. Then $\psi(t) = F(\varphi_1(t), \ldots, \varphi_n(t))$ is continuously differentiable on (a, b), and*

$$\psi'(t) = \sum_{j=1}^{n} \frac{\partial F}{\partial x_j} (\varphi_1(t), \ldots, \varphi_n(t))\, \varphi_j'(t), \quad t \in (a, b) .$$

Curves in R_n. If $\varphi_1(t), \ldots, \varphi_n(t)$ are real-valued continuous functions, then

$$x(t) = (\varphi_1(t), \ldots, \varphi_n(t)), \quad t \in (a, b) = I ,$$

is called a curve in R_n. Here I is an interval. $x(t)$ is called a differentiable curve if the components $\varphi_j(t)$ are differentiable functions on I. Here the tangent vector is of interest, being defined as

$$v(t) = \lim_{h \to 0} \left(\ldots, \frac{\varphi_j(t+h) - \varphi_j(t)}{h}, \ldots \right), \quad t \in (a, b), \quad |h| < \varepsilon$$

(cf. Fig. 8.2).

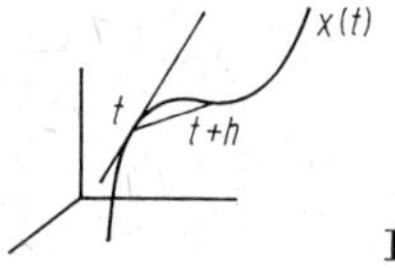

Fig. 8.2

Lemma. *If $x(t)$ is a differentiable curve, then*

$$v(t) = (\varphi_1'(t), \ldots, \varphi_n'(t)) \quad \textit{for} \quad t \in (a, b) .$$

Remark 1. The line passing through the points $x(t)$ and $x(t+h)$, $h \neq 0$, is called secant line. As $h \to 0$, the secant lines approach the tangent line, whose direction is given by $v(t)$.

Surfaces in R_n. If $F(x) = F(x_1, \ldots, x_n)$ is a given real-valued function, then the set of all points $x \in R_n$ where $F(x) = 0$ is called a surface in R_n. If a_j, $j = 1, \ldots, n$, and c are real, then

$$F(x) = \sum_{j=0}^{n} a_j x_j - c = 0$$

is a plane in R_n; $\sum_{j=1}^{n} |a_j| > 0$. If $r > 0$, then

$$F(x) = \sum_{j=1}^{n} (x_j - x_j^0)^2 - r^2 = 0$$

is a sphere in R_n, with centre x^0 and radius r. By analogy with the curves, we look for tangent planes of surfaces and vectors orthogonal to them.

Definition 1. *Two vectors $x = (x_1, \ldots, x_n)$ and $y = (y_1, \ldots, y_n)$ in R_n are said to be orthogonal if* $\sum_{j=1}^{n} x_j y_j = 0$.

Remark 2. From analytical geometry it is known that, for $n = 2$ and $n = 3$, two vectors x and y are orthogonal if and only if they are (in a graphic, geometrical sense) perpendicular to each other (provided that $|x|\ |y| > 0$).

Normal vector of a plane. A vector $v=(v_1, \ldots, v_n)$ is called normal vector of the plane $F(x)=\sum_{j=1}^{n} a_h x_j - c = 0$ if v is orthogonal to all lines (or, more strictly speaking, to the directions of all lines) extending within the plane. If x^0 and x are, respectively, a fixed and a variable point in the plane, then all the directions in question are given by $x-x^0$ (Fig. 8.4). This implies that $\sum_{j=1}^{n} v_j (x_j - x_j^0)=0$ must hold. On the other hand it is immediately obvious that $\sum_{j=1}^{n} a_j (x_j - x_j^0)=0$. Hence $(a_1, \ldots, a_n)$ is a normal vector. It is easily shown that every other normal vector v has the form $v=(ca_1, \ldots, ca_n)$, c real, and hence is proportional to $(a_1, \ldots, a_n)$. Our object is now to extend these considerations to surfaces.

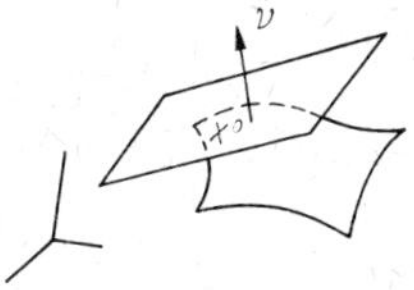

Fig. 8.3

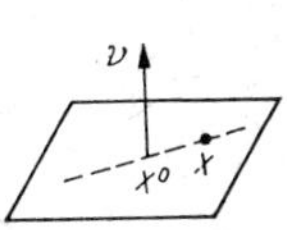

Fig. 8.4

Definition 2. *Let x^0 be a point of the surface $F(x)=0$. Then $\sum_{j=1}^{n} a_j (x_j - x_j^0)=0$, with $\sum_{j=1}^{n} |a_j| > 0$, is called tangent plane (of the surface at the point x^0) if, in a neighbourhood of x^0,*

$$F(x) - \sum_{j=1}^{n} a_j (x_j - x_j^0) = O(|x-x^0|^2) \tag{1}$$

Then $v=(a_1, \ldots, a_n)$ is called normal vector (of the surface at the point x^0) (Fig. 8.3).

Remark 3. Hence, tangent planes are defined by approximation properties. This is the generalization of the considerations described in Subsec. 3.1.1. For the following theorem, see also Lemma 3.1.1/1.

Theorem 2. *If $F(x)$ has continuous partial derivatives of first and second order in a neighbourhood of the point $x^0 \in R_n$, and if $F(x^0)=0$ and $\sum_{j=1}^{n} \left| \frac{\partial F}{\partial x_j}(x^0) \right| > 0$, then there exists exactly one tangent plane (of the surface $F(x)=0$ at the point x^0). It is given by*

$$\sum_{j=1}^{n} \frac{\partial F}{\partial x_j}(x^0)\,(x_j - x_j^0) = 0 .$$

$v=(\operatorname{grad} F)(x^0)=\left(\frac{\partial F}{\partial x_1}(x^0), \ldots, \frac{\partial F}{\partial x_n}(x^0)\right)$ *is the associated normal vector.*

8.1.6. Geometric Interpretation of the Taylor Polynomial

Suppose that the function $f(x_1, \ldots, x_n)$ has continuous partial derivatives of first and second order in an open domain $U=D(f)$ in R_n. We represent f as a surface in R_{n+1} (Fig. 8.5),

$$x_{n+1} = f(x_1, \ldots, x_n) ,$$

or

$$F(x_1, \ldots, x_{n+1}) = f(x_1, \ldots, x_n) - x_{n+1} = 0 .$$

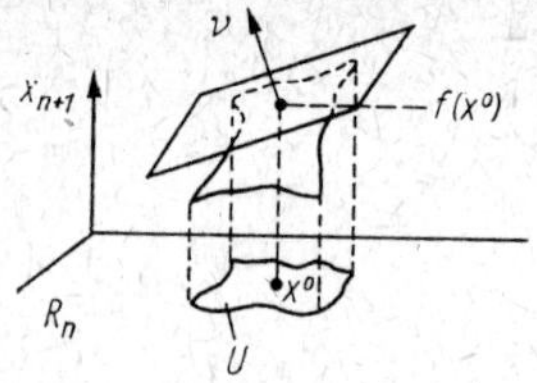

Fig. 8.5

F satisfies the assumptions of Theorem 8.1.5/2, where one has to substitute $n+1$ for n. Thus

$$\nu = \left(\frac{\partial f}{\partial x_1}, \frac{\partial f}{\partial x_2}, \ldots, \frac{\partial f}{\partial x_n}, -1 \right) \neq 0$$

is the normal vector and

$$\sum_{j=1}^{n} \frac{\partial f}{\partial x_j} (x^0) (x_j - x_j^0) - (x_{n+1} - f(x_0)) = 0$$

is the associated tangent plane. Consequently (in the sense of Theorem 8.1.5/2 and Eq. (8.1.5/1)),

$$x_{n+1} = f(x^0) + \sum_{j=1}^{n} \frac{\partial f}{\partial x_j} (x^0) (x_j - x_j^0)$$

is the best approximation to f at the point x^0 by a polynomial of degree one. This polynomial coincides, however, with the Taylor polynomial of degree one, as given in Subsection 8.1.3. If f has higher derivatives, one obtains corresponding geometric interpretations for the higher Taylor polynomials of Eq. (8.1.3/1). For $m=2$ one obtains the best approximation to $f(x)$ at the point x^0 by surfaces of second order, i.e., by ellipsoids, hyperboloids etc.

8.1.7. Directional Derivative

Let $\nu = (\nu_1, \ldots, \nu_n)$, where $|\nu| = \sqrt{\nu_1^2 + \ldots + \nu_n^2} = 1$.

Definition. *If $f(x)$ is continuously differentiable in the neighbourhood of a point $x^0 \in R_n$, then*

$$\frac{\partial f}{\partial \nu} (x^0) = \sum_{j=1}^{n} \frac{\partial f}{\partial x_j} (x^0) \nu_j$$

is called directional derivative.

Remark. Without loss of generality, let $x^0 = 0$. As is well known from analytical geometry, rotations in R_n can be described by

$$x_j = \sum_{k=1}^{n} a_{j,k} y_k, \quad j = 1, \ldots, n,$$

where $(a_{j,k})_{j,k=1}^{n}$ is a proper orthogonal matrix, i.e.,

$$\sum_{k=1}^{n} a_{j,k} a_{l,k} = \begin{cases} 1 & \text{for } j = l, \\ 0 & \text{for } j \neq l. \end{cases}$$

In order that the y_1-axis coincides with the ν-direction, it is necessary that $a_{k,1} = \nu_k$ (Fig. 8.6). Setting now $f(x_1, \ldots, x_n) = f\left(\ldots, \sum_{k=1}^{n} a_{j,k} y_k, \ldots\right) = g(y_1, \ldots, y_n)$, one obtains

$$\frac{\partial g}{\partial y_1} = \sum_{l=1}^{n} \frac{\partial f}{\partial x_l} (\ldots, \sum \ldots, \ldots)\, a_{l,1} = \frac{\partial f}{\partial \nu}$$

at the point $x = y = 0$. Thus the directional derivative proves to be an ordinary partial derivative with respect to y_1.

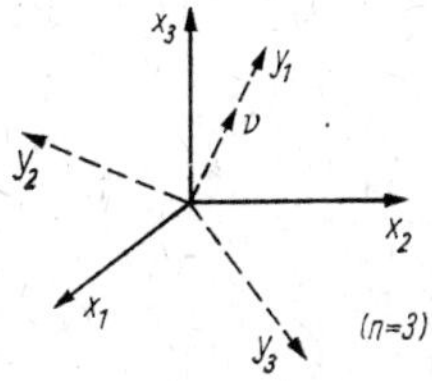

Fig. 8.6

8.2. Implicit Functions and Theorems of Implicit Functions

8.2.1. Formulation of the Problem

In Subsec. 3.1.4. we had discussed inverse functions $g(y) = f^{-1}(y)$ of continuously differentiable functions $f(x)$ in the neighbourhood of a point $x^0 \in R_1$. The decisive requirement was that $f'(x^0) \neq 0$. If this requirement is satisfied, f is a one-to-one mapping of a neighbourhood U of x^0 onto a neighbourhood V of $y^0 = f(x^0)$ (Fig. 8.7). Let us now extend these considerations to the case of n dimensions. Let $y = f(x) = (f_1(x), \ldots, f_n(x))$ be a vector function, where the $f_j(x)$ are assumed to be functions continuous in a neighbourhood U of the point $x^0 \in R_n$. We look for conditions which make $y = f(x)$ a one-to-one mapping of U onto a neighbourhood V of the point $y^0 = f(x^0)$ (Fig. 8.8). If the functions $f_k(x)$ are twice continuously differentiable on U, then it follows from Theorem 8.1.3 that

$$f_k(x) = f_k(x^0) + \sum_{j=1}^{n} \frac{\partial f_k}{\partial x_j} (x^0)\, (x_j - x_j^0) + O\, (|x - x^0|^2)\,, \quad k = 1, \ldots, n\,. \tag{1}$$

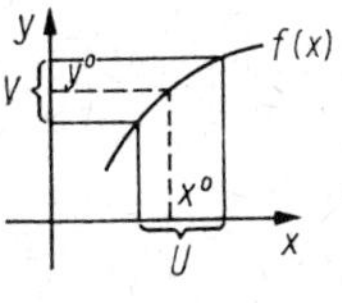

Fig. 8.7

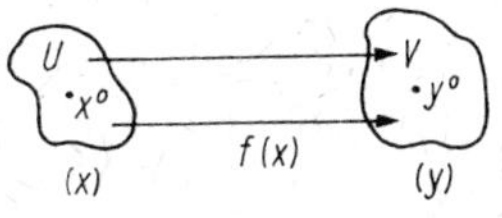

Fig. 8.8

From analytical geometry it is known that the first two terms on the right-hand side of Eq. (1) give a one-to-one (and linear) mapping of R_n onto itself if and only if the determinant of the coefficients $\dfrac{\partial f_k}{\partial x_j}(x^0)$, $j = 1, \ldots, n$ and $k = 1, \ldots, n$, is different from zero. One may now hope that the "perturbation term" $O\,(|x - x^0|^2)$ for small values of $|x - x^0|$ does not alter this property. The above determinant of coefficients plays a fundamental role and is therefore given a particular name.

Definition. *If $y(x)=f(x)=(f_1(x), \dots, f_n(x))$ has partial derivatives of first order, then*

$$\frac{\partial(f_1, \dots, f_n)}{\partial(x_1, \dots, x_n)}=\frac{\partial(y_1, \dots, y_n)}{\partial(x_1, \dots, x_n)}=\begin{vmatrix} \frac{\partial f_1}{\partial x_1} & \cdots & \frac{\partial f_1}{\partial x_n} \\ \vdots & & \vdots \\ \frac{\partial f_n}{\partial x_1} & \cdots & \frac{\partial f_n}{\partial x_n} \end{vmatrix}$$

is called functional determinant (or Jacobian determinant). $\left(\frac{\partial(y_1, \dots, y_n)}{\partial(x_1, \dots, x_n)}\right)_{j,k}$ *denotes the minor determinant obtained by deleting the j-th row and the k-th column in* $\frac{\partial(y_1, \dots, y_n)}{\partial(x_1, \dots, x_n)}$ $\left(\text{minor of the element } \frac{\partial f_j}{\partial x_k}\right)$.

8.2.2. Theorem of Implicit Functions, Curvilinear Co-Ordinates

Theorem 1. *If the functions $f_j(x)$, $j=1, \dots, n$, are twice continuously differentiable in a neighbourhood U of $x^0 \in R_n$ and if $\frac{\partial(f_1, \dots, f_n)}{\partial(x_1, \dots, x_n)}(x^0) \neq 0$, then there exists a neighbourhood W of x^0 such that $f(x)=(f_1(x), \dots, f_n(x))$ is a one-to-one mapping of W onto a neighbourhood V of $y^0=f(x^0)$ (Fig. 8.9). The inverse mapping $x=g(y)=$ $=(g_1(y), \dots, g_n(y))$ is twice continuously differentiable in V, and for $k=1, \dots, n$ and $j=1, \dots, n$ the following relation holds:*

$$\frac{\partial g_k}{\partial y_j}(y^0)=\left(\frac{\partial(f_1, \dots, f_n)}{\partial(x_1, \dots, x_n)}\right)_{k,j}(x^0)\left[\frac{\partial(f_1, \dots, f_n)}{\partial(x_1, \dots, x_n)}(x^0)\right]^{-1}$$

Remark 1. For $n=1$ one has the situation described at the beginning of Subsec. 8.2.1.; see also Subsec. 3.1.4.

Remark 2. According to the considerations made in Subsec. 8.2.1., the theorem is plausible. Its proof, however, which is based on Banach's fixed point theorem 2.5.3, is rather complicated.

Theorem 2. *The assumptions shall be the same as in Theorem 1. Let the functions $f_j(x)$ be m times continuously differentiable in U. Then the functions $g_j(y)$, with $j=1, \dots, n$, are also m times continuously differentiable in V.*

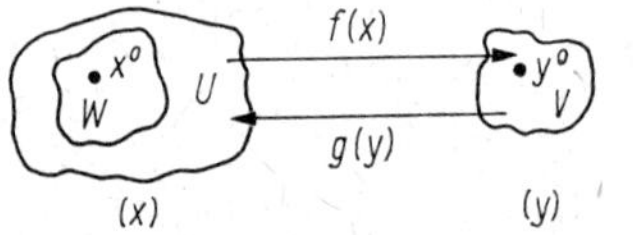

Fig. 8.9

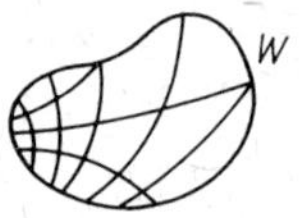

Fig. 8.10

Remark 3. If the above assumptions are fulfilled, then for a suitable choice of c_j one obtains surfaces in W by

$$f_j(x_1, \dots, x_n) \equiv c_j, \quad j=1, \dots, n$$

(Fig. 8.10). For that purpose it is only necessary to choose $(c_1, \dots, c_n) \in V$. Then it follows from Theorem 1 that every point $x \in W$ can be identified in a one-to-one manner by indicating the parameter values $c_1, \dots, c_n$. $c_1, \dots, c_n$ are called curvilinear co-ordinates.

8.2.3. Theorem of Parameter-Depending Implicit Functions

We consider functions of the form $f_j(x, \lambda) = f_j(x_1, \ldots, x_n, \lambda_1, \ldots, \lambda_r)$, which depend on $x = (x_1, \ldots, x_n)$ as well as on the parameters $\lambda = (\lambda_1, \ldots, \lambda_r)$ where $j = 1, \ldots, n$. Here λ varies within a ball, $K = \{\lambda \mid |\lambda - \lambda^0| < \eta\} \subset R_r$.

Theorem. (a) *If, for a fixed $\lambda \in K$, the assumptions of Theorem 8.2.2/1 are satisfied and if $f_j(x, \lambda)$ and $\frac{\partial f_j}{\partial x_k}(x, \lambda)$ are continuous in an $(n+r)$-dimensional neighbourhood of (x^0, λ^0), then the inverse functions $g_k(y, \lambda)$ of Theorem 8.2.2/1 are also continuous in an $(n+r)$-dimensional neighbourhood (y^0, λ^0), where $y^0 = f(x^0, \lambda^0)$.*

(b) *If, moreover, the functions $f_j(x, \lambda)$ are m times continuously differentiable with respect to $\lambda_1, \ldots, \lambda_r$ in a neighbourhood of λ^0, then the inverse functions $g_k(y, \lambda)$ also have partial derivatives up to the order m with respect to $\lambda_1, \ldots, \lambda_r$, in a neighbourhood of λ^0. Here the $\frac{\partial g_k}{\partial \lambda_\varrho}$ are calculated from*

$$\frac{\partial f_j}{\partial \lambda_s} + \sum_{t=1}^{n} \frac{\partial f_j}{\partial x_t} \frac{\partial g_t}{\partial \lambda_s} = 0 \quad \textit{for} \quad j = 1, \ldots, n \quad \textit{and} \quad s = 1, \ldots, r\,. \tag{1}$$

Remark. As the Jacobian determinant $\frac{\partial(f_1, \ldots, f_n)}{\partial(x_1, \ldots, x_n)}(x_0)$ is different from zero, (1) can be resolved uniquely for $\frac{\partial g}{\partial \lambda_s}$.

8.2.4. Implicit Functions

The question is whether $F(x_1, \ldots, x_n) = 0$ can be resolved for one of the variables, e.g., $x_1 = f(x_2, \ldots, x_n)$. Let $\tilde{x} = (x_2, \ldots, x_n)$, and correspondingly $\tilde{x}^0 = (x_2^0, \ldots, x_n^0)$ for $x^0 = (x_1^0, \ldots, x_n^0)$. The surface in R_n which is described by $F(x) = 0$ shall be denoted by F.

Theorem 1. (a) *Let $F(x)$ be twice continuously differentiable in an n-dimensional neighbourhood U of $x^0 \in R_n$. Let $F(x^0) = 0$ and $\frac{\partial F}{\partial x_1}(x^0) \neq 0$. Then F can be represented by $x_1 = f(\tilde{x})$ in an $(n-1)$-dimensional neighbourhood V of $\tilde{x}^0$. $f(\tilde{x})$ is continuous in a neighbourhood of $\tilde{x}^0$* (Fig. 8.11).

(b) *If, moreover, $F(x)$ is m times continuously differentiable in U, then $f(\tilde{x})$ is likewise m times continuously differentiable in V.*

Remark 1. The theorem follows from the implicit function theorems of Subsecs. 8.2.2. and 8.2.3.

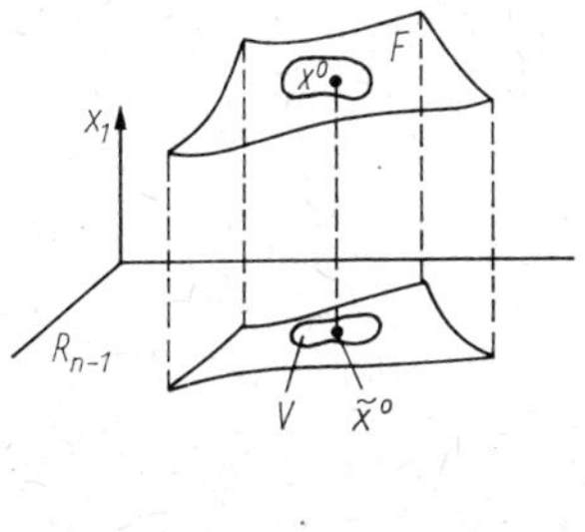

Fig. 8.11

Remark 2. If $(\text{grad } F)(x^0) = \left(\frac{\partial F}{\partial x_1}(x^\circ), \ldots, \frac{\partial F}{\partial x_n}(x^0)\right) \neq 0$, then the above theorem can be applied, where one possibly has to substitute another co-ordinate for x_1.

Generalization: Let us now consider surfaces $F_1, \ldots, F_l$ in R_n which are given by $F_j(x) = 0$ for $j = 1, \ldots, l$. Suppose that $1 \leq l < n$. Further let $\tilde{x} = (x_{l+1}, \ldots, x_n)$ for $x = (x_1, \ldots, x_n)$, and correspondingly $\tilde{x}^0 = (x^0_{l+1}, \ldots, x^0_n)$ for $x^0 = (x^0_1, \ldots, x^0_n)$ (Fig. 8.12).

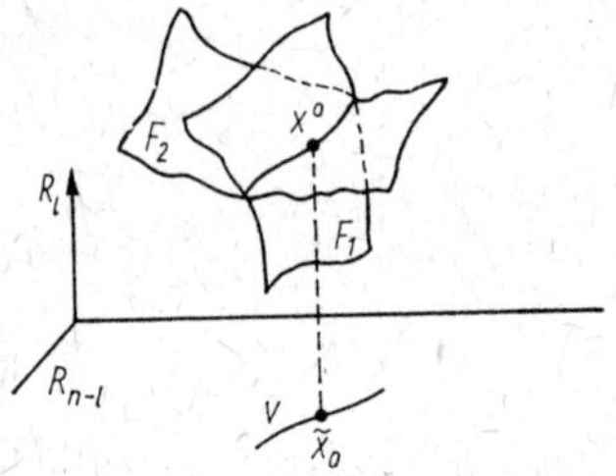

Fig. 8.12

Theorem 2. (**a**) *If the functions $F_1(x), \ldots, F_l(x)$ are twice continuously differentiable in an n-dimensional neighbourhood U of $x^0 \in R_n$, and if $F_j(x^0) = 0$ for $j = 1, \ldots, l$ and $\frac{\partial(F_1, \ldots, F_l)}{\partial(x_1, \ldots, x_l)}(x^0) \neq 0$, then $\bigcap_{j=1}^{l} F_j$ can be represented as $x_j = f_j(\tilde{x})$, with $j = 1, \ldots, l$, in an $(n-l)$-dimensional neighbourhood V of $\tilde{x}^0$. Here the $f_j(\tilde{x})$ are continuous in V.*

(**b**) *If, moreover, the functions $F_1(x), \ldots, F_l(x)$ are m times continuously differentiable in U, then the functions $f_1(\tilde{x}), \ldots, f_l(\tilde{x})$ are likewise m times continuously differentiable in V.*

Remark 3. Again the proof is based on the implicit function theorems of Subsecs. 8.2.2. and 8.2.3.

8.3. Extreme Values of Functions

8.3.1. The One-Dimensional Case

If $y = f(x)$ is three times continuously differentiable on the interval $[a, b]$, with $-\infty < a < b < \infty$ (Fig. 8.13), then according to Theorem 5.4.4/1 one has

$$f(x) = f(x^0) + f'(x^0)(x - x^0) + \frac{f''(x^\circ)}{2!}(x - x^0)^2 + O(|x - x^0|^3)\,. \tag{1}$$

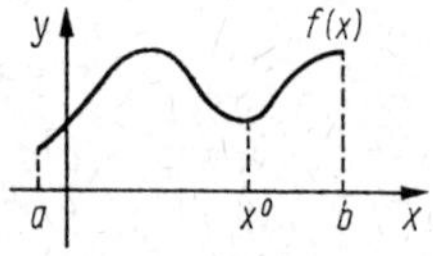

Fig. 8.13

If x^0 is a relative maximum or minimum of $f(x)$, then $f'(x^0)$ must be zero, because otherwise $f(x)$ would be strictly monotone in a neighbourhood of x^0. Conversely, if $f'(x^0) = 0$ and $f''(x^0) > 0$, then Eq. (1) shows that $f(x)$ has a relative minimum at x^0. If $f'(x^0) = 0$ and $f''(x^0) < 0$, then a relative maximum exists at x^0. If $f'(x^0) = f''(x^0) = 0$, then a relative extremum need not necessarily exist at the point x^0, as is shown by the example $f(x) = x^3$ near the origin.

8.3.2. The n-Dimensional Case

Definition 1. *Let $f(x)=f(x_1, \ldots, x_n)$ be continuous in a neighbourhood of the point $x^0 \in R_n$.*

(**a**) *$f(x^0)$ is a relative maximum if $f(x) \leqq f(x^0)$ for all x in a suitable neighbourhood of x^0.*

(**b**) *$f(x^0)$ is a relative minimum if $f(x) \geqq f(x^0)$ for all x in a suitable neighbourhood of x^0.*

Definition 2. *Let $a_{j,k}$ be real numbers. The quadratic form $\sum_{j,k=1}^{n} a_{j,k}\xi_j\xi_k$ is called positive definite if there exists a positive number c such that $\sum_{j,k=1}^{n} a_{j,k}\xi_j\xi_k \geqq c\,|\xi|^2$ for all $\xi \in R_n$; it is called negative definite if there exists a positive number c such that $\sum_{j,k=1}^{n} a_{j,k}\xi_j\xi_k \leqq -c\,|\xi|^2$ for all $\xi \in R_n$, where $\xi=(\xi_1, \ldots, \xi_n)$.*

Remark. $\sum_{j=1}^{n} c_j\xi_j^2$ with $c_j>0$ is positive definite.

Theorem. *Let $f(x)$ be three times continuously differentiable in a neighbourhood of $x^0 \in R_n$.*

(**a**) *If $f(x^\circ)$ is a relative extremum (minimum or maximum), then $\frac{\partial f}{\partial x_j}(x^0)=0$ for $j=1, \ldots, n$.*

(**b**) *Let $\frac{\partial f}{\partial x_j}(x^0)=0$ for $j=1, \ldots, n$. If $\sum_{j,k=1}^{n} \frac{\partial^2 f}{\partial x_j \partial x_k}(x^0)\,\xi_j\xi_k$ is positive definite, then $f(x^0)$ is a relative minimum. If $\sum_{j,k=1}^{n} \frac{\partial^2 f}{\partial x_j \partial x_k}(x^0)\,\xi_j\xi_k$ is negative definite, then $f(x^0)$ is a relative maximum.*

9. Integral Calculus in R_n

9.1. Definitions and Properties

9.1.1. Q-Domains and I-Domains

Let $Q=\{x \mid x \in R_n,\ |x-x_j^0|<a_j \text{ for } j=1, \ldots, n\}$ be an n-dimensional interval in R_n (Fig. 9.1), and let $|Q|=2^n a_1 \ldots a_n$ be its volume. Here the a_j are positive numbers. A Q-domain Ω is a finite union of n-dimensional intervals (Fig. 9.2),

$$\Omega=\bigcup_{j=1}^{N} \bar{Q}_j-\partial\left(\bigcup_{j=1}^{N} \bar{Q}_j\right). \tag{1}$$

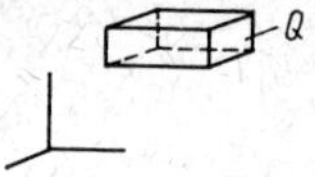

Fig. 9.1

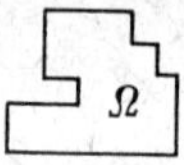

Fig. 9.2

Here $\bar{Q}_j$ is the closure of the n-dimensional open interval Q_j (i.e. the n-dimensional interval including its boundary surface). Further let $\partial\omega$ be the boundary of a set ω in R_n (Fig. 9.3): $y \in \partial\omega$ if and only if every neighbourhood of y contains points which belong to ω, but also points which do not belong to ω. We shall use the term "domain" whenever the set concerned is open. In this sense, Ω of Eq. (1) is a domain.

Definition. *A bounded domain Ω in R_n is called an I-domain if for every positive number ε there exists a finite covering of $\partial\Omega$ by n-dimensional intervals Q_r^ε,*

$$\partial\Omega \subset \sum_{r=1}^{N_\varepsilon} Q_r^\varepsilon, \quad \textit{such that} \quad \sum_{r=1}^{N_\varepsilon} |Q_r^\varepsilon| \leq \varepsilon \quad \text{(Fig. 9.4)}.$$

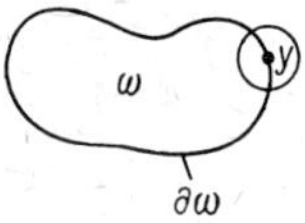

Fig. 9.3

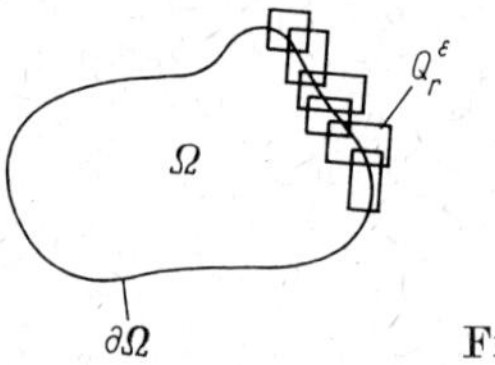

Fig. 9.4

Lemma. *Let Ω be a bounded domain in R_n, with $\partial\Omega = \bigcup_{j=1}^{N} F_j$, where F_j is an interior patch of a surface G_j in R_n which is represented by $G_j(x) \equiv 0$* (Fig. 9.5 and Fig. 9.6). *If $G_j(x)$ has continuous partial derivatives of first order and if $\sum_{l=1}^{n} \left| \frac{\partial G_j}{\partial x_l}(x) \right| > 0$ for $x \in G_j$, then Ω is an I-domain.*

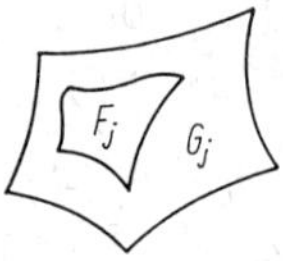

Fig. 9.5

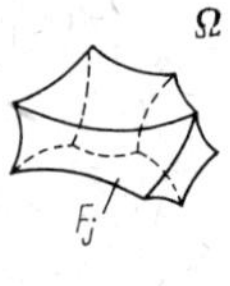

Fig. 9.6

Remark. *I*-domains are a relatively general concept. They include bounded domains with smooth boundaries (also with holes), but also certain domains with edges and corners.

9.1.2. Integrals over Q-Domains

Let Ω be a Q-domain, then we consider arbitrary finite decompositions Z into n-dimensional intervals: $\bar{\Omega} = \bigcup_{j=1}^{N} \bar{Q}_j$ (Fig. 9.7).

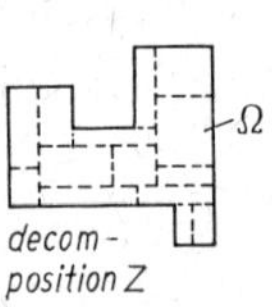

Fig. 9.7

Definition. *Let $f(x)$ be a real-valued bounded function with the domain of definition $D(f)=\bar{\Omega}$, where Ω is a Q-domain.*

(**a**) $\underline{\int} f(x)\,dx = \sup\limits_Z \sum_Z^*$, *where* $\sum_Z^* = \sum\limits_{j=1}^N |Q_j| \inf\limits_{y\in Q_j} f(y)$,

$\overline{\int} f(x)\,dx = \inf\limits_Z \sum_Z^{**}$, *where* $\sum_Z^{**} = \sum\limits_{j=1}^N |Q_j| \sup\limits_{y\in Q_j} f(y)$.

(**b**) *$f(x)$ is said to be integrable if $\underline{\int} f(x)\,dx = \overline{\int} f(x)\,dx$.*

Remark 1. This is the generalization of Def. 3.2.1. As in that definition, $\sup\limits_Z$ (or $\inf\limits_Z$) denotes the supremum (or infimum) taken over all permissible decompositions of the kind indicated above, where N is variable.

Remark 2. The statements made in Subsec. 3.2.1. can be transferred immediately: Lemma 3.2.1/1 remains valid. With a suitable modification, Lemma 3.2.1/2 is also applicable.

Remark 3. If f is integrable, we write $\int f(x)\,dx$ or $\int\limits_\Omega f(x)\,dx$ instead of $\underline{\int} f(x)\,dx$ or $\overline{\int} (f(x)\,dx$.

9.1.3. Properties

The theorems of this subsection are analogues to corresponding propositions made in Subsecs. 3.2.2. and 3.2.3.

Theorem 1. *Let $f(x)$ and $g(x)$ be functions integrable over the Q-domain Ω.*
(**a**) *Let λ and μ be real numbers, then $\lambda f(x)+\mu g(x)$ is integrable over Ω, and*

$$\int\limits_\Omega (\lambda f(x)+\mu g(x))\,dx = \lambda \int\limits_\Omega f(x)\,dx + \mu \int\limits_\Omega g(x)\,dx\,.$$

(**b**) *$|f(x)|$ is integrable, and $\left|\int\limits_\Omega f(x)\,dx\right| \leqq \int\limits_\Omega |f(x)|\,dx$.*

(**c**) *If, moreover, $f(x)\leqq g(x)$, then $\int\limits_\Omega f(x)\,dx \leqq \int\limits_\Omega g(x)\,dx$.*

Definition. *If $\{f_j(x)\}_{j=1}^\infty$ and $f(x)$ are real-valued functions defined on $\bar{\Omega}$, then $f_j \rightrightarrows f$ on Ω (uniform convergence) means that for every positive number ε there exists a natural number $j_0(\varepsilon)$ such that $\sup\limits_{x\in\Omega} |f(x)-f_j(x)| \leqq \varepsilon$ for all j with $j\geqq j_0(\varepsilon)$.*

Theorem 2. *Let the functions $f_j(x)$, $j=1, 2, 3, \ldots$, be integrable over the Q-domain Ω, and let $f_j \rightrightarrows f$ on Ω. Then $f(x)$ is integrable over Ω, and*

$$\int\limits_\Omega f(x)\,dx = \lim_{j\to\infty} \int\limits_\Omega f_j(x)\,dx\,.$$

9.1.4. Integrable Functions

Theorem. (**a**) *Let Ω be a Q-domain, then every function that is continuous on $\bar{\Omega}$ is integrable.*

(**b**) *Let ω be an I-domain, and let $f(x)$ be continuous on $\bar{\omega}$* (Fig. 9.8). *If Ω is a Q-domain and $\bar{\omega}\subset\Omega$, then*

$$g(x) = \begin{cases} f(x) & \text{for} \quad x\in\bar{\omega} \\ 0 & \text{for} \quad x\in\bar{\Omega}-\bar{\omega} \end{cases}$$

is integrable over Ω.

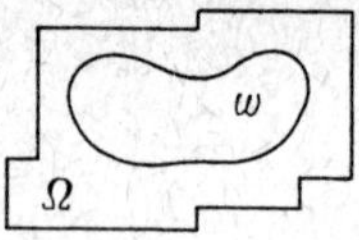

Fig. 9.8

Remark. ω and Ω are open. From $\bar{\omega} \subset \Omega$ it follows that ω has a positive distance from $\partial\Omega$. Further it is seen that $\int_\Omega g(x)\,dx$ is independent of Ω.

9.1.5. Integrals over *I*-Domains

Definition. *Let $f(x)$ be a function continuous on $\bar{\omega}$, where ω is an I-domain. Let Ω be a Q-domain with $\bar{\omega} \subset \Omega$, then one sets*

$$\int_\omega f(x)\,dx = \int_\Omega g(x)\,dx, \quad \textit{where} \quad g(x) = \begin{cases} f(x) & \textit{for} \quad x \in \bar{\omega}\,, \\ 0 & \textit{for} \quad x \in \bar{\Omega} - \bar{\omega}\,. \end{cases}$$

Remark. From Theorem 9.1.4 and Remark 9.1.4 it follows that the definition is meaningful and that $\int_\omega f(x)\,dx$ is independent of Ω.

Theorem 1. *If $f(x)$, $g(x)$ are assumed to be continuous, then Theorems* 9.1.3/1 *and* 9.1.3/2 *remain valid for I-domains, too.*

Theorem 2. *Let ω, ω_1 and ω_2 be three I-domains, and let $\bar{\omega} = \bar{\omega}_1 \cup \bar{\omega}_2$* and $\omega_1 \cap \omega_2 = \emptyset$ (Fig. 9.9), *then, for every function $f(x)$ that is continuous on $\bar{\omega}$,*

$$\int_\omega f(x)\,dx = \int_{\omega_1} f(x)\,dx + \int_{\omega_2} f(x)\,dx\,.$$

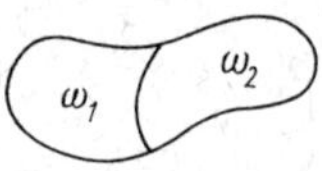

Fig. 9.9

9.1.6. Iteration Theorem for *n*-Dimensional Integrals

For a domain ω in R_n, let $\omega_c = \omega \cap \{x \mid x_n = c\}$ be the intersection of ω with the plane $x_n = c$ (Fig. 9.10). If ω is an n-dimensional Q-domain, then ω_c is either empty or an $(n-1)$-dimensional Q-domain.

Theorem. *If ω is a Q-domain, and if $f(x)$ is continuous on $\bar{\omega}$, then*

$$\int_\omega f(x)\,dx = \int_a^b \left(\int_{\omega_{x_n}} f(x^*, x_n)\,dx^* \right) dx_n\,, \tag{1}$$

where $x^* = (x_1, \dots, x_{n-1})$ and $dx^* = dx_1 \dots dx_{n-1}$ (Fig. 9.11).

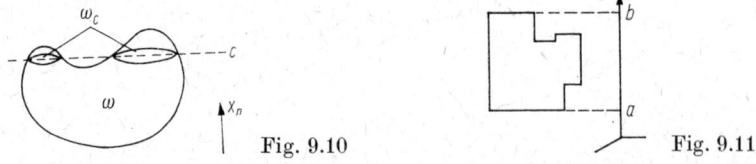

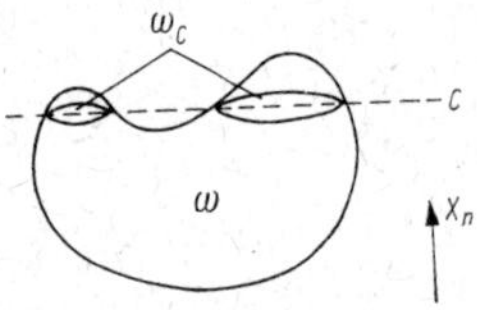

Fig. 9.10

Fig. 9.11

Remark 1. According to the above considerations, the inner, $(n-1)$-dimensional integral over ω_{x_n} is meaningful. It gives a function that is integrable over the interval $[a, b]$.

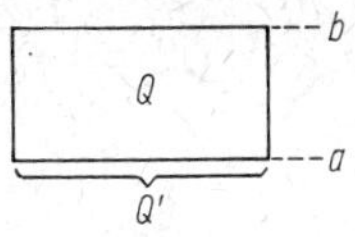

Fig. 9.12

Remark 2. If $\omega = Q$ is an n-dimensional interval (Fig. 9.12) and if $f(x) = g(x^*)\, h(x_n)$, then one obtains

$$\int_Q f(x)\, dx = \int_{Q'} g(x^*)\, dx^* \int_a^b h(x_n)\, dx_n \,.$$

Remark 3. If $\omega = Q = \{x \mid a_j < x_j < b_j\}$ is an n-dimensional interval, then an iterative application of Eq. (1) gives

$$\int_Q f(x)\, dx = \int_{a_n}^{b_n} \left(\int_{a_{n-1}}^{b_{n-1}} \cdots \left(\int_{a_1}^{b_1} f(x_1, \ldots, x_{n-1}, x_n)\, dx_1 \right) \ldots dx_{n-1} \right) dx_n \,.$$

Remark 4. The theorem is not fully satisfactory; in particular the restriction to Q-domains is inconvenient. Later on we shall generalize this theorem substantially (Fubini's theorem).

9.2. Transformation Formulas, Volume and Area Measurement

9.2.1. Volume Measurement

Definition. *Let ω be an I-domain in R_n, then $|\omega| = \int_\omega 1 dx$ is the volume of ω.*

Remark 1. This definition is compatible with the statement made in Subsec. 9.1.1., saying that the n-dimensional interval $Q = \{x \mid |x_j - x_j^0| < a_j\}$ has the volume $|Q| = 2^n a_1 \ldots a_n$. To prove this, we consider the more general case of a parallelepipedon. Let $a^j = (a_1^j, \ldots, a_n^j)$, $j = 1, \ldots, n$ be linearly independent vectors in R_n, then they span a parallelepipedon with the 2^n corners $\sum_{j=1}^{n} \varepsilon_j a^j$ (Fig. 9.13). Here $\varepsilon_j = 0$ or $\varepsilon_j = 1$. A parallelepipedon is an I-domain; its volume is denoted by $|(a^1, \ldots, a^n)|$.

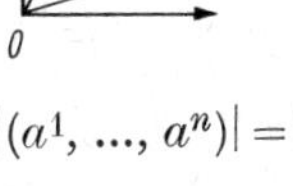

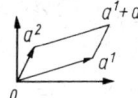

Fig. 9.13

Lemma. $|(a^1, \ldots, a^n)| = |\det (a^1, \ldots, a^n)|$, where

$$\det (a^1, \ldots, a^n) = \begin{vmatrix} a_1^1 & \ldots & a_n^1 \\ \vdots & & \vdots \\ a_1^n & \ldots & a_n^n \end{vmatrix} .$$

Remark 2. It is clear that n-dimensional intervals are special parallelepipedons and that the volume formula indicated above is valid for n-dimensional intervals.

Remark 3. Let $y_j = \sum_{k=1}^{n} d_{j,k} x_k$ be a rotation in R_n, then from the lemma as well as from elementary calculating rules for determinants it follows that the volume of a parallelepipedon is rotation-invariant. This implies that the volume of an I-domain is rotation-invariant, too.

9.2.2. Transformation Formulas

Lemma. *Let $y=f(x)=(f_1(x), \ldots, f_n(x))$ with $y^0=f(x^0)$, and let $f_j(x)$ be twice continuously differentiable in a neighbourhood of $x^0 \in R_n$. Let $z=g(y)=(g_1(y), \ldots, g_n(y))$ with $z^0=g(y^0)$, and let $g_j(y)$ be twice continuously differentiable in a neighbourhood of $y^0 \in R_n$. If $z=g(f(x))=h(x)=(h_1(x), \ldots, h_n(x))$, then $z^0=h(x^0)$, where the $h_j(x)$ are twice continuously differentiable in a neighbourhood of x^0, and* (Fig. 9.14)

$$\frac{\partial(z_1, \ldots, z_n)}{\partial(x_1, \ldots, x_n)}(x^0)=\frac{\partial(z_1, \ldots, z_n)}{\partial(y_1, \ldots, y_n)}(y^0) \cdot \frac{\partial(y_1, \ldots, y_n)}{\partial(x_1, \ldots, x_n)}(x^0) .$$

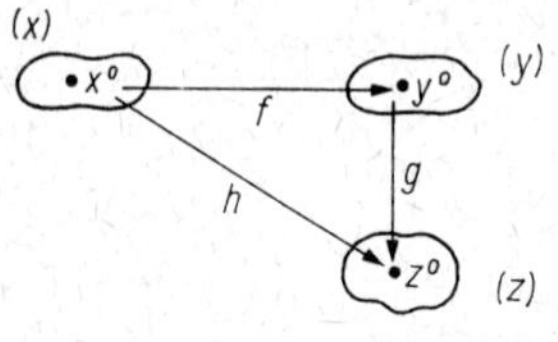

Fig. 9.14

If, moreover, $\frac{\partial(y_1, \ldots, y_n)}{\partial(x_1, \ldots, x_n)}(x^0) \neq 0$, *then for the inverse mapping $x=x(y)$ the following relation holds:*

$$\frac{\partial(y_1, \ldots, y_n)}{\partial(x_1, \ldots, x_n)}(x^0) \cdot \frac{\partial(x_1, \ldots, x_n)}{\partial(y_1, \ldots, y_n)}(y^0)=1 .$$

Remark 1. If $\frac{\partial(y_1, \ldots, y_n)}{\partial(x_1, \ldots, x_n)}(x^0) \neq 0$, then it follows from Theorem 8.2.2./1 that $y(x)$ is a one-to-one mapping of a neighbourhood ω of x^0 onto a neighbourhood Ω of y^0. In particular, there exists an inverse mapping.

Theorem. *Let $y=y(x)$ be a one-to-one mapping of the I-domain $\omega \subset R_n$ onto the I-domain $\Omega \subset R_n$, with* $\frac{\partial(y_1, \ldots, y_n)}{\partial(x_1, \ldots, x_n)}(x) \neq 0$ *for $x \in \bar{\omega}$. If $f(x)$ is continuous on $\bar{\omega}$ and if $x=x(y)$ is the inverse mapping of $y(x)$, then $g(y)=f(x(y))$ is continuous on $\bar{\Omega}$, and*

$$\int_{\omega} f(x)\, \mathrm{d}x=\int_{\Omega} g(y)\left|\frac{\partial(x_1, \ldots, x_n)}{\partial(y_1, \ldots, y_n)}(y)\right| \mathrm{d}y .$$

Remark 2. Note that it is the absolute value of the Jacobian determinant which appears in the last formula.

9.2.3. Arc Length of Curves

Differentiable curves $x(t)=(\varphi_1(t), \ldots, \varphi_n(t))$ in R_n have been dealt with in Subsec. 8.1.5. Here we assume that the tangent $v(t)=(\varphi_1'(t), \ldots, \varphi_n'(t)) \neq 0$, where the functions $\varphi_j(t)$ are assumed to be continuously differentiable on $[a, b]$, with $-\infty<a<b<\infty$. Further let $a=t_0<t_1<\ldots<t_{l+1}=b$. The points $x(t_j)$ and $x(t_{j+1})$, where $j=0, \ldots, l$, are connected by line segments (Fig. 9.15). This gives a polygon P of length $L_P=\sum_{j=0}^{l}|x(t_{j+1})-x(t_j)|$.

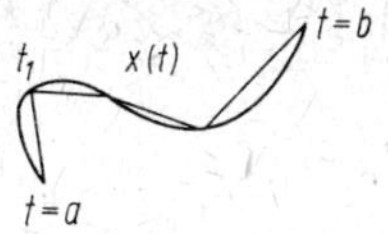

Fig. 9.15

Definition. $L = \sup L_P$ *is the length of the curve, where the supremum has to be taken over all polygons of the kind described above.*

Theorem. $L = \int_a^b \sqrt{\sum_{j=1}^{n} \varphi_i'^2(t)}\, dt.$

9.2.4. Area Measurement

We want to measure the area of surfaces F which are represented in R_{n+1} by $F(x_1, ..., x_{n+1}) = 0$ (Fig. 9.16). According to Subsec. 8.2.4. we may assume without loss of generality that $F(x_1, ..., x_n, x_{n+1}) = f(x) - x_{n+1}$, where $x = (x_1, ..., x_n)$. Hence F is represented by $x_{n+1} = f(x)$, $x \in \omega \subset R_n$. Let ω be an n-dimensional I-domain. The method is analogous to that described in Subsec. 9.2.3. For $n = 2$ we perform a triangulation: points are marked on F and interconnected by triangles. Thus the surface is approximated by a web whose segments are triangles. For $n > 2$ the web consists of patches Δ_j of n-dimensional planes. As these patches are n-dimensional I-domains, the volumes $|\Delta_j|$ can be determined according to Subsec. 9.2.1.

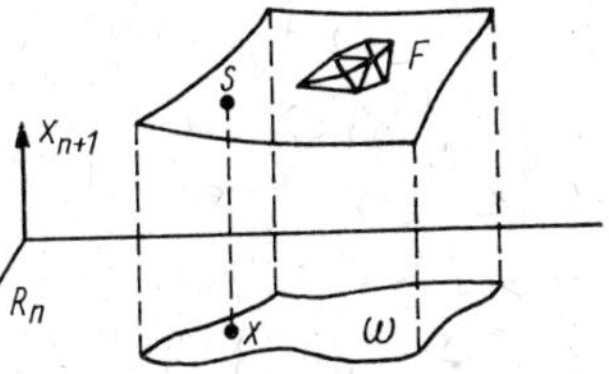

Fig. 9.16

Definition $|F| = \sup\limits_{Z} \sum\limits_{j} |\Delta_j|$ *is the area of the surface F, where the supremum has to be taken over all permissible decompositions of the kind indicated above.*

Remark 1. For $n = 2$ the definition is satisfactory, because the triangulation is easy to visualize. For $n > 2$, imagination will fail, and a more precise formulation of the above definition will be required. We shall not, however, go into greater detail here.

Theorem. *If $f(x)$ has continuous partial derivatives of first order in a neighbourhood of $\bar{\omega}$, then*

$$|F| = \int_\omega |\text{grad } F|\, dx = \int_\omega \sqrt{1 + \sum_{j=1}^{n} \left|\frac{\partial f}{\partial x_j}\right|^2}\, dx\,. \tag{1}$$

Remark 2. One has grad $F = \left(\frac{\partial f}{\partial x_1}, ..., \frac{\partial f}{\partial x_n}, -1\right) \neq 0$. The expression

$$ds = |\text{grad } F|\, dx \tag{2}$$

is usually called surface element on F. ds as well as the formula in the above theorem can be given an intuitively geometric interpretation by referring to "infinitesimal" surface elements (instead of the above plane segments). From Fig. 9.17 it is observed that $\frac{\mathrm{d}x}{\mathrm{d}s} = \frac{|\nu_{n+1}|}{|\nu|}$, where $\nu = \operatorname{grad} F$ is the normal vector and ν_{n+1} is the $(n+1)$-th component of ν. Substituting now ν and $|\nu_{n+1}| = 1$, one obtains Eq. (2), which also gives an intuitively geometric interpretation of (1).

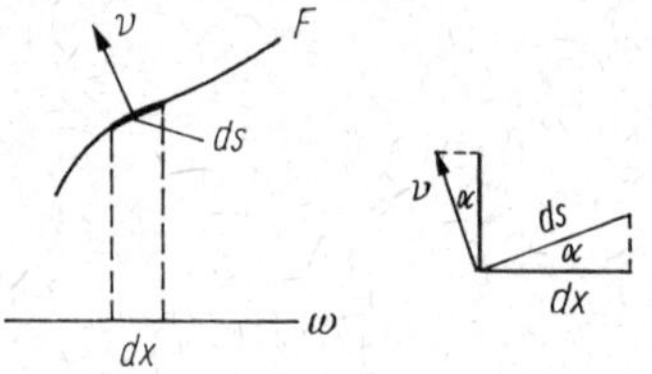

Fig. 9.17

9.2.5. Surface Integrals

We consider the same surface F as in Subsec. 9.2.4. The general point on F is denoted by s. Every $s \in F$ is assigned a unique $x \in \omega$ by $s = s(x)$. If a function $g(s)$ is defined on F, then we obtain a corresponding function $G(x) = g(s(x))$ defined on $\bar{\omega}$. If $G(x)$ is continuous on $\bar{\omega}$, $g(s)$ is said to be continuous on F.

Definition. *Let $g(s)$ be a real-valued function that is defined and continuous on F, then we set*

$$\int_F g(s)\,\mathrm{d}s = \int_\omega G(x)\,|\operatorname{grad} F|\,\mathrm{d}x\,. \tag{1}$$

Remark 1. The geometric interpretation of this formula follows from Remark 9.2.4/2. In particular, the left-hand side of Eq. (1) is independent of the specific choice of the co-ordinates $x_1, \ldots, x_n$. It is also possible to define the left-hand side of Eq. (1) directly on F by extending Def. 9.1.2 to F. Note that, for Eq. (1) to be valid, F must have the form described at the beginning of Subsec. 9.2.4.

Remark 2. Let F be an arbitrary surface $F(x_1, \ldots, x_{n+1}) = 0$ in R_{n+1}, with $(\operatorname{grad} F)(x) \neq 0$ on F. Then F can be decomposed into partial surfaces $F_1, F_2, \ldots$, where each of the partial surfaces has the standard shape indicated in Subsec. 9.2.4. (where possibly x_{n+1} must be replaced by $x_1, \ldots, x_n$; cf. Subsec. 8.2.4.) (Fig. 9.18). Now it is possible to calculate $\int_F g(s)\,\mathrm{d}s$ by adding together the integrals $\int_{F_j} g(s)\,\mathrm{d}s$.

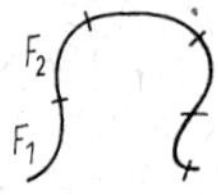

Fig. 9.18

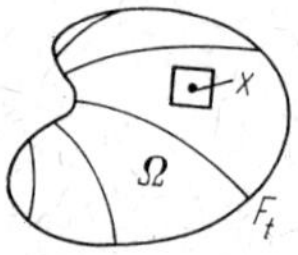

Fig. 9.19

Remark 3. A possible application of surface integrals is as follows. Let Ω be a smooth, bounded domain in R_n, being covered by smooth surfaces F_t, with $F(x) \equiv t$ (Fig. 9.19). Here t is a parameter, $t \in (a, b)$, and $F(x)$ is continuously differentiable on $\bar{\Omega}$, with $|\operatorname{grad} F(x)| \neq 0$ for $x \in \bar{\Omega}$. If

$$\mathrm{d}x = \varrho(s_t, t)\,\mathrm{d}s_t\mathrm{d}t, \quad \varrho(s_t, t) > 0\,,$$

where $\mathrm{d}s_t$ is the surface element on F_t, then one obtain s

$$\int_\Omega g(x)\mathrm{d}x = \int_a^b \Big(\int_{F_t} \tilde{g}(s_t, t)\,\varrho(s_t, t)\,\mathrm{d}s_t\Big)\,\mathrm{d}r\,.$$

Here $g(x)$ is continuous on $\bar{\Omega}$, $s_t \in F_t$, and $\tilde{g}(s_t, t) = g(x)$ if x corresponds to the point $(s_t, t) \in F_t$. The factor $\varrho(s_t, t)$ must be determined in each particular case. We consider an important example. Let ds_r be the surface element of an n-dimensional sphere of radius $r > 0$ centered at the origin (Fig. 9.20). Then $dx = ds_r dr$ (i.e., in this case $\varrho \equiv 1$ because, roughly speaking, ds_r and dr are orthogonal). If $ds = ds_1$ is the surface element of the unit sphere, then $ds_r = r^{n-1} ds$, and hence $dx = r^{n-1} ds dr$.

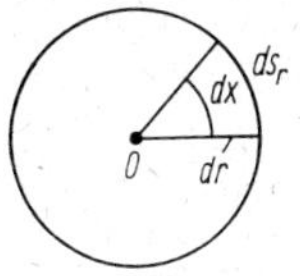

Fig. 9.20

9.2.6. The Unit Ball, $\Gamma\left(\frac{n}{2}\right)$

$|V_n|$ shall denote the volume of the unit ball $V_n = \{x \mid |x| < 1\}$ in R_n, and $|\omega_n|$ shall denote the area of the surface $\omega_n = \{x \mid |x| = 1\}$ of V_n. Well-known values are $|\omega_2| = 2\pi$ (definition of π), $|V_2| = \pi$ as well as $|\omega_3| = 4\pi$ and $|V_3| = \frac{4}{3}\pi$.

Theorem.

(a)

$$\Gamma\left(\frac{1}{2}\right) = \int_{-\infty}^{\infty} e^{-t^2}\, dt = \sqrt{\pi}\,.$$

(b)

$$\Gamma\left(\frac{n}{2}\right) = \begin{cases} \left(\frac{n}{2} - 1\right)! & \text{for } n = 2, 4, 6, \ldots \quad (0! = 1)\,, \\ \left(\frac{n}{2} - 1\right)\left(\frac{n}{2} - 2\right) \cdots \frac{3}{2} \cdot \frac{1}{2} \cdot \sqrt{\pi} & \text{for } n = 3, 5, 7, \ldots \end{cases}$$

(c)

$$|\omega_n| = \frac{2\pi^{\frac{n}{2}}}{\Gamma\left(\frac{n}{2}\right)} = \frac{(2\pi)^{\frac{n}{2}}}{(n-2)(n-4)\cdots} \cdot \begin{cases} 1 & \text{for } n = 2, 4, 6, \ldots \\ \sqrt{\frac{2}{\pi}} & \text{for } n = 3, 5, 7, \ldots, \end{cases}$$

$$|V_n| = \frac{|\omega_n|}{n}\,.$$

Remark 1. The proof is based on a trick. One calculates the n-dimensional improper Riemann integral $\int_{R_n} e^{-|x^2|}\, dx$ in two ways:

$$\left(\int_{-\infty}^{\infty} e^{-t^2}\, dt\right)^n = \int_{R_n} e^{-|x|^2}\, dx = \int_0^{\infty}\int_{\omega_n} e^{-r^2} r^{n-1}\, dr ds = \frac{|\omega_n|}{2}\, \Gamma\left(\frac{n}{2}\right),$$

where we have used the formula stated at the end of Subsec. 9.2.5. From this one obtains all the propositions of the theorem relatively easily; cf. [66], Appendix 2.

Remark 2. $|\omega_n|$ attains its maximum for $n = 7$. Further, $|\omega_n| \to 0$ as $n \to \infty$.

9.2.7. Improper Integrals

By analogy with Subsec. 7.2.1., where improper one-dimensional integrals have been considered, it is also possible to study improper n-dimensional Riemann integrals. Again we shall restrict ourselves to two types.

(1) Let ω be an I-domain in R_n, then $\omega_\varepsilon = \omega - \{x \mid |x - x^0| < \varepsilon\}$ is also an I-domain, provided that $x^0 \in \omega$ and $\varepsilon > 0$ is sufficiently small (Fig. 9.21). If $f(x)$ is continuous on $\bar{\omega} - \{x^0\}$, one may ask whether $\lim\limits_{\varepsilon \downarrow 0} \int\limits_{\omega_\varepsilon} f(x)\,dx$ exists. If this is the case, the limit is denoted by $\int\limits_{\omega} f(x)\,dx$, and called improper (Riemann) integral.

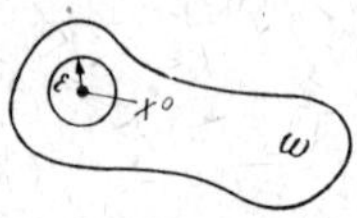

Fig. 9.21

(2) $\Omega_N = \{x \mid 1 < |x| < N\}$ is an I-domain. If $f(x)$ is continuous for $|x| \geqq 1$, then, by analogy with (1), one can investigate the limit

$$\lim_{N \to \infty} \int\limits_{\Omega_N} f(x)\,dx = \int\limits_{|x|>1} f(x)\,dx\,.$$

Lemma. (**a**) $\int\limits_{|x|<1} |x|^\alpha\,dx$ *exists if and only if* $\alpha > -n$.

(**b**) $\int\limits_{|x|>1} |x|^\alpha\,dx$ *exists if and only if* $\alpha < -n$.

Remark. This is the analogue to Example 7.2.1/2. In the case (a), $x^0 = 0$ in the sense indicated above.

9.3. Integral Theorems

9.3.1. Gauss' Theorem

Domains which are columnar in the direction of x_n are bounded by a finit number of smooth surfaces, where the boundary surfaces have the properties of the surfaces F_j of Lemma 9.1.1 (including the differentiability properties of $G_j(x)$) (Fig. 9.22). Let $x^0 = (x_1, \ldots, x_{n-1}, x_n^0)$ and $x^1 = (x_1, \ldots, x_{n-1}, x_n^1)$ be two boundary points that differ only in their x_n-co-ordinates, $x_n^0 > x_n^1$, then the line segment $\{x \mid x = (x_1, \ldots, x_{n-1}, x_n),\ x_n^1 < x_n < x_n^0\}$ is assumed to be included in the domain. Columnar domains are bounded. Riemann standard domains are domains composed of finitely many domains which are columnar with respect to all of the directions $x_1, \ldots, x_n$ (Fig. 9.23). Let $\nu = (\nu_1, \ldots, \nu_n)$ be the normed outer normal, hence $|\nu| = 1$ $\left(\text{cf. Subsec. 8.1.5.: if the surface is given by } F(x) = 0, \text{ then } \nu = \pm \frac{\text{grad } F}{|\text{grad } F|}\right)$. This normal exists at every boundary point, except for possi-

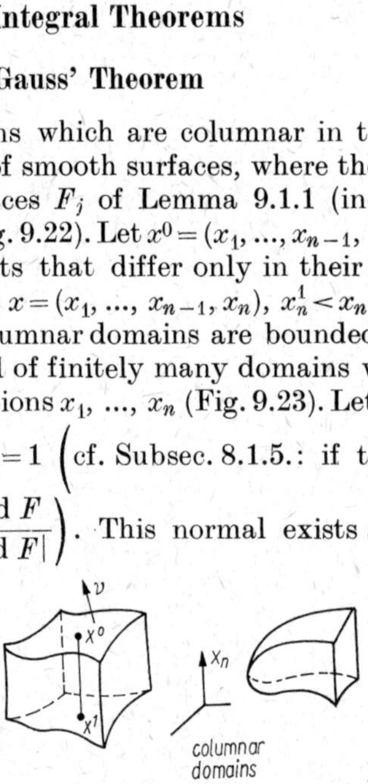

Fig. 9.22

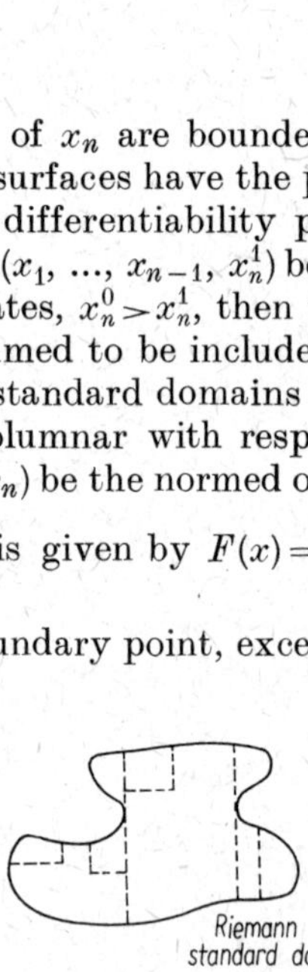

Fig. 9.23

ble ridge points, which are of no importance. Let $a=(a_1, \ldots, a_n)$ and $b=(b_1, \ldots, b_n)$ be two vectors in R_n, then $(a, b)=\sum_{j=1}^{n} a_j b_j$ is their scalar product. For $|a|=1$ and $|b|=1$ one obtains:

$$|(a, b)| \leqq \sum_{j=1}^{n} |a_j|\,|b_j| \leqq \left(\sum_{j=1}^{n} a_j^2\right)^{\frac{1}{2}} \left(\sum_{j=1}^{n} b_j^2\right)^{\frac{1}{2}} = 1 .$$

Hence we can set $(a, b)=\cos\alpha$ with $0 \leqq \alpha \leqq \pi$. α is called the angle included by the normed vectors a and b. We do not risk confusion when writing $\cos(a, b)$ instead of $\cos\alpha$, where it is understood that (a, b) in $\cos(a, b)$ does not mean the scalar product (a, b). Let $e_j=(0, \ldots, 0, 1, 0, \ldots, 0)$ be the unit vector in the direction of x_j, then $\nu_j=(\nu, e_j)=\cos(\nu, x_j)$, the latter being a common (though not quite correct) notation. Hence,

$$\nu=(\cos(\nu, x_1), \ldots, \cos(\nu, x_n)) .$$

Theorem 1 (*Gauss' integral formula*). *Let Ω be a Riemann standard domain, and let $f(x)$ be continuously differentiable on $\overline{\Omega}$, then, for $j=1, \ldots, n$,*

$$\int_{\Omega} \frac{\partial f}{\partial x_j}(x)\,\mathrm{d}x = \int_{\partial\Omega} f(s) \cos(\nu_s, x_j)\,\mathrm{d}s .$$

Here ds *is the surface element of $\partial\Omega$ and ν_s is the normed outer normal at the point $s \in \partial\Omega$.*

Definition. *If the functions $f_j(x)$, $j=1, \ldots, n$, are differentiable at the point $x^0 \in R_n$, then one denotes*

$$(\operatorname{div} f)(x^0)=\sum_{j=1}^{n} \frac{\partial f_j}{\partial x_j}(x^0), \quad \textit{where} \quad f=(f_1, \ldots, f_n) .$$

Theorem 2. *If Ω is a Riemann standard domain, and if the functions $f_j(x)$, $j=1, \ldots, n$, are continuously differentiable on $\overline{\Omega}$, then*

$$\int_{\Omega} (\operatorname{div} f)(x)\,\mathrm{d}x = \int_{\partial\Omega} (f, \nu)\,\mathrm{d}s ,$$

where $f=(f_1, \ldots, f_n)$ and $(f, \nu)=\sum_{j=1}^{n} f_j(s) \cos(\nu_s, x_j)$. Here ds *is the surface element of $\partial\Omega$ and ν_s is the normed outer normal at the point $s \in \partial\Omega$.*

9.3.2. Green's Formulas

Definition. *Let $f(x)$ be twice differentiable at the point $x^0 \in R_n$, then one sets $(\Delta f)(x^0)=\sum_{j=1}^{n} \frac{\partial^2 f}{\partial x_j^2}(x^0)$ (Laplacian differential operator).*

Theorem 1 (*First Green formula*). *If Ω is a Riemann standard domain, and if $f(x)$ is twice and $g(x)$ is once continuously differentiable on $\overline{\Omega}$, then*

$$\int_{\Omega} g(x)\,(\Delta f)(x)\,\mathrm{d}x = -\int_{\Omega} (\operatorname{grad} f, \operatorname{grad} g)\,\mathrm{d}x + \int_{\partial\Omega} g \frac{\partial f}{\partial \nu}\,\mathrm{d}s .$$

Here $\nu=\nu_s$ is the normed outer normal to $\partial\Omega$ and $\frac{\partial f}{\partial \nu}$ is the directional derivative of Subsec. 8.1.7.

Remark 1. Again $\mathrm{d}s$ is the surface element of $\partial\Omega$, and

$$(\text{grad } f, \text{grad } g) = \sum_{j=1}^{n} \frac{\partial f}{\partial x_j} \frac{\partial g}{\partial x_j}.$$

Theorem 2 (*Second Green formula*). *If Ω is a Riemann standard domain, and if $f(x)$ and $g(x)$ are twice continuously differentiable on $\bar{\Omega}$, then*

$$\int_{\Omega} (g(x)\,(\Delta f)\,(x) - f(x)\,(\Delta g)\,(x))\,\mathrm{d}x = \int_{\partial\Omega} \left(g \frac{\partial f}{\partial \nu} - f \frac{\partial g}{\partial \nu} \right) \mathrm{d}s\,.$$

Here $\nu = \nu_s$ is the normed outer normal to $\partial\Omega$, and $\frac{\partial f}{\partial \nu}$ is the directional derivative of Subsec. 8.1.7.

Remark 2. The differential and integral calculus in R_n developed in Chapters 8. and 9. frequently appeals to intuitive power. This is especially true for the integral theorems of Sec. 9.3. For a modern, rigorous but yet largely elementary introduction into this subject, the reader is referred to [60].

10. Ordinary Differential Equations (Methods of Solution)

10.1. Separable, Homogeneous, and Exact Differential Equations

10.1.1. Formulation of the Problem

In Subsecs. 4.1.2. and 4.2.3., we had formulated the typical initial value problem for ordinary differential equations of order n: desired are real functions $f(x)$ that are defined and n times continuously differentiable in a neighbourhood of x_0, solve the differential equation $f^{(n)}(x) = h(x, f(x), \ldots, f^{(n-1)}(x))$, and assume the initial values $f^{(j)}(x_0) = c_j$, where $j = 0, \ldots, n-1$. A corresponding problem has to be solved for systems of differential equations. In Chapter 4, the existence and uniqueness of the solutions of such problems has been established, provided that h exhibits certain smoothness properties. In Chapter 5 elementary functions have been defined by means of ordinary differential equations. Now we choose the reverse approach: desired are explicit solutions for special classes of ordinary differential equations. Here we utilize the theory of elementary functions and the one-dimensional integral calculus.

10.1.2. Separable Differential Equations

Theorem. *Let* $-\infty < a < b < \infty$ *and* $-\infty < c < d < \infty$ (Fig. 10.1.). *Further let $f_1(t)$ and $f_2(t)$ be continuously differentiable on $[a, b]$ and $[c, d]$, respectively, and let $f_2(t) \neq 0$ for $t \in [c, d]$. If $x_0 \in (a, b)$ and $y_0 \in (c, d)$, then the unique solution of the initial value*

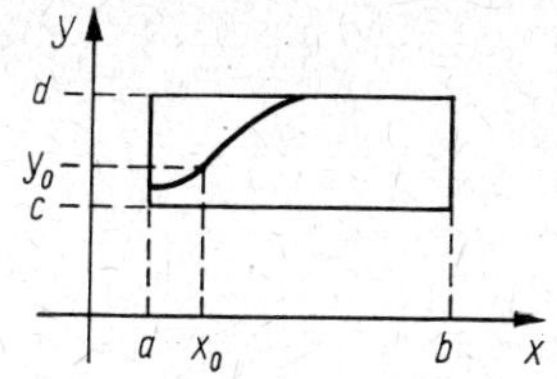

Fig. 10.1

problem,

$$y' = f_1(x)\, f_2(y), \quad y(x_0) = y_0\,, \tag{1}$$

can be calculated from

$$F(x, y) = \int_{y_0}^{y} \frac{\mathrm{d}u}{f_2(u)} - \int_{x_0}^{x} f_1(v)\,\mathrm{d}v = 0. \tag{2}$$

Remark 1. Theorem 4.2.3 implies the existence and uniqueness of the solution (at least locally in a neighbourhood of $\{x_0, y_0\}$). As $\frac{\partial F}{\partial y}(x_0, y_0) = \frac{1}{f_2(y_0)} \neq 0$, Theorem 8.2.4/1 shows that $F(x, y) = 0$ can be represented as $y = y(x)$ in a neighbourhood of x_0. One has $y_0 = y(x_0)$. This is the desired solution of Eq. (1). Whether the integrals in (2) can be calculated explicitly, and whether the resulting equation can be resolved explicitly for $y = y(x)$ by means of elementary functions, depends on f_1 and f_2.

Remark 2. Formally, one obtains (2) as follows.

$$\frac{\mathrm{d}y}{\mathrm{d}x} = f_1(x)\, f_2(x) \quad \text{is written in the form} \quad \frac{\mathrm{d}y}{f_2(y)} = f_1(x)\,\mathrm{d}x\,,$$

and the last expression is integrated:

$$\int^{y} \frac{\mathrm{d}u}{f_2(u)} = \int^{x} f_1(v)\,\mathrm{d}v + c, \quad c \text{ real}. \tag{3}$$

Here the lower limit of integration in (3) is omitted. $\int^{y} \frac{\mathrm{d}u}{f_2(u)}$ and $\int^{x} f_1(v)\,\mathrm{d}v$ are arbitrary primitives of $\frac{1}{f_2(u)}$ and $f_1(v)$, respectively, in the sense of Subsec. 3.2.5. Then the arbitrary parameter c is given a value that makes the curve pass through $\{x_0, y_0\}$.

Remark 3 (Examples). 1. Find solutions of $y' = \frac{y}{x}$. One obtains

$$\frac{\mathrm{d}y}{y} = \frac{\mathrm{d}x}{x}, \quad \ln y = \ln x + \ln c \quad \text{and} \quad y = cx\,.$$

These are straight lines passing through the origin (Fig. 10.2). If it is required that the line passes through a given point $\{x_0, y_0\}$, then c must be determined accordingly.

2. Find solutions of $y' = -\frac{x}{y}$. One obtains

$$y\mathrm{d}y = -x\mathrm{d}x, \quad \frac{1}{2}\, y^2 = -\frac{1}{2}\, x^2 + \frac{c}{2} \quad \text{and} \quad x^2 + y^2 = c\,.$$

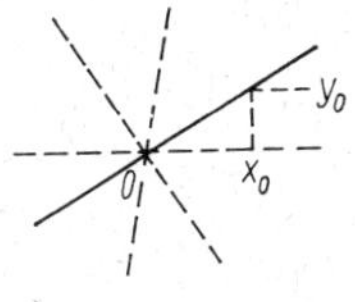

Fig. 10.2

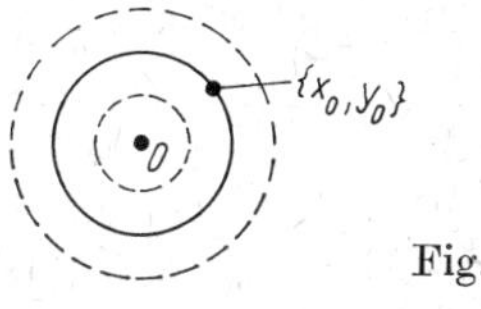

Fig. 10.3

These are concentric circles centered at the origin (Fig. 10.3). These two examples also show that it is not always desirable to resolve $F(x, y)=0$ of Eq. (2) for y.

Remark 4 (Special cases). 1. If $y'=f(x)=f_1(x)$ (i.e., if $f_2(y)\equiv 1$), then $y=y_0+\int_{x_0}^{x} f(v)\,dv$. The integral curves are shifted in parallel with the y-axis (Fig. 10.4). 2. If $y'=f(y)=f_2(y)$ (i.e. if $f_1(x)\equiv 1$), then $x-x_0=\int_{y_0}^{y} \frac{du}{f(u)}$. The integral curves are shifted in parallel with the x-axis (Fig. 10.5).

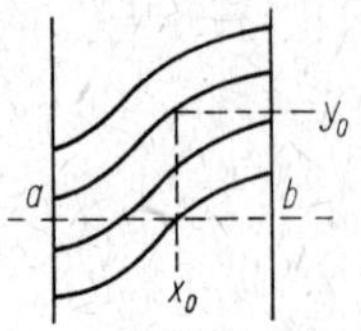

Fig. 10.4

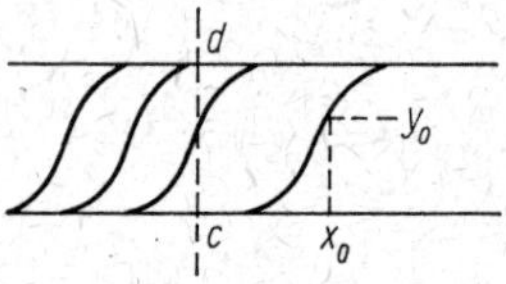

Fig. 10.5

10.1.3. Homogeneous Differential Equations

Theorem. *The setup* $y(x)=xz(x)$ *reduces the homogeneous differential equation* $y'=f\left(\frac{y}{x}\right)$ *to the separable differential equation* $z=\frac{f(z)-z}{x}$.

Remark. In the sense of Eq. (10.1.2/2), one has $f_1(x)=\frac{1}{x}$, and hence $\int^{x} f_1(v)\,dv=\ln x$. Naturally, f must have the necessary differentiability properties in order that Theorem 10.1.2 be applicable.

10.1.4. Exact Differential Equations

Theorem. *Let* $-\infty<a<b<\infty$ *and* $-\infty<c<d<\infty$ (Fig. 10.6). *Let the functions* $P(x, y)$ *and* $Q(x, y)$ *be continuously differentiable on the rectangle* $R=(a, b)\times(c, d)$, *and let*

$$\frac{\partial P}{\partial y}(x, y)=\frac{\partial Q}{\partial x}(x, y) \quad \textit{for} \quad \{x, y\}\in R. \tag{1}$$

Further let $Q(x_0, y_0)\neq 0$ *for a fixed point* $\{x_0, y_0\}\in R$. *If*

$$M(x, y)=\int_{x_0}^{x} P(u, y_0)\,du+\int_{y_0}^{y} Q(x, v)\,dv \quad \textit{for} \quad \{x, y\}\in R, \tag{2}$$

then the unique solution of

$$P(x, y)+Q(x, y)y'=0, \quad y(x_0)=y_0, \tag{3}$$

can be determined from $M(x, y)=0$ *in a neighbourhood of* $\{x_0, y_0\}$.

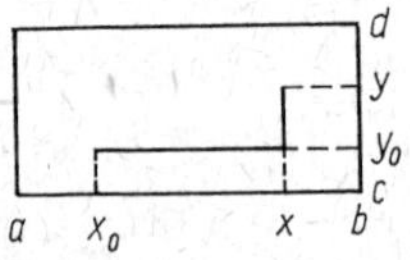

Fig. 10.6

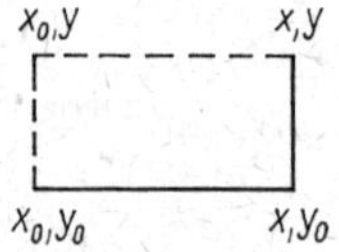

Fig. 10.7

Remark 1. In Eq. (2) we have assumed that $x \geqq x_0$ and $y \geqq y_0$. If $x < x_0$, one has to interpret (as usual) $\int_{x_0}^{x}$ as $-\int_{x}^{x_0}$. The y-integral is interpreted analogously. From Subsec. 4.2.3. it follows that Eq. (3) has a unique solution in a neighbourhood of $\{x_0, y_0\}$. Further, with the use of Eq. (1) one obtains

$$\frac{\partial M}{\partial x}(x, y) = P(x, y), \quad \frac{\partial M}{\partial y}(x, y) = Q(x, y) . \tag{4}$$

In particular one has $\frac{\partial M}{\partial y}(x, y) \neq 0$ in a neighbourhood of $\{x_0, y_0\}$. Theorem 8.2.4/1 then shows that $M(x, y) = 0$ can locally be represented as $y = y(x)$, where $y_0 = y(x_0)$. This is the solution of Eq. (3).

Remark 2. The path of integration in (2) is indicated by the solid line in Fig. 10.7. Choosing the broken line, one obtains

$$\widetilde{M}(x, y) = \int_{y_0}^{y} Q(x_0, v)\, dv + \int_{x_0}^{x} P(u, y)\, du .$$

With the use of Eq. (1) this gives $M(x, y) = \widetilde{M}(x, y)$. In other words: the two paths are completely equivalent, as it should also be expected.

Remark 3. Differential equations of the type (3), (1) are called exact differential equations. According to (4) they have the form

$$\frac{\partial M}{\partial x} + \frac{\partial M}{\partial y} y' = 0 . \tag{5}$$

Conversely, if for two continuously differentiable functions $P(x, y)$ and $Q(x, y)$ one wants to find a function $M(x, y)$ that is twice continuously differentiable and satisfies (4), then it follows from Theorem 8.1.2 that

$$\frac{\partial P}{\partial y} = \frac{\partial^2 M}{\partial x \partial y} = \frac{\partial Q}{\partial x} .$$

Hence (1) is necessary and sufficient for the differential equation in (3) to be written in the form (5).

Remark 4. If $M(x, y)$ is continuously differentiable, then

$$dM = \frac{\partial M}{\partial x} dx + \frac{\partial M}{\partial y} dy \tag{6}$$

is called the total differential of M. If $M(x, y)$ is the function of (2), then according to (4) one has

$$dM = P\, dx + Q\, dy . \tag{7}$$

Thus the approach to the solution of exact differential equations can be described as follows. Desired are functions $M(x, y)$ which show that $P\, dx + Q\, dy$ is a total differential in the sense of (6), (7). Then functions $y(x)$ for which $M(x, y(x)) \equiv c$ are solutions of $P + Qy' = 0$.

Remark 5. The differential equation $x + yy' = 0$ referred to in Remark 10.1.2/3 is of the exact type. One has $P = x$ and $Q = y$, and hence $\frac{\partial P}{\partial y} = \frac{\partial Q}{\partial x} = 0$. According to (4), $M(x, y) = \frac{1}{2}(x^2 + y^2)$. This gives the solutions. $x^2 + y^2 = c$

10.1.5. Integrating Factor

In Subsec. 10.1.4. we dealt with differential equations of the type $P + Qy' = 0$ by representing them in the form (10.1.4/5). The necessary and sufficient condition for this approach was that (10.1.4/1) was true. If this condition is not fulfilled, it is possible to look for an exact differential equation which has the same solution $y(x)$ as $P+Qy'=0$. If, for example, $\mu(x, y)>0$, then every solution of $P+Qy'=0$ is also a solution of

$$\mu(x, y)\, P(x, y)+\mu(x, y)\, Q(x, y)\, y'=0 \tag{1}$$

and vice versa. (1) becomes an exact differential equation if

$$0=\frac{\partial}{\partial y}(\mu P)-\frac{\partial}{\partial x}(\mu Q)=\frac{\partial \mu}{\partial y} P-\frac{\partial \mu}{\partial x} Q+\mu\left(\frac{\partial P}{\partial y}-\frac{\partial Q}{\partial x}\right). \tag{2}$$

For given P and Q this is a partial differential equation for $\mu(x, y)$. In general, partial differential equations are much more complicated than ordinary ones. A systematic theory of solution for (2) cannot be stated. But in special cases it is possible to guess solutions $\mu(x, y)$. Then $\mu(x, y)$ is called integrating factor. It reduces the original problem to an exact differential equation.

10.2. Linear Differential Equations of First Order

10.2.1. The Equation $y'=f(x)y$

Theorem. *Let* $-\infty<a<b<\infty$ *and* $x_0\in[a, b]$. *If* $f(x)$ *is continuous on* $[a, b]$, *then*

$$y(x)=y_0 e^{\int_{x_0}^{x} f(u)\mathrm{d}u}, \quad a\leqq x\leqq b, \tag{1}$$

is the unique solution of the problem

$$y'=f(x)\, y, \quad y(x_0)=y_0 .$$

Remark 1. As usual, one has to interpret $\int_{x_0}^{x}$ as $-\int_{x}^{x_0}$ if $x<x_0$.

Remark 2. The conditions of problem 4.1.2/1 are satisfied. This, according to Theorem 4.2.2, ensures the existence and uniqueness of the solution. $y'=f(x)y$ is a special separable differential equation as described in Subsec. 10.1.2. In this case, however, it is easy to find an explicit solution of $F(x, y)=0$ of Eq. (10.1.2/2) (where Theorem 8.2.4/1 with its additional differentiability assumptions is not required).

10.2.2. The Inhomogeneous Linear Differential Equation

Theorem. *Let* $-\infty<a<b<\infty$ *and* $x_0\in[a, b]$. *If* $f(x)$ *and* $g(x)$ *are continuous on* $[a, b]$, *then*

$$y(x)=y_0 e^{\int_{x_0}^{x} f(u)\mathrm{d}u}+e^{\int_{x_0}^{x} f(u)\mathrm{d}u}\int_{x_0}^{x} e^{-\int_{x_0}^{z} f(u)\mathrm{d}u} g(z)\,\mathrm{d}z, \quad a\leqq x\leqq b, \tag{1}$$

is the unique solution of the problem

$$y'=f(x)\, y+g(x), \quad y(x_0)=y_0 . \tag{2}$$

Remark 1. As regards the interpretation, existence and uniqueness, see the remarks of Subsec. 10.2.1.

Remark 2. What is more interesting than the complicated formula for $y(x)$ is its derivation by means of a method of variation of the constant. The general solution of the homogeneous equation $y' = f(x)\,y$ is obtained from Eq. (10.2.1/1) by regarding $y_0 = c$ as an arbitrary real parameter. To solve the differential equation in (2), one uses the setup

$$y(x) = c(x)\, e^{\int_{x_0}^{x} f(u)du},$$

i.e., the constant c in the solution of the homogeneous equation is varied, $c = c(x)$. Substitution of this setup into the equation $y' = f(x)y + g(x)$ gives the following differential equation for $c(x)$:

$$c'(x) = g(x)\, e^{-\int_{x_0}^{x} f(u)du}.$$

The solution of this equation is (1). When the functions $f(x)$ and $g(x)$ are given explicitly, it is often preferable to follow this straightforward way of solution instead of using the given formula (1): this will simplify the calculation.

10.3. Linear Systems of Differential Equations of First Order

10.3.1. Fundamental Systems and Wronski's Determinant

If $f_{j,k}(x)$ are real-valued functions continuous on $[a, b]$, then

$$y_j'(x) = \sum_{k=1}^{n} f_{j,k}(x)\, y_k(x), \quad y_j(x_0) = y_j^{(0)}, \tag{1}$$

where $j = 1, \ldots, n$, has a unique solution. Here $-\infty < a < b < \infty$ and $a \leqq x \leqq b$. Further, $y_1^{(0)}, \ldots, y_n^{(0)}$ are given real numbers and $x_0 \in [a, b]$ (Fig. 10.8). This follows from Subsec. 4.1.2. (problem 2) and Theorem 4.2.1. In contrast to the general case discussed there, (1) is linear with respect to $y_1, \ldots, y_n$ (this explains the name "linear systems of differential equations").

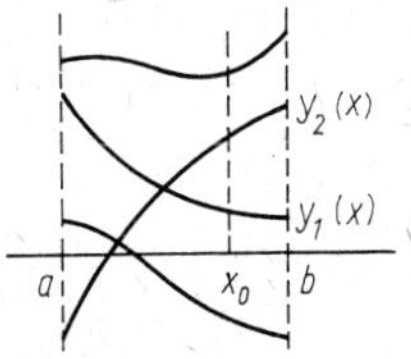

Fig. 10.8

Theorem 1. *Let* $(y_{1,1}(x), \ldots, y_{1,n}(x))$ *and* $(y_{2,1}(x), \ldots, y_{2,n}(x))$ *be solutions of*

$$y_j'(x) = \sum_{k=1}^{n} f_{j,k}(x)\, y_k(x), \quad j = 1, \ldots, n. \tag{2}$$

If c_1 *and* c_2 *are real numbers, then* $(c_1 y_{1,1}(x) + c_2 y_{2,1}(x), \ldots, c_1 y_{1,n}(x) + c_2 y_{2,n}(x))$ *is also a solution of* (2) (*linear combination*).

Remark 1. What is meant is that (2) shall hold with $y_{1,k}$ instead of y_k, or with $y_{2,k}$ instead y_k (corresponding to $y_{1,j}'$ or $y_{2,j}'$). The proof is simple. For abbreviation, from now on we shall write $\vec{y}(x) = (y_1(x), \ldots, y_n(x))$. The linear combination in the above theorem is then written as $c_1\vec{y}_1(x) + c_2\vec{y}_2(x)$, where it is understood that $\vec{y}_1$ and $\vec{y}_2$ are the two solutions. Now the following problem arises: can every solution of (2) be represented as a linear combination of special solutions?

Definition 1. *The solutions* $\vec{y}_j(x)=(y_{j,1}(x), \dots, y_{j,n}(x))$ *of* (2), *with* $j=1, \dots, n$, *are called fundamental system if, for all* $x\in[a, b]$,

$$W(x)=\begin{vmatrix} y_{1,1}(x) & \dots & y_{1,n}(x) \\ \vdots & & \vdots \\ y_{n,1}(x) & \dots & y_{n,n}(x) \end{vmatrix} \neq 0$$

($W(x)$ *is called Wronski's determinant*).

Remark 2. Hence for the $\vec{y}_j$ one has

$$y'_{j,k}(x)=\sum_{l=1}^{n} f_{k,l}(x)\, y_{j,l}(x) \tag{2'}$$

where $k=1, \dots, n$ and $j=1, \dots, n$.

Theorem 2. (a) *There exist fundamental systems for* (2).

(b) *If* $W(x_0)\neq 0$ *for a point* $x_0\in[a, b]$ *with respect to the system* $\vec{y}_1, \dots, \vec{y}_n$ *of solutions of* (2), *then* $W(x)\neq 0$ *for all* x *with* $a\leq x\leq b$.

(c) *If* $\vec{y}_1(x), \dots, \vec{y}_n(x)$ *is a fundamental system of* (2), *then* $\vec{y}(x)=\sum_{j=1}^{n} c_j\vec{y}_j(x)$ *is the general solution of* (2). *Here* $c_1, \dots, c_n$ *are arbitrary real constants.*

Remark 3. To prove (b), one shows that

$$W'(x)=\left(\sum_{l=1}^{n} f_{l,l}(x)\right) W(x) \quad \text{for} \quad a\leq x\leq b\,.$$

If $W(x_0)=0$ for a point $x_0\in[a, b]$, then it follows from the uniqueness theorem that $W(x)\equiv 0$ for $a\leq x\leq b$. This leads to the following alternative: either $W(x)\neq 0$ for all $x\in[a, b]$ or $W(x)\equiv 0$ for all $x\in[a, b]$. Now the proof of (a) is straightforward. The solutions $\vec{y}_1, \dots, \vec{y}_n$ of (2), for instance, are a fundamental system if $y_{j,k}(x_0)=\delta_{j,k}$ for a point $x_0\in[a, b]$.

Definition 2. *The solutions* $\vec{y}_j(x)=(y_{j,1}(x), \dots, y_{j,n}(x))$ *of* (2) (*or* (2′)), *where* $j=1, \dots, m$, *are said to be linearly dependent if there exist real numbers* $c_1, \dots, c_m$, *with* $\sum_{j=1}^{m} |c_j|>0$, *such that*

$$\sum_{j=1}^{m} c_j\vec{y}_j(x)\equiv 0 \quad \textit{for all} \quad x\in[a, b]\,. \tag{3}$$

If such numbers $c_1, \dots, c_m$ *do not exist, so that* (3) *holds for all* $x\in[a, b]$, *then the solutions* $\vec{y}_1, \dots, \vec{y}_m$ *are said to be linearly independent.*

Theorem 3. *The solutions* $\vec{y}_1(x), \dots, \vec{y}_n(x)$ *of* (2) (*or* (2′)) *are linearly dependent if and only if* $W(x)\equiv 0$ *for all* $x\in[a, b]$.

Remark 4. Consequently, according to Remark 3, $\vec{y}_1, \dots, \vec{y}_n$ are linearly independent if and only if there exists a point $x_0\in[a, b]$ such that $W(x_0)\neq 0$.

10.3.2. Inhomogeneous Systems of Differential Equations

In Remark 10.2.2/2 we had described the method of variation of the constants for the solution of inhomogeneous differential equations. The method can be transferred to systems. Desired are solutions of the inhomogeneous linear system of differential equations of first order

$$y'_j(x)=\sum_{k=1}^{n} f_{j,k}(x)\, y_k(x)+g_j(x), \quad j=1, \dots, n\,. \tag{1}$$

Here $f_{j,k}(x)$ and $g_j(x)$ are real functions that are continuous on $[a, b]$, with $-\infty < a < b < \infty$. As regards the existence and uniqueness of solutions, the statements made in Subsec. 10.3.1. are also true here (cf. Theorem 4.2.1). Let $\vec{y}_j = (y_{j,1}, \ldots, y_{j,n})$, $j = 1, \ldots, n$, be a fundamental system for (10.3.1/2) in the sense of Def. 10.3.1/1, then (by analogy with Subsec. 10.2.2.) we make the setup

$$\vec{y}(x) = \sum_{l=1}^{n} c_l(x)\, \vec{y}_l(x)\,. \tag{2}$$

If this setup is introduced in (1), one obtains the following differential equations for the functions $c_1(x), \ldots, c_n(x)$:

$$\sum_{l=1}^{n} c_l'(x)\, y_{l,j}(x) = g_j(x), \quad j = 1, \ldots, n \quad \text{and} \quad x \in [a, b]\,. \tag{3}$$

The determinant of the $(y_{l,j}(x))_{l,j=1}^{n}$ is $W(x)$, and hence is different from zero for $x \in [a, b]$. Consequently, (3) can be resolved for $c_1'(x), \ldots, c_n'(x)$, and integrated. Substitution of the result into (2) gives a special solution $\vec{y}_0(x)$ of (1). The general solution of (1) is then given by

$$\vec{y}(x) = \vec{y}_0(x) + \sum_{j=1}^{n} b_j \vec{y}_j(x), \quad x \in [a, b]\,,$$

where $b_1, \ldots, b_n$ are arbitrary real numbers. Hence, if a fundamental system for the homogeneous system of equations is known, it can be used to construct the general solution of the inhomogeneous system.

10.3.3. Special Systems of Differential Equations

Suppose that all the assumptions made in Subsecs. 10.3.1. and 10.3.2. are fulfilled. In addition, let $f_{j,k}(x) \equiv 0$ for $k > j$. Consequently, (10.3.2./1) has the form

$$\begin{aligned}
y_1' &= f_{1,1}y_1 && + g_1\\
y_2' &= f_{2,1}y_1 + f_{2,2}y_2 && + g_2\\
y_3' &= f_{3,1}y_1 + f_{3,2}y_2 + f_{3,3}y_3 && + g_3\\
&\vdots \qquad\qquad \ddots && \vdots\\
y_n' &= f_{n,1}y_1 + \cdots \qquad + f_{n,n}y_n && + g_n\,.
\end{aligned}$$

Here $f_{j,k} = f_{j,k}(x)$, $g_j = g_j(x)$ and $y_j = y_j(x)$. Such triangular systems can be solved explicitly. The solution $y_1(x)$ is found according to Subsec. 10.2.2. Substituting $y_1(x)$ into the second row, one finds the solution $y_2(x)$ likewise according to Subsec. 10.2.2. Iteration shows that in this way all the $y_j(x)$ can be determined step by step. By this method it is possible to fulfil given initial conditions $y_j(x_0) = a_j$, $j = 1, \ldots, n$.

Remark. In general it is not possible to find explicit solutions for arbitrary systems of the form (10.3.2/1).

10.3.4. Systems of Differential Equations with Constant Coefficients

From the considerations set out in Subsecs. 10.3.2. and 10.3.3., it follows that systems of the form (10.3.2/1) can be solved explicitly if they can be reduced to triangular systems as described in Subsec. 10.3.3. This is not generally possible, except when the functions $f_{j,k}(x)$ of (10.3.2/1) are constants. Further it follows from Subsec. 10.3.2. that we may restrict ourselves to the homogeneous case $g_j(x) \equiv 0$. Setting

$$F = \begin{pmatrix} f_{1,1} & \dots & f_{1,n} \\ \vdots & & \vdots \\ f_{n,1} & \dots & f_{n,n} \end{pmatrix}, \quad f_{j,k} \text{ real numbers },$$

we can write the homogeneous system (10.3.2/1) with constant coefficients $f_{j,k}(x) = f_{j,k}$ as

$$\vec{y}'(x) = F\vec{y}(x) \tag{1}$$

where, as previously, $\vec{y} = (y_1, \dots, y_n)$ and $F\vec{y} = \left(\sum_{j=1}^{n} f_{1,j}y_j, \dots, \sum_{j=1}^{n} f_{n,j}y_j \right)$ (matrix multiplication). With the setup $\vec{z} = A\vec{y}$, where $A = (a_{j,k})_{j,k=1}^{n}$ is a matrix with constant coefficients, Eq. (1) becomes

$$\vec{z}'(x) = AFA^{-1}\vec{z}(x)\,. \tag{2}$$

Here we assume that the determinant of $(a_{j,k})_{j,k=1}^{n}$ is different from zero. From the matrix calculus it is known that A can be so chosen that AFA^{-1} has a triangular form (with zeros above the main diagonal). This system (and hence also the original system of equations) can be solved in the manner described in Subsec. 10.3.3. The details, which shall not be discussed here, are rather complicated. A special case corresponding to a differential equation of order n shall be studied more closely later on.

10.4. Linear Differential Equations of Higher Order

10.4.1. Formulation of the Problem

In $[a, b]$, with $-\infty < a < b < \infty$, we consider the linear differential equation of order n,

$$y^{(n)}(x) = \sum_{j=0}^{n-1} f_j(x)\, y^{(j)}(x) + g(x)\,, \tag{1}$$

where $f_0(x), \dots, f_{n-1}(x)$ and $g(x)$ are real-valued functions continuous on $[a, b]$. In particular, the assumptions of Problem 1 of Subsec. 4.1.2. are fulfilled. Then, according to Theorem 4.2.2., there exists a unique solution of (1) for given initial conditions $y^{(j)}(x_0) = c_j$, $j = 0, \dots, n-1$, where $x_0 \in [a, b]$ (Fig. 10.9). As has been stated in Remark 4.2.2, Eq. (1) reduces to a special system of first order. If $z_j(x) = y^{(j)}(x)$ for $j = 0, 1, \dots, n-1$, then

$$\begin{aligned} z_0' &= z_1 \\ z_1' &= \quad z_2 \;\; \ddots \\ &\vdots \\ z_{n-2}' &= \qquad\qquad z_{n-1} \\ z_{n-}' &= f_0(x)\, z_0 + \dots + f_{n-1}(x)\, z_{n-1} + g(x)\,. \end{aligned}$$

This is a special system of equations, but unfortunately it has not the form dealt with in Subsec. 10.3.3. The conclusions of 10.3.1., 10.3.2. and 10.3.4. can, however, be transferred to this case, which shall be done in the next two subsections.

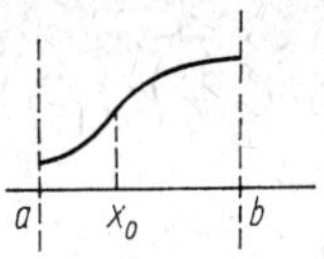

Fig. 10.9

10.4.2. Fundamental Systems and Wronski's Determinant

The functions $f_j(x)$ have the same meaning as in Subsec. 10.4.1.

Definition. *The solutions* $y_1(x), \ldots, y_n(x)$ *of the homogeneous differential equation*

$$y^{(n)}(x) = \sum_{j=0}^{n-1} f_j(x)\, y^{(j)}(x) \tag{1}$$

are called a fundamental system if, for all $x \in [a, b]$,

$$W(x) = \begin{vmatrix} y_1(x) & y_1'(x) & \ldots & y_1^{(n-1)}(x) \\ \vdots & \vdots & & \vdots \\ y_n(x) & y_n'(x) & \ldots & y_n^{(n-1)}(x) \end{vmatrix} \neq 0$$

(*Wronski's determinant*).

Remark 1. $W(x)$ coincides with $W(x)$ of Def. 10.3.1/1 with respect to the system of Subsec. 10.4.1.

Remark 2. As in Def. 10.3.1/2, solutions $y_1(x), \ldots, y_m(x)$ of (1) are called linearly dependent if there exist real numbers $c_1, \ldots, c_m$ such that $\sum_{j=1}^{m} |c_j| > 0$ and $\sum_{j=1}^{m} c_j y_j(x) \equiv 0$ for $x \in [a, b]$. If such numbers do not exist, then $y_1(x), \ldots, y_m(x)$ are called linearly independent.

Theorem. (**a**) *If* $y_1(x), \ldots, y_m(x)$ *are solutions of* (1), *and if* $c_1, \ldots, c_m$ *are real numbers, then* $\sum_{j=1}^{m} c_j y_j(x)$ *is also a solution of* (1).

(**b**) *If* $y_1(x), \ldots, y_m(x)$ *is a fundamental system, then every solution* $y(x)$ *of* (1) *can be represented as* $y(x) = \sum_{j=1}^{n} c_j y_j(x)$, *with suitable real numbers* $c_1, \ldots, c_n$.

(**c**) *There exist fundamental systems for* (1).

(**d**) *The solutions* $y_1(x), \ldots, y_n(x)$ *of* (1) *are linearly dependent if and only if* $W(x) \equiv 0$ *for* $x \in [a, b]$.

(**e**) *If* $y_1(x), \ldots, y_n(x)$ *are solutions of* (1), *then either* $W(x) \neq 0$ *for all* $x \in [a, b]$ *or* $W(x) \equiv 0$ *for all* $x \in [a, b]$.

Remark 3. The above theorem follows from Subsec. 10.4.1. and the considerations set out in Sec. 10.3. for systems of differential equations.

10.4.3. Differential Equations with Constant Coefficients

For real numbers $f_1, \ldots, f_n$, where $f_n = 1$, we consider the linear differential equation of order n

$$\sum_{j=0}^{n} f_j y^{(j)}(x) \equiv 0 \tag{1}$$

with constant coefficients. It is seen immediately that $y(x)=e^{\lambda x}$ is a solution of (1) if and only if λ is a (generally complex) zero of the polynomial $P(\lambda)=\sum_{j=0}^{n} f_j\lambda^j$. (For the moment, we also admit complex functions as solutions.) Thus it is clear that pairwise different zeros $\lambda_1, \ldots, \lambda_l$ with their multiplicities $m_1, \ldots, m_l$ play a decisive role. According to Subsec. 5.3.3. one has

$$P(\lambda)=\sum_{j=0}^{n} f_j\lambda^j=(\lambda-\lambda_1)^{m_1}\ldots(\lambda-\lambda_l)^{m_l}.$$

Theorem.

$$\begin{matrix} e^{\lambda_1 x}, & xe^{\lambda_1 x}, & \ldots, & x^{m_1-1}e^{\lambda_1 x}, \\ e^{\lambda_2 x}, & xe^{\lambda_2 x}, & \ldots, & x^{m_2-1}e^{\lambda_2 x}, \\ \vdots & \vdots & & \vdots \\ e^{\lambda_l x}, & xe^{\lambda_l x}, & \ldots, & x^{m_l-1}e^{\lambda_l x} \end{matrix} \tag{2}$$

is a (complex) fundamental system for (1).

Remark. These are n functions, because $n=m_1+\ldots+m_l$. According to Theorem 5.3.3/3, $\bar{\lambda}_1$ is likewise a zero of $P(\lambda)$, with multiplicity m_1 (here we can assume that $\operatorname{Im}\lambda_1 \neq 0$). If, for example, $\lambda_2=\bar{\lambda}_1$, then it is easy to obtain $2m_1=m_1+m_2$ real linearly independent solutions of (1) by linear combination of the first two rows of (2). By pairwise combination it is thus possible to construct a real fundamental system. On the other hand, (2) is also a fundamental system if the coefficients f_j are complex (and in this case a real combination is neither generally possible nor desirable).

10.5. Continuous Dependence on Initial Values

10.5.1. Systems of Differential Equations of First Order

We consider first-order systems

$$y_j'(x)=h_j(x, y_1(x), \ldots, y_n(x)), \quad y_j(x_0)=c_j, \tag{1}$$

$j=1, \ldots, n$, on $[a, b]$. Here $-\infty<a<b<\infty$ and $x_0\in[a, b]$. The functions $h_j(x_1, \ldots, x_n)$ are assumed to fulfil the conditions of Problem 4.1.2/2. According to Theorem 4.2.1, the system (1) has a unique solution. The question to be answered is how this solution $y_j(x, c_1, \ldots, c_n)$, $j=1, \ldots, n$, depends on the initial values $c_1, \ldots, c_n$.

Theorem. *There exist a number $\delta>0$ and a number $L=L(\delta)>0$ such that*

$$\sum_{j=1}^{n} |y_j(x, c_1, \ldots, c_n)-y_j(x, \tilde{c}_1, \ldots, \tilde{c}_n)| \le L \sum_{l=1}^{n} |c_l-\tilde{c}_l|$$

for all real numbers $c_1, \ldots, c_n, \tilde{c}_1, \ldots, \tilde{c}_n$ and all $x\in[x_0-\delta, x_0+\delta]\cap[a, b]$ (continuous dependence on the initial values).

Remark. This is a local statement. Accordingly, local assumptions concerning the functions h_j, for example in the sense of Subsec. 4.2.3., will also be sufficient.

10.5.2. Differential Equations of Order n

According to Subsec. 10.4.1., differential equations of order n can be reduced to systems of first-order differential equations. Consequently there is also an analogue to Theorem 10.5.1 for differential equations of order n. Let us consider the problem

$$y^{(n)}(x)=h(x, y(x), y'(x), \ldots, y^{(n-1)}(x)), \quad y^{(j)}(x_0)=c_j\,, \tag{1}$$

where $j=0, \ldots, n-1$, $x\in[a, b]$. Here, $-\infty<a<b<\infty$ and $x_0\in[a, b]$. The function $h(x, x_1, \ldots, x_n)$ is assumed to fulfil the conditions of Problem 4.1.2/1. According to Theorem 4.2.2, Eq. (1) has a unique solution $y(x, c_0, \ldots, c_{n-1})$.

Theorem. *There exist a number $\delta>0$ and a number $L=L(\delta)>0$ such that*

$$\sum_{j=0}^{n-1} |y^{(j)}(x, c_0, \ldots, c_{n-1})-y^{(j)}(x, \tilde{c}_0, \ldots, \tilde{c}_{n-1})|\leqq L \sum_{l=0}^{n-1} |c_l-\tilde{c}_l|$$

for all real numbers $c_0, \ldots, c_{n-1}, \tilde{c}_0, \ldots, \tilde{c}_{n-1}$ and all $x\in[x_0-\delta, x_0+\delta]\cap[a, b]$ (continuous dependence on initial values).

Remark. The statement made in Remark 10.5.1 applies analogously.

10.5.3. Continuous Dependence on the Right-Hand-Side

We shall restrict ourselves to differential equations of first order. Corresponding assertions are, however, also true for systems for first-order equations and for differential equations of higher order. As in Subsec. 4.2.3., let Ω be an open domain in the x,y-plane. Let $\{x_0, y_0\}\in\Omega$ and $\{x_0, \tilde{y}_0\}\in\Omega$. Further let $h(x, y)$ and $\tilde{h}(x, y)$ be real-valued functions that are continuous and bounded on Ω and fulfil the Lipschitz condition

$$|h(x, y)-h(x, y')|\leqq M\,|y-y'|, \quad |\tilde{h}(x, y)-\tilde{h}(x, y')|\leqq M\,|y-y'|$$

for all $\{x, y\}\in\Omega$ and $\{x, y'\}\in\Omega$ (cf. Subsec. 4.2.3.). According to Theorem 4.2.3, the differential equations

$$y'(x)=h(x, y(x)), \quad y(x_0)=y_0\,,$$

and

$$\tilde{y}'(x)=\tilde{h}(x, \tilde{y}(x)), \quad \tilde{y}(x_0)=\tilde{y}_0\,,$$

have unique solutions (this holds at least locally on an interval $[x_0-\delta, x_0+\delta]$) (Fig. 10.10). The problem now is to compare these two solutions.

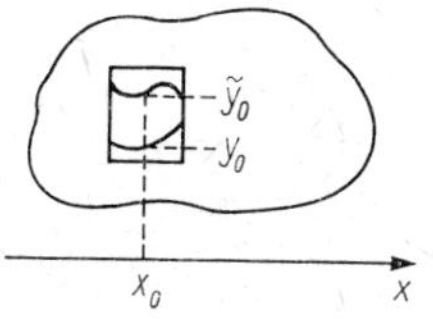

Fig. 10.10

Theorem. *There exist a number $\delta>0$ and a number $L=L(\delta)>0$ such that*

$$|\tilde{y}(x)-y(x)|\leqq L\,(\delta \sup_{\{u,v\}\in\Omega} |h(u, v)-\tilde{h}(u, v)|+|y_0-\tilde{y}_0|)$$

for all admissible $x_0, y_0, \tilde{y}_0$ and $x\in[x_0-\delta, x_0+\delta]$.

Remark 1. "Admissible" means that the rectangles of Fig. 4.5 in Subsec. 4.2.3. are included in Ω.

Remark 2. The theorems stated in Subsecs. 10.5.1. and 10.5.2. as well as the above theorem are useful for numerical and physical applications: small perturbations of the initial values and of the differential equation will have little effect on the solution.

Remark 3. Qualitative aspects of ordinary differential equations play an essential role in modern mathematics. The books [4, 5, 27] give an introduction into these fields of the theory.

These barbarians measure everything with the same yardstick, the theorem as well as the epigram ... May they distrust their senses narcotized by the opium of differential and integral calculus. (Friedrich II, King of Prussia, on Euler and d'Alembert, 1762)

11. Calculus of Variations

11.1. Fundamental Equations of the Calculus of Variations

11.1.1. Formulation of the Problems

To illustrate the problems to be solved, let us consider a simple case that will be generalized later on. Let $L(t, u_1, v_1)$ be a real-valued function that is continuous on $[a, b]\times R_2$. For a function $x(t)$ which is continuously differentiable on $[a, b]$, one may consider the integral

$$L(x(t))=\int_a^b L(t, x(t), \dot{x}(t))\,\mathrm{d}t \quad \text{where} \quad \dot{x}(t)=\frac{\mathrm{d}x}{\mathrm{d}t}(t)\,. \tag{1}$$

The problem is to find functions continuously differentiable on $[a, b]$, such that $L(x(t))$ attains an extremum.

1st Problem (*variational problems with free boundary values*). Admitted to competition are all functions $x(t)$ that are continuously differentiable on $[a, b]$.

2nd Problem (*variational problems with fixed boundary values*). Admitted to competition are only those functions continuously differentiable on $[a, b]$ which assume given boundary values at a and b: $x(a)=c_a$ and $x(b)=c_b$ (Fig. 11.1).

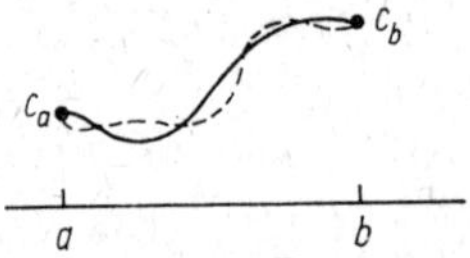

Fig. 11.1

Neighbouring curves. We shall now describe the method by which extremal curves, or extremals, are determined. For a given $x(t)$, let

$$I(\varphi, \varepsilon)=\int_a^b L(t, x(t)+\varepsilon\varphi(t), \dot{x}(t)+\varepsilon\dot{\varphi}(t))\,\mathrm{d}t\,,$$

where $\varphi(t)$ is an arbitrary function which is continuously differentiable on $[a, b]$ (with $\varphi(a)=\varphi(b)=0$ in the case of the second problem). ε is a real parameter. $x(t)$ is called extremal if $\frac{\mathrm{d}}{\mathrm{d}\varepsilon} I(\varphi, \varepsilon)\,|_{\varepsilon=0}=0$ for all admissible $\varphi(t)$. This does not quite correspond to the usual approach in the calculus of variations, but will suffice for our purposes. In this way, variational problems for functions have been reduced to the problem of finding extreme values of functions; cf. Sec. 8.3. In Sec. 8.3., points had been varied in order to determine extreme values of functions. Now it is functions that are varied.

11.1.2. Preparatory Considerations

We need two preparatory lemmas, which are also of interest on their own: one concerning the differentiation of parameter-dependent integrals, and the fundamental lemma of the calculus of variations.

Lemma 1. *Let $L(t, \lambda)$ be a real-valued function that is continuous on $[a, b]\times R_1$, with $-\infty<a<b<\infty$, and differentiable with respect to λ, with $t\in[a, b]$ being fixed. Let $\frac{\partial}{\partial\lambda} L(t, \lambda)$ be continuous on $[a, b]\times R_1$. If $a(\lambda)$ and $b(\lambda)$ are two real-valued functions being continuously differentiable on R_1, with $a<a(\lambda)<b(\lambda)<b$, then $\int_{a(\lambda)}^{b(\lambda)} L(t, \lambda)\,\mathrm{d}t$ is continuously differentiable on R_1, and*

$$\frac{\mathrm{d}}{\mathrm{d}\lambda}\int_{a(\lambda)}^{b(\lambda)} L(t, \lambda)\,\mathrm{d}\lambda = L(b(\lambda), \lambda)\, b'(\lambda) - L(a(\lambda), \lambda)\, a'(\lambda) + \int_{a(\lambda)}^{b(\lambda)} \frac{\partial}{\partial\lambda} L(t, \lambda)\,\mathrm{d}t\,.$$

Lemma 2 (*fundamental lemma of the calculus of variations*). *Let $y(t)$ be continuous on $[a, b]$, and let*

$$\int_a^b y(t)\,\varphi(t)\,\mathrm{d}t = 0$$

for all functions $\varphi(t)$ that are continuously differentiable on $[a, b]$ and vanish in a neighbourhood on the right of a as well as in a neighbourhood on the left of b, (Fig. 11.2), *then $y(t)=0$ on $[a, b]$.*

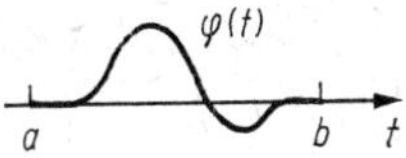

Fig. 11.2

11.1.3. The Euler-Lagrange Equations

Definition. *Suppose that $L(t, u_1, \ldots, u_n, v_1, \ldots, v_n)$ is real-valued and continuous on $[a, b]\times R_{2n}$, with $-\infty<a<b<\infty$, where it has continuous partial derivatives of first and second order with respect to all of its variables. If ε is a real number, and if $x_1(t), \ldots, x_n(t), \varphi_1(t), \ldots, \varphi_n(t)$ are real-valued functions that are continuously differentiable on $[a, b]$, then let*

$$I(\varphi_1, \ldots, \varphi_n, \varepsilon) = \int_a^b L(t, x_1(t)+\varepsilon\varphi_1(t), \ldots, x_n(t)+\varepsilon\varphi_n(t), \dot{x}_1(t)+\varepsilon\dot{\varphi}_1(t), \ldots, \dot{x}_n(t)+\varepsilon\dot{\varphi}_n(t))\,\mathrm{d}t.$$

(**a**) *$x(t)=(x_1(t), \ldots, x_n(t))$ is called extremal of the variational problem with free boundary values if*

$$\frac{\mathrm{d}}{\mathrm{d}\varepsilon} I(\varphi_1, \ldots, \varphi_n, \varepsilon)\,|_{\varepsilon=0} = 0 \tag{1}$$

for all $\varphi(t)=(\varphi_1(t), \ldots, \varphi_n(t))$ as defined above.

(**b**) *$x(t)$ is called extremal of the variational problem with fixed boundary values if* (1) *holds for all of the above $\varphi(t)$ that satisfy $\varphi_j(a)=\varphi_j(b)=0$ for $j=1, \ldots, n$.*

Remark 1. This is a generalization (and more precise definition) of the situation described in Subsec. 11.1.1.

Theorem 1 (*Euler-Lagrange equations*). *$x(t)$ is an extremal of the variational problem with fixed boundary values if and only if*

$$\frac{\partial L}{\partial x_j}-\frac{\mathrm{d}}{\mathrm{d}t}\left(\frac{\partial L}{\partial \dot{x}_j}\right)=0 \quad \textit{for} \quad j=1, \ldots, n \quad \textit{and} \quad t\in[a, b]\,. \tag{2}$$

Remark 2. The equations (2) are to be understood as follows. The functions

$$\frac{\partial L}{\partial x_j}=\frac{\partial L}{\partial u_j}(t, x_1(t), \ldots, \dot{x}_n(t)), \quad \frac{\partial L}{\partial \dot{x}_j}=\frac{\partial L}{\partial v_j}(t, x_1(t), \ldots, \dot{x}_n(t))=g_j(t)$$

depend on $t\in[a, b]$ alone. Here u_j and v_j have the same meaning as in the above definition. So one first has to take the partial derivatives in (2), then insert $x(t)$, and thereafter form $\frac{\mathrm{d}}{\mathrm{d}t}g_j(t)=\frac{\mathrm{d}}{\mathrm{d}t}\left(\frac{\partial L}{\partial \dot{x}_j}\right)$.

Theorem 2. *$x(t)$ is an extremal of the variational problem with free boundary values if (2) holds and, in addition,*

$$\frac{\partial L}{\partial \dot{x}_j}(b, x_1(b), \ldots, \dot{x}_n(b))=\frac{\partial L}{\partial \dot{x}_j}(a, x_1(a), \ldots, \dot{x}_n(a))=0 \tag{3}$$

is satisfied for $j=1, \ldots, n$.

Remark 3. In the case of the variational problem with free boundary values one automatically gets certain boundary conditions for an extremal (so-called natural boundary conditions).

Remark 4. Taking the derivative with respect to t in (2), one obtains

$$\frac{\partial L}{\partial x_j}-\frac{\partial^2 L}{\partial t\partial \dot{x}_j}-\sum_{l=1}^{n}\frac{\partial^2 L}{\partial x_l\partial \dot{x}_j}\dot{x}_l-\sum_{l=1}^{n}\frac{\partial^2 L}{\partial \dot{x}_l\partial \dot{x}_j}\ddot{x}_l=0 \tag{4}$$

for $j=1, \ldots, n$. Here $\ddot{x}_l=\frac{\mathrm{d}^2 x_l}{\mathrm{d}t^2}$. This is a system of ordinary differential equations of second order, which is generally non-linear, since $L=L(t, x_1(t), \ldots, \dot{x}_n(t))$. Using the same trick as in, say, Subsec. 10.4.1., one can reduce this system to a system of first order. If the determinant of $\left(\frac{\partial^2 L}{\partial \dot{x}_l\partial \dot{x}_j}\right)_{l,j=1}^{n}$ is different from zero, then it is possible to resolve (4) for the highest-order derivatives.

11.2. Examples

11.2.1. A Preliminary Physical Remark

The considerations presented in Subsec. 11.1.3. are of great importance for theoretical physics. One attempts to describe physical processes (e.g. in classical mechanics, electrodynamics, or in the special or general theory of relativity) by variational principles. Such a physical situation is assigned a Lagrangian density, which corresponds closely the function L of Def. 11.1.3. The physical process in question is then found to take place so that $L(x, (t))$, as defined in (11.1.1/1), assumes an extreme value. If L coincides with the function of Def. 11.1.3, then this means that the process is described by the Euler-Lagrange equations (11.1.3/2). Many physical processes are described by partial differential equations rather than by ordinary ones. If one wishes to stick to the concept of variational principles, the considerations of Sec. 11.1. must be substantially generalized. In Chap. 24 we will investigate more closely the mathematical and physical aspects of this principle. For classical point mechanics, however, the results of Sec. 11.1. are fully sufficient. A more detailed description, especially of the physical aspect of this matter, will be given at the beginning of Chap. 12. In Sec. 11.2. we consider a few examples which will, however give an impression of the physical importance of variational principles.

11.2.2. The Brachistochrone

Consider a mass point (e.g., a steel ball) that moves under the action of gravity on a certain path line (e.g., a rail) from A to B (Fig. 11.3). Suppose that the motion takes place in a plane perpendicular to the earth's surface. Further suppose that initially the ball is at rest at the point A. For which path (or paths) will the ball require minimum time to get from A to B? If the ball moves in the straight line from A to B (broken line), then this is the shortest path indeed, but it will take some time for the ball to accelerate. If, on the other hand, a path like that indicated by the dash-dot line is chosen, then the ball will rapidly reach high velocities, but the path has now become comparatively long. The physical compromise might be the path indicated by the solid line. Let A be the origin of a y,x-co-ordinate system (Fig. 11.4). Let $y(x)$ be the path of the ball, i.e., $y(0)=0$, and let $v=v(x)$ be its velocity. If m is the mass of the ball, then its total energy at the point P is given by

$$\mathrm{E}=\frac{m}{2}v^2-mgy=\text{kinetic energy}+\text{potential energy}.$$

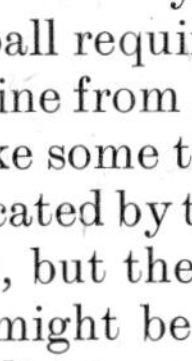

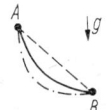

Fig. 11.3

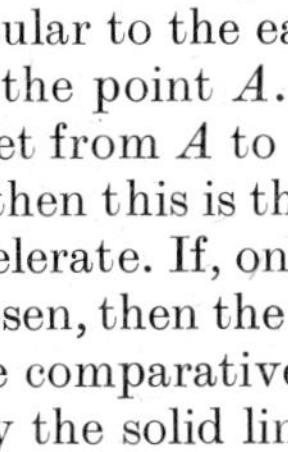

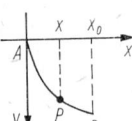

Fig. 11.4

Here g denotes the acceleration of gravity, $\mathrm{g}\approx 9.81\,\frac{\mathrm{m}}{\mathrm{s}^2}$. Using the physical fact that $\mathrm{E}(x)=\mathrm{E}$ is a constant independent of x (energy conservation law), and the initial conditions $v(0)=y(0)=0$, one obtains $v(x)=\sqrt{2\mathrm{g}y(x)}$. If $s(x)$ denotes the length of the curve $y(x)$ from A to P, then in the light of Subsec. 9.2.3. one obtains

$$s(x)=\int_0^x \sqrt{1+y'^2(z)}\,\mathrm{d}z,\quad y'=\frac{\mathrm{d}y}{\mathrm{d}z}.$$

If t is the time required for the ball to move from A to P, then t is a strictly increasing function, and for the inverse function $s=s(t)$ one has $v(t)=\frac{\mathrm{d}s}{\mathrm{d}t}$. Then, by Theorem 3.1.4, $\frac{\mathrm{d}t}{\mathrm{d}s}=\frac{1}{v}$. If T is the time required for the ball to move from A to B on the path $y(x)$, one now obtains

$$T=\int_0^T \mathrm{d}t=\int_0^{x_0}\frac{\mathrm{d}t}{\mathrm{d}s}\cdot\frac{\mathrm{d}s}{\mathrm{d}x}\,\mathrm{d}x=\int_0^{x_0}\frac{1}{v}\sqrt{1+y'^2}\,\mathrm{d}x=\int_0^{x_0}\sqrt{\frac{1+y'^2(x)}{2\mathrm{g}y(x)}}\,\mathrm{d}x.$$

Thus the problem of minimizing T is an extreme value problem in the sense of Def. 11.1.3, where

$$L(x, y, y') = \sqrt{\frac{1+y'^2}{2gy}}\,. \tag{1}$$

Paths for which T attains a minimum are called brachistochrones. The petty difficulty that L becomes singular for $y=0$ can be avoided. For that purpose it is convenient to consider the curve in a parametric representation, $y=y(u)$ and $x=x(u)$.

Theorem. *The brachistochrones are cycloids,*

$$x(u)=c(u-\sin u), \quad y(u)=c(1-\cos u) \tag{2}$$

where $0 \leq u \leq 2\pi$, *and* $c>0$ *is a parameter.*

Remark 1. The curve (2) is obtained by solving the Euler-Lagrange equation (11.1.3/2). Here $n=1$, and one has to substitute x for t and y for x. One obtains an ordinary differential equation of second order, where it has been taken into account in (2) that the only solutions of interest are those passing through A. $x(u)$ is strictly monotonic, and hence there is an inverse function, and (2) can be represented as $y=y(x)$. By variation of c one obtains a schlicht and gapless covering of the whole quadrant $\{x>0, y>0\}$. Hence, if B is given, there exists exactly one $c>0$ such that (2) passes through A and B (Fig. 11.5).

Remark 2. The cycloid allows of the following nice geometric interpretation. Suppose a wheel is rolling on a road. Find the trajectory of the valve of this wheel (Fig. 11.6). If c is the radius of the wheel, then Figs. 11.7 and 11.8 show that the cycloid (2) describes the trajectory of the valve.

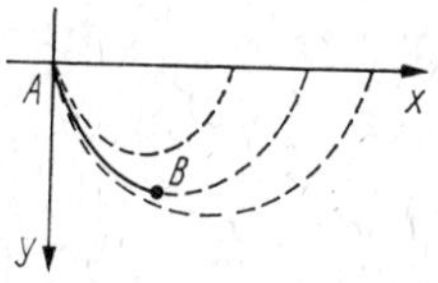

Fig. 11.5

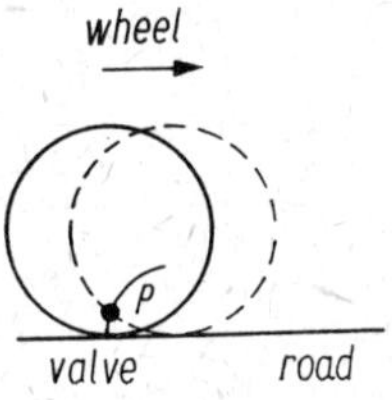

Fig. 11.6

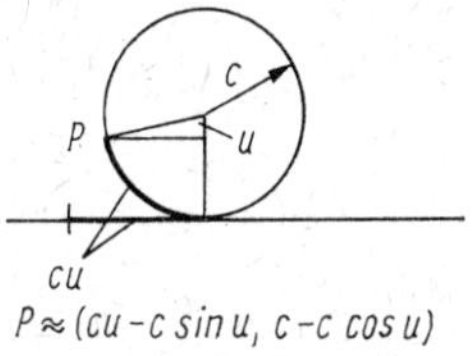

Fig. 11.7

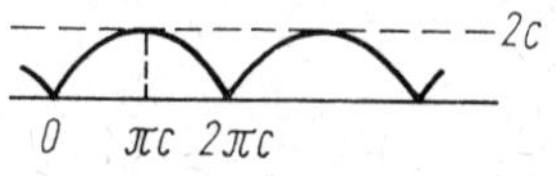

Fig. 11.8

11.2.3. The Problem of the Straight Line as the Shortest Line Connecting Two Points

(This is the heading of the 4th problem of the famous lecture by D. Hilbert on "Mathematische Probleme", Paris, 1900 [85].) It deals with curves $x(t)=(x_1(t), \ldots, x_n(t))$, $t_0 \leq t \leq t_1$, which are continuously differentiable in R_n and pass through two given points A and B, with $x(t_0)=A$ and $x(t_1)=B$. It is further supposed that $\dot{x}(t)=(\dot{x}_1(t), \ldots, \dot{x}_n(t)) \neq 0$ for $t_0 \leq t \leq t_1$. According to Theorem 9.2.3,

$$L=\int_{t_0}^{t_1} \sqrt{\dot{x}_1^2(t)+\ldots+\dot{x}_n^2(t)}\, dt \tag{1}$$

is the length of the curve from A to B. What is desired are curves of minimum length.

Theorem. *The straight line from A to B is the unique curve of minimum length between these two points.*

Remark 1. This is a variational problem as defined in Subsec. 11.1.3., with $L(v_1, \ldots, v_n) = \sqrt{v_1^2 + \ldots + v_n^2}$. As regards the condition that L be twice differentiable, there is a minor difficulty at 0, which can, however, be avoided by choosing the parameter to be, say, $t = x_1$, provided that A and B have different co-ordinate values x_1 (the total number of comparable curves will then be somewhat restricted, but this is of no consequence). If the above formulation is retained, the solution of the differential equation (11.1.3/2) gives the desired result.

Remark 2. We will study problems of this kind in detail later on in the fields of differential geometry and general relativity. Thus one may look for curves that lie in a given surface and pass through two given points. Considerations of this kind are necessary, for example, in calculating the paths of light rays in four-dimensional curved space-times by the general theory of relativity.

11.2.4. Rotation-Symmetric Minimal Surfaces

Take a closed wire sling and dip it into soap solution. After removing it from the solution, you may have the luck to find a thin liquid film spanning the sling; it realizes a surface of minimum area that passes through the sling. In other words, one considers two-dimensional surfaces that pass through a given closed curve in R_3. Desired are surfaces of minimum area in the sense of Subsec. 9.2.4. This is Plateau's problem; the corresponding surfaces are called minimal surfaces. The general theory of these surfaces is relatively complex, but a special case is within our reach: given two points A and B in the upper half-plane. Let $y(x)$ be a curve in the upper half-plane which is twice continuously differentiable and connects these two points. If this curve is allowed to rotate about the x-axis, it generates a two-dimensional surface of revolution in R_3 (Fig. 11.9).

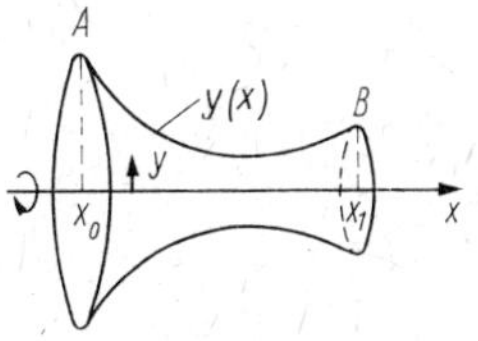

Fig. 11.9

Lemma. *The area of the surface of revolution just described is*

$$L = 2\pi \int_{x_0}^{x_1} y(x) \sqrt{1 + y'^2(x)}\, dx .$$

Remark 1. What is now desired are curves $y(x)$ that minimize the area L of this surface. We treat this question as a variational problem in the sense of Def. 11.1.3, with $L(u, v) = 2\pi u \sqrt{1 + v^2}$. Solving the associated Euler-Lagrange differential equation (11.1.3/2), one obtains the following result.

Theorem. *If a is a positive parameter and b a real one, then*

$$y(x) = a \cosh \frac{x - b}{a} \tag{1}$$

is the generating line of a rotation-symmetric minimal surface. Here cosh *denotes the hyperbolic cosine defined in Subsec.* 7.1.4.

Remark 2. The generated rotation-symmetric minimal surface is called catenoid. We shall not discuss here how a and b must be chosen (provided that this is possible) in order to make the curve (1) pass through the given points A and B.

> ... the task is suggested to us ... to treat those physical disciplines axiomatically in which already today mathematics plays a prominent role; above all, these fields are the theory of probability and mechanics.
> (D. Hilbert, "Mathematische Probleme", Paris, 1900, [85])

12. Principles of Classical Mechanics

12.1. Modelling in Physics

12.1.1. On the Relationship between Mathematics and Physics

When we speak of physics here, we always mean theoretical physics. A mathematician takes a different view of physics than does a physicist. The latter is interested in including experimental facts into existing theories, in the (qualitative and quantitative) prediction of new effects, and in the development of new theories if the old ones do no longer perform satisfactorily. For him, mathematics is a tool. (In circles of well-meaning physicists, they like to compare physics to a queen served by mathematics in a double role: bearing the trail behind and the torch ahead.) New theories (or new propositions within existing theories) must carry conviction, the mathematics used must be credible. Here it is quite customary (and perhaps even necessary, if one breaks new ground) that mathematical deduction and physical argumentation go hand in hand. This presupposes a sure physical feeling. In the history of physics there are many very impressive examples of this. Perhaps a mathematician would feel reminded of the instinctive certainty with which once Euler handled divergent infinite series. It is just this physical feeling that even physically interested mathematicians (and other ones we are not speaking of here) are often lacking in. For a mathematician, the mathematics used must be not only credible but definitely logically founded: and this makes an enormous difference. Further, he will try to separate mathematics carefully from its physical interpretation. This leads to axiomatics and modelling. The heart of such an axiomatic approach is a mathematical theory which, per se, has nothing to do with physics. Interconnecting mechanisms and interpretations are required to translate the physical problems into the mathematical language used. The situation resembles that of a computer which, taken separately, is capable of functioning and has its internal logics, but is of no use until a whole set of input and output units have been connected. By their experimental colleagues, theoretical physicists are often cross-examined about the sense or use of their theories. They take their revenge for that on the mathematicians who deal with the axiomatics of physical disciplines. The question they ask is: "What will this bring about physically?" There are two possible answers: (1) nothing, and (2) certainty. The first answer has been given from the questioner's point of view, for axiomatics cannot make any contribution to those problems he is interested in (see above): it does not create any physically new theories, nor explain any new effects. After all, axiomatics seems to be possible (or even desirable?) not until the physical discipline in question has developed to a relatively advanced state. (Perhaps like a black hole as the decline of life of a one-time shining star? This would also account for the magic attraction.) The second answer hits the gist of the matter: axiomatics creates mathematical certainty.[1] It is like the construction of wide highways that open up a wonderful physical landscape for mathematical tourism. (After a period of modesty, the physical footsloggers will also tramp along these ways.) Fortification of the base is of advantage also for those pioneers who cut their way with the cutlass through the jungle at the frontline of physics. In this sense, the axiomatics of physical disciplines is not only mathematically desirable, but also physically necessary.

Remark. We quote a few relevant opinions of outstanding physicists and mathematicians. The Nobel laureate S. Weinberg writes in [71]: "Physics is not a finished logical system. Rather, at any moment it spans a great confusion of ideas, some that survive like folk epics from the heroic periods of the past, and others that arise like utopian novels from our dim premonitions of a future grand synthesis". In R. K. Sachs, H. Wu [53], one finds the following sentences: "Physical theories are guessed, not deduced; if only deduction were required, every competent hack could be an Einstein or a Feynman . . . No wonder a mathematician, accustomed to regarding a mathematical framework as primary, finds the plethora of alternatives confusing. The physics view is that nature is primary, and that even the most careful, mathematically rigorous mathematical models are at best approximations to nature. So the physicist finds nothing odd in having to switch models occasionally, and indeed revels in the intuitive arguments needed to decide which model is least inaccurate physically". The Nobel laureate E. Segrè writes in [57]: "Mathematics is the natural language of physics, as Galileo pointed out, and altough Volta and Faraday wrote great physics without using a formal mathematical language, they thought mathematically, and their ignorance of standard mathematics makes them less, not more, intelligible".

In "Physics and Philosophy" the Nobel laureate W. Heisenberg remarks: "The conception of objective reality . . . evaporated into the . . . mathematics that represents no longer the behaviour of elementary particles but rather our knowledge of this behaviour". In "The Character of Physical Law" the Nobel laureate R. Feynman writes: "Mathematicians like to make their reasoning as general as possible. If you say to them 'I want to talk about ordinary three dimensional space' they say 'If you have a space of n dimensions, then here are the theorems'. 'But I only want the case 3'. 'Well substitute n=3' . . . So a certain amount of reducing is necessary, because the mathematicians have prepared these things for a wide range of problems. This is very useful, and later on it always turns out that the poor physicist has to come back and say, 'Excuse me, when you wanted to tell me about four dimensions . . .' . . . The mathematical rigour of great precision is not very useful in physics".

---

[1] It was noted by Bertrand Russell that the axiomatic method has many advantages, similar to the advantages of shirking honest work.

Moreover, it is an error to believe that rigor of argumentation would be adverse to simplicity.
(D. Hilbert, "Mathematische Probleme", Paris, 1900 [85])

12.1.2. Mathematical Models

A mathematical model for a physical discipline can be subdivided into five components (Fig. 12.1).

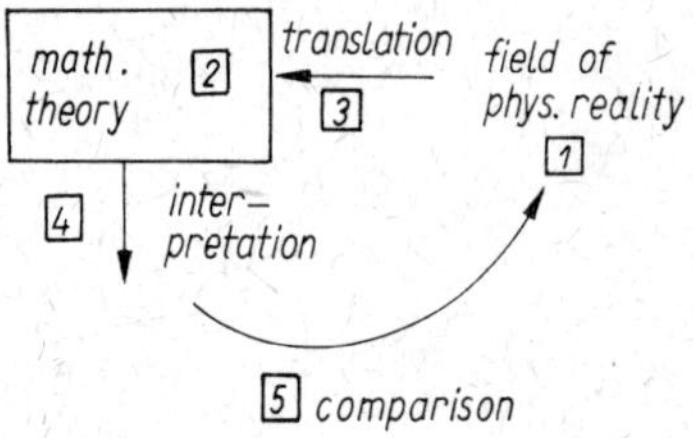

Fig. 12.1

1. A given field of physical reality, such as mechanics, electrodynamics, thermodynamics, relativity theory, quantum mechanics etc. What is meant here is the whole set of physical processes classified under the respective keyword. This is not a very precise description, but in concrete cases the meaning will be clear.
2. An intramathematical theory which (as far as its internal logical structures are concerned) has nothing to do with physics.
3. A rule of translation that relates physical quantities to corresponding mathematical ones.
4. When translated, the theory yields data (real numbers) that must be interpreted as physical quantities.
5. Finally, the theoretically obtained, interpreted data must be compared with corresponding experimental material. If the result of this comparison is not fully satisfactory, this means a criticism of the translation, the interpretation, and of the expediency of using the particular mathematical theory concerned. It is not, however, a criticism of the internal structure of this theory, which can at best be criticized from an intramathematical, but not from a physical point of view.

Physical laws should have mathematical beauty
(P. A. M. Dirac)

12.1.3. Criteria for Models

1. A model is good if it works. In this sense there are no true or false models but only suitable and unsuitable ones. Here the term "suitable" means that, in the sense of Subsec. 12.1.2., all effects of a defined field of physical reality can be described within a given accuracy of measurement. Consequently a model is characterized not only by the subject it describes but also by the accuracy of

this description. To give an example, classical point mechanics is an excellent theory, provided that the velocities involved are small compared with the velocity of light. It is only for high velocities and high accuracies that it must be replaced by the special theory of relativity.

2. Mathematically, a model should be as simple as possible without making concessions with respect to mathematical rigor. Hence there must be substantial physical arguments in order that a theory be replaced by a new, more complex one. At the time of Newton, special relativity (in competition with classical mechanics) would have violated this postulate.

3. There is a hierarchy of theories and associated models. A new theory covers a larger field of reality with a higher accuracy than did the corresponding old theory. History shows that the inventory of mathematical means used in these theories becomes more and more complex. Very likely the full potential of a theory can be surveyed not until the next better theory is made out in outlines. During the development of point mechanics, our ancestors in physics believed that it would give a definite description of everything that happens in the world. They had been dazzled by the glory of this theory, and hugely overestimated its potential (hindsight is easier than foresight[1])). At least the physicists have not made this mistake a second time: theories and models always give only approximative descriptions.

4. An essential justification of dealing with theories and models lies in the ability of the latter ones to predict new effects. There are impressive examples of this ability: Einstein's prediction that light rays are deflected by heavy masses, or Dirac's prediction of the existence of positrons. Formulas are indeed wiser than men.[2])

Remark. We quote a few sentences from R. Feynman, The Character of Physical Law: "The moons of Jupiter appeared to get sometimes eight minutes ahead of time and sometimes eight minutes behind time, where the time is the calculated value according to Newton's Law . . . O. Roemer (1644–1710), having confidence in the Law of Gravitation, came to the interesting conclusion that it takes light some time to travel from the moons of Jupiter to the earth, . . . In this way he was able to determine the velocity of light. This was the first demonstration that light was not an instantaneously propagating material. I bring this particular matter to your attention because it illustrates that when a law is right it can be used to find another one. If we have confidence in a law, then if something appears to be wrong it can suggest to us another phenomenon".

12.1.4. An Example

In this book we shall deal with several physical theories, following closely the model thinking described above. To illustrate this, let us consider here a simple example.

---

[1]) "Given for one instant an intelligence which could comprehend all the forces of nature and the respective situation of the things that compose it . . . for it nothing would be uncertain and the future, as the past, would be present to its eyes" (Laplace, Philosophical Essay on Probabilities, 1812–1820).

[2]) Heinrich Hertz: "One cannot escape the feeling that these mathematical formulas have an independent existence and an intelligence of their own, that they are wiser than we are, wiser even than their discoverers, that we got more out of them than was originally put into them".

General scheme	Application to mechanics	Special example (12.2.2.)
1. field of physical reality	classical point mechanics	falling stone
2. mathematical theory	extremal principle, Eulerian equations	$\frac{\partial L}{\partial x} - \frac{\mathrm{d}}{\mathrm{d}t}\frac{\partial L}{\partial \dot{x}} = 0$, $\ddot{x} = \mathrm{g}$, $x = \frac{\mathrm{g}}{2}t^2 + c_1 t + c_2$
3. translation	Lagrangian function $L(t, x_j, \dot{x}_j)$	$L(t, x, \dot{x}) = \frac{m}{2}\dot{x}^2 + mgx$
4. interpretation	$x_j(t)$ trajectories of particles	$x(t)$: way
5. comparison		Galilei on the leaning tower of Pisa

12.2. The Model for Point Mechanics

12.2.1. Hamilton's Principle

The field of physical reality to be considered here is a system of n mass points moving in the three-dimensional Euclidean space. These mass points are acted upon by internal and external forces. To give an example, gravitational forces are internal, whereas electromagnetic fields acting upon charged particles are external forces. Suppose that the position of the j-th particle at the time t is $(x_{3j-2}(t), x_{3j-1}(t), x_{3j}(t)) \in R_3$. Hence the total system is described at the time t by a point $(x_1(t), \ldots, x_{3n}(t)) \in R_{3n}$, with the interpretation given above. To every system there is associated a Lagrangian function $L(t, x_j(t), \dot{x}_j(t))$, which must be determined by means of physical considerations.

Axiom (*Hamilton's principle*). *A system of n particles with the Lagrangian function $L(t, x_1(t), \ldots, x_{3n}(t), \dot{x}_1(t), \ldots, \dot{x}_{3n}(t))$ moves so that $x(t) = (x_1(t), \ldots, x_{3n}(t))$ is an extremal of the associated variational problem with fixed boundary values.*

Remark 1. This in effect determines the cycle pointed out in Subsec. 12.1.2. The mathematical theory is the calculus of variations referred to in Subsec. 11.1.3. Translation is accomplished by finding a suitable Lagrangian, $L(t, x_1, \ldots, x_{3n}, \dot{x}_1, \ldots, \dot{x}_{3n})$. The axiom and the theory of Subsec. 11.1.3. lead to the Euler-Lagrange equations (11.1.3/2) (here it is always assumed that L has the necessary differentiability properties). Their solution gives trajectories $(x_1(t), \ldots, x_{3n}(t))$ in R_{3n}. The interpretation then identifies $(x_{3j-2}(t), x_{3j-1}(t), x_{3j}(t))$ as the position of the j-th particle in the physical three-dimensional space.

Theorem. *The equations of motion of a system of n particles with the (twice continuously differentiable) Lagrangian function $L(t, x_1, \ldots, x_{3n}, \dot{x}_1, \ldots, \dot{x}_{3n})$ are the Euler-Lagrange equations*

$$\frac{\partial L}{\partial x_j} - \frac{\mathrm{d}}{\mathrm{d}t}\left(\frac{\partial L}{\partial \dot{x}_j}\right) = 0, \quad j = 1, \ldots, 3n. \tag{1}$$

Remark 2. The theorem is an immediate consequence of Theorem 11.1.3/1. The equations (1) are also called Lagrange's equations of the second kind. The above considerations show that the problem is reduced to two questions: (i) determination of L on the basis of pysical principles, and (ii) solution of the system of differential equations (1). If the determinant of $\left(\frac{\partial^2 L}{\partial \dot{x}_l \partial \dot{x}_k}\right)_{l,k=1}^{n}$ is always different from zero, then (1) can be resolved for $\ddot{x}_l$, as has been stated in Remark 11.1.3/4. The theorems of Chap. 4 then show that (1) has a local unique solution which assumes given values $x_j(t_0) = a_j$ and $\dot{x}_j(t_0) = b_j$, $j = 1, \ldots, 3n$, at a given time t_0 (positions and velocities at t_0).

12.2.2. An Example (Free Fall)

Let us explain the example of Subsec. 12.1.4. more closely. We consider the free fall with $x = 0$ for $t = 0$. Let

$$L(t, x(t), \dot{x}(t)) = \frac{m}{2} \dot{x}^2 + mgx = T - V$$
$$= \text{kin. energy} - \text{pot. energy}\,.$$

Here m is the mass, g is the acceleration of gravity, and $V = -mgx$ is the potential energy. (12.2.1/1) then reads $\ddot{x} = g$, giving $x(t) = \frac{g}{2} t^2 + c_1 t + c_2$. From $x(0) = 0$ and $c_1 = \dot{x}(0) = v_0$ (initial velocity) one obtains $x(t) = \frac{g}{2} t^2 + v_0 t$.

12.2.3. The First Integral of Motion

Theorem. *If the Lagrangian function*

$$L(t, x_1, \ldots, x_{3n}, \dot{x}_1, \ldots, \dot{x}_{3n}) = L(x_1, \ldots, x_{3n}, \dot{x}_1, \ldots, \dot{x}_{3n})$$

does not explicitly depend on t, then for every trajectory

$$L - \sum_{k=1}^{3n} \frac{\partial L}{\partial \dot{x}_k} \dot{x}_k \equiv \mathrm{E} = \text{const} \quad (\textit{first integral of motion}). \tag{1}$$

Remark. Substituting a solution of (12.2.1/1) into the left-hand side of (1), one obtains a function of t. The theorem says that this function of t is constant. (1) is interpreted as an energy conservation law. The example of Subsec. 12.2.2. yields $(T + V)(t) \equiv \mathrm{E}$.

> ... the felicitous elucidation, deduced from this hypothesis, of most of the phenomena of nature furnishes sufficient evidence that it be absolutely true.
> (Euler on Newton's law of gravitation, 1760)

12.3. Systems of n Mass Points

12.3.1. The Basic Model

Consider n point-shaped particles of masses $m_1, \ldots, m_n$ in the three-dimensional physical space (e.g. celestial bodies). Suppose that

$$r_{j,k} = \sqrt{(x_{3j-2} - x_{3k-2})^2 + (x_{3j-1} - x_{3k-1})^2 + (x_{3j} - x_{3k})^2}$$

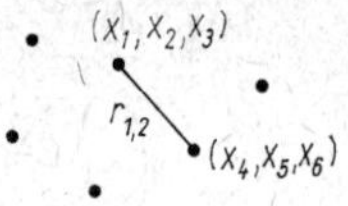

Fig. 12.2

be the distance between the j-th and the k-th particle (Fig. 12.2). Here the coordinates are numbered as stipulated in Subsec. 12.2.1. The gravitational potential of the system is

$$V_0 = V_0(x_1, \ldots, x_{3n}) = -\frac{1}{2}\gamma \sum_{j \neq k} \frac{m_j m_k}{r_{j,k}},$$

where γ is the constant of gravitation.[1]) Further suppose that an external potential $V_1 = V_1(t, x_1, \ldots, x_{3n}, \dot{x}_1, \ldots, \dot{x}_{3n})$ is given (e.g. an electromagnetic potential). On the basis of physical experience and physical principles, which are exterior to mathematical axiomatics, the following set-up is made for the Lagrangian function:

$$\begin{aligned} L(t, x_1, \ldots, \dot{x}_{3n}) &= \sum_{k=1}^{n} \frac{m_k}{2} (\dot{x}_{3k-2}^2 + \dot{x}_{3k-1}^2 + \dot{x}_{3k}^2) - V \qquad (1) \\ &= T - V = \text{kin. energy} - \text{pot. energy} \end{aligned}$$

where $V = V_0 + V_1$ is the potential energy. The trajectories of this system are solutions of (12.2.1/1).

Theorem. *If $V = V(t, x_1, \ldots, x_{3n})$ is independent of $\dot{x}_k$, then the trajectories of the system are solutions of*

$$m_k \ddot{x}_{3k-r}(t) = -\frac{\partial V}{\partial x_{3k-r}} \quad \textit{with} \quad k = 1, \ldots, n \quad \textit{and} \quad r = 0, 1, 2\,. \qquad (2)$$

Remark. This follows immediately from (12.2.1/1) and the above set-up for L. The interpretation shows that $\vec{x}_k = (x_{3k-2}, x_{3k-1}, x_{3k})$ is the position of the k-th particle, and that $\dot{\vec{x}}_k = (\dot{x}_{3k-2}, \dot{x}_{3k-1}, \dot{x}_{3k})$ is its velocity. $\vec{K}_k = \left(-\frac{\partial V}{\partial x_{3k-2}}, -\frac{\partial V}{\partial x_{3k-1}}, -\frac{\partial V}{\partial x_{3k}}\right)$ is the force acting upon the k-th particle. Then formula (2) above is Newton's second law of motion:

force = mass $\times$ acceleration.

12.3.2. Force-Free Systems

The system of Subsec. 12.3.1. is called force-free if $V(t, x_1, \ldots, \dot{x}_{3n}) \equiv 0$ in Eq. (12.3.1/1). In other words: the internal forces are neglected, $V_0 \equiv 0$, and external forces are absent, $V_1 \equiv 0$.

Theorem. *The trajectories of a force-free system are*

$$x_k(t) = a_k t + b_k\,, \quad k = 1, \ldots, 3n\,, \qquad (1)$$

where the a_k and b_k, are real numbers.

[1]) "This law has been called 'the greatest generalization achieved by a human mind' . . . Therefore our main concentration will not be on how clever we are to have found it all out, but on how clever nature is to pay attention to it" (R. Feynman, The Character of Physical Law, with respect to Newton's law of gravitation).

Remark. The theorem follows from (12.3.1/2). According to the interpretation in Subsec. 12.2.1., the k-th particle moves in the straight line $x(t)=\vec{a}_k t+\vec{b}_k$, where $\vec{a}_k=(a_{3k-2}, a_{3k-1}, a_{3k})$ and $\vec{b}_k=(b_{3k-2}, b_{3k-1}, b_{3k})$. The velocity of the particle is $\frac{\mathrm{d}x(t)}{\mathrm{d}t}=\vec{a}_k$. So the particles in a force-free system move uniformly in a straight line.

Historical note. We quote the famous three "Axioms, or Laws of Motion" of I. Newton (1686), cf. [46, p. 13]:
Lex prima (Law I). Every body continues in its state of rest, or of uniform motion in a right line, unless it is compelled to change that state by forces impressed upon it.
Lex secunda (Law II). The change of motion is proportional to the motive force impressed; and is made in the direction of the right line in which that force is impressed.
Lex tertia (Law III). To every action there is always opposed an equal reaction: or, the mutual actions of two bodies upon each other are always equal, and directed to contrary parts.

12.3.3. Conservative Systems

The system of Subsec. 12.3.1. is called conservative if $V(t, x_1, \ldots, \dot{x}_{3n})=V(x_1, \ldots, x_{3n})$ of Eq. (12.3.1/1) is independent of $t, \dot{x}_1, \ldots, \dot{x}_{3n}$. The kinetic energy T has the same meaning as in that formula.

Theorem. *For the trajectories of a conservative system,*

$$\sum_{k=1}^{3n} \frac{m_k}{2} (\dot{x}_{3k-2}^2+\dot{x}_{3k-1}^2+\dot{x}_{3k}^2) + V(x_1, \ldots, x_{3n}) = T+V \equiv \mathrm{E} = \text{const}$$

(*energy conservation law*).

Remark. The theorem follows from Theorem 12.2.3. So the total energy of a conservative system remains constant in time.

12.3.4. Particle in a Potential Well, Harmonic Oscillator

Consider a particle of mass m that moves in R_1. The particle is acted upon by a force $K(x)=-V'(x)$, $x\in R_1$, which is generated by a potential $V(x)$. According to Subsec. 12.3.1. one has $L(t, x, \dot{x})=\frac{m}{2}\dot{x}^2-V(x)$ as the associated Lagrangian function. This is a one-dimensional conservative system as defined in Subsec. 12.3.3., and one obtains

$$\frac{m}{2}\dot{x}^2+V(x)=\mathrm{E}=\text{const} \tag{1}$$

for the trajectory $x(t)$ of the particle. E is a characteristic quantity of the system (energy), and (1) gives

$$\dot{x}^2=\frac{2}{m}(\mathrm{E}-V(x)). \tag{2}$$

This implies that $V(x) \leq E$. Hence the particle cannot leave the potential well (Fig. 12.3). Integration of (2) gives

$$t - t_0 = \pm \sqrt{\frac{m}{2}} \int_{x_0}^{x} \frac{du}{\sqrt{E - V(u)}}. \tag{3}$$

If (without loss of generality) the positive sign is chosen in (3), then this means that the particle moves from x_0 to the right. The question is whether the particle will arrive at the point x_1 within a finite time.

Theorem 1. *Let $V(x)$ be twice continuously differentiable on R_1.*

(a) *If $V'(x_1) \neq 0$, then the particle arrives at $x_1 = x(t_1)$ after a finite time, and*

$$t_1 - t_0 = \sqrt{\frac{m}{2}} \int_{x_0}^{x_1} \frac{du}{\sqrt{E - V(u)}} \quad \textit{(convergent improper integral).}$$

(b) *If $V'(x_1) = 0$, then the particle does not arrive at x_1 (creeping approach), and*

$$\lim_{x \uparrow x_1} \int_{x_0}^{x} \frac{du}{\sqrt{E - V(u)}} = \infty \quad \textit{(divergent improper integral).}$$

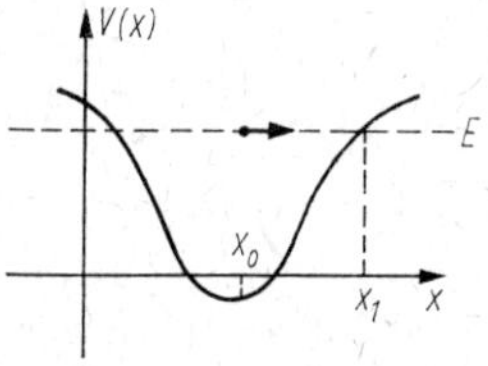

Fig. 12.3

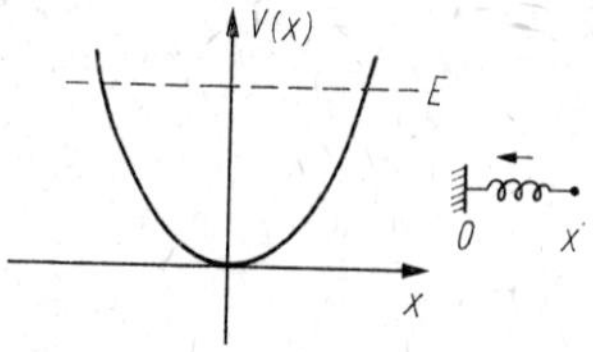

Fig. 12.4

Harmonic oscillator. The special case of the harmonic oscillator, where $V(x) = \frac{k}{2} x^2$ (Fig. 12.4), is of particular interest. The force $K(x) = -kx$ acts on the particle to pull it towards the origin from every nonzero position (restoring force). An elongated steel spring attached to the origin is a technical realization of such a restoring force.

Theorem 2. *If $x(0) = 0$ (initial position) and $x'(0) = v_0$ (initial velocity), then the harmonic oscillator oscillates about the origin:*

$$x(t) = v_0 \sqrt{\frac{m}{k}} \sin \sqrt{\frac{k}{m}}\, t. \tag{4}$$

Remark. (4) is the solution of the equation of motion $m\ddot{x} = -kx$ (cf. Eq. (12.3.1/2)), taking into account the above initial conditions. According to (4), $x_{max} = v_0 \sqrt{\frac{m}{k}}$ is the maximum deflection and $T = 2\pi \sqrt{\frac{m}{k}}$ is the period of the oscillation.

Look to the Heavens, and learn from them
How one should really honor the Master.
The stars in their courses extol Newton's laws –
In silence eternal.
(A. Einstein on the three-hundredth anniversary of I. Newton's birth, 1942)

12.4. Planetary Motion

12.4.1. Formulation of the Problem and Basic Model

Kepler's laws found empirically for the planets' motion about the sun read as follows.

Kepler's 1st law: The orbit of a planet is an ellipse that lies in a fixed plane. The position of the sun is in one of the two focuses.

Kepler's 2nd law (*principle of conservation of areas*): In equal periods, the radius vector drawn from the sun to the planet describes equal areas. Let F_1 and F_2 be the areas of the shaded (curvilinear) triangles (Fig. 12.5). If $t_2-t_1=t_4-t_3$, then $F_1=F_2$. Here the t_k are the times at which the planet passes through the respective positions.

Kepler's 3rd law: Let a_1 be the major semi-axis of the ellipse, and let T_1 be the period of revolution of a first planet. Let a_2 and T_2 be the corresponding parameters of a second planet. Then $\frac{T_1^2}{T_2^2}=\frac{a_1^3}{a_2^3}$.

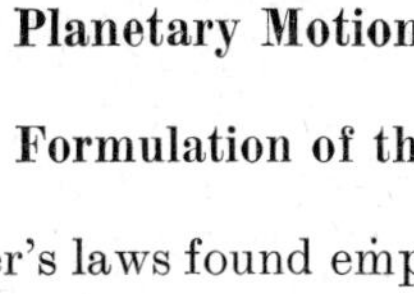

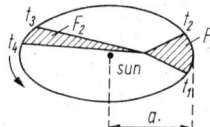

Fig. 12.5

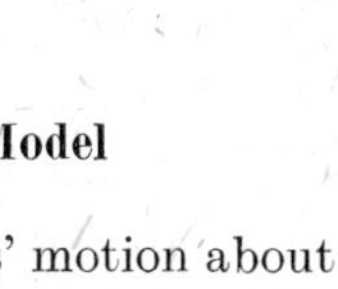

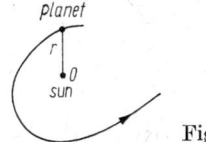

Fig. 12.6

The basic model. It is our aim to deduce Kepler's laws from the previous considerations. The problem is treated as a one-body problem. The sun, with the mass M, is fixed to the origin of the R_3, and the celestial body of mass m moves in an orbit $(x(t), y(t), z(t))$ about the sun (Fig. 12.6). t is the time. If $r(t)=\sqrt{x^2+y^2+z^2}$ is the distance from the origin, we make the following set-up for the Lagrangian function:

$$L(t, x, y, z, \dot{x}, \dot{y}, \dot{z})=\frac{m}{2}(\dot{x}^2+\dot{y}^2+\dot{z}^2)+\gamma\frac{Mm}{r}. \tag{1}$$

Here $-\gamma\frac{Mm}{r}$ is the gravitational potential, and γ is the constant of gravitation.

This agrees with the set-up of Subsec. 12.3.1., except that one mass point, namely the sun, is regarded as being fixed: the physical justification of this is that the mass of the sun is much greater than the mass of all of the planets, so that the influence the planets have on the sun can be neglected.

Polar co-ordinates. We use the spherical polar co-ordinates known from analytical geometry (Fig. 12.7):

$$x=r\sin\vartheta\cos\varphi,\quad 0<r<\infty,$$
$$y=r\sin\vartheta\sin\varphi,\quad 0<\vartheta<\pi,$$
$$z=r\cos\vartheta,\quad 0\leq\varphi<2\pi .$$

The singularities at $r=0$, $\vartheta\to 0$ and $\vartheta\to\pi$ will not disturb our considerations. Transformation of (1) to r, ϑ, and φ gives

$$L(t, x, y, z, \dot{x}, \dot{y}, \dot{z}) = \frac{m}{2}(\dot{r}^2 + r^2\dot{\vartheta}^2 + r^2\dot{\varphi}^2 \sin^2\vartheta) + \gamma\frac{Mm}{r} \tag{2}$$
$$= \tilde{L}(t, r, \vartheta, \varphi, \dot{r}, \dot{\vartheta}, \dot{\varphi}) .$$

According to the axiom of Subsec. 12.2.1., the desired trajectory is an extremal of

$$\int_{t_0}^{t_1} L(t, x, y, z, \dot{x}, \dot{y}, \dot{z})\, \mathrm{d}t = \int_{t_0}^{t_1} \tilde{L}(t, r, \vartheta, \varphi, \dot{r}, \dot{\vartheta}, \dot{\varphi})\, \mathrm{d}t .$$

As the value of the integral is invariant under the transformation from Cartesian to spherical polar co-ordinates, the extremal of the transformed integral is identical with the original one. Thus, according to Theorem 11.1.3/1, the desired trajectory is a solution of the Euler-Lagrange equations associated to $\tilde{L}$ of formula (2) above:

$$0 = \frac{\partial\tilde{L}}{\partial r} - \frac{\mathrm{d}}{\mathrm{d}t}\left(\frac{\partial\tilde{L}}{\partial\dot{r}}\right) = mr(\dot{\vartheta}^2 + \dot{\varphi}^2\sin^2\vartheta) - \gamma\frac{Mm}{r^2} - \frac{\mathrm{d}}{\mathrm{d}t}(m\dot{r}) ; \tag{3}$$

$$0 = \frac{\partial\tilde{L}}{\partial\varphi} - \frac{\mathrm{d}}{\mathrm{d}t}\left(\frac{\partial\tilde{L}}{\partial\dot{\varphi}}\right) = -\frac{\mathrm{d}}{\mathrm{d}t}(mr^2\dot{\varphi}\sin^2\vartheta), \tag{4}$$

$$0 = \frac{\partial\tilde{L}}{\partial\vartheta} - \frac{\mathrm{d}}{\mathrm{d}t}\left(\frac{\partial\tilde{L}}{\partial\dot{\vartheta}}\right) = mr^2\dot{\varphi}^2\sin\vartheta\cos\vartheta - m\frac{\mathrm{d}}{\mathrm{d}t}(r^2\dot{\vartheta}) . \tag{5}$$

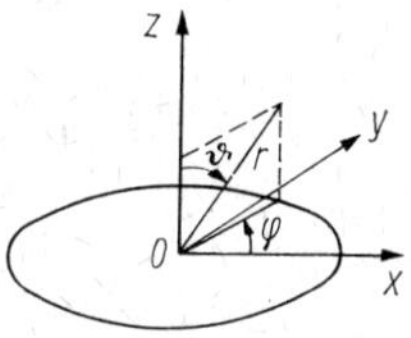

Fig. 12.7

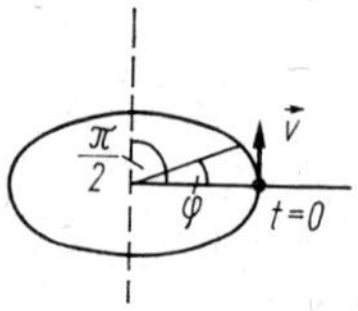

Fig. 12.8

Initial values: Suppose that at $t=0$ the celestial body has its shortest distance from the sun (at least compared with $t\in(-\delta, \delta)$, $\delta>0$). This only excludes the case that the celestial body falls into the sun. Let $\vec{v}(t) = (\dot{x}(t), \dot{y}(t), \dot{z}(t))$ be the velocity vector tangential to the trajectory, then we choose the x,y,z-coordinate system so that $y(0) = z(0) = \dot{z}(0) = 0$ (Fig. 12.8). In spherical polar coordinates this gives

$$\vartheta(0) = \frac{\pi}{2}, \quad \varphi(0) = 0, \quad r(0) = r_0 > 0, \tag{6}$$
$$\dot{\vartheta}(0) = 0, \quad \dot{\varphi}(0) = v, \quad \dot{r}(0) = 0 .$$

The system of equations (3)–(5) can be resolved for $\ddot{r}$, $\ddot{\vartheta}$, and $\ddot{\varphi}$. The theorems of Chap. 4 then show that (3)–(5) has a unique solution with the initial values (6). (This has been established at least locally in a neighbourhood of $t=0$, but will turn out to hold also globally for all t-values.) This then is the desired trajectory.

12.4.2. Plane Orbits, Kepler's Second Law

Theorem 1. *If $r(t)$ and $\varphi(t)$ are a solution of*

$$0 = mr\dot{\varphi}^2 - \frac{\gamma Mm}{r^2} - m\ddot{r}, \quad r(0) = r_0, \quad \dot{r}(0) = 0, \tag{1}$$

$$0 = m \frac{\mathrm{d}}{\mathrm{d}t} (r^2\dot{\varphi}), \quad \varphi(0) = 0, \quad \dot{\varphi}(0) = v\,, \tag{2}$$

then $r(t)$, $\varphi(t)$, $\vartheta(t) \equiv \frac{\pi}{2}$ are the solution, unique in a neighbourhood of $t=0$, of (12.4.1/3)–(12.4.1/6).

Remark 1. The theorem is astonishing and trivial: substitution of $\vartheta(t) \equiv \frac{\pi}{2}$ into (12.4.1/3)–(12.4.1/6) gives the above system of equations. The rest follows from the uniqueness theorem. The problem is thus reduced to (1), (2).

Theorem 2. *The orbit of a planet (or of another celestial body) lies in a fixed plane, obeying Kepler's second law.*

Remark 2. The first part of the theorem follows from the fact that $\vartheta(t) \equiv \frac{\pi}{2}$. Further, $\mathrm{d}F = \frac{1}{2} r r \mathrm{d}\varphi$, and hence $\dot{F} = \frac{\mathrm{d}F}{\mathrm{d}t} = \frac{1}{2} r^2\dot{\varphi}$ (Fig. 12.9). From (2) it is now seen that $\dot{F}(t) \equiv c =$ const. But this is the principle of conservation of areas.

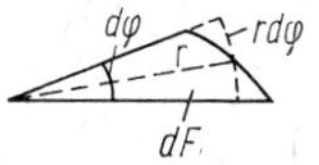

Fig. 12.9

12.4.3. Kepler's First Law

From Theorem 12.4.2/1 it follows that $r^2\dot{\varphi} = r_0^2 v = h \neq 0$ (with $h \neq 0$ we only exclude the case that the planet falls into the sun, $\varphi(t) \equiv 0$). Here h is the constant in the law of areas, which is assumed to be known (observational data). So $\varphi(t)$ is strictly monotone, and we can write $r(t) = r(t(\varphi)) = \frac{1}{\varrho(\varphi)}$ as a function of φ. Eq. (12.4.2/1) then gives $\frac{\mathrm{d}^2\varrho}{\mathrm{d}\varphi^2} = \frac{\gamma M}{h^2} - \varrho$. Taking into account the initial values, one obtains the solution

$$\varrho(\varphi) = \frac{\gamma M}{h^2} + \left(\frac{1}{r_0} - \frac{\gamma M}{h^2}\right) \cos \varphi\,.$$

Hence

$$r(\varphi) = \frac{p}{1 + \varepsilon \cos \varphi} \quad \text{where} \quad p = \frac{h^2}{\gamma M} \quad \text{and} \quad \frac{\varepsilon}{p} = \frac{1}{r_0} - \frac{\gamma M}{h^2}\,. \tag{1}$$

Here it is possible to determine p and ε from astronomical data.

Theorem (*Kepler's first law*). *The orbit of a celestial body as described by* (1) *is a curve of second degree (that lies in a fixed plane). The position of the sun is in one of the focuses. The curve of second degree is*

an ellipse (planetary orbit) for $|\varepsilon| < 1$ (Fig. 12.10),
a parabola for $\varepsilon = 1$,
a hyperbola (cometary orbit) for $\varepsilon > 1$ (Fig. 12.11).

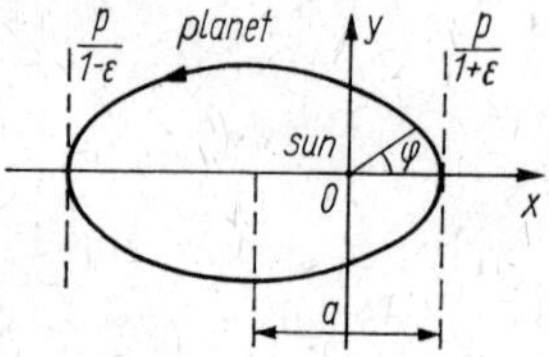

Fig. 12.10

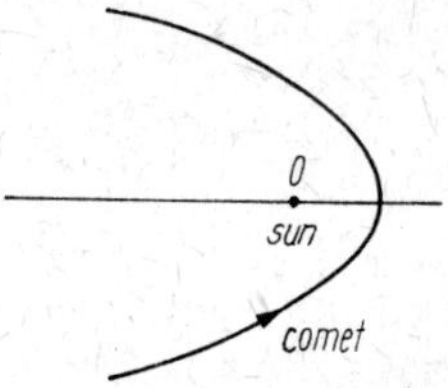

Fig. 12.11

Remark. The theorem is proved by returning to x,y-co-ordinates in the above formula (1). According to our convention that $\varphi = 0$ defines the orbital point nearest to the sun, we may assume that $0 \leq \varepsilon$. In the case of the ellipse one obtains $a = \frac{p}{1-\varepsilon^2}$ for the major semi-axis.

12.4.4. Kepler's Third Law

Theorem. *Let a be the major semi-axis of the ellipse of a planetary orbit, and let T be the period of revolution of that planet, then* $\frac{T^2}{a^3} = \frac{4\pi^2}{\gamma M}$.

Remark 1. From the law of areas it follows that $\dot{F} = \frac{h}{2}$. If b is the minor semi-axis of the ellipse, then πab is its area. T can be calculated from

$$\pi ab = \int_0^T \frac{\mathrm{d}F}{\mathrm{d}t}\,\mathrm{d}t = \frac{h}{2}\,T\,.$$

With the use of $b = \sqrt{\frac{a}{\gamma M}}\,h$, one obtains the assertion of the theorem.

Remark 2. Since $\frac{T^2}{a^3}$ does not depend on the particular planet, this leads to Kepler's third law.

13. Measure Theory

13.1. Classes of Sets

13.1.1. Algebras and σ-Algebras

In Chap. 9 we considered the Riemann integral in R_n. At many points there an intuitive argumentation was necessary. It is now the aim of Chaps. 13 and 14 to present a substantial generalization of the Riemann integral. The approach will be abstract, and will do without intuitive aids.

Sets. We recall some concepts of set theory (Fig. 13.1). Let X be an arbitrary (abstract) set. A set E with $E \subset X$ is called a subset of X. If $E \subset X$ and $F \subset X$, then one defines

the union $E \cup F = \{x \mid x \in E \quad \text{or} \quad x \in F\}$,
the intersection $E \cap F = \{x \mid x \in E \quad \text{and} \quad x \in F\}$,
the complement $E' = X \backslash E = \{x \mid x \notin E\}$, and
the difference $E \backslash F = \{x \mid x \in E \quad \text{and} \quad x \notin F\} = E \cap F'$.

We make the convention that $\emptyset \subset X$ ($\emptyset$ being the empty set).

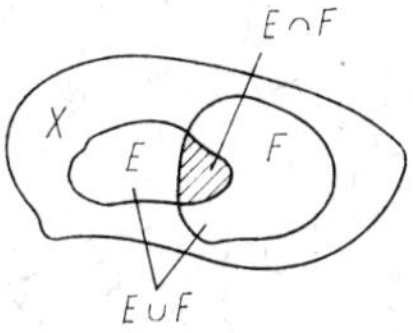

Fig. 13.1

Fig. 13.2

Definition. *Let X be an arbitrary non-empty set. Let $\mathfrak{A}$ be a non-empty class* (*or set*) *of subsets of X.*

(**a**) *$\mathfrak{A}$ is called an algebra if it has the following properties:*
if $E \in \mathfrak{A}$, then $E' \in \mathfrak{A}$, and
if $E \in \mathfrak{A}$ and $F \in \mathfrak{A}$, then $E \cup F \in \mathfrak{A}$.

(**b**) *$\mathfrak{A}$ is called a σ-algebra if it has the following properties:*
if $E \in \mathfrak{A}$, then $E' \in \mathfrak{A}$, and
if $E_j \in \mathfrak{A}$ for $j = 1, 2, 3, \ldots$, then $\bigcup_{j=1}^{\infty} E_j \in \mathfrak{A}$.

Remark 1. The class $P(X)$ of all subsets of a given set X (power set of X) is an algebra as well as a σ-algebra.

Remark 2. Let $X = R_1$ (Fig. 13.2). Let a subset E of R_1 belong to the class $\mathfrak{A}$ if and only if E is a finite union of half-open intervals $[a, b)$. Here the intervals $(-\infty, b)$ and $[a, \infty)$ as well as R_1 and $\emptyset$ are admitted to competition. $\mathfrak{A}$ is an algebra. But $(0, 1) = \bigcup_{n=2}^{\infty} \left[\frac{1}{n}, 1\right)$ shows that $\mathfrak{A}$ is not a σ-algebra.

Theorem. (**a**) *Every σ-algebra is an algebra.*

(**b**) *Let $\mathfrak{A}$ be an algebra. Then $\emptyset \in \mathfrak{A}$ and $X \in \mathfrak{A}$. Further it follows from $E_j \in \mathfrak{A}$ for $j = 1, 2, \ldots, N$ that*

$$\bigcup_{j=1}^{N} E_j \in \mathfrak{A} \quad \text{and} \quad \bigcap_{j=1}^{N} E_j \in \mathfrak{A} \quad (N = 1, 2, 3, \ldots).$$

(**c**) *If $\mathfrak{A}$ is a σ-algebra, then it follows from $E_j \in \mathfrak{A}$ for $j = 1, 2, 3, \ldots$ that $\bigcup_{j=1}^{\infty} E_j \in \mathfrak{A}$.*

Remark 3. Consequently, algebras are closed under complementation and under the formation of finite intersections and finite unions. σ-algebras are also closed under the formation of countably infinite intersections and unions.

13.1.2. Extension Theorems

Let $\mathfrak{A}$ and $\mathfrak{B}$ be non-empty classes of subsets of the given set X, then $\mathfrak{B} \supset \mathfrak{A}$ means that every set of $\mathfrak{A}$ also belongs to $\mathfrak{B}$. For a given $\mathfrak{A}$, we ask for the smallest class $\mathfrak{B}$ that includes $\mathfrak{A}$ and is an algebra or a σ-algebra. In other words, if $\mathfrak{C}$ is an algebra or a σ-algebra such that $\mathfrak{C} \supset \mathfrak{A}$, then $\mathfrak{C} \supset \mathfrak{B}$ should also be true.

Theorem. *Let $\mathfrak{A}$ be a non-empty class of subsets of X.*
(a) *There exists a smallest algebra that includes $\mathfrak{A}$. It is denoted by $R(\mathfrak{A})$.*
(b) *There exists a smallest σ-algebra that includes $\mathfrak{A}$. It is denoted by $S(\mathfrak{A})$.*

Remark. There arises the question as to whether $R(\mathfrak{A})$ and $S(\mathfrak{A})$ can be constructed. For $R(\mathfrak{A})$ there is the following possibility. Let $\mathfrak{A}_0 = \mathfrak{A}$ and $\mathfrak{A}_{k+1} = \left\{E \mid E' \in \mathfrak{A}_k \text{ or } E = \bigcup_{j=1}^{N} E_j \text{ with } E_j \in \mathfrak{A}_k\right\}$, $j = 0, 1, 2, \ldots$.

Lemma. $R(\mathfrak{A}) = \bigcup_{k=0}^{\infty} \mathfrak{A}_k$.

13.1.3. Borel Sets in R_n

Let us extend the example of Remark 13.1.1/2 and generalize it to the n-dimensional case. Let $X = R_n$, and let $Q = \{x \mid x \in R_n,\ a_j < x_j < b_j \text{ for } j = 1, \ldots, n\}$ be a so-called n-dimensional open interval. As usual, we set $x = (x_1, \ldots, x_n)$.

Definition. *If $\mathfrak{A}$ is the class of n-dimensional open intervals in R_n, then the elements of $S(\mathfrak{A})$ are called Borel sets.*

Remark 1. The Borel sets are said to be generated by the n-dimensional open intervals. With the aid of Lemma 13.1.2, it is not so difficult to imagine the sets of $R(\mathfrak{A})$, whereas general Borel sets have a more complicated structure.

Theorem. *The Borel sets form the smallest σ-algebra that includes one of the following classes:*
(a) *the class of all n-dimensional open intervals $\{x \mid a_j < x_j < b_j\}$,*
(b) *the class of all n-dimensional closed intervals $\{x \mid a_j \leq x_j \leq b_j\}$,*
(c) *the class of all n-dimensional half-open intervals $\{x \mid a_j \leq x_j < b_j\}$,*
(d) *the class of all open sets in R_n,*
(e) *the class of all closed sets in R_n.*

Remark 2. In particular, the set $\{x\}$ consisting of the point $x \in R_n$ alone is a Borel set. Consequently, $\bigcup_{j=1}^{\infty} \{x_j\}$ is also a Borel set. Hence, for $n = 1$, $\{x \mid x \text{ rational}\}$ is a Borel set. Then the complement $\{x \mid x \text{ irrational}\}$ is also a Borel set in R_1.

13.2. Elementary Measures and Measures

13.2.1. Definitions

Let X be an arbitrary set, and let $\mathfrak{A}$ be an algebra of subsets of X. We consider non-negative set functions μ defined on $\mathfrak{A}$: to every $E \in \mathfrak{A}$ there is assigned a number $\mu(E)$ such that $0 \leq \mu(E) \leq \infty$. Here it is admissible that $\mu(E) = \infty$. As regards calculation with the symbol ∞, the following relations are adopted: for any real number a, one has $a + \infty = \infty + \infty = \infty$.

Definition. (a) *A non-negative set function μ on an algebra $\mathfrak{A}$ is called an elementary measure if it has the following properties.*

1. *If* $E_j \in \mathfrak{A}$, $j=1, 2, 3, \ldots$, *and* $E_j \cap E_k = \emptyset$ *for* $j \neq k$, *and if* $\bigcup_{j=1}^{\infty} E_j \in \mathfrak{A}$, *then* $\mu\left(\bigcup_{j=1}^{\infty} E_j\right) = \sum_{j=1}^{\infty} \mu(E_j)$ (*σ-additivity, or countable additivity*).

2. $\mu(\emptyset) = 0$.

3. *There exists a sequence of sets* $E_j \in \mathfrak{A}$, *with* $E_1 \subset E_2 \subset E_3 \subset, \ldots$, *and* $\mu(E_j) < \infty$ *for* $j=1, 2, 3, \ldots$, *such that* $X = \bigcup_{j=1}^{\infty} E_j$ (*σ-finiteness*).

(**b**) *An elementary measure that is defined on a σ-algebra is called a measure.*

Remark 1. If $\mathfrak{A}$ is a σ-algebra, then $\bigcup_{j=1}^{\infty} E_j \in \mathfrak{A}$ for $E_j \in \mathfrak{A}$. For an algebra $\mathfrak{A}$, $\bigcup_{j=1}^{\infty} E_j \in \mathfrak{A}$ had to be postulated explicitly in the above definition. Frequently σ-finiteness is not included in the definition of the elementary measure. For our purposes, however, the above formulation will be sufficient.

Remark 2. An elementary measure μ is called finite if $\mu(X) < \infty$.

Remark 3. If $\mathfrak{A}$ is the σ-algebra of the Borel sets in R_n, then

$$\mu(E) = \begin{cases} 1 & \text{if} \quad 0 \in E \\ 0 & \text{if} \quad 0 \notin E \end{cases}, \quad E \in \mathfrak{A}, \text{ is a finite measure.}$$

It is called Dirac's measure. Let $a_1, \ldots, a_k$ be fixed points in R_n, and let $\mu(E)$, with $E \in \mathfrak{A}$, be the number of those points $a_1, \ldots, a_k$ which lie in E. Then $\mu(E)$ is a measure.

13.2.2. Properties

Since measures are also elementary measures, it will suffice to formulate propositions for elementary measures.

Theorem 1. *Let* μ *be an elementary measure on the algebra* $\mathfrak{A}$.

(**a**) *If* $E_j \in \mathfrak{A}$ *for* $j=1, \ldots, N$, *and if* $E_j \cap E_k = \emptyset$ *for* $j \neq k$, *then* $\mu\left(\bigcup_{j=1}^{N} E_j\right) = \sum_{j=1}^{N} \mu(E_j)$ (*additivity*).

(**b**) *If* $E \in \mathfrak{A}$, $F \in \mathfrak{A}$, *and* $E \subset F$, *then* $\mu(E) \leqq \mu(F)$ (*monotonicity*).

(**c**) *If, for a finite or countably infinite number of* $E_j \in \mathfrak{A}$ *and for* $E \in \mathfrak{A}$, $E \subset \bigcup E_j$, *then* $\mu(E) \leqq \sum \mu(E_j)$ (*subadditivity*).

(**d**) *If, for a finite or countably infinite number of* $E_j \in \mathfrak{A}$ *with* $E_j \cap E_k = \emptyset$ *for* $j \neq k$ *and for* $E \in \mathfrak{A}$, $\bigcup E_j \subset E$, *then* $\sum \mu(E_j) \leqq \mu(E)$.

(**e**) *If* $E_j \in \mathfrak{A}$, *with* $E_1 \subset E_2 \subset E_3 \subset \ldots$, *and if* $\bigcup_{j=1}^{\infty} E_j \in \mathfrak{A}$, *then*

$$\mu\left(\bigcup_{k=1}^{\infty} E_k\right) = \lim_{j \to \infty} \mu(E_j) .$$

(**f**) *If* $E_j \in \mathfrak{A}$, *with* $E_1 \supset E_2 \supset E_3 \supset \ldots$, *and if* $\bigcap_{j=1}^{\infty} E_j \in \mathfrak{A}$ *and if there exists a* j_0 *such that* $\mu(E_{j_0}) < \infty$, *then*

$$\mu\left(\bigcap_{k=1}^{\infty} E_k\right) = \lim_{j \to \infty} \mu(E_j) .$$

Remark 1. The proof is relatively simple. The condition that $\mu(E_{j_0}) < \infty$ for a suitable j_0 in part (f) of the theorem cannot be dispensed with. (e) and (f) are continuity properties.

Remark 2. If $E \in \mathfrak{A}$, $F \in \mathfrak{A}$, $E \supset F$ and $\mu(E) < \infty$, then the following subtraction property holds: $\mu(E \backslash F) = \mu(E) - \mu(F)$.

Remark 3. If $\mathfrak{A}$ is an algebra, then a non-negative set function μ on $\mathfrak{A}$ is said to be finite if $\mu(E)<\infty$ for all $E\in\mathfrak{A}$. A non-negative set function μ on $\mathfrak{A}$ is said to be additive if $\mu(\emptyset)=0$ and $\mu(E\cup F)=\mu(E)+\mu(F)$ for all $E\in\mathfrak{A}$ and $F\in\mathfrak{A}$ with $E\cap F=\emptyset$.

Theorem 2. *Let μ be a non-negative, finite and additive set function on the algebra $\mathfrak{A}$.*

(a) *μ is an elementary measure if $\mu\left(\bigcup_{j=1}^{\infty} E_j\right)=\lim_{j\to\infty}\mu(E_j)$ for all $E_j\in\mathfrak{A}$ with $\bigcup_{j=1}^{\infty} E_j\in\mathfrak{A}$ and $E_1\subset E_2\subset E_3\subset \dots$.*

(b) *μ is an elementary measure if $\mu(E_j)\downarrow 0$ for all $E_j\in\mathfrak{A}$ with $E_1\supset E_2\supset \dots$ and $\bigcap_{j=1}^{\infty} E_j=\emptyset$.*

Remark 4. Consequently a non-negative, finite and additive set function is an elementary measure if and only if either of the two continuity properties (e) or (f) of Theorem 1 holds.

13.2.3. Heine-Borel's Theorem

This is not a theorem of measure theory but one that has been omitted in the chapters on the differential and integral calculus.

Theorem. *Let E be a closed, bounded set in R_n, and let $F_1, F_2, F_3, \dots$ be open sets in R_n, such that $E\subset\bigcup_{j=1}^{\infty} F_j$, then there exists a natural number N such that $E\subset\bigcup_{j=1}^{N} F_j$* (Fig. 13.3).

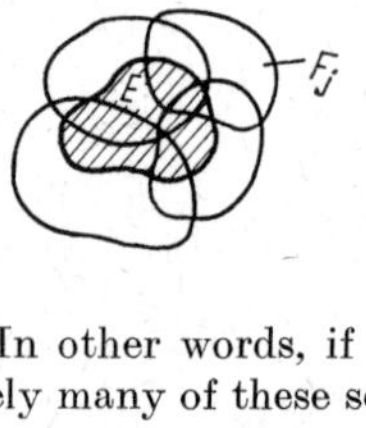

Fig. 13.3

Remark. In other words, if E is covered by a (countably) infinite number of open sets, then finitely many of these sets suffice to cover E.

13.2.4. Elementary Borel Measures in R_1

It is our aim to characterize measures on the Borel sets in R_1. Let $\mathfrak{A}$ be the algebra referred to in Remark 13.1.1/2, then it follows from Theorem 13.1.3 that the Borel sets form the smallest σ-algebra $S(\mathfrak{A})$ that includes $\mathfrak{A}$. We call $\mathfrak{A}$ the algebra of the pre-Borelian sets. First of all, let us describe all elementary measures on $\mathfrak{A}$. We consider increasing, left-continuous real-valued functions $\varphi(t)$ on R_1, i.e., $\varphi(t_0)=\lim_{t\uparrow t_0}\varphi(t)$ (Fig. 13.4). Here it is allowed that $\varphi(t)$ be constant on intervals (i.e., not strictly monotone). If

$$\varphi(\infty)=\lim_{t\to\infty}\varphi(t),\quad \varphi(-\infty)=\lim_{t\to-\infty}\varphi(t)\,,$$

then the values $\varphi(\infty)=\infty$ and $\varphi(-\infty)=-\infty$ are permissible (but $\varphi(\infty)$ and/or $\varphi(-\infty)$ may also be finite). For the calculation with ∞ we adopt the usual relations: $a+\infty=\infty$, $a-\infty=-\infty$ for any real a, $-(-\infty)=\infty$.

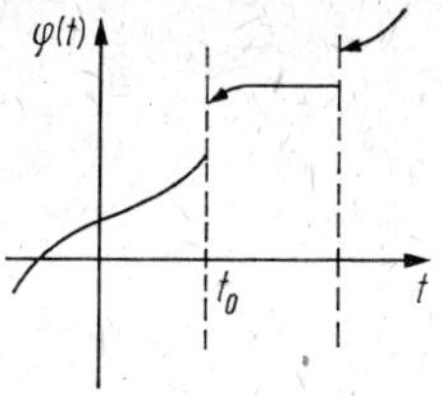

Fig. 13.4

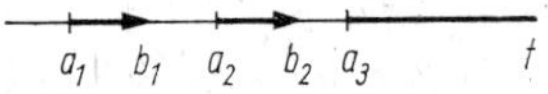

Fig. 13.5

Theorem 1. *Let* $a_1 < b_1 < a_2 < b_2 < \ldots < a_n < b_n$ *(where* $a_1 = -\infty$ *and/or* $b_n = \infty$ *is allowable), and let* $E = \bigcup_{j=1}^{n} [a_j, b_j)$ (Fig. 13.5). *If* $\varphi(t)$ *is the function described above, then*

$$\mu(E) = \sum_{j=1}^{n} (\varphi(b_j) - \varphi(a_j)) \quad \textit{where} \quad \mu(\emptyset) = 0 \tag{1}$$

is an elementary measure on the pre-Borelian sets.

Remark 1. The above theorem is plausible. In the case $a_1 = -\infty$ one must of course substitute $(-\infty, b_1)$ for $[a_1, b_1)$. The proof of σ-additivity is, however, somewhat tedious.

Remark 2. If c is a real number, then $\varphi(t) + c$ generates the same elemantary measure μ as $\varphi(t)$ does. Hence $\varphi(t)$ can be normalized by $\varphi(0) = 0$.

Theorem 2. *If* μ *is an elementary measure on the algebra of pre-Borelian sets, then there exists a real-valued, increasing, left-continuous function* $\varphi(t)$ *that generates* μ *in the sense of Theorem* 1. $\varphi(t)$ *is unique apart from an additive constant* (Fig. 13.6).

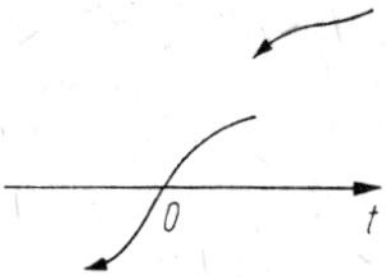

Fig. 13.6

Remark 3. The function

$$\varphi(t) = \begin{cases} \mu[(0, t)) & \text{for } t > 0 \\ 0 & \text{for } t = 0 \\ -\mu([t, 0)) & \text{for } t < 0 \end{cases} \tag{2}$$

is the desired generating function.

Remark 4. The two theorems (and the construction formulas (1) and (2)) provide a one-to-one correspondence between increasing functions on R_1 and the elementary measures on the pre-Borelian sets in R_1.

13.2.5. Elementary Lebesgue Measure in R_1

Setting $\varphi(t) = t$ in (13.2.4/1), one obtains the so-called elementary Lebesgue measure on the algebra of the pre-Borelian sets in R_1. One has $\mu([a, b)) = b - a$. We shall now generalize this elementary measure to the n-dimensional case. The algebra of pre-Borelian sets in R_n is constructed as follows. Let

$$Q = \{x \mid x = (x_1, \ldots, x_n) \in R_n,\ a_j \leqq x_j < b_j \quad \text{for} \quad j = 1, \ldots, n\} \tag{1}$$

be an n-dimensional half-open interval. Here $-\infty \leqq a_j < \infty$ and $-\infty < b_j \leqq \infty$. If $a_j = -\infty$, then $a_j \leqq x_j < b_j$ must be read as $-\infty < x_j < b_j$.)

Lemma. *$\mathfrak{A}=\{E \mid E=\bigcup_{k=1}^{N} Q_k$, Q_k being an n-dimensional interval as defined in (1)$\}$ is an algebra of subsets of R_n (the algebra of pre-Borelian sets in R_n* (Fig. 13.7)).

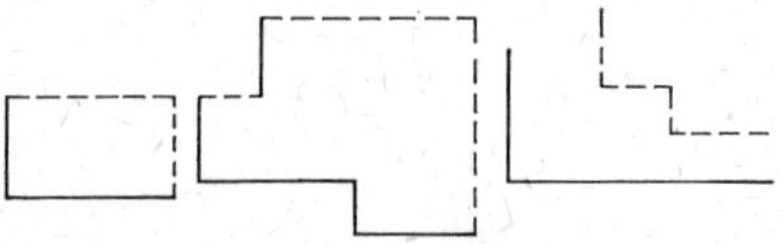

Fig. 13.7

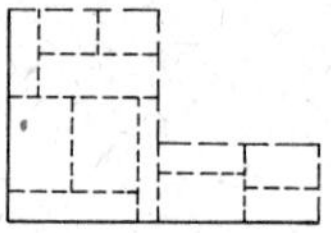

Fig. 13.8

Remark 1. N is any natural number. $\emptyset \in \mathfrak{A}$ (to show this, one may choose $a_j=b_j$ in (1); otherwise one naturally assumes that always $a_j<b_j$). Hence a pre-Borelian set E is a finite union of (possibly unbounded) n-dimensional half-open intervals. The "left" and the "lower" ends of E belong to E, whereas the "right" and the "upper" ones do not. For $n=1$ one obtains the previously defined algebra of pre-Borelian sets.

Remark 2. According to Subsec. 13.1.3., the class of Borel sets is the smallest σ-algebra $S(\mathfrak{A})$ that includes the algebra of pre-Borelian sets.

Remark 3. We generalize the elementary Lebesgue measure in R_1 as follows. If $E=Q$ has the form (1), then let $\mu(Q)=\prod_{j=1}^{n}(b_j-a_j)$ (here we assume that $b_j>a_j$, $a\cdot\infty=\infty$ for $a>0$). If $E\in\mathfrak{A}$ is an arbitrary set, then there exists a representation (not a unique one)

$$E=\bigcup_{k=1}^{N} Q_k, \quad Q_k\cap Q_l=\emptyset \quad \text{for} \quad k\neq l\,, \tag{2}$$

where the Q_k's have the form (1) (Fig. 13.8). We set

$$\mu(E)=\sum_{k=1}^{N}\mu(Q_k)\,.$$

The question now is whether $\mu(E)$ is independent of the kind of representation.

Theorem. (a) *$\mu(E)$ is independent of the kind of the representation* (2).

(b) *$\mu(E)$ (with $\mu(\emptyset)=0$) is an elementary measure on the algebra of pre-Borelian sets. (It is called elementary Lebesgue measure.)*

13.3. The Outer Measure, Extension of Elementary Measures

13.3.1. The Outer Measure

The aim of the following considerations is to construct measures on σ-algebras. Here the σ-algebra of the Borel sets in R_n is of special interest. For the time being, the constructions of Subsecs. 13.2.4. and 13.2.5. only provide elementary measures on the algebra of the pre-Borelian sets in R_n. Now the question arises whether these elementary measures can be extended in such a way that they become measures on the Borel sets. This will be realized in several steps, where we consider the abstract case.

Definition. *Let $\mathfrak{A}$ be an algebra of subsets of the set X, and let μ be an elementary measure on $\mathfrak{A}$. If E is any set of X, then*

$$\mu^*(E)=\inf\sum_{j=1}^{\infty}\mu(E_j), \quad E\subset\bigcup_{j=1}^{\infty}E_j, \quad E_k\in\mathfrak{A}\,, \tag{1}$$

is the outer measure (corresponding to μ*). The infimum is taken over all systems* $\{E_j\}_{j=1}^{\infty} \subset \mathfrak{A}$ *with* $E \subset \bigcup_{j=1}^{\infty} E_j$.

Remark. In contrast to μ, μ^* is defined on every subset of X.

Theorem. *The outer measure* μ^* *has the following properties.*

(a) $0 \leqq \mu^*(E) \leqq \infty, \quad \mu^*(\emptyset) = 0.$

(b) $\mu^*(E) \leqq \mu^*(F)$ *for* $E \subset F$ *(monotonicity).*

(c) $\mu^*(E) \leqq \sum_{j=1}^{\infty} \mu(E_j)$ *for* $E \subset \bigcup_{j=1}^{\infty} E_j$ *(subadditivity)*

(d) *There exists a sequence* $E_1 \subset E_2 \subset E_3 \subset \ldots$ *such that* $\bigcup_{j=1}^{\infty} E_j = X$ *and* $\mu^*(E_k) < \infty$ *for* $k = 1, 2, 3, \ldots$ *(σ-finiteness).*

13.3.2. The Induced Measure

We have the same situation as in Subsec. 13.3.1.: an algebra $\mathfrak{A}$ of subsets of the set X, an elementary measure μ on $\mathfrak{A}$, and a corresponding outer measure μ^*.

Definition 1. *A set* E, $E \subset X$, *is called* μ^**-measurable if, for all subsets* A *of* X,

$$\mu^*(A) = \mu^*(A \cap E) + \mu^*(A \cap E'). \tag{1}$$

Remark 1. It is our aim to construct measures. Measures have the property (1). This suggests to pick out those subsets which have the property (1) from the class of all subsets of E.

Definition 2. *A measure* μ *on a σ-algebra* $\mathfrak{C}$ *is called complete if the conditions* $E \in \mathfrak{C}$, $\mu(E) = 0$ *and* $F \subset E$ *imply that* $F \in \mathfrak{C}$.

Remark 2. Then it follows from $\mu(F) \leqq \mu(E)$ that $\mu(F) = 0$.

Theorem. (a) *The class* $\mathfrak{B}$ *of all* μ^**-measurable sets is a σ-algebra.*

(b) *If* $\mu^*(E) = 0$ *for* $E \subset X$, *then* E *belongs to* $\mathfrak{B}$.

(c) *The restriction* $\bar{\mu}(E) = \mu^*(E)$ *of* μ^* *to the sets* $E \in \mathfrak{B}$ *induces a complete measure* $\bar{\mu}$ *on* $\mathfrak{B}$.

Remark 3. $\bar{\mu}$ is called the measure induced by μ on $\mathfrak{B}$.

13.3.3. The Extension Theorem

What has been attained so far can be described as follows. The starting point is $[X, \mathfrak{A}, \mu]$: a set X, an algebra $\mathfrak{A}$, and an elementary measure μ on $\mathfrak{A}$. In Subsec. 13.3.1. we have constructed the corresponding outer measure μ^* on the class $P(X)$ of all subsets of X:$[X, P(X), \mu^*]$. By restriction of this measure to the σ-algebra $\mathfrak{B}$ of the μ^*-measurable sets we obtained the induced measure $\bar{\mu}$ on $\mathfrak{B}$ in Subsec. 13.3.2: $[X, \mathfrak{B}, \bar{\mu}]$. What is now of interest is the relationship between $\mathfrak{B}$ and $\mathfrak{A}$ and between $\bar{\mu}$ and μ. As usual, $S(\mathfrak{A})$ is the smallest σ-algebra over $\mathfrak{A}$.

Theorem. (a) $\mathfrak{B} \supset S(\mathfrak{A})$.

(b) *The restriction of* $\bar{\mu}$ *to* $S(\mathfrak{A})$ *is an extension of* μ, *i.e., if* $E \in \mathfrak{A}$, *then* $\bar{\mu}(E) = \mu(E)$.

(c) $\bar{\mu}$ *is a complete extension of* μ *on* $\mathfrak{B}$, *i.e.,* $\bar{\mu}$ *is an extension of* μ *in the sense of (b), and* $\bar{\mu}$ *is a complete measure on* $\mathfrak{B}$.

Remark. So we have reached the goal set out at the beginning of Subsec. 13.3.1. What remains is the question for the uniqueness of these extensions. We shall discuss this point in Subsec. 13.3.5.

13.3.4. Borel, Lebesgue and Dirac Measures

In Subsec. 13.2.4. we considered elementary Borel measures on the algebra $\mathfrak{A}$ of the pre-Borelian sets in R_1. In Subsec. 13.2.5. we introduced the elementary Lebesgue measure on the algebra $\mathfrak{A}$ of the pre-Borelian sets in R_n.

Definition. (a) *A Borel measure in R_1 is the extension (in the sense of Theorem* 13.3.3(b)) *of an elementary Borel measure from the algebra $\mathfrak{A}$ of the pre-Borelian sets in R_1 to the σ-algebra $S(\mathfrak{A})$ of the Borel sets in R_1.*

(b) *The Lebesgue measure is the complete extension of the elementary Lebesgue measure from the algebra $\mathfrak{A}$ of the pre-Borelian sets in R_n to the σ-algebra $\mathfrak{B}$ in the sense of Theorem* 13.3.3(c).

Remark 1. According to Subsec. 13.3.2., $\mathfrak{B}$ consists of the μ^*-measurable sets, where μ is the elementary Lebesgue measure on the pre-Borelian sets $\mathfrak{A}$ in R_n. The sets of $\mathfrak{B}$ are called Lebesgue-measurable sets, the sets of $S(\mathfrak{A})$ (i.e. the Borel sets) are also called Borel-measurable sets.

Remark 2. $[X, \mathfrak{C}, \nu]$ is called a measure space if X is an arbitrary set, $\mathfrak{C}$ is a σ-algebra of subsets of X, and ν is a measure on $\mathfrak{C}$. According to the above definition we have two examples: $[R_1$, σ-algebra of the Borel-measurable sets in R_1, Borel measure], and $[R_n$, σ-algebra of the Lebesgue-measurable sets in R_n, Lebesgue measure].

Remark 3. Consider an increasing step function $\varphi(t)$ on R_1, which has a finite or countably infinite number of jump discontinuities t_j (Fig. 13.9). The magnitude of the discontinuity is given by

$$\sigma_j = \lim_{t \downarrow t_j} \varphi(t) - \varphi(t_j)$$

where it has been assumed that $\varphi(t)$ is again left-continuous). If E is a pre-Borelian set, then

$$\mu(E) = \sum_{t_j \in E} \sigma_j \tag{1}$$

(where $\mu = \infty$ is not excluded) is the value of the corresponding elementary Borel measure (Theorem 13.2.4/1). Here the sum is to be taken over those σ_j-values for which t_j belongs to E. The corresponding Borel measure $\bar{\mu}$, as per Definition (a), is called Dirac measure. The formula (1) (with $\bar{\mu}$ instead of μ) holds for arbitrary Borel-measurable sets. In particular, $\bar{\mu}(\{t_j\}) = \sigma_j$ and $\bar{\mu}(R_1) = \sum \sigma_j$. If none of the jump discontinuities t_j belongs to the Borel-measurable set E, then $\mu(E) = 0$. One says that the "mass" of the Dirac measure $\bar{\mu}$ concentrates at the points t_j.

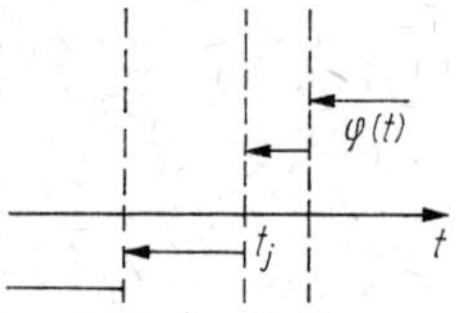

Fig. 13.9

13.3.5. Uniqueness Theorems

We consider now the problem of the uniqueness of the extensions $\bar{\mu}$ referred to in Theorem 13.3.3.

Theorem 1. *The extended measure* $\bar{\mu}$ *of Theorem* 13.3.3(b) *is unique: if* ν *is a measure on* $S(\mathfrak{A})$ *such that* $\mu(E)=\nu(E)$ *for* $E\in\mathfrak{A}$, *then* $\nu=\bar{\mu}$.

Remark 1. In particular, the Borel measures are thus uniquely determined by the elementary Borel measures. Every measure on the Borel-measurable sets is a Borel measure in the sense of Definition 13.3.4.

Theorem 2. *If* $[X, \mathfrak{C}, \varrho]$ *is a measure space, then there exists a smallest complete extension* $[X, \overline{\mathfrak{C}}, \bar{\varrho}]$.

Remark 2. $[X, \overline{\mathfrak{C}}, \bar{\varrho}]$ is a measure space with $\overline{\mathfrak{C}} \supset \mathfrak{C}$ and $\bar{\varrho}(E)=\varrho(E)$ for $E\in\mathfrak{C}$. Further, $\bar{\varrho}$ is a complete measure. This is the definition of a complete extension. The term "smallest complete extension" means: if $[X, \mathfrak{B}, \nu]$ is a complete extension of $[X, \mathfrak{C}, \varrho]$, then $[X, \mathfrak{B}, \nu]$ is also an extension of $[X, \overline{\mathfrak{C}}, \bar{\varrho}]$.

Remark 3. It can be shown that $[X, \mathfrak{C}, \bar{\mu}]$ of Theorem 13.3.3(c) is the smallest complete extension of $[X, S(\mathfrak{A}), \bar{\mu}]$ of Theorem 13.3.3(b). This also includes a uniqueness theorem for the Lebesgue measure of Def. 13.3.4: it is the smallest complete measure that coincides with the elementary Lebesgue measure on the pre-Borelian sets in R_n.

13.4. Measurable Functions

13.4.1. Definition

As stated above, a measure space $[X, \mathfrak{B}, \mu]$ consists of a set X, a σ-algebra $\mathfrak{B}$ of subsets of X, and a measure μ on $\mathfrak{B}$. We consider real-valued functions $f(x)$ on X. If M is a subset of R_1, then

$$f^{-1}(M)=\{x \mid x\in X, f(x)\in M\}$$

is called the inverse image of M (Fig. 13.10).

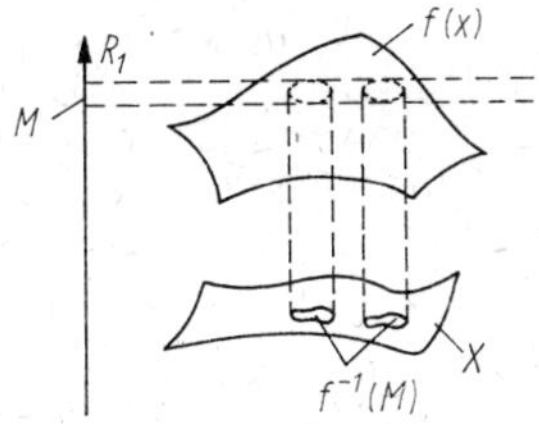

Fig. 13.10

Definition. $f(x)$ *is called measurable* (*or a measurable function*) *if* $f^{-1}(M)\in\mathfrak{B}$ *for every Borel set* M *in* R_1.

Remark 1. Consequently the measurability of a function only depends on $\mathfrak{B}$, not on μ.

Remark 2. If $X=R_n$, then there are two important σ-algebras: $\mathfrak{B}_B$, the σ-algebra of the Borel-measurable sets, and $\mathfrak{B}_L$, the σ-algebra of the Lebesgue-measurable sets; cf. Subsec. 13.3.4. Since $\mathfrak{B}_B\subset\mathfrak{B}_L$, every Borel-measurable function (i.e., $\mathfrak{B}=\mathfrak{B}_B$ in the above definition) is also Lebesgue-measurable (i.e., $\mathfrak{B}=\mathfrak{B}_L$ in the above definition).

Lemma. *Let $[X, \mathfrak{B}, \mu]$ be a measure space. A function $f(x)$ is measurable if and only if the inverse image $f^{-1}((-\infty, c))$ belongs to $\mathfrak{B}$ for every real number c.*

Remark 3. $(-\infty, c)$ is a Borel set in R_1. Hence $f^{-1}((-\infty, c)) \in \mathfrak{B}$ if $f(x)$ is measurable. The lemma says that it suffices to check whether or not the inverse images of the special Borel sets $(-\infty, c)$ belong to $\mathfrak{B}$.

Remark 4. Let $[X, \mathfrak{B}, \mu]$ be a measure space. If $f(x)$ is measurable, then

$$\{x \mid x \in X, c_1 \underset{(=)}{<} f(x) \underset{(=)}{<} c_2\} \in \mathfrak{B} \tag{1}$$

for $-\infty \leqq c_1 \leqq c_2 \leqq \infty$. Here $\underset{(=)}{<}$ means that both $<$ and $\leqq$ is possible in the above formula. So (1) includes four cases. If $c_1 = -\infty$ and/or $c_2 = \infty$, then one has to read $\underset{(=)}{<}$ as $<$. As a special case it is seen from (1) that $f^{-1}(\{c\}) = \{x \mid f(x) = c\}$ is measurable for every real number c.

Remark 5. Sometimes it is practical to admit $f(x) = \infty$ and $f(x) = -\infty$. The above definition must then be supplemented by $\{x \mid f(x) = \infty\} \in \mathfrak{B}$ and $\{x \mid f(x) = -\infty\} \in \mathfrak{B}$.

13.4.2. Properties of Measurable Functions

Theorem 1. *Let $[X, \mathfrak{B}, \mu]$ be a measure space, and let $f(x)$ be a measurable function.*

(a) *$|f(x)|$, $f^+(x) = \max(f(x), 0)$ (positive part of $f(x)$) and $f^-(x) = -\min(f(x), 0)$ (negative part of $f(x)$) are measurable.*

(b) *If*

$$\left(\frac{1}{f}\right)(x) = \begin{cases} \dfrac{1}{f(x)} & \text{for} \quad f(x) \neq 0, \\ 0 & \text{for} \quad f(x) = 0, \end{cases}$$

then $\left(\frac{1}{f}\right)(x)$ is measurable.

(c) *For any real number λ, $(\lambda f)(x) = \lambda f(x)$ is measurable.*

(d) *If $f(x) \geqq 0$ and $\alpha \geqq 0$, then $f^\alpha(x)$ is measurable.*

Remark. Consequently one has $f = f^+ - f^-$ and $|f| = f^+ + f^-$.

Theorem 2. *Let $[X, \mathfrak{B}, \mu]$ be a measure space, and let $f(x)$ and $g(x)$ be measurable functions.*

(a) *For any real number c,*

$$\{x \mid f(x) < g(x) + c\} \in \mathfrak{B}, \ \{x \mid f(x) \leqq g(x) + c\} \in \mathfrak{B},$$
$$\{x \mid f(x) = g(x) + c\} \in \mathfrak{B}.$$

(b) *$(f+g)(x)$, $\max(f(x), g(x))$, $\min(f(x), g(x))$, and $(f \cdot g)(x)$ are measurable.*

(c) *If*

$$\left(\frac{f}{g}\right)(x) = \begin{cases} \dfrac{f(x)}{g(x)} & \text{for} \quad g(x) \neq 0, \\ 0 & \text{for} \quad g(x) = 0, \end{cases}$$

then $\left(\frac{f}{g}\right)(x)$ is measurable.

13.4.3. Sequences of Measurable Functions

If $[X, \mathfrak{B}, \mu]$ is a measure space, and if $\{f_j(x)\}_{j=1}^{\infty}$ is a sequence of measurable functions that converges for every $x \in X$, then one may ask whether $f(x) = \lim_{j \to \infty} f_j(x)$ is also a measurable function. It is practical to modify this question. A subset

E of X is called a set of measure 0 if E is measurable (i.e., if it belongs to $\mathfrak{B}$) and $\mu(E)=0$.

Convention. A property (e.g., the convergence of a sequence) is said to hold almost everywhere in X (abbreviated: a.e.) if this property holds for all $x\in X$ except for a set of measure zero.

Remark 1. Since $\emptyset$ is also a set of measure zero, the convention "a.e." does not exclude that the property holds for all $x\in X$.

Examples. The statement that $\lim\limits_{j\to\infty} f_j(x)$ exists a.e. means that there is a set $E\in\mathfrak{B}$ with $\mu(E)=0$, such that $\lim\limits_{j\to\infty} f_j(x)$ exists for all $x\in X\setminus E$. $|f(x)|<\infty$ a.e. means that there is a set $E\in\mathfrak{B}$ with $\mu(E)=0$, such that $|f(x)|<\infty$ for $x\in X\setminus E$.

Theorem 1. *Let $[X, \mathfrak{B}, \mu]$ be a measure space, and let $\{f_j(x)\}_{j=1}^{\infty}$ be a sequence of measurable functions.*

(**a**) *If $f_0(x)=\sup\limits_j f_j(x)<\infty$ a.e., then*

$$f(x)=\begin{cases} f_0(x) & \text{for } f_0(x)<\infty\,, \\ 0 & \text{for } f_0(x)=\infty \end{cases}$$

is measurable.

(**b**) *If $f_0(x)=\inf\limits_j f_j(x)>-\infty$ a.e., then*

$$f(x)=\begin{cases} f_0(x) & \text{for } f_0(x)>-\infty\,, \\ 0 & \text{for } f_0(x)=-\infty \end{cases}$$

is measurable.

(**c**) *If $f_0(x)=\overline{\lim\limits_j}\, f_j(x)$ and $|f_0(x)|<\infty$ a.e., then*

$$f(x)=\begin{cases} f_0(x) & \text{for } |f_0(x)|<\infty\,, \\ 0 & \text{for } |f_0(x)|=\infty \end{cases}$$

is measurable.

(**d**) *If $f_0(x)=\underline{\lim\limits_j}\, f_j(x)$ and $|f_0(x)|<\infty$ a.e., then*

$$f(x)=\begin{cases} f_0(x) & \text{for } |f_0(x)|<\infty \\ 0 & \text{for } |f_0(x)|=\infty \end{cases}$$

is measurable.

Remark 2. $\overline{\lim}$ (upper limit) and $\underline{\lim}$ (lower limit) have been explained in Subsec. 5.4.1. If $\lim\limits_{j\to\infty} f_j(x)$ exists a.e., then $\overline{\lim}=\underline{\lim}=\lim$ a.e. If E is the exceptional set with $\mu(E)=0$, then it follows from (c) and (d) that

$$f(x)=\begin{cases} \lim\limits_{j\to\infty} f_j(x) & \text{for } x\in X\setminus E \\ 0 & \text{for } x\in E \end{cases}$$

is measurable.

Definition. *Let $[X, \mathfrak{B}, \mu]$ be a measure space. $f(x)$ is called step function if*

$$f(x)=\sum_{j=1}^{N} \alpha_j\chi_j(x) \quad \textit{where} \quad \chi_j(x)=\begin{cases} 1 & \text{for } x\in E_j \\ 0 & \text{for } x\in X\setminus E_j\,, \end{cases} \quad E_j\in\mathfrak{B}.$$

Here the α_j are real numbers.

Remark 3. χ_j is called the characteristic function of the set E_j. It is immediately observed that a step function is measurable.

Theorem 2. *If $[X, \mathfrak{B}, \mu]$ is a measure space, and if $f(x)$ is a measurable function, then there exists a sequence of step functions $f_j(x)$ such that*

$$f(x) = \lim_{j\to\infty} f_j(x) \quad for \quad x \in X .$$

Remark 4. The sequence

$$f_j(x) = \begin{cases} j & \text{for} \quad f(x) \geqq j, \\ \dfrac{l-1}{2^j} & \text{for} \quad \dfrac{l-1}{2^j} \leqq f(x) < \dfrac{l}{2^j}, \\ -j & \text{for} \quad f(x) < -j \end{cases} \tag{1}$$

will serve the above purpose, with $l = -j2^j + 1, \ldots, 0, 1, \ldots, j2^j$.

Remark 5. If $f(x) \geqq 0$, then (1) shows that $f_j(x) \uparrow f(x)$ is a monotonically increasing sequence of step functions (Fig. 13.11).

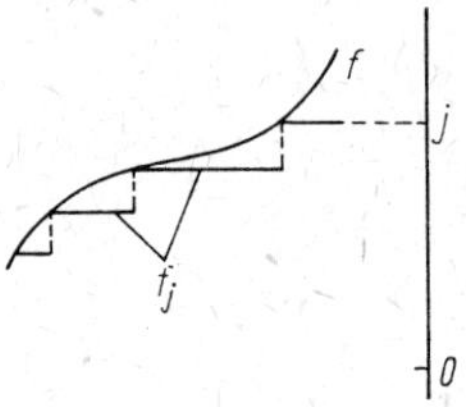

Fig. 13.11

13.4.4. Convergence Almost Everywhere, Convergence in Measure

Definition. *Let $[X, \mathfrak{B}, \mu]$ be a measure space, and let $\{f_j(x)\}_{j=1}^{\infty}$ be a sequence of measurable functions.*

(**a**) $f_j \xrightarrow[a.e.]{} f$ *(convergence a.e.) means that there exists a set $E \in \mathfrak{B}$ with $\mu(E) = 0$, such that $f_j(x) \to f(x)$ for all $x \in X \setminus E$ (pointwise convergence a.e.).*

(**b**) $\{f_j(x)\}_{j=1}^{\infty}$ *is called an a.e. Cauchy sequence (or a.e. fundamental sequence) if there exists a set $E \in \mathfrak{B}$ with $\mu(E) = 0$, such that, for every $\varepsilon > 0$ and all $x \in X \setminus E$, there exists a natural number $m_0 = m_0(\varepsilon, x)$ with*

$$|f_n(x) - f_m(x)| \leqq \varepsilon$$

for all $n \geqq m \geqq m_0$.

(**c**) *Let $f(x)$ be a measurable function. $f_j \xrightarrow[\mu]{} f$ (μ-convergence or convergence in measure) means that, for every $\varepsilon > 0$,*

$$\lim_{j\to\infty} \mu(\{x \mid |f(x) - f_j(x)| \geqq \varepsilon\}) = 0 .$$

(**d**) $\{f_j(x)\}_{j=1}^{\infty}$ *is called a μ-Cauchy sequence (or μ-fundamental sequence) if for every $\varepsilon > 0$ and every $\delta > 0$ there exists a natural number $m_0 = m_0(\varepsilon, \delta)$ such that*

$$\mu(\{x \mid |f_k(x) - f_l(x)| \geqq \varepsilon\}) \leqq \delta \tag{1}$$

for all $l \geqq k \geqq m_0$.

Remark 1. The following generalization of (c) and (d) is of interest. If $A \in \mathfrak{B}$, then $f_j \xrightarrow[\mu|A]{} f$ (μ-convergence in A) means that, for every $\varepsilon > 0$,

$$\lim_{j\to\infty} \mu(\{x \mid x \in A, |f(x) - f_j(x)| \geqq \varepsilon\}) = 0 .$$

By analogy, for μ-Cauchy sequences in A one has to replace $\{x \mid \ldots$ in (1) by $\{x \mid x \in A \ldots$.

Theorem 1. *Let $[X, \mathfrak{B}, \mu]$ be a measure space, and let $\{f_j(x)\}_{j=1}^{\infty}$ be a sequence of measurable functions.*

(**a**) $f_j \xrightarrow[\text{a.e.}]{} f$ *implies* $f_j \xrightarrow[\mu|A]{} f$ *for every set* $A \in \mathfrak{B}$ *with* $\mu(A) < \infty$.

(**b**) *An a.e. Cauchy sequence is also a μ-Cauchy sequence in A for every $A \in \mathfrak{B}$ with $\mu(A) < \infty$.*

(**c**) *If* $f_j \xrightarrow[\mu]{} f$, *then there exists a sequence of natural numbers* $j_1 < j_2 < j_3 < \dots$ *such that* $f_{j_k} \xrightarrow[\text{a.e.}]{} f$ *as* $k \to \infty$.

(**d**) *If* $\{f_j\}_{j=1}^{\infty}$ *is a μ-Cauchy sequence, then there exists a subsequence* $\{f_{j_k}\}_{k=1}^{\infty}$ *which is an a.e. Cauchy sequence.*

Remark 2. If $f_j \xrightarrow[\text{a.e.}]{} f$, then it follows from Remark 13.4.3/2 that $f(x)$ is measurable. Thus proposition (a) of the above theorem is meaningful.

Remark 3. The propositions (a) and (b) cannot be extended to $A = X$ if $\mu(X) = \infty$. It is easy to find counter-examples.

Remark 4. It is also possible to construct $\{f_{j_k}(x)\}_{k=1}^{\infty}$ of proposition (c) so that there exists a sequence $E_1 \supset E_2 \supset E_3 \supset \dots$ of sets in $\mathfrak{B}$ with $\mu(E_l) \leqq 2^{-l}$ and

$$\sup_{x \in X \setminus E_l} |f_{j_k}(x) - f(x)| \to 0 \quad \text{as} \quad k \to \infty$$

for $l = 1, 2, 3, \dots$ In other words, for every $\varepsilon > 0$ there exists a set $E_\varepsilon \in \mathfrak{B}$ with $\mu(E_\varepsilon) \leqq \varepsilon$, such that $\{f_{j_k}(x)\}$ converges uniformly to $f(x)$ on $X \setminus E_\varepsilon$. The exceptional set in proposition (c) may be chosen to be, say, $\bigcap_{l=1}^{\infty} E_l$.

Remark 5. If $\mu(X) < \infty$, then one may set $A = X$ in (a) and (b). The convergence a.e. then implies μ-convergence. Generally the converse is not true, but for a suitable subsequence one has proposition (c).

Theorem 2. *Let $[X, \mathfrak{B}, \mu]$ be a measure space, and let $\{f_j(x)\}_{j=1}^{\infty}$ be a sequence of measurable functions.*

(**a**) $f_j \xrightarrow[\text{a.e.}]{} f$ *implies that* $\{f_j(x)\}_{j=1}^{\infty}$ *is an a.e. Cauchy sequence.*

(**b**) $f_j \xrightarrow[\mu]{} f$ *implies that* $\{f_j(x)\}_{j=1}^{\infty}$ *is a μ-Cauchy sequence.*

(**c**) *If* $\{f_j(x)\}_{j=1}^{\infty}$ *is an a.e. Cauchy sequence, then there is a measurable function* $f(x)$ *such that* $f_j \xrightarrow[\text{a.e.}]{} f$. *If* $f_j \xrightarrow[\text{a.e.}]{} g$, *then* $f(x) = g(x)$ *a.e.*

(**d**) *If* $\{f_j(x)\}_{j=1}^{\infty}$ *is a μ-Cauchy sequence, then there is a measurable function* $f(x)$ *such that* $f_j \xrightarrow[\mu]{} f$. *If* $f_j \xrightarrow[\mu]{} g$, *then* $f(x) = g(x)$ *a.e.*

14. Integration Theory

14.1. Integrable Functions, Properties of Integrals

14.1.1. Integrable Step Functions

Definition. *Let* $[X, \mathfrak{B}, \mu]$ *be a measure space.*

(a) *A step function* $f(x)=\sum_{j=1}^{N} \alpha_j \chi_{E_j}(x)$, *where the* α_j *are real,* $E_j \in \mathfrak{B}$, $\bigcup_{j=1}^{N} E_j = X$, $E_k \cap E_l = \emptyset$ *for* $k \neq l$, *and* $\chi_{E_j}(x)$ *is the characteristic function of* E_j (*cf. Def.* 13.4.3), *is called integrable if* $\mu(E_j) < \infty$ *for* $\alpha_j \neq 0$. *One sets*

$$\int f(x)\, \mathrm{d}\mu = \sum_{j=1}^{N} \alpha_j \mu(E_j) \quad (\textit{with } 0 \cdot \infty = 0)\,. \tag{1}$$

(b) *If* $f(x)$ *is an integrable step function, and if* $A \in \mathfrak{B}$, *then let*

$$\int_A f(x)\, \mathrm{d}\mu = \int f(x)\, \chi_A(x)\, \mathrm{d}\mu\,,$$

where χ_A *is the characteristic function of* A.

Remark 1. It is easily shown that (1) is independent of the representation: if $f(x)=\sum_{k=1}^{M} \beta_k \chi_{F_k}(x)$, with real β_k, $F_k \in \mathfrak{B}$, $\bigcup_{k=1}^{M} F_k = X$, $F_k \cap F_l = \emptyset$ for $k \neq l$, is the step function referred to in (a), then

$$\sum_{j=1} \alpha_j \mu(E_j) = \sum_{k=1}^{M} \beta_k \mu(F_k)\,.$$

Remark 2. If $f(x)$ is an integrable step function, and if $A \in \mathfrak{B}$, then $f(x)\, \chi_A(x)$ is also an integrable step function. This justifies part (b) of the above definition.

Lemma. *Let* $[X, \mathfrak{B}, \mu]$ *be a measure space.*

(a) *If* $f(x)$ *and* $g(x)$ *are integrable step functions, and if* α *and* β *are real numbers, then* $\alpha f(x) + \beta g(x)$ *is also an integrable step function, and*

$$\int (\alpha f(x) + \beta g(x))\, \mathrm{d}\mu = \alpha \int f(x)\, \mathrm{d}\mu + \beta \int g(x)\, \mathrm{d}\mu\,.$$

(b) *If* $f(x)$ *is an integrable step function, then* $|f(x)|$ *is also an integrable step function, and*

$$\left|\int f(x)\, \mathrm{d}\mu\right| \leqq \int |f(x)|\, \mathrm{d}\mu\,.$$

(c) *If* $f(x)$ *and* $g(x)$ *are integrable step functions such that* $f(x) \leqq g(x)$, *then* $\int f(x)\, \mathrm{d}\mu \leqq \int g(x)\, \mathrm{d}\mu$.

(d) *If* $f(x)$ *is an integrable step function, and if* $A \in \mathfrak{B}$, *then*

$$\int f(x)\, \mathrm{d}\mu = \int_A f(x)\, \mathrm{d}\mu + \int_{X \setminus A} f(x)\, \mathrm{d}\mu.$$

Remark 3. All the above properties are easily deduced from the definition. From (c) one obtains

$$\int g(x)\, \mathrm{d}\mu \geqq 0\,,$$

if $g(x) \geqq 0$ is an integrable step function.

14.1.2. Integrable Functions

Definition. *Let $[X, \mathfrak{B}, \mu]$ be a measure space, and let $f(x)$ be a measurable function. $f(x)$ is called integrable if there is a sequence $\{f_j(x)\}_{j=1}^{\infty}$ of integrable step functions with $f_j \underset{\mu}{\rightarrow} f$, such that for all $\varepsilon > 0$ there exists a natural number $k_0 = k_0(\varepsilon)$ with the property that*

$$\int |f_l(x) - f_k(x)| \, \mathrm{d}\mu \leqq \varepsilon \quad \textit{for all} \quad l \geqq k \geqq k_0 \, . \tag{1}$$

We set

$$\int f(x) \, \mathrm{d}\mu = \lim_{j \to \infty} \int f_j(x) \, \mathrm{d}\mu \, . \tag{2}$$

Remark. The limit in (2) exists, because

$$\begin{aligned} \left| \int f_k(x) \, \mathrm{d}\mu - \int f_l(x) \, \mathrm{d}\mu \right| &= \left| \int (f_k(x) - f_l(x)) \, \mathrm{d}\mu \right| \\ &\leqq \int |f_k(x) - f_l(x)| \, \mathrm{d}\mu \leqq \varepsilon \end{aligned}$$

for $l \geqq k \geqq k_0(\varepsilon)$. It is more difficult to show that this limit does not depend on the sequence $\{f_j(x)\}_{j=1}^{\infty}$: if $g_j \underset{\mu}{\rightarrow} f$ is a sequence of integrable step functions, and if the analogue to (1) (with g instead of f) is true, then it follows that

$$\lim_{j \to \infty} \int g_j(x) \, \mathrm{d}\mu = \lim_{j \to \infty} \int f_j(x) \, \mathrm{d}\mu \, .$$

This justifies (2).

14.1.3. Properties of Integrable Functions

The properties stated in Lemma 14.1.1 can be transferred to integrable functions.

Theorem *Let $[X, \mathfrak{B}, \mu]$ be a measure space.*

(a) *If $f(x)$ is integrable, then $|f(x)|$ is also integrable, and*

$$\left| \int f(x) \, \mathrm{d}\mu \right| \leqq \int |f(x)| \, \mathrm{d}\mu \, .$$

(b) *If $f(x)$ and $g(x)$ are integrable, and if α and β are real numbers, then $\alpha f(x) + \beta g(x)$ is also integrable, and*

$$\int (\alpha f(x) + \beta g(x)) \, \mathrm{d}\mu = \alpha \int f(x) \, \mathrm{d}\mu + \beta \int g(x) \, \mathrm{d}\mu \, .$$

(c) *If $f(x)$ and $g(x)$ are integrable, and if $f(x) \leqq g(x)$, then*

$$\int f(x) \, \mathrm{d}\mu \leqq \int g(x) \, \mathrm{d}\mu \, .$$

In particular, $\int g(x) \, \mathrm{d}\mu \geqq 0$ if $g(x) \geqq 0$ is integrable.

(d) *Let $f(x)$ be integrable. Then $\int |f(x)| \, \mathrm{d}\mu = 0$ if and only if $f(x) = 0$ a.e.*

(e) *Let $f(x)$ be integrable, and let $g(x)$ be measurable. If $f(x) = g(x)$ a.e., then $g(x)$ is integrable, and $\int f(x) \, \mathrm{d}\mu = \int g(x) \, \mathrm{d}\mu$.*

Remark 1. The properties (a), (b) and (c) are also valid for Riemann integrals; cf. Theorem 9.1.3/1 and Theorem 9.1.5/1. On the other hand, (d) and (e) are typical new porperties.

Remark 2. (a) implies that $f^+(x) = \max(f(x), 0) = \frac{1}{2}(|f(x)| + f(x))$ is integrable too.

14.1.4. Properties of Integrals

Definition. *Let $[X, \mathfrak{B}, \mu]$ be a measure space, and let $f(x)$ be an integrable function. If $E \in \mathfrak{B}$, and if $\chi_E(x)$ is the characteristic function of E, then one sets*

$$\int_E f(x)\,\mathrm{d}\mu = \int f(x)\,\chi_E(x)\,\mathrm{d}\mu\,.$$

Remark 1. This is the generalization of Def. 14.1.1(b). If $f(x)$ is integrable, and if $E \in \mathfrak{B}$, then $f(x)\,\chi_E(x)$ is also integrable. So the above definition is meaningful.

Theorem. *Let $[X, \mathfrak{B}, \mu]$ be a measure space.*

(a) *If $f(x)$ is integrable, and if $E \in \mathfrak{B}$, then*

$$\int f(x)\,\mathrm{d}\mu = \int_E f(x)\,\mathrm{d}\mu + \int_{X \setminus E} f(x)\,\mathrm{d}\mu\,.$$

(b) *Let $f(x)$ be integrable, and let $f(x) > 0$ for $x \in X$. If $\int_E f(x)\,\mathrm{d}\mu = 0$, then $\mu(E) = 0$.*

(c) *If $f(x)$ is integrable, and if $X = \bigcup_{j=1}^{\infty} E_j$, with $E_j \in \mathfrak{B}$ and $E_k \cap E_l = \emptyset$ for $k \neq l$, then*

$$\int f(x)\,\mathrm{d}\mu = \sum_{k=1}^{\infty} \int_{E_k} f(x)\,\mathrm{d}\mu$$

(absolutely convergent series).

Remark 2. The following proposition can be deduced from (c). If $f(x)$ is integrable, and if $E_1 \subset E_2 \subset E_3 \subset \ldots$, with $X = \bigcup_{j=1}^{\infty} E_j$ and $E_j \in \mathfrak{B}$, then

$$\lim_{j \to \infty} \int_{E_j} f(x)\,\mathrm{d}\mu = \int f(x)\,\mathrm{d}\mu\,.$$

Remark 3. Another implication of (c) reads as follows. If $f(x)$ is integrable and $f(x) \geqq 0$, then

$$\nu(E) = \int_E f(x)\,\mathrm{d}\mu, \quad E \in \mathfrak{B}\,,$$

is a measure.

14.2. Fundamental Theorems of Integration Theory

14.2.1. Convergence in L_1

Definition. *Let $[X, \mathfrak{B}, \mu]$ be a measure space, and let $f(x)$ and $f_j(x)$, $j = 1, 2, 3, \ldots$, be integrable functions.*

(a) *$f_j \underset{L_1}{\to} f$ (convergence in L_1, or convergence in the mean) means that*

$$\lim_{j \to \infty} \int |f_j(x) - f(x)|\,\mathrm{d}\mu = 0\,.$$

(b) *$\{f_j(x)\}_{j=1}^{\infty}$ is called a fundamental sequence in L_1 (or Cauchy sequence in L_1) if for all $\varepsilon > 0$ there exists a natural number $m_0 = m_0(\varepsilon)$ such that*

$$\int |f_n(x) - f_m(x)|\,\mathrm{d}\mu \leqq \varepsilon \quad \textit{for all} \quad n \geqq m \geqq m_0\,.$$

Lemma. *Let $[X, \mathfrak{B}, \mu]$ be a measure space.*

(a) *$f_j \underset{L_1}{\to} f$ implies $f_j \underset{\mu}{\to} f$.*

(b) *If $\{f_j(x)\}_{j=1}^{\infty}$ is a fundamental sequence in L_1, then $\{f_j(x)\}_{j=1}^{\infty}$ is also a μ-fundamental sequence.*

Remark. If $\mu(X) < \infty$, then according to Subsec. 13.4.4 one has the following situation; convergence a.e. implies μ-convergence, and convergence in L_1 likewise implies μ-convergence. μ-convergence of a sequence implies the convergence a.e. of a suitable subsequence. In particular, it is thus possible to select from a sequence convergent in L_1 a subsequence that is convergent a.e. (Fig. 14.1).

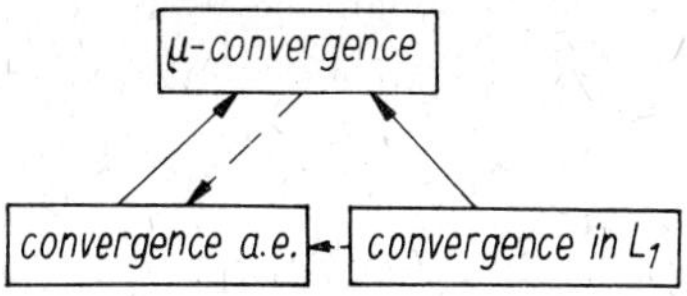

Fig. 14.1

14.2.2. Lebesgue's Bounded Convergence Theorem

Theorem. *Let $[X, \mathfrak{B}, \mu]$ be a measure space, let $\{f_j(x)\}_{j=1}^{\infty}$ be a sequence of integrable functions, and let $g(x)$ be an integrable function such that*

$$|f_j(x)| \leqq g(x) \quad \text{for} \quad j = 1, 2, 3, \ldots \quad \text{and} \quad x \in X .$$

If either $f_j \underset{\mu}{\rightarrow} f$ or $f_j \xrightarrow[a.e.]{} f$ or $f_j \underset{L_1}{\rightarrow} f$, then $f(x)$ is integrable, and

$$\int f(x) \, d\mu = \lim_{j \to \infty} \int f_j(x) \, d\mu . \tag{1}$$

Remark. Formally, one may write (1) as

$$\int (\lim_{j \to \infty} f_j(x)) \, d\mu = \lim_{j \to \infty} \int f_j(x) \, d\mu .$$

Hence Lebesgue's theorem expresses the commutativity of limit and integration. $g(x)$ is called an integrable majorant. If such an integrable majorant does not exist, the theorem is no longer valid, as is shown by the following example (Fig. 14.2): in R_1, let $f_j(x) = j$ for $0 \leqq x \leqq \frac{1}{j}$, and $f_j(x) = 0$ otherwise. Then $\int f_j(x) \, dx = 1$ for $j = 1, 2, \ldots$ On the other hand, $f_j \to 0$ a.e. and $\int 0 \, dx = 0$. This shows that Lebesgue's theorem cannot be valid in this case.

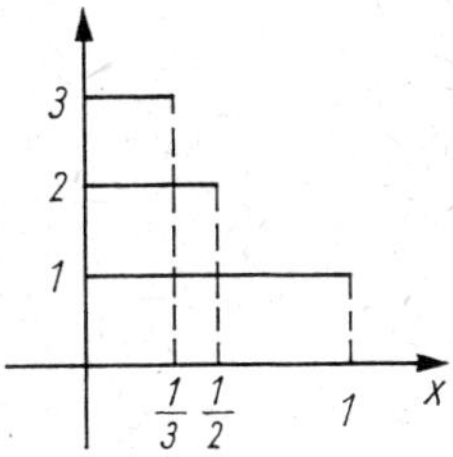

Fig. 14.2

14.2.3. Other Properties of Integrable Functions

There are some simple conclusions that can be drawn from Lebesgue's theorem of bounded convergence as well as from previous considerations.

Theorem 1. *Let $[X, \mathfrak{B}, \mu]$ be a measure space, and let $f(x) \geqq 0$ be integrable. If the functions $f_j(x)$ are as given in (13.4.3/1), then*

$$\int f(x) \, d\mu = \lim_{j \to \infty} \int f_j(x) \, d\mu . \tag{1}$$

Remark 1. The theorem is an immediate consequence of Theorem 14.2.2, if it is taken into account that $0 \leq f_j(x) \leq f(x)$ and $f_j \xrightarrow[\text{a.e.}]{} f$. Thus the integral over a non-negative integrable function can be represented as the limit of integrals over step functions. This reminds us of the Riemann integral. The methods are, however, different. Let, for instance, $X = R_1$, and let μ be the Lebesgue measure on R_1. For the construction of approximating step functions, one subdivides the x-axis for the Riemann integral (cf. Subsec. 3.2.1.), whereas for the Lebesgue integral the y-axis is subdivided (Fig. 14.3).

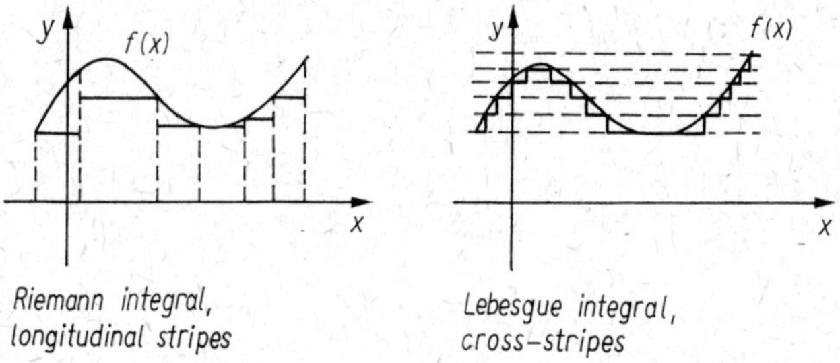

Fig. 14.3

Theorem 2. *Let* $[X, \mathfrak{B}, \mu]$ *be a measure space.* $f(x)$ *is integrable if and only if* $|f(x)|$ *is integrable.*

Remark 2. There is no analogue to this theorem in Riemann's theory of integration. Compare this with Theorem 3.2.2/2(b) and with the counter-example of Subsec. 3.2.4. The function considered there is Lebesgue-integrable.

Theorem 3. *Let* $[X, \mathfrak{B}, \mu]$ *be a measure space. If* $f(x)$ *is measurable and* $g(x)$ *is integrable, and if* $|f(x)| \leq |g(x)|$ *for* $x \in X$, *then* $f(x)$ *is also integrable.*

14.2.4. The Banach Space $L_1(X, \mathfrak{B}, \mu)$

If $[X, \mathfrak{B}, \mu]$ is a measure space, then we introduce an equivalence relation in the set of all measurable functions: $f \sim g$ if $f(x) = g(x)$ a.e. It is easily seen that $\sim$ satisfies the requirements placed on an equivalence relation: $f \sim f$ (reflexive); $f \sim g$ if and only if $g \sim f$ (symmetric); $f \sim g$ and $g \sim h$ implies that $f \sim h$ (transitive). So the measurable functions can be subdivided into equivalence classes: let $[f]$ be the equivalence class that contains $f(x)$ as a representative.

Definition. $L_1 = L_1(X, \mathfrak{B}, \mu)$ *is the set of all equivalence classes of measurable functions for which*

$$\|[f]\|_{L_1} = \int |h(x)| \, d\mu < \infty \quad \textit{where} \quad h(x) \in [f] \, . \tag{1}$$

Remark 1. It is easily seen that (1) is independent from the particular selection of the representatives $h(x) \in [f]$. Hence (1) is meaningful.

Theorem. L_1 *is a Banach space with the norm* $\|[\cdot]\|_{L_1}$.

Remark 2. There is no corresponding theorem for Riemann integrals. This shows a decisive advantage of the present, more general concept.

Remark 3 (convention). In our future calculations we shall use representatives $f \in L_1$ instead of equivalence classes $[f] \in L_1$. We also write $\|f\|_{L_1}$ instead of $\|[f]\|_{L_1}$. In the sense of L_1, two functions are equal if they coincide a.e.

Remark 4. $f_j \xrightarrow[L_1]{} f$ implies that $\int |f_j(x)| \, d\mu \to \int |f(x)| \, d\mu$.

14.2.5. The Theorems of B. Levi and Fatou

Theorem 1 (*B. Levi*). *Let* $[X, \mathfrak{B}, \mu]$ *be a measure space, and let* $0 \leqq f_1(x) \leqq f_2(x) \leqq \leqq f_3(x) \leqq \ldots$ *be a monotone sequence of integrable functions. Let* $\sup_j \int f_j(x)\,\mathrm{d}\mu < \infty$. *Then* $f(x) = \lim_{j\to\infty} f_j(x)$ *is integrable, and* $\int f(x)\,\mathrm{d}\mu = \lim_{j\to\infty} \int f_j(x)\,\mathrm{d}\mu$.

Remark 1. In particular the theorem says that $f_j(x)$ converges almost everywhere. The theorem also remains valid for $\sup_j \int f_j(x)\,\mathrm{d}\mu = \infty$, provided that $\int f(x)\,\mathrm{d}\mu = \infty$ is interpreted to express that $f(x)$ is measurable but not integrable. (Here it is also possible that $f(x) = \infty$ on a set of positive measure.)

Theorem 2 (*Fatou's lemma*). *Let* $[X, \mathfrak{B}, \mu]$ *be a measure space. Let* $\{f_j(x)\}_{j=1}^{\infty}$ *be a sequence of integrable functions,* $f_j(x) \geqq 0$ *for* $j = 1, 2, 3, \ldots$ *If* $\underline{\lim}_j \int f_j(x)\,\mathrm{d}\mu < \infty$, *then* $f(x) = \underline{\lim}_j f_j(x)$ *is integrable, and*

$$\int f(x)\,\mathrm{d}\mu \leqq \underline{\lim}_j \int f_j(x)\,\mathrm{d}\mu\,.$$

Remark 2. $\underline{\lim}_j = \underline{\lim}$ has been defined in Subsec. 5.4.1.

14.3. Transformation Formulas

14.3.1. Measurable Mappings and Image Measures

We will consider a measure space $[X, \mathfrak{B}, \mu]$ and a measurable space $(Y, \mathfrak{C})$ consisting of a basic set Y and a σ-algebra $\mathfrak{C}$ of subsets of Y (Fig. 14.4).

Definition. *A single-valued mapping* $T(x)$ *of* X *into* Y *is said to be measurable if*

$$T^{-1}(M) = \{x \mid x \in X,\ T(x) \in M\} \in \mathfrak{B} \tag{1}$$

for every set $M \in \mathfrak{C}$.

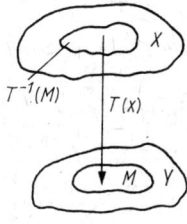

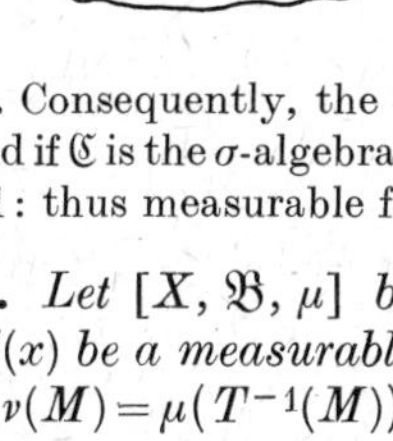

Fig. 14.4

Remark 1. Consequently, the inverse image $T^{-1}(M)$ of every set M of $\mathfrak{C}$ belongs to $\mathfrak{B}$. If $Y = R_1$, and if $\mathfrak{C}$ is the σ-algebra of the Borel sets in R_1, then the above definition agrees with Def. 13.4.1: thus measurable functions are special measurable mappings.

Theorem. *Let* $[X, \mathfrak{B}, \mu]$ *be a measure space, let* $(Y, \mathfrak{C})$ *be a measurable space, and let* $T(x)$ *be a measurable mapping of* X *into* Y *in the sense of the above definition. Let* $\nu(M) = \mu(T^{-1}(M))$ *for* $M \in \mathfrak{C}$. *If there exists a sequence* $F_1 \subset F_2 \subset F_3 \subset \ldots$ *such that* $F_j \in \mathfrak{C}$, $\nu(F_j) < \infty$ *and* $Y = \bigcup_{j=1}^{\infty} F_j$, *then* ν *is a measure.*

Remark 2. Since $T(x)$ is measurable, $\nu(M)$ is defined for every set $M \in \mathfrak{C}$. To prove the above theorem, one has to check the properties stated in Def. 13.2.1: The σ-additivity follows from the setup, and σ-finiteness has been required by us. It is possible to state examples which show that σ-finiteness cannot automatically be derived from the setup for ν.

Remark 3. One writes $\nu = \mu T^{-1}$, calling ν the image measure (with respect to μ and T). We make the convention that the term "image measure" shall imply the existence of a sequence of sets $F_1 \subset F_2 \subset F_3 \subset \dots$ with the properties referred to in the above theorem.

14.3.2. A Special Transformation Formula

Consider a measure space $[X, \mathfrak{B}, \mu]$, a measurable space $(Y, \mathfrak{C})$, and a measurable mapping $T(x)$ of X into Y (Fig. 14.5). According to Remark 14.3.1/3, let $\nu = \mu T^{-1}$ be the associated image measure. Then $[Y, \mathfrak{C}, \nu]$ is a measure space. Let x be the general point in X, and y, the general point in Y. Functions f on X are written as $f(x)$, and functions f on Y, as $f(y)$. Measurability, integrability etc. are then always to be understood as being referred to the associated measure space.

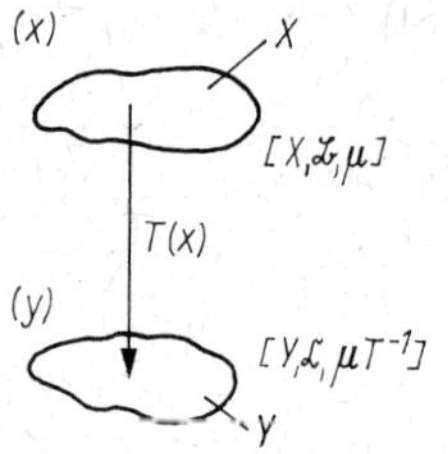

Fig. 14.5

Theorem. (a) *If $f(y)$ is measurable, then $h(x) = f(T(x))$ is also measurable.*

(b) *If $f(y)$ is integrable, then $h(x) = f(T(x))$ is also integrable, and*

$$\int f(y)\, d(\mu T^{-1}) = \int f(T(x))\, d\mu .$$

14.3.3. Absolutely Continuous Measures, Theorem of Radon-Nikodým

Definition. *Let $[X, \mathfrak{B}, \mu]$ be a measure space. A measure ν on $\mathfrak{B}$ is called absolutely continuous (with respect to μ) if $\mu(E) = 0$, $E \in \mathfrak{B}$, always implies that $\nu(E) = 0$.*

Remark 1. If ν is absolutely continuous with respect to μ, one writes $\nu \ll \mu$. Examples can easily be stated. If $f(x) \geqq 0$ is integrable, then

$$\nu(E) = \int_E f(x)\, d\mu, \quad E \in \mathfrak{B}, \tag{1}$$

is absolutely continuous with respect to μ; cf. Remark 14.1.4/3.

Remark 2. If $f(x) \geqq 0$ is measurable but not integrable, we write $\int f(x)\, d\mu = \infty$. Then Eq. (1) is meaningful for any measurable functions $f(x) \geqq 0$. If ν is a measure (and hence, in particular, is σ-finite), then it is absolutely continuous with respect to μ. The following theorem is to be understood in this sense.

Theorem (*Radon-Nikodým*). *Let $[X, \mathfrak{B}, \mu]$ be a measure space. If $\nu \ll \mu$, then there exists a measurable function $f(x) \geqq 0$ such that*

$$\nu(E) = \int_E f(x)\, d\mu \quad \textit{for all} \quad E \in \mathfrak{B} .$$

If $g(x) \geqq 0$ is a measurable function such that $\nu(E) = \int_E g(x)\, d\mu$ for all $E \in \mathfrak{B}$, then $f(x) = g(x)$ almost everywhere.

Remark 3. So $f(x)$ is in effect uniquely determined, and one writes $f(x) = \frac{d\nu}{d\mu}$: Radon-Nikodým derivative.

14.3.4. The General Transformation Formula

Consider two measure spaces $[X, \mathfrak{B}, \mu]$ and $[Y, \mathfrak{C}, \varrho]$ and a measurable mapping $T(x)$ of X into Y. Let μT^{-1} be the image measure with respect to μ and T in the sense of Remark 14.3.1/3. As regards the designation by x or y, the conventions made in Subsec. 14.3.2. shall apply.

Theorem. *Let $\mu T^{-1} \ll \varrho$. If $f(y)$ is measurable, and if $f(y) \frac{d(\mu T^{-1})}{d\varrho}(y)$ is integrable, then $h(x) = f(T(x))$ is integrable, and*

$$\int f(T(x))\, d\mu = \int f(y) \frac{d(\mu T^{-1})}{d\varrho}(y)\, d\varrho\,.$$

Remark 1. This theorem is a combination of the theorems of Subsecs. 14.3.2. and 14.3.3.

14.4. Product Measures, Fubini's Theorem

14.4.1. The σ-Algebra in a Product Space, Measurable Intersections

Let $\mathfrak{B}$ be a σ-algebra of subsets of the set X, and let $\mathfrak{C}$ be a σ-algebra of subsets of the set Y. Let

$$X \times Y = \{(x, y) \mid x \in X, y \in Y\}, \quad E \times F = \{(x, y) \mid x \in E, y \in F\}\,,$$

where E is a subset of X and F is a subset of Y (Fig. 14.6).

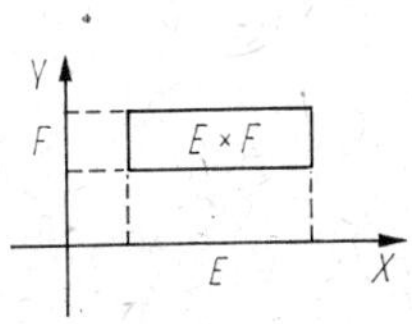

Fig. 14.6

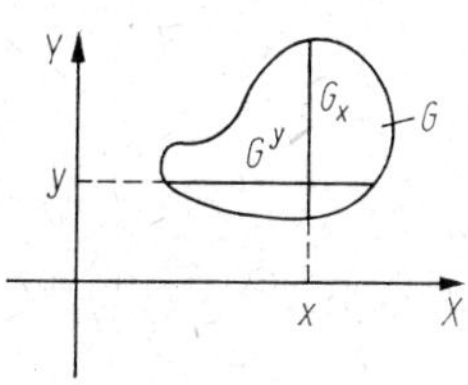

Fig. 14.7

Definition. (a) *$\mathfrak{B} \times \mathfrak{C}$ is the smallest σ-algebra of subsets of the set $X \times Y$, containing all sets of the form $E \times F$, with $E \in \mathfrak{B}$ and $F \in \mathfrak{C}$ (product σ-algebra).*

(b) *If $G \subset X \times Y$, and $x \in X$, then $G_x = \{y \mid y \in Y, (x, y) \in G\}$ is called an intersection of G. Accordingly, $G^y = \{x \mid x \in X, (x, y) \in G\}$, with $y \in Y$, is called an intersection of G* (Fig. 14.7).

(c) *If $f(x, y)$ is a real-valued function on $X \times Y$, then $f_x(y) = f(x, y)$ is called an intersection of f, and $f^y(x) = f(x, y)$ is called an intersection of f.*

Remark 1. One has $G_x \subset Y$ and $G^y \subset X$. Further $f_x(y)$ means that, with $x \in X$ being fixed, $f(x, y)$ is considered a function of y, and analogously, $f^y(x)$ is considered a function of x, with $y \in Y$ being fixed.

Theorem. (a) *If $G \in \mathfrak{B} \times \mathfrak{C}$, then $G_x \in \mathfrak{C}$ for all $x \in X$, and $G^y \in \mathfrak{B}$ for all $y \in Y$.*

(b) *If $f(x, y)$ is measurable with respect to $\mathfrak{B}\times\mathfrak{C}$, then $f_x(y)$ is measurable with respect to $\mathfrak{C}$ for all $x\in X$, and $f^y(x)$ is measurable with respect to $\mathfrak{B}$ for all $y\in Y$.*

Remark 2. Intersections of measurable sets are also measurable sets, and intersections of measurable functions are also measurable functions.

14.4.2. The Product Measure

Definition. *Let $[X, \mathfrak{B}, \mu]$ and $[Y, \mathfrak{C}, \nu]$ be two measure spaces. A measure λ on the product σ-algebra $\mathfrak{B}\times\mathfrak{C}$ is called product measure if $\lambda(E\times F)=\mu(E)\cdot\nu(F)$ for all $E\in\mathfrak{B}$ and all $F\in\mathfrak{C}$.*

Remark 1. $\mu(E)$ and/or $\nu(F)$ may be ∞. We recall the previous convention that $0\cdot\infty=0$ and $a\cdot\infty=\infty$ for $a>0$. Further we set $\int f(x)\,d\mu=\infty$ (as we did in Remark 14.3.3/2) if $f(x)\geqq 0$ is measurable but not integrable. Here it is expedient for the formulation of the following theorem that $f(x)=\infty$ is allowed on an arbitrary measurable set; cf. Remark 13.4.1/5. If the set with $f(x)=\infty$ has a positive measure, then $\int f(x)\,d\mu=\infty$.

Theorem. *In the sense of the above definition, there exists exactly one product measure. Let λ be such product measure, and let $G\in\mathfrak{B}\times\mathfrak{C}$, then*

$$\lambda(G)=\int \nu(G_x)\,d\mu=\int \mu(G^y)\,d\nu\,. \tag{1}$$

Remark 2. According to Remark 1, it is possible that $\lambda(G)=\infty$. The expressions formulated in (1) are meaningful. $G_x\in\mathfrak{C}$ by Theorem 14.4.1, and hence one may form $\nu(G_x)$, considering it a function of x. It has to be shown that $\nu(G_x)$ is measurable with respect to $\mathfrak{B}$. Since $\nu(G_x)\geqq 0$, the first integral in (1) is then meaningful in view of Remark 1. A corresponding statement can be made for the second integral. The product measure is also written as $\lambda=\mu\times\nu$.

14.4.3. Fubini's Theorem for non-negative Functions

Theorem. *Let $[X, \mathfrak{B}, \mu]$ and $[Y, \mathfrak{C}, \nu]$ be two measure spaces. Let $[X\times Y, \mathfrak{B}\times\mathfrak{C}, \mu\times\nu]$ be the associated product measure space. If $f(x, y)\geqq 0$ is a function measurable with respect to $\mathfrak{B}\times\mathfrak{C}$, then $\int f(x, y)\,d\mu$ is a $\mathfrak{C}$-measurable function and $\int f(x, y)\,d\nu$ is a $\mathfrak{B}$-measurable function. Further,*

$$\int f(x, y)\,d(\mu\times\nu)=\int\left(\int f(x, y)\,d\mu\right)d\nu=\int\left(\int f(x, y)\,d\nu\right)d\mu\,. \tag{1}$$

Remark 1. In accordance with Remark 14.4.2/1, the integrals encountered may also be infinite. In any case, $\int f(x, y)\,d\mu=\int f^y(x)\,d\mu$ as well as $\int f(x, y)\,d\nu=\int f_x(y)\,d\nu$ are meaningful. The first assertion of the above theorem states that $\int f^y(x)\,d\mu$ is measurable with respect to the σ-algebra $\mathfrak{C}$, and an analogous statement is made on $\int f_x(y)\,d\nu$. As these functions are non-negative, all of the integrals in (1) are meaningful. (1) is meant to include the following assertion: if one of the three integrals is finite, then the other two integrals are finite too, and all three integrals are then equal.

14.4.4. Fubini's Theorem for Arbitrary Functions

Theorem. *Let $[X, \mathfrak{B}, \mu]$ and $[Y, \mathfrak{C}, \nu]$ be two measure spaces. Let $[X\times Y, \mathfrak{B}\times\mathfrak{C}, \mu\times\nu]$ be the associated product measure space. Let $f(x, y)$ be $\mu\times\nu$-integrable. Then $f_x(y)$ is a ν-integrable function for almost all $x\in X$, and $f^y(x)$ is a μ-integrable function for almost all $y\in Y$,*

$$\int f_x(y)\,d\nu \text{ is } \mu\text{-integrable}, \int f^y(x)\,d\mu \text{ is } \nu\text{-integrable}, \tag{1}$$

and

$$\int f(x, y)\, d(\mu \times \nu) = \int \left(\int f(x, y)\, d\mu\right) d\nu = \int \left(\int f(x, y)\, d\nu\right) d\mu\,. \tag{2}$$

Remark 1. Consequently $f_x(y)$ belongs to $L_1(Y, \mathfrak{C}, \nu)$ for all $x \in X \backslash E$, with $E \in \mathfrak{B}$ and $\mu(E) = 0$. For these x-values the first integral in (1) exists. For $x \in E$ one may set, say, $\int f_x(y)\, d\nu = 0$. Analogous statements can be made on $f^y(x)$.

Remark 2. This theorem does not include Theorem 14.4.3 as a special case. Theorem 14.4.3 has an independent meaning, because (unlike the above theorem) it is applicable to arbitrary non-negative measurable functions.

Riemann's integral is a special case of Lebesgue's integral
(mathematical folklore)

14.5. Comparison between Riemann's and Lebesgue's Integrals

14.5.1. Integrable Functions

In Def. 9.1.2(b) it had been stated when a real-valued bounded function $f(x)$ defined on a Q-domain Ω in R_n is Riemann-integrable. We shall now denote the corresponding Riemann integral by $\mathrm{R}\int f(x)\, dx$. If Ω is an I-domain, the Riemann integral for functions being continuous on $\bar{\Omega}$ had been introduced by Def. 9.1.5. We shall likewise denote it by $\mathrm{R}\int f(x)\, dx$. In accordance with Subsec. 13.3.4., let $[R_n, \mathfrak{B}_{\mathrm{L}}, \mu_{\mathrm{L}}]$ be the measure space consisting of the basic set $X = R_n$, the σ-algebra $\mathfrak{B}_{\mathrm{L}}$ of the Lebesgue-measurable sets in R_n, and the Lebesgue measure μ_{L}. We shall now denote the corresponding Lebesgue integral in the sense of Def. 14.1.2. by $\mathrm{L}\int f(x)\, dx$, where we write dx instead of $d\mu$. If $\Omega \in \mathfrak{B}_{\mathrm{L}}$, we set $\int_\Omega f(x)\, d\mu_{\mathrm{L}} = \mathrm{L}\int_\Omega f(x)\, dx$, in accordance with Def. 14.1.4.

Theorem. (**a**) *Every I-domain (and hence also every Q-domain) belongs to $\mathfrak{B}_{\mathrm{L}}$.*

(**b**) *If $f(x)$ is Riemann-integrable over the Q-domain Ω, then $f(x)$ is Lebesgue-integrable, and*

$$\mathrm{R}\int_\Omega f(x)\, dx = \mathrm{L}\int_\Omega f(x)\, dx\,. \tag{1}$$

(**c**) *If Ω is an I-domain, and if $f(x)$ is continuous on $\bar{\Omega}$, then $f(x)$ is Lebesgue-integrable, and Eq. (1) holds.*

Remark 1. Part (a) can be strengthened to read: every I-domain is a Borel set. In (b) and (c), one may imagine $f(x)$ as extended beyond Ω, being zero outside of Ω, so that $f(x)$ is defined on the entire R_n.

Remark 2. The theorem shows that the Lebesgue integral is a substantial generalization of the Riemann integral. In the case of the Riemann integral we considered only bounded domains and bounded functions. For the Lebesgue integral, arbitrary (Lebesgue-measurable) sets are admitted, and the functions may also be unbounded. Absolutely convergent improper Riemann integrals, as referred to in Subsec. 9.2.7., are special cases of Lebesgue integrals. In this sense, Lemma 9.2.7 makes assertions about the existence of the special Lebesgue integrals $\mathrm{L}\int_{|x|<1} |x|^\alpha\, dx$ and $\mathrm{L}\int_{|x|>1} |x|^\alpha\, dx$. Reference should also be made to Examples 2 and 3 in Subsec. 7.2.1. The Γ-function referred to in Subsec. 7.2.3. is a Lebesgue integral. If improper Riemann integrals do not converge absolutely, the Lebesgue integrals in question need not exist. This is indicated by Example 1 of Subsec. 7.2.1. There one has

$\mathrm{L}\int\limits_{R_1}\left|\frac{\sin x}{x}\right|\mathrm{d}x=\infty$, and hence $\mathrm{L}\int\limits_{R_1}\frac{\sin x}{x}\mathrm{d}x$ does not exist either. It is, however, possible to introduce improper Lebesgue integrals following the proven Riemannian model.

Remark 3. From now on we shall write Lebesgue integrals in the form

$$\int f(x)\,\mathrm{d}x \quad \text{instead of} \quad \mathrm{L}\int f(x)\,\mathrm{d}x \quad \text{or} \quad \int f(x)\,\mathrm{d}\mu_{\mathrm{L}}\,.$$

14.5.2. Lebesgue's and Fubini's Theorems

Lebesgue's theorem: Theorems 9.1.3/2 and 9.1.5/1 are precursors to Theorem 14.2.2. The uniform convergence as well as the boundedness required there for the functions immediately ensure the existence of an integrable majorant function $g(x)$ in the sense of Theorem 14.2.2.

Fubini's theorem: Theorem 9.1.6 now is a very modest special case of Theorem 14.4.4. Let $X=R_n$, let $\mathfrak{B}=\mathfrak{B}_{\mathrm{L}}^n$ be the σ-algebra of the Lebesgue-measurable sets in R_n, and let $\mu=\mu_{\mathrm{L}}^n$ be the Lebesgue measure on R_n, with analogous assumptions concerning $Y=R_m$, $\mathfrak{C}=\mathfrak{B}_{\mathrm{L}}^m$, and $\nu=\mu_{\mathrm{L}}^m$. Then $[X\times Y, \mathfrak{B}\times\mathfrak{C}, \mu\times\nu]=$ $=[R_{n+m}, \mathfrak{B}_{\mathrm{L}}^{n+m}, \mu_{\mathrm{L}}^{n+m}]$ is the product measure space in the sense of Theorem 14.4.4. If $f(x, y)$, with $x\in R_n$ and $y\in R_m$, is Lebesgue-integrable in R_{n+m}, then

$$\int\limits_{R_{n+m}} f(x, y)\,\mathrm{d}x\mathrm{d}y=\int\limits_{R_n}\Big(\int\limits_{R_m} f(x, y)\,\mathrm{d}y\Big)\,\mathrm{d}x=\int\limits_{R_m}\Big(\int\limits_{R_n} f(x, y)\,\mathrm{d}x\Big)\,\mathrm{d}y\,.$$

If $f(x)$, $x=(x_1, \ldots, x_n)$, is Lebesgue-integrable in R_n, then by iteration one finds that

$$\int\limits_{R_n} f(x)\,\mathrm{d}x=\int\limits_{R_1}\Big(\int\limits_{R_1}\Big(\ldots\int\limits_{R_1} f(x_1, \ldots, x_{n-1}, x_n)\,\mathrm{d}x_1\ldots\Big)\,\mathrm{d}x_{n-1}\Big)\,\mathrm{d}x_n\,,$$

where the parentheses are usually omitted. If $f(x)$ is defined only on a Lebesgue-measurable set Ω in R_n, then one continues $f(x)$ to $R_n\backslash\Omega$, setting it zero there. If the result is a Lebesgue-integrable function on R_n, then the above formula can be applied. In this sense one has, for example,

$$\int\limits_{\Omega} f(x)\,\mathrm{d}x=\int\limits_{\Omega^*}\int\limits_{a(x^*)}^{b(x^*)} f(x_1, x^*)\,\mathrm{d}x_1\mathrm{d}x^*$$

where $x^*=(x_2, \ldots, x_n)$, the meaning of the other quantities being as indicated in Fig. 14.8. If the functions considered are non-negative, then Theorem 14.4.3 including its special cases is also applicable in the sense indicated above.

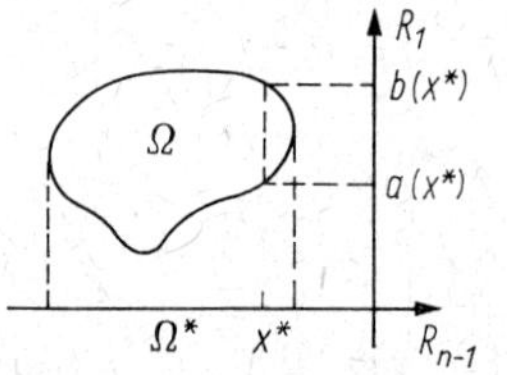

Fig. 14.8

14.5.3. Transformation Formulas

Let Ω be a Lebesgue-measurable set in R_n, and let $f(x)$ be a real-valued function defined on Ω, then $f(x)$ is continued to $R_n\backslash\Omega$ by setting it zero there. We say that $f(x)$ is Lebesgue-measurable or Lebesgue-integrable on Ω if the function continued in this way is Lebesgue-measurable or Lebesgue-integrable, respectively, on R_n. By Theorem 14.5.1, in particular I-domains are Lebesgue-measurable on R_n. Now Theorems 9.2.2 and 14.3.4 can be combined, where the symbols used have the same meaning as indicated there.

Theorem. *Let $y=y(x)$ be a one-to-one mapping of the I-domain ω in R_n onto the I-domain Ω in R_n, with $\frac{\partial(y_1, \ldots, y_n)}{\partial(x_1, \ldots, x_n)}(x) \neq 0$ for $x \in \bar{\omega}$. If $f(x)$ is Lebesgue-integrable on ω, and if $x=x(y)$ is the inverse function of $y(x)$, then $g(y)=f(x(y))$ is Lebesgue-integrable on Ω, and*

$$\int_\omega f(x)\,\mathrm{d}x = \int_\Omega g(y) \left| \frac{\partial(x_1, \ldots, x_n)}{\partial(y_1, \ldots, y_n)}(y) \right| \mathrm{d}y\,.$$

Remark. In the sense of Subsec. 14.3.4., both μ and ϱ are the Lebesgue measures on R_n, and $T(x)=y(x)$ (in this case T is a one-to-one mapping) and $f(x)=g(y(x))=g(T(x))$. Comparison of the formulas in Theorems 9.2.2 and 14.3.4 gives

$$\frac{\mathrm{d}(\mu T^{-1})}{\mathrm{d}\varrho} = \left| \frac{\partial(x_1, \ldots, x_n)}{\partial(y_1, \ldots, y_n)}(y) \right|. \tag{1}$$

This also presents a concrete example of a Radon-Nikodým derivative.

Polar co-ordinates: To illustrate (1), we consider polar co-ordinates. In R_2 one obtains

$$\begin{aligned} x &= r\cos\varphi, \\ y &= r\sin\varphi, \end{aligned} \qquad 0<r<\infty, \quad 0 \leqq \varphi < 2\pi \quad \text{and} \quad \left| \frac{\partial(x, y)}{\partial(r, \varphi)} \right| = r.$$

As to R_3, the polar co-ordinates r, ϑ, φ have been described in Subsec. 12.4.1. One obtains

$$\left| \frac{\partial(x, y, z)}{\partial(r, \vartheta, \varphi)} \right| = r^2 \sin\vartheta\,.$$

14.6. L_p Spaces

14.6.1. Definition

In the measure space $[X, \mathfrak{B}, \mu]$, we shall consider equivalence classes $[f]$ of measurable functions as introduced in Subsec. 14.2.4.

Definition. (a) *For $1 \leqq p < \infty$, $L_p = L_p(X, \mathfrak{B}, \mu)$ is the system of all equivalence classes of measurable functions such that*

$$\|[f]\|_{L_p} = \left(\int |h(x)|^p \,\mathrm{d}y\right)^{\frac{1}{p}} < \infty \quad \textit{with} \quad h(x) \in [f]\,. \tag{1}$$

(b) *$L_\infty = L_\infty(X, \mathfrak{B}, \mu)$ is the system of all equivalence classes of measurable functions such that*

$$\|[f]\|_{L_\infty} = \operatorname{ess\,sup} |h(x)| < \infty \quad \textit{with} \quad h(x) \in [f]\,. \tag{2}$$

Here ess sup $|h(x)| = \inf N$, *where the infimum is taken over all non-negative numbers N such that* $\mu(\{x \mid h(x) > N\}) = 0$ (Fig. 14.9).

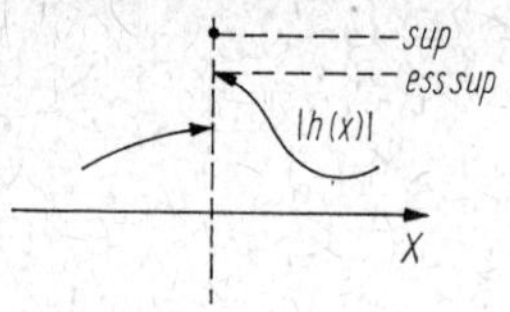

Fig. 14.9

Remark 1. For $p=1$ this agrees with Def. 14.2.4. It is easily seen that (1) and (2) are independent of the particular selection of the representatives $h(x)\in[f]$. In (2) one only takes the "essential supremum" instead of the supremum. Otherwise (2) would not be independent of the selection of the representatives. The values of a given function can be altered at choice on a set of measure zero.

Remark 2 (Convention). As indicated in Remark 14.2.4/3 for L_1, from now on we shall use functions $f\in L_p$ instead of equivalence classes $[f]\in L_p$ in our calculations. We shall also write $\|f\|_{L_p}$ instead of $\|[f]\|_{L_p}$.

14.6.2. Hölder's and Minkowski's Inequalities

Theorem. *Let $[X, \mathfrak{B}, \mu]$ be a measure space, and let $L_p=L_p(X, \mathfrak{B}, \mu)$ be the spaces defined in Def.* 14.6.1.

(a) *If $f(x)\in L_p$ and $g(x)\in L_{p'}$, with $1<p<\infty$ and $\frac{1}{p}+\frac{1}{p'}=1$, then $f(x)\cdot g(x)\in L_1$, and the two functions satisfy Hölder's inequality,*

$$\int |f(x)\cdot g(x)|\,d\mu \leqq \|f\|_{L_p}\|g\|_{L_{p'}}\,.$$

(b) *If $f(x)\in L_p$ and $g(x)\in L_p$, with $1\leqq p\leqq\infty$, then $f(x)+g(x)\in L_p$, and the two functions satisfy Minkowski's inequality*

$$\|f+g\|_{L_p}\leqq \|f\|_{L_p}+\|g\|_{L_p}\,.$$

Remark. Corresponding inequalities for complex numbers have been stated in Theorem 6.2.1. The above theorem is proved by reducing it to Theorem 6.2.1.

14.6.3. The Spaces $L_p(X, \mathfrak{B}, \mu)$

Definition. *A set M in a Banach space B is said to be dense if for every element $b\in B$ and for every $\varepsilon>0$ there exists an element $m\in M$ such that $\|b-m\|\leqq\varepsilon$.*

Remark 1. Banach spaces have been considered in Chap. 6; $\|\cdot\|$ denotes the norm. The definition says that every element of B can be approximated in the norm by elements of M to any accuracy desired.

Theorem. *Let $[X, \mathfrak{B}, \mu]$ be a measure space, and let $L_p=L_p(X, \mathfrak{B}, \mu)$ be the spaces defined in Def.* 14.6.1.

(a) *L_p with the norm $\|f\|=\|f\|_{L_p}$ is a Banach space. Here $1\leqq p\leqq\infty$.*

(b) *For $1\leqq p<\infty$, the integrable step functions constitute a set dense in L_p.*

Remark 2. Part (a) shall include the assertion that $\|f\|_{L_p}$ is a norm. As regards integrable step functions, we refer to Def. 14.1.1. To prove (b), one may use Theorem 14.2.3/1 and the step functions of (13.4.3/1).

14.6.4. The Spaces $L_p(R_n)$ and $L_p(\Omega)$

In this section we shall specialize the measure space $[X, \mathfrak{B}, \mu]$: $X = R_n$; $\mathfrak{B} = \mathfrak{B}_L$ is the σ-algebra of the Lebesgue measurable sets in R_n, and $\mu = \mu_L$ is the Lebesgue measure.

Definition 1. *Let* $1 \leqq p \leqq \infty$.

(a) $L_p(R_n) = L_p(R_n, \mathfrak{B}_L, \mu_L)$ *in the sense of Def.* 14.6.1.

(b) *If* Ω *is a Lebesgue-measurable set in* R_n, *then*

$$L_p(\Omega) = \{f \mid f \in L_p(R_n),\ f(x) = 0 \quad \text{for} \quad x \in R_n \backslash \Omega\}\,. \tag{1}$$

Remark 1. We recall the previous convention that we write $f \in L_p$ instead of (more precisely) $[f] \in L_p$: (1) has to be understood in this sense. Thus, like $L_p(R_n)$, $L_p(\Omega)$ consists of equivalence classes of functions that are equal almost everywhere (in the Lebesgue sense). Likewise following previous conventions, we write

$$\|f\|_{L_p(\Omega)} = \Big(\int_\Omega |f(x)|^p \,\mathrm{d}x\Big)^{\frac{1}{p}} \quad \text{for} \quad 1 \leqq p < \infty$$

and

$$\|f\|_{L_\infty(\Omega)} = \operatorname*{ess\,sup}_{x \in \Omega} |f(x)| \quad \text{for} \quad p = \infty\,.$$

Definition 2. *Let* Ω *be a domain in* R_n *(i.e., an open set in* R_n*). A Lebesgue-measurable function* $f(x)$ *is called a function with bounded support (or a finitely nonzero function) (with respect to* Ω*) if* $\operatorname{supp} f = \overline{\{x \mid f(x) \neq 0\}}$ *is bounded and lies in* Ω (Fig. 14.10).

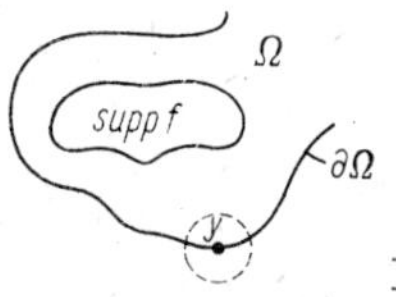

Fig. 14.10

Remark 2. $\operatorname{supp} f$ means the support of f. Since Ω is open and $\operatorname{supp} f$ is closed and bounded, $\operatorname{supp} f$ has a positive distance from the boundary $\partial\Omega$ of Ω. Here the boundary is defined as $\partial\Omega = \bar{\Omega} \backslash \Omega$.

Theorem 1. *Let* Ω *be a domain in* R_n.

(a) *For* $1 \leqq p \leqq \infty$, $L_p(\Omega)$ *with the norm* $\|f\|_{L_p(\Omega)}$ *is a Banach space.*

(b) *If* $1 \leqq p < \infty$, *then the functions which are continuous and have bounded support on* Ω *constitute a dense set in* $L_p(\Omega)$.

Remark 3. Part (a) is also true if Ω is an arbitrary Lebesgue-measurable set. Part (b) is proved using Theorem 14.6.3(b). An arbitrary integrable step function is approximated by functions of the form $\sum_{j=1}^{N} \alpha_j \chi_{Q_j}(x)$, where $\chi_{Q_j}(x)$ is the characteristic function of an (n-dimensional) rectangular parallelepiped Q_j, with $\bar{Q}_j \subset \Omega$. Subsequently the characteristic function of an n-dimensional rectangular parallelepiped Q, with $\bar{Q} \subset \Omega$, is approximated in an elementary way by continuous functions with bounded support.

Theorem 2. *If* Ω *is a domain in* R_n, *and if* $1 \leqq p < \infty$, *then the functions with bounded support that are differentiable on* Ω *up to arbitrary order constitute a dense set in* $L_p(\Omega)$.

Remark 4. This important theorem is proved using Sobolev's mollification method. Let

$$\omega(x) = \begin{cases} c e^{\frac{-1}{1-|x|^2}} & \text{for } |x| < 1, \\ 0 & \text{for } |x| \geqq 1, \end{cases} \qquad \int_{R_n} \omega(x)\, dx = 1 .$$

Using Theorem 5.4.5/3 it can easily be shown that $\omega(x)$ has derivatives up to arbitrary order in R_n. Further, supp $\omega = \{y \mid |y| \leqq 1\}$. For $h > 0$ one sets $\omega_h(x) = h^{-n}\omega\left(\frac{x}{n}\right)$ (Fig. 14.11). One has supp $\omega_h = \{y \mid |y| \leqq h\}$. By Theorem 1(b) it will suffice to approximate continuous functions with bounded support. If $f(x)$ is such a function, then one sets

$$f_h(x) = \int_{R_n} f(z)\, \omega_h(x-z)\, dz, \quad h > 0 .$$

It can now be shown that $f_h(x)$ has derivatives up to arbitrary order on R_n. Further, supp $f_h \subset \Omega$ for small positive h, and $f_h \to f$ in $L_p(\Omega)$ (Fig. 14.12). For details, cf. [66], pp. 37–40.

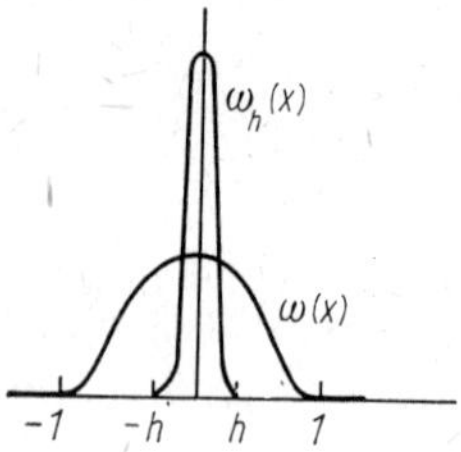

Fig. 14.11

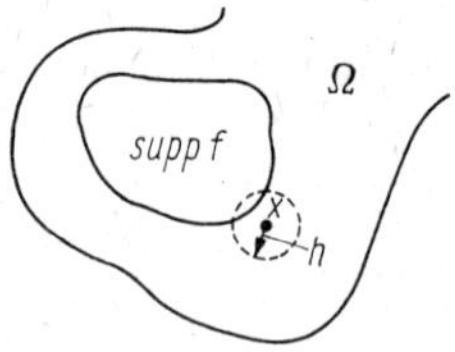

Fig. 14.12

15. Complex Function Theory

15.1. Holomorphic Functions

15.1.1. The Complex Number Plane *C*

A complex number $z \in C_1$ can be written as $z = x + iy$ (normal representation introduced in Subsec. 1.2.4.) or as $z = re^{i\varphi}$ (representation in polar co-ordinates as introduced in Subsec. 5.2.4.) (Fig. 15.1). For the following considerations it will be convenient to close C_1 by a point at infinity. To this end we imagine a sphere placed on the complex number plane C_1, touching C_1 at the point 0 (Fig. 15.2).

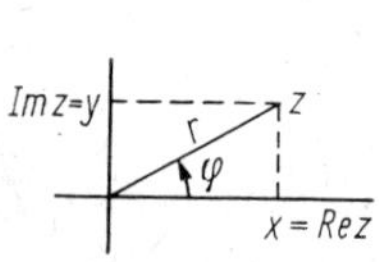

Fig. 15.1

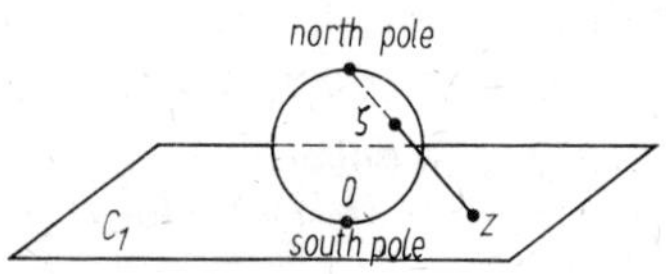

Fig. 15.2

If this point of contact is the south pole of the sphere, then we draw from its north pole a ray to the point $z \in C_1$, which pierces the sphere at the point ζ. The correspondence is one-to-one: every ζ on the sphere, except for the north pole, is assigned exactly one $z \in C_1$. The method is called stereographic projection, and the sphere is called Riemann's sphere. Now this sphere is closed by including the north pole. In the z-plane, the north pole is assigned the point ∞ by the stereographic projection, and one writes $C = C_1 \cup \{\infty\}$ (closure of the complex number plane C_1 by ∞, the point at infinity). Any concepts related to ∞ can then be made transparent by reference to Riemann's sphere, e.g., the phrase that $\{z \mid z \in C_1,\ |z| > N\} \cup \{\infty\}$ are neighbourhoods of ∞. Neighbourhoods of $z \in C_1$ have the usual meaning: e.g., circles with the centre z. If $z = re^{i\varphi}$, then $w = \frac{1}{z} = \frac{1}{r} e^{-i\varphi}$, in particular then $\{z \mid |z| > N\} \cup \{\infty\}$ becomes $\left\{w \mid |w| < \frac{1}{N}\right\}$. $\left(\text{Here one has } 0 = \frac{1}{\infty}.\right)$ Thus neighbourhoods of ∞ become neighbourhoods of 0. (Following the convention made in Subsec. 6.2.2., we set $|z| = \sqrt{x^2 + y^2}$, with $x = \operatorname{Re} z$ and $y = \operatorname{Im} z$). This suggests the following method: (i) Concepts of the complex function theory are defined initially for points z with $|z| < \infty$. (ii) For $z \neq \infty$ one defines the concept in question by starting from the already defined concept for $w = 0$ and substituting $\frac{1}{z}$ for w.

15.1.2. Holomorphic Functions

We shall consider functions $f(z)$ defined on the plane $C = C_1 \cup \{\infty\}$, whose values are elements of C, i.e. $f(z) \in C$. If C is imagined as being mapped onto the Riemann sphere in a one-to-one way, then this means that we consider functions defined on the Riemann sphere, with their values lying on the Riemann sphere too. It is, however, more convenient to refer to the plane C, even though in this case the point ∞ must sometimes be given a preferential treatment. Let G be a domain (i.e. an open set) in C (Fig. 15.3). If $\infty \in G$, then $\{z \mid |z| > N\} \subset G$ must be true for sufficiently large values of N.

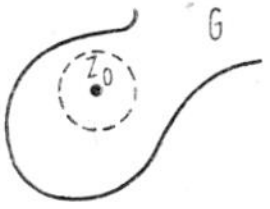

Fig. 15.3

Definition 1. *Let $f(z)$ be defined on the domain G in C, i.e., $D(f) = G$, and let $f(z) \in C_1$ for $z \in G$ (i.e., $f(z) \neq \infty$ for $z \in G$). Then $f(z)$ is said to be continuous at the point $z_0 \in G$, $z_0 \neq \infty$, if for every $\varepsilon > 0$ there exists a sufficiently small positive number $\delta = \delta(\varepsilon)$ such that $|f(z) - f(z_0)| \leq \varepsilon$ for $|z - z_0| < \delta$. $f(z)$ is said to be continuous at the point ∞ (provided that $\infty \in G$) if for every $\varepsilon > 0$ there exists a sufficiently large number $N = N(\varepsilon)$ such that $|f(z) - f(\infty)| \leq \varepsilon$ for $|z| > N$.*

Remark 1. Here $\delta > 0$ must be chosen so small that $\{z \mid |z - z_0| < \delta\} \subset G$. If $\infty \in G$, then N must be chosen so large that $\{z \mid |z| > N\} \subset G$. One observes that the continuity at the point ∞ is shown following the principle set out at the end of Subsec. 15.1.1.

Definition 2. *Let $f(z)$ be defined on a domain G in C, $D(f)=G$; and let $f(z)\in C_1$ for $z\in G$.*

(**a**) *$f(z)$ is said to be differentiable at the point $z_0\in G$, with $z_0\neq\infty$, if the C_1-limit, $\lim\limits_{C_1\ni h\to 0}\frac{1}{h}(f(z_0+h)-f(z_0))$, exists, which is then denoted by $f'(z_0)$. $f(z)$ is said to be differentiable at the point ∞ (provided that $\infty\in G$) if $w(z)=f\left(\frac{1}{z}\right)$ is differentiable at the point $z_0=0$. In this case one sets $f'(\infty)=0$.*

(**b**) *$f(z)$ is called holomorphic on G if $f(z)$ is differentiable at every point $z\in G$.*

Remark 2. Consequently, $f(z)$ is differentiable at the point $z_0\neq\infty$, the value of its derivative being $f'(z_0)$, if for every $\varepsilon>0$ there exists a $\delta=\delta(\varepsilon)$ such that

$$\left|\frac{f(z_0+h)-f(z_0)}{h}-f'(z_0)\right|\leqq\varepsilon \quad\text{for}\quad h\in C_1 \quad\text{and}\quad |h|\leqq\delta\,.$$

Here $f'(z_0)\in C_1$ (i.e. that $f'(z_0)=\infty$ is not admitted). This is the complex analogue to the real differentiability considered in Subsec. 3.1.1. The rules stated in Subsec. 3.1.2. for real differentiation are also applicable to complex differentiation. In particular, the chain rule of Theorem 3.1.2/2 also applies to complex differentiation.

Remark 3. The calculating rules for ∞ are as follows: $z+\infty=z-\infty=z\cdot\infty=\infty$, and $\frac{z}{\infty}=0$, $\frac{z}{0}=\infty$ for $z\neq 0$ and $z\neq\infty$. (Other than in the case of R_1, no distinction is made between ∞ and $-\infty$.) Setting $f'(\infty)=0$ in part (b) of the definition is natural because, if $w(z)=f\left(\frac{1}{z}\right)$ is differentiable on a neighbourhood of 0, then applying the chain rule one obtains

$$f'(z)=-w'\left(\frac{1}{z}\right)\frac{1}{z^2}\to -w'(0)\frac{1}{\infty}=0 \quad\text{for}\quad z\to\infty\,.$$

Lemma. *If $f(z)$ is differentiable at the point $z_0\in G$, then $f(z)$ is continuous at the point z_0.*

Remark 4. The lemma holds especially for $\infty=z_0\in G$.

15.1.3. Examples of Holomorphic Functions

Polynomials: Let $a_0, \ldots, a_N$ be any complex numbers, then the polynomial $f(z)=P(z)=\sum\limits_{k=0}^{N}a_kz^k$ is holomorphic on $G=C_1=C\setminus\{\infty\}$. One has $f'(z)=\sum\limits_{k=1}^{N}ka_kz^{k-1}$. In the special case $f(z)\equiv a_0$, $f(z)$ is holomorphic on C, and $f'(z)\equiv 0$.

The function $\frac{1}{z}$: The function $f(z)=\frac{1}{z}$ is holomorphic on $G=C\setminus\{0\}$, and one has $f'(z)=-\frac{1}{z^2}$ (including $z=\infty$).

Power series: Complex power series have been treated in Subsec. 5.4.1.; the radius of convergence has the meaning given in (5.4.1/2).

Theorem. *If the power series $P(z)=\sum\limits_{k=0}^{\infty}a_k(z-z_0)^k$, with $z_0\in C_1$, has the radius of convergence R, then $P(z)$ is holomorphic on $\{z\mid |z-z_0|<R\}$, and $P'(z)=$ $=\sum\limits_{k=1}^{\infty}ka_k(z-z_0)^{k-1}$.*

Remark. This is the complex analogue to Theorem 5.4.3/3. The polynomials are a special case where the radius of convergence is $R=\infty$.

The function e^z: The function e^z has been treated in Subsec. 5.4.6. According to Theorem 5.4.6 the above theorem is applicable, and it follows that e^z is holomorphic on $G=C_1$. One has $(e^z)'=\sum_{k=0}^{\infty}\frac{z^k}{k!}=e^z$. Consequently, the functions $a^z=e^{z\ln a}$, with $a>0$, are also holomorphic on C_1, and application of the chain rule gives $(a^z)'=\ln a\cdot a^z$.

15.1.4. The Cauchy-Riemann Differential Equations, Harmonic Functions

If $f(z)=f(x, y)$ is defined on a domain G in C_1, and if $f(z)\in C_1$, then we set $f(z)=\operatorname{Re} f(z)+\mathrm{i}\operatorname{Im} f(z)=u(x, y)+\mathrm{i}\, v(x, y)$, where $u(x, y)=\operatorname{Re} f(z)$ and $v(x, y)=\operatorname{Im} f(z)$ are regarded as real-valued functions on R_2 (Fig. 15.4). If $f(z)$ is holomorphic on G, then it is easily verified that $u(x, y)$ and $v(x, y)$ are continuous on G and have partial derivatives of first order on G with respect to x and y. From

$$f'(z_0)=\lim_{h\to 0}\frac{f(x_0+h_1, y_0+h_2)-f(x_0, y_0)}{h_1+\mathrm{i}h_2}, \quad h=h_1+\mathrm{i}h_2\,,$$

one observes that

$$\frac{\partial f}{\partial x}=\frac{\partial u}{\partial x}+\mathrm{i}\frac{\partial v}{\partial x}=f'(z_0), \quad \frac{\partial f}{\partial y}=\frac{\partial u}{\partial y}+\mathrm{i}\frac{\partial v}{\partial y}=\mathrm{i}f'(z_0)\,.$$

By comparison of the real and imaginary parts of the last two equations one obtains the Cauchy-Riemann differential equations

$$\frac{\partial u}{\partial x}=\frac{\partial v}{\partial y}, \quad \frac{\partial u}{\partial y}=-\frac{\partial v}{\partial x}, \quad (x, y)\in G\,. \tag{1}$$

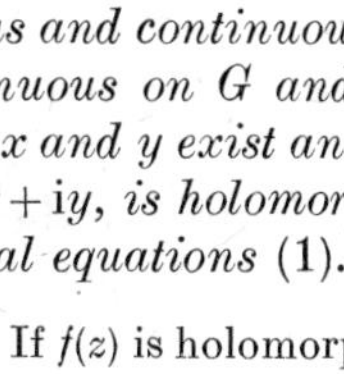

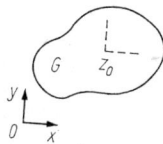

Fig. 15.4

Theorem 1. *Let G be a domain in C_1 $(=R_2)$. Further let $u(x, y)$ and $v(x, y)$ be continuous and continuously differentiable on G (this means that $u(x, y)$ and $v(x, y)$ are continuous on G and that the partial derivatives of $u(x, y)$ and $v(x, y)$ with respect to x and y exist and are also continuous on G). Then $f(z)=u(x, y)+\mathrm{i}\, v(x, y)$, with $z=x+\mathrm{i}y$, is holomorphic on G if and only if it satisfies the Cauchy-Riemann differential equations* (1).

Remark 1. If $f(z)$ is holomorphic on G, then the above preparatory considerations show that $u(x, y)$ and $v(x, y)$ are continuous on G, that the partial derivatives $\frac{\partial u}{\partial x}, \frac{\partial u}{\partial y}, \frac{\partial v}{\partial x}$, and $\frac{\partial v}{\partial y}$ exist on G and that the Eqs. (1) are satisfied. Consequently, except for the continuity of these partial derivatives, the conditions of the theorem are necessary for $f(z)$ to be holomorphic on G.

Remark 2 (*The functions* ln z *and* z^w). Let G be a connected domain in C_1, and let $0 \notin G$ (thus any two points z_1 and z_2 can be connected by a smooth curve that lies entirely in G) (Fig. 15.5). Further we require that there be no closed path within G on which it is possible to move round the point 0. Then, by Def. 5.3.1,

$$f(z) = \ln z = \ln r + i\varphi \tag{2}$$

is continuous on G. One has $u(x, y) = \ln r = \ln \sqrt{x^2+y^2}$ and $v(x, y) = \arctan \frac{y}{x}$ (Fig. 15.6). It is now easily verified that the conditions of the theorem are met by $u(x, y)$ and $v(x, y)$. Hence $\ln z$ is holomorphic on G, and $(\ln z)' = \frac{1}{z}$. The concept of holomorphy can be extended

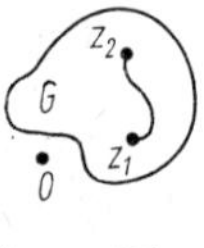

Fig. 15.5

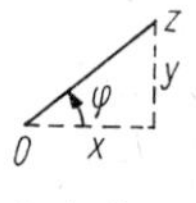

Fig. 15.6

immediately to Riemann surfaces. Then $\ln z$ is holomorphic on the Riemann surface described in Subsec. 5.3.1. If $w \neq 0$ is a complex number, then, using the chain rule, it can be shown that $z^w = e^{w \ln z}$ is holomorphic on G. According to Subsec. 5.3.2. the holomorphy can again be extended to Riemann surfaces.

Definition. *Let G be a domain in C_1. The real-valued function $u(x, y)$ is called harmonic on G if $u(x, y)$ is continuous on G, if it has first- and second-order partial derivatives on G with respect to x and y, and satisfies on G the differential equation*

$$\Delta u = \frac{\partial^2 u}{\partial x^2} + \frac{\partial^2 u}{\partial y^2} \equiv 0 .$$

Theorem 2. *Let G be a domain in C_1, and let $f(z)$ be holomorphic on G. If $u(x, y) = \operatorname{Re} f(z)$ and $v(x, y) = \operatorname{Im} f(z)$ have continuous partial derivatives of second order on G with respect to x and y, then $u(x, y)$ and $v(x, y)$ are harmonic on G.*

Remark 3. The theorem is a simple implication of (1):

$$\frac{\partial^2 u}{\partial x^2} + \frac{\partial^2 u}{\partial y^2} = \frac{\partial^2 v}{\partial x \partial y} - \frac{\partial^2 v}{\partial y \partial x} = 0$$

(cf. Theorem 8.1.2). As we shall see later on, $u(x, y) = \operatorname{Re} f(z)$ and $v(x, y) = \operatorname{Im} f(z)$ have partial derivatives of arbitrary order if the function $f(z)$ is holomorphic. So the additional assumption on the second-order derivatives is not needed in the above theorem.

15.2. Integral Theorems

15.2.1. Complex Curvilinear Integrals

Following the lines given in Subsec. 8.1.5., we consider curves $\mathfrak{C}$ in the complex plane C_1 which are given by $z(t) = x(t) + iy(t)$, where $\alpha \leq t \leq \beta$. Further, $x(t)$ and $y(t)$ are assumed to be functions continuously differentiable on $[\alpha, \beta]$. We write

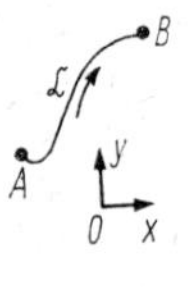

Fig. 15.7

$z'(t)=x'(t)+iy'(t)$. The orientation of the curve is given by the direction of increasing t-values, i.e. $A=z(\alpha)$ and $B=z(\beta)$ (Fig. 15.7). The curve may be closed, i.e. $z(\alpha)=z(\beta)$, but shall not have any other double points, i.e., $z(t_1)\neq z(t_2)$ for $\alpha\leqq t_1<t_2<\beta$ and $\alpha<t_1<t_2\leqq\beta$.

Definition. *If $f(z)$ is a continuous complex-valued function on $\mathfrak{C}$, then*

$$\int_{\mathfrak{C}} f(z)\,\mathrm{d}z=\int_{\alpha}^{\beta} f(z(t))\,z'(t)\,\mathrm{d}t$$
$$=\int_{\alpha}^{\beta} \mathrm{Re}\,[f(z(t))\,z'(t)]\,\mathrm{d}t+\mathrm{i}\int_{\alpha}^{\beta} \mathrm{Im}\,[f(z(t))\,z'(t)]\,\mathrm{d}t\,.$$

Remark 1. Complex curvilinear integrals $\int_{\mathfrak{C}} f(z)\,\mathrm{d}z$ are thus reduced to real integrals over intervals $[\alpha, \beta]$. Consequently the properties of these real integrals are transferred to the complex ones, e.g.,

$$\int_{\mathfrak{C}} (\lambda_1 f_1(z)+\lambda_2 f_2(z))\,\mathrm{d}z=\lambda_1\int_{\mathfrak{C}} f_1(z)\,\mathrm{d}z+\lambda_2\int_{\mathfrak{C}} f_2(z)\,\mathrm{d}z\,,$$

where λ_1 and λ_2 are complex numbers and $f_1(z)$ and $f_2(z)$ are complex-valued functions continuous on $\mathfrak{C}$.

Remark 2. $\int_{\mathfrak{C}} f(z)\,\mathrm{d}z$ is independent of the choice of the parameter t. If $\tau=\tau(t)$ is continuously differentiable on $[\alpha, \beta]$, if $\tau'(t)>0$ for $t\in[\alpha, \beta]$, and if $\gamma=\tau(\alpha)$ and $\delta=\tau(\beta)$ (Fig. 15.8), then

$$\int_{\alpha}^{\beta} f(z(t))\,z'(t)\,\mathrm{d}t=\int_{\gamma}^{\delta} f(w(\tau))\,\frac{\mathrm{d}w}{\mathrm{d}\tau}\,\mathrm{d}\tau \quad\text{with}\quad w(\tau)=z(t(\tau))\,.$$

Remark 3. It is also possible to determine $\int_{\mathfrak{C}} f(z)\,\mathrm{d}z$ directly as a limit of Riemann sums, by analogy to Def. 3.2.1. To this end one chooses any points $z_k\in\mathfrak{C}$, with $z_1=A$ and $z_N=B$ (Fig. 15.9). Then

$$\int_{\mathfrak{C}} f(z)\,\mathrm{d}z=\lim\sum_{k=1}^{N-1} f(z_k)\,(z_{k+1}-z_k)\,,$$

where the limit is understood in the sense of increasing fineness of subdivision (in a way analogous to that of the Riemann integral).

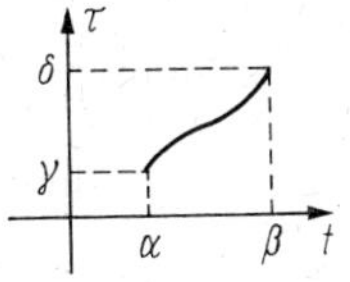

Fig. 15.8

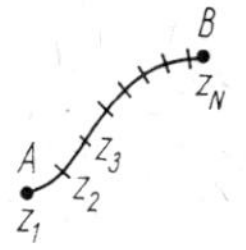

Fig. 15.9

Properties of curvilinear integrals. 1. (Fig. 15.10) If $\mathfrak{C}$ is a smooth curve, and if $\mathfrak{C}=\mathfrak{C}_1\cup\mathfrak{C}_2$, then

$$\int_{\mathfrak{C}} f(z)\,\mathrm{d}z=\int_{\mathfrak{C}_1} f(z)\,\mathrm{d}z+\int_{\mathfrak{C}_2} f(z)\,\mathrm{d}z\,.$$

If $\mathfrak{C}'$ coincides with the curve $\mathfrak{C}$, but has the reverse orientation, then

$$\int_{\mathfrak{C}'} f(z)\,\mathrm{d}z=-\int_{\mathfrak{C}} f(z)\,\mathrm{d}z\,.$$

We point out that $\mathfrak{C}$ denotes the curve including its orientation.

2. If $f_j(z) \Rightarrow f(z)$ (uniform convergence of $f_j(z)$ to $f(z)$ on $\mathfrak{C}$), then

$$\int_{\mathfrak{C}} f_j(z)\,\mathrm{d}z \to \int_{\mathfrak{C}} f(z)\,\mathrm{d}z\,.$$

This can, for instance, be concluded from Theorem 3.2.3.

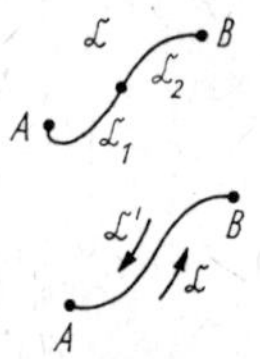

Fig. 15.10

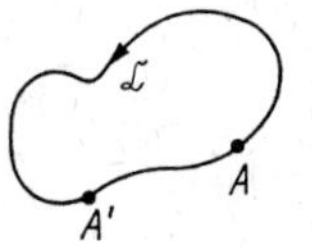

Fig. 15.11

3. If $\mathfrak{C}$ is a smooth closed curve (Fig. 15.11), then $\int_{\mathfrak{C}} f(z)\,\mathrm{d}z$ does not depend on whether the integral $\int_{\mathfrak{C}}$ is taken from A to A or from A' to A'.

4. If $\mathfrak{C} = \mathfrak{C}_1 \cup \mathfrak{C}_2 \cup \mathfrak{C}_3 \cup \ldots$ is a piecewise smooth curve with a finite number of smooth curve segments $\mathfrak{C}_1$, $\mathfrak{C}_2$, $\mathfrak{C}_3$ etc. (Fig. 15.12), then we set

$$\int_{\mathfrak{C}} f(z)\,\mathrm{d}z = \int_{\mathfrak{C}_1} f(z)\,\mathrm{d}z + \int_{\mathfrak{C}_2} f(z)\,\mathrm{d}z + \int_{\mathfrak{C}_3} f(z)\,\mathrm{d}z + \ldots$$

The properties stated above remain valid for curves of this type.

5. Let $f(z)$ be a function holomorphic on a domain G in C_1. If $\mathfrak{C}$ is a smooth curve in G (Fig. 15.13), then

$$\int_{\mathfrak{C}} f'(z)\,\mathrm{d}z = f(B) - f(A)\,.$$

Hence in this case the integral is independent of the path of integration.

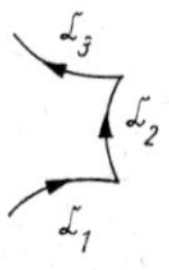

Fig. 15.12

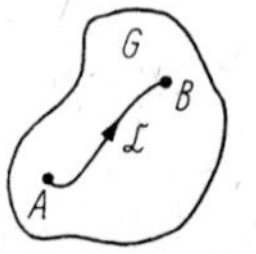

Fig. 15.13

6. Under the conditions of the above definition, one has

$$\left| \int_{\mathfrak{C}} f(z)\,\mathrm{d}z \right| \leq \max_{z \in \mathfrak{C}} |f(z)| \int_{\alpha}^{\beta} |z'(t)|\,\mathrm{d}t\,.$$

According to Subsec. 9.2.3., $\int_{\alpha}^{\beta} |z'(t)|\,\mathrm{d}t = \int_{\alpha}^{\beta} \sqrt{x'^2(t) + y'^2(t)}\,\mathrm{d}t$ is the length of the curve $\mathfrak{C}$.

15.2.2. Cauchy's Integral Theorem

A domain G in C_1 is called simply connected if every closed curve $\mathfrak{C}$ in G can continuously be contracted within G into a point (Fig. 15.14). The curve $\mathfrak{C}'$ in Fig. 15.15 cannot continuously be contracted within G to a point: hence in this case G is not simply connected.

Theorem (*Cauchy's integral theorem*). *Let G be a bounded, simply connected domain in C_1. Further let $\mathfrak{C}_1$ and $\mathfrak{C}_2$ be two piecewise smooth curves in G, extending*

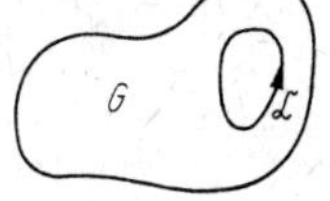

Fig. 15.14

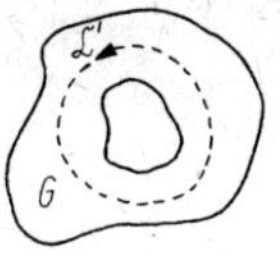

Fig. 15.15

from A to B (Fig. 15.16). *If $f(z)$ is holomorphic on G, then*

$$\int_{\mathfrak{C}_1} f(z)\,\mathrm{d}z = \int_{\mathfrak{C}_2} f(z)\,\mathrm{d}z\,.$$

Remark 1. Mathematics is art. Surely this is an exaggerated and somewhat biased statement. But it emphasizes an essential criterion of mathematics, that is the aesthetics of its theorems. There are not few mathematicians who regard Cauchy's integral theorem as the most beautiful theorem of mathematics.

Remark 2. The theorem can be reformulated as follows. Let G again be a bounded simply connected domain in C_1, and let $\mathfrak{C}$ be a piecewise smooth, closed curve in G. If $f(z)$ is holomorphic on G, then $\int_{\mathfrak{C}} f(z)\,\mathrm{d}z = 0$.

Remark 3. The requirement that G be simply connected is essential for the theorem. If $G = \left\{z \mid \frac{1}{2} < |z| < 2\right\}$, then G is not simply connected (Fig. 15.17). $f(z) = \frac{1}{z}$ is holomorphic on G. If $\mathfrak{C} = \{\mathrm{e}^{\mathrm{i}t},\ 0 \leq t < 2\pi\}$ is the unit circle, then one obtains

$$\int_{\mathfrak{C}} \frac{1}{z}\,\mathrm{d}z = \int_0^{2\pi} \mathrm{e}^{-\mathrm{i}t}(\mathrm{i}\mathrm{e}^{\mathrm{i}t})\,\mathrm{d}t = \mathrm{i}2\pi.$$

Hence in this case Cauchy's integral theorem does not apply.

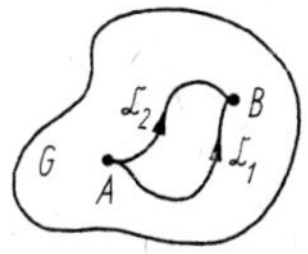

Fig. 15.16

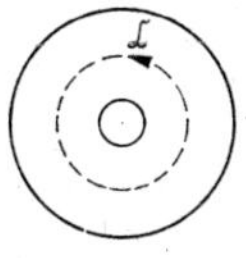

Fig. 15.17

Remark 4 (*Fresnel's integrals and Cornu's spiral*). The theorem can also be used effectively in the calculation of real integrals. In geometrical optics, Fresnel's integrals

$$z(t) = x(t) + \mathrm{i}y(t) = \int_0^t \cos v^2\,\mathrm{d}v + \mathrm{i}\int_0^t \sin v^2\,\mathrm{d}v$$

play an important role (Fig. 15.18). What is of interest is their behaviour as $t \to \infty$. One obtains the picture of Cornu's spiral (Fig. 15.19). For the calculation of the limits,

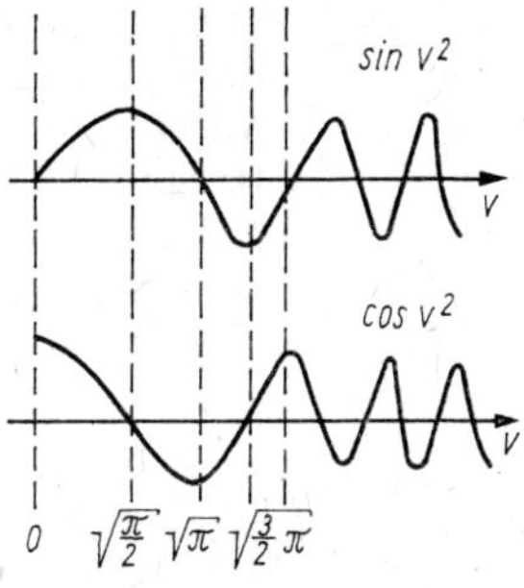

Fig. 15.18

Cauchy's integral theorem gives the starting formula

$$\int_{\mathfrak{C}_1 \cup \mathfrak{C}_2 \cup \mathfrak{C}_3} e^{iz^2}\, dz = 0 \,. \tag{1}$$

Here $\mathfrak{C}_1$ and $\mathfrak{C}_3$ are straight line segments, while $\mathfrak{C}_2$ is a circular arc (Fig. 15.20). One has

$$\int_{\mathfrak{C}_1} e^{iz^2} dz = \int_0^t e^{ix^2} dx = z(t)\,, \tag{2}$$

$$\int_{\mathfrak{C}_3} e^{iz^2} dz = -\int_0^t e^{-r^2} e^{i\frac{\pi}{4}} dr\,. \tag{3}$$

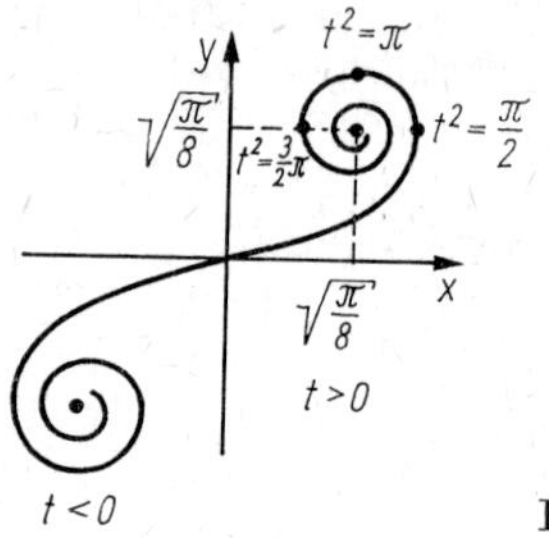

Fig. 15.19

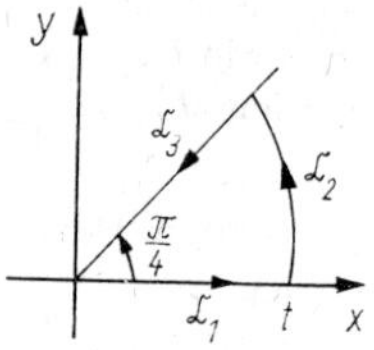

Fig. 15.20

On $\mathfrak{C}_3$ we set $z(r) = r \cdot e^{i\frac{\pi}{4}}$. Then $z^2 = ir^2$, and we obtain (3). Letting $t \to \infty$ one gets $\int_0^t e^{-r^2} dr \to$ $\to \int_0^\infty e^{-r^2} dr = \frac{\sqrt{\pi}}{2}$ (cf. Subsec. 9.2.6.). It can also be shown that $\lim_{t\to\infty} \int_{\mathfrak{C}_2} e^{iz^2} dz = 0$. As $t \to \infty$ in (1), this gives

$$z(\infty) = \frac{\sqrt{\pi}}{2}\, e^{\frac{i\pi}{4}} = \frac{\sqrt{\pi}}{2}\, \frac{1+i}{\sqrt{2}} = \sqrt{\frac{\pi}{8}}\,(1+i)\,.$$

15.2.3. Cauchy's Integral Formula

Let G be a connected, bounded domain in C_1, its boundary consisting of a finite number of components. We choose smooth, closed oriented curves $\mathfrak{C}_0$, $\mathfrak{C}_1$, ..., $\mathfrak{C}_n$ which enclose the respective boundary components (see Fig. 15.21). We recall that, according to Subsec. 15.1.2., curves must not have any double points. Now let $\mathfrak{C}_0$ be the outer curve. We connect the curves $\mathfrak{C}_1$, ..., $\mathfrak{C}_n$ by smooth curve segments with $\mathfrak{C}_0$ (Fig. 15.22), so that altogether we obtain a piecewise smooth, oriented curve $\mathfrak{C}$ consisting of $\mathfrak{C}_0$, $\mathfrak{C}_1$, ..., $\mathfrak{C}_n$ and the connecting segments, which are passed through in both directions (by way of exception we admit double points here, because the connecting segments have to be counted doubly). If $f(z)$ is holomorphic on G, then, by Theorem 15.2.2, $\int_{\mathfrak{C}} f(z)\, dz = 0$. Here the integrals

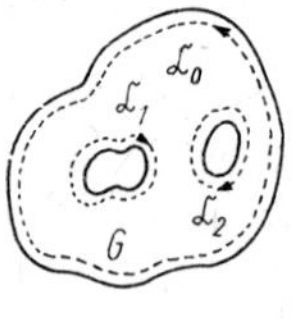

Fig. 15.21

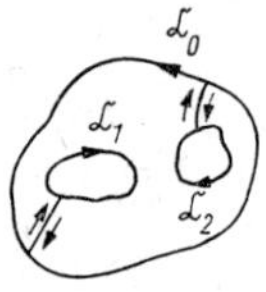

Fig. 15.22

over the connecting segments cancel each other, and one obtains

$$\int_{\mathfrak{C}_0} f(z)\,\mathrm{d}z + \sum_{k=1}^{n} \int_{\mathfrak{C}_k} f(z)\,\mathrm{d}z = 0\,. \tag{1}$$

This is a generalization of Theorem 15.2.2 for multiply connected domains.

Theorem (*Cauchy's integral formula*). *Let G be a bounded, simply connected domain in C_1, and let $\mathfrak{C}$ be a smooth curve in G* (Fig. 15.23). *If $f(z)$ is holomorphic on G, and if z_0 is a point lying inside the curve $\mathfrak{C}$, then*

$$f(z_0) = \frac{1}{2\pi \mathrm{i}} \int_{\mathfrak{C}} \frac{f(z)}{z - z_0}\,\mathrm{d}z\,. \tag{2}$$

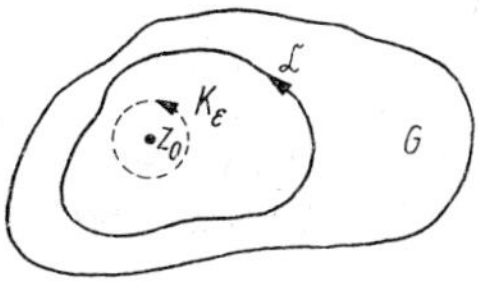

Fig. 15.23

Remark 1. $\dfrac{f(z)}{z - z_0}$ is holomorphic on $G \backslash \{z_0\}$. If K_ε is a circle of radius ε around z_0, then (1) can be applied to $\dfrac{f(z)}{z - z_0}$, giving

$$\int_{\mathfrak{C}} \frac{f(z)}{z - z_0}\,\mathrm{d}z = \int_{K_\varepsilon} \frac{f(z)}{z - z_0}\,\mathrm{d}z \underset{\varepsilon \downarrow 0}{\rightarrow} f(z_0)\,\mathrm{i}2\pi\,,$$

where Remark 15.2.2/3 has been used. In other words, the theorem is obtained from (1) by passing to the limit $\varepsilon \downarrow 0$.

Remark 2. Formula (2) means that $f(z_0)$ can be calculated if the values of $f(z)$ are known on a curve that encloses z_0. This is a very remarkable situation, to which there is no analogue in the case of real functions. It also shows that holomorphy is something quite extraordinary.

15.3. Properties of Holomorphic Functions

15.3.1. Differentiability and Derivative Formulas

Theorem. *If $f(z)$ is holomorphic on a domain G in C_1, then $f(z)$ has derivatives (with respect to the complex variable z) up to arbitrary order on G, and the n-th derivative $f^{(n)}(z)$ is also holomorphic on G. Let $\mathfrak{C}$ be a closed, piecewise smooth curve in G, and let the interior of $\mathfrak{C}$ solely contain points of G* (Fig. 15.24). *If z is a point inside of $\mathfrak{C}$, then*

$$f^{(n)}(z) = \frac{n!}{2\pi \mathrm{i}} \int_{\mathfrak{C}} \frac{f(\zeta)}{(\zeta - z)^{n+1}}\,\mathrm{d}\zeta, \quad n = 0, 1, 2, \ldots \tag{1}$$

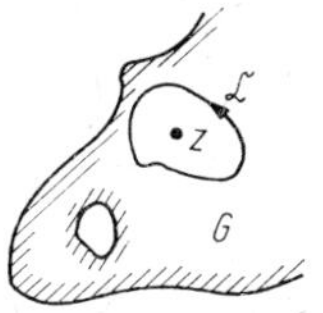

Fig. 15.24

Remark 1. For a holomorphic function $f(z)$, the derivative $f'(z)$ has been defined in Def. 15.1.2/2. If $f'(z)$ is also holomorphic, then one can form $f^{(2)}(z)=(f'(z))'$, etc. $f^{(n)}(z)$ is the n-th iteration. Since the complex differentiation with respect to $z=x+iy$ includes real differentiation with respect to x and y as special cases, the above theorem implies that $f(z)=$ $=f(x+iy)=g(x, y)$ has derivatives up to arbitrary order with respect to x and y. If $f(z)$ is once differentiable with respect to the complex variable (and hence holomorphic), then, by the above theorem, $f(z)$ has derivatives up to arbitrary order with respect to the complex variable. Also for this property there is no analogue in real analysis.

Remark 2. Formula (1) is derived formally from (15.2.3/2) by differentiation under the sign of integration. For $n=0$ one has $f^{(0)}(z)=f(z)$, and (1) coincides with (15.2.3/2).

15.3.2. Taylor Series

Theorem. *If $f(z)$ is holomorphic on a domain G in C_1, then $f(z)$ can be expanded in an absolutely and uniformly convergent power series in the neighbourhood of an arbitrary point $z_0 \in G$:*

$$f(z)=\sum_{k=0}^{\infty} a_k(z-z_0)^k \quad \textit{with} \quad a_k=\frac{1}{2\pi \mathrm{i}} \int_{\mathfrak{C}} \frac{f(\zeta)}{(\zeta-z_0)^{k+1}}\,\mathrm{d}\zeta\,. \tag{1}$$

Here $\mathfrak{C}$ is a piecewise smooth curve in G that encloses z_0, and the interior of $\mathfrak{C}$ contains points of G alone (Fig. 15.25).

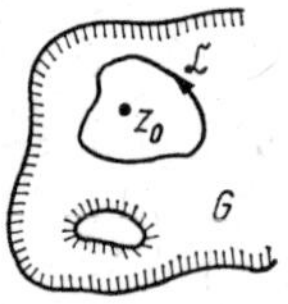

Fig. 15.25

Remark 1. In Theorem 15.1.3 we had seen that complex power series with a positive radius of convergence, R, are holomorphic functions. The above theorem states (at least locally) the converse of this assertion: if one chooses any point $z_0 \in G$, then $f(z)$ can be represented as a power series with a positive radius of convergence R in a neighbourhood of z_0. R has the meaning of (5.4.1/2). Thus holomorphic functions can also be defined as those functions which can be expanded in a complex power series with a positive radius of convergence at every point of their domain of definition. This is Weierstrass' approach, whereas the previous variant is referred to as Riemann's approach.

Remark 2. A point of interest is the magnitude of R, the radius of convergence of the series (1). It turns out that (1) converges on every disk around z_0 that lies entirely in G. Thus the radius of convergence R of (1) is at least equal to the radius of the largest disk around z_0 that lies in G (Fig. 15.26).

Remark 3. From (1) and (15.3.1/1) it follows that

$$f(z)=\sum_{k=0}^{\infty} \frac{f^{(k)}(z_0)}{k!}(z-z_0)^k\,. \tag{2}$$

This is the complex analogue to (5.4.4/1).

Remark 4. As an example, let us expand $\ln z$ at the point 1 (Fig. 15.27). According to Remark 2 and Remark 15.1.4/2, the power series for $\ln z$ at the point $z_0=1$ converges on a disk of radius 1 around the point 1. From (2), $\ln 1=0$, $(\ln z)'=\dfrac{1}{z}$ and the expressions for the higher derivatives one obtains

$$\ln z = \sum_{k=1}^{\infty} \frac{(-1)^{k+1}}{k} (z-1)^k.$$

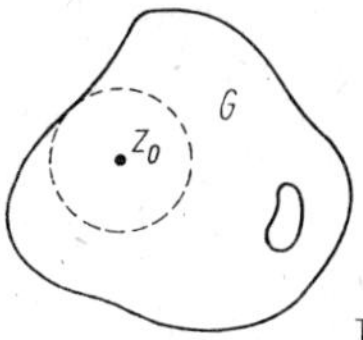

Fig. 15.26

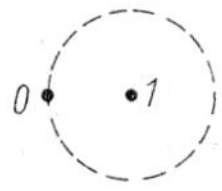

Fig. 15.27

15.3.3. The Identity Theorem

Theorem. *Let G be a connected domain in C_1, and let $\{z_k\}_{k=1}^{\infty}$ be a sequence of points of G, with $z_k \neq z_l$ for $k \neq l$, which has at least one accumulation point in G* (Fig. 15.28). *If $f(z)$ and $g(z)$ are any two functions holomorphic on G, with $f(z_k)=g(z_k)$ for $k=1, 2, 3, \ldots$, then $f(z) \equiv g(z)$ on G.*

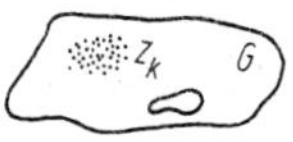

Fig. 15.28

Remark. Thus a holomorphic function $f(z)$ on G is uniquely determined by the values $f(z_k)$. The behaviour in the small (e.g., in the neighbourhood of the accumulation point of $\{z_k\}_{k=1}^{\infty}$) already implies the global behaviour.

15.3.4. The Maximum Principle

Theorem. *Let $f(z)$ be holomorphic on a connected domain G in C_1. Let G_0 be a subdomain of G, and let $z_0 \in G_0$* (Fig. 15.29). *If $|f(z_0)| = \max_{z \in G_0} |f(z)|$, then $f(z) \equiv c$ is constant on G.*

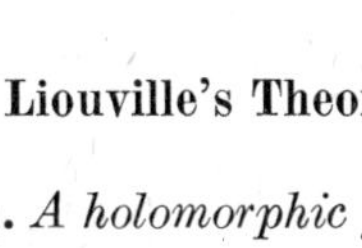

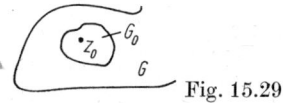

Fig. 15.29

Remark. Hence the constants are the only holomorphic functions $f(z)$ on G for which $|f(z)|$ may have a local maximum on G. A corresponding assertion is also true for local minima of $|f(z)|$ (where, of course, $|f(z_0)|=0$ must be excluded).

15.3.5. Liouville's Theorem

Theorem. *A holomorphic function that is bounded on the complex plane is constant.*

Remark. The theorem is a consequence of (15.3.2/1). If one sets $z_0=0$ and $\mathfrak{C}=\{\zeta \mid |\zeta|=R\}$ there, one obtains

$$|a_k| \leq (\sup_{z \in C_1} |f(z)|) \frac{1}{R^k} \to 0 \quad \text{as} \quad R \to \infty \quad \text{and} \quad k=1, 2, \ldots$$

Hence $f(z) \equiv a_0$.

15.3.6. The Fundamental Theorem of Algebra

We prove the assertion made in Remark 5.3.3/4: the polynomial $P(z)=\sum_{k=0}^{n} a_k z^k$ with the complex coefficients $a_0, \ldots, a_n$, $a_n \neq 0$, $n \geqq 1$, has at least one complex zero. If we suppose that $P(z)$ has no zero, then $\frac{1}{P(z)}$ is holomorphic on C_1. It is easily seen that $\left|\frac{1}{P(z)}\right|$ is also bounded. Then, by Theorem 15.3.5, we find that $\frac{1}{P(z)} \equiv c$, i.e. that $P(z)$ is constant, which contradicts the assumption. So simple is the matter if adequate mathematical means are at hand.

15.4. Theory of Singularities

15.4.1. Laurent's Series

Theorem. *If $f(z)$ is holomorphic on the circular ring $\{z \mid R_1 < |z-z_0| < R_2\}$, with $z_0 \in C_1$ and $0 \leqq R_1 < R_2 < \infty$ (Fig. 15.30), then*

$$f(z)=\sum_{k=-\infty}^{\infty} a_k (z-z_0)^k \quad \textit{with} \quad a_k=\frac{1}{2\pi i}\int_{\mathfrak{C}} \frac{f(\zeta)}{(\zeta-z_0)^{k+1}}\, d\zeta . \tag{1}$$

Here $\mathfrak{C}=\{w \mid |w|=R\}$, $R_1<R<R_2$, with the orientation indicated in the figure. For a sufficiently small $\varepsilon>0$, the series (1) converges absolutely and uniformly on the circular ring $\{z \mid R_1+\varepsilon<|z-z_0|<R_2-\varepsilon\}$.

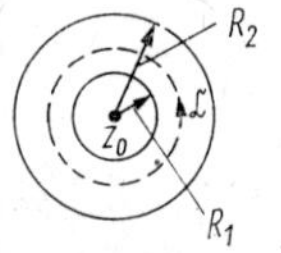

Fig. 15.30

Remark 1. The representation (1) is called a Laurent series. One has

$$f(z)=\underbrace{\sum_{k=0}^{\infty} a_k (z-z_0)^k}_{\text{regular part}}+\underbrace{\sum_{k=-\infty}^{-1} a_k (z-z_0)^k}_{\text{main part}} .$$

The regular part converges for $|z-z_0|<R_2$, whereas the main part converges for $|z-z_0|>R_1$.

Remark 2. The following assertion can be derived from the identity theorem 15.3.3: if $f(z)$, as given in the above theorem, can be represented as $f(z)=\sum_{k=-\infty}^{\infty} b_k (z-z_0)^k$ on the circular ring $\{z \mid R_1<|z-z_0|<R_2\}$, then $b_k=a_k$ for $k=0, \pm 1, \pm 2, \ldots$ Hence the coefficients a_k in (1) are unique. In particular this is then also true for power series.

15.4.2. Singularities

Definition. *Let $f(z)$ be holomorphic on the punctured disk $\{z \mid 0<|z-z_0|<\varepsilon\}$, $z_0 \in C_1$. Let $f(z)=\sum_{k=-\infty}^{\infty} a_k (z-z_0)^k$ be the corresponding Laurent series.*

(**a**) *z_0 is called a removable singularity if $a_k=0$ for $k=-1, -2, -3, \ldots$*

(**b**) *z_0 is called a pole of order m if $a_{-m} \neq 0$ and $a_k = 0$ for $k = -m-1, -m-2, \ldots$, where $m = 1, 2, 3, \ldots$*

(**c**) *z_0 is called an essential singularity if z_0 is neither a removable singularity nor a pole.*

Remark 1. In the case of a removable singularity one simply sets $f(z_0) = a_0$. Then $f(z)$ is holomorphic on the disk $\{z \mid |z-z_0| < \varepsilon\}$. The singularity has disappeared, and $f(z) = \sum_{k=0}^{\infty} a_k (z-z_0)^k$ is a normal expansion in a power series on $\{z \mid |z-z_0| < \varepsilon\}$. In the case of a pole at z_0 one has

$$f(z) = \frac{a_{-m}}{(z-z_0)^m} + \frac{a_{-m+1}}{(z-z_0)^{m-1}} + \ldots + \frac{a_{-1}}{z-z_0} + a_0 + a_1 (z-z_0) + \ldots$$

If z_0 is an essential singularity, then the main part (cf. Remark 15.4.1/1) contains infinitely many terms with non-vanishing coefficients a_k.

Theorem. *Let $f(z)$ be a function holomorphic on the punctured disk $\{z \mid 0 < |z-z_0| < \varepsilon\}$, $z_0 \in C_1$.*

(**a**) *z_0 is a removable singularity if and only if $|f(z)|$ is bounded in a neighbourhood of z_0.*

(**b**) *z_0 is a pole if and only if $f(z) \to \infty$ as $z \to z_0$.*

(**c**) *(Theorem of Casorati-Weierstrass). z_0 is an essential singularity if and only if for every $\varepsilon > 0$, every $\delta > 0$, and every $w \in C_1$ there exists a $z \in C_1$ such that $0 < |z-z_0| < \varepsilon$ and $|f(z) - w| < \delta$.*

Remark 2. In the older literature, the Casorati-Weierstrass theorem is also referred to as the theorem with the three "arbitraries": let $w \in C_1$ be an arbitrary point, then in an arbitrary neighbourhood of z_0 there exist points z whose images $f(z)$ lie arbitrarily close to the point w. The values of the function $f(z)$ are thus terribly jumbled up in the neighbourhood of z_0. If, on the other hand, z_0 is a pole, then it is natural to set $f(z_0) = \infty$. Then z_0 is a harmless singularity that can be included in the domain of definition of f.

Remark 3. So far we assumed that $z_0 \in C_1$. If $z_0 = \infty$, then, following the lines of Subsec. 15.1.1., we set $g(z) = f\left(\frac{1}{z}\right)$, considering $g(z)$ in a neighbourhood of 0. If $f(z)$ is holomorphic for $|z| > N$, then $g(z)$ is holomorphic for $0 < |z| < \frac{1}{N}$. Then

$$f\left(\frac{1}{z}\right) = g(z) = \sum_{k=-\infty}^{\infty} a_k z^k = \sum_{k=-\infty}^{\infty} a_{-k} \left(\frac{1}{z}\right)^k,$$

and hence

$$f(z) = \sum_{k=-\infty}^{\infty} a_{-k} z^k = \underbrace{\sum_{k=-\infty}^{0} a_{-k} z^k}_{\text{regular part}} + \underbrace{\sum_{k=1}^{\infty} a_{-k} z^k}_{\text{main part}}.$$

Thus a pole at $z_0 = \infty$ has the form $a_{-m} z^m + a_{-m+1} z^{m-1} + \ldots$

15.4.3. Systematic Complex Function Theory, Rational Functions

If G is a domain in C (that is the complex plane including ∞), then one may ask for the set of all functions $f(z)$ that are holomorphic on G. Or one may ask for functions $f(z)$ that have nothing but poles as singularities on G (according to

Remark 15.4.2/2, one sets $f(z_0)=\infty$ at a pole z_0). Investigations of this kind are the subject of systematic complex function theory. In some simple cases it is possible to give final answers.

Theorem. (**a**) *The constants are the only holomorphic functions on C.*

(**b**) *The rational functions are the only functions that have only poles as singularities on C.*

Remark 1. Naturally, the functions admitted to competition in (b) are only those which are either holomorphic in the neighbourhood of every point $z_0 \in C$ or have a pole in the sense of Remark 15.4.2/1 there. If $z_0=\infty$, then see Def. 15.1.2/2 and Remark 15.4.2/3 for comparison. Part (a) is in effect a consequence of Theorem 15.3.5. The fact that the rational functions $R(z)=\left(\sum_{k=0}^{n} a_k z^k\right)\left(\sum_{l=0}^{m} b_l z^l\right)^{-1}$, with complex coefficients a_k, b_l, and with $b_m \neq 0$ and $n=0, 1, 2, \ldots$ and $m=0, 1, 2, \ldots$, have only poles as singularities on C follows from the partial fraction decomposition in Theorem 7.1.2, which is also applicable to $z \in C_1$ instead of $x \in R_1$.

Remark 2. Functions $f(z)$ which are holomorphic on $C_1 = C\backslash\{\infty\}$ are called entire functions. Functions $f(z)$ which have only poles as singularities on C_1 (and are holomorphic in the neighbourhood of all other points of C_1) are called meromorphic functions.

15.5. Theory of Residues

15.5.1. Cauchy's Residue Theorem

Theorem. *Let G be a simply connected domain in C_1, and let $\mathfrak{C}$ be a smooth closed curve in G* (Fig. 15.31). *If $z_1, \ldots, z_N$ are points lying inside of $\mathfrak{C}$, and if $f(z)$ is holomorphic on $G\backslash\{z_1\}\cup\{z_2\}\cup\ldots\cup\{z_N\}$, then let*

$$f(z)=\sum_{k=-\infty}^{\infty} c_{l,k}(z-z_l)^k, \quad l=1, \ldots, N,$$

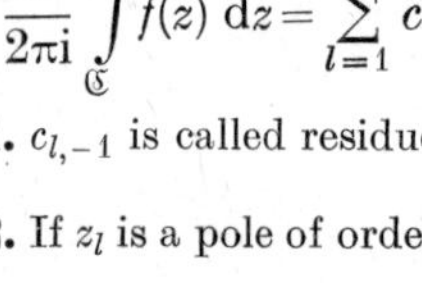

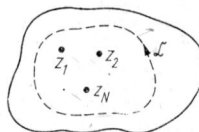

Fig. 15.31

be the local expansions in Laurent series. Then

$$\frac{1}{2\pi i}\int_{\mathfrak{C}} f(z)\,dz=\sum_{l=1}^{N} c_{l,-1} \quad (\textit{residue theorem})\,. \tag{1}$$

Remark 1. $c_{l,-1}$ is called residue at the point z_l.

Remark 2. If z_l is a pole of order 1, i.e., $f(z)=\frac{c_{l,-1}}{z-z_l}+c_{l,0}+\ldots$, then it is possible to calculate $c_{l,-1}$ from $c_{l,-1}=\lim_{z\to z_l}(z-z_l)\,f(z)$. This is the same method as in the partial fraction decomposition used in Subsec. 7.1.2.

Remark 3. As in many integral theorems of complex function theory, it can be used in a tricky way for the calculation of real integrals (see for instance Remark 15.2.2/4). Let us outline here how the improper Riemann integral $\int_{-\infty}^{\infty}\frac{\sin x}{x}\,dx=2\int_{0}^{\infty}\frac{\sin x}{x}\,dx$ mentioned in

Subsec. 7.2.1. can be calculated. $\frac{1}{2iz}(e^{iz}-e^{-iz})$ is holomorphic on $C_1=C\setminus\{\infty\}$ (also at the point $z=0$). If $z=x$ is real, one obtains

$$\frac{e^{iz}-e^{-iz}}{2iz}=\frac{e^{ix}-e^{-ix}}{2ix}=\frac{\sin x}{x}.$$

(The holomorphic function e^z has been treated in Subsec. 15.1.3.) From Cauchy's integral theorem, Subsec. 15.2.2., it follows that

$$\int_{-N}^{N}\frac{\sin x}{x}\,dx=\int_{\mathfrak{C}}\frac{e^{iz}-e^{-iz}}{2iz}\,dz=\frac{1}{2i}\int_{\mathfrak{C}}\frac{e^{iz}}{z}\,dz-\frac{1}{2i}\int_{\mathfrak{C}}\frac{e^{-iz}}{z}\,dz\,, \tag{2}$$

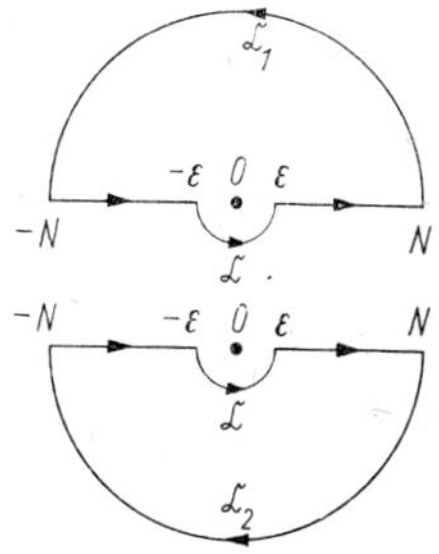

Fig. 15.32

Fig. 15.33

where $\mathfrak{C}$ is the path of integration indicated in Fig. 15.33. If $\mathfrak{C}_1$ and $\mathfrak{C}_2$ are the semicircles indicated in Fig. 15.32, then it follows from the above theorem that

$$\int_{\mathfrak{C}}\frac{e^{iz}}{z}\,dz=-\int_{\mathfrak{C}_1}\frac{e^{iz}}{z}\,dz+2\pi i\cdot 1,\qquad \int_{\mathfrak{C}}\frac{e^{-iz}}{z}\,dz=-\int_{\mathfrak{C}_2}\frac{e^{-iz}}{z}\,dz\,,$$

where 1 is the residue of $\frac{e^{iz}}{z}$ at the point 0. One can now estimate that the integrals over $\mathfrak{C}_1$ and $\mathfrak{C}_2$ approach 0 as $N\to\infty$. Then from (2) one obtains

$$\int_{-\infty}^{\infty}\frac{\sin x}{x}\,dx=\pi\,.$$

15.5.2. The Logarithmic Residue

Theorem. *Let $z_1, \ldots, z_n$ be any points in the simply connected domain G in C_1. Let the function $f(z)$ be holomorphic on $G\setminus(\{z_1\}\cup\ldots\cup\{z_n\})$, and let $f(z)$ have a pole of order N_k at the point z_k, $k=1, \ldots, n$. Further let $f(z)$ have a finite number of zeros in G, $w_1, \ldots, w_m$, where w_l is a zero of order M_l, $l=1, \ldots, m$. If the smooth curve $\mathfrak{C}$ encloses all zeros and poles of $f(z)$* (Fig. 15.34), *then*

$$\frac{1}{2\pi i}\int_{\mathfrak{C}}\frac{f'(z)}{f(z)}\,dz=\sum_{l=1}^{m}M_l-\sum_{k=1}^{n}N_k\,. \tag{1}$$

Remark 1. Consequently, $f(z)$ can be expanded at z_k in a local Laurent series having the form

$$f(z)=\frac{1}{(z-z_k)^{N_k}}\left(c_{k,-N_k}+c_{k,-N_k+1}(z-z_k)+\ldots\right),\quad c_{k,-N_k}\neq 0\,.$$

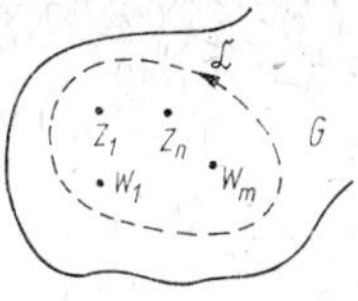

Fig. 15.34

Here $N_k = 1, 2, \ldots$ Thus $\sum_{k=1}^{n} N_k$ is the sum of the poles, where their multiplicities have been taken into account. By analogy, in a neighbourhood of w_l one has

$$f(z) = (z - w_l)^{M_l}(d_{l,M_l} + d_{l,M_l+1}(z - w_l) + \ldots), \quad d_{l,M_l} \neq 0 .$$

Here $M_l = 1, 2, \ldots$, and $\sum_{l=1}^{m} M_l$ is the sum of the zeros, with their multiplicities having been taken into account. From these expansions one obtains

$$\frac{f'(z)}{f(z)} = \begin{cases} -\dfrac{N_k}{z - z_k}(1 + \ldots) & \text{near } z_k , \\ \dfrac{M_l}{z - w_l}(1 + \ldots) & \text{near } w_l . \end{cases} \tag{2}$$

At all other points $\frac{f'(z)}{f(z)}$ is holomorphic. Then (1) follows immediately from (15.5.1/1). The term "logarithmic residue" has its origin in (2), since the residues of $(\ln f(z))' = \frac{f'(z)}{f(z)}$ are calculated.

Remark 2. With the use of (1), the fundamental theorem of algebra can be proved on the basis of Theorem 5.3.3/2 (and Subsec. 15.3.6.) as follows. Initially one estimates that all zeros of $f(z) = \sum_{k=0}^{n} a_k z^k$ lie inside a circle $\mathfrak{C} = \{z \mid |z| = R\}$. Further one has $\frac{f'(z)}{f(z)} = \frac{n}{z}(1 + \ldots)$ for $|z| > R$. $f(z)$ has no poles. Now it follows from (1) that

$$n = 2\pi i \frac{n}{2\pi i} = \frac{1}{2\pi i} \int_{\mathfrak{C}} \frac{n}{z}(1 + \ldots)\, dz = \frac{1}{2\pi i} \int_{\mathfrak{C}} \frac{f'(z)}{f(z)}\, dz = \sum_{l=1}^{m} M_l ,$$

which is the desired result.

15.5.3. Mapping Properties of Holomorphic Functions

Theorem 1. *Let $f(z)$ be holomorphic on a neighbourhood of a point $z_0 \in C_1$, and let*

$$f(z) = w_0 + c_n(z - z_0)^n + c_{n+1}(z - z_0)^{n+1} + \ldots$$

in that neighbourhood, with $c_n \neq 0$. Then there are a neighbourhood U of z_0 and a neighbourhood V of w_0, such that $f(U) = V$ (Fig. 15.35). *If $w_1 \in V$, $w_1 \neq w_0$, then there are exactly n different points $z_1, \ldots, z_n$ in U, such that $f(z_k) = w_1$ for $k = 1, \ldots, n$. On the other hand there is no point $z \in U$ such that $z \neq z_0$ and $f(z) = w_0$.*

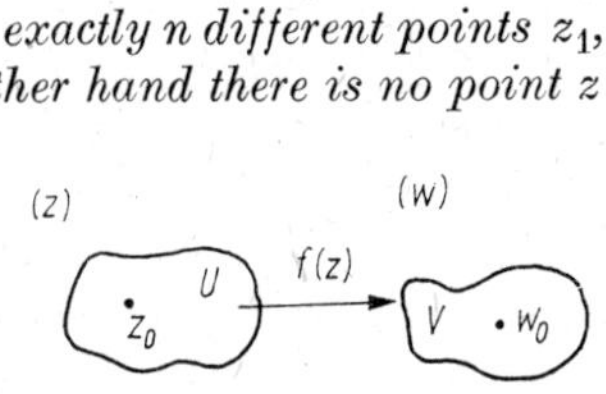

Fig. 15.35

Remark 1. One has $f(U) = \{w \mid \exists z \in U$ such that $f(z) = w\}$. The theorem can be proved in an elegant way by means of Theorem 15.5.2: if one substitutes $f(z) - w_1$ for $f(z)$ in (15.5.2/1), then the integral counts the number of points where $f(z) = w_1$. Unfortunately the (necessary) details cannot be dealt with here.

Theorem 2 (*theorem on domain invariance*). *Let $f(z)$ be holomorphic on a domain G in C_1. If $f(z)$ is not constant, then $f(G) = \{w \mid \exists z \in G$ such that $f(z) = w\}$ is also a domain in C_1* (Fig. 15.36).

Remark 2. This theorem is an immediate consequence of Theorem 1.

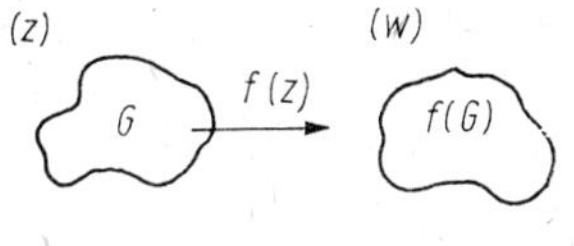

Fig. 15.36

15.5.4. Inverse Functions

Theorem. *Let $f(z)$ be holomorphic in a neighbourhood of a point $z_0 \in C_1$. Let $f(z_0) = w_0$ and $f'(z_0) \neq 0$* (Fig. 15.37). *Then there exist a neighbourhood U of z_0 and a neighbourhood V of w_0, such that $f(z)$ maps U onto V in a one-to-one way, the inverse function $z = g(w)$ is holomorphic on V, and the relation*

$$g'(w) = \frac{1}{f'(z)}, \quad \textit{with} \quad w = f(z)$$

holds on V.

Remark. The first part of the theorem is inferred immediately from Theorem 15.5.3/1. The second part is the complex analogue to Theorem 3.1.4.

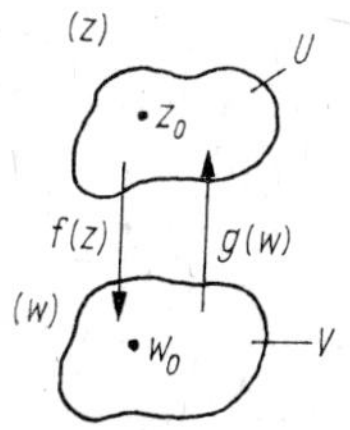

Fig. 15.37

15.6. Holomorphic Continuation

15.6.1. The Disk-Chain Method

If $f_0(z)$ is holomorphic on the disk K_0, and if $z_1 \in K_0$, then $f_0(z)$ can be expanded in a power series at the point z_1:

$$f_0(z) = \sum_{k=0}^{\infty} a_k (z - z_1)^k . \tag{1}$$

If K_1 is a disk with centre z_1, then the absolute and uniform convergence of (1) on K_1 is ensured if $K_1 \subset K_0$ (Theorem 15.3.2 and Remark 15.3.2/2). However, it may happen that the right-hand side of (1) converges absolutely on a disk K_1 that extends beyond K_0 (Fig. 15.38). By Theorem 15.1.3, the right-hand side of (1) is a holomorphic function on K_1 also in this case. Thus we obtain a holomorphic continuation of $f_0(z)$, namely

$$f(z)=\begin{cases} f_0(z) & \text{for} \quad z\in K_0\,, \\ \sum\limits_{k=0}^{\infty} a_k(z-z_1)^k & \text{for} \quad z\in K_1\,. \end{cases}$$

The iteration of this method is called disk-chain method: along a given smooth curve $\mathfrak{C}$ in C_1, one attempts to continue the starting function $f_0(z)$ holomorphically to a point $w\in\mathfrak{C}$ (Fig. 15.39). The continued holomorphic function is again denot-

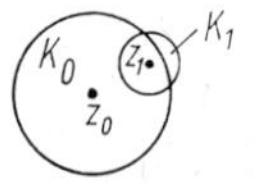

Fig. 15.38

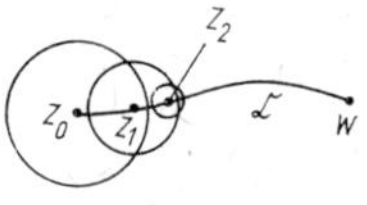

Fig. 15.39

ed by $f(z)$. Whether or not the point w will be reached cannot be stated generally. Further the question arises as to whether the value $f(w)$ will depend on the particular choice of the disks in the disk-chain method.

Theorem. *The disk-chain method applied along the curve $\mathfrak{C}$ yields a value $f(w)$ that is independent of the particular choice of the disks.*

Remark. More precisely, one should write $f_{\mathfrak{C}}(w)$ instead of $f(w)$, because in general $f(w)$ depends on $\mathfrak{C}$ (but not on the disks chosen along $\mathfrak{C}$).

15.6.2. The Monodromy Theorem

Theorem (*monodromy theorem*). *Suppose that $f(z)$ is holomorphic on the disk K and that it can be holomorphically continued to w along the curves $\mathfrak{C}_0$ and $\mathfrak{C}_1$, in the sense of Subsec.* 15.6.1. (Fig. 15.40). *Let $f_{\mathfrak{C}_0}(w)$ and $f_{\mathfrak{C}_1}(w)$ be the corresponding values. If there exists a family of smooth curves, $\{\mathfrak{C}_s\}_{0\leq s\leq 1}$, which cover the domain between $\mathfrak{C}_0$ and $\mathfrak{C}_1$ in a schlicht, gapless, and continuous manner, and if $f(z)$ can be continued holomorphically to w along each of the curves $\mathfrak{C}_s$ by means of a disk-chain method, then $f_{\mathfrak{C}_0}(w)=f_{\mathfrak{C}_1}(w)$.*

Remark 1. "Schlicht, gapless and continuous" means the following: for $t\neq s$, $\mathfrak{C}_s$ and $\mathfrak{C}_t$ only have the points z_0 and w in common (this applies especially also for $\mathfrak{C}_0$ and $\mathfrak{C}_1$); every point in the domain between $\mathfrak{C}_0$ and $\mathfrak{C}_1$ belongs to one of the curves $\mathfrak{C}_s$; and $\mathfrak{C}_s$ approaches $\mathfrak{C}_t$ as $|s-t|$ becomes small.

Remark 2. If $f(z)$ can be continued holomorphically from the disk K into the simply connected domain G by means of the disk-chain method, then the continuation is unique (and especially also independent of the path of continuation) (Fig. 15.41). This is a direct implication of the above theorem.

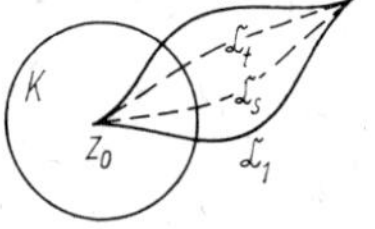

Fig. 15.40

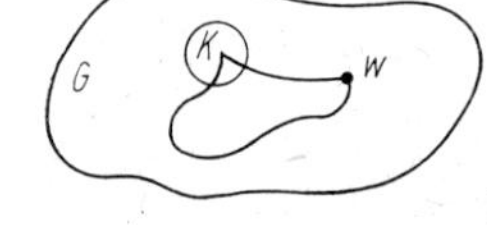

Fig. 15.41

15.6.3. Riemann Surfaces

If the domain G referred to in Remark 15.6.2/2 is not simply connected, then the continuation by means of the disk-chain method generally depends on the path. In particular one may arrive at new values of the function along closed paths. We continue $f(z)=\sqrt{z}$ along the unit circle $\mathfrak{C}$ (Fig. 15.42). The surface of existence of $\sqrt{z}$ has been described at the end of Subsec. 5.3.2. On the unit circle one has $z=e^{i\varphi}$ and $\sqrt{z}=e^{i\frac{\varphi}{2}}$. Starting from, say, $\sqrt{1}=1$ (i.e. $\varphi=0$), after a full traverse of the unit circle by means of the disk-chain method one obtains the value $e^{i\frac{2\pi}{2}}=e^{i\pi}=-1$. This many-valuedness is avoided by introducing the Riemann surface for $\sqrt{z}$ in the sense of Subsec. 5.3.2.: two sheets with the "branch points" 0 and ∞ and a slit that connects these two branch points and along which the two sheets are glued to one another (Fig. 15.43). (The path of this slit is irrelevant.)

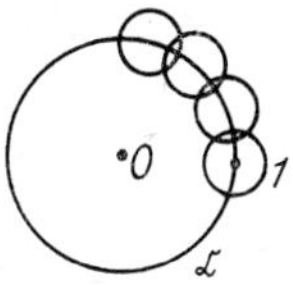

Fig. 15.42

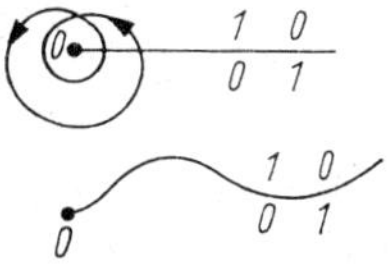

Fig. 15.43

Other than in the case of the unit circle, the disk-chain method on the Riemann surface for $\sqrt{z}$ is independent of the path. This opens the possibility to construct Riemann surfaces for arbitrary holomorphic functions $f(z)$. Let $f(z)$ be any function holomorphic on a disk K. By means of an optimum utilization of the disk-chain method one then obtains the Riemann surface (surface of existence) for $f(z)$. If these procedures are carried out for $f(z)=\ln z$, starting with a disk around $z=1$ and the value $f(1)=\ln 1=0$, one obtains the Riemann surface for $\ln z$ as given in Subsec. 5.3.1. Another example is

$$f(z)=\sqrt{(z-z_1)\,(z-z_2)\,(z-z_3)\,(z-z_4)}, \quad z_j \neq z_k\,.$$

It turns out that the corresponding surface of existence consists of 2 sheets with the 4 branch points z_1, z_2, z_3 and z_4 and the glued boundaries indicated in Fig. 15.44. It is easy to check that the disk-chain method on this Riemann surface for $f(z)$ is path-independent. There is still a lot to tell about Riemann surfaces; in particular the inexact formulations given above should be made more precise. A fine representation has been given in [6].

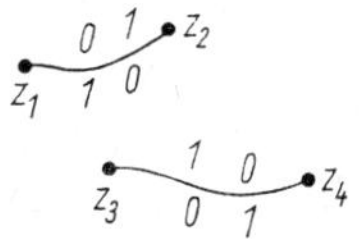

Fig. 15.44

15.7. Conformal Mappings

15.7.1. Fundamental Properties

Definition. *A continuously differentiable, one-to-one mapping $u = u(x, y)$, $v = v(x, y)$ of a domain G in $R_2 = C_1$ onto a domain G' in R_2 is called conformal if it leaves the angles including their orientations unchanged.*

Remark 1. The meaning is as follows (Fig. 15.45). The continuously differentiable mapping $u = u(x, y)$, $v = v(x, y)$ maps G onto G' in a one-to-one way. Consider two continuously differentiable curves $\mathfrak{C}_1$ and $\mathfrak{C}_2$ in G, the tangents of which shall include an angle α. Here α is considered positive if measured from $\mathfrak{C}_1$ to $\mathfrak{C}_2$ (orientation). Now the mapping is called conformal if the tangents of the image curves $\mathfrak{C}_1'$ and $\mathfrak{C}_2'$ include the same angle α with the same orientation.

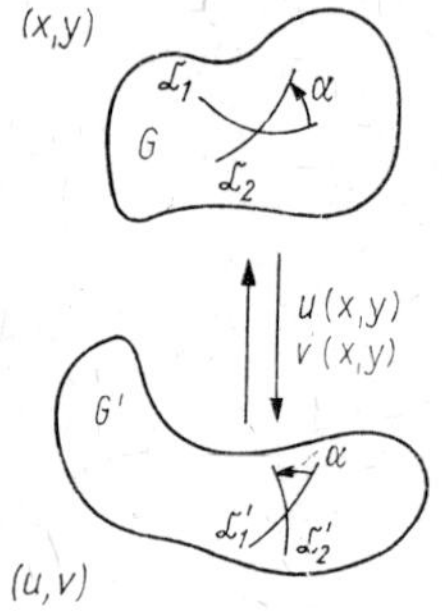

Fig. 15.45

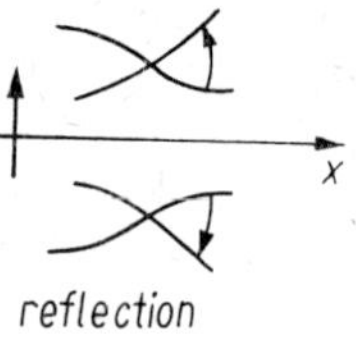

Fig. 15.46

Remark 2. If a and b are real numbers, then the translation $u = x + a$, $v = y + b$ in R_2 is a conformal mapping. On the other hand, the reflection $u = x$, $v = -y$ is not conformal: while the magnitude of the angles remains unchanged, their orientation does not (Fig. 15.46).

Theorem 1. *If the non-constant holomorphic function $f(z)$ maps the domain G in C_1 onto $G' = f(G)$ in a one-to-one way, then $f(z)$ is a conformal mapping. Here G' is again a domain.*

Remark 3. The fact that G' is again a domain follows from Theorem 15.5.3/2. Further, since the mapping $f(z)$ is assumed to be one-to-one, it follows from Theorem 15.5.3/1 that $f'(z) \neq 0$ for every point $z \in G$.

Theorem 2. *Let $\mathfrak{C}$ be a piecewise smooth, closed curve that encloses a bounded domain G in C_1, let $z_1, \ldots, z_N$ be any points on $\mathfrak{C}$, especially including all of the corner points of $\mathfrak{C}$. Let $f(z)$ be holomorphic on $\overline{G} \setminus \bigcup_{k=1}^{N} \{z_k\}$ and continuous on $\overline{G}$. If the mapping $z \in \mathfrak{C} \to f(z) \in \mathfrak{C}' = f(\mathfrak{C})$ is one-to-one, then $f(z)$ effects a conformal mapping of G onto the domain G' enclosed by the curve $\mathfrak{C}'$.*

Remark 4. Consequently $\{z_1, \ldots, z_N\}$ contains all corner points of $\mathfrak{C}$ at which smooth curve segments coincide, as well as possible other points on the smooth curve segments. Smooth means continuously differentiable. Further we allow $\mathfrak{C}$ to consist (as indicated in Fig. 15.47) of several connected boundary components. The assumption that $f(z)$ be holomorphic on $\overline{G} \setminus \bigcup_{k=1}^{N} \{z_k\}$ means that $f(z)$ is holomorphic in a neighbourhood $U(z_0)$ of every point $z_0 \in \overline{G} \setminus \bigcup_{k=1}^{N} \{z_k\}$.

Remark 5. The value of the theorem consists in that one has only to consider the images of the boundaries.

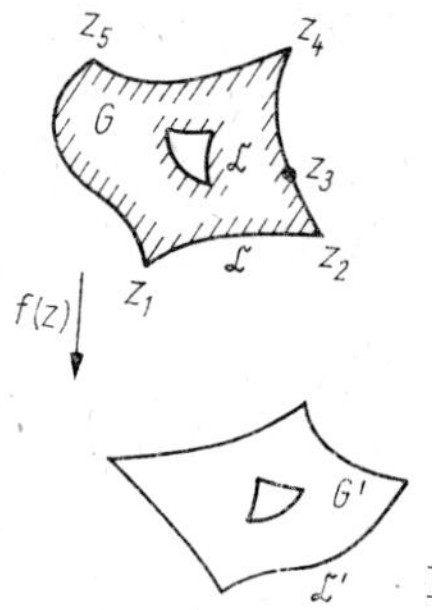

Fig. 15.47

15.7.2. Riemann's Mapping Theorem

Lemma. *If $f(z)$ is a holomorphic function that effects a conformal mapping of the unit disk $\{z \mid |z|<1\}$ onto itself, and if $f(0)=0$ and $f'(0)>0$, then $f(z)$ is the identical mapping.*

Remark 1. What is required above is that $f(z)$ maps the unit disk $K=\{z \mid |z|<1\}$ conformally onto $f(K)=K$, satisfying the additional conditions that $f(0)=0$ and $f'(0)>0$. The assertion is that $f(z)=z$ is the only mapping that has these properties. This is a first, essential step towards determining all conformal mappings $f(z)$ that map K onto itself; cf. Theorem 15.8.7.

Remark 2. In connection with the above lemma, the following problem arises: which domains G in C can be mapped one-to-one onto the unit disk by means of holomorphic functions? Now let $\infty \in G$ be admitted. If G is a domain in C_1, then Theorem 15.7.1/1 shows that this mapping is conformal in the sense indicated above. Therefore in the case of arbitrary domains G in C one also speaks of conformal mappings, which means one-to-one mappings by holomorphic functions.

Theorem (*Riemann's mapping theorem*). *Let G be a simply connected domain in C that has at least two different boundary points. Further let $z_0 \in G$, $z_0 \neq \infty$, and $0 \leqq \varphi < 2\pi$. Then there exists exactly one conformal mapping $f(z)$ of G onto $K=\{z \mid |z|<1\}$, such that $f(z_0)=0$ and $f'(z_0)=|f'(z_0)|\, e^{i\varphi}$.*

Remark 3. Consequently, $f(G)=K$, where one may additionally prescribe which point $z_0 \neq \infty$ of G shall be mapped onto 0 and what shall be the value of $\arg f'(z_0)$. The uniqueness assertion is easily derived from the above lemma. On the other hand, the proof that there exists a conformal mapping with the desired properties proves difficult.

Remark 4. Since G has at least two different boundary points z_0 and z_1 and is simply connected, there is also a curve from z_0 to z_1 that consists of nothing but boundary points. If, for instance, $G=C\backslash\mathfrak{C}$, where $\mathfrak{C}$ is a curve from z_0 to z_1 (see Fig. 15.48), then this is a simply connected domain, because every closed curve in G can be contracted to a point, if necessary, via the point ∞ (for reference, see the corresponding situation on the Riemann sphere (Fig. 15.49)). There are only two types of simply connected domains which are not covered,

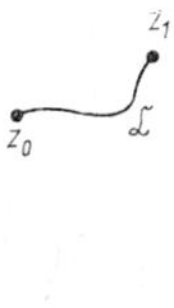

Fig. 15.48

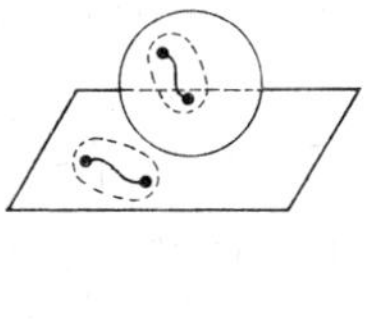

Fig. 15.49

namely the extended plane C and the punctured extended plane $C\setminus\{z_0\}$. It is easily seen that $C\setminus\{z_0\}$ can be mapped conformally onto $C\setminus\{\infty\}=C_1$. Further it is relatively easy to see that neither C nor C_1 can be mapped conformally onto $K=\{z \mid |z|<1\}$. After all, it is not possible either to map C conformally onto C_1. Thus the class of all simply connected domains in C can be subdivided into three conformal types: K, C_1, and C.

15.8. Linear Transformations

15.8.1. Conformal Mappings of C and C_1

At the end of Subsec. 15.7.2. we stated that for simply connected domains in C there are three conformal types, namely the unit disk K, the complex plane C_1, and the extended plane C. Thus it is of interest to determine the class of all conformal mappings that map C onto itself, C_1 onto itself, and K onto itself. We shall return to the case of the unit disk in Subsec. 15.8.7.

Theorem. (a) *$f(z)$ is a conformal mapping of C onto itself if and only if either $f(z)=\frac{a}{z-b}+c$ or $f(z)=az+b$. Here a, b, and c are complex numbers, and $a\neq 0$.*

(b) *$f(z)$ is a conformal mapping of C_1 onto itself if and only if $f(z)=az+b$. Here a and b are complex numbers, and $a\neq 0$.*

15.8.2. The Group of Linear Transformations

A conformal mapping of C onto itself is called linear transformation (more precisely: bilinear transformation). From Theorem 15.8.1 (a) it is concluded that $L(z)$ is a linear transformation if and only if

$$L(z)=\frac{az+b}{cz+d}, \quad \text{with} \quad ad-bc\neq 0\,. \tag{1}$$

If $L(z)=\frac{az+b}{cz+d}$, with $ad-bc=0$ and $|c|+|d|>0$, then $L(z)$ is constant and of no interest for our purposes.

Theorem. (a) *The linear mappings constitute a group with the identical mapping as its identity element and the law of composition L_1L_2: $(L_1L_2)(z)=L_1(L_2(z))$, $z\in C$, L_1 and L_2 linear transformations.*

(b) *Every linear transformation can be represented as a finite product of special linear transformations of the form*

$$L_1(z)=z+b, \quad L_2(z)=az \quad \text{and} \quad L_3(z)=\frac{1}{z}.$$

Here a and b are complex numbers, and $a\neq 0$.

Remark. In general, $L_1L_2\neq L_2L_1$, which means that the group is not commutative.

15.8.3. Invariance of Circles

Theorem. *The class of all straight lines and circles in C is mapped onto itself by a linear transformation.*

Remark 1. If one regards straight lines as special circles that pass through ∞, then the above theorem can also be formulated as follows: under a linear transformation, a circle is mapped onto another circle. The transformation $L(z)=\frac{1}{z}$ maps a circle passing through the origin onto a straight line, and vice versa.

Remark 2. The theorem needs only to be proved for the three standard types of Theorem 15.8.2(b); the rest is done by Theorem 15.8.2(b).

15.8.4. Mapping Properties and Cross Ratios

Theorem 1. *Let z_1, z_2, z_3 be three different points in C. Further let w_1, w_2, w_3 be three different points in C. Then there exists exactly one linear transformation $L(z)$ such that $L(z_k)=w_k$ for $k=1, 2, 3$.*

Remark 1. Thus one is free to prescribe 3 counter images and 3 image points. If all 6 points are different from ∞, then $w=L(z)$, with

$$\frac{z-z_1}{z-z_2}\cdot\frac{z_3-z_2}{z_3-z_1}=\frac{w-w_1}{w-w_2}\cdot\frac{w_3-w_2}{w_3-w_1} \tag{1}$$

is the desired linear transformation.

Theorem 2 (*invariance of the cross ratio*). *If $L(z)$ is a linear transformation, and if $w_k=L(z_k)$ for $k=1, 2, 3, 4$, where the points z_k are pairwise different, then*

$$\frac{w_4-w_1}{w_4-w_2}\cdot\frac{w_3-w_2}{w_3-w_1}=\frac{z_4-z_1}{z_4-z_2}\cdot\frac{z_3-z_2}{z_3-z_1}. \tag{2}$$

Remark 2. (2) is obtained from (1) and the uniqueness proposition of Theorem 1. If one of the points is ∞, then ∞ is treated in the usual way in the calculation, setting $\frac{\infty}{\infty}=1$.

15.8.5. Fixed Points and Types of Mappings

Theorem. *A linear transformation other than the identical one has either exactly one fixed point or exactly two fixed points.*

Remark. What is desired are points $z\in C$ such that $\frac{az+b}{cz+d}=L(z)=z$. Such a point is called fixed point. Here ∞ is admitted to competition. For the identical transformation $L(z)\equiv z$, of course, every point $z\in C$ is a fixed point. If $L(z)$ is not the identical transformation, then the equation $\frac{az+b}{cz+d}=z$ has either one or two solutions.

Classification by types: If the linear transformation $L_0(z)$ has exactly one fixed point and if this fixed point is ∞, then $L_0(z)=z+b$, where b is a complex number, $b\neq 0$. If the linear transformation $L_1(z)$ has exactly two fixed points and if 0 and ∞ are these fixed points, then $L_1(z)=az$, where a is a complex number, $a\neq 0$, and $a\neq 1$. Hence the two standard types are

$$L_0(z)=z+b \quad \text{and} \quad L_1(z)=az, \quad \text{where} \quad b\neq 0, \quad a\neq 0, \quad a\neq 1. \tag{1}$$

If $L(z)$ is an arbitrary linear transformation that is not the identical one, then there exists a linear transformation l such that

$$\text{either} \quad L=lL_0l^{-1} \quad \text{or} \quad L=lL_1l^{-1}. \tag{2}$$

l^{-1} is the transformation inverse to l. L has exactly one fixed point in the first case and exactly two fixed points in the second (L_0 and L_1 have the meaning of (1)).

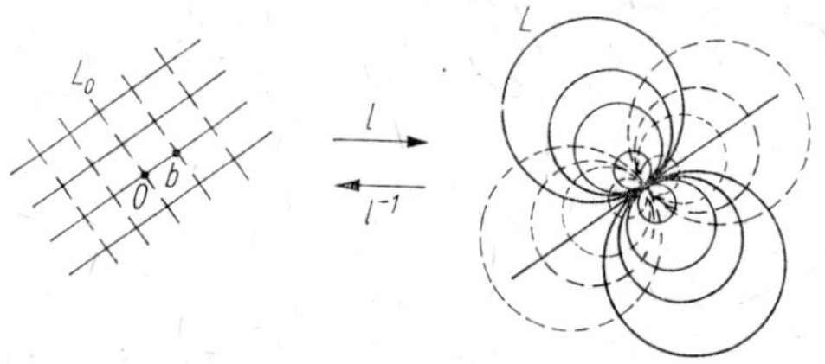

Fig. 15.50

Parabolic pencil of circles (Fig. 15.50): $L_0(z) = z + b$ is a translation. When L_0 is applied iteratively, the points z in C move along the solid lines, where the orthogonal dashed lines indicate the distances between counter image and image points. Hence the solid lines are the trajectories, and the dashed lines indicate the step interval in an iterative application of L_0. When l is applied, this picture is transformed into a parabolic pencil of circles: this follows from the invariance of circles and the conformity. If L has exactly one fixed point, then the first formula in (2) can be used: the solid circles are then the trajectories, and the broken circles indicate the step interval in an iterative application of L. The pencil point at which all the circles coincide is the fixed point.

Elliptic-hyperbolic pencil of circles (Fig. 15.51): If one considers $L_1(z) = az$, where $a > 0$, one obtains a dilatation. The trajectories are the straight lines passing through the origin, and the concentric circles around the origin indicate the step interval. If $a = e^{i\varphi}$, φ real, then L_1 is a rotation. One obtains the same picture, except that now the circles are the trajectories, while the straight lines indicate the step interval. Application of a linear transformation l gives an elliptic-hyperbolic pencil of circles. The hyperbolic pencil of circles consists of all circles passing through the points z_0 and z_1 (corresponding to the straight lines passing through 0). The elliptic pencil of circles consists of all circles orthogonal to the former ones (corresponding to the circles with centre 0). If L has exactly two fixed points, then the second formula of (2) can be used. z_0 and z_1 then are the fixed points. If in the corresponding standard transformation either $a > 0$ or $a = e^{i\varphi}$, then the circles of the elliptic-hyperbolic pencil are the corresponding trajectories. In the general case, $L_1(z) = az$, with $a = |a|\, e^{i\varphi}$, is a rotation and stretching. Such mappings are also called loxodromic transformations.

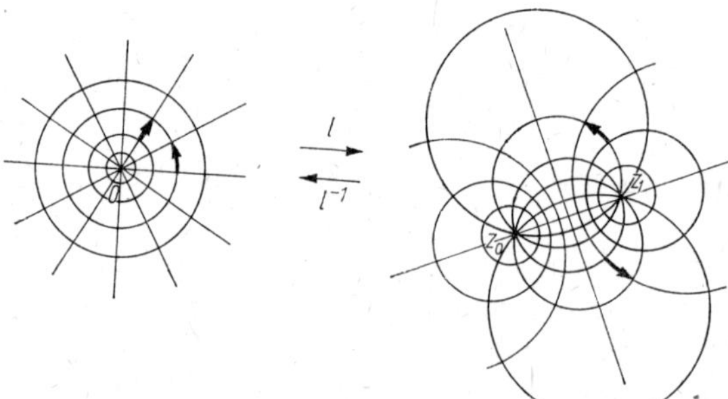

Fig. 15.51

15.8.6. Schwarz's Reflection Principle

If K is a circle with the centre $a \in C_1$ and the radius $R > 0$, then the points z_0 and z_1 are called mirror points (with respect to K) if $z_0 - a = R_0 e^{i\varphi}$, $z_1 - a = R_1 e^{i\varphi}$, and $R_0 R_1 = R^2$ (Fig. 15.52). For straight lines, mirror points have the usual meaning.

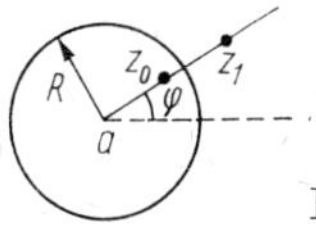

Fig. 15.52

Theorem. *Under linear transformations, mirror points are transformed into mirror points.*

Remark. Since circles are transformed into circles under linear transformations, the assertion of the theorem is meaningful. (As usual, we consider straight lines to be special circles.)

15.8.7. Conformal Mappings of the Unit Disk

To supplement Theorem 15.8.1, let us determine the conformal mappings of the unit disk $K = \{z \mid |z| < 1\}$ onto itself.

Theorem. *$f(z)$ is a conformal mapping of K onto itself if and only if* $f(z) = e^{i\varphi} \cdot \dfrac{z-a}{1-z\bar{a}}$, *where* $a \in C_1$, $|a| < 1$, $0 \leqq \varphi < 2\pi$.

Remark 1. It is relatively easy to reduce the theorem to Theorem 15.8.6 and Lemma 15.7.2: let $f(z)$ be a conformal mapping of K onto itself, with $f(a) = 0$ and $a \neq 0$. If $h(z) = \dfrac{z-a}{1-z\bar{a}}$, then $h(a) = 0$, $h\left(\dfrac{1}{\bar{a}}\right) = \infty$. Since a and $\dfrac{1}{\bar{a}}$ are mirror points with respect to the unit circle, 0 and ∞ must be mirror points with respect to the image of the unit circle. But this is only possible if this image is a circle with centre 0. From $|h(1)| = 1$ it follows that this image is again the unit circle. Hence $h(z)$ maps the unit disk onto itself. If $g(z) = f(h^{-1}(z))$, then $g(z)$ is a conformal mapping of K onto itself, and $g(0) = 0$. The rest now follows from Lemma 15.7.2.

Remark 2. The transformations of the unit disk onto itself constitute a subgroup of the group of all linear transformations.

15.9. Special Functions

15.9.1. The Functions e^z and $\ln z$

The holomorphic function e^z has been dealt with in Subsecs. 15.1.3. and 5.4.6. For the holomorphic function $\ln z$ we refer to Subsecs. 15.1.4., 5.3.1. and 15.3.2. The known properties of the real-valued functions e^x and $\ln x$ can be transferred to e^z and $\ln z$: for arbitrary complex numbers z_1 and z_2 and $z \neq 0$ one has

$$e^{z_1+z_2} = e^{z_1} e^{z_2} \quad \text{and} \quad e^{\ln z} = z . \tag{1}$$

Here it is irrelevant from which branch the value of $\ln z$ is taken: one has $(\ln z)_k = (\ln z)_0 + 2\pi i k$ and $e^{2\pi i k} = 1$. Relations of the form (1) can be proved in a simple way: $f(z)=e^{\ln z}-z$ is holomorphic, and $f(x)=0$ for x real, $x\neq 0$. Then the identity theorem 15.3.3 implies that $f(z)\equiv 0$. In other words, identities for real functions can be extended to complex functions, provided that the functions considered are holomorphic.

Mapping properties (Fig. 15.53): From $e^z=e^x e^{iy}$ for $z=x+iy$, it is easy to infer the following mapping properties: e^z maps the strip region $\{z \mid 0<\operatorname{Im} z<2\pi\}$ conformally onto the plane slit along the positive semi-axis. e^z maps the strip region $\{z \mid 0<\operatorname{Im} z<\pi\}$ conformally onto the upper half-plane. Here the marked straight lines are transformed into the corresponding straight lines and circles.

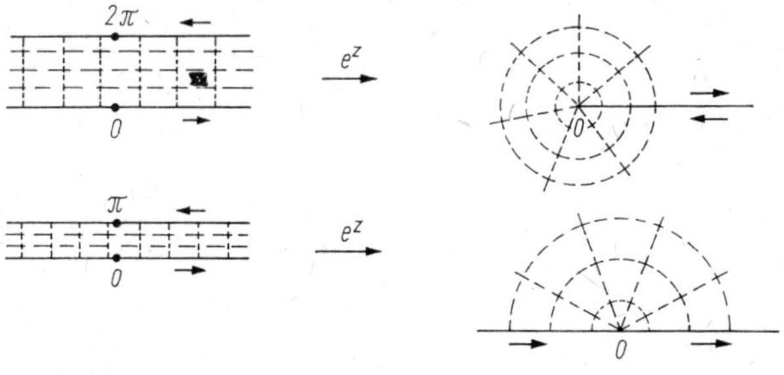

Fig. 15.53

15.9.2. The Functions sin z, cos z, tan z, and cot z

We extend the expansions in power series for sin x and cos x, as described in Subsec. 5.4.5., from R_1 to the complex plane. Thus, for $z\in C_1$ let

$$\sin z=z-\frac{z^3}{3!}+\frac{z^5}{5!}-\frac{z^7}{7!}+\ldots \quad \text{and} \quad \cos z=1-\frac{z^2}{2!}+\frac{z^4}{4!}-\frac{z^6}{6!}+\ldots,$$

as well as $\tan z=\dfrac{\sin z}{\cos z}$ and $\cot z=\dfrac{\cos z}{\sin z}$. The radius of convergence of both series is ∞ (cf. Subsec. 5.4.1.). Then it follows from Theorem 15.1.3 that sin z and cos z are holomorphic functions on C_1.

Theorem. *All the propositions of Theorem* 5.2.1 *remain true if one substitutes* $z\in C_1$ *for* $x\in R_1$ *and* $w\in C_1$ *for* $y\in R_1$.

Remark. As was indicated in Subsec. 15.9.1., the theorem is proved by using Theorem 15.3.3 and the corresponding propositions for real z and w. Further, for all $z\in C_1$ one has

$$\sin z=\frac{1}{2i}(e^{iz}-e^{-iz}), \quad \cos z=\frac{1}{2}(e^{iz}+e^{-iz}).$$

Thus, for instance, $\sin i=\dfrac{1}{2i}(e^{-1}-e)$.

Mapping properties (Fig. 15.54): From $\tan z=-i\dfrac{e^{iz}-e^{-iz}}{e^{iz}+e^{-iz}}$ and the mapping properties of e^z it follows that tan z maps the strip region $\left\{z \mid |\operatorname{Re} z|<\dfrac{\pi}{4}\right\}$ conformally onto the unit disk.

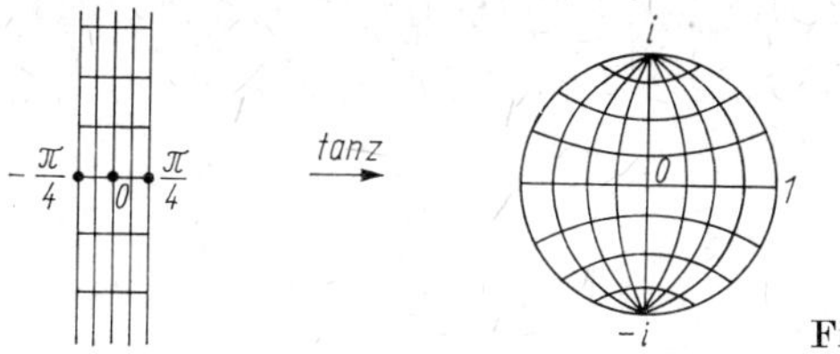

Fig. 15.54

15.9.3. Partial Fraction Decomposition for cot z

From the power series expansions for sin z and cos z it follows that $\cot z = \frac{\cos z}{\sin z}$ has the Laurent expansion

$$\cot z = \frac{1 - \frac{z^2}{2} + \dots}{z\left(1 - \frac{z^2}{6} + \dots\right)} = \frac{1}{z} - \frac{z}{3} + \dots \tag{1}$$

at the point 0. Since cot z has the period π, the term $\frac{1}{z - k\pi}$ is the main part of the Laurent series at the point $k\pi$. Here k is an integer. Apart from these poles of order one, cot z has no other poles.

Theorem 1. $\cot z = \frac{1}{z} + \sum_{k=1}^{\infty} \left[\frac{1}{z - k\pi} + \frac{1}{z + k\pi}\right]$, *and this series converges uniformly in the neighbourhood of every point z that is not a pole.*

Remark 1. This is an infinite analogue to the partial fraction decomposition of Subsec. 7.1.2. Thus one simply collects all the main parts, where convergence is achieved by combining those main parts which contain $k\pi$ and $-k\pi$.

Theorem 2. $$\sum_{k=1}^{\infty} \frac{1}{k^2} = \frac{\pi^2}{6}. \tag{2}$$

Remark 2. If $\sum_{k=1}^{\infty} \left[\frac{1}{z - k\pi} + \frac{1}{z + k\pi}\right] = \sum_{k=1}^{\infty} \frac{2z}{z^2 - k^2\pi^2}$ is expanded in a power series at 0, then this gives

$$\cot z = \frac{1}{z} - 2z \sum_{k=1}^{\infty} \frac{1}{\pi^2 k^2} + \dots \tag{3}$$

Comparison with (1) leads immediately to (2). In the same way it is possible to compare the coefficients of z^3, z^5, ... in (1) and (3). Thus one obtains

$$\sum_{k=1}^{\infty} \frac{1}{k^4} = \frac{\pi^4}{90}, \quad \sum_{k=1}^{\infty} \frac{1}{k^6} = \frac{\pi^5}{945}, \dots$$

16. Principles of Hydrodynamics of Plane Flows

16.1. The Fundamental Equations of Hydrodynamics

16.1.1. Preliminary Remarks on Modeling

In the next three subsections we shall take the following line to formulate a mathematical model for the physical problem of plane flows.

1. By heuristic considerations we attempt to find mathematical propositions which characterize the physical problem. In our case this is tantamount to finding formulas that describe when a plane flow is source-free and circulation-free (Subsec. 16.1.2.).

2. An intramathematical formulation of the problem resulting from these investigations: the real formulation of the fundamental equations of hydrodynamics (Subsec. 16.1.3.).

3. Intramathematical reformulation of the problem, looking out for as efficient a mathematical set of tools as possible: complex formulation of the fundamental equations of hydrodynamics (Subsec. 16.1.3.).

4. Axiomatization in the sense of Subsec. 12.1.2. The mathematical theory takes the lead (in our case it is very simple); physics is connected to it by translation and interpretation (Subsec. 16.1.4.).

16.1.2. Source-Free and Circulation-Free Flows

We shall consider flows that can approximately be regarded as plane flows, e.g. flows at a water surface or atmospheric flows in which one dimension can be neglected (Fig. 16.1). The flowing particles move at a certain velocity. Their trajectories are called streamlines. In the x, y-plane they can be characterized by the velocity vector $w(x, y) = (p(x, y), q(x, y))$ (Fig. 16.2). Then $|w| = \sqrt{p^2 + q^2}$ is

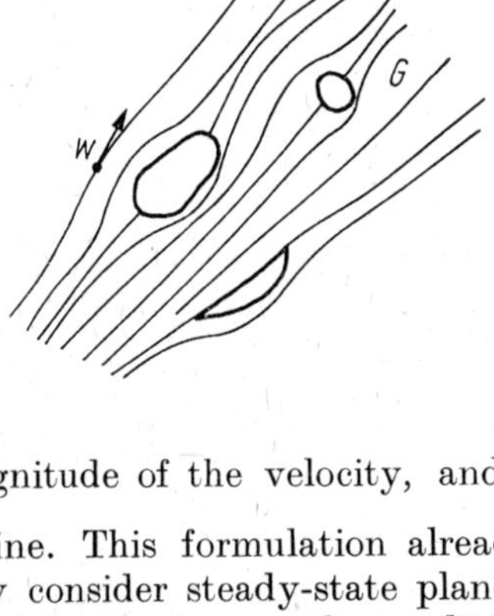

Fig. 16.1

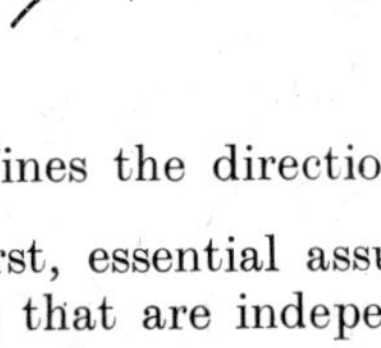

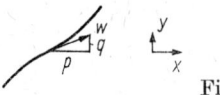

Fig. 16.2

the magnitude of the velocity, and $\frac{\mathrm{d}y}{\mathrm{d}x} = \frac{q(x, y)}{p(x, y)}$ defines the direction of the streamline. This formulation already contains a first, essential assumption: we only consider steady-state plane flows, i.e. flows that are independent of time. Further we impose the conditions that these flows have no sources, sinks, or vortices. Consider a simply connected domain Ω with a smooth boundary $\partial\Omega$ that is fully included in the flow region G (Fig. 16.3). The domain Ω is said to be free of sources and sinks if the amount of liquid (or air etc.) that flows into Ω per unit of time is equal to the amount flowing out of Ω within this unit of time. Let $\mathrm{d}s$ be the element of length of $\partial\Omega$, and let ν be the exterior normal. Then the amount flowing through $\mathrm{d}s$ per unit time is $c \cdot \mathrm{d}s \cdot \cos(\nu, w)\,|w| = c\,\mathrm{d}s\nu \cdot w$, where c is a factor of proportionality (Fig. 16.4). $w_\nu = \nu \cdot w$ is the normal component of w. Hence freedom of sources and sinks means that $\int_{\partial\Omega} w_\nu \mathrm{d}s = 0$ for every domain Ω of this type. One has $\nu = (\cos(\nu, x), \cos(\nu, y))$, and hence $w_\nu = p\cos(\nu, x) + q\cos(\nu, y)$. Then it follows from Theorem 9.3.1/2 that $\int_\Omega \left(\frac{\partial p}{\partial x} + \frac{\partial q}{\partial y}\right) \mathrm{d}x\,\mathrm{d}y = 0$. As this equation must hold for every admissible domain Ω, one finally obtains

$$\frac{\partial p(x, y)}{\partial x} + \frac{\partial q(x, y)}{\partial y} \equiv 0 \quad \text{in the flow region } G. \tag{1}$$

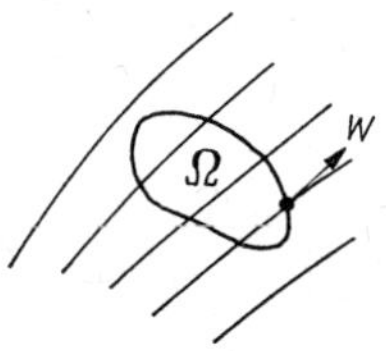

Fig. 16.3

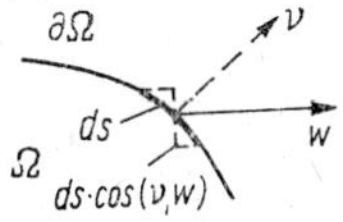

Fig. 16.4

The vector $t = (-\cos(\nu, y), \cos(\nu, x))$ is orthogonal to ν, and hence it is the normalized tangent vector to the curve $\partial\Omega$. Then $w_t = w \cdot t = -p\cos(\nu, y) + q\cos(\nu, x)$ is the tangential component of w. Hence the property of local freedom of circulation means that $\int_{\partial\Omega} w_t \mathrm{d}s = 0$ for every admissible domain Ω. From the above representation of w_t and from Theorem 9.3.1/1 it then follows that $\int_\Omega \left(\frac{\partial p}{\partial y} - \frac{\partial q}{\partial x}\right) \mathrm{d}x\,\mathrm{d}y = 0$. As this equation must hold for every admissible domain Ω, one finally obtains

$$\frac{\partial p(x, y)}{\partial y} - \frac{\partial q(x, y)}{\partial x} \equiv 0 \quad \text{in the flow region } G. \tag{2}$$

16.1.3. Real and Complex Fundamental Equations

Fundamental equations (*real variant*). Let $p(x, y)$ and $q(x, y)$ be real functions that are continuous and continuously differentiable on a domain G in R_2, where they satisfy the conditions

$$\frac{\partial p(x, y)}{\partial x} + \frac{\partial q(x, y)}{\partial y} = 0 \quad \text{and} \quad \frac{\partial p(x, y)}{\partial y} - \frac{\partial q(x, y)}{\partial x} = 0\,. \tag{1}$$

The streamlines of the corresponding source-, sink-, and circulation-free plane flow are the solutions of the differential equation $p(x, y)\,\mathrm{d}y - q(x, y)\,\mathrm{d}x = 0$. The velocity vector of the flow at the point $(x, y) \in G$ is $w(x, y) = (p(x, y), q(x, y))$.

Remark 1. If $p(x, y) \neq 0$, then $\frac{dy}{dx} = \frac{q(x, y)}{p(x, y)}$. According to the existence and uniqueness theorems for ordinary differential equations as stated in Chap. 4, then for every point $(x_0, y_0) \in G$ there exists (at least locally) a unique streamline that passes through this point. From (1) it follows that $p\,dy - q\,dx = 0$ is an exact differential equation as defined in Subsec. 10.1.4.

The stream potential. The above formulation reduces the problem to the solution of ordinary exact differential equations. Here the role played by the conditions (1) is not quite clear. It shall be attempted to find a formulation in which the conditions (1) are taken into consideration in a natural way. Let again G be a domain in R_2, and let P_0 be a point fixed in G (Fig. 16.5). If γ is a continuously differentiable curve in G that connects P_0 with $P \sim (x, y) \in G$, then we set

$$u(x, y) = \int_\gamma \left(p(\xi, \eta) \frac{d\xi}{ds} + q(\xi, \eta) \frac{d\eta}{ds} \right) ds\,, \tag{2}$$

$$v(x, y) = \int_\gamma \left(p(\xi, \eta) \frac{d\eta}{ds} - q(\xi, \eta) \frac{d\xi}{ds} \right) ds\,, \tag{3}$$

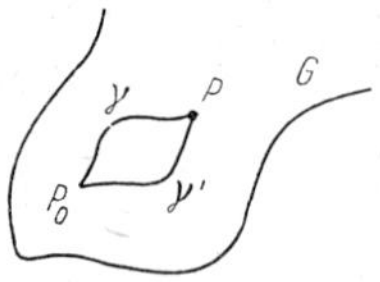

Fig. 16.5

where γ is given by $(\xi(s), \eta(s))$, with the arc length s as a parameter. $u(x, y)$, $v(x, y)$ and $f(z) = f(x + iy) = u(x, y) + iv(x, y)$ are called velocity potential, stream function, and stream potential, respectively.

Theorem 1. *Let G be a simply connected domain in R_2. Then $u(x, y)$ and $v(x, y)$ are path-independent and satisfy the Cauchy-Riemann differential equations on G:*

$$\frac{\partial u}{\partial x} = \frac{\partial v}{\partial y} = p(x, y) \quad \textit{and} \quad \frac{\partial u}{\partial y} = -\frac{\partial v}{\partial x} = q(x, y)\,. \tag{4}$$

$f(z)$ is holomorphic on G.

Remark 2. $u(x, y)$ and $v(x, y)$ are independent of the path γ. Hence, if in Eqs. (2) and (3) one replaces the curve γ by a different curve γ' (see Fig. 16.5), then the values of $u(x, y)$ and $v(x, y)$ are left unchanged. This also justifies the notation used (otherwise in (2) and (3) we should have written, say, u_γ and v_γ instead of u and v, respectively). For the proof of the path-independence, Eqs. (1) are required in a natural way. A problem of the same kind has been dealt with in Subsec. 10.1.4. The fact that $f(z)$ is holomorphic if the conditions (4) are satisfied is a consequence of Theorem 15.1.4/1.

Theorem 2. *With the assumptions of Theorem 1, $w(x, y) = \overline{f'(z)}$ is the velocity vector. The streamlines are determined by $v(x, y) \equiv c$.*

Remark 3. The first proposition follows from

$$\overline{f'(z)} = \frac{\partial u}{\partial x} - i\,\frac{\partial v}{\partial x} = p + iq = w\,,$$

where $w(x, y)$ is now represented as a complex number whose real and imaginary parts are the respective components of the vector $w(x, y)$. The second proposition can be deduced from Theorem 10.1.4, where the function denoted by $M(x, y)$ there coincides with the above function $v(x, y)$ (cf. Remark 1).

Fundamental equations (complex variant): Let G be a simply connected domain in R_2, and let $p(x, y)$ and $q(x, y)$ be the functions considered in the real variant of the fundamental equations. If $v(x, y)$ and $f(z)$ have the meaning indicated above, then the streamlines can be calculated from $v(x, y) \equiv c$, and the velocity vector is given by $w(x, y) = \overline{f'(z)}$.

Remark 4. What is annoying in this version is the restrictive condition that G be simply connected.

16.1.4. The Mathematical Model

Axiom. *If G is an arbitrary domain in R_2, then a source-, sink-, and locally circulation-free flow is described by a function $f(z)$ that is holomorphic on G. The streamlines can be calculated from* Im $f(z) \equiv c$, *and the velocity vector is given by $w(x, y) = \overline{f'(z)}$.*

Remark 1. Hence, if $f(z) = u(x, y) + iv(x, y)$, then $w(x, y) = (p(x, y), q(x, y))$, where $p = \frac{\partial u}{\partial x} = \frac{\partial v}{\partial y}$ and $q = -\frac{\partial v}{\partial x} = \frac{\partial u}{\partial y}$. From this it follows that

$$\frac{\partial p}{\partial x} + \frac{\partial q}{\partial y} = \frac{\partial p}{\partial y} - \frac{\partial q}{\partial x} = 0 \quad \text{in } G .$$

Thus the conditions (16.1.3/1) are satisfied. From $v(x, y) \equiv c$ one obtains $0 = \frac{\partial v}{\partial x}\,dx + \frac{\partial v}{\partial y}\,dy = p\,dy - q\,dx$. Thus we have achieved a full correspondence to the real variant of the fundamental equations of Subsec. 16.1.3. Moreover we have got rid of the awkward constraint of the complex variant of the fundamental equations, namely that G be simply connected.

The model: In the sense of Subsec. 12.1.2., we thus have the following situation (Fig. 16.6): the mathematical theory consists in determining the level curves $c \equiv v(x, y) = \text{Im } f(z)$ of holomorphic functions $f(z)$. Given an actual flow, then there arises the following translational problem: find (or guess) the holomorphic stream potential $f(z)$. The interpretation is then clear: the level curves of Im $f(z)$ are the streamlines, and $w(x, y) = \overline{f'(z)}$ is the velocity.

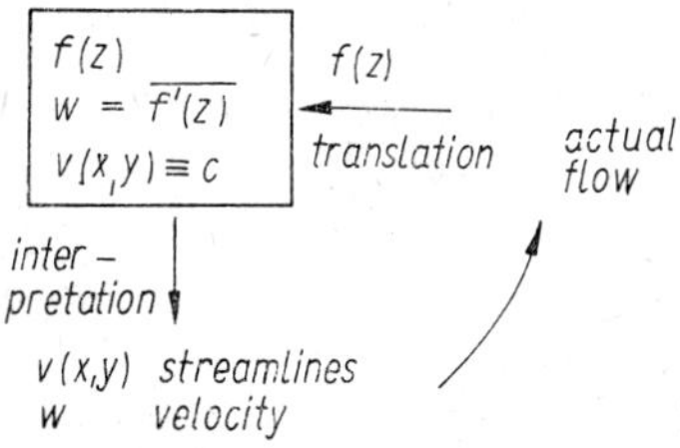

Fig. 16.6

Remark 2. We proceed here as follows. First we investigate simple holomorphic functions, thus providing for a stock of elementary flow patterns. By combining such basic flow patterns we will then arrive at more complicated flow types that give a mathematical description of actual flows which are of physical interest.

16.2. Flow Regimes

16.2.1. Stagnation Point Flow and Multipole Current Flow

Parallel flow (Fig. 16.7): For $a=\alpha+i\beta$ complex, $f(z)=az$ is regarded as a stream potential. One has $\overline{f'(z)}=\bar{a}$ and $\operatorname{Im} f(z)=\beta x+\alpha y$. The flow in question is a parallel flow with the streamlines $\beta x+\alpha y\equiv c$ and the velocity $\overline{f'(z)}=\bar{a}$. The flow region can be chosen to be $G=C_1$.

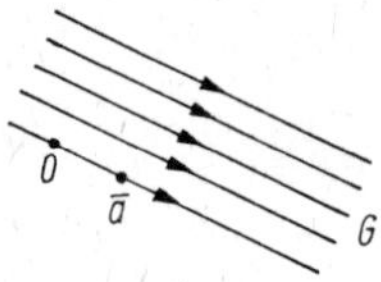

Fig. 16.7

Stagnation point flow (Fig. 16.8): Let the stream potential be $f(z)=z^n$, $n=2, 3, \ldots$ Then in polar co-ordinates one has $f(z)=r^n e^{in\varphi}$ and $\overline{f'(z)}=nr^{n-1}e^{-(n-1)i\varphi}$. The streamlines are given by $r^n \sin n\varphi\equiv c$. One obtains a stagnation point flow with $G=C_1\backslash\{0\}$ as its flow region.

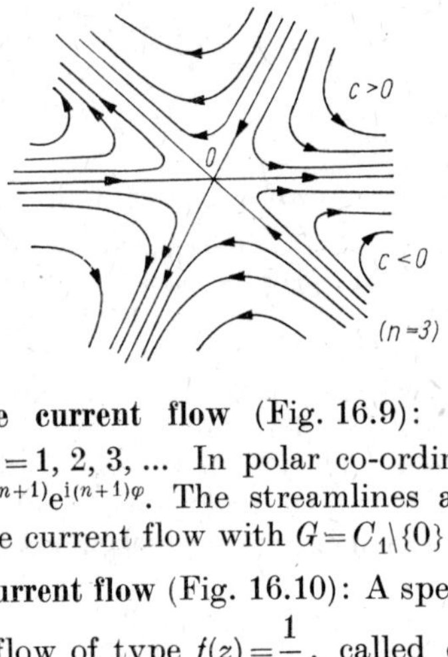

Fig. 16.8

Multipole current flow (Fig. 16.9): Let the stream potential be $f(z)=z^{-n}$, where $n=1, 2, 3, \ldots$ In polar co-ordinates one has $f(z)=r^{-n}e^{-in\varphi}$, and $\overline{f'(z)}=$ $=-nr^{-(n+1)}e^{i(n+1)\varphi}$. The streamlines are given by $r^n=c \sin n\varphi$. One obtains a multipole current flow with $G=C_1\backslash\{0\}$ as its flow region.

Dipole current flow (Fig. 16.10): A special case of interest is that of a multipole current flow of type $f(z)=\dfrac{1}{z}$, called dipole current flow. The streamlines are given by $r=c \sin \varphi$. One obtains $x^2+y^2=r^2=cr \sin \varphi=cy$. Hence the streamlines are a parabolic pencil of circles (cf. Subsec. 15.8.5.).

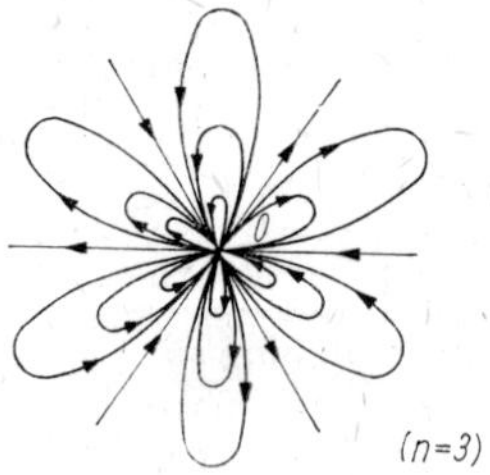

Fig. 16.9

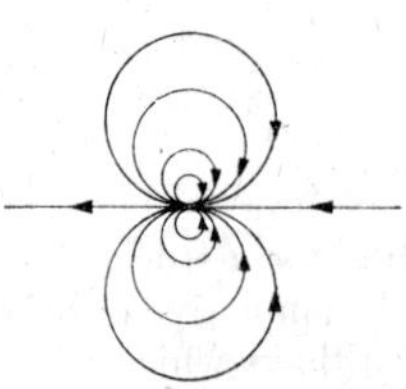
Fig. 16.10

16.2.2. Swirl Flow Regimes

Sources and sinks: Let the stream potential be $f(z)=\ln z$. In polar co-ordinates this means $f(z)=\ln r+i\varphi$ and $\overline{f'(z)}=\frac{1}{r}e^{i\varphi}$. The streamlines are given by $\varphi\equiv c$, i.e., the straight lines passing through the origin. One obtains a source (Fig. 16.11) with $G=C_1\setminus\{0\}$ as its flow region. If one starts from $f(z)=-\ln z$, then $\overline{f'(z)}=-\frac{1}{r}e^{i\varphi}$. One obtains a sink (Fig. 16.12) with $G=C_1\setminus\{0\}$ as its flow region.

Fig. 16.11

Fig. 16.12

Remark. The flow regimes considered have been assumed to be free of sources and sinks. This is only a local statement that applies to a neighbourhood of an arbitrary point $z\in G$ and does not contradict the above examples (0 does not belong to the flow region).

Vortices (Fig. 16.13): Let the stream potential be $f(z)=i\,a\ln z$, where a is a real number. In polar co-ordinates one has $f(z)=i\,a\ln r-a\varphi$ and $\overline{f'(z)}=-\frac{ia}{r}e^{i\varphi}$. The streamlines are given by $r\equiv c$, i.e., they are circles with centre 0.

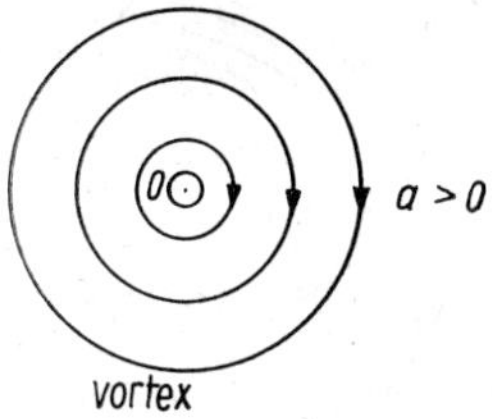

Fig. 16.13

Swirls (Fig. 16.14): It is possible to combine sources and sinks with vortices. Here the actual combination corresponds to the combination of the stream potential. Let the stream potential be $f(z)=(\alpha+i\beta)\ln z$, where α and β are real numbers, $\beta\neq 0$. Then the streamlines are given by $\beta\ln r+\alpha\varphi\equiv c$, which gives $\ln r=\frac{c}{\beta}-\frac{\alpha}{\beta}\varphi$, $\varphi\in R_1$, that means logarithmic spirals.

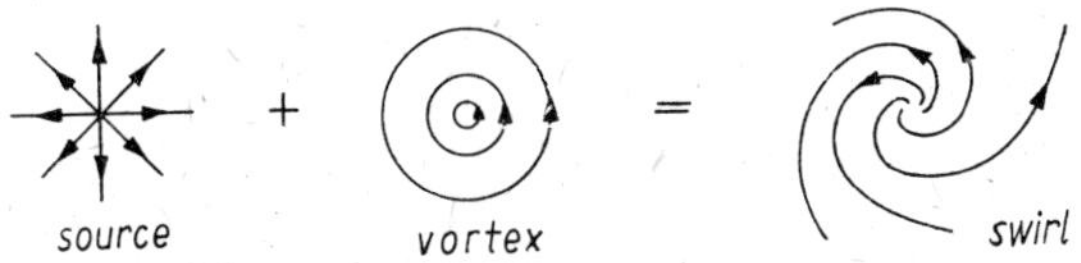

Fig. 16.14

16.2.3. Profile Flow Regimes

The hitherto considered elementary flow regimes exhibit only points as singularities. The question as to how does a flow respond to an obstacle, or what is the pattern of flow around a given profile, is of much greater interest (from the physical point of view). This is closely connected with the following question: what happens when additional sources or sinks are put into a given flow? The latter problem can be treated by the method indicated at the end of Subsec. 16.2.2., that is by combination of suitable elementary flow regimes and of the corresponding stream potentials.

Parallel flow with a source (Fig. 16.15): Consider a parallel flow (in C_1) which is disturbed by a source (at the point 0), for instance the discharge of chemical sewage into our Saale river, which might become too clear otherwise. Intuitively, one should expect something like the situation shown in the figure. As the parallel flow and the source are known to have the stream potentials z and $\ln z$, respectively, it suggests itself to set the combined stream potential to be $f(z) = z + \ln z$. If x, y are Cartesian co-ordinates and r, φ are polar co-ordinates, then for the streamlines one obtains $\operatorname{Im}(z + \ln z) = y + \varphi \equiv c$. Let $-\pi < \varphi \leqq \pi$. One obtains the pattern of Fig. 16.15. The interesting limiting curve is $y + \varphi = \pi$, or $y + \varphi = -\pi$. It joins the x-axis at -1 at right angles. One has $f'(z) = 1 + \dfrac{1}{z}$, and hence $f'(-1) = 0$, which means that -1 is a stagnation point. But the above pattern can also be regarded as a profile flow, namely the flow around the profile described by the curve $y + \varphi = \pi$, or $y + \varphi = -\pi$. The stream potential is $f(z) = z + \ln z$, and it is only the external part of the pattern that is of interest in this case.

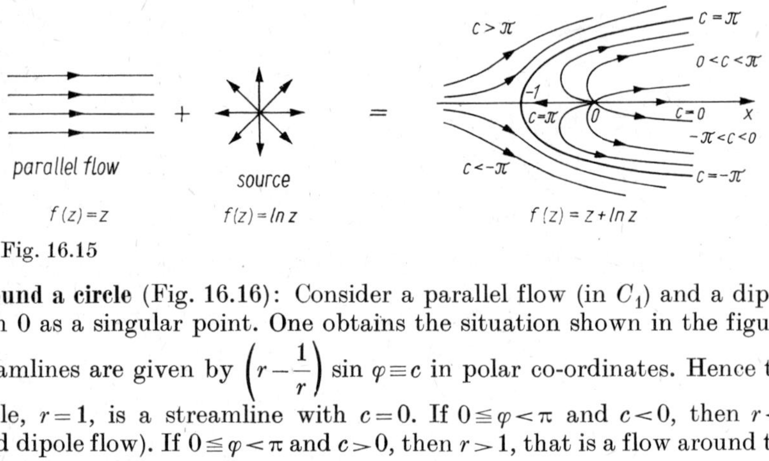

Fig. 16.15

Flow around a circle (Fig. 16.16): Consider a parallel flow (in C_1) and a dipole flow with 0 as a singular point. One obtains the situation shown in the figure. The streamlines are given by $\left(r - \dfrac{1}{r}\right) \sin \varphi \equiv c$ in polar co-ordinates. Hence the unit circle, $r = 1$, is a streamline with $c = 0$. If $0 \leqq \varphi < \pi$ and $c < 0$, then $r < 1$ (distorted dipole flow). If $0 \leqq \varphi < \pi$ and $c > 0$, then $r > 1$, that is a flow around the unit circle. The result for $-\pi < \varphi < 0$ is analogous. From $f'(z) = 1 - \dfrac{1}{z^2} = 0$ one obtains $z = \pm 1$ as stagnation points.

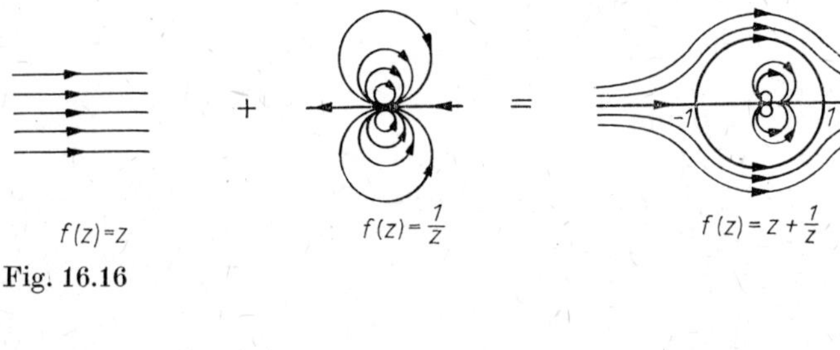

Fig. 16.16

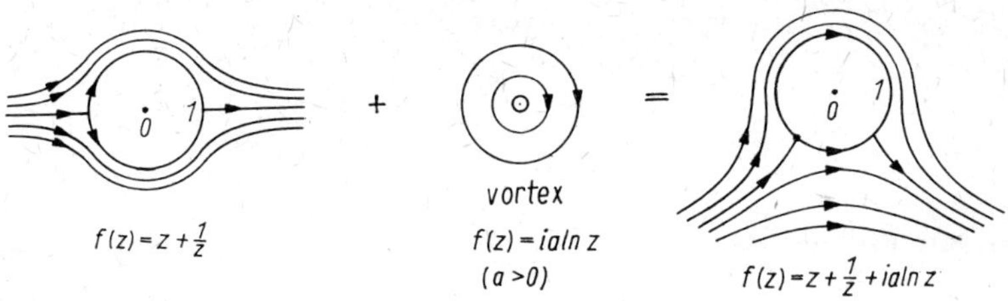

Fig. 16.17

Stagnation point flows (Fig. 16.17): We consider the flow around the unit circle mentioned in the last example (where now the streamlines inside the circle are of no interest), and combine it with a vortex (with 0 as a singular point). The stream potential is $f(z)=z+\frac{1}{z}+\mathrm{i}a\ln z$, where $a>0$. In polar coordinates, the streamlines are given by $\left(r-\frac{1}{r}\right)\sin\varphi+a\ln r\equiv c$. In particular, the unit circle $r=1$ is a streamline. Thus one again obtains a flow around the unit circle. The stagnation points resulting from $f'(z)=0$ are of interest. From $f'(z)=1-\frac{1}{z^2}+\frac{\mathrm{i}a}{z}=0$ one obtains

$$z=-\frac{\mathrm{i}a}{z}\pm\sqrt{-\frac{a^2}{4}+1}\,.$$

Here three cases can be distinguished, where it is clearly seen that the effect of the vortex increases with increasing a (Fig. 16.18).

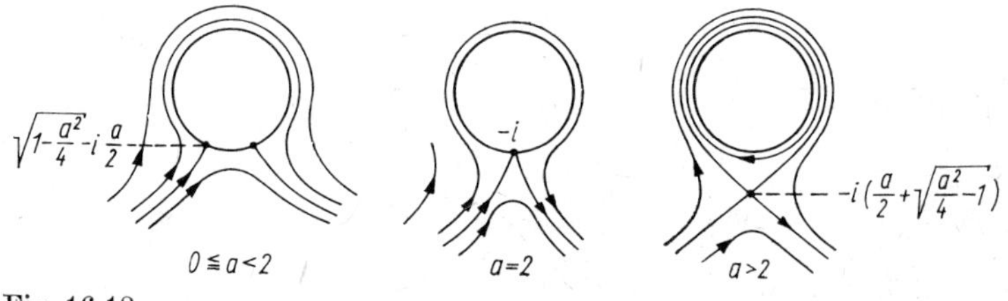

Fig. 16.18

16.2.4. Conformal Mappings in Hydrodynamics, Žukovskij Profiles

Conformal Mapping: Let $f(z)$ be the stream potential in a flow region G (Fig. 16.19). The streamlines are given by $\operatorname{Im} f(z)\equiv c$, the stagnation points can be obtained from $f'(z)=0$. If $w=w(z)$ is a conformal mapping of G onto G', then we consider the transformed stream potential $g(w)=f(z(w))$. From $\operatorname{Im} g(w)=\operatorname{Im} f(z)$ it follows that the images of streamlines are again streamlines. From $g'(w)=$ $=f'(z(w))\cdot z'(w)$ and $z'(w)\neq 0$ it follows that the images of stagnation points are also stagnation points. In other words, conformal mappings transform the flow pattern as a whole. It is clear that on this basis it is possible to construct a great number of new flow patterns from the examples treated so far.

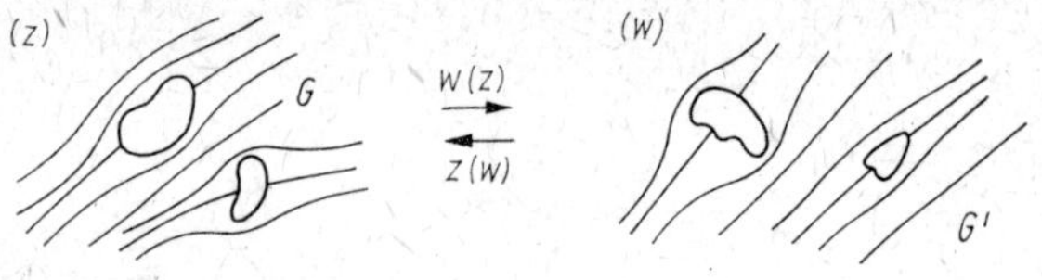

Fig. 16.19

Žukovskij profiles: Our aim is to describe the flow around lifting surfaces (e.g. of airplanes). For that purpose we use the principle of conformal mapping as described above. Before doing so, let us investigate some conformal mappings (Fig. 16.20). The linear mapping $v = L(z) = \dfrac{z-1}{z+1}$ maps the exterior of the

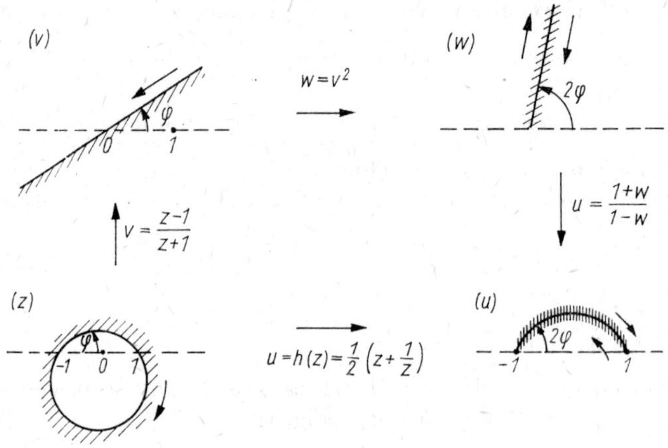

Fig. 16.20

circle in the z-plane, as indicated in the figure, onto a half-plane, which in its turn is mapped onto a slit plane by $w = v^2$. Finally, the linear mapping $u = \dfrac{1+w}{1-w}$ yields an image which is a plane slit by a circular arc. The combined mapping is $u = \dfrac{1}{2}\left(z + \dfrac{1}{z}\right)$. We now consider a flow around a circle K as described in Subsec. 16.2.3., where K touches the broken circle at the point 1 (Fig. 16.21). The image of K obtained by application of $u = \dfrac{1}{2}\left(z + \dfrac{1}{z}\right)$ is drop-shaped. Here the angle φ and the circle K can be varied. One then obtains slender and thick

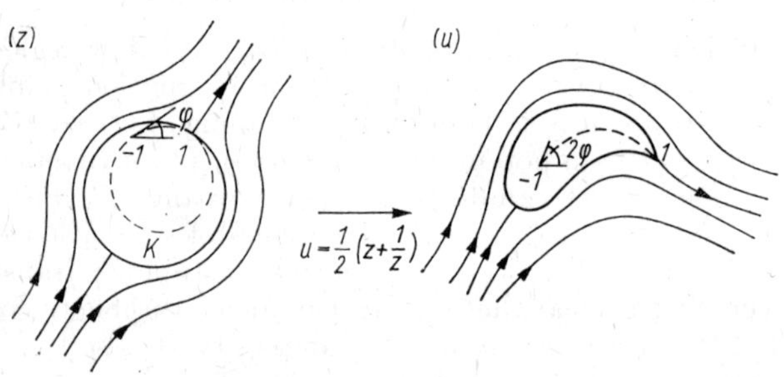

Fig. 16.21

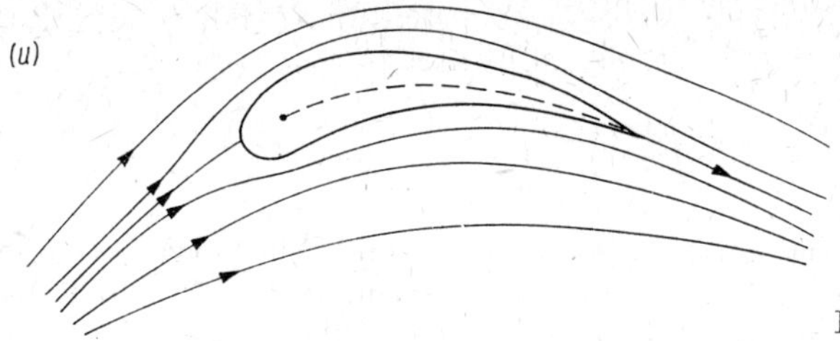

Fig. 16.22

profiles, as well as prolate and greatly curved ones. According to the principle of conformal mapping, $u=\frac{1}{2}\left(z+\frac{1}{z}\right)$ maps streamlines onto streamlines. Thus one obtains patterns of flow around lifting surfaces (Fig. 16.22). As the stream potential $f(z)$ is explicitly known for the flow around a circle $\left(\text{a suitable linear transformation of } z+\frac{1}{z}\right)$, the stream potential $g(u)=f(z)$, with $u=\frac{1}{2}\left(z+\frac{1}{z}\right)$, is also explicitly known for the profile of the lifting surface. For lifting surfaces, the question for the lift $A=X+\mathrm{i}Y$ is of special interest. It obeys Blasius' formula

$$X-\mathrm{i}Y=c\mathrm{i}\int_{\mathfrak{C}} g'^2(u)\,\mathrm{d}u\,,$$

where $\mathfrak{C}$ is the curve flown around and c is a material constant.

17. Elements of Geometry

17.1. Geometry of Space Curves in R_3

17.1.1. The Moving Trihedral

Curves in R_n have been dealt with in Subsec. 8.1.5. Now let $n=3$. We consider the space curves

$$x(u)=(x_1(u),\,x_2(u),\,x_3(u))\,,$$

where u is a parameter, $u\in[a,b]$, $-\infty<a<b<\infty$.

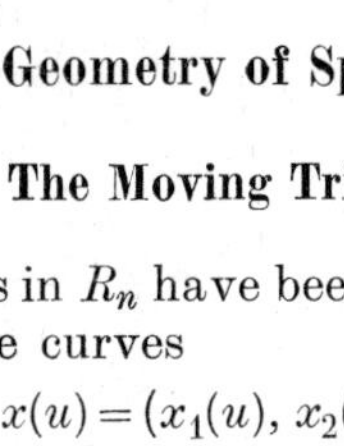

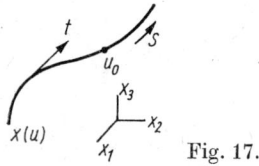

Fig. 17.1

For functions $x_k(u)$ that are continuously differentiable, we had determined the arc length of $x(u)$ in Subsec. 9.2.3. We found that (Fig. 17.1)

$$s = \int_{u_0}^{u} \sqrt{\left(\frac{dx_1}{dv}\right)^2 + \left(\frac{dx_2}{dv}\right)^2 + \left(\frac{dx_3}{dv}\right)^2}\, dv .$$

Apart from the initial point and the orientation of the curve, s is unique. Hence s is a natural parameter that we shall use from now on. We shall denote derivatives with respect to s by a prime "′". In particular one has

$$x_k'(s) = \frac{dx_k}{du} \left(\left(\frac{dx_1}{du}\right)^2 + \left(\frac{dx_2}{du}\right)^2 + \left(\frac{dx_3}{du}\right)^2\right)^{-\frac{1}{2}}, \quad k = 1, 2, 3 .$$

Thus, according to Subsec. 8.1.5., for the tangent $t = t(s)$ to the curve $x(s)$ one obtains

$$t = (x_1'(s), x_2'(s), x_3'(s)) \quad \text{with} \quad |t| = 1 .$$

If $t'(s) \neq 0$, then

$$n(s) = \frac{t'(s)}{|t'(s)|}$$

is the normal, and

$$b(s) = t(s) \times n(s)$$

is the binormal. From $t \cdot t' = \frac{1}{2}(t \cdot t)' = \frac{1}{2}(1)' = 0$ it follows that n is orthogonal to t. b is orthogonal to t and n, and further $|n| = |t| = 1$. Combining the three vectors, we have an orthonormal right-handed trihedral $(t, n\, b,)$, called the moving trihedral (of the curve) (Fig. 17.2). Here we are using well-known notations of analytical geometry: if $a = (a_1, a_2, a_3)$ and $b = (b_1, b_2, b_3)$ are vectors in R_3, then $a \cdot b = a_1b_1 + a_2b_2 + a_3b_3$ denotes the scalar product, and $a \times b = (a_2b_3 - a_3b_2,\ a_3b_1 - a_1b_3,\ a_1b_2 - a_2b_1)$ denotes the vector product, which is orthogonal to both a and b. (Fig. 17.3). We recall the mnemonic rule that the components of $a \times b$ are the subdeterminants of $\begin{vmatrix} * & * & * \\ a_1 & a_2 & a_3 \\ b_1 & b_2 & b_3 \end{vmatrix}$ (including the signs).

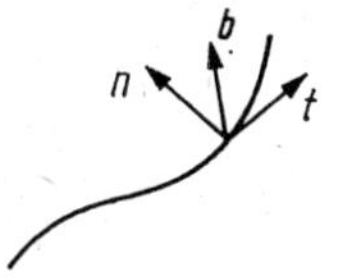

Fig. 17.2

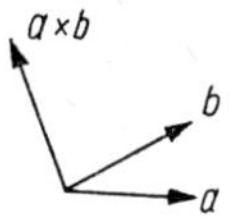

Fig. 17.3

17.1.2. Frenet's Formulae

Theorem. *Let $x(s)$ be a space curve in R_3 that is three times continuously differentiable, with s as a parameter, $s_0 < s < s_1$. Further let $x''(s) \neq 0$ for $s_0 < s < s_1$. Then there exist a positive function $\varkappa(s)$ and a real function $\tau(s)$ such that*

$$\begin{aligned} t' &= \quad\ \varkappa n, \\ n' &= -\varkappa t \quad + \tau b, \qquad s_0 < s < s_1 \quad \textit{(Frenet's formulae)} \\ b' &= \quad -\tau n, \end{aligned}$$

Remark 1. As $t(s)$, $n(s)$, $b(s)$ span the R_3, one has, for example, $t'(s) = a_1(s)\,t(s) + a_2(s)\,n(s) + a_3(s)\,b(s)$, with analogous relations for $n'(s)$ and $b'(s)$. From the orthogonality of t, n, b and the above construction it then follows that the coefficients must be as given in the theorem.

Remark 2. If $x''(s) \neq 0$ does not hold for the entire curve, then it is decomposed into curve segments for which either $x''(s) \neq 0$ for all s or $x''(s) \equiv 0$ for all s. If $x''(s) \equiv 0$, then $x(s)$ is a straight line segment.

Remark 3. $\varkappa(s)$ is called the (first) curvature, $\tau(s)$ the torsion (second curvature) of a space curve. From

$$\varkappa(s_2)\,n(s_2) = t'(s_2) = \lim_{s_1 \to s_2} \frac{t(s_2) - t(s_1)}{s_2 - s_1}$$

it follows that $\varkappa(s)$ is a measure of the deviation of the curve $x(s)$ from a straight line (Fig. 17.4). t and n span the "osculating plane" (or plane of curvature). As we shall see in Subsec. 17.1.3., $\tau(s) \equiv 0$ if and only if $x(s)$ extends within a fixed plane, which is then the osculating plane for all s. In the general case, $\tau(s)$ indicates to what extent the curve turns off the osculating plane.

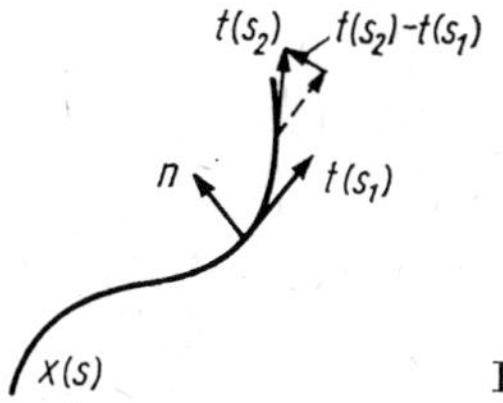

Fig. 17.4

Remark 4. Consider a circle

$$x_1(s) = R\cos\frac{s}{R}, \quad x_2(s) = R\sin\frac{s}{R},$$

$$x_3(s) \equiv 0, \text{ where } s \text{ is the arc length.}$$

It is readily verified that $\varkappa(s) = \dfrac{1}{R}$ and $\tau(s) \equiv 0$.

17.1.3. Plane Curves

Theorem. *A curve $x(s)$ in R_3 which is three times continuously differentiable and for which $x''(s) \neq 0$ extends in a fixed plane if and only if $\tau(s) \equiv 0$. (Here s is the arc length, and $s_0 < s < s_1$).*

Remark 1. Confer Remark 17.1.2/3.

Remark 2 (evolute of a plane curve; Figs. 17.5 and 17.6). For plane curves $x(s)$, Frenet's formulae reduce to

$$t' = \varkappa n \quad \text{and} \quad n' = -\varkappa t\,.$$

In accordance with Remark 17.1.2/4 we call $R(s) = \dfrac{1}{\varkappa(s)}$ the radius of curvature. The evolute $y(s)$ of the curve $x(s)$,

$$y(s) = x(s) + R(s)\,n(s) = x(s) + \frac{n(s)}{\varkappa(s)},$$

is the locus of the so-called centres of curvature. From Frenet's formulae it follows that $y'(s) = x'(s) + \dfrac{1}{\varkappa}n' + \dfrac{\varkappa'}{\varkappa^2}n = -\dfrac{\varkappa'}{\varkappa^2}n$. This implies that $y(s)$ is also the envelope of the normal, provided that $\varkappa'(s) \neq 0$.

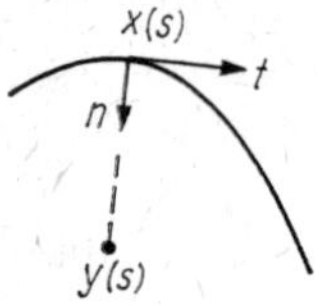

Fig. 17.5

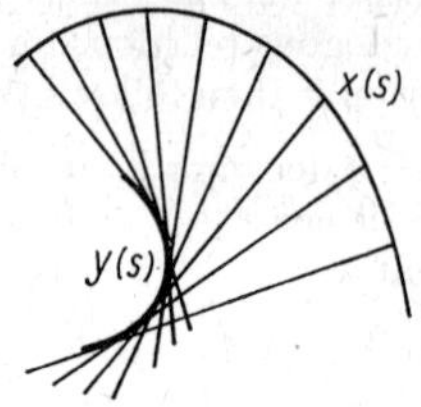

Fig. 17.6

17.1.4. Existence and Uniqueness Theorem

Theorem. *Let $\varkappa(s)$ and $\tau(s)$ be functions continuously differentiable for $s_0 < s < s_1$. Further let $\varkappa(s) > 0$. Then there exists a space curve $x(s)$ in R_3 such that s is the arc length, $\varkappa(s)$ is the curvature, and $\tau(s)$ the torsion of the curve. This space curve $x(s)$ is unique in R_3, except for congruent transformations.*

Remark. If any two space curves $x(s)$ and $\tilde{x}(s)$ have identical curvature and identical torsion, then $\tilde{x}(s)$ can be transformed into $x(s)$ by a translation combined with a rotation in R_3. The intrinsic geometry of the curve is thus well-determined by $\varkappa(s)$ and $\tau(s)$. Therefore $\varkappa(s)$ and $\tau(s)$ are also called the natural equations of the curve.

17.2. Hyperbolic Geometry

17.2.1. Principles of Axiomatic Geometries

About 325 B.C., Euclid, in his "Elementa", made an attempt at an axiomatic formulation of plane geometry, including the concepts of points, straight lines, and circles. Later on these axiomatic geometries were substantially refined, including new aspects (by D. Hilbert about the turn of our century). The aim is to put a few propositions (axioms) at the top and derive all other propositions from them by purely logical deduction, without having recourse to intuition. Even the fundamental concepts are defined axiomatically, i.e. with no content. This reads, for instance, as follows. The elements P of a set M are called points. Certain subsets g of M are called straight lines. Now definitional statements are made on the interrelations between points and straight lines as well as between straight lines and straight lines, etc. We shall give three examples below.

1. If g_1 and g_2 are two straight lines, then either $g_1 = g_2$ or $g_1 \cap g_2 = \emptyset$, or there exists exactly one point P such that $P \in g_1 \cap g_2$.

2. If P_1 and P_2 are two different points, $P_1 \neq P_2$, then there exists exactly one straight line g such that $P_1 \in g$ and $P_2 \in g$.

3. (Euclid's parallel axiom). If g_1 is a straight line and P is a point such that $P \notin g_1$, then there exists exactly one straight line g_2 such that $P \in g_2$ and $g_1 \cap g_2 = \emptyset$.

This is not an exhaustive enumeration, but a few typical examples. The parallel axiom played a decisive role in the history of mathematics. It is much more complicated than Euclid's other axioms. It has been attempted to derive this axiom from the other ones. Not earlier than in the first half of the 19th century, Gauss, J. Bolyai, and Lobačevskij found out that Euclid's parallel axiom was not derivable from the other axioms of Euclid. For it can be replaced by either of the two following parallel axioms 3′ or 3″ (keeping all the other axioms of Euclid), which yields a consistent geometry in either case:

3′. If g_1 is a straight line, and if P is a point not of g_1, then there does not exist any straight line g_2 such that $P \in g_2$ and $g_1 \cap g_2 = \emptyset$ (elliptic geometry).

3″. If g_1 is a straight line, and if P is a point not of g_1, then there exist infinitely many straight lines g_2 such that $P \in g_2$ and $g_1 \cap g_2 = \emptyset$ (hyperbolic geometry).

What is desired are models of such geometries, and the purpose of the considerations in the following subsections will be to describe the so-called conformal model of hyperbolic geometry.

Another problem involves the measurable quantities in a goemetry. In Euclid's plane geometry, these quantities are, for example, lengths, areas, angles etc. These quantities are invariant under congruent transformations of the plane, i.e., under translations and rotations. This simple statement has been generalized substantially by F. Klein in his "Erlanger Programm" in 1872. It defines geometry as the theory of invariants of a transformation group. Hence, what is given is a group of transformations defined on a certain fundamental set M. The only quantities to be regarded as geometrically admissible are those which are left unchanged under every transformation of the group. We shall pursue here this aspect, too.

17.2.2. A Model of Hyperbolic Geometry

Following the notation of Subsec. 17.2.1., we define: $M = \{z \mid |z| < 1\}$ is the unit disk in the complex plane. The points P are the usual points of the unit disk. The straight lines g are the circles orthogonal to the unit circle, i.e., those circles which intersect the unit circle at right angles (Fig. 17.7). The transformation group consists of all conformal mappings of the unit circle onto itself. These are the linear transformations referred to in Theorem 15.8.7. According to Theorem 15.8.3, linear transformations map circles onto circles. Since the unit circle is mapped onto itself and angles are left unchanged under the linear transformations of Theorem 15.8.7, orthogonal circles are transformed into orthogonal circles. In other words, the transformation group maps the above points P again onto points and the above straight lines onto straight lines (as it should also be expected of a reasonable geometry with a transformation group).

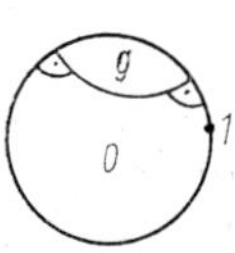

Fig. 17.7

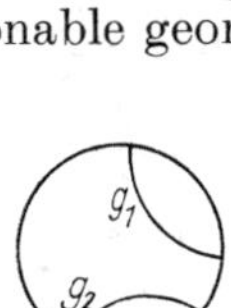

Fig. 17.8

Theorem. (**a**) *If g_1 and g_2 are any two straight lines, then either $g_1 = g_2$ or $g_1 \cap g_2 = \emptyset$, or there exists exactly one point P such that $P \in g_1 \cap g_2$* (Fig. 17.8).

(**b**) *If P_1 and P_2 are any two different points, then there exists exactly one straight line g such that $P_1 \in g$ and $P_2 \in g$.*

(**c**) *If g_1 is a straight line, and if P is a point not of g_1, then there exist infinitely many straight lines g_2 such that $P \in g_2$ and $g_1 \cap g_2 = \emptyset$ (parallels)* (Fig. 17.9).

Remark. These are Axioms 1, 2 and 3″ of Subsec. 17.2.1.

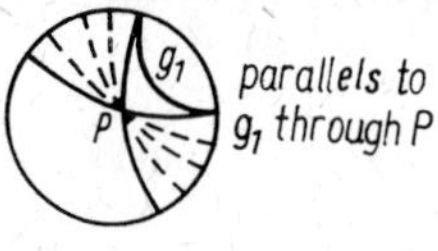

Fig. 17.9

17.2.3. Distances, Angles, and Triangles

Definition. (a) *If z_0 and z_1 are any two different points of M, and if ζ_0 and ζ_1 are the points of intersection of the connecting straight line with the unit circle, then*

$$d(z_0, z_1) = \frac{1}{2} \ln \frac{\zeta_0 - z_1}{\zeta_0 - z_0} \cdot \frac{\zeta_1 - z_0}{\zeta_1 - z_1} \tag{1}$$

is the (hyperbolic) distance of z_0 and z_1 (Fig. 17.10).

(b) *The (hyperbolic) angle between two straight lines g_1 and g_2 is equal to the Euclidean angle between the associated tangents* (Fig. 17.11).

(c) *Triangles are formed by three straight line segments* (Fig. 17.12).

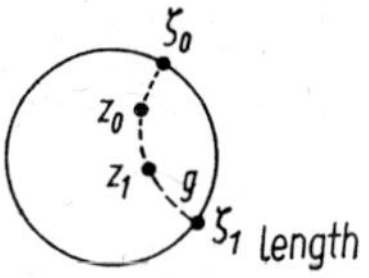

Fig. 17.10

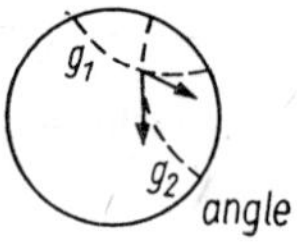

Fig. 17.11

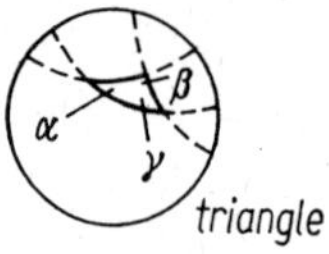

Fig. 17.12

Remark 1. $d(z_0, z_1)$ is $\frac{1}{2}$ ln of a cross ratio in the sense of Theorem 15.8.4/2. That theorem also shows that the cross ratio is invariant under linear transformations (and hence especially under the above transformations). It can now easily be shown that $\frac{\zeta_0 - z_1}{\zeta_0 - z_0} \cdot \frac{\zeta_1 - z_0}{\zeta_1 - z_1} > 1$. Thus $d(z_0, z_1)$ is a positive number that is invariant under the transformation group admitted. We set $d(z_0, z_1) = 0$ for $z_0 = z_1$. Since the mappings of the transformation group are conformal, part (b) of the definition is also meaningful.

Remark 2. It is seen immediately that $d(z_0, z_1) = d(z_1, z_0)$. If we regard z_1 as fixed and let $z_0 \to \zeta_0$, it follows that $d(z_0, z_1) \to \infty$: so the straight lines are of infinite length. Consequently, an "inhabitant" of the unit disk who commands a hyperbolic perception lives in an infinite world.

Theorem. (a) *If z_0, z_1 and z_2 are three points on a straight line, where z_1 lies between z_0 and z_2, then $d(z_0, z_2) = d(z_0, z_1) + d(z_1, z_2)$* (Fig. 17.13).

(b) $\lim_{z \to 0} \frac{d(0, z)}{|z|} = 1$.

(c) *If α, β and γ are the interior angles of a triangle, then* $0 \leqq \alpha + \beta + \gamma < \pi$ (Fig. 17.14).

Fig. 17.13

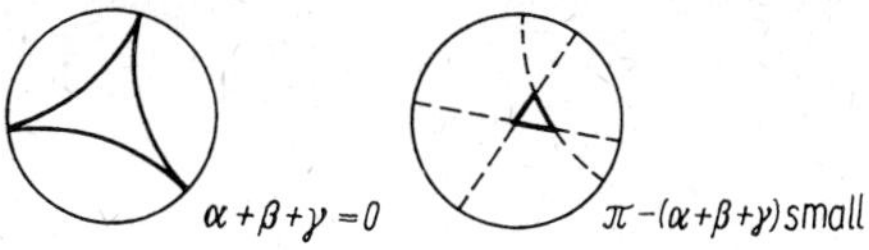

Fig. 17.14

Remark 3. (b) means that at the point 0 the conditions are approximately Euclidean.

Remark 4. Other than in the Euclidean case, the sum of the three angles α, β, and γ of a triangle can assume every value from (and including) 0 to (and excluding) π.

17.2.4. Circles

Definition. *If* $z_0 \in M$ *and* $r > 0$, *then* $K = \{z \mid z \in M,\ d(z_0, z) = r\}$ *is a (hyperbolic) circle.* z_0 *is called the centre of* K.

Remark 1. As M is an infinite world, one obtains a circle for every $z_0 \in M$ and every $r > 0$.

Theorem. *The (hyperbolic) circles with centre* z_0 *are the common Euclidean circles in* M *which belong to the elliptic pencil of circles with the centres* z_0 *and* $\frac{1}{\bar{z}_0}$ (Fig. 17.15).

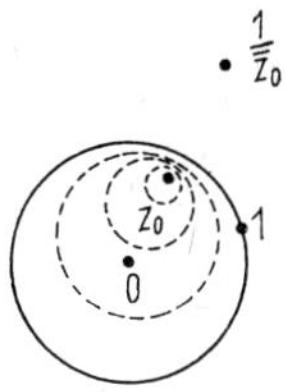

Fig. 17.15

Remark 2. The elliptic pencil of circles (with the centres z_0 and z_1) has been described in Subsec. 15.8.5. z_0 and $\frac{1}{\bar{z}_0}$ are mirror points with respect to the unit circle in the sense of Theorem 15.8.6. Then it is relatively easy to show that, in particular, the unit circle belongs to the elliptic pencil of circles with the centres z_0 and $\frac{1}{\bar{z}_0}$.

17.2.5. Arc Length and Area

The (hyperbolic) length of continuously differentiable curves $\mathfrak{C}$ in M is determined by the same method as in Subsec. 9.2.3.: approximation of the curve by polygons, where a polygon consists of (hyperbolic) straight line segments (i.e., of segments of circles that are orthogonal to the unit circle), the ends of which lie on $\mathfrak{C}$. The supremum of the lengths of these polygons is the length L of the curve $\mathfrak{C}$ (Fig. 17.16).

Theorem 1. *A continuously differentiable curve in* M *with the parametric representation* $z(s)$, $s_0 \leqq s \leqq s_1$, *has the (hyperbolic) length* $L = \int_{s_0}^{s_1} \frac{ds}{1-|z(s)|^2}$, *where* s *is the Euclidean length of arc.*

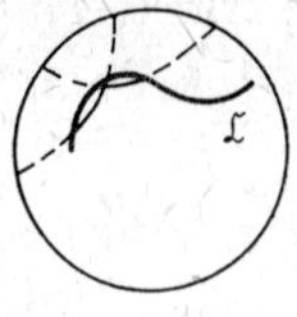

Fig. 17.16

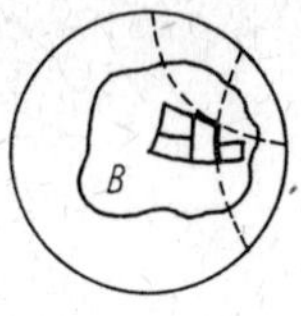

Fig. 17.17

Remark 1. The formula suggests defining $\frac{dxdy}{(1-|z|^2)^2}$ as an infinitesimal element of area in M.

Theorem 2. *If B is a (hyperbolic) bounded domain in M, the boundary of which consists of piecewise smooth curves, then the (hyperbolic) area of this domain* (Fig. 17.17) *is given by*

$$|B| = \int_B \frac{dxdy}{(1-|z|^2)^2}.$$

Remark 2. This formula can also be obtained by first defining the area of (hyperbolic) rectangles (formed by elliptic-hyperbolic pencils of circles), and then exhausting an arbitrary surface by rectangles of this kind.

Remark 3. The area of M is infinite, because

$$\int_{0\leq|z|<1} \frac{dxdy}{(1-|z|^2)^2} = 2\pi \int_0^1 \frac{rdr}{(1-r^2)^2} = \infty.$$

17.2.6. Area of Triangles

Theorem. *If B is a triangle with the angles α, β and γ, then* $|B| = \frac{1}{4}(\pi-\alpha-\beta-\gamma)$ (Fig. 17.18).

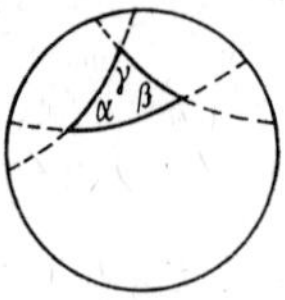

Fig. 17.18

Remark 1. In Theorem 17.2.3 we had already found that $\alpha+\beta+\gamma<\pi$.

Remark 2. Many properties in hyperbolic geometry are analogous to corresponding properties in Euclidean geometry. The theorem shows, however, that there may also be entirely new effects which have no counterpart in Euclidean geometry.

17.3. Geometry of the Hilbert Space

17.3.1. Hilbert Spaces

Definition 1. *Let H be a complex linear space. A scalar product on H assigns a complex number (x_1, x_2) with the following properties to ervery ordered pair $x_1 \in H$ and $x_2 \in H$:*

1. $(\lambda_1 x_1 + \lambda_2 x_2, x_3) = \lambda_1(x_2, x_3) + \lambda_2(x_2, x_3)$.
2. $(x_1, x_2) = \overline{(x_2, x_1)}$.
3. $(x, x) > 0$ *for* $x \neq 0$, *where one sets* $\|x\| = \sqrt{(x, x)}$.

Here x_1, x_2, x_3 *are elements of* H, *and* λ_1, λ_2 *are complex numbers.*

Remark 1. The concept of the complex linear space has been explained in Subsec. 6.1.1.

Theorem. (**a**) $(x, 0) = (0, x) = 0$ *for all* $x \in H$ (*in particular*, $\|0\| = 0$).

(**b**) *If* x_1, x_2, x_3 *are elements of* H, *and if* λ_1, λ_2 *are complex numbers, then*

$$(x_3, \lambda_1 x_1 + \lambda_2 x_2) = \bar{\lambda}_1(x_3, x_1) + \bar{\lambda}_2(x_3, x_2) .$$

(**c**) (*Schwarz's inequality*). *For* $x_1 \in H$ *and* $x_2 \in H$,

$$|(x_1, x_2)| \leqq \|x_1\| \, \|x_2\| .$$

Here $|(x_1, x_2)| = \|x_1\| \, \|x_2\|$ *if and only if* $x_1 = 0$ *or* $x_2 = 0$ *or* $x_1 = \lambda x_2$.

(**d**) $\|x\|$ *is a norm.*

(**e**) *If* $\|x_n - x\| \to 0$ *as* $n \to \infty$, *then* $\|x_n\| \to \|x\|$. *If, moreover* $\|y_n - y\| \to 0$ *as* $n \to \infty$, *then* $(x_n, y_n) \to (x, y)$.

Remark 2. Thus $\|x\|$ satisfies the conditions of a norm in the sense of Definition 6.1.1/1. Therefore a complex linear space with a scalar product is also a normed linear space. From now on we will use the terminology for normed linear spaces, which will always refer to $\|x\|$ in the sense of the above definition.

Definition 2. *A complex linear space* H *with a scalar product is called Hilbert space if it is complete and if there exists a countably infinite, dense subset in* H.

Remark 3. Consequently, in order that a complex linear space H with a scalar product be called a Hilbert space, two conditions must be satisfied:

1. H must be a complex Banach space in the sense of Def. 6.1.1/2.
2. There must be a set $\{x_j\}_{j=1}^{\infty} \subset H$ that is dense in H.

The conditions for a set to be dense in a Banach space have been described in Def. 14.6.3. It is customary not to include the last requirement in the definition of a Hilbert space; accordingly, a Hilbert space is a special complex Banach space whose norm $\|x\|$ is derived from a scalar product in the sense described above. A Banach space B is called separable if there exists a dense set $\{x_j\}_{j=1}^{\infty} \subset B$. In the sense of the usual terminology, we have thus defined by Def. 2 under which conditions a complex linear space H is a separable Hilbert space. However, since Hilbert spaces other than separable ones are of no interest for us here, we directly chose the above formulation.

Remark 4. We restrict ourselves to complex spaces here. All the definitions are, however, also valid for real linear spaces. In particular one may also speak of real (separable) Hilbert spaces.

17.3.2. Examples of Hilbert Spaces

The space C_n: If $x = (x_1, ..., x_n) \in C_n$ and $y = (y_1, ..., y_n) \in C_n$, then one sets

$$(x, y) = \sum_{j=1}^{n} x_j \bar{y}_j .$$

This is a scalar product, and one has $C_n = l_{2,C}^n$ in the sense of Subsec. 6.2.2. It is easily seen that $\{x \mid x = (x_1, ..., x_n), \operatorname{Re} x_k \text{ and } \operatorname{Im} x_k \text{ rational}\}$ is a countable, dense set in C_n. Hence C_n is a Hilbert space.

The space l_2: We set $l_2 = l_{2,C}$ in the sense of subsec. 6.2.2. If $x = (x_1, x_2, ...) \in l_2$ and $y = (y_1, y_2, ...) \in l_2$, then

$$(x, y) = \sum_{j=1}^{\infty} x_j \bar{y}_j$$

is a scalar product. According to Schwarz's inequality one has $|(x, y)| \leq \|x\|_{l_2} \|y\|_{l_2} < \infty$, so that the product formation is meaningful. (In this case Schwarz's inequality can also be derived from Hölder's inequality referred to in Subsec. 6.2.1.) The vectors $x = (x_1, ..., x_N, 0, 0, ...)$, where Re x_k and Im x_k are rational, are dense in l_2. Hence l_2 is a Hilbert space.

The space $L_2(\Omega)$: For Ω being a domain in R_n, we had introduced the real space $L_2(\Omega)$ in Subsec. 14.6.4. In the same way it is possible to define the complex space $L_2(\Omega)$: It is the system of all complex-valued functions $f(x)$ (more precisely: of equivalence classes of functions) such that $|f(x)|$ belongs to the real space $L_2(\Omega)$. As in Subsec. 14.6.4. one has

$$\|f\|_{L_2(\Omega)} = \left(\int_\Omega |f(x)|^2 \, dx\right)^{\frac{1}{2}}. \tag{1}$$

We make the convention that from now on $L_2(\Omega)$ shall always denote the complex space $L_2(\Omega)$. The properties of the real space $L_2(\Omega)$ are also valid for the (complex) space $L_2(\Omega)$. In particular, $L_2(\Omega)$ with the norm (1) is a Banach space. (1) is derived here from the scalar product

$$(f, g)_{L_2(\Omega)} = \int_\Omega f(x) \overline{g(x)} \, dx .$$

Theorem 14.6.2 shows that $(f, g)_{L_2(\Omega)}$ is meaningful. With the use of Theorem 14.6.3 it can be shown that the set

$$\left\{ f \mid f = \sum_{j=1}^{N} r_j \chi_{Q_j}(x),\ r_j \text{ complex rational}, \right. \tag{2}$$
$$\left. Q_j \text{ being a cube with rational corners}, \bar{Q}_j \subset \Omega \right\}$$

is dense in $L_2(\Omega)$. "r_j complex rational" means that Re r_j as well as Im r_j are rational. $\chi_{Q_j}(x)$ is the characteristic function of the cube Q_j. "Rational corners" means that the corners of Q_j have the form $x = (x_1, ..., x_n)$, x_k rational. This set is countable. Hence $L_2(\Omega)$ is a Hilbert space.

Remark. In the same way it is possible to define complex spaces $L_p(\Omega)$ with $1 \leq p \leq \infty$. If $1 \leq p < \infty$, then (2) is a dense set, and one obtains separable complex Banach spaces. Correspondingly, the real spaces $L_p(\Omega)$ of Subsec. 14.6.4. are real separable Banach spaces.

17.3.3. Orthogonal Systems

Definition. *Let H be a Hilbert space.*

(**a**) *Two elements $u \in H$ and $v \in H$ are called orthogonal if $(u, v) = 0$; $u \in H$ is said to be normalized if $\|u\| = 1$.*

(**b**) *A (finite or countably infinite) system $\{u_j\}_j$ is called orthonormal if $(u_j, u_k) = 0$ for $j \neq k$ and $\|u_j\| = 1$. If $x \in H$, then $\alpha_k = (x, u_k)$, $k = 1, 2, ...$, are called the Fourier coefficients (of x with respect to $\{u_j\}$).*

(c) *An orthonormal system* $\{u_j\}_{j=1,2,\dots}$ *is called complete* (*or closed*) *if* $(x, u_j)=0$ *for* $j=1, 2, \dots$ *if and only if* $x=0$.

Remark 1. $H=C_n$ is a special, finite-dimensional Hilbert space. In this case $u=(u_1, \dots, u_n)$ and $v=(v_1, \dots, v_n)$ are orthogonal if and only if the (complex) scalar product $\sum_{j=1}^{n} u_j\bar{v}_j=0$. If $\{u_j\}_{j=1}^{n}$ is orthonormal, then in this case every element $x\in H$ can be represented as

$$x=\sum_{j=1}^{n} \alpha_j u_j .$$

From this it follows that $(x, u_k)=\sum_{j=1}^{n} \alpha_j(u_j, u_k)=\alpha_k$, which means that the α_k's are the Fourier coefficients. A special complete orthonormal system in C_n is $u_j=(0, \dots, 0, \underset{j}{1}, 0, \dots, 0)$, $j=1, \dots, n$. In other words, the above concepts represent the beginning of an analytical geometry in C_n. The Hilbert spaces to be considered here will in general be infinite-dimensional, i.e., different from C_n. Thus the considerations to follow can be regarded as an analytical geometry in the infinite-dimensional (complex) Hilbert space.

Remark 2. If $x=0$, then $(x, u_j)=0$, i.e., in one direction the condition in part (c) of the above definition is always satisfied. This raises the question as to whether there exist complete orthonormal systems in an arbitrary Hilbert space. In the case $H=C_n$ this is clearly shown by Remark 1.

Theorem. *Let* $\{u_j\}$ *be an orthonormal system in the Hilbert space* H. *Further let* $\alpha_k=(x, u_k)$ *be the Fourier coefficients of* $x\in H$.

(a) *For every* $x\in H$,

$$\sum_j |\alpha_j|^2\leqq\|x\|^2 \quad (\textit{Bessel's inequality}) .$$

Further, $\left\{\sum_{j=1}^{n} \alpha_j u_j\right\}_{n=1,2,\dots}$ *is a Cauchy sequence in* H.

(b) $\sum_{j=1}^{n} \alpha_j u_j$ *converges to* x *if and only if*

$$\sum_j |\alpha_j|^2=\|x\|^2 \quad (\textit{Parseval's equation}) .$$

(c) $\{u_j\}$ *is complete if and only if Parseval's equation holds for every* $x\in H$.

Remark 3. If $\{u_j\}=\{u_j\}_{j=1}^{\infty}$ is an infinite orthonormal system (which is the normal case), then $\sum_j=\sum_{j=1}^{\infty}$ in Bessel's inequality as well as in Parseval's equation. Then the convergence in (b) must be understood in the sense of letting $n\to\infty$. It is clear how the propositions are to be understood in the case of finite systems (and especially in the case of finite-dimensional Hilbert spaces).

Remark 4. In the space l_2, $u^{(j)}=(0, \dots, 0, \underset{j}{1}, 0, \dots)$, $j=1, 2, 3, \dots$, is a complete orthonormal system. If $x=(x_1, x_2, \dots)\in l_2$, then $x_j=(x, u^{(j)})$ are the Fourier coefficients of x with respect to this system.

Remark 5. From (b) and (c) one obtains the following proposition. The orthonormal system $\{u_j\}$ is complete if and only if the (finite) linear combinations of the elements $u_1, u_2, \dots$ form a dense set in H.

17.3.4. Schmidt's Orthogonalization Process

Theorem. *In every Hilbert space there exist complete orthonormal systems.*

Remark. The proof of this theorem is based on Schmidt's orthogonalization process. One starts with a countable, dense set $\{x_j\}_{j=1}^{\infty}$, where it is allowed to assume that $x_j \neq 0$, and sets $v_1 = \frac{x_1}{\|x_1\|}$. If $v'_2 = x_2 - (x_2, v_1)v_1 \neq 0$, then one sets $v_2 = \frac{v'_2}{\|v'_2\|}$, and $v_2 = 0$ otherwise. Then one sets $v'_3 = x_3 - (x_3, v_1)v_1 - (x_3, v_2)v_2$, and $v_3 = \frac{v'_3}{\|v'_3\|}$ for $v'_3 \neq 0$, $v_3 = 0$ otherwise. Iteration yields a system $v_1, v_2, v_3, \dots$ Finally, omitting those v_j's which are zero, one obtains a complete orthonormal system.

17.3.5. Orthogonal Decompositions

The space C_n is spanned by the orthonormal elements $(0, \dots, 0, \underset{j}{1}, 0, \dots, 0)$. Further, C_n can be represented as an orthogonal sum $C_n = C_{n_1} \oplus C_{n_2}$, with $n = n_1 + n_2$. This raises the question as to whether this is feasible in arbitrary Hilbert spaces. Theorem 17.3.4 provides the necessary basis for investigations of this kind.

Definition 1. *A set M in a Hilbert space is called convex if $x \in M$, $y \in M$ and $0 \leq t \leq 1$ implies that $tx + (1-t)y \in M$* (Fig. 17.19).

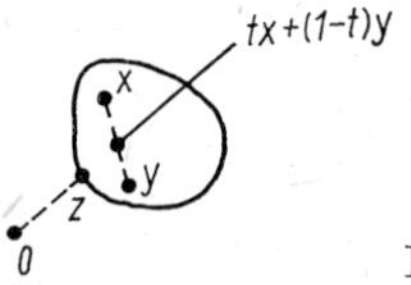

Fig. 17.19

Lemma. *Every nonvoid closed convex set in a Hilbert space contains exactly one element with a minimum norm.*

Remark 1. If M is convex and closed, then the lemma says that there exists a $z \in M$ such that $\|z\| = \inf_{x \in M} \|x\|$. This is plausible but not trivial.

Definition 2. (a) *A linear subset of a Hilbert space is called a subspace.*

(b) *If H_1 is a subspace of a Hilbert space H, then $H_1^{\perp} = \{y \mid y \in H, (x, y) = 0$ for all $x \in H_1\}$ is called the orthogonal complement (to H_1).*

Remark 2. Thus H_1 is a subspace if $\lambda x + \mu y \in H_1$ for every $x \in H_1$ and $y \in H_1$ and arbitrary complex numbers λ and μ. In particular, one thus has $0 \in H_1$. From the linearity of the scalar product it follows immediately that $H_1^{\perp}$ is a subspace, too. By analogy with the case of C_n (or R_n), the question that now arises is whether H_1 and $H_1^{\perp}$ span the entire space H.

Theorem. *If H_1 is a closed subspace of a Hilbert space H, then every element $x \in H$ can be uniquely represented as*

$$x = x_1 + x_1^{\perp}, \quad \textit{where} \quad x_1 \in H_1 \quad \textit{and} \quad x_1^{\perp} \in H_1^{\perp}$$

(Fig. 17.20).

Remark 3. The hyperplane $M = \{y \mid y = x + z, z \in H_1\}$ is a closed convex set (Fig. 17.21). Let $x_1^{\perp}$ be the element with the minimum norm as referred to in the above lemma. Then $x = x_1 + x_1^{\perp}$ yields the desired decomposition. A corresponding conclusion in R_3 is intuitively clear.

Many propositions in the Hilbert space are simply the counterpart of corresponding geometric facts in R_3. A minor difficulty is involved by the fact that in general Hilbert spaces are not finite-dimensional. Thus, for instance, the above theorem is not true if H_1 is not closed. (In a finite-dimensional space, every subspace is closed, so that this difficulty does not occur.)

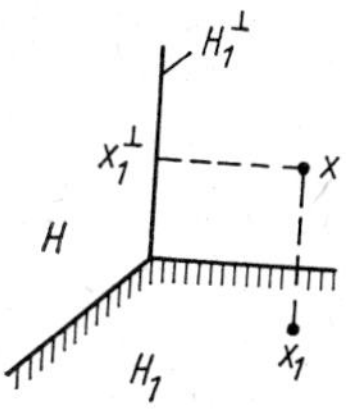

Fig. 17.20

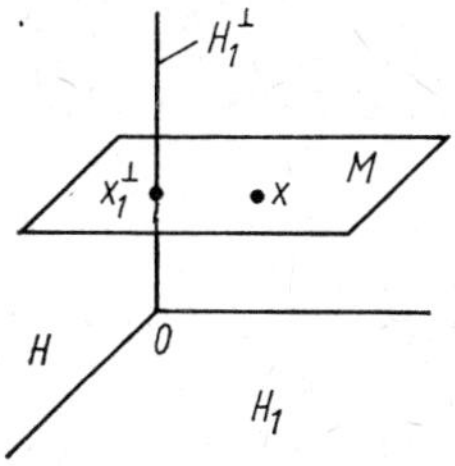

Fig. 17.21

Remark 4. The theorem can also be written as

$$H = H_1 \oplus H_1^\perp \quad \text{(orthogonal sum)}\,,$$

and H is said to be spanned by H_1 and $H_1^\perp$. It is easily seen that H_1 and $H_1^\perp$ are again Hilbert spaces (according to our definition, it is necessary to prove their separability for that purpose). Now the above procedure can be repeated to represent $H_1^\perp$ as an orthogonal sum of two Hilbert spaces H_2 and H_3: $H_1^\perp = H_2 \oplus H_3$. Combination with the above formula gives

$$H = H_1 \oplus H_2 \oplus H_3.$$

Iteration yields

$$H = H_1 \oplus H_2 \oplus \ldots \oplus H_n = \sum_{k=1}^{n} \oplus H_k\,.$$

To this there corresponds a unique decomposition $x = \sum_{k=1}^{n} x_k$, and in particular one obtains $(x_k, x_l) = 0$ for $k \neq l$. Conversely, if $H_1, \ldots, H_n$ are closed, pairwise orthogonal subspaces of a Hilbert space H, then it is clear what is meant by $\sum_{k=1}^{n} \oplus H_k$, even if the space spanned is not the whole space H: the set of all elements $x \in H$ that can be represented by $x = \sum_{k=1}^{n} x_k$, where $x_k \in H_k$. If $H_1, H_2, \ldots$ are closed, pairwise orthogonal subspaces of a Hilbert space H, then let $\sum_{k=1}^{\infty} \oplus H_k$ be the smallest closed subset of H that includes all spaces $\sum_{k=1}^{n} \oplus H_k$, $n = 1, 2, 3, \ldots$

Remark 5. If H_1 is a closed subspace of a Hilbert space H, then $(H_1^\perp)^\perp = H_1$. Besides, $H_1^\perp$ is always a closed subspace (even if H_1 is not closed).

18. Orthogonal Series

18.1. n-Dimensional Trigonometric Functions

18.1.1. Orthonormal Systems

If Ω is a domain in R_n, then, according to Subsec. 17.3.2., $L_2(\Omega)$ is a (complex separable) Hilbert space. Theorem 17.3.4 then shows that there exist complete orthonormal systems $\{f_j(x)\}_{j=1}^{\infty}$ in $L_2(\Omega)$. These functions $f_1(x)$, $f_2(x)$, ... span the space $L_2(\Omega)$, in the sense set out in Subsec. 17.3.5.: they play the role of the co-ordinate axes in R_n or C_n. It is plausible that it will be attempted to construct explicit complete orthonormal systems in $L_2(\Omega)$. For general domains Ω, however, this is hardly possible (at least if well-interpretable, simple systems are desired). The situation is improved decisively if, say, the cube

$$\Omega = Q = \{x \mid x = (x_1, \ldots, x_n) \in R_n, \ |x_j| < \pi \text{ for } j = 1, \ldots, n\} \tag{1}$$

is chosen as a fundamental domain (Fig. 18.1). Let $Z_n = \{m \mid m = (m_1, \ldots, m_n) \in R_n;$ m_j integers$\}$ be the lattice of the points of R_n with integer co-ordinates. As usual, $xm = \sum_{j=1}^{n} x_j m_j$ for $x = (x_1, \ldots, x_n) \in R_n$ and $m = (m_1, \ldots, m_n) \in R_n$.

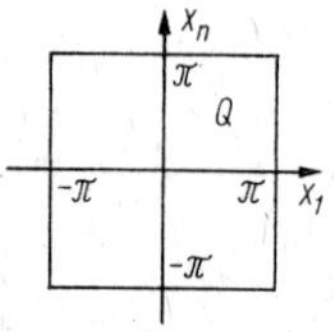

Fig. 18.1

Lemma 1. $\{(2\pi)^{-\frac{n}{2}} e^{ixm}\}_{m \in Z_n}$ *is an orthonormal system in* $L_2(Q)$.

Remark 1. This can be proved by simple integration. A less simple problem raised by this lemma is whether this system is complete.

Remark 2. The system stated in the above lemma can be rewritten as follows. One has

$$e^{ixm} = (\cos m_1x_1 + i \sin m_1x_1)(\cos m_2x_2 + i \sin m_2x_2) \ldots (\cos m_nx_n + i \sin m_nx_n),$$

and conversely, cos ... and sin ... can be represented as linear combinations of $e^{i\cdots}$. As cos and sin are even and odd functions, respectively, it will suffice to restrict our considerations to $N_n = \{m \mid m = (m_1, \ldots, m_n) \in Z_n, \ m_j \geq 0\}$.

Lemma 2. *In the system* $\left\{\varrho_m^{-1} \pi^{-\frac{n}{2}} \begin{Bmatrix}\cos \\ \sin\end{Bmatrix} m_1x_1 \ldots \begin{Bmatrix}\cos \\ \sin\end{Bmatrix} m_nx_n\right\}_{m \in N_n}$, *let* $\varrho_m = (\sqrt{2})^k$, *where* k *is the number of zero co-ordinates,* $m_j = 0$, *in* $m = (m_1, \ldots, m_n) \in N_n$. *Here* $\begin{Bmatrix}\cos \\ \sin\end{Bmatrix} m_1x_1$ *means that either* $\cos m_1x_1$ *or* $\sin m_1x_1$ *can be taken (all combinations being admissible). If the identically vanishing terms in the above system are omitted, one obtains an orthonormal system in* $L_2(Q)$.

Remark 3. The somewhat clumsy formulation is explicable by the different values of

$$\int_{-\pi}^{\pi} \cos^2 lt\,dt = \int_{-\pi}^{\pi} \sin^2 lt\,dt = \pi, \quad \int_{-\pi}^{\pi} \cos 0t\,dt = \int_{-\pi}^{\pi} dt = 2\pi\,,$$

$l=1, 2, \dots$ Anyway, it is easily shown that Lemma 2 is a reformulation of Lemma 1.

Remark 4. In the case $n=1$, i.e., $Q=(-\pi, \pi)$, the systems of Lemma 1 and Lemma 2 are

$$\left\{\frac{1}{\sqrt{2\pi}}\,e^{imx}\right\}_{m=-\infty}^{\infty} \quad \text{and} \quad \left\{\frac{1}{\sqrt{2\pi}},\ \frac{1}{\sqrt{\pi}}\cos mx,\ \frac{1}{\sqrt{\pi}}\sin mx\right\}_{m=1}^{\infty}.$$

These are periodic functions in the interval $(-\pi, \pi)$.

Remark 5. If the domain considered is not the special n-cube Q of (1) but a general n-dimensional interval (n-cuboid) $q=\{x \mid x\in R_n,\ |x_j-a_j|<b_j \text{ for } j=1, \dots, n\}$, then the systems of Lemma 1 and Lemma 2 can easily be converted to this case (Fig. 18.2). Then Lemma 1 reads:

$$\left\{\frac{1}{\sqrt{2b_1\cdot 2b_2 \dots 2b_n}}\prod_{k=1}^{n} e^{im_k(x_k-a_k)\frac{\pi}{b_k}}\right\}_{m\in Z_n} \tag{2}$$

is an orthonormal system in $L_2(q)$. Lemma 2 has an analogous formulation.

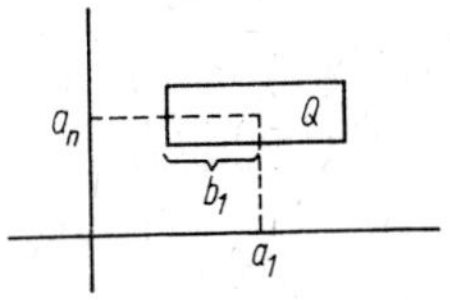

Fig. 18.2.

18.1.2. Fourier Coefficients and Absolute Convergence

Consider the Hilbert space $H=L_2(Q)$ of Subsec. 18.1.1. and the orthonormal system of Lemma 18.1.1/1, then

$$a_m = (f, (2\pi)^{-\frac{n}{2}}e^{imx})_{L_2(Q)} = (2\pi)^{-\frac{n}{2}}\int_Q f(x)\,e^{-imx}dx \tag{1}$$

are the corresponding Fourier coefficients in the sense of Def. 17.3.3 (b). Here $m\in Z_n$ and $f\in L_2(Q)$. We recall that, by Def. 14.6.4/2, $f(x)$ is called a function with bounded support (or finitely nonzero) if $\operatorname{supp} f \subset Q$.

Theorem. *Let Q be the n-cube of* (18.1.1/1), *and let a_m be the Fourier coefficients of* (1).

(a) *If $f\in L_2(Q)$, then*

$$\sum_{m\in Z_n} |a_m|^2 \leqq \|f\|^2_{L_2(Q)} \quad (\textit{Bessel's inequality})\,. \tag{2}$$

(b) *If $f(x)$ is k times continuously differentiable and a function with bounded support on Q, then*

$$|a_m| \leqq |m|^{-k}|b_m| \quad \textit{where} \quad |m|\neq 0 \quad \textit{and} \quad \sum_{m\in Z_n}|b_m|^2<\infty\,. \tag{3}$$

Here $k=1, 2, 3, \dots$ and $|m|^2=\sum_{k=1}^{n} m_k^2$.

(c) *If $f(x)$ is k times continuously differentiable and a function with bounded sup-*

port on Q, and if $k>\frac{n}{2}$, then

$$g(x)=\sum_m \frac{a_m}{(2\pi)^{\frac{n}{2}}} \mathrm{e}^{imx} \tag{4}$$

converges absolutely and uniformly on $\bar{Q}$. In particular, $g(x)$ is continuous on $\bar{Q}$.

Remark. Formula (2) can be directly inferred from Theorem 17.3. 3(a) and Lemma 18.1.1/1. The same theorem also implies that the series (4) converges at least in $L_2(Q)$. The absolute convergence of (4) results from (3):

$$\sum_{\substack{m\in Z_n\\ m\neq 0}} |a_m| \leq \Big[\sum_{m\neq 0} (|a_m|\,|m|^k)^2\Big]^{\frac{1}{2}} \left(\sum_{m\neq 0} \frac{1}{|m|^{2k}}\right)^{\frac{1}{2}} < \infty, \tag{5}$$

because $\sum_{m\neq 0} \frac{1}{|m|^{2k}}$ can be essentially estimated by

$$\sum_{m_l\geq 1} (m_1^2+\ldots+m_n^2)^{-k} \leq \sum_{m_1=1}^{\infty} \frac{1}{m_1^{2\frac{k}{n}}} \cdot \sum_{m_2=1}^{\infty} \frac{1}{m_2^{2\frac{k}{n}}} \cdots < \infty.$$

Thus part (c) can also be reworded: with the assumptions of part (c), the series $\sum_{m\in Z_n} |a_m|$ converges.

18.1.3. Periodic Trigonometric Series in $L_2(Q)$

Theorem. *If Q is the n-cube of* (18.1.1/1), *then the orthonormal systems of Lemma* 18.1.1/1 *and Lemma* 18.1.1/2 *are complete.*

Remark 1. A relatively concise proof of this important theorem has been given in [36].

Remark 2. Thus we have found the connection to the general theory of Subsec. 17.3. Theorem 17.3.3 implies that, for all $f\in L_2(Q)$,

$$f(x)=\sum_{m\in Z_n} \frac{a_m}{(2\pi)^{\frac{n}{2}}} \mathrm{e}^{imx} \quad \text{in } L_2(Q) \tag{1}$$

and $\sum_{m\in Z_n} |a_m|^2=\|f\|^2_{L_2}$ (Parseval's equation). Here a_m are the Fourier coeffients of (18.1.2/1). This is the expansion of functions $f(x)\in L_2(Q)$ in Fourier series. Accordingly, an expansion can be written in sin/cos terms. If the assumptions of Theorem 18.1.2(c) are satisfied, then $g(x)=f(x)$ in (18.1.2/4), and (1) converges not only in $L_2(Q)$ but also absolutely and uniformly.

18.1.4. Semiperiodic Trigonometric Series in $L_2(Q)$

The numbers ϱ_m have the same meaning as in Lemma 18.1.1/2. Further let $\bar{N}_n=\{m \mid m=(m_1, \ldots, m_n)\in Z_n,\ m_j\geq 1\}$, and let

$$q=\{x \mid x=(x_1, \ldots, x_n)\in R_n,\quad 0<x_j<\pi \quad \text{for} \quad j=1, \ldots, n\} \quad \text{(Fig. 18.3)}.$$

Theorem. $\left\{\left(\frac{2}{\pi}\right)^{\frac{n}{2}} \prod_{k=1}^{n} \sin m_k x_k\right\}_{m\in \bar{N}_n}$ *and* $\left\{\left(\frac{2}{\pi}\right)^{\frac{n}{2}} \frac{1}{\varrho_m} \prod_{k=1}^{n} \cos m_k x_k\right\}_{m\in N_n}$ *are complete orthonormal systems in $L_2(q)$.*

Remark 1. For $n=1$, $q=(0, \pi)$, and the systems have the form

$$\left\{\sqrt{\frac{2}{\pi}}\sin mx\right\}_{m=1}^{\infty} \quad \text{and} \quad \left\{\frac{1}{\sqrt{\pi}}, \sqrt{\frac{2}{\pi}}\cos mx\right\}_{m=1}^{\infty} \quad \text{(Fig. 18.4).}$$

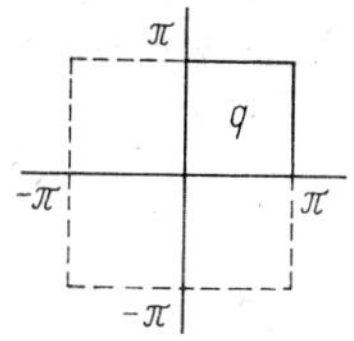

Fig. 18.3

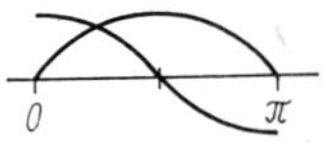

Fig. 18.4

Hence these systems comprise the semiperiodic functions on the interval $(0, \pi)$. The number of sin and cos functions admitted to competition has now increased (as compared with the periodic functions of Subsecs. 18.1.2. and 18.1.3.; note that $(0, \pi)$ is the fundamental interval now), but it will suffice to consider either the sin or the cos functions alone.

Remark 2. The theorem is proved by reducing it to Theorem 18.1.3. If $f \in L_2(q)$, then f is continued to the n-cube Q of (18.1.1/1) as an even or odd function, and the function continued in this way is expanded in the sense of (18.1.3/1). If this expansion is reduced to $x \in q$, one obtains the completeness of the systems referred to in the above theorem. The fact that the systems are orthonormal can be verified directly.

Remark 3. By analogy with Remark 18.1.3/2, it is now possible to expand functions $f \in L_2(q)$ in terms of the systems of the above theorem. If q is the general n-dimensional interval (n-cuboid) of Subsec. 18.1.1., then, by analogy with Remark 18.1.1/5, a corresponding modification must be applied.

18.1.5. Example

We develop $f(x)=x^2$ on the interval $(-\pi, \pi)$ in terms of the (complete orthonormal) second system of Remark 18.1.1/4. The Fourier coefficients are easy to calculate explicitly, and one obtains

$$x^2 = \frac{\pi^2}{3} + \sum_{m=1}^{\infty} 4\frac{(-1)^m}{m^2}\cos mx, \quad x \in (-\pi, \pi)\,. \tag{1}$$

The theory ensures the convergence of the series in $L_2(Q)$, with $Q=(-\pi, \pi)$. It is, however, immediately seen that the series also converges absolutely and uniformly. Then the series must also converge pointwise to x^2. Setting $x=\pi$ and $x=0$, one obtains

$$\sum_{m=1}^{\infty}\frac{1}{m^2}=\frac{\pi^2}{6} \quad \text{and} \quad \sum_{m=1}^{\infty}\frac{(-1)^{m+1}}{m^2}=\frac{\pi^2}{12}\,.$$

The first result is already known to us from Theorem 15.9.3/2. It is possible to obtain other results of this kind: by twice integrating (1) one obtains a representation of x^4, from which a formula for $\sum_{m=1}^{\infty}\frac{1}{m^4}$ can be derived. It is worth while playing a little with such expansions.

18.2. Orthogonal Polynomials

18.2.1. Approximation Theorems

If δ is the small positive number from the Figs. 18.5 and 18.6, then $x_k = \cos y_k$, $k = 1, \ldots, n$ is a mapping of Q onto q, with derivatives up to an arbitrary order. A function $f(x_1, \ldots, x_n)$ which is k times continuously differentiable is

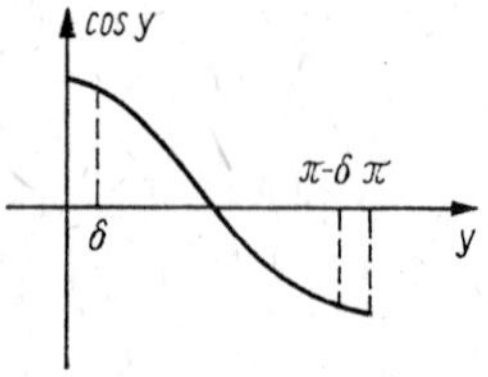

Fig. 18.5

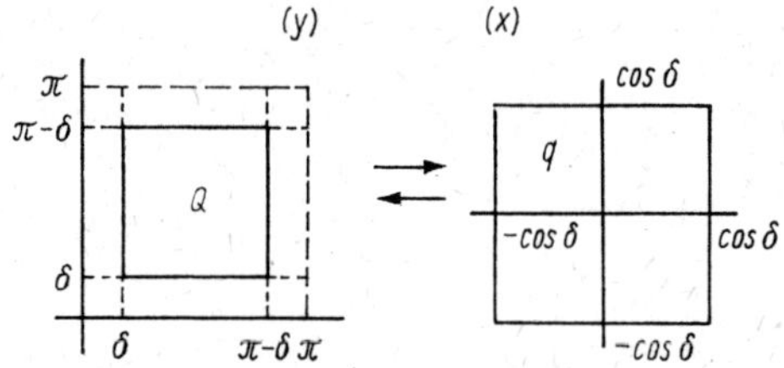

Fig. 18.6

transformed into a function $g(y_1, \ldots, y_n) = f(x_1(y_1), \ldots, x_n(y_n))$ with the same property, and vice versa.

Theorem 1. *If $f(x)$ is a function with bounded support on q which is k times continuously differentiable there, where $k > \frac{n}{2}$, then for every positive number ε there exists a polynomial $P_\varepsilon(x) = P_\varepsilon(x_1, \ldots, x_n)$ such that* $\sup\limits_{x \in q} |f(x) - P_\varepsilon(x)| \leq \varepsilon$.

Remark 1. The theorem is reduced to Theorem 18.1.2(c): by the mapping $x_k = \cos y_k$, $f(x)$ is transformed into a function on Q, and the new function $g(y)$ is approximated in the same way as in Theorem 18.1.2(c). Here, however, one chooses the cos system of Theorem 18.1.4, which gives

$$\left| g(y) - \sum_{|m| \leq M} a_m \prod_{k=1}^{n} \cos m_k y_k \right| \leq \varepsilon$$

with a suitable choice of M and a_m. In view of Moivre's formulas of Subsec. 5.2.4. one has $\cos my = P(\cos y)$ (polynomial in $\cos y$). By insertion of this polynomial and inverse transformation one obtains the above theorem. Thus the theorem is derived from Theorem 18.1.2(c) by a few simple transformations and modifications.

Remark 2. Application of simple linear transformations shows that the theorem remains true if q is an arbitrary n-dimensional rectangular parallelepiped in R_n. Further, by applying Sobolev's mollification method referred to in Subsec. 14.6.4., one observes that the theorem is also true for arbitrary n-dimensional rectangular parallelepipeds q and for arbitrary functions that are continuous on $\bar{q}$. This is Weierstrass' approximation theorem.

Theorem 2. *If Ω is a boundet domain in R_n, and if $1 \leq p < \infty$, then the polynomials in $L_p(\Omega)$ constitute a dense set.*

Remark 3. The theorem follows from Theorem 14.6.4/2 and Theorem 1.

Remark 4. Now a complete orthonormal system in $L_2(\Omega)$ can be constructed using Schmidt's orthogonalization as described in Subsec. 17.3.4.

18.2.2. Legendre's Polynomials

Let us consider the one-dimensional case, with $\Omega=(-1, 1)$.

Theorem. (a) *With a suitable choice of the real constants* c_k, *the Legendre polynomials*

$$L_k(x)=c_k \frac{\mathrm{d}^k}{\mathrm{d}x^k}[(1-x^2)^k] \quad with \quad k=0, 1, 2, \dots \quad and \quad -1\leqq x\leqq 1 \tag{1}$$

constitute a complete orthonormal system in $L_2(\Omega)$, *with* $\Omega=(-1, 1)$.

(b) *If* $\{P_k(x)\}_{k=0}^{\infty}$ *is an orthogonal system of polynomials in* $L_2(\Omega)$, *with* $\Omega=(-1, 1)$, *and if* $P_k(x)$ *has the degree* k, *then* $P_k(x)=d_k L_k(x)$, *where the* $L_k(x)$ *are the Legendre polynomials of* (1) *and* d_k *are suitable numbers.*

Remark 1. The truth of $(L_k, L_l)_{L_2(\Omega)}=0$ for $k\neq l$ is verified by partial integration. The completeness of the system is inferred from Theorem 18.2.1/2.

Remark 2. Thus every function $f\in L_2(\Omega)$, with $\Omega=(-1, 1)$, can be expanded in terms of Legendre polynomials:

$$f(x)=\sum_{k=0}^{\infty} L_k(x) \int_{-1}^{1} f(y)\, L_k(y)\, \mathrm{d}y\,. \tag{2}$$

19. Partial Differential Equations

19.1. Types of Partial Differential Equations and Physical Examples

19.1.1. Types

In Chaps. 4 and 10 we considered ordinary differential equations of first and higher orders,

$$F(x, y')=0, \quad F(x, y', y'', \dots, y^{(n)})=0\,.$$

Qualitative propositions (Chap. 4) and explicit solutions (Chap. 10) were a primary consideration (Fig. 19.1). Now we are looking for functions $u(x)=u(x_1, \dots, x_n)$ (Fig. 19.2) which are solutions of partial differential equations of first and higher orders,

$$F\left(x, \frac{\partial u}{\partial x_1}, \dots, \frac{\partial u}{\partial x_n}\right)=0, \quad F\left(x, \frac{\partial u}{\partial x_j}, \dots, \frac{\partial^2 u}{\partial x_j \partial x_k}, \dots\right)=0\,.$$

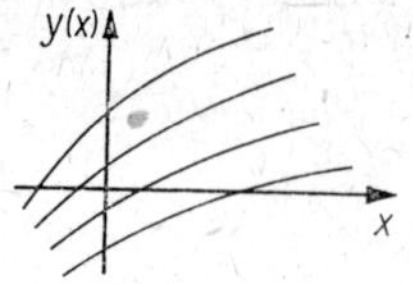

Fig. 19.1

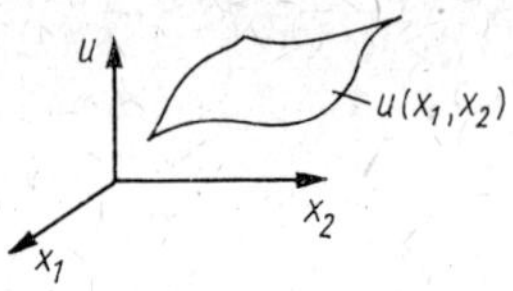

Fig. 19.2

In this case there is no counterpart to Chap. 10: apart from very special partial differential equations it is not possible to state the set of all solutions of a given partial differential equation (of higher order) explicitly. There is, however, an established qualitative theory. Here it turns out that some essential parts of the theory of first-order partial differential equations can be developed largely on the model of the theory of ordinary differential equations. Virtually new effects will however occur in partial differential equations of higher order. Partial differential equations of second order are of special interest here, because many problems of physics can be reduced to such kind of differential equations. In this chapter we shall discuss the three physically interesting basic types of linear partial differential equations of second order: if $x=(x_1, \ldots, x_n) \in R_n$, then $\Delta = \sum_{j=1}^{n} \frac{\partial^2}{\partial x_j^2}$ is called Laplace's differential operator (Laplacian). $(x, t) \in R_{n+1}$ means that $x=(x_1, \ldots, x_n) \in R_n$ and $t \in R_1$. Thus in particular one has $u(x, t) = u(x_1, \ldots, x_n, t)$. The three basic types are

$$\Delta u = f(x) \qquad \textit{Laplace-Poisson equation}\,, \tag{1}$$

$$\Delta u - \frac{\partial^2 u}{\partial t^2} = f(x, t) \qquad \textit{wave equation}\,, \tag{2}$$

$$\Delta u - \frac{\partial u}{\partial t} = f(x, t) \qquad \textit{heat conduction equation}\,. \tag{3}$$

In (1), $u = u(x)$ depends on $x \in R_n$ alone, whereas in (2) and (3) it depends on $(x, t) \in R_{n+1}$. In any case, however, the Laplacian applied to $u(x, t)$ is defined as $\Delta u(x, t) = \sum_{k=1}^{n} \frac{\partial^2 u}{\partial x_k^2}$, that is, the derivatives with respect to t are written separately in (2) and (3). It turns out that a great number of physically interesting linear partial differential equations of second order can (at least qualitatively) be reduced to these three basic types. Their treatment thus gives a representative insight into the theory of linear partial differential equations of second order.

19.1.2. Physical Examples

The vibrating string (Fig. 19.3): Consider a string that extends over the interval $[0, l]$ of the x-axis, being fixed at its end points $x=0$ and $x=l$. If the string is deflected and then allowed to move, it will vibrate. When the string is let loose, it can be given a push to realize an initial velocity in this way. If $u(x, t)$, with $0 \leqq x \leqq l$ and $t \geqq 0$, is the deflection of the string from the normal position at

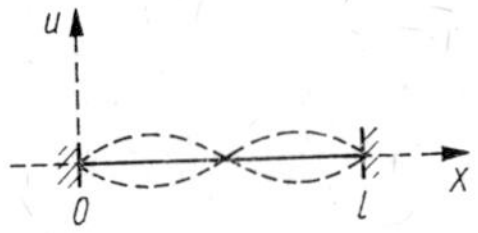

Fig. 19.3

the time t and the position x, then for small deflections the vibration process is described by the wave equation (19.1.1/2), with $f(x, t) \equiv 0$ (for normalized matter constants). Altogether this leads to the following problem: desired a solution of the differential equation

$$\Delta u - \frac{\partial^2 u}{\partial t^2} = 0 \quad \text{for} \quad 0 \leqq x \leqq l \quad \text{and} \quad t \geqq 0, \quad \text{where}$$

$$u(x, 0) = u_0(x) \quad \text{and} \quad \frac{\partial u}{\partial t}(x, 0) = u_1(x) \quad \text{(initial conditions)},$$

$$u(0, t) = u(l, t) = 0 \quad \text{(boundary conditions)}.$$

Here $u_0(x)$ (initial deflection) and $u_1(x)$ (initial velocity) are given functions.

Heat conduction: Consider a bounded body Ω in R_3, with a smooth boundary $\partial\Omega$. Suppose that at the time $t = 0$ the body exhibits the temperature distribution $w(x)$, $x \in \Omega$. Further suppose that on the surface $\partial\Omega$ of the body a time-independent temperature distribution $\varphi(y)$, $y \in \partial\Omega$, is given. Find the temperature distribution $u(x, t)$ at the time t and the position $x \in \Omega$ in the interior of the body. For normalized matter constants this leads to the solution of the problem

$$\Delta u(x, t) - \frac{\partial u(x, t)}{\partial t} = 0 \quad \text{for} \quad x \in \Omega \quad \text{and} \quad t \geqq 0, \quad \text{where}$$

$$u(x, 0) = w(x) \quad \text{for} \quad x \in \Omega \quad \text{(initial condition)},$$

$$u(y, t) = \varphi(y) \quad \text{for} \quad y \in \partial\Omega \quad \text{and} \quad t \geqq 0 \qquad \text{(boundary condition)}.$$

Stationary distributions: In the problem of heat conduction just investigated, consider the case that $t \to \infty$: what will be the temperature reached in the interior of the body Ω after a long time? It will be presumed that the initial temperature distribution will have no influence on this final temperature, whereas the (time-independent) temperature distribution on the surface will influence it. The final temperature distribution $u(x)$ does not depend on t. If (rather daringly) $u(x, t) = u(x)$ is inserted into the above equation, then one has a plausible argument for the following statement: the temperature distribution $u(x)$ in the interior of the body Ω after a long time, with the time-independent temperature distribution $\varphi(y)$ being given on the surface $\partial\Omega$, is the solution of the problem

$$\Delta u(x) \equiv 0 \qquad \text{for} \quad x \in \Omega,$$

$$u(y) = \varphi(y) \quad \text{for} \quad y \in \partial\Omega \qquad \text{(boundary condition)}.$$

Remark. Thus we have three typical physical problems which lead to three typical problems for the three basic types. Physics gives indications as to what kind of mathematical problems might be fruitful.

19.2. The Laplace-Poisson Equation

19.2.1. Fundamental Solutions and Integral Representations

Singularity solutions: In $R_n \backslash \{0\}$, we are looking for twice continuously differentiable functions $u(x) = u(x_1, \ldots, x_n)$ with $\Delta u(x) \equiv 0$ in $R \backslash \{0\}$, which depend on $r = \sqrt{x_1^2 + \ldots + x_n^2}$ alone, that is, $u(x_1, \ldots, x_n) = v(r)$. If one introduces $\Delta u(x) = 0$ into

this setup, one obtains the ordinary differential equation $\frac{d^2v}{dr^2}+\frac{n-1}{r}\frac{dv}{dr}=0$.
Besides the solution $v(r)\equiv c$ (which is of no interest here) one obtains

$$v(r)=\frac{1}{r^{n-2}} \quad \text{for} \quad n>2 \quad \text{and} \quad v(r)=\ln r \quad \text{for} \quad n=2 \tag{1}$$

(apart from a constant factor). Here we always assume that $n\geqq 2$.

Notation: In this subsection we consider connected standard domains Ω in R_n in the sense of Subsec. 9.3.1. (Fig. 19.4). In particular Ω is then bounded, the integral theorems of Sec. 9.3. are applicable and it makes sense to speak of the exterior normal ν with respect to the boundary $\partial\Omega$ of Ω (except for some corners and edges). If on an arbitrary domain Ω (which need not necessarily be a standard domain) the real-valued function $f(x)$ has all its partial derivatives up to and including the order k, and if all of these derivatives are continuous on Ω, then we write $f\in C^k(\Omega)$. If, in addition, all these partial derivatives can be continued by continuity to $\bar{\Omega}$, then we write $f\in C^k(\bar{\Omega})$. Here $k=0, 1, 2, \ldots$, and $f(x)$ is identified with the zeroth derivative. Instead of $C^0(\Omega)$ or $C^0(\bar{\Omega})$ we also write $C(\Omega)$ or $C(\bar{\Omega})$, respectively.

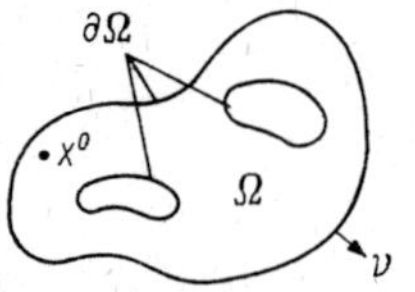

Fig. 19.4

Definition. *Let Ω be a connected standard domain in R_n, and let $x^0\in\Omega$. Then*

$$\gamma_{x^0}(x)=\begin{cases}\dfrac{1}{(n-2)\,|\omega_n|}\,\dfrac{1}{|x-x^0|^{n-2}}+\Phi(x) & \text{for} \quad n>2\,,\\[2mm] -\dfrac{1}{2\pi}\ln|x-x^0|+\Phi(x) & \text{for} \quad n=2\end{cases} \tag{2}$$

is called a fundamental solution if $\Phi\in C^2(\bar{\Omega})$ and $\Delta\Phi(x)\equiv 0$ on Ω.

Remark 1. $|\omega_n|$ is the volume of the unit sphere in R_n as defined in Subsec. 9.2.6.

Remark 2. (1) are solutions of the Laplace equation $\Delta u(x)=0$. Then $cu(x-x^0)$ are solutions, too. Thus one has $\Delta\gamma_{x^0}(x)\equiv 0$ on $\Omega\backslash\{x^0\}$. Hence the fundamental solution is a normalized singularity solution, where $\Phi(x)$ is still artibrary.

Theorem. *Let Ω be a connected standard domain in R_n. Further let $u\in C^2(\bar{\Omega})$ and $\Delta u(x)=f(x)$ for $x\in\Omega$. If $x^0\in\Omega$, then*

$$u(x^0)=\int_{\partial\Omega}\left[\gamma_{x^0}\frac{\partial u}{\partial\nu}-u\,\frac{\partial\gamma_{x^0}}{\partial\nu}\right]ds-\int_{\Omega}\gamma_{x^0}(x)\,f(x)\,dx\,. \tag{3}$$

Remark 3. Here $\gamma_{x^0}(x)$ is an arbitrary fundamental solution as per the above definition. The first integral is a surface integral in the sense of Subsec. 9.2.5.

Remark 4. (3) is proved by applying the second Green formula as stated in Subsec. 9.3.2., where the notation Ω used there is replaced by $\Omega\backslash K_\varepsilon$, and g and f are replaced by u and γ_{x^0}, respectively (Fig. 19.5). Here K_ε is a small ball of radius ε about x^0. Some estimates lead to the desired result for $\varepsilon\downarrow 0$. So the formula is less intricate than it looks like.

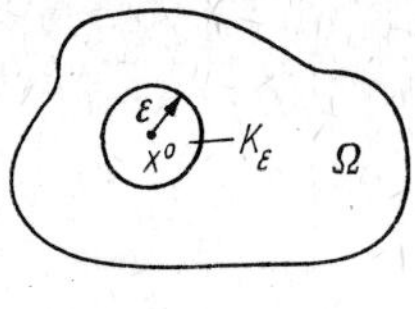

Fig. 19.5

19.2.2. Green Functions

Definition 1. *Let Ω be a connected standard domain in R_n. A real-valued function $u(x)$ is called harmonic on Ω if $u \in C^2(\Omega)$ and $\Delta u(x) \equiv 0$ for $x \in \Omega$.*

Remark 1. There exist functions $u(x)$ which are not continuous at all points of Ω, but nevertheless satisfy the equation $\Delta u(x) = 0$ for all $x \in \Omega$. According to the definition, such functions are not harmonic.

Definition 2. *If Ω is a connected standard domain in R_n, then $g(x^0, x)$ is called Green function (of the first kind) if, for the fixed $x^0 \in \Omega$, the function $g(x^0, x)$ is a fundamental solution in the sense of Def. 19.2.1 and $g(x^0, y) = 0$ for all $y \in \partial\Omega$.*

Remark 2. Every $x^0 \in \Omega$ is admitted to competition, and $g(x^0, x)$ is regarded as a function of x on Ω. A comparison with Def. 19.2.1 shows the problem that arises here: are there any harmonic functions Φ such that $\gamma_{x^0}(x)$ of (19.2.1/2) becomes a Green function?

Remark 3. (19.2.1/3) takes a simpler form if $\gamma_{x^0}(x) = g(x^0, x)$ is a Green function. With the assumptions of Theorem 19.2.1 one has

$$u(x^0) = -\int_{\partial\Omega} u \frac{\partial g(x^0, y)}{\partial \nu} \, ds_y - \int_{\Omega} g(x^0, x) f(x) \, dx . \tag{1}$$

If $u(x) \in C^2(\bar{\Omega})$ is harmonic on Ω, then one obtains

$$u(x^0) = -\int_{\partial\Omega} u(y) \frac{\partial g(x^0, y)}{\partial \nu} \, ds_y . \tag{2}$$

This is an interesting formula we shall make use of later on.

Theorem. *Let $\Omega = K_R = \{x \mid |x| < R\}$.*
(**a**) *If $n > 2$ and $x^0 \in K_R$, then*

$$g(x^0, x) = \frac{1}{(n-2)\,|\,\omega_n|} \left[\frac{1}{|x - x^0|^{n-2}} - \left(\frac{R}{|x^0|}\right)^{n-2} \frac{1}{\left| x - \dfrac{R^2}{|x^0|^2} x^0 \right|^{n-2}} \right] \quad \text{for } x^0 \neq 0 ,$$

$$g(0, x) = \frac{1}{(n-2)\,|\omega_n|} \left[\frac{1}{|x|^{n-2}} - \frac{1}{R^{n-2}} \right] \tag{3}$$

is a Green function on K_R.
(**b**) *If $u(x) \in C^2(\bar{K}_R)$ is harmonic on K_R, if $n \geqq 2$, and if $x^0 \in K_R$, then*

$$u(x^0) = \frac{R^2 - |x^0|^2}{R\,|\omega_n|} \int_{\partial K_R} \frac{u(y)}{|y - x^0|^n} \, ds_y . \tag{4}$$

Remark 4. The point $y^0 = x^0 \dfrac{R^2}{|x^0|^2}$ is the mirror image of $x^0 \in K_R$ with respect to the sphere (cf. Subsec. 15.8.6; Fig. 19.6). In particular, y^0 is exterior to K_R. Then it is clear that

$g(x^0, x)$ is a fundamental solution in the sense of Def. 19.2.1. It remains to show that $g(x^0, x) = 0$ for $|x| = R$. One obtains (4) from (2) by insertion of $g(x^0, x)$. (4) is also valid for $n = 2$, whereas $g(x^0, x)$ has to be modified for $n = 2$.

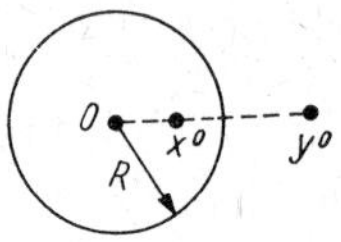

Fig. 19.6

Remark 5. By insertion of $u(x) \equiv 1$ into (4) one obtains, for $|x^0| < R$ and $n = 2, 3, \ldots$,

$$1 = \frac{R^2 - |x^0|^2}{R|\omega_n|} \int\limits_{\partial K_R} \frac{ds_y}{|y - x^0|^n} . \tag{5}$$

Marginal note of philosophical nature: if you should have forgotten the magnitude of 1, then you can evaluate the integral in (5) numerically, which will give the desired value (at least approximately).

Remark 6. For general standard domains it is not possible to state the Green function explicitly. It is however possible to make the following important statement (which is equally important for physics and mathematics): if Ω is a connected standard domain in R_n and $x^0 \in \Omega$, then there exists a unique Green function on Ω with x^0 as singular point.

19.2.3. Properties of Harmonic Functions

If Ω is a bounded domain in R_n, then $u(x) \in C(\bar{\Omega})$ is said to have the mean value property on Ω if, for every point $x^0 \in \Omega$ and every ball $K = \{x \mid |x - x^0| < \varepsilon\}$ with $K \subset \Omega$,

$$u(x^0) = \frac{1}{|\partial K|} \int\limits_{\partial K} u(y) \, ds .$$

Here ∂K is the boundary of K and $|\partial K|$ is its volume (see also Subsec. 9.2.6.).

Theorem. *Let Ω be a bounded domain in R_n, and let $u(x) \in C(\bar{\Omega})$ be harmonic on Ω. Then*

(**a**) *$u(x)$ has the mean value property on Ω.*

(**b**) *$u(x)$ is differentiable on Ω up to arbitrary order.*

(**c**) *(Maximum-minimum principle): $u(x)$ assumes its maximum and its minimum on $\partial\Omega$.*

Remark 1. All the propositions of the theorem are rather simple implications of the representation formula (19.2.2/4).

Remark 2. For $n = 2$ there is a relationship with holomorphic functions; cf. Subsecs. 15.1.4. and 15.3.4.

Remark 3. Part (a) of the theorem characterizes harmonic functions: $u(x) \in C(\bar{\Omega})$ has the mean value property if and only if $u(x)$ is harmonic on Ω.

Remark 4. By analogy with Subsecs. 15.3.4. and 15.3.5., the following statements can be made: (a) A function that is harmonic and bounded on R_n is constant. (b) If Ω is a connected bounded domain in R_n, if $u(x) \in C(\bar{\Omega})$ is harmonic on Ω, and if $u(x^0) = \max_{y \in \bar{\Omega}} u(y)$ for some point $x^0 \in \Omega$, then $u(x)$ is constant on Ω. An analogous proposition holds for the minimum.

19.2.4. Dirichlet's Boundary Value Problem

Definition. (*Dirichlet's boundary value problem*). *Let Ω be a connected standard domain in R_n, and let $\varphi(y)$ be a continuous function on the boundary $\partial\Omega$. Desired a function $u(x) \in C(\bar{\Omega})$ such that $u(y) = \varphi(y)$ for $y \in \partial\Omega$, which is harmonic on Ω.*

Theorem 1. *Dirichlet's boundary value problem has at most one solution.*

Remark 1. The theorem is a simple implication of the maximum-minimum principle stated in Theorem 19.2.3(c).

Theorem 2. *If $\Omega = K_R = \{x \mid |x| < R\}$, and if $\varphi(y)$ is continuous on ∂K_R, then*

$$u(x) = \frac{R^2 - |x|^2}{R\,|\omega_n|} \int\limits_{\partial K_R} \frac{\varphi(y)}{|x-y|^n}\, ds_y \tag{1}$$

is the unique solution of Dirichlet's boundary value problem for the ball K_R.

Remark 2. If there is a solution at all of Dirichlet's boundary value problem for $\Omega = K_R$, then it must necessarily be of the form (1). This follows from (19.2.2/4). One must, however, check explicitly whether (1) is indeed a solution.

Remark 3. It can be shown that Dirichlet's boundary value problem as defined above (i.e., for arbitrary connected standard domains) has exactly one solution. The existence theorem is, however, rather complicated; see for instance [48].

Remark 4 (stability). Let $\varphi_1(y)$ and $\varphi_2(y)$ be continuous functions on $\partial\Omega$, and let $u_1(x)$ and $u_2(x)$ be the associated solutions of Dirichlet's boundary value problem in the sense of the above definition. Then.

$$\max_{x\in\Omega} |u_1(x) - u_2(x)| \leq \max_{y\in\partial\Omega} |\varphi_1(y) - \varphi_2(y)|\,. \tag{2}$$

This expresses stability: small changes of the boundary values will also cause only small changes in the solutions. (2) is an implication of Theorem 19.2.3(c).

Remark 5 (physical interpretation). The physical interpretation of Dirichlet's boundary value problem has been given in Subsec. 19.1.2. (stationary distributions) for the case $n = 3$.

19.2.5. The Poisson Equation

Usually $\Delta u(x) = 0$ is called the Laplace equation and $\Delta u(x) = f(x)$ is called the Poisson equation (or inhomogeneous Laplace equation).

Theorem 1. *Let $f(x) \in C^2(R_n)$ be a function with compact support in R_n. If*

$$u(x) = -\frac{1}{(n-2)\,|\omega_n|} \int\limits_{R_n} \frac{f(y)}{|x-y|^{n-2}}\, dy \quad (\textit{Newton's potential})\,, \tag{1}$$

with $x \in R_n$ and $n \geqq 3$, then $u(x) \in C^2(R_n)$, and $\Delta u(x) = f(x)$ for $x \in R_n$.

Remark 1. If we are unscrupulous enough to differentiate

$$u(x) = -\frac{1}{(n-2)\,|\omega_n|} \int\limits_{R_n} \frac{f(x-z)}{|z|^{n-2}}\, dz$$

under the sign of integration, then by Theorem 19.2.1 we obtain

$$\Delta u(x) = -\frac{1}{(n-2)\,|\omega_n|} \int_{R_n} \frac{(\Delta f)(x-z)}{|z|^{n-2}}\, dz = f(x)\,.$$

This is the desired proposition, where however the differentiability properties of $u(x)$ must be carefully proved.

Remark 2. The theorem is also true for $f \in C^1(R_n)$ (provided that f has a compact support in R_n), but the proof is more complicated.

Theorem 2. *Let $\Omega = K_R = \{x \mid x \in R_n,\ |x| < R\}$, and let $n \geqq 3$. If $\varphi(y)$ is continuous on ∂K_R, and if $f(x) \in C^2(R_n)$ with a compact support in R_n, then Dirichlet's boundary value problem for the Poisson equation*

$$\Delta u(x) = f(x) \quad \text{for} \quad x \in K_R\,, \qquad u(x) \in C^2(\Omega)\,,$$
$$u(y) = \varphi(y) \quad \text{for} \quad y \in \partial K_R\,, \qquad u(x) \in C(\overline{\Omega})\,,$$

has a unique solution.

Remark 3. It is rather easy to reduce this theorem to Theorem 1 and Theorem 19.2.4/2.

Remark 4. The theorem remains true if Ω is a connected standard domain, $\varphi(y)$ is continuous on $\partial\Omega$, and $f \in C^1(\overline{\Omega})$. But the proof is then more complicated; cf. Remark 2 and Remark 19.2.4/3.

19.3. The Wave Equation

19.3.1. Uniqueness Theorems

If $x = (x_1, \ldots, x_n) \in R_n$ and $t \in R_1$, then we set $(x, t) \in R_{n+1}$. Let

$$R^+_{n+1} = \{(x, t) \mid (x, t) \in R_{n+1},\ t > 0\}, \quad n = 1, 2, 3, \ldots$$

Then $\overline{R^+_{n+1}} = \{(x, t) \mid (x, t) \in R_{n+1},\ t \geqq 0\}$.

Definition 1 (*initial value problem or Cauchy's problem*). *Let $u_0(x) \in C(R_n)$, $u_1(x) \in C(R_n)$, and let $f(x, t) \in C(R^+_{n+1})$. Desired a function $u(x, t) \in C^2(R^+_{n+1})$ such that*

$$\Delta u(x, t) - \frac{\partial^2 u}{\partial t^2}(x, t) = f(x, t) \qquad \text{for} \quad (x, t) \in R^+_{n+1}\,,$$

$$u(x, t) \in C(\overline{R^+_{n+1}}) \quad \text{and} \quad u(x, 0) = u_0(x) \quad \text{for} \quad x \in R_n\,,$$

$$\frac{\partial u}{\partial t}(x, t) \in C(\overline{R^+_{n+1}}) \quad \text{and} \quad \frac{\partial u}{\partial t}(x, 0) = u_1(x) \quad \text{for} \quad x \in R_n\,.$$

Remark 1. The spaces $C^k(R_n)$ etc. have the same meaning as in Subsec. 19.2.1. According to the convention made in Subsec. 19.1.1., $\Delta u(x, t) = \sum_{k=1}^{n} \frac{\partial^2 u}{\partial x_k^2}(x, t)$, that is, Δ always applies to the space co-ordinates $x_1, \ldots, x_n$ alone.

Theorem 1. *The initial value problem as per Definition 1 has at most one solution (in the class of sufficiently smooth functions $u(x, t)$).*

Remark 2. The proof is based on the derivation of an integral identity for sufficiently smooth functions $u(x, t)$. One considers a circular cone K whose base D_0 is a ball in the plane

$t=0$ and whose lateral surface includes an angle of $\frac{\pi}{4}$ with the t-axis (Fig. 19.7). Let D_t be the surface of intersection with the plane $t=\text{const}$, as indicated in the figure, and let S_t be the lateral surface between D_0 and D_t. Further let ν be the normal vector of S_t $\left(\text{then one has } \cos(\nu, t)=\frac{1}{\sqrt{2}}\right)$. Finally, let K_t be the truncated cone between D_0 and D_t. Then, for sufficiently high-order differentiability properties, the following identity holds:

$$2\int_{K_t}\left(\frac{\partial^2 u}{\partial t^2}-\Delta u\right)\frac{\partial u}{\partial t}\,dx\,dt+\int_{D_0}\left[\left(\frac{\partial u}{\partial t}\right)^2+\sum_{k=1}^{n}\left(\frac{\partial u}{\partial x_k}\right)^2\right]dx \tag{1}$$

$$=\int_{D_t}\left[\left(\frac{\partial u}{\partial t}\right)^2+\sum_{k=1}^{n}\left(\frac{\partial u}{\partial x_k}\right)^2\right]dx+\int_{S_t}\frac{1}{\cos(\nu,t)}\left[\sum_{k=1}^{n}\left(\frac{\partial u}{\partial x_k}\cos(\nu,t)-\frac{\partial u}{\partial t}\cos(\nu,x_k)\right)^2\right]ds\,.$$

To prove the theorem, one can assume that $\frac{\partial^2 u}{\partial t^2}-\Delta u\equiv 0$ and $u(x,0)=\frac{\partial u}{\partial t}(x,0)\equiv 0$. Then the left-hand side of (1) is zero and the right-hand side is $\geqq 0$. This then implies that $u(x,t)\equiv \equiv 0$. The cone considered above plays a fundamental role in the theory of the wave equation.

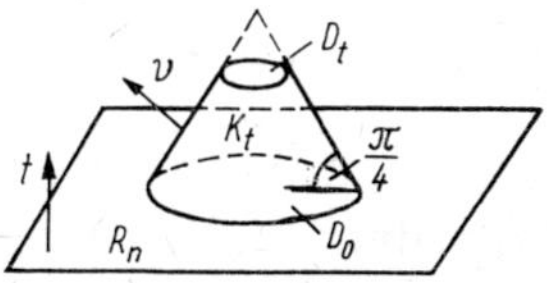

Fig. 19.7

Definition 2 (*boundary and initial value problem*). *Let Ω be a connected standard domain in R_n, and let $Z=\Omega\times(0,\infty)$ be the cylinder over Ω, as shown in Fig. 19.8. Let $u_0(x)\in C(\bar{\Omega})$, $u_1(x)\in C(\bar{\Omega})$, and let $f(x,t)\in C(Z)$. Further let $\varphi(y)$ be a function continuous on $\partial\Omega$. Desired a function $u(x,t)\in C^2(Z)$ such that*

$$\Delta u(x,t)-\frac{\partial^2 u}{\partial t^2}(x,t)=f(x,t) \quad \text{for} \quad (x,t)\in Z\,,$$

$$u(x,t)\in C(\bar{Z}) \quad \text{and} \quad u(x,0)=u_0(x) \quad \text{for} \quad x\in\Omega\,,$$

$$\frac{\partial u}{\partial t}(x,t)\in C(\bar{Z}) \quad \text{and} \quad \frac{\partial}{\partial t}u(x,0)=u_1(x) \quad \text{for} \quad x\in\Omega \quad \text{as well as}$$

$$u(y,t)=\varphi(y) \quad \text{for} \quad y\in\partial\Omega \quad \text{and} \quad t>0\,.$$

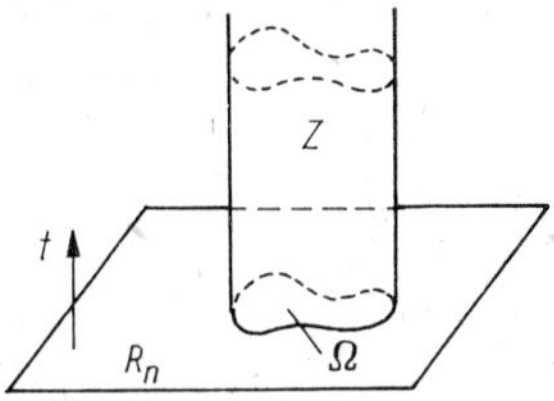

Fig. 19.8

Theorem 2. *The boundary and initial value problem as per Definition 2 has at most one solution (in the class of sufficiently smooth functions).*

Remark 3. The kind of the boundary and initial value realization leads to difficult problems we shall not discuss here. This also accounts for the vague formulations of Theorems 1 and 2. For a more profound discussion on the basis of generalized solutions in Sobolev spaces, the reader is referred to [66].

19.3.2. The Wave Equation in One Dimension

Now let $n=1$. In other words, we consider the (homogeneous) wave equation $\frac{\partial^2 u}{\partial x^2}=\frac{\partial^2 u}{\partial t^2}$, where $u=u(x, t)$ and $x \in R_1$, $t \in R_1$.

Theorem 1. *$u(x, t) \in C^2(R_2)$ is a solution of $\frac{\partial^2 u}{\partial x^2}=\frac{\partial^2 u}{\partial t^2}$ if and only if*

$$u(x, t)=v(x+t)+w(x-t), \quad \text{where} \quad v \in C^2(R_1) \quad \text{and} \quad w \in C^2(R_1). \tag{1}$$

Here $x \in R_1$ and $t \in R_1$.

Remark 1. It is immediately verified that (1) is a solution. To show the converse is not difficult either. If $v(x) \in C^2(R_1)$, then, as $t \to \infty$, $v(x+t)$ is a wave travelling to the left, and $v(x-t)$ is a wave travelling to the right (Fig. 19.9). Thus (1) is a superposition of two waves travelling to the right and left, respectively.

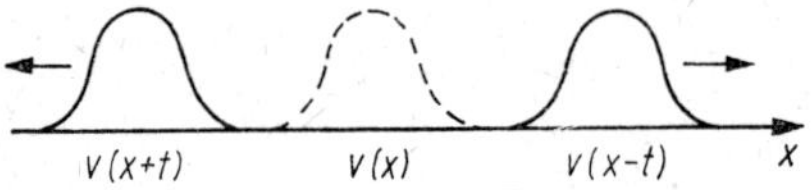

Fig. 19.9

Theorem 2 (*vibrating string of infinite length*). *If $u_0(x) \in C^2(R_1)$ and $u_1(x) \in C^1(R_1)$, then*

$$u(x, t)=\frac{1}{2}\left(u_0(x+t)+u_0(x-t)\right)+\frac{1}{2} \int_{x-t}^{x+t} u_1(\tau)\, d\tau, \quad x \in R_1, \quad t \geqq 0, \tag{2}$$

is the unique solution of the initial value problem

$$\frac{\partial^2 u}{\partial x^2}=\frac{\partial^2 u}{\partial t^2}, \quad u(x, 0)=u_0(x) \quad \text{and} \quad \frac{\partial u}{\partial t}(x, 0)=u_1(x) \tag{3}$$

in the sense of Def. 19.3.1/1.

Remark 2. $u(x, t)$ has the structure stated in Theorem 1. Further it is easily seen that the initial values are correctly assumed.

Remark 3 (physical interpretation). In the sense of Subsec. 19.1.2. it is possible to interpret (3) as the vibration of an infinitely long string. Then $u_0(x)$ is the initial deflection and $u_1(x)$ is the initial velocity. The travelling waves referred to in Remark 1 now have a real physical meaning.

Remark 4 (domain of dependence). In connection with the physical interpretation given in Remark 1, the following statement is of interest, too. From (2) it is seen that for calculating the value of $u(x, t)$ at the point (x, t), with $x \in R_1$ and $t>0$, it suffices to know the values of $u_0(y)$ and $u_1(y)$ in the interval $[x-t, x+t]$ (domain of dependence; Fig. 19.10). If $u_0(y)$ and $u_1(y)$ are zero outside the interval I, then the effect of the perturbation introduced by the values of $u_0(y)$ and $u_1(y)$ in I becomes perceptible at the point x only after a finite time, that is, when $[x-t, x+t] \cap I \neq \emptyset$.

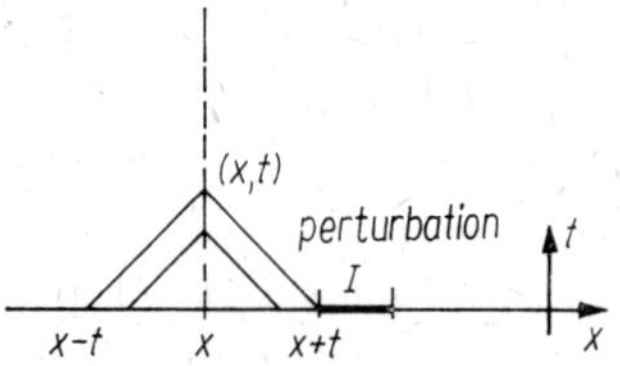

Fig. 19.10

Theorem 3 *(fixed-ended string). Let $\Omega=(0, l)$, where $l>0$. If $u_0(x)\in C^2(\bar{\Omega})$, $u_1(x)\in C^1(\bar{\Omega})$, and if $u_0(0)=u_0(l)=u_1(0)=u_1(l)=u_0''(0)=u_0''(l)=0$, then the boundary and initial value problem*

$$\frac{\partial^2 u}{\partial x^2}=\frac{\partial^2 u}{\partial t^2} \quad \textit{for} \quad x\in\bar{\Omega} \quad \textit{and} \quad t\geqq 0\,,$$

$$u(x, 0)=u_0(x), \quad \frac{\partial u}{\partial t}(x, 0)=u_1(x) \quad \textit{for} \quad x\in\bar{\Omega}\,,$$

$$u(0, t)=u(l, t)=0 \quad \textit{for} \quad t\geqq 0\,,$$

has a unique solution $u(x, t)\in C^2(\bar{Z})$, where $Z=(0, l)\times(0, \infty)$.

Remark 5. The solution $u(x, t)$ can be constructed as follows: $u_0(x)$ and $u_1(x)$ are extended to the interval $(-l, 0)$ as odd functions. Then the functions obtained in this way are extended to become periodic functions, with the period $2l$, on the whole R_1. If the result is again denoted by $u_0(x)$ and $u_1(x)$, then $u_0\in C^2(R_1)$ and $u_1\in C^1(R_1)$. Now formula (2), with $0\leqq x\leqq l$ and $t\geqq 0$, represents the desired solution.

Remark 6. The physical interpretation has been described in Subsec. 19.1.2.

19.3.3. Initial Value Problems for the Wave Equation in Two and Three Dimensions

Spherical mean values in R_3. If $u(x)\in C(R_3)$, then for $t>0$ we set

$$(M(t)\,u)\,(x)=\frac{1}{4\pi}\int_{\omega} u(x+\nu t)\,\mathrm{d}\nu=\frac{1}{4\pi t^2}\int_{|y-x|=t} u(y)\,\mathrm{d}s_y\,. \tag{1}$$

Here ω is the unit sphere in R_3. Further let $\mathrm{d}\nu$ be surface element on ω, and let $\nu\in\omega$ (Fig. 19.11). The second integral is taken over the sphere of radius t centered at x, where $\mathrm{d}s_y$ is the surface element. (Surface integrals have been discussed in Subsec. 9.2.5.) It is easy to see that the two integrals in (1) are equal. Since $4\pi t^2$ is the surface area of a sphere of radius t in R_3, (1) represents a mean value.

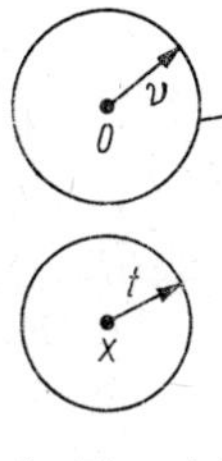

Fig. 19.11

Theorem 1. *If $u_0(x)\in C^3(R_3)$ and $u_1(x)\in C^2(R_3)$, then*

$$u(x, t)=t(M(t)\,u_1)\,(x)+\frac{\partial}{\partial t}\,[t(M(t)\,u_0)\,(x)], \quad \textit{where} \quad x\in R_3 \quad \textit{and} \quad t>0 \tag{2}$$

is the unique solution of the initial value problem

$$\Delta u=\frac{\partial^2 u}{\partial t^2} \quad \textit{for} \quad x\in R_3 \quad \textit{and} \quad t>0\,,$$

$$u(x, 0)=u_0(x), \quad \frac{\partial u}{\partial t}(x, 0)=u_1(x) \quad \textit{for} \quad x\in R_3, \quad \textit{with} \quad u(x, t)\in C^2(\overline{R_4^+}).$$

Remark. It is rather troublesome to prove this theorem, and especially to show that $u(x, t)$ is a solution of the wave equation. $u \in C^2(\overline{R_4^+})$ is to be understood as saying that u as well as its first- and second-order partial derivatives are continuously extendable from R_4^+ to $\overline{R_4^+}$.

The two-dimensional wave equation. The two-dimensional wave equation is treated using the method of reduction of dimensions. If $u_0(x_1, x_2)$ and $u_1(x_1, x_2)$ are the given initial data, then one adds a third dimension: $u_0(x) = u_0(x_1, x_2)$ and $u_1(x) = u_1(x_1, x_2)$, where $x = (x_1, x_2, x_3)$. It is not difficult to see that then the three-dimensional solution $u(x, t)$ is also independent of x_3. The problem then is to convert (1) and (2) into two-dimensional expressions. The two-dimensional counterpart to (1) is

$$(\overline{M}_t u)(x_1, x_2) = \frac{1}{2\pi} \int_K \frac{u(y_1, y_2)}{\sqrt{t^2 - (x_1 - y_1)^2 - (x_2 - y_2)^2}} \, dy_1 \, dy_2 , \tag{3}$$

where the integral is taken over the disk $K = \{(y_1, y_2) \mid (x_1 - y_1)^2 + (x_2 - y_2)^2 < t^2\}$ with centre (x_1, x_2) and radius t.

Theorem 2. *If* $u_0(x_1, x_2) \in C^3(R_2)$ *and* $u_1(x_1, x_2) \in C^2(R_2)$, *then*

$$u(x_1, x_2, t) = (\overline{M}_t u_1)(x_1, x_2) + \frac{\partial}{\partial t} [(\overline{M}_t u_0)(x_1, x_2)], \quad \textit{with} \quad (x_1, x_2) \in R_2$$

$$\textit{and} \quad t \geqq 0$$

is the unique solution of the initial value problem

$$\frac{\partial^2 u}{\partial x_1^2} + \frac{\partial^2 u}{\partial x_2^2} = \frac{\partial^2 u}{\partial t^2} \quad \textit{for} \quad (x_1, x_2) \in R_2 \quad \textit{and} \quad t > 0 ,$$

$$u(x_1, x_2, 0) = u_0(x_1, x_2), \quad \frac{\partial u}{\partial t}(x_1, x_2, 0) = u_1(x_1, x_2) \quad \textit{for} \quad (x_1, x_2) \in R_2 ,$$

where $u(x_1, x_2, t) \in C^2(\overline{R_3^+})$.

19.3.4. Physical Interpretations, Huyghenian Property, Spherical Waves

Vibrating membrane: By analogy with the infinitely long vibrating string discussed in Subsec. 19.3.2., one may consider a membrane of infinite extension which, in its rest position, extends over the whole R_2 plane. If the membrane is locally deflected at the time $t = 0$, then $u_0(x_1, x_2)$ and $u_1(x_1, x_2)$ are assumed to be the initial deflection and the initial velocity, respectively. The (small) vibrations of this membrane are then described (for normalized matter constants) by the initial value problem for the two-dimensional wave equation as stated in Theorem 19.3.3/2, where $u(x_1, x_2, t)$ is the deflection at the position $(x_1, x_2) \in R_2$ at the time $t \geqq 0$. By analogy with Remark 19.3.2/4 it is possible to determine the domain of dependence: the value of $u(x_1, x_2, t)$ depends only on the values of $u_0(x_1, x_2)$ and $u_1(x_1, x_2)$ in $\{(y_1, y_2) \mid (x_1 - y_1)^2 + (x_2 - y_2)^2 \leqq t^2\}$. If u_0 and u_1 vanish outside the perturbation indicated in Fig. 19.12, then this perturbation comes into effect not until the cones indicated in the figure touch the domain of perturbation for the first time.

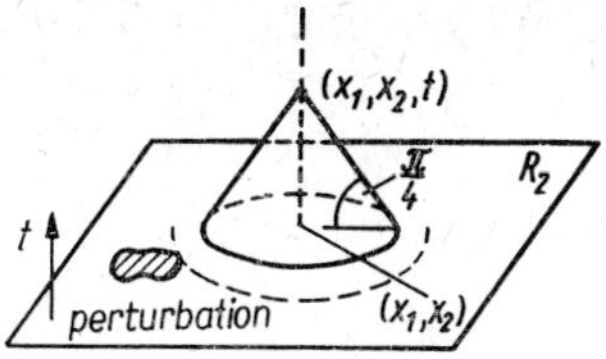

Fig. 19.12

Propagation of sound: Sound waves propagate in the R_3 as pressure waves. If $u(x, t)$ is the deviation from the (constant) normal pressure at the position $x \in R_3$ and at the time $t \geqq 0$, then $u(x, t)$ satisfies the three-dimensional wave equation $\Delta u = \frac{\partial^2 u}{\partial t^2}$. If one now assumes that at the time $t=0$ a (local) pressure fluctuation $u_0(x)$ and an initial variation $u_1(x)$ of this pressure fluctuation are generated, then $u(x, t)$ is the solution of the initial value problem for the three-dimensional wave equation as stated in Theorem 19.3.3/1. The domain of dependence now exhibits a remarkable peculiarity: for calculating u at the point (x, t) one needs nothing but the knowledge of $u_0(x)$ and $u_1(x)$ in a small neighbourhood of the sphere $\{y \mid y \in R_3, |y-x| = t\}$. This follows from Theorem 19.3.3/1 and from (19.3.3/1). Now we regard u_0 and u_1 as a local perturbation, e.g. an explosion, assuming that u_0 and u_1 vanish outside the perturbation indicated in Fig. 19.13. An observer at the point $x \in R_3$ will perceive the explosion only after a finite time. At the time t_1 the sound will set in suddenly, and it will end suddenly at the time t_2 (Fig. 19.13). If the domain of dependence has the form just described, then the differential equation is said to have the Huyghenian property. It turns out that the wave equations in 1, 2, 4, 6, 8, ... dimensions do not have the Huyghenian property, whereas the wave equations in 3, 5, 7, . . . dimensions have it. What a mercy we live in an odd-dimensional space (apart from the R_1, which does not offer any comfort anyway)!

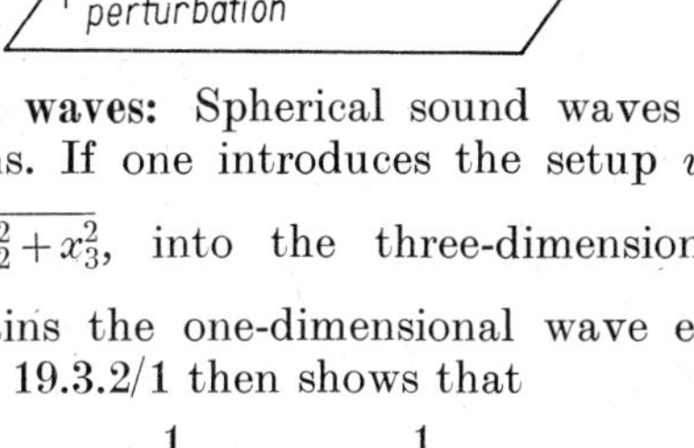

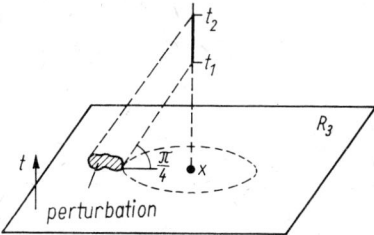

Fig. 19.13

Spherical waves: Spherical sound waves are of interest in connection with explosions. If one introduces the setup $u(x, t) = v(r, t)$, with $x \in R_3$, $t \geqq 0$, $r = \sqrt{x_1^2 + x_2^2 + x_3^2}$, into the three-dimensional wave equation $\Delta u = \frac{\partial^2 u}{\partial t^2}$, then one obtains the one-dimensional wave equation $\frac{\partial^2}{\partial r^2}(rv) = \frac{\partial^2}{\partial t^2}(rv)$ for $rv(r, t)$. Theorem 19.3.2/1 then shows that

$$u(x, t) = \frac{1}{r}\, v_1(r+t) + \frac{1}{r}\, v_2(r-t)\,.$$

$\frac{1}{r}\, v_1(r+t)$ describes an incoming spherical wave, and $\frac{1}{r}\, v_2(t-r)$ describes a propagating spherical wave.

19.3.5. The Inhomogeneous Wave Equation, Retarded Potentials

In this subsection we shall confine ourselves to the case $n=3$, i.e., to the three-dimensional wave equation. If $f(x, t)\in C^2(\overline{R_4^+})$, then

$$(Rf)(x, t)=\frac{1}{4\pi}\int\limits_{|x-y|\leq t}\frac{f(y, t-|x-y|)}{|x-y|}\,dy\,, \tag{1}$$

where $x\in R_3$, $t>0$, is called retarded potential. If $x\in R_3$ and $t>0$ are given, then in (1) only those values of $f(y, \tau)$ are required which lie on the lateral surface of the cone (see Fig. 19.14).

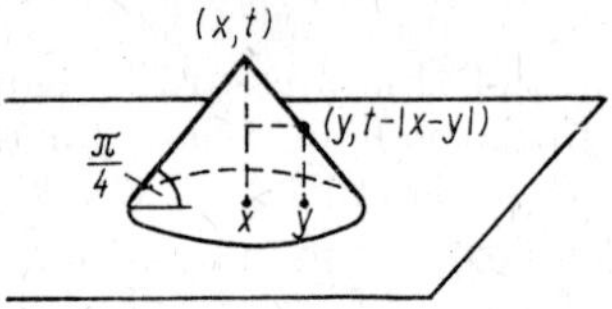

Fig. 19.14

Theorem 1. *If* $f(x, t)\in C^2(\overline{R_4^+})$, *then* $u(x, t)=(Rf)(x, t)$, *with* $x\in R_3$ *and* $t>0$, *is the unique solution of the initial value problem*

$$\frac{\partial^2 u}{\partial t^2}-\Delta u=f(x, t) \quad \textit{for} \quad x\in R_3 \quad \textit{and} \quad t>0\,,$$

$$u(x, 0)=\frac{\partial u}{\partial t}(x, 0)=0 \quad \textit{for} \quad x\in R_3, \quad \textit{where} \quad u(x, t)\in C^2(\overline{R_4^+})\,.$$

Remark. Thus the retarded potential is a solution of the inhomogeneous wave equation with vanishing initial data. $u\in C^2(\overline{R_4^+})$ has again the same meaning as in Remark 19.3.3. If we now combine this theorem with Theorem 19.3.3/1, then we obtain the following result.

Theorem 2. *If* $f(x, t)\in C^2(\overline{R_4^+})$, *and if* $u_0(x)\in C^3(R_3)$ *and* $u_1(x)\in C^2(R_3)$, *then*

$$u(x, t)=(Rf)(x, t)+t(M(t)u_1)(x)+\frac{\partial}{\partial t}[t(M(t)u_0)(x)],\ x\in R_3,\ t>0\,,$$

is the unique solution of the initial value problem

$$\frac{\partial^2 u}{\partial t^2}-\Delta u=f(x, t) \quad \textit{for} \quad x\in R_3 \quad \textit{and} \quad t>0\,,$$

$$u(x, 0)=u_0(x),\ \frac{\partial}{\partial t}u(x, 0)=u_1(x) \quad \textit{for} \quad x\in R_3, \quad \textit{where} \quad u(x, t)\in C^2(\overline{R_4^+})\,.$$

19.4. The Heat Conduction Equation

19.4.1. The Singularity Solution

Again let $x=(x_1, \ldots, x_n)\in R_n$, $t\in R_1$, and $(x, t)\in R_{n+1}$. Further let R_{n+1}^+ have the same meaning as in Subsec. 19.3.1.

Lemma. $s(x, t)=(4\pi t)^{-\frac{n}{2}}\,e^{-\frac{|x|^2}{4t}}$ *is a solution of the heat conduction equation* $\dfrac{\partial s}{\partial t}=\Delta s$ *in* R_{n+1}^+.

Remark. For the heat conduction equation, $s(x, t)$ plays the same role as r^{-n+2} in Subsec. 19.2.1. for the Laplace equation. We recall that $\Delta s = \sum_{k=1}^{n} \frac{\partial^2 s}{\partial x_k^2}$. For a fixed t, $s(x, t)$ describes a bell-shaped surface (error distribution curve for $n=1$; Fig. 19.15).

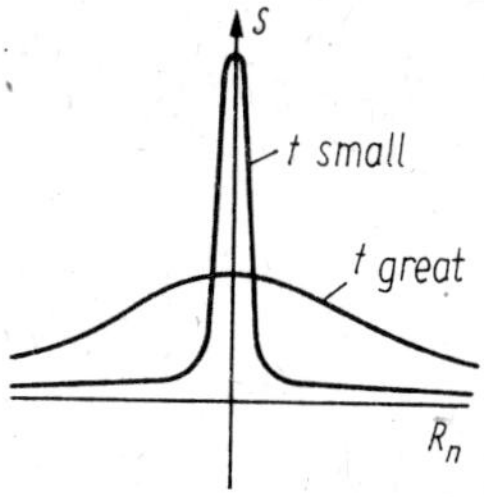

Fig. 19.15

19.4.2. The Maximum-Minimum Principle

Theorem. *Let Ω be a bounded domain in R_n with the boundary $\partial\Omega$* (Fig. 19.16). *Further let $Z = \Omega \times (0, T]$ with $T > 0$ be the cylinder over Ω (including the top surface). If $u(x, t) \in C^2(Z)$, $u(x, t) \in C(\bar{Z})$, and if $\frac{\partial u}{\partial t} = \Delta u$ on Z, then $u(x, t)$ assumes its maximum and its minimum either on the bottom surface Ω or on the lateral surface $\partial\Omega \times [0, T]$.*

Remark. This is the counterpart to the maximum-minimum principle for the Laplace equation as stated in Theorem 19.2.3(c).

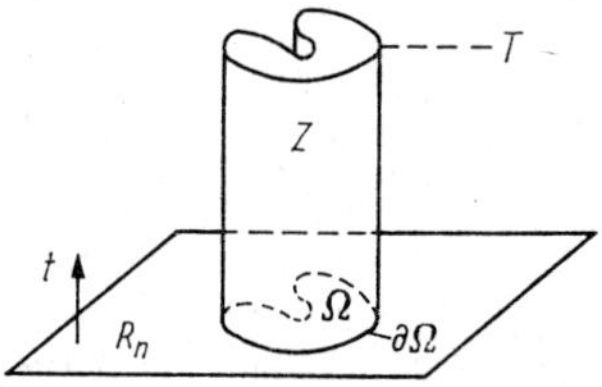

Fig. 19.16

19.4.3. The Initial Value Problem

Definition. (*initial value problem or Cauchy problem*). *If $\varphi(x) \in C(R_n)$, then a function $u(x, t) \in C^2(R_{n+1}^+)$ is desired, such that*

$$\frac{\partial u}{\partial t}(x, t) = (\Delta u)(x, t) \quad \textit{for} \quad (x, t) \in R_{n+1}^+ ,$$

$$u(x, t) \in C(\overline{R_{n+1}^+}) \quad \textit{and} \quad u(x, 0) = \varphi(x) \quad \textit{for} \quad x \in R_n .$$

Remark 1. This is the counterpart to Def. 19.3.1/1.

Theorem. *If $\varphi(x) \in C(R_n)$ is a bounded function, then in $\overline{R_{n+1}^+}$ there exists a unique bounded solution of the initial value problem in the sense of the above definition.*

This solution is

$$u(x,t)=(4\pi t)^{-\frac{n}{2}}\int_{R_n} e^{-\frac{|x-y|^2}{4t}}\varphi(y)\,dy \quad \textit{for} \quad x\in R_n \quad \textit{and} \quad t>0\,, \tag{1}$$

$$u(x,0)=\varphi(x) \quad \textit{for} \quad x\in R_n\,.$$

Remark 2. (1) is called Poisson's integral, being obtained from $\varphi(y)$ and the singularity solution referred to in Subsec. 19.4.1. If the requirement that $u(x,t)$ be bounded in $\overline{R^+_{n+1}}$ is dropped, then the uniqueness is no longer ensured.

Remark 3. This theorem as well as Theorem 19.4.2 show that the heat conduction equation has some properties resembling those of the Laplace equation, and other properties that are analogous to those of the wave equation.

Remark 4 (physical interpretation). Consider an infinitely long rod (e.g., of iron), whose lateral dimensions are neglected, the rod being identified with R_1 (Fig. 19.17). If $\varphi(x)$, $x\in R_1$, is the heat distribution in this rod at the time $t=0$, then (1) describes the heat distribution at the time $t>0$. For example, if $\varphi(x)=\frac{1}{\sqrt{4\pi\varepsilon}}\,e^{-\frac{x^2}{4\varepsilon}}$, where ε is a small positive number (heat source at $x=0$), then

$$u(x,t)=\frac{1}{\sqrt{4\pi\,(t+\varepsilon)}}\,e^{-\frac{x^2}{4(\varepsilon+t)}}$$

for $t>0$ (singularity solution according to Subsec. 19.4.1.). It is seen how the heat is dissipated.

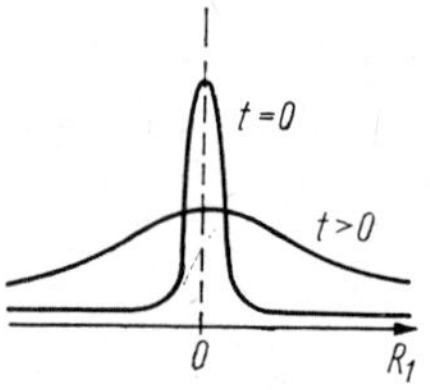

Fig. 19.17

Remark 5. The physical interpretation of (1) is not fully satisfactory: if $\varphi(y)$ is a local perturbation (e.g., like that described in Remark 19.3.2/4), then this perturbation comes into effect at all positions $x\in R_1$ immediately. This contradicts physical principles, according to which information propagates with finite velocity. The corresponding remarks for the wave equation have shown that in this case perturbations propagate with finite velocity. Nevertheless the heat conduction equation gives a satisfactory description of a great number of physical effects (heat conduction, diffusion etc.).

19.5. Separation Setups

19.5.1. Introductory Note

Separation setups for boundary value problems as well as for boundary and initial value problems of partial differential equations are of fundamental importance for mathematics and physics. Frequently they allow informative physical interpretations. Since the methods are constructive, they are also of numerical interest. They can frequently be used to calculate solutions of physical problems by numerical approximation. On the other hand a rigorous mathematical foundation of these methods is rather complicated. Within classical mathematics this foundation is governed by many complicated constraints. We refer, for example, to [48]. The theory of Sobolev spaces and of generalized solutions offers a natural access to these problems; see for instance [66]. We shall not, however, develop this theory here. In the following subsections we shall present the fundamental ideas, where we shall frequently adopt a formal procedure, dispensing with the precise formulation of assumptions etc.

19.5.2. The Fixed Loaded Plate

In the x_1,x_2-plane we consider the square elastic plate $Q=\{x \mid x\in R_2,\ 0<x_1<\pi,\ 0<x_2<\pi\}$, which is fixed at its edges (Fig. 19.18). A load $f(x)\in L_2(Q)$ deflects the plate. If $u(x)$ is the deflection from the normal position, $x\in Q$, then $u(x)$ turns out to be the solution of the Dirichlet boundary value problem

$$\Delta u(x)=f(x) \quad \text{for} \quad x\in Q \quad \text{and} \quad u(y)=0 \quad \text{for} \quad y\in\partial Q\,. \tag{1}$$

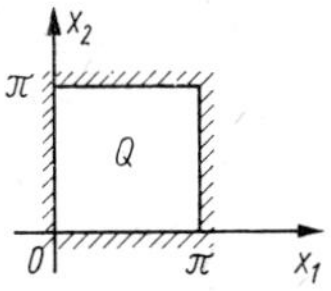

Fig. 19.18

As $f(x)\in L_2(Q)$ (a very general assumption), by Theorem 18.1.4 it is possible to expand $f(x)$ in terms of semiperiodic sine functions,

$$f(x)=\sum_{n,m=1}^{\infty} a_{n,m}\sin nx_1 \sin mx_2, \qquad \sum_{n,m=1}^{\infty} |a_{n,m}|^2=\|f\|^2_{L_2(Q)}\,. \tag{2}$$

Using the setup

$$u(x)=-\sum_{n,m=1}^{\infty}\frac{a_{n,m}}{n^2+m^2}\sin nx_1\cdot\sin mx_2\,, \tag{3}$$

for $u(x)$, one obtains by (formal) differentiation

$$\begin{aligned}(\Delta u)\,(x)&=-\sum_{n,m=1}^{\infty}\frac{a_{n,m}}{n^2+m^2}\Delta\,(\sin nx_1\cdot\sin mx_2)\\ &=\sum_{n,m=1}^{\infty} a_{n,m}\sin nx_1\cdot\sin mx_2=f(x)\,,\end{aligned} \tag{4}$$

$$u(y)=0 \quad \text{for} \quad y\in\partial\Delta\,. \tag{5}$$

Hence (3) is the desired (unique) solution. The elegance of the method is clear, but it is even clearer that the foundation is imperfect. A satisfactory explanation of (4) and (5) can be given within the theory of Sobolev spaces. Then (5) has to be understood in the sense of the boundary values for functions in Sobolev spaces. For details we refer to [66]. $u(x)$ in (3) is called a generalized solution of (1). If $f(x)$ is sufficiently smooth (for instance if $f(x)$ is differentiable on Q up to arbitrary order and has a compact support there), then $u(x)$ is a classical solution and (5) holds in the usual sense. One may also formulate the following statement. If $f(x)\in L_2(Q)$, then the above method is permissible, numerically practicable, and (within a new theory) rigorously justifiable.

19.5.3. The Separation Setup for the Laplace Equation

We now generalize the problem stated in Subsec. 19.5.2. to arbitrary connected standard domains Ω in R_n. (Standard domains have been described in Subsec. 9.3.1. They are always bounded.) Thus, given $f(x) \in L_2(\Omega)$, and desired a function $u(x)$ such that

$$\Delta u(x) = f(x) \quad \text{for} \quad x \in Q \quad \text{and} \quad u(y) = 0 \quad \text{for} \quad y \in \partial\Omega \,. \tag{1}$$

First of all one will ask for an analogue to the (complete orthonormal) system of semiperiodic sine functions of Subsec. 19.5.2. A real-valued function $v(x)$ is called a (normalized) eigenfunction of the operator Δ (with zero boundary values) if there exists a real number λ such that

$$\Delta v(x) = \lambda v(x) \quad \text{for} \quad x \in \Omega \,, \tag{2}$$

$$\|v\|_{L_2(\Omega)} = 1 \quad \text{and} \quad v(y) = 0 \quad \text{for} \quad y \in \partial\Omega \,. \tag{3}$$

It is immediately seen that for $\Omega = Q$ the sine functions of Subsec. 19.5.2. are eigenfunctions. λ is called eigenvalue. If Ω is an arbitrary connected standard domain in R_n, then the following statement is true:

There exists an orthonormal system $\{v_j(x)\}_{j=1}^{\infty}$ of eigenfunctions in the sense of (2), (3) *that is complete in* $L_2(\Omega)$.

If λ_j are the corresponding eigenvalues, i.e., if $\Delta v_j = \lambda_j v_j$, then $0 > \lambda_1 \geqq \lambda_2 \geqq \lambda_3 \geqq \dots \to -\infty$ as $j \to \infty$. Here Δv_j as well as $v_j \,|_{\partial\Omega} = 0$ have to be understood in the sense of the theory of Sobolev spaces, which shall not be discussed here (cf. [66]). If the boundary of Ω is sufficiently smooth (for example, if it is differentiable up to arbitrary order), then $v_j(x) \in C^2(\overline{\Omega})$, and Δv_j as well as $v_j \mid \partial\Omega = 0$ are to be understood in the classical sense. But this allows us to transfer the method of Subsec. 19.5.2. to the present, more general case. $f(x) \in L_2(\Omega)$ is expanded in terms of the complete orthonormal system $\{v_k(x)\}_{k=1}^{\infty}$ (cf. Theorem 17.3.3),

$$f(x) = \sum_{k=1}^{\infty} a_k v_k(x) \quad \text{for} \quad x \in \Omega, \qquad \sum_{k=1}^{\infty} |a_k|^2 = \|f\|^2_{L_2(\Omega)} \,.$$

Using the setup

$$u(x) = \sum_{k=1}^{\infty} \frac{a_k}{\lambda_k} v_k(x) \quad \text{for} \quad x \in \Omega$$

for $u(x)$, one obtains by formal differentiation

$$\Delta u(x) = \sum_{k=1}^{\infty} \frac{a_k}{\lambda_k} \Delta v_k(x) = \sum_{k=1}^{\infty} a_k v_k(x) = f(x) \quad \text{for} \quad x \in \Omega$$

and $u(y) = 0$ for $y \in \partial\Omega$. Hence $u(x)$ is the (unique) solution of (1). Again the same remarks as in Subsec. 19.5.2. are applicable: the calculations can be given a rigorous foundation within the theory of Sobolev spaces; $u(x)$ is a generalized solution. If Ω is a sufficiently smooth domain and if $f(x)$ is sufficiently smooth, then $u(x)$ is a classical solution and $u \,|_{\partial\Omega} = 0$ holds in the usual sense. The numerical practicability of the method depends on whether or not the eigenfunctions $u_j(x)$ and the eigenvalues λ_j can be calculated explicitly or by numerical approximation.

19.5.4. The Fourier Method for the Wave Equation

Consider the initial and boundary value problem as per Def. 19.3.1/2 for the homogeneous wave equation with zero boundary conditions: let Ω be a connected standard domain in R_n, and let $Z = \Omega \times (0, \infty)$ (Fig. 19.19). Desired a function $u(x, t)$ such that

$$\frac{\partial^2 u}{\partial t^2} = \Delta u \quad \text{for} \quad (x, t) \in Z, \quad u(y, t) = 0 \quad \text{for} \quad y \in \partial\Omega \quad \text{and} \quad t > 0\,, \tag{1}$$

$$u(x, 0) = u_0(x) \quad \text{and} \quad \frac{\partial u}{\partial t}(x, 0) = u_1(x) \quad \text{for} \quad x \in \Omega\,. \tag{2}$$

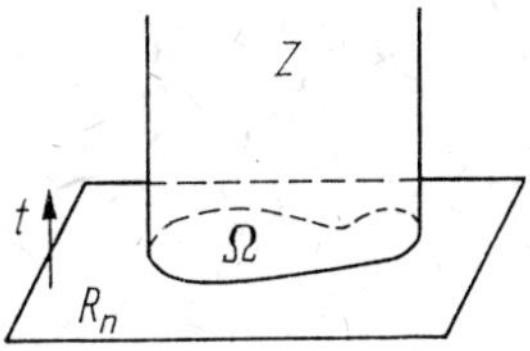

Fig. 19.19

By Theorem 19.3.1/2 the uniqueness is ensured (at least for smooth functions). We ask for the existence and the construction of the solution. First of all we look for non-trivial solutions $v(x, t) = X(x)\, T(t)$, such that

$$\frac{\partial^2 v}{\partial t^2} = \Delta v \quad \text{for} \quad (x, t) \in Z \quad \text{and} \quad X(y) = 0 \quad \text{for} \quad y \in \partial\Omega\,. \tag{3}$$

Insertion of $v = XT$ gives

$$\frac{\Delta X(x)}{X(x)} = \frac{T''(t)}{T(t)} = \lambda\,,$$

where λ must be a constant. Together with (3) this shows that $X(x)$ is an eigenfunction in the sense of Subsec. 19.5.3., i.e., that $X(x) = v_k(x)$ and $\lambda = \lambda_k$. Then $T''(t) = \lambda_k T(t)$. Since $\lambda_k < 0$, one obtains

$$v_k(x, t) = v_k(x)\, (a_k e^{i\sqrt{|\lambda_k|}t} + b_k e^{-i\sqrt{|\lambda_k|}t})\,, \tag{4}$$

where we have written $v_k(x, t)$ instead of $v(x, t)$. a_k and b_k are arbitrary (complex) constants. If one makes the setup

$$u(x, t) = \sum_{k=1}^{\infty} v_k(x)\, (a_k e^{i\sqrt{|\lambda_k|}t} + b_k e^{-i\sqrt{|\lambda_k|}t})\,, \tag{5}$$

then (1) is formally satisfied (i.e.) without regard to convergence problems). If $u_0(x) \in L_2(\Omega)$ and $u_1(x) \in L_2(\Omega)$, then u_0 and u_1 can be expanded in terms of the complete orthonormal system $\{v_k(x)\}_{k=1}^{\infty}$,

$$u_0(x) = \sum_{k=1}^{\infty} c_k v_k(x), \quad u_1(x) = \sum_{k=1}^{\infty} d_k v_k(x)\,. \tag{6}$$

On the other hand (5) (formally) implies that

$$u(x, 0) = \sum_{k=1}^{\infty} (a_k + b_k)\, v_k(x), \quad \frac{\partial u}{\partial t}(x, 0) = \sum_{k=1}^{\infty} i\sqrt{|\lambda_k|}\, (a_k - b_k)\, v_k(x)\,. \tag{7}$$

If a_k and b_k are so chosen that $a_k + b_k = c_k$ and $\mathrm{i}\sqrt{|\lambda_k|}\,(a_k - b_k) = d_k$, then (2) is also satisfied. Further one has $u(y, t) = 0$ for $y \in \partial\Omega$ and $t > 0$. All the considerations (problems of convergence, differentiability etc.) can be justified on the basis of the theory of Sobolev spaces. We refer to [66]. In this sense $u(x, t)$ is the (generalized) solution of (1), (2).

Physical interpretation: For $n = 2$, the problem (1), (2) describes the vibrations of a membrane with fixed edges which at rest extends over the domain Ω in the x_1,x_2-plane. $u(x, t)$ is the deflection from the position of rest at the time t and the position $x \in \Omega$. According to (5) $u(x, t)$ is the superposition of the fundamental modes

$$v_k(x)\,\mathrm{e}^{\pm \mathrm{i}\sqrt{|\lambda_k|}t}, \quad x \in \Omega, \quad t \in R_1 .$$

Such a fundamental mode has the period $T_k = \frac{2\pi}{\sqrt{|\lambda_k|}}$ and the frequency $\nu_k = \frac{1}{T_k} = \frac{\sqrt{|\lambda_k|}}{2\pi}$. Thus the fundamental tone corresponding to $|\lambda_1|$ is superposed by overtones which correspond to $|\lambda_2|$, $|\lambda_3|$, ...

19.5.5. Vibrating Membrane, Vibrating String

Vibrating membrane: We refer immediately to the last statements made in Subsec. 19.5.4., considering the square membrane with fixed edges, $\Omega = Q = \{(x_1, x_2) \mid 0 < x_1 < \pi,\ 0 < x_2 < \pi\}$ (Fig. 19.20). In this case the eigenfunctions $\{v_k(x)\}_{k=1}^{\infty}$ are explicitly known:

$$\left\{\frac{2}{\pi}\sin n x_1 \sin m x_2\right\}_{n,m=1}^{\infty}, \quad \text{Theorem 18.1.4}$$

(see also Subsec. 19.5.2.). The eigenvalues are $\lambda_{n,m} = -n^2 - m^2$. Hence the general vibration is

$$u(x, t) = \sum_{n,m=1}^{\infty} \left(a_{n,m}\mathrm{e}^{\mathrm{i}\sqrt{n^2+m^2}t} + b_{n,m}\mathrm{e}^{-\mathrm{i}\sqrt{n^2+m^2}t}\right) \sin n x_1 \cdot \sin m x_2 .$$

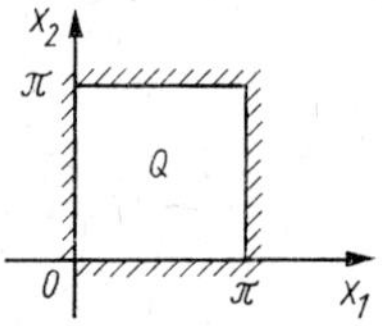

Fig. 19.20

Fig. 19.21

Vibrating string: We return to the problem of the fixed-ended vibrating string as referred to in Theorem 19. 3.2/3 (Fig. 19.21). The remarks made at the end of Subsec. 19.5.4. are of course also valid for $n = 1$ and $\Omega = (0, \pi)$. By Theorem 18.1.4, $v_k(x) = \frac{\sqrt{2}}{\pi}\sin kx$ are the eigenfunctions and $\lambda_k = -k^2$ are the eigenvalues. Hence the general vibration of a fixed-ended string of length π is described by

$$u(x, t) = \sum_{k=1}^{\infty} \left(a_k \mathrm{e}^{\mathrm{i}kt} + b_k \mathrm{e}^{-\mathrm{i}kt}\right) \sin kx .$$

The frequencies of the fundamental modes are $\nu_k = \frac{k}{2\pi}$.

19.5.6. The Fourier Method for the Heat Conduction Equation

By analogy with Subsec. 19.5.4., we now consider the initial and boundary value problem for the homogeneous heat conduction equation with zero boundary conditions: let Ω be a connected standard domain in R_n, and let $Z = \Omega \times (0, \infty)$. Desired a function $u(x, t)$ such that

$$\frac{\partial u}{\partial t} = \Delta u \quad \text{for} \quad (x, t) \in Z, \quad u(y, t) = 0 \quad \text{for} \quad y \in \partial\Omega \quad \text{and} \quad t > 0\,, \tag{1}$$

$$u(x, 0) = u_0(x) \quad \text{for} \quad x \in \Omega\,. \tag{2}$$

Again it can be shown that this problem has at most one solution (at least for sufficiently smooth functions). We ask for the existence and the construction of this solution. As in Subsec. 19.5.4., one first looks for non-trivial functions $v(x, t) = X(x)T(t)$ such that

$$\frac{\partial v}{\partial t} = \Delta v \quad \text{for} \quad (x, t) \in Z \quad \text{and} \quad X(y) = 0 \quad \text{for} \quad y \in \partial\Omega\,.$$

The functions must satisfy the condition $\dfrac{\Delta X(x)}{X(x)} = \dfrac{T'(t)}{T(t)} = \lambda$, where λ is a constant. Hence, as in Subsec. 19.5.4., $X(x)$ is an eigenfunction $v_k(x)$ in the sense of Subsec. 19.5.3., with the eigenvalue λ_k. For $T(t)$ one then has $T(t) = a_k e^{\lambda_k t}$. If we now, by analogy with (19.5.4/5), make the setup

$$u(x, t) = \sum_{k=1}^{\infty} a_k v_k(x)\, e^{\lambda_k t}\,, \tag{3}$$

then (1) is formally satisfied. If $u_0(x) \in L_2(\Omega)$, then $u_0(x)$ can be expanded in the same way as in (19.5.4/6). If we now insert $a_k = c_k$ into (3), then (2) is also satisfied, and we obtain the desired solution. Again it is possible to justify all calculations within the theory of Sobolev spaces (we refer to [66]). In this sense $u(x, t)$ is a generalized solution of (1), (2).

Physical interpretation: If $n = 3$ and $\varphi(y) = 0$, then $u(x, t)$ is the solution of the "heat conduction" problem of Subsec. 19.1.2. Since $\lambda_k < 0$, one has $u(x, t) \to 0$ as $t \to \infty$, as it should also be expected, because the temperature on $\partial\Omega$ is always zero. If $n = 1$ and $\Omega = (0, \pi)$, then by analogy with Subsec. 19.5.5. we obtain

$$u(x, t) = \sum_{k=1}^{\infty} a_k e^{-k^2 t} \sin kx\,.$$

This can be interpreted as a temperature distribution in a rod of length π which is heat-insulated in the lateral directions and whose ends have the temperature zero.

20. Operators in Banach Spaces

20.1. Banach Spaces

20.1.1. Separable Banach Spaces

Real and complex Banach spaces have been dealt with in Chap. 6. From now on a Banach space shall always be a complex Banach space if not explicitly stated otherwise.

Definition. *A Banach space B is called separable if there exists a countably infinite set $\{x_k\}_{k=1}^{\infty} \subset B$ that is dense in B.*

Remark 1. By Def. 17.3.1/2 (and Remark 17.3.1/3), Hilbert spaces are special separable Banach spaces. Denseness of a set in a Banach space is defined in Def. 14.6.3.

Remark 2 (examples). The Banach spaces $l_p^n[=l_{p,C}^n]$ and $l_p[=l_{p,C}]$ referred to in Subsec. 6.2.2., with $1 \leq p < \infty$, are separable: the set described in Subsec. 17.3.2. with respect to l_2 is also dense in l_p, with $1 \leq p < \infty$. If Ω is a domain in R_n, then, by Remark 17.3.2, $L_p(\Omega)$ is also a separable Banach space for $1 \leq p < \infty$. If Ω is a bounded domain in R_n, then it can be shown that the countable set of polynomials $\sum a_m x_1^{m_1} \dots x_n^{m_n}$ with (complex) rational coefficients a_m is dense in $L_p(\Omega)$ for $1 \leq p < \infty$. This is rather easily deduced from the considerations described in Subsec. 18.2.1.

20.1.2. Special Sets in Banach Spaces

Definition 1. *Let B be a Banach space, and let E be a nonvoid set in B.*

(**a**) *E is said to be closed if $\{x_k\}_{k=1}^{\infty} \subset E$ and $x_k \to x$ as $k \to \infty$ implies that $x \in E$.*

(**b**) *E is said to be bounded if $\sup\limits_{x \in E} \|x\| < \infty$.*

(**c**) *E is said to be precompact if from every sequence $\{x_k\}_{k=1}^{\infty} \subset E$ it is possible to select a convergent subsequence $\{x_{k_l}\}_{l=1}^{\infty}$.*

(**d**) *E is said to be compact if E is precompact and closed.*

Remark 1. We recall that $x_k \to x$ as $k \to \infty$ means that $\lim\limits_{k \to \infty} \|x_k - x\| = 0$. Hence E is said to be closed if the limit of convergent sequences in E also lies in E.

Theorem 1. *Every precompact set in a Banach space is bounded.*

Remark 2. In general the converse is not true; see also Theorem 20.1.4(a).

Definition 2. *Let N and M be any two sets in a Banach space B. If $\varepsilon > 0$, then N is called an ε-net for M if $M \subset \bigcup\limits_{x \in N} \{y \mid x \in B, \|x-y\| < \varepsilon\}$.*

Remark 3. $K_\varepsilon(x) = \{y \mid y \in B, \|x-y\| < \varepsilon\}$ is a ball with centre x and radius ε. The definition says that such balls cover M if their centres belong to N.

Theorem 2. (**a**) *A set in a Banach space B is precompact if and only if for every $\varepsilon > 0$ there exists a finite ε-net.*

(**b**) *A set in a Banach space B is precompact if and only if for every $\varepsilon > 0$ there exists a precompact ε-net.*

Remark 4. It would be wrong to suspect that proposition (b) is of no use.

20.1.3. The Space $C(\bar{\Omega})$

In Subsecs. 2.5.2. and 6.1.2. we considered the real Banach space $C[a, b]$. In Subsec. 19.2.1. we further introduced $C(\bar{\Omega})$ as an abbreviating notation. Now we consider complex-valued functions, but retain the previous notations.

Definition 1. *If Ω is a bounded domain in R_n, then $C(\bar{\Omega})=\{f(x) \mid f(x)$ complex-valued and continuous on $\bar{\Omega}\}$, with*

$$\|f\|_{C(\bar{\Omega})}=\sup_{x\in\bar{\Omega}}|f(x)| . \tag{1}$$

Remark 1. As $f(x)$ is continuous on the closed set $\bar{\Omega}$, $f(x)$ is also bounded. Hence (1) is meaningful.

Theorem 1. *If in $C(\bar{\Omega})$ the addition of functions is defined by $(f+g)(x)=f(x)+g(x)$, and if the multiplication of functions $f(x)$ by complex numbers λ is defined by $(\lambda f)(x)=\lambda f(x)$, then $C(\bar{\Omega})$ is a Banach space with respect to the norm* (1).

Remark 2. The case $n=1$ and $\Omega=(a, b)$ is largely in accordance with Theorem 2.5.2.

Definition 2. *If M is an arbitrary (not necessarily countable) index set, then a family of functions $\{f_\mu(x)\}_{\mu\in M}$ in $C(\bar{\Omega})$ is equicontinuous if for every $\varepsilon>0$ there exists a number $\delta=\delta(\varepsilon)$ such that $|f_\mu(x)-f_\mu(y)|\leqq\varepsilon$ for all $\mu\in M$ and for all $x\in\bar{\Omega}$ and $y\in\bar{\Omega}$ with $|x-y|\leqq\delta$.*

Remark 3. From Theorem 2.3.2/4 (and from its n-dimensional counterpart) it follows that every function $f\in C(\bar{\Omega})$ is uniformly continuous on $\bar{\Omega}$. Thus the meaning of the definition does not consist in the independence of δ with respect to x, y, but rather in the independence of μ, x, y.

Theorem 2 (*theorem of Arzelà-Ascoli*). *A family of functions $\{f_\mu(x)\}_{\mu\in M}$ in $C(\bar{\Omega})$ is precompact if and only if $\{f_\mu(x)\}_{\mu\in M}$ is a bounded and equicontinuous set in $C(\bar{\Omega})$.*

20.1.4. Finite-Dimensional Banach Spaces

A complex linear vector space B as defined in Subsec. 6.1.1. is called n-dimensional if there exist n linearly independent vectors $x_1, \dots, x_n$ in B, whereas every set of $n+1$ vectors in B is linearly dependent. Here $y_k\in B$, $k=1, \dots, m$, are called linearly dependent if there exist complex numbers λ_k, $k=1, \dots, m$, such that $\sum_{k=1}^{m}|\lambda_k|>0$, and $\sum_{k=1}^{m}\lambda_k y_k=0$. If there are no such numbers $\lambda_1, \dots, \lambda_m$, then $y_1, \dots, y_m$ are called linearly independent. The spaces C_n of Subsec. 6.1.2. and l_p^n $[=l_{p,C}^n]$ of Subsec. 6.2.2. are n-dimensional. If $x_1, \dots, x_n$ have the above meaning and if $x\in B$, then there exist complex numbers $\lambda, \lambda_1, \dots, \lambda_n$ such that $\lambda x+\sum_{k=1}^{n}\lambda_k x_k=0$ and $|\lambda|+\sum_{k=1}^{n}|\lambda_k|>0$. Since λ must be different from zero, one obtains

$$x=\sum_{k=1}^{n}\varrho_k x_k , \quad \varrho_k \text{ complex} ,$$

which means that $x_1, \dots, x_n$ form a basis in B. Thus the "co-ordinate spaces" C_n and l_p^n are already the most general n-dimensional (complex) linear vector spaces. In Def. 6.2.2/2 we had defined when two norms on a linear vector space are called equivalent. A proposition on equivalent norms in l_p^n has been made in Theorem 6.2.2. It turns out that this theorem can be substantially generalized as follows.

Theorem. (a) *A Banach space is finite-dimensional if and only if every bounded set is precompact.*

(b) *All the norms defined on an n-dimensional linear vector space are equivalent.*

Remark 1. Confer Remark 20.1.2/2.

Remark 2. A normed finite-dimensional linear vector space is always complete, i.e., it is a Banach space. Part (b) then shows that in effect there is only one way to define a norm on C_n.

20.1.5. Completion of Normed Spaces

In Remark 20.1.4/2 we just stated that normed finite-dimensional linear vector spaces are automatically complete. This is not true for general (infinite-dimensional) normed spaces. The polynomials $P(x)$ over $\Omega=(-1, 1)$ form a linear vector space that can be normed by $\|P(x)\|_{L_2(\Omega)}$. But this space is not complete, as can be deduced, for example, from Theorem 18.2.2. This raises the problem of completing a given (incomplete) normed linear vector space to obtain a Banach space.

Definition. *If M_1 is a linear vector space with the norm $\|x\|_1$, and if M_2 is a linear vector space with the norm $\|x\|_2$, then M_1 and M_2 are called isometric-isomorphic if there exists a linear mapping I of M_1 onto M_2 such that $\|Ix\|_2=\|x\|_1$ for all $x\in M_1$.*

Remark 1. Linearity of the mapping means that $I(\lambda x+\mu y)=\lambda I(x)+\mu I(y)$ for all $x\in M_1$, $y\in M_1$, λ and μ complex. That I maps M_1 *onto* M_2 means that the range of I is the entire M_2. From $\|Ix\|_2=\|x\|_1$ it follows that the mapping is one to one. In particular the definition is symmetrical in M_1 and M_2 (one only needs to replace I by the inverse operator I^{-1} in order to interchange the roles of M_1 and M_2).

Theorem. *If M_1 is a linear normed vector space, then there exist a Banach space B and a linear subset M_2 that is dense in B, such that M_1 can be mapped onto M_2 in an isometric-isomorphic way* (Fig. 20.1).

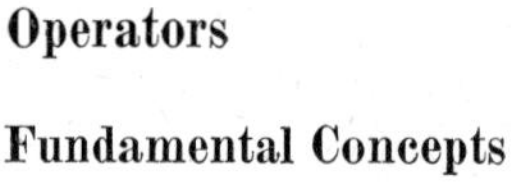

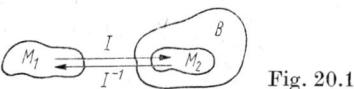

Fig. 20.1

Remark 2. Thus in the sense of the theorem it is possible to complete every linear normed vector space to a Banach space. (In fact, it cannot be observed from the figure that all spaces are linear.)

20.2. Operators

20.2.1. Fundamental Concepts

Definition. *Let B_1 and B_2 be Banach spaces, and let D be a linear subspace (vector space) of B_1.*

(a) *A mapping A of D into B_2 is called linear if, for all $x\in D$ and $y\in D$ and for* *all complex numbers λ and μ, $A(\lambda x+\mu y)=\lambda Ax+\mu Ay$. Here $D=D(A)$ is called the domain of definition of A.*

(b) *A linear mapping A of D into B_2, with $D=D(A)=B_1$, is called continuous if $x_k \to x$ in B_1 as $k\to\infty$ always implies that $Ax_k \to Ax$ in B_2.*

(c) *A linear mapping A of D into B_2, with $D=D(A)=B_1$, is called bounded if there exists a positive number c such that $\|Ax\|_{B_2} \leqq c\|x\|_{B_1}$ for all $x \in B_1$.*

(d) *A linear mapping A of D into B_2, with $D=D(A)=B_1$, is called compact if A maps every bounded set in B_1 into a precompact set in B_2.*

Remark 1. We write Ax instead of $A(x)$ if there is no fear of confusion. In (b), (c) and (d) we assume that the domain of definition is $D(A)=B_1$. It is only operators of this kind that shall be discussed in this chapter. For the later investigations, however, the weaker formulation (a) will be of use. A mapping *into* B_2 is one whose range lies in B_2 but is not necessarily identical with the entire B_2. If the range is the entire B_2, then we speak of a mapping *onto* B_2.

Theorem. (a) *A mapping A is bounded if and only if it is continuous.*

(b) *A compact mapping A is continuous.*

(c) *A mapping A is compact if and only if from every bounded sequence $\{x_k\}_{k=1}^{\infty} \subset B_1$ it is possible to select a subsequence $\{x_{k_l}\}_{l=1}^{\infty}$ such that $\{Ax_{k_l}\}_{l=1}^{\infty}$ converges.*

Remark 2. All operators (=mappings) considered by us are linear, so that we will not indicate this explicitly.

20.2.2. The Space $L(B_1, B_2)$

Definition 1. *If B_1 and B_2 are any two Banach spaces, then $L(B_1, B_2)$ is the class of all continuous operators that map $B_1=D(A)$ into B_2.*

Remark 1. $L(B_1, B_2)$ will become a linear vector space if the addition of continuous operators A_1 and A_2 in $L(B_1, B_2)$ and the multiplication by complex numbers λ_1 and λ_2 are defined by

$$(\lambda_1 A_1+\lambda_2 A_2)(x)=\lambda_1 A_1 x+\lambda_2 A_2 x \quad \text{for all} \quad x\in B_1 .$$

It is easily seen that $\lambda_1 A_1+\lambda_2 A_2$ again belongs to $L(B_1, B_2)$. Further

$$\|A\| = \sup_{\|x\|_{B_1}\leqq 1} \|Ax\|_{B_2} \tag{1}$$

is a norm on $L(B_1, B_2)$. One has

$$\|A\| = \sup_{\|x\|_{B_1}=1} \|Ax\|_{B_2} \quad \text{and} \quad \|A\|=\inf c ,$$

where the infimum is taken over all numbers c with $\|Ax\|_{B_2} \leqq c\|x\|_{B_1}$ for all $x\in B_1$.

Theorem 1. *$L(B_1, B_2)$ with the norm (1) is a Banach space. The class of all compact operators in $L(B_1, B_2)$ forms a closed subspace in $L(B_1, B_2)$.*

Remark 2. A subspace is a linear subset. With the norm (1) the compact operators in $L(B_1,B_2)$ thus also form a Banach space.

Remark 3. If $A_1\in L(B_1, B_2)$ and $A_2\in L(B_2, B_3)$, then $A_2A_1\in L(B_1, B_3)$, where A_2A_1 is defined by $(A_2A_1)(x)=A_2(A_1x)$. One has $\|A_2A_1\|\leqq\|A_2\|\cdot\|A_1\|$.

Definition 2. *Let B_1 and B_2 be any two Banach spaces. Further let A and $\tilde{A}$ be two linear operators, $D(A)\subset D(\tilde{A})\subset B_1$, which map into B_2. Then $\tilde{A}$ is called extension of A if $\tilde{A}x=Ax$ for all $x\in D(A)$.*

Theorem 2. *Let B_1 and B_2 be any two Banach spaces. Let A be a linear operator mapping into B_2, whose domain of definition $D(A)$ is dense in B_1. If there exists a positive number c such that $\|Ax\|_{B_2} \leq c\|x\|_{B_1}$ for all $x \in D(A)$, then there exists exactly one operator $A \in L(B_1, B_2)$ that is an extension of A.*

Linear functionals: If $B_2 = C_1$ **is the** complex plane, then one writes $L(B, C_1) = B'$, where $B_1 = B$ is a Banach space. $f \in B'$ is called a linear continuous functional. Thus $f \in B'$ maps the Banach space B into the complex plane. If $B_2 = C_1$, then it is possible to strengthen Theorem 2 substantially. For that purpose we need the concept of a continuous linear form. Let B be a Banach space, and let f be a linear operator in the sense of Def. 20.2.1(a) that maps $D(f) \subset B$ into the complex plane C_1. Then f is called a continuous linear form if there exists a positive number c such that $|f(x)| \leq c\|f\|_B$ for all $x \in D(f)$.

Theorem 3 (*Hahn-Banach theorem*). *If f is a continuous linear form with $D(f) \subset B$, then there exists a linear continuous functional $\tilde{f} \in B'$ that is an extension of f. Here it is possible to achieve that*

$$\|f\| = \sup_{\substack{\|x\|=1 \\ x \in D(f)}} |f(x)| = \sup_{\substack{\|x\|=1 \\ x \in B}} |\tilde{f}(x)| = \|\tilde{f}\| .$$

Remark 4. In contrast to Theorem 2 it is not required that $D(f)$ be dense in B. On the other hand the uniqueness proposition of Theorem 2 is missing here. The proof of Theorem 3 is much more complex than the (simple) proof of Theorem 2.

20.2.3. Spectrum and Resolvents

If B_1 and B_2 are any two Banach spaces, and if $A \in L(B_1, B_2)$, then

$$R(A) = \{y \mid y \in B_2, \text{ there exists an } x \in B_1 \text{ such that } y = Ax\}$$

denotes the range of A. If $A \in L(B_1, B_2)$ is a one to one mapping of B_1 onto $R(A)$, then A^{-1}, with $x = A^{-1}y$ for $y = Ax$, is called inverse operator (to A). One has $D(A^{-1}) = R(A) \subset B_2$. It is easily seen that A^{-1} is a linear operator. (In any case $R(A)$ is a linear subspace of B_2.) If $B_1 = B_2 = B$, then we write $L(B) = L(B, B)$. By E we denote the identity operator in B.

Definition. *If B is a Banach space, and if $A \in L(B)$, then*

$$M_A = \{\lambda \mid \lambda \in C_1, \text{ there exists } (A - \lambda E)^{-1}, \text{ and } (A - \lambda E)^{-1} \in L(B)\}$$

is called the resolvent set (of A), and $S_A = C_1 \backslash M_A$ is called the spectrum (of A).

Remark 1. In the complex plane C_1 we thus ask for those complex numbers λ for which $A - \lambda E$ is a one to one mapping of B onto B, $R(A - \lambda E) = B$, and for which the inverse operator $(A - \lambda E)^{-1}$ (which then exists) is continuous. The set of all these numbers is called resolvent set, the difference is called spectrum. For $\lambda \in M_A$, $(A - \lambda E)^{-1}$ is called resolvent.

Theorem. *If B is a Banach space, and if $A \in L(B)$, then*

$$\{\lambda \mid \lambda \in C_1, |\lambda| > \|A\|\} \subset M_A .$$

For $|\lambda| > \|A\|$, $\left\{-\sum_{k=0}^{N} \frac{A^k}{\lambda^{k+1}}\right\}_{N=1}^{\infty}$ is a Cauchy sequence (or fundamental sequence) in $L(B)$, and the limit of this sequence is $(A - \lambda E)^{-1} = -\sum_{k=0}^{\infty} \frac{A^k}{\lambda^{k+1}}$ (Neumann series).

Remark 2. Here one has $A^0 = E$, $A^2 = AA$, $A^{k+1} = A^k A$ for $k = 2, 3, \ldots$. Hence the spectrum of an operator $A \in L(B)$ lies in a disk of radius $\|A\|$ about the origin.

20.2.4. The Space $(l_p)'$

If $1 \leqq p < \infty$, then let $l_p\,[=l_{p,C}]$ be the (complex) Banach space referred to in Subsec. 6.2.2. We ask for $(l_p)'$, the space of the linear continuous functionals over l_p in the sense of Subsec. 20.2.2. If $f \in (l_p)'$, then, as previously, $\|f\|_{(l_p)'} = \sup\limits_{\|x\|_{l_p} \leqq 1} |f(x)|$.

Theorem. *Let* $1 < p < \infty$ *and* $\frac{1}{p} + \frac{1}{p'} = 1$.

(a) *If* $y = (y_1, y_2, \ldots) \in l_{p'}$, *then* $(l_p)'$ *contains every* f *with*

$$f(x) = \sum_{k=1}^{\infty} x_k y_k \quad \text{for} \quad x = (x_1, x_2, \ldots) \in l_p\,. \tag{1}$$

Furthermore, $\|f\|_{(l_p)'} = \|y\|_{l_{p'}}$.

(b) *If* $f \in (l_p)'$, *then there exists exactly one element* $y \in l_{p'}$ *such that* $f(x)$ *can be represented by* (1).

Remark 1. The fact that the element f in (1) belongs to $(l_p)'$ is in effect an implication of Hölder's inequality (6.2.1/1). The theorem also remains true for $p=1$ if one sets $p'=\infty$ and

$$l_\infty = \{x \mid x = (x_1, x_2, \ldots),\quad \|x\|_{l_\infty} = \sup_k |x_k| < \infty\}\,.$$

Remark 2. It is easy to deduce from the above theorem that $(l_p)'$ and $l_{p'}$ are isometric-isomorphic in the sense of Def. 20.1.5. Therefore one also writes $(l_p)' = l_{p'}$, where it must, however, be kept in mind that this is to be understood in the sense of (1).

Remark 3. If Ω is a domain in R_n, then the (complex) Banach space $L_p(\Omega)$ has the same meaning as in Remark 17.3.2 and in Remark 20.1.1/2. If $1 < p < \infty$, and if $g \in L_{p'}(\Omega)$, then $(L_p(\Omega))'$ contains every f with

$$f(h) = \int_\Omega h(x)\, g(x)\, dx \quad \text{for} \quad h \in L_p(\Omega)\,. \tag{2}$$

Furthermore, $\|f\|_{(L_p(\Omega))'} = \|g\|_{L_{p'}(\Omega)}$. Conversely, every $f \in (L_p(\Omega))'$ can be represented in the form (2) in a unique way, with $g \in L_{p'}(\Omega)$. In this sense one has $(L_p(\Omega))' = L_{p'}(\Omega)$.

20.2.5. Integral Operators

If Ω is a bounded domain in R_n, then we consider the integral operator K,

$$(Kf)(x) = \int_\Omega K(x, y)\, f(y)\, dy, \quad x \in \Omega\,, \tag{1}$$

by which every function $f(x)$ in Ω is assigned a function $(Kf)(x)$ in Ω (provided that the integral exists). $K(x, y)$ is defined over $\Omega \times \Omega$ and is called the kernel of the integral operator. K is a linear operator (provided that it exists).

Theorem 1. *If* $K(x, y) \in C(\overline{\Omega \times \Omega})$, *then* K *is a compact operator in* $L(C(\bar{\Omega}))$.

Remark 1. The space $C(\bar{\Omega})$ has been considered in Subsec. 20.1.3. Further $L(C(\bar{\Omega}))$ has the same meaning as $L(B)$ in Subsec. 20.2.3., with $B = C(\bar{\Omega})$.

Remark 2. If $K(x, y) \in L_2\,(\Omega \times \Omega)$, then from Fubini's theorem in Subsec. 14.4.4. it follows that, for a fixed $x \in \Omega$, $K(x, y)$ as a function of y belongs to $L_2(\Omega)$ for almost all $x \in \Omega$. For such $x \in \Omega$, $K(x, y)\, f(y)$ as a function of y is then integrable, provided that $f(y) \in L_2(\Omega)$. It is then possible to form (1). The following theorem is to be understood in this sense.

Theorem 2. *If* $K(x, y) \in L_2(\Omega \times \Omega)$, *then* K *is a compact operator in* $L(L_2(\Omega))$.

21. Operators in Hilbert Spaces

21.1. Classes of Continuous Operators

21.1.1. Isomorphy of Hilbert Spaces

Hilbert spaces have been introduced in Sec. 17.3. All notations shall have the meaning indicated there. By Def. 17.3.1/2, in a Hilbert space there exists a countably infinite dense subset. Hence Hilbert spaces are special separable Banach spaces in the sense of Def. 20.1.1.

Definition. *Two Banach spaces B_1 and B_2 are isomorphic if there exists an operator $A \in L(B_1, B_2)$ that has an inverse operator $A^{-1} \in L(B_1, B_2)$.*

Remark 1. What is required is the existence of a continuous operator A, with $D(A) = B_1$ and $R(A) = B_2$, which provides a one to one mapping of B_1 onto B_2, while its (hence existing) inverse operator A^{-1} provides a continuous mapping of B_2 onto B_1. If $y = Ax$, then it follows from $\|y\|_{B_2} \leqq \|A\| \, \|x\|_{B_1}$ and $\|x\|_{B_1} \leqq \|A^{-1}\| \, \|y\|_{B_2}$ that there exist two positive numbers c_1 and c_2 such that for all $x \in B_1$

$$c_1\|y\|_{B_2} \leqq \|x\|_{B_1} \leqq c_2\|y\|_{B_2} \quad \text{with} \quad y = Ax \,. \tag{1}$$

The above definition is thus a generalization of the concept of equivalent norms as per Def. 6.2.2/2 and a generalization of the concept of isometric-isomorphic Banach spaces as per Def. 20.1.5.

Remark 2. According to Subsecs. 17.3.2. and 20.1.4., C_n is an n-dimensional Hilbert space. Now it is easy to deduce from Theorem 20.1.4(b) that every finite-dimensional Hilbert space is isomorphic to C_n with a suitably chosen n (the dimension of the space). If a Hilbert space is called infinite-dimensional if it is not finite-dimensional, then the following statement can be made.

Theorem. *An infinite-dimensional Hilbert space is isomorphic to l_2.*

21.1.2. Linear Functionals

Theorem (*F. Riesz, Fischer*). *Let H be a Hilbert space, and let H' be the space of the linear continuous functionals over H.*

(a) *If $y \in H$, then $f(x) = (x, y)$ is a linear continuous functional over H, and $\|y\|_H = \|f\|_{H'}$.*

(b) *If $f \in H'$, then there exists exactly one $y \in H$ such that $f(x) = (x, y)$ for all $x \in H$.*

Remark. In the sense of this theorem one has $H = H'$. If $H = l_2$ or $H = L_2(\Omega)$, then the theorem (largely) agrees with Theorem 20.2.4 and with Remark 20.2.4/3.

21.1.3. Bilinear Forms

Definition. *If H is a Hilbert space, then $L(x, y)$ is called a bounded bilinear form if to every ordered pair $x \in H$ and $y \in H$ a complex number $L(x, y)$ with the following properties is assigned:*

(a) *If* x_1, x_2, y *are elements of* H, *and if* λ_1, λ_2 *are complex numbers, then*

$$L(\lambda_1 x_1 + \lambda_2 x_2, y) = \lambda_1 L(x_1, y) + \lambda_2 L(x_2, y) ,$$
$$L(y, \lambda_1 x_1 + \lambda_2 x_2) = \bar{\lambda}_1 L(y, x_1) + \bar{\lambda}_2 L(y, x_2) .$$

(b) *There exists a positive number c such that, for all* $x \in H$ *and* $y \in H$,

$$|L(x, y)| \leq c\|x\|_H \|y\|_H . \qquad (1)$$

Remark 1. $L(x, y) = (x, y)$ is a bounded bilinear form. If $A \in L(H)$, then (Ax, y) and (x, Ay) are likewise bounded bilinear forms.

Remark 2. By analogy with the norm of operators and functionals, we set

$$\|L\| = \sup_{\|x\| = \|y\| = 1} |L(x, y)| = \sup_{\substack{\|x\| \leq 1 \\ \|y\| \leq 1}} |L(x, y)| = \inf c ,$$

where the infimum is taken over all numbers c that satisfy (1).

Theorem. *If* $L(x, y)$ *is a bounded bilinear form in the Hilbert space* H, *then there exist exactly one operator* $A_1 \in L(H)$ *and exactly one operator* $A_2 \in L(H)$ *such that*

$$L(x, y) = (A_1 x, y) = (x, A_2 y) \quad \text{for all} \quad x \in H \quad \text{and} \quad y \in H .$$

Further, $\|L\| = \|A_1\| = \|A_2\|$.

Remark 3. So the examples stated in Remark 1 are already the most general bounded bilinear forms.

21.1.4. Adjoint Operators

Definition. *If* H *is a Hilbert space, and if* $A \in L(H)$, *then* $A^* \in L(H)$ *is called the adjoint operator of* A *if, for all* $x \in H$ *and all* $y \in H$,

$$(Ax, y) = (x, A^* y) .$$

Remark. $A \in L(H)$ implies that $L(x, y) = (Ax, y)$ is a bounded bilinear form. Hence, by Theorem 21.1.3, there exists exactly one operator $A^* \in L(H)$ such that $(Ax, y) = (x, A^* y)$ for all $x \in H$ and all $y \in H$. This justifies the definition.

Theorem. *Let* H *be a Hilbert space.*

(a) *If* $A \in L(H)$, *then* $\|A\| = \|A^*\|$ *and* $(A^*)^* = A$.

(b) *If* $A_1 \in L(H)$ *and* $A_2 \in L(H)$, *and if* λ_1, λ_2 *are complex numbers, then*

$$(\lambda_1 A_1 + \lambda_2 A_2)^* = \bar{\lambda}_1 A_1^* + \bar{\lambda}_2 A_2^* \quad \text{and} \quad (A_1 A_2)^* = A_2^* A_1^* .$$

(c) *If* $A \in L(H)$, *and if* $A^{-1} \in L(H)$ *exists, then* $(A^*)^{-1} \in L(H)$ *exists too, and* $(A^*)^{-1} = (A^{-1})^*$.

21.1.5. Projection Operators

Definition. *Let* H *be a Hilbert space, and let* H_1 *be a closed subspace of* H. *If* $x = x_1 + x_1^\perp$ *is the representation of* x *in the sense of Theorem* 17.3.5, *then the operator* P *with* $Px = x_1$ *is called projection operator (or projector).* P *projects* H *onto* H_1.

Remark 1. It is easily seen that the correspondence $x \to x_1$, that means P, is linear, and that $D(P) = H$ and $\|Px\| \leq \|x\|$. Hence $P \in L(H)$. If $x \in H_1$, then $Px = x$. Hence it follows that $\|P\| = 1$, provided that H_1 is at least one-dimensional.

Theorem. *If H is a Hilbert space, then $P \in L(H)$ is a projection operator if and only if $P = P^* = P^2$.*

Remark 2. The corresponding projection space is $H_1 = \{x \mid x \in H,\ Px = x\}$.

21.1.6. Isometric and Unitary Operators

Definition. *Let H_1 and H_2 be any two Hilbert spaces.*

(a) *If V is a linear operator in the sense of Def.* 20.2.1, *with $D(V) \subset H_1$ and $R(V) \subset H_2$, then V is called isometric if $\|Vx\|_{H_2} = \|x\|_{H_1}$ for all $x \in D(V)$.*

(b) *An isometric operator in the sense of* (a) *is called unitary if $D(V) = H_1$ and $R(V) = H_2$.*

Remark. A unitary operator belongs to $L(H_1, H_2)$. If for any two Hilbert spaces H_1 and H_2 there exists a unitary operator, then these spaces are isometric-isomorphic in the sense of Def. 20.1.5 (and hence also isomorphic in the sense of Def. 21.1.1).

Theorem. *If H_1 and H_2 are any two Hilbert spaces, and if V is an isometric operator in the above sense, then*

$$(Vx, Vy)_{H_2} = (x, y)_{H_1} \quad \text{for all} \quad x \in D(V) \quad \text{and} \quad y \in D(V).$$

The inverse operator V^{-1} is an isometric operator which maps $D(V^{-1}) = R(V)$ in H_2 onto $R(V^{-1}) = D(V)$ in H_1. If V is unitary, then V^{-1} is unitary, too.

21.1.7. Compact and Degenerate Operators

The definition of a compact operator has been given in Def. 20.2.1(d).

Definition. *If H is a Hilbert space, then $A \in L(H)$ is called degenerate if the dimension of its range $R(A)$ is finite.*

Remark 1. It is easy to see that every degenerate operator is also compact.

Lemma. *If H is a Hilbert space, and if $A \in L(H)$ is degenerate, then A can be represented in the form*

$$Ax = \sum_{k=1}^{m} (x, e_k^*)\, e_k \quad \text{for all} \quad x \in H\,.$$

Here $\{e_k\}_{k=1}^{m}$ is an orthonormal system and m is the dimension of the (finite-dimensional linear) space $R(A)$. Further, A^* is also degenerate, the dimension of $R(A^*)$ is m, and

$$A^*x = \sum_{k=1}^{m} (x, e_k)\, e_k^* \quad \text{for all} \quad x \in H\,.$$

Remark 2. The dimension of a linear vector space was explained in Subsec. 20.1.4.

Theorem. *Let H be a Hilbert space.*

(a) *$A \in L(H)$ is compact if and only if there exists a sequence $\{A_k\}_{k=1}^{\infty} \subset L(H)$ of degenerate operators such that $\lim_{k\to\infty} \|A_k - A\| = 0$.*

(b) *$A \in L(H)$ is compact if and only if A^* is compact.*

21.2. The Theory of Riesz and Schauder

21.2.1. Formulation of the Problem

The solution theory of algebraic equations, $\sum_{k=1}^{m} a_{l,k}x_k = y_l$, with $l=1, 2, \ldots, m$, is well known. Here $(a_{l,k})_{l,k=1}^{m}$ is a given matrix. The properties of this matrix as well as of the adjoint matrix (rank, subdeterminants etc.) determine the theory of these equations. We look for the infinite-dimensional analogue to this theory. If H is a Hilbert space, and if $A \in L(H)$, then one may ask for which elements $y \in H$ the equation $Ax=y$ has a solution, and what can be said about the set of all solutions. Here it is expedient to consider the family of equations $(A-\lambda E)x = Ax - \lambda x = y$ rather than $Ax=y$ alone. Here E is the identity operator in H, and λ is an arbitrary complex number. If $\lambda \in M_A$, then, by Def. 20.2.3, $(A-\lambda E)x = y$ has a unique solution $x=(A-\lambda E)^{-1}y$ for every $y \in H$. Hence a solution theory for the equation $(A-\lambda E)x=y$ is closely related to the determination of the resolvent set M_A and of the spectrum S_A as per Def. 20.2.3.

Definition. *Let H be a Hilbert space, and let $A \in L(H)$.*

(a) *$R(A)=\{y \mid y\in H$, there exists an $x \in H$ such that $Ax=y\}$ is called the range of A, and $N(A)=\{y \mid Ay=0\}$ is called null space.*

(b) *$\lambda \in C_1$ is called eigenvalue of A if there exists an element $x \neq 0$ such that $Ax=\lambda x$. The dimension of $N(A-\lambda E)$ is called multiplicity of the eigenvalue λ.*

Remark 1. It is clear that $R(A)$ and $N(A)$ are linear subspaces of H. If $N(A-\lambda E)$ is not finite-dimensional, then λ is said to be an eigenvalue of infinite multiplicity. Further it is easy to see that in any case $N(A-\lambda E)$ is a closed subspace. If λ is an eigenvalue, then $\lambda \in S_A$.

Remark 2. If $R(A-\lambda E)=H$, then $Ax-\lambda x=y$ has at least one solution for every $y\in H$. If $N(A-\lambda E)=\{0\}$ (the set consisting of the zero element alone), then for a given $y \in H$ the equation $Ax-\lambda x=y$ has at most one solution.

21.2.2. Decomposition Theorems

If H_1 is a (linear) subspace of the Hilbert space H, then

$$\overline{H}_1 = \{x \mid x \in H, \text{ there exists a sequence } \{x_k\}_{k=1}^{\infty} \subset H_1 \text{ such that } x_k \to x\}$$

is the closure of H_1. It is easy to see that $\overline{H}_1$ is a closed (linear) subspace of H.

Theorem. *Let H be a Hilbert space, let $A \in L(H)$, and let λ be a complex number.*

(a) $H=\overline{R(A-\lambda E)} \oplus N(A^*-\bar{\lambda}E) = \overline{R(A^*-\bar{\lambda}E)} \oplus N(A-\lambda E)$.

(b) *If A is compact and if $\lambda \neq 0$, then*

$$H=R(A-\lambda E)\oplus N(A^*-\bar{\lambda}E) = R(A^*-\bar{\lambda}E)\oplus N(A-\lambda E)\,. \tag{1}$$

Remark. $\oplus$ has the meaning indicated in Subsec. 17.3.5. The essential proposition of the theorem is contained in (1): $R(A-\lambda E)$ as well as $R(A^*-\bar{\lambda}E)$ are closed subspaces if A is compact and $\lambda \neq 0$. According to (1), $Ax-\lambda x=y$ has a solution if and only if $(y, z)=0$ for all $z \in N(A^*-\bar{\lambda}E)$, that means if $A^*z=\bar{\lambda}z$.

21.2.3. The Spectrum of Compact Operators

We had already stated that an eigenvalue λ of an operator $A \in L(H)$ belongs to the spectrum S_A of the operator A. In contrast to the finite-dimensional case, it may however happen that there are complex numbers $\lambda \in S_A$ which are not eigenvalues. In this case the inverse operator $(A-\lambda E)^{-1}$ exists, but it does not belong to $L(H)$. The following (rather far-reaching) theorem now shows that compact operators (unlike any other operator in $L(H)$) largely behave as one is accustomed to from the finite-dimensional case.

Theorem 1. *Let H be an infinite-dimensional Hilbert space. The spectrum S_A of the compact operator $A \in L(H)$ consists of the point $\lambda = 0$ and of at most countably many eigenvalues that can accumulate only at the origin. Every nonzero eigenvalue is of finite multiplicity. Further, $S_{A^*} = \{\lambda \mid \lambda \in C_1, \bar{\lambda} \in S_A\}$.*

Remark 1. This theorem together with Theorem 21.2.2 enable a remarkable conclusion to be drawn: let $A \in L(H)$ be compact, and let $\lambda \neq 0$. Then the equation $Ax - \lambda x = y$ has a unique solution for every $y \in H$ if and only if λ is not an eigenvalue; further one has $N(A^* - \bar{\lambda}E) = \{0\}$ if and only if $N(A - \lambda E) = \{0\}$. The last proposition can even be strengthened as follows.

Theorem 2. *If H is a Hilbert space, and if $\lambda \neq 0$ is an eigenvalue of a compact operator $A \in L(H)$, then $N(A-\lambda E)$ and $N(A^* - \bar{\lambda}E)$ are of equal (finite) dimension.*

Remark 2. The theorems stated in this section are the abstract formulation of the so-called Fredholm alternatives for integral equations we shall discuss in the following section.

21.3. Fredholm's Integral Equations

21.3.1. The Adjoint Integral Operator

If Ω is a bounded domain in R_n, then the operator K,

$$(Kf)(x) = \int_\Omega K(x, y)\, f(y)\, dy\,,$$

is the integral operator of Subsec. 20.2.5., where we now assume that the kernel $K(x, y) \in L_2(\Omega \times \Omega)$. Then, by Theorem 20.2.5/2, $K \in L(L_2(\Omega))$ is a compact operator. Hence it is possible to apply the theory of Riesz and Schauder as described in Sec. 21.2. The first problem is to determine the adjoint operator $K^* \in L(L_2(\Omega))$.

Theorem. *With the above assumtions, one has*

$$(K^*f)(x) = \int_\Omega \overline{K(y, x)}\, f(y)\, dy \quad \textit{for} \quad f \in L_2(\Omega)\,.$$

Remark. The equations of Sec. 21.2. now take the form

$$\int_\Omega K(x, y)\, f(y)\, dy - \lambda f(x) = h(x)\,. \tag{1}$$

Here $h(x) \in L_2(\Omega)$ and $K(x, y) \in L_2(\Omega \times \Omega)$ are given, and $\lambda \neq 0$ is a complex number. Desired is a function $f(x) \in L_2(\Omega)$ such that (1) is satisfied. Equations of this kind are called Fredholm integral equations of the second kind. The theory of Sec. 21.2. and the above theorem show that, in addition to (1), the adjoint equation

$$\int_\Omega \overline{K(y, x)}\, g(y)\, dy - \bar{\lambda} g(x) = h(x) \tag{2}$$

is of interest too.

21.3.2. Fredholm's Alternatives

We now transfer the abstract theory of Sec. 21.2. to the operators and equations of Subsec. 21.3.1. All notations shall have the same meaning as in Subsec. 21.3.1.

Theorem 1. *The spectrum S_K of the compact integral operator $K \in L(L_2(\Omega))$ of Subsec. 21.3.1. consists of the point $\lambda=0$ and of an at most countably infinite number of nonzero eigenvalues of finite multiplicities which can accumulate only at the point $\lambda=0$.*

Remark 1. The theorem is an implication of Theorem 21.2.3/1 and Subsec. 21.3.1.

Theorem 2. *If $\lambda \neq 0$ is not an eigenvalue of the integral operator K, then the Fredholm integral equations of second kind (21.3.1/1) and (21.3.1/2) have a unique solution $f(x) \in L_2(\Omega)$ and $g(x) \in L_2(\Omega)$ for every function $h(x) \in L_2(\Omega)$.*

Remark 2. The theorem is an implication of Theorem 21.2.2(b) and Remark 21.2.3/1.

Theorem 3. *If $\lambda \neq 0$, then the following alternative is true: either the Fredholm integral equation of second kind (21.3.1/1) has a unique solution in $L_2(\Omega)$ for every function $h(x) \in L_2(\Omega)$ or the homogeneous integral equation*

$$\int_\Omega K(x, y)\, f(y)\, \mathrm{d}y - \lambda f(x) = 0 \tag{1}$$

has a non-trivial solution in $L_2(\Omega)$.

Remark 3. This theorem has also been transferred from Remark 21.2.3/1.

Theorem 4. *If $\lambda \neq 0$, then the homogeneous integral equations* (1) *and*

$$\int_\Omega \overline{K(y, x)}\, g(y)\, \mathrm{d}y - \bar{\lambda} g(x) = 0 \tag{2}$$

have an equal (finite) number N of linearly independent solutions in $L_2(\Omega)$. Let $f_1(x), \ldots, f_N(x)$ be linearly independent solutions of (1), *and let $g_1(x), \ldots, g_N(x)$ be linearly independent solutions of* (2). *Then* (21.3.1/1), *with $h(x) \in L_2(\Omega)$, has a solution in $L_2(\Omega)$ if and only if*

$$\int_\Omega h(x)\, \overline{g_k(x)}\, \mathrm{d}x = 0 \quad \text{for} \quad k = 1, \ldots, N\,.$$

Accordingly, (21.3.1/2) *with $h(x) \in L_2(\Omega)$ has a solution in $L_2(\Omega)$ if and only if*

$$\int_\Omega h(x)\, \overline{f_k(x)}\, \mathrm{d}x = 0 \quad \text{for} \quad k = 1, \ldots, N\,.$$

Remark 4. The theorem is an implication of Theorem 21.2.2 and Theorem 21.2.3/2. The theorem is of interest only for those $\lambda \neq 0$ which are eigenvalues of K. For other numbers λ the theorem is also true, but it does not give any new information as compared with the previous theorems.

Remark 5. The theorems of this section are called Fredholm's alternatives for integral equations.

22. Distributions

22.1. Fundamental Concepts

22.1.1. Introduction

A continuous function in R_n does not necessarily have continuous partial derivatives of first order, a continuously differentiable function does not necessarily have continuous partial derivatives of second order, and so on. These statements are elementary and can be easily supported by examples. But from the physical point of view these statements are regrettable. On the one hand many physical theories are based on partial differential equations, but on the other hand the mathematical idealizations of physical problems in a natural way lead to functions that are not differentiable or even discontinuous. This can be demonstrated by many examples in electrodynamics and quantum mechanics. There are two possibilities to escape this dilemma. One may either reformulate the fundamental laws of corresponding physical theories or look for a new basis, an extended concept of function, a new concept of differentiability etc. Both ways are practicable. In the first case one may, for example, replace partial differential equations by integral identities that are also valid for more general functions. But here one frequently encounters new difficulties, not to mention the fact that mathematicians and physicists prefer to calculate with partial differential equations rather than with integral identities. The other way leads to the theory of distributions (generalized functions): the concepts of a function, of differentiability etc. are extended, while the partial differential equations are retained. Many technical difficulties and (from the physical point of view) artificial constraints on the partial differential equations of theoretical physics are avoided by this new calculus. Anyway, the physicists hadn't bothered so much about such "mathematical subtleties" (and very rightly so, as we know today). After all, it seems to be a sort of quality characteristic of good theoretical physicists that even on unexplored mathematical terrain they instinctively avoid or jump over the pitfalls (including the well masked ones) that are abundant there. Less competent theoretical physicists collect in these pitfalls, but will deny this vehemently. Here it is recommendable to wait until mathematics will reclaim these areas with heavy clearing tools and mark "popular" hiking trails. A mathematician may either loftily turn up his nose, or observe with astonishment and admiration (though with slight discomfort and a trace of jealousy) that results which are equally striking and spectacular from both the physical and the mathematical point of view come out at the end of a way which, for the present, lacks full mathematical support. In this chapter we shall develop the fundamentals of the theory of distributions, while applications of this theory to partial differential equations will be described in the following Chap. 23.

22.1.2. The Spaces $D(\Omega)$ and $D'(\Omega)$

If Ω is a domain in R_n, then we denote by $D(\Omega)$ the class of all complex functions with compact support that are differentiable on Ω up to arbitrary order (another common notation is $C_0^\infty(\Omega)$). The definition of a function with compact support has been given in Def. 14.6.4/2. Theorem 14.6.4/2 has shown that $D(\Omega)$ is dense in $L_p(\Omega)$, with $1 \leqq p < \infty$.

Definition 1. *A sequence of functions $\{\varphi_k(x)\}_{k=1}^\infty \subset D(\Omega)$ converges in $D(\Omega)$ to $\varphi(x) \in D(\Omega)$ if all the supports of the functions $\varphi_k(x)$ are included in a compact subset of Ω and if $D^\alpha\varphi_k(x)$ converges uniformly to $D^\alpha\varphi(x)$ for every multiple index α.*

Remark 1. For the convergence in $D(\Omega)$ we also write $\varphi_k \underset{D(\Omega)}{\longrightarrow} \varphi$. Thus, what is required is that there exists a compact set ω, with $\omega \subset \Omega$ (depending on the sequence $\{\varphi_k\}$) such that supp $\varphi_k \subset \omega$ (Fig. 22.1). The concept of a support, supp, has been explained in Def. 14.6.4/2 and in Remark 14.6.4/2. Since ω is closed and bounded and Ω is open, ω has a positive distance from the boundary $\partial\Omega$ of Ω. Since $\varphi(x)$ is the limit approached by uniform convergence of $\varphi_k(x)$, one also has supp $\varphi \subset \omega$ (to which there corresponds the multiple index $\alpha = (0, ..., 0)$). The notation $D^\alpha\varphi$ for a multiple index α was introduced in Remark 8.1.2.

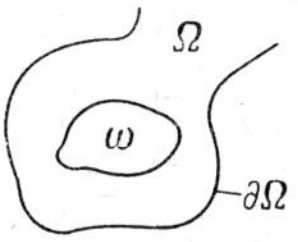

Fig. 22.1

Definition 2. *A distribution is a complex continuous linear form T over $D(\Omega)$. That means, every $\varphi \in D(\Omega)$ is assigned a complex number $T(\varphi)$ with the following two properties:*

(1) *For $\varphi \in D(\Omega)$, $\psi \in D(\Omega)$, and the complex numbers λ and μ,*
$T(\lambda\varphi + \mu\psi) = \lambda T(\varphi) + \mu T(\psi)$ *(linearity).*

(2) $\varphi_k \underset{D(\Omega)}{\longrightarrow} \varphi$ *implies that $T(\varphi_k) \to T(\varphi)$ (continuity).*

The class of all distributions over $D(\Omega)$ is denoted by $D'(\Omega)$. Here $T_1 = T_2$ means that $T_1(\varphi) = T_2(\varphi)$ for all $\varphi \in D(\Omega)$.

Remark 2. This is the fundamental concept in the theory of distributions, the elements of which were developed by S. L. Sobolev in 1936, while L. Schwartz completed the theory systematically in the late forties and fifties. Since that time distributions play a fundamental role in modern analysis as well as in some parts of theoretical physics.

22.1.3. Examples of Distributions

If Ω is a domain in R_n, then $L_1^{\text{loc}}(\Omega)$ denotes the class of all locally integrable complex functions on Ω. This means: if ω is a compact subset of Ω, and if $\chi_\omega(x)$ is the associated characteristic function, then $f(x)$ belongs to $L_1^{\text{loc}}(\Omega)$ if and only if $\chi_\omega(x)\, f(x) \in L_1(\Omega)$ for every set ω of this kind. For example, one has $\frac{1}{t} \in L_1^{\text{loc}}(\Omega)$, where $\Omega = (0, 1)$, but $\frac{1}{t} \notin L_1(\Omega)$. The fact that $f(x)$ belongs to $L_1^{\text{loc}}(\Omega)$ does not include any information about the boundary behaviour. As for the spaces $L_p(\Omega)$, we make the convention that two functions $f(x)$ and $g(x)$ are considered to be identical elements of $L_1^{\text{loc}}(\Omega)$ if they differ from one another only on a set of the Lebesgue measure zero. For this point, refer to Subsec. 14.6.1.

Lemma. *If* $f(x) \in L_1^{\text{loc}}(\Omega)$, *and if*

$$\int_\Omega f(x)\,\varphi(x)\,\mathrm{d}x = 0 \quad \textit{for all} \quad \varphi(x) \in D(\Omega)\,, \tag{1}$$

then $f(x) = 0$ *for almost all* $x \in \Omega$ *(i.e., the zero element in* $L_1^{\text{loc}}(\Omega)$*).*

Remark. Since $\varphi(x)$ has a compact support, the integral in (1) is meaningful. The lemma is a generalization of the fundamental lemma of the calculus of variations as given in Subsec. 11.1.2.

Regular distributions (or *distributions of the function type*): If Ω is an arbitrary domain in R_n, and if $f(x) \in L_1^{\text{loc}}(\Omega)$, then it is easy to see that

$$T_f(\varphi) = \int_\Omega f(x)\,\varphi(x)\,\mathrm{d}x \quad \text{for} \quad \varphi(x) \in D(\Omega) \tag{2}$$

fulfils the conditions of Def. 22.1.2/2. Hence $T_f \in D'(\Omega)$. On the other hand, if $g \in L_1^{\text{loc}}(\Omega)$, and if $T_f(\varphi) = T_g(\varphi)$ for all $\varphi \in D(\Omega)$, then it follows from the above lemma that $f = g$ in $L_1^{\text{loc}}(\Omega)$. Thus, according to the above convention, the correspondence between $f \in L_1^{\text{loc}}(\Omega)$ and $T_f \in D'(\Omega)$ is one to one. In this sense we identify $L_1^{\text{loc}}(\Omega)$ with the corresponding distributions (Fig. 22.2). Instead of $T_f \in D'(\Omega)$ we then write for abbreviation $f(x) \in D'(\Omega)$. Distributions of this kind are called regular.

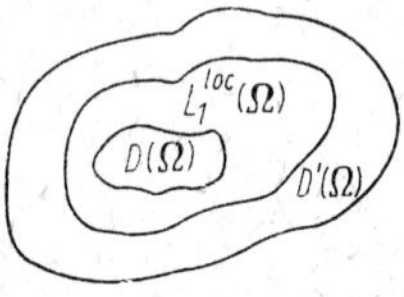

Fig. 22.2

δ-Distribution (or *Dirac's distribution*): A distribution which is not regular in the above sense is called singular. If $a \in \Omega$, then δ_a,

$$\delta_a(\varphi) = \varphi(a) \quad \text{for all} \quad \varphi \in D(\Omega)\,,$$

is a singular distribution. If $0 \in \Omega$, then one also writes δ instead of δ_0.

Other examples: If $a \in \Omega$, and if $f(x) \in L_1^{\text{loc}}(\Omega)$, then it is easily seen that T, with

$$T(\varphi) = D^\alpha \varphi(a) \quad \text{and} \quad T(\varphi) = \int_\Omega f(x)\,D^\alpha \varphi(x)\,\mathrm{d}x \quad \text{for} \quad \varphi \in D(\Omega)\,,$$

are distributions in $D'(\Omega)$ for every multiple index α.

22.1.4. Operations with Distributions

Definition 1. *Let* Ω *be a domain in* R_n.

(**a**) (*Differentiation*). *For* $T \in D'(\Omega)$ *and a multiple index* α,

$$(D^\alpha T)(\varphi) = (-1)^{|\alpha|}\,T(D^\alpha \varphi) \quad \textit{for} \quad \varphi \in D(\Omega)\,.$$

(**b**) (*Multiplication*). *For a complex function* $a(x)$ *that is differentiable on* Ω *up to arbitrary order, and for* $T \in D'(\Omega)$,

$$(aT)(\varphi) = (a(x)T)(\varphi) = T(a(x)\,\varphi(x)) \quad \textit{for} \quad \varphi \in D(\Omega)\,.$$

(c) (*Addition*). *For* $T_1 \in D'(\Omega)$, $T_2 \in D'(\Omega)$, *and for the complex numbers* λ_1 *and* λ_2,

$$(\lambda_1 T_1 + \lambda_2 T_2)(\varphi) = \lambda_1 T_1(\varphi) + \lambda_2 T_2(\varphi) \quad \text{for} \quad \varphi \in D(\Omega) .$$

Theorem 1. *If all quantities have the meaning stated in the above definition, then*

$$D^\alpha T \in D'(\Omega), \qquad aT \in D'(\Omega) \quad \text{and} \quad \lambda_1 T_1 + \lambda_2 T_2 \in D'(\Omega) .$$

Further (in the sense of the above definition) the following relation holds

$$\frac{\partial}{\partial x_j}(a(x)\ T) = \frac{\partial a}{\partial x_j}\ T + a\frac{\partial T}{\partial x_j} \quad \text{for} \quad j = 1, \ldots, n \tag{1}$$

and $D^{\alpha+\beta}T = D^\alpha(D^\beta T) = D^\beta(D^\alpha T)$. *Here* $\alpha = (\alpha_1, \ldots, \alpha_n)$ *and* $\beta = (\beta_1, \ldots, \beta_n)$ *are multiple indices, and* $\alpha + \beta = (\alpha_1 + \beta_1, \ldots, \alpha_n + \beta_n)$.

Remark 1. Hence it is possible to differentiate distributions up to arbitrary order, and to multiply them by smooth functions. If $f \in L_1^{\text{loc}}(\Omega)$, and if $T_f \in D'(\Omega)$ is the corresponding distribution, then $T_{af} = aT_f$, where $a(x)$ is differentiable up to arbitrary order on Ω. If $f(x) \in L_1^{\text{loc}}(\Omega)$ has classical partial derivatives up to the order $|\alpha|$, which shall be denoted by $D^\alpha f(x)$, then $D^\alpha T_f = T_{D^\alpha f(x)}$. In the sense of the identification of $L_1^{\text{loc}}(\Omega)$ with the regular distributions, the previous concepts (multiplication by functions, differentiation) thus agree with the newly introduced ones.

Remark 2. It follows from (1) that for $D^\alpha(a(x)\ T)$ the usual Leibniz formula for the derivative of a product is applicable.

Remark 3 (examples). If Ω is a domain in R_n and if $a \in \Omega$, then

$$(D^\alpha \delta_a)(\varphi) = (-1)^{|\alpha|} D^\alpha \varphi(a) \quad \text{for} \quad \varphi \in D(\Omega) .$$

Moreover, if $p(x)$ is a function differentiable on Ω up to arbitrary order, then

$$(p(x)\delta_a)(\varphi) = p(a)\ \varphi(a) .$$

We consider another example. Let $\Omega = R_1$, and let

$$y(t) = \begin{cases} 1 & \text{for} \quad 0 < t < \infty \\ 0 & \text{for} \quad -\infty < t \leqq 0 \end{cases} \quad \text{be the Heaviside function} .$$

If $y(t) \in D'(R_1)$ is regarded as a distribution over R_1, then $y' = \delta$ (where y' is the distribution derivative).

Lemma (*partition of unity*). *Let* B *be a bounded closed set in* R_n, *and let* $\{O_k\}_{k=1}^N$ *be a system of bounded open sets in* R_n, *such that* $B \subset \bigcup_{k=1}^N O_k$ (Fig. 22.3). *Then there exist real-valued functions* $\varphi_k \in D(O_k)$ *such that* $\sum_{k=1}^N \varphi_k(x) = 1$ *for* $x \in B$. (*Here the functions* $\varphi_k(x)$ *are continued to vanish outside of* O_k.)

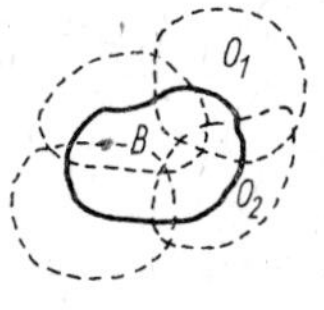

Fig. 22.3

Remark 4. Thus the function $f(x) \equiv 1$ for $x \in B$ can be decomposed into smooth partial functions, each of which is different from zero only in a small domain. For the above functions it is also possible to achieve that $0 \leqq \varphi_k(x) \leqq 1$. The meaning of this important lemma is that it is possible to reduce global properties (of functions and distributions) to local ones. This is exemplified by the following theorem.

Theorem 2. *Let Ω be a domain in R_n, and let $\{O_k\}_{k=1}^{\infty}$ be a system of nonvoid bounded domains in R_n, such that $\bigcup_{k=1}^{\infty} O_k = \Omega$. If the restrictions of two distributions $T_1 \in D'(\Omega)$ and $T_2 \in D'(\Omega)$ coincide on O_k, $k = 1, 2, 3, \ldots$, then $T_1 = T_2$.*

Remark 5. If $T \in D'(\Omega)$, then also $T \in D'(O_k)$, because the restriction $T(\varphi)$ to $\varphi \in D(O_k)$ yields a distribution over O_k. In this sense, the restriction of a distribution over Ω to a subdomain of Ω again yields a distribution. The theorem then says that (the global property) $T_1 = T_2 \in D'(\Omega)$ holds if and only if (the local property) $T_1 = T_2 \in D'(O_k)$ for $k = 1, 2, \ldots$ The theorem is proved by means of the above lemma.

Definition 2. *Let Ω be a domain in R_n, and let $T \in D'(\Omega)$. Then* supp T *(the support of T) is the set of all points $x \in \bar{\Omega}$ for which the restriction of T to $\Omega \cap \{y \mid |x-y| < \delta\}$ is different from the zero distribution for every positive number δ.*

Remark 6. Hence the method of checking $x \in \operatorname{supp} T$ is as follows. One asks whether for every positive number δ there exists a function $\varphi_\delta \in D(\Omega \cap \{y \mid |x-y| < \delta\})$ such that $T(\varphi_\delta) \neq 0$. If this is the case, then x belongs to supp T, otherwise not. supp T may also contain points $x \in \partial\Omega$ that are not in Ω. It is easily seen that supp T is a closed set in R_n. The restriction of T to $\Omega \setminus (\Omega \cap \operatorname{supp} T)$ is the zero distribution. This follows from Theorem 2. Moreover it is seen that $\Omega \setminus (\Omega \cap \operatorname{supp} T)$ is the largest domain with this property.

Remark 7. If $f(x) \in L_1^{\text{loc}}(\Omega)$, then we can form supp f by Def. 14.6.4/2 and supp $f = \operatorname{supp} T_f$ by Def. 2. One has $\operatorname{supp} T_f \subset \operatorname{supp} f$, but it may happen that supp f with f as a function includes $\operatorname{supp} T_f = \operatorname{supp} f$, with f as a distribution, as a proper subset. In the first case every point x with $f(x) \neq 0$ is considered, whereas in the second case points of a set of measure zero do not play any role. For continuous functions the two concepts coincide.

Remark 8. If Ω is a domain in R_n, and if $a \in \Omega$, then $\operatorname{supp} D^\alpha \delta_a = \{a\}$ for every multiple index α. If $y(t)$ is the Heaviside function referred to in Remark 3, then $\operatorname{supp} y = [0, \infty)$.

22.1.5. The Space $E'(\Omega)$

Definition. *If Ω is a domain in R_n, then $E'(\Omega)$ is the class of all distributions in $D'(\Omega)$ whose supports are bounded and lie in Ω.*

Remark 1. As $\operatorname{supp} T \subset \bar{\Omega}$ for arbitrary $T \in D'(\Omega)$, the above requirement that $\operatorname{supp} T \subset \Omega$ means an essential restriction. Since, moreover, supp T is bounded and closed for $T \in E'(\Omega)$, in this case supp T has a positive distance from the boundary $\partial\Omega$ (Fig. 22.4). This is also the essential property required for the proof of the following theorem.

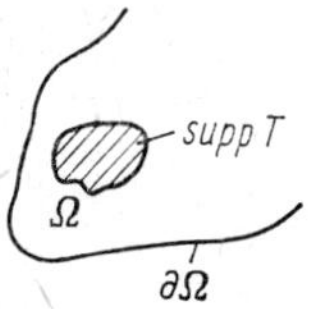

Fig. 22.4

Theorem 1. *If Ω is a domain in R_n, and if $T \in E'(\Omega)$, then there exist a natural number N and a positive number C such that, for all $\varphi \in D(\Omega)$,*

$$|T(\varphi)| \leq C \sup_{x \in \Omega} \sum_{|\alpha| \leq N} |D^\alpha \varphi(x)| . \tag{1}$$

Remark 2. Conversely, if an arbitrary linear form T over $D(\Omega)$ can be estimated in the form (1), then it is immediately obvious that $T \in D'(\Omega)$.

Theorem 2. *If Ω is a domain in R_n, and if $a \in \Omega$ and* supp $T=\{a\}$, *then there exist a natural number N and complex numbers a_α such that*

$$T = \sum_{|\alpha| \leqq N} a_\alpha D^\alpha \delta_a \,. \tag{2}$$

Remark 3. If T has the form (2), then T is either the zero distribution, or supp $T=\{a\}$. In other words, with the exception of the zero distribution, supp $T=\{a\}$ if and only if T has the form (2).

22.2. The Fourier Transform and the Spaces $S(R_n)$ and $S'(R_n)$

22.2.1. The Space $S(R_n)$ and the Fourier Transform

Definition 1. *The space $S(R_n)$ consists of all complex functions $\varphi(x)$ that are differentiable on R_n up to arbitrary order and have the property that*

$$\|\varphi\|_{k,l} = \sup_{x \in R_n} (1+|x|^k) \sum_{|\alpha| \leqq l} |D^\alpha \varphi(x)| < \infty \tag{1}$$

for $k=1, 2, 3, \ldots$ and $l=1, 2, \ldots$ If $\{\varphi_m\}_{m-1}^{\infty} \subset S(R_n)$ and $\varphi \in S(R_n)$, then we write $\varphi_m \xrightarrow[S(R_n)]{} \varphi$ (convergence in $S(R_n)$) if, for all k and l, $\|\varphi_m - \varphi\|_{k,l} \to 0$ as $m \to \infty$.

Remark 1. $\|\varphi\|_{k,l}$ is a norm. Thus $S(R_n)$ is characterized by a countably infinite number of norms. Spaces of this kind are also called locally convex. $S(R_n)$ is called the space of rapidly decreasing functions. The meaning of the designation will be clear from a look at (1). One has $D(R_n) \subset S(R_n)$. A function that belongs to $S(R_n)$ but not to $D(R_n)$ is $e^{-|x|^2}$.

Definition 2. *If $\varphi \in S(R_n)$, then $F\varphi$, with*

$$(F\varphi)(x) = (2\pi)^{-\frac{n}{2}} \int_{R_n} e^{-i\langle x,\xi\rangle} \varphi(\xi)\, d\xi, \quad x \in R_n \,, \tag{2}$$

is the Fourier transform of φ.

Remark 2. If $\xi = (\xi_1, \ldots, \xi_n)$ and $x = (x_1, \ldots, x_n)$ then we now set (in order to avoid confusion) $\langle x, \xi \rangle = \sum_{k=1}^{n} x_k \xi_k$. Since $\varphi \in S(R_n)$, it is easily seen that (2) exists for every $x \in R_n$.

22.2.2. Properties of the Fourier Transform

As previously we write $x^\alpha = x_1^{\alpha_1} \ldots x_n^{\alpha_n}$, where $x = (x_1, \ldots, x_n) \in R_n$ *and* $\alpha = (\alpha_1, \ldots, \alpha_n)$ is a multiple index.

Lemma 1. (a) *If $\varphi \in S(R_n)$, then also $F\varphi \in S(R_n)$, and*

$$D^\alpha(F\varphi) = (-i)^{|\alpha|} F(x^\alpha \varphi) \quad \text{as well as} \quad \xi^\alpha F\varphi = (-i)^{|\alpha|} F(D^\alpha \varphi) \,. \tag{1}$$

(b) *If $\{\varphi_k\}_{k=1}^{\infty} \subset S(R_n)$ and if $\varphi_k \xrightarrow[S(R_n)]{} \varphi$, then $F\varphi_k \xrightarrow[S(R_n)]{} F\varphi$.*

Remark 1. According to (1) the Fourier transform interchanges derivatives D^α with multiplications x^α. This already suggests that it is possible to reduce partial differential equations to algebraic equations by means of Fourier transforms.

Lemma 2. (a) $F(e^{-\frac{|\xi|^2}{2}})=e^{-\frac{|x|^2}{2}}$.

(b) *If* $\varphi \in S(R_n)$ *and* $\varepsilon > 0$, *then*

$$F(\varphi(\varepsilon\xi))(x)=\varepsilon^{-n}F(\varphi(\xi))(x\varepsilon^{-1}) .$$

Remark 2. Consequently, $e^{-\frac{|x|^2}{2}}$ is an eigenfunction of F.

Theorem. *The Fourier transform F provides a one to one mapping of $S(R_n)$ onto $S(R_n)$. The inverse mapping F^{-1} is called inverse Fourier transform, being given by*

$$(F^{-1}\varphi)(x)=(2\pi)^{-\frac{n}{2}}\int_{R_n} e^{i\langle x,\xi\rangle}\varphi(\xi)\,d\xi, \quad \varphi\in S(R_n) . \tag{2}$$

Remark 3. A mapping "onto" means that the range of F is the total $S(R_n)$, i.e., $F(S(R_n))= =S(R_n)$. The formula (2) differs from (22.2.1/2) only by a sign in the exponent. For $\varphi\in S(R_n)$, $F(F^{-1}\varphi)=F^{-1}(F\varphi)=\varphi$.

22.2.3. The Space $S'(R_n)$

Definition. *A tempered (or slowly increasing) distribution is a complex continuous linear form T over $S(R_n)$. This means that every $\varphi\in S(R_n)$ is assigned a complex number $T(\varphi)$ with the following two properties:*

(1) *For $\varphi\in S(R_n)$, $\psi\in S(R_n)$ and the complex numbers λ and μ,*

$$T(\lambda\varphi+\mu\psi)=\lambda T(\varphi)+\mu T(\psi) \qquad (\textit{linearity}) .$$

(2) $\varphi_k \xrightarrow[S(R_n)]{} \varphi$ *implies that* $T(\varphi_k)\to T(\varphi)$ *(continuity).*

The class of all tempered distributions is denoted by $S'(R_n)$. Here $T_1=T_2$ means that $T_1(\varphi)=T_2(\varphi)$ for all $\varphi\in S(R_n)$.

Remark 1. This is the analogue to Def. 22.1.2/2. $\varphi_k \xrightarrow[D(R_n)]{} \varphi$ implies that $\varphi_k \xrightarrow[S(R_n)]{} \varphi$. Hence it follows that the restriction of $T\in S'(R_n)$ to $D(R_n)$ belongs to $D'(R_n)$. In this sense one thus has $S'(R_n)\subset D'(R_n)$. On the other hand, if $T\in E'(R_n)$ and if $\psi\in D(R_n)$ is a function such that $\psi(x)=1$ in a neighbourhood of supp T, then $T(\varphi)=T(\varphi\psi)$ is meaningful for $\varphi\in S(R_n)$. It can be verified that $T(\varphi)$ is independent of ψ and that $T\in S'(R_n)$. Hence

$$E'(R_n)\subset S'(R_n)\subset D'(R_n)$$

in the sense of the above interpretations.

Theorem. *If $T\in S'(R_n)$, then there exist natural numbers k and l as well as a positive number C, such that for all $\varphi\in S(R_n)$*

$$|T(\varphi)|\leq C\,\|\varphi\|_{k,l} . \tag{1}$$

Remark 2. This is the analogue to Theorem 22.1.5/1. Conversely, if an arbitrary linear form T over $S(R_n)$ can be estimated in the form (1), then it is immediately seen that $T\in S'(R_n)$ is a tempered distribution.

Remark 3 (examples). T_f with

$$T_f(\varphi)=\int_{R_n} f(x)\,\varphi(x)\,dx, \quad \varphi\in S(R_n) , \tag{2}$$

$f\in L_p(R_n)$, $1\leq p\leq\infty$, belongs to $S'(R_n)$. Here $L_p(R_n)$ has the same meaning as in Subsec. 14.6.4., where $f(x)\in L_p(R_n)$ can now also be complex-valued (cf. Remark 20.1.1/2 and Remark 17.3.2). By analogy with the interpretation that $L_1^{\mathrm{loc}}(R_n)\subset D'(R_n)$, as stated in Subsec. 22.1.3., we now have $L_p(R_n)\subset S'(R_n)$ for $1\leqq p\leqq\infty$. If in formula (2) one chooses polynomials in R_n, then one likewise obtains tempered distributions. Moreover it is observed that the increase of $|f(x)|$ as $|x|\to\infty$ must not be too rapid if T_f is to belong to $S'(R_n)$ (this accounts for the terminology in the above definition).

22.2.4. The Fourier Transform in $S'(R_n)$

Definition. *If $T\in S'(R_n)$, then FT defined by*

$$(FT)(\varphi)=T(F\varphi) \quad \textit{for} \quad \varphi\in S(R_n)$$

is called the Fourier transform of T, and $F^{-1}T$ defined by

$$(F^{-1}T)(\varphi)=T(F^{-1}\varphi) \quad \textit{for} \quad \varphi\in S(R_n)$$

is called the inverse Fourier transform of T.

Remark 1. $(FT)(\varphi)$ and $(F^{-1}T)(\varphi)$ are linear. If $|T(\varphi)|\leqq c\,\|\varphi\|_{k,l}$, then it follows that

$$|(FT)(\varphi)|=|T(F\varphi)|\leqq c\,\|F\varphi\|_{k,l}\leqq c'\,\|\varphi\|_{l+n+1,k}\,.$$

Hence $FT\in S'(R_n)$, and analogously $F^{-1}T\in S'(R_n)$. Thus one again obtains tempered distributions.

Remark 2. If $\varphi\in S(R_n)$, then we have two definitions for $F\varphi$, namely Def. 22.2.1/2 and the above definition, provided that $\varphi=T_\varphi$ is interpreted as an element of $S'(R_n)$. It turns out that both definitions lead to the same result. In other words, the above definition extends the Fourier transform and the inverse Fourier transform from $S(R_n)$ to $S'(R_n)$.

Theorem. *The Fourier transform F provides a one to one mapping of $S'(R_n)$ onto $S'(R_n)$. Further,*

$$F^{-1}FT=FF^{-1}T=T \quad \textit{for} \quad T\in S'(R_n)\,.$$

Remark 3. This is the extension of Theorem 22.2.2 to $S'(R_n)$. In particular F^{-1} also provides a one to one mapping of $S'(R_n)$ onto $S'(R_n)$.

Remark 4. If $f(x)\in L_1(R_n)$, then by Remark 22.2.3/3 it is possible to regard $f(x)$ as a regular distribution in $S'(R_n)$. Then Ff is also a regular distribution in $S'(R_n)$, and

$$(Ff)(x)=(2\pi)^{-\frac{n}{2}}\int_{R_n} e^{-i\langle x,\xi\rangle}f(\xi)\,d\xi\,.$$

Remark 5. The following example is of interest for the three-dimensional wave equation. Let

$$V_R(\varphi)=\int_{|x|=R}\varphi(x)\,ds \quad \text{for} \quad \varphi\in S(R_3) \quad \text{and} \quad R>0\,.$$

One has $V_R\in E'(R_3)$, and hence also $V_R\in S'(R_3)$. The support of the (singular) distribution V_R is the sphere $\{x\mid |x|=R\}$, $R>0$. The Fourier transform FV_R is a regular distribution, and

$$FV_R(x)=\sqrt{\frac{2}{\pi}}\,R\,\frac{\sin R\,|x|}{|x|}\,.$$

22.2.5. Other Properties of Fourier Transforms

Theorem 1. (a) *If T is a distribution with* supp $T=\{0\}$, *then FT is regular and equals a polynomial.*

(b) *If $T\in S'(R_n)$, then for every multiple index*

$$F(D^\alpha T)=\mathrm{i}^{|\alpha|}x^\alpha FT \quad \text{and} \quad F(x^\alpha T)=\mathrm{i}^{|\alpha|}D^\alpha(FT)\,.$$

Remark 1. By Theorem 22.1.5/2, part (a) of the above theorem follows from

$$F\Big(\sum_{|\alpha|\leqq N} a_\alpha D^\alpha\delta\Big)=\sum_{|\alpha|\leqq N}(2\pi)^{-\frac{n}{2}}\mathrm{i}^{|\alpha|}a_\alpha x^\alpha\,.$$

Hence in particular one has $F(\delta)=(2\pi)^{-\frac{n}{2}}$.

Remark 2. Part (b) of the theorem is a generalization of Lemma 22.2.2/1. If $T\in S'(R_n)$, then $x^\alpha T$ and $D^\alpha T$ also belong to $S'(R_n)$.

Theorem 2. *The Fourier transform F and the inverse Fourier transform F^{-1} are unitary operators that map the Hilbert space $L_2(R_n)$ onto itself.*

Remark 3. This theorem is of fundamental interest. The concept of the unitary operator has been described in Def. 21.1.6. In the notation used there, one has $V=F$ (or $V=F^{-1}$) and $D(V)=R(V)=L_2(R_n)$. Thus, what is meant here are the restrictions of F and F^{-1} (which are originally defined on the total $S'(R_n)$) to $L_2(R_n)$.

22.3. Tensor Products and Convolutions

22.3.1. Tensor Products

Let $x=(x_1, \ldots, x_n)\in R_n$, $y=(y_1, \ldots, y_m)\in R_m$, and let $(x, y)=(x_1, \ldots, x_n, y_1, \ldots, y_m)\in R_{n+m}$. We consider distributions $T\in D'(R_n)$ and $S\in D'(R_m)$. To characterize the dependence on the variables, we sometimes also write T_x instead of T and S_y instead of S.

Lemma. *If $S_y\in D'(R_m)$, and if $\varphi(x, y)\in D(R_{n+m})$, then $\psi(x)=S_y(\varphi(x, y))\in D(R_n)$, and*

$$D^\alpha\psi(x)=S_y(D_x^\alpha\varphi(x, y)) \quad \textit{for every multiple index } \alpha.$$

$$\varphi_k(x, y)\xrightarrow[D(R_{n+m})]{}\varphi(x, y) \quad \textit{implies that}$$

$$S_y(\varphi_k(x, y))=\psi_k(x)\xrightarrow[D(R_n)]{}\psi(x)=S_y(\varphi(x, y))\,.$$

Remark 1. Here $D_x^\alpha\varphi(x, y)=\dfrac{\partial^{|\alpha|}\varphi(x, y)}{\partial x_1^{\alpha_1}\ldots\partial x_n^{\alpha_n}}$, hence the index x indicates that the differentiation is carried out only with respect to $x_1, \ldots, x_n$, but not with respect to $y_1, \ldots, y_m$. An analogous statement holds for D_y^β. If there is no fear of confusion, we prefer to write D^α instead of D_x^α or D_y^α, which then means that the differentiation can be carried out with respect to all variables that come into question.

Theorem. *If $T\in D'(R_n)$ and $S\in D'(R_m)$, then there exists exactly one distribution $U\in D'(R_{n+m})$ such that, for all $\varphi(x)\in D(R_n)$ and all $\psi(y)\in D(R_m)$,*

$$U(\varphi(x)\,\psi(y))=T(\varphi(x))\,S(\psi(y))\,. \tag{1}$$

For $\varrho(x, y) \in D(R_{n+m})$, this distribution U can be calculated from

$$U(\varrho(x, y)) = S_y(T_x(\varrho(x, y))) = T_x(S_y(\varrho(x, y))) \,. \tag{2}$$

Remark 2. The lemma shows that the terms formed in (2) are meaningful. The linearity of U is obvious, and so is property (1). What remains to be shown is the continuity of U and its uniqueness.

Definition. *If $T \in D'(R_n)$ and $S \in D'(R_m)$, then $U \in D'(R_{n+m})$, as given in the above theorem, is called the tensor product of T and S. We write $U = T \otimes S$.*

22.3.2. Properties of Tensor Products

Theorem. *If $T \in D'(R_n)$, $S \in D'(R_m)$, and $V \in D'(R_l)$, then*

$$T_x \otimes S_y = S_y \otimes T_x \quad (\textit{commutativity})\,, \tag{1}$$

$$(T \otimes S) \otimes V = T \otimes (S \otimes V) \quad (\textit{associativity})\,, \tag{2}$$

$$D_x^\alpha (T_x \otimes S_y) = (D^\alpha T_x) \otimes S_y \quad \textit{and} \tag{3}$$

$$\operatorname{supp} T \otimes S = \{(x, y) \mid (x, y) \in R_{n+m},\ x \in \operatorname{supp} T,\ y \in \operatorname{supp} S\} \,.$$

Further, the tensor product is continuous: from $T_k \in D'(R_n)$, $T \in D'(R_n)$, and $T_k(\varphi) \to T(\varphi)$ for all $\varphi \in D(R_n)$ as $k \to \infty$, it follows that

$$(T_k \otimes S)(\varrho) \to (T \otimes S)(\varrho) \quad \textit{for all} \quad \varrho \in D(R_{n+m}) \quad \textit{and} \quad k \to \infty \,.$$

Remark 1. Formula (1) is a reformulation of (22.3.1/2); the associative law as well as (3) are also easily obtained from this representation.

Remark 2. Let $f(x) \in L_1^{\text{loc}}(R_n)$, and let $g(y) \in L_1^{\text{loc}}(R_m)$. If f and g are interpreted as regular distributions, then $f \otimes g$ is also a regular distribution, and

$$(f \otimes g)(x, y) = f(x)\, g(y) \quad \text{for} \quad x \in R_n \quad \text{and} \quad y \in R_m \,.$$

Thus the tensor product of distributions is a generalization of the point-by-point product of functions.

22.3.3. Convolutions

Lemma. *If $T \in D'(R_n)$, $S \in D'(R_n)$, and if $\varphi \in D(R_n)$, then*

$$\operatorname{supp} [\varphi\,(x+y)\,(T \otimes S)]$$
$$\subset \{(x, y) \mid (x, y) \in R_{2n},\ x \in \operatorname{supp} T,\ y \in \operatorname{supp} S,\ x+y \in \operatorname{supp} \varphi\} \,.$$

Remark 1. supp T has been introduced in Def. 22.1.4/2. According to Remark 22.1.4/7 it is of no account whether supp φ is understood in the sense of Def. 14.6.4/2 (as a function) or in the sense of Def. 22.1.4/2 (as a distribution).

Definition. *Let $T \in D'(R_n)$ and $S \in D'(R_n)$. If for every function $\varphi \in D(R_n)$ the set*

$$\{(x, y) \mid (x, y) \in R_{2n}\ x \in \operatorname{supp} T,\ y \in \operatorname{supp} S,\ x+y \in \operatorname{supp} \varphi\} \tag{1}$$

is bounded in R_{2n}, then $T * S$ defined by

$$(T * S)(\varphi) = [\varphi(x+y)(T \otimes S)](\varrho(x, y)) \quad \textit{for} \quad \varphi \in D(R_n) \tag{2}$$

is the convolution of T and S. Here $\varrho(x, y) \in D(R_{2n})$, with $\varrho(x, y) = 1$ in a neighbourhood of the set (1).

Remark 2. The above lemma shows that $M = \operatorname{supp}[\varphi(x+y)(T \otimes S)]$ is included in the set (1). Hence M is a closed and bounded set in R_{2n}. But then there exist functions $\varrho(x, y) \in D(R_{2n})$ such that $\varrho(x, y) = 1$ in a neighbourhood Ω (domain) of M, i.e., $M \subset \Omega$. (This is meant by the formulation in the above definition.) It is easily verified that (2) is independent of the choice of such functions ϱ.

22.3.4. Properties of Convolutions

Theorem 1. (a) *Let the assumptions of Def. 22.3.3 be satisfied. Then* $T*S \in D'(R_n)$. *Further* $S*T$ *exists (and belongs to* $D'(R_n)$ *too), and* $T*S = S*T$ *(commutativity). If* α *is a multiple index, then the convolutions* $D^\alpha T*S$ *and* $T*D^\alpha S$ *exist too (and belong to* $D'(R_n)$*), and*

$$D^\alpha(T*S) = D^\alpha T*S = T*D^\alpha S \,. \tag{1}$$

(b) *If the convolutions* $T*S_1$ *and* $T*S_2$ *exist in the sense of Def. 22.3.3, and if* λ_1 *and* λ_2 *are complex numbers, then* $T*(\lambda_1 S_1 + \lambda_2 S_2)$ *exists, and*

$$T*(\lambda_1 S_1 + \lambda_2 S_2) = \lambda_1(T*S_1) + \lambda_2(T*S_2) \,.$$

Remark 1. One has $\operatorname{supp} D^\alpha T \subset \operatorname{supp} T$. Hence it is clear that all distributions in (1) exist

Theorem 2. (a) *If* $T \in D'(R_n)$ *and* $S \in E'(R_n)$, *then* $T*S$ *exists (in the sense of Def. 22.3.3), and*

$$(T*S)(\varphi) = S_y(\eta(y)\, T_x(\varphi(x+y))) = T_x(S_y(\varphi(x+y))) \,, \tag{2}$$

where $\eta \in D(R_n)$ *is a function such that* $\eta(x) = 1$ *in a neighbourhood of* $\operatorname{supp} S$.

(b) *For every distribution* $T \in D'(R_n)$,

$$T*\delta = \delta*T = T \,. \tag{3}$$

Remark 2. The space $E'(R_n)$ has been introduced in Subsec. 22.1.5. In particular, $\delta \in E'(R_n)$, where δ has the same meaning as in Subsec. 22.1.3. Since $\operatorname{supp} S$ is bounded for $S \in E'(R_n)$, it is easily seen that, independently of $T \in D'(R_n)$, the set (22.3.3/1) is bounded. Thus one can form $T*S = S*T$, and all the terms formed in (2) are meaningful. If one sets $S_y = \delta$ in the last term in (2), then one obtains (3).

Theorem 3. *For* $f(x) \in L_1^{\text{loc}}(R_n)$ *and* $g(x) \in L_1^{\text{loc}}(R_n)$, *let* $\operatorname{supp} f$ *and* $\operatorname{supp} g$ *be the supports of these functions in the sense of Def. 14.6.4/2. If for every positive number* N *the set*

$$\{(x, y) \mid (x, y) \in R_{2n}, |x+y| \leq N, x \in \operatorname{supp} f, y \in \operatorname{supp} g\}$$

is bounded, then

$$h(x) = \int_{R_n} f(x-y)\, g(y)\, dy = \int_{R_n} f(y)\, g(x-y)\, dy \,, \tag{4}$$

exists for almost all $x \in R_n$, *and* $h(x) \in L_1^{\text{loc}}(R_n)$. *If* f *and* g *are interpreted as distributions in* $D'(R_n)$, *then the convolution* $f*g$ *exists. It is a regular distribution, and one has* $(f*g)(x) = h(x)$ *almost everywhere.*

Remark 3. Formula (4) was the starting point for the theory of convolutions of functions. Convolutions of distributions have to be regarded as a generalization of this original concept.

23. Partial Differential Equations and Distributions

23.1. Fundamental Solutions

23.1.1. Basic Properties

If $a_\alpha(x)$ are complex functions that are differentiable on R_n up to arbitrary order, and if $U \in D'(R_n)$, then we ask for distributions $T \in D'(R_n)$ such that

$$\sum_{|\alpha| \leqq m} a_\alpha(x)\, D^\alpha T = U\,. \tag{1}$$

In other words, desired are distribution solutions of partial differential equations.

Lemma. *If U and T are regular distributions, and if T has (classical) partial derivatives up to the order m, then T is a classical solution of* (1).

Remark 1. Since $T = t(x) \in C^m(R_n)$, it follows from (1) that $U = u(x) \in C(R_n)$. A classical solution then means that (1) is satisfied point by point. Conversely, every classical solution of (1) is also a distribution solution. So the above problem is the generalization of problems like those treated in Chap. 19.

Definition. *If $a_\alpha(x)$ are complex functions that are differentiable on R_n up to arbitrary order, then $G \in D'(R_n)$ is called a fundamental solution if it satisfies*

$$\sum_{|\alpha| \leqq m} a_\alpha(x)\, D^\alpha G = \delta\,.$$

Remark 2. Speaking more strictly, one should say: a fundamental solution with respect to the differential expression on the left-hand side of (1). Here δ is the δ-distribution defined in Subsec. 22.1.3.

Theorem. *Let the differential expression $\sum_{|\alpha| \leqq m} a_\alpha D^\alpha T$, with constant coefficients a_α, have a fundamental solution G.*

(a) *If for $U \in D'(R_n)$ the convolution $T = G * U$ exists, then T is a solution of*

$$\sum_{|\alpha| \leqq m} a_\alpha D^\alpha T = U\,.$$

(b) *In the class of the distributions that allow a convolution with G, $\sum_{|\alpha| \leqq m} a_\alpha D^\alpha T = U$ has at most one solution.*

Remark 3. Convolutions were described in Subsecs. 22.3.3. and 22.3.4. From the properties stated there it follows that

$$\sum_{|\alpha| \leqq m} a_\alpha D^\alpha (G * U) = \Big(\sum_{|\alpha| \leqq m} a_\alpha D^\alpha G\Big) * U = \delta * U = U\,,$$

which proves (a). The theorem also shows the importance of fundamental solutions. Of interest is the proof of the existence of fundamental solutions and their explicit construction.

23.1.2. The Laplace Equation

Theorem 1. *The regular distribution*

$$G(x)=\begin{cases}-\dfrac{1}{(n-2)\,|\omega_n|\,|x|^{n-2}} & \textit{for} \quad n>2\,,\\ \dfrac{1}{2\pi}\ln|x| & \textit{for} \quad n=2\end{cases} \tag{1}$$

is a fundamental solution for Δ, *which means that* $\Delta G=\delta$.

Remark 1. Confer Def. 19.2.1. As stated there, $\Delta=\sum_{k=1}^{n}\frac{\partial^2}{\partial x_k^2}$ is the Laplace differential operator. Here we always assume that $n\geqq 2$.

Remark 2. We thus can apply Theorem 23.1.1 (a). For the following theorem it is of interest to point out that G belongs not only to $D'(R_n)$ but even to $S'(R_n)$. A polynomial $P(x)=\sum_{|\alpha|\leqq m} b_\alpha x^\alpha$ is called harmonic if $\Delta P(x)\equiv 0$. Here (as previously) we set $x^\alpha=x_1^{\alpha_1}\dots x_n^{\alpha_n}$. Examples of harmonic polynomials are $1, x_1, \dots, x_n$, but also $x_1^2-x_2^2$ for $n=2$, etc.

Theorem 2. (**a**) *If* $U\in E'(R_n)$ *then* $U*G\in S'(R_n)$ *is a solution of* $\Delta(U*G)=U$. *Conversely, if* $T\in S'(R_n)$ *is a solution of* $\Delta T=U$, *then* T *can be represented as*

$$T=U*G+P(x)\,, \tag{2}$$

where $P(x)$ *is a harmonic polynomial.*

(**b**) *If, in addition,* $U=u(x)\in E'(R_n)$ *is a regular distribution, then* (2) *can be written as*

$$T=t(x)=\int_{R_n} G(x-y)\,u(y)\,\mathrm{d}y+P(x)\,, \tag{3}$$

where $T=t(x)$ *is also a regular distribution.*

Remark 3. Since $U\in E'(R_n)$, $U*G\in D'(R_n)$ exists according to Theorem 22.3.4/2. As stated in the above theorem, it is possible to prove a much stronger assertion, namely that $U*G\in S'(R_n)$. (2) or (3) then give a complete survey of all solutions of $\Delta T=U$ that belong to $S'(R_n)$. (b) says that $u(x)\in L_1(R_n)$ has a bounded support, that the integral in (2) exists for almost all $x\in R_n$, and that $t(x)\in L_1^{\mathrm{loc}}(R_n)\cap S'(R_n)$. Compare (2) with the Newton potential referred to in Theorem 19.2.5/1. Differentiability assumptions of, say, the form $u(x)\in C^2(R_n)$ have now become superfluous.

23.1.3. The Heat Conduction Equation

Theorem 1. *The regular distribution*

$$G(x,t)=\begin{cases}2^{-n}\pi^{-\frac{n}{2}}t^{-\frac{n}{2}}\mathrm{e}^{-\frac{|x|}{4t}} & \textit{for} \quad t>0,\quad x\in R_n\,,\\ 0 & \textit{for} \quad t\leqq 0,\quad x\in R_n\,,\end{cases}$$

is a fundamental solution of the heat conduction equation, that means $\frac{\partial G}{\partial t}-\Delta G=\delta$.

Remark 1. One has $(x,t)\in R_{n+1}$, where $x=(x_1,\dots,x_n)\in R_n$ and $t\in R_1$. Further $G(x,t)\in L_1^{\mathrm{loc}}(R_{n+1})\subset D'(R_{n+1})$. As stated in Chap. 19., $\Delta=\sum_{k=1}^{n}\frac{\partial^2}{\partial x_k^2}$ does not contain any derivatives with respect to t. Compare $G(x,t)$ with $s(x,t)$ in Lemma 19.4.1. We are now able to apply Theorem 23.1.1 (a).

Theorem 2. *If* $U \in E'(R_n)$, *and if* $G(x, t)$ *is the function given in Theorem* 1, *then* $T = U*G$ *is a solution of the differential equation* $\frac{\partial T}{\partial t} - \Delta T = U$. *If additionally* $U = u(x, t)$ *is a regular distribution, then* $T = T(x, t)$ *is also a regular distribution, and*

$$T(x, t) = 2^{-n}\pi^{-\frac{n}{2}} \int_{-\infty}^{t} \int_{R_n} \frac{e^{-\frac{|x-y|^2}{4(t-\tau)}}}{(t-\tau)^{\frac{n}{2}}} u(y, \tau)\, dy\, d\tau . \tag{1}$$

Remark 2. According to Theorem 22.3.4/2 it is clear that $U * G$ exists. The last part of the theorem means that for almost all $(x, t) \in R_{n+1}$ the integral in (1) exists and $T(x, t) \in L_1^{\text{loc}}(R_{n+1})$, provided that $u(x, t) \in L_1(R_{n+1})$ has a bounded support. Moreover (1) shows that for the calculation of $T(x, t)$ one only needs τ-values with $\tau \leqq t$. This is physically plausible if t is interpreted as the time variable.

23.1.4. The Wave Equation

In Sec. 19.3. we discussed the wave equation in 1, 2, and 3 dimensions. Here we shall largely confine ourselves to the physically interesting three-dimensional case.

Theorem (**a**) *For* $x = (x_1, x_2, x_3) \in R_3$ *and* $t \in R_1$, *the singular distribution* $G \in D'(R_4)$,

$$G(\varphi) = \frac{1}{4\pi} \int_0^{\infty} \frac{1}{t} \int_{|x|=t} \varphi(x, t)\, ds_x dt, \quad \varphi \in D(R_4) , \tag{1}$$

is a fundamental solution of the three-dimensional wave equation, that means $\frac{\partial^2 G}{\partial t^2} - \Delta G = \delta$.

(**b**) *If* $U \in D'(R_4)$, *with* supp $U \subset \{(x, t) \mid (x, t) \in R_4, t \geqq 0\} = \overline{R_4^+}$, *then* $T = U*G$ *exists and is a solution of the differential equation* $\frac{\partial^2 T}{\partial t^2} - \Delta T = U$. *One has* supp $T \subset \overline{R_4^+}$. *If additionally* $U = u(x, t)$ *is a regular distribution, then* $T = T(x, t)$ *is also a regular distribution, and*

$$T(x, t) = \begin{cases} \frac{1}{4\pi} \int_{|x-y| \leqq t} \frac{u(y, t-|x-y|)}{|x-y|}\, dy & \text{for } t > 0 , \\ 0 & \text{for } t \leqq 0 . \end{cases} \tag{2}$$

Remark 1. Formula (1) can also be written as

$$G(\varphi) = \frac{1}{4\pi} \int_{R_3} \frac{\varphi(x, |x|)}{|x|}\, dx, \quad \varphi \in D(R_4) . \tag{3}$$

Hence it follows that supp $G = \{(x, t) \mid (x, t) \in R_4, t = |x|\}$. Since this is a set of the four-dimensional measure 0, it follows that G in (1) must be singular. In proving that G is a fundamental solution, one largely makes use of Remark 22.2.4/5. It is now easily seen that the convolution $U * G$ as per Def. 22.3.3 exists if supp $U \subset \overline{R_4^+}$. Then by Theorem 23.1.1 (a) $T = U * G$ is a solution of $\frac{\partial^2 T}{\partial t^2} - \Delta T = U$, where again $\Delta T = \sum_{k=1}^{n} \frac{\partial^2 T}{\partial x_k^2}$. (2) largely coincides with the retarded potentials defined by (19.3.5/1). In contrast to the considerations described there, here we do not need any assumptions on the differentiability of $u(x, t)$.

Remark 2. In the same way it is possible to indicate fundamental solutions for the one-dimensional and the two-dimensional wave equation. We refer to [66]. For the one-dimensional wave equation, the regular distribution (Fig. 23.1)

$$G(x,t)=\begin{cases}\frac{1}{2} & \text{for} \quad t>|x| \\ 0 & \text{for} \quad t\leq|x|,\end{cases}$$

$x\in R_1$, $t\in R_1$, is a fundamental solution, i.e., $\frac{\partial^2 G}{\partial t^2}-\frac{\partial^2 G}{\partial x^2}=\delta$.

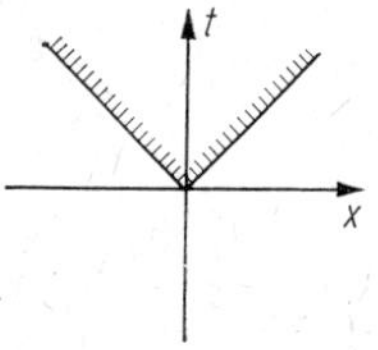

Fig. 23.1

23.2. Initial Value Problems

23.2.1. Formulation of the Problem

We already discussed the initial value problem for the wave equation in Sec. 19.3., and for the heat conduction equation in Subsec. 19.4.3. The problems and solutions described there shall now be referred to as "classical" ones. In this section we are going to consider generalizations within the theory of distributions. We again make use of the terminology of tensor products and convolutions as introduced in Sec. 22.3. As we did there, we shall write $U_{0,x}\in D'(R_n)$ instead of $U_0\in D'(R_n)$, where $x=(x_1, \ldots, x_n)$ is the general point in R_n. Further, in the following definition, let δ be the δ-distribution in R_1, and let δ' be its first derivative. If t is the general point in R_1, then we also write δ_t instead of δ (there should be no fear of confusion with the terminology used in Subsec. 22.1.3.). By analogy we write δ_t' instead of δ'. If $(x, t)\in R_{n+1}$, $x\in R_n$, $t\in R_1$, then in this sense $U_0\otimes\delta'= =U_{0,x}\otimes\delta_t'$ is the tensor product of $U_0\in D'(R_n)$ and $\delta'\in D'(R_1)$. Finally, R_{n+1}^+ and $\overline{R_{n+1}^+}$ have the same meaning as in Subsec. 19.3.1., and $\Delta=\sum_{k=1}^{n}\frac{\partial^2}{\partial x_k^2}$, which means that Δ does not contain any derivatives with respect to t.

Definition. (a) *If $U_0\in D'(R_n)$, $U_1\in D'(R_n)$, and if $F\in D'(R_{n+1})$ with* supp $F\subset\overline{R_{n+1}^+}$, *then $U\in D'(R_{n+1})$, with* supp $U\subset\overline{R_{n+1}^+}$ *and*

$$\frac{\partial^2 U}{\partial t^2}-\Delta U=F+U_0\otimes\delta'+U_1\otimes\delta \tag{1}$$

is said to be a solution of the initial value problem for the wave equation with the right-hand side F and the initial values U_0 and U_1.

(b) *If $U_0\in D'(R_n)$, and if $F\in D'(R_{n+1})$ with* supp $F\subset\overline{R_{n+1}^+}$, *then $U\in D'(R_{n+1})$, with* supp $U\subset\overline{R_{n+1}^+}$ *and*

$$\frac{\partial U}{\partial t}-\Delta U=F+U_0\otimes\delta \tag{2}$$

is said to be a solution of the initial value problem for the heat conduction equation with the right-hand side F and the initial value U_0.

Remark 1. As was already indicated, (x, t) with $x\in R_n$ and $t\in R_1$ is the general point in R_{n+1}. Further we have $U_0\otimes\delta'=U_{0,x}\otimes\delta'_t\in D'(R_{n+1})$ and $U_1\otimes\delta=U_{1,x}\otimes\delta_t\in D'(R_{n+1})$. We also have $\operatorname{supp}(U_0\otimes\delta')\subset\{(x,t)\mid (x,t)\in R_{n+1}, t=0\}$, and analogous relations for $U_1\otimes\delta$ and $U_0\otimes\delta$.

Remark 2. There arises the problem of how the above definition is related to the "classical" Definitions 19.3.1/1 and 19.4.3.

Theorem. (a) *Let U be a solution of the initial value problem for the wave equation as per part* (a) *of the above definition. Let $U_0=u_0(x)\in C^1(R_n)$, $U_1=u_1(x)\in C(R_n)$, $F=f(x,t)\in C(\overline{R^+_{n+1}})$ and $U=u(x,t)\in C^2(R^+_{n+1})\cap C^1(\overline{R^+_{n+1}})$ be regular distributions in $D'(R_n)$ and $D'(R_{n+1})$, respectively. Then $u(x, t)$ is a "classical" solution of the initial value problem for the wave equation in the sense of Def.* 19.3.1/1.

(b) *Let U be a solution of the initial value problem for the heat conduction equation as per part* (b) *of the above definition. Let $U_0=u_0(x)\in C(R_n)$, $F=f(x,t)\in C(\overline{R^+_{n+1}})$ and $U=u(x,t)\in C^2(R^+_{n+1})\cap C(\overline{R^+_{n+1}})$. Then $u(x, t)$ is a "classical" solution of the initial value problem for the heat conduction equation*

$$\frac{\partial u}{\partial t}(x,t)-\Delta u(x,t)=f(x,t) \quad \textit{for} \quad (x,t)\in R^+_{n+1}\,,$$

$$u(x,0)=u_0(x) \quad \textit{for} \quad x\in R_n\,.$$

Remark 3. In the formulation of the above theorem we have allowed ourselves some minor incorrectnesses. Instead of $F=f(x,t)\in C(\overline{R^+_{n+1}})$ we should formulate more correctly: let the restriction of F to R^+_{n+1} be a regular distribution in $D'(R^+_{n+1})$ that belongs to $C(\overline{R^+_{n+1}})$. The wording for $U=u(x,t)$ should be analogous. But since by definition we have $\operatorname{supp} F\subset\overline{R^+_{n+1}}$, misinterpretations should hardly be possible. The notations $C^2(R^+_{n+1})$ etc. were introduced in Subsec. 19.2.1.

Remark 4. There is a converse to this theorem. For $u_0(x)\in C^1(R_n)$, $u_1(x)\in C(R_n)$ and $f(x,t)\in C(\overline{R^+_{n+1}})$, let $u(x,t)\in C^2(R^+_{n+1})\cap C^1(\overline{R^+_{n+1}})$ be a classical solution of the initial value problem for the wave equation in the sense of Def. 19.3.1/1. If one sets $U_0=u_0(x)$, $U_1=u_1(x)$,

$$F=\begin{cases} f(x,t) & \text{for } x\in R_n\,,\ t\geqq 0,\\ 0 & \text{for } x\in R_n\,,\ t<0,\end{cases} \qquad U=\begin{cases} u(x,t) & \text{for } x\in R_n\,,\ t\geqq 0\,,\\ 0 & \text{for } x\in R_n\,,\ t<0\,,\end{cases}$$

then U is a solution of the initial value problem as per part (a) of the above definition. For $u_0(x)\in C(R_n)$ and $f(x,t)\in C(\overline{R^+_{n+1}})$, let $u(x,t)\in C^2(R^+_{n+1})\cap C(\overline{R^+_{n+1}})$ be a classical solution of the initial value problem for the heat conduction equation in the sense of part (b) of the above theorem. If U_0, F and U have the same meaning as above, then U is a solution of the initial value problem as per part (b) of the above definition. These remarks show that the above definition is an elegant generalization of classical initial value problems within the theory of distributions.

23.2.2. The Wave Equation

Theorem. *The initial value problem for the three-dimensional wave equation has a unique solution in the sense of Def.* 23.2.1(a). *If U_0, U_1 and F have the meaning indicated there, then*

$$U=G*(F+U_0\otimes\delta'+U_1\otimes\delta) \tag{1}$$

represents this solution, where G is the fundamental solution stated in Theorem 23.1.4. If $U_0 = u_0(x)$, $U_1 = u_1(x)$ *and* $F = f(x, t)$ *are regular distributions in* $D'(R_3)$ *and* $D'(R_4)$, *respectively (with* supp $F \subset \overline{R_4^+}$), *then* (1) *can be represented as*

$$U = \frac{\chi(t)}{4\pi} \int\limits_{|x-y| \leq t} \frac{f(y, t-|x-y|)}{|x-y|} \, dy + \frac{\chi(t)}{4\pi t} \int\limits_{|x-y|=t} u_1(y) \, ds_y + \frac{1}{4\pi} \frac{\partial}{\partial t} \left[\frac{\chi(t)}{t} \int\limits_{|y-x| \leq t} u_0(y) \, ds_y \right]. \tag{2}$$

Here $\chi(t) = 1$ *for* $t \geq 0$ *and* $\chi(t) = 0$ *for* $t < 0$. *If* $u_0(x) \in C^3(R_3)$, $u_1(x) \in C^2(R_3)$, *and if* $f(x, t) \in C^2(\overline{R_4^+})$, *then the regular distribution* $U = u(x, t)$ *is a classical solution of the initial value problem for the wave equation in the sense of Def.* 19.3.1/1.

Remark 1. The last part of the theorem again is to be understood in the sense of Remarks 23.2.1/3 and 23.2.1/4.

Remark 2. Since supp $(F + U_0 \otimes \delta' + U_1 \otimes \delta) \subset \overline{R_4^+}$, it follows from Theorem 23.1.4 that (1) is a solution of the problem. The derivative $\frac{\partial}{\partial t}$ in (2) is to be understood as a distribution derivative; thus it is not asserted that the right-hand side of (2) is a regular distribution (in contrast to the last part of the theorem, where rather strong assumptions are made on u_0, u_1 and f).

Remark 3. The last part of the theorem is identical with Theorem 19.3.5/2. Thus in particular $U = u(x, t) \in C^2(\overline{R_4^+})$ in the sense of Remark 1. So the above theorem is an essential generalization of the previous classical considerations presented in Chap. 19.

Remark 4. The solution (1) continuously depends on the initial data in the following sense. For $\{U_0^k\}_{k=1}^\infty \subset D'(R_n)$, $\{U_1^k\}_{k=1}^\infty \subset D'(R_n)$ and $\{F^k\}_{k=1}^\infty \subset D'(R_{n+1})$, with supp $F^k \subset \overline{R_{n+1}^+}$ for $k = 1, 2, 3, \ldots$, let

$$U_0^k(\varphi) \to U_0(\varphi), \qquad U_1^k(\varphi) \to U_1(\varphi) \quad \text{for} \quad \varphi \in D(R_n)\,,$$
$$F^k(\psi) \to F(\psi) \quad \text{for} \quad \psi \in D(R_{n+1}) \quad \text{and} \quad k \to \infty\,.$$

U, U_0, U_1 and F have the previous meaning. If U^k is the solution (1) with F^k, U_0^k and U_1^k instead of F, U_0 and U_1, then $U^k(\psi) \to U(\psi)$ as $k \to \infty$, where $\psi \in D(R_{n+1})$. Stability propositions of this kind are physically meaningful: small perturbations of the initial data cause only small perturbations of the solutions.

Remark 5. Analogous theorems and remarks can be presented for the one- and two-dimensional wave equation; cf. Par. 15 in [66]. In the case of the homogeneous wave equation (i.e., with $F = 0$) one then obtains generalizations of Theorems 19.3.2/2 and 19.3.3/2.

23.2.3. The Heat Conduction Equation

Theorem. *If* $U_0 \in E'(R_n)$, $F \in E'(R_{n+1})$, *and if* supp $F \subset \overline{R_{n+1}^+}$, *then*

$$U = G * (F + U_0 \otimes \delta) \tag{1}$$

is a solution of the initial value problem for the (n-dimensional) heat conduction equation as per Def. 23.2.1(b). *Here G is the fundamental solution given in Theorem* 23.1.3/1. *If* $U_0 = u_0(x) \in E'(R_n)$ *and* $F = f(x, t) \in E'(R_{n+1})$, *with* supp $F \subset \overline{R_{n+1}^+}$, *are* *regular distributions, then $U = u(x, t)$ in* (1) *is also a regular distribution, and*

$$u(x,t)=\frac{\chi(t)}{2^n\pi^{\frac{n}{2}}}\int_0^t\int_{R_n}\frac{e^{-\frac{|x-y|^2}{4(t-\tau)}}}{(t-\tau)^{\frac{n}{2}}}f(y,\tau)\,dy\,d\tau+\frac{\chi(t)}{2^n\pi^{\frac{n}{2}}t^{\frac{n}{2}}}\int_{R_n}e^{-\frac{|x-y|^2}{4t}}u_0(y)\,dy\,. \qquad (2)$$

Here $\chi(t)$ has the same meaning as in Theorem 23.2.2.

Remark 1. One has supp $(F+U_0\otimes\delta)\subset\overline{R^+_{n+1}}$ and $F+U_0\otimes\delta\in E'(R_{n+1})$. Theorem 23.1.3/2 now shows that U in (1) is a solution of the problem (it is readily verified that supp $U\subset\overline{R^+_{n+1}}$). The first summand in (2) is identical with (23.1.3/1) (apart from the modified notation). The second summand in (2) is analogous to (19.4.3/1). A comparison with Theorem 19.4.3 shows that here $u_0(x)=\varphi(x)\in L_1(R_n)$ is generally not bounded but, on the other hand, has a bounded support.

Remark 2. In contrast to Theorem 23.2.2 the above theorem does not contain a uniqueness proposition. The solution of the initial value problem for the heat conduction equation is not unique (cf. [18], Chap. I, Par. 9). Uniqueness can however be enforced by making additional demands upon the initial data and the solutions (limitations of growth as $|x|\to\infty$ and $t\to\infty$). Theorem 19.4.3 represents an example of this.

Remark 3. Stronger requirements for the regular distributions $u_0(x)$ and $f(x, t)$ ensure that $u(x, t)$ is a classical solution. Further, by analogy with Remark 23.2.2/4, it can be shown that (1) depends continuously on the data U_0 and F.

24. Fundamental Concepts of Classical Field Theory

24.1. Tensors

24.1.1. Introductory Remark

The axiom of classical point mechanics stated in Subsec. 12.2.1. is a variational principle. A problem of mechanics was assigned a Lagrangian function $L(t, x_k(t), \dot{x}_k(t))$, and we asked for extremals of the corresponding variational problem in the sense of Def. 11.1.3. This led to the Euler-Lagrange equations stated in Theorem 11.1.3/1. The solutions of this system of ordinary differential equations (to which initial conditions had been added) then yielded the trajectories of the mechanical problem. Thus the idea to describe the dynamics of natural processes by variational principles has proved highly successful in mechanics. Therefore it suggests itself to ask whether this principle can be extended to other areas of physics. We had made some remarks on this subject in Subsec. 11.2.1. In particular we had mentioned that the theory developed in Chap. 11 (calculus of variations) will not suffice for such an intent. On this basis it is only possible to develop theories whose fundamental equations are ordinary differential equations. But many theories of physics are governed by partial differential equations. If we want to keep to the line of obtaining these partial differential equations as a result of variational principles, then the basic ideas of the calculus of variations as presented in Chap. 11 must be decisively generalized. Here first of all it is our aim to present the fundamental ideas of the special theory of relativity, including electrodynamics. This is a so-called absolute space-time theory: the geometry of space is given (the four-dimensional Minkowskian space), and physical processes are described in this four-space. On the other hand, in general space-time theories such as the general theory of relativity, physical processes will (at least partially) determine the geometry of space. These few indications already suggest that the set of mathematical tools of general space-time theories will be much more complex than that of absolute space-time theories. For the presentation of the general theory of relativity we will, later on, need the theory of C^∞-manifolds as a basis. Starting from this point, we shall consider tensors and other geometric objects on manifolds (Chap. 29). For the moment, however, we can permit ourselves the luxury to postpone this rather complex theory to a later time. For the special theory of relativity and the electrodynamics associated with it, it will suffice for the present to consider tensor fields that are defined over a fixed domain in R_n. This will essentially facilitate the understanding of the basic ideas of this theory and of its interpretations. The purely mathematical aspect of this matter is not essentially different in these two cases: it is tantamount to a more or less masterly application of the chain rule of the differential calculus. In other words, in this chapter we reduce the mathematical foundations to the minimum required here and in the following Chap. 25. Those who want to know more must also invest more: for understanding the general theory of relativity to be presented in Chaps. 30 and 31 it will be necessary to study Chap. 29; Chap. 32 will be required for understanding Chap. 33, but for understanding Chap. 25 it will just suffice to read Chap. 24.

24.1.2. The Fundamental Tensor

Euclidean line element: If one considers the continuously differentiable curve $x(t) = (x^1(t), \ldots, x^n(t))$ in R_n, then, according to Subsec. 9.2.3.,

$$s(t) = \int_{t_0}^{t} \sqrt{(\dot{x}^1(\tau))^2 + \ldots + (\dot{x}^n(\tau))^2}\, \mathrm{d}\tau, \quad \dot{x}^k = \frac{\mathrm{d}x^k}{\mathrm{d}\tau}, \tag{1}$$

is the arc length. Within the theory we are going to develop here, it is expedient (and common practice) to use superscripts for the co-ordinates, i.e., $x = (x^1, \ldots, x^n) \in R_n$. From (1) it follows that

$$\left(\frac{\mathrm{d}s}{\mathrm{d}t}\right)^2 = \sum_{k=1}^{n} (\dot{x}^k(t))^2 = \sum_{k=1}^{n} \dot{x}^k(t)\, \dot{x}^k(t)\,. \tag{2}$$

It is usual (and, as we shall see, highly expedient) to write

$$\mathrm{d}s^2 = \sum_{k=1}^{n} \mathrm{d}x^k \mathrm{d}x^k = \mathrm{d}x^k \mathrm{d}x^k \tag{3}$$

instead of (2). In the last expression we have used Einstein's summation convention: take the sum over all doubly occurring indices, where from one case to another it is clear in which limits the sum has to be taken (in our case over $k=1, \ldots, n$). Thus (3) is an abbreviating notation for (2), which indicates that the parameter t is of no interest. Besides the Cartesian co-ordinates $x^1, \ldots, x^n$ in R_n (or in a domain in R_n) we now consider curvilinear co-ordinates $x'^1, \ldots, x'^n$ in the sense of Subsec. 8.2.2., i.e., $x'^k = x'^k(x^l)$ and $x^l = x^l(x'^k)$, where from now on we shall always assume that all co-ordinate transformations of this kind are differentiable up to arbitrary order (Fig. 24.1). We denote the functional determinant (or Jacobian) by $\dfrac{\partial(x'^l)}{\partial(x^k)} = \dfrac{\partial(x'^1, \ldots, x'^n)}{\partial(x^1, \ldots, x^n)}$, requiring that $\dfrac{\partial(x'^l)}{\partial(x^k)} > 0$ (in R_n or in the domain where the co-ordinate transformation is considered). If we now change over to x'^k in (3), then we obtain

$$ds^2 = g_{kl} dx'^k dx'^l, \quad \text{where} \quad g_{kl} = \frac{\partial x^m}{\partial x'^k} \cdot \frac{\partial x^m}{\partial x'^l}. \tag{4}$$

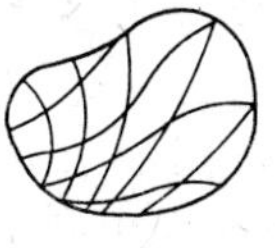

Fig. 24.1

Here we have to take account of the above summation convention. It is easily seen that

$$\det(g_{kl}) = \left(\frac{\partial(x^m)}{\partial(x'^p)}\right)^2 \neq 0 \quad \text{and} \quad g_{kl} = g_{lk}. \tag{5}$$

Here $\det(g_{kl})$ is the determinant of the positive definite matrix $(g_{kl})_{k,l=1}^n$. Hence $(g_{kl})_{k,l=1}^n$ is a positive definite, symmetric matrix. If one fixes a point $P \sim x$, then it is possible to find a non-singular matrix $A = A(P) = (a_{kl})_{k,l=1}^n$ such that

$$A^t (g_{kl})_{k,l=1}^n A = \begin{pmatrix} 1 & & 0 \\ & \ddots & \\ 0 & & 1 \end{pmatrix} \quad \text{(identity matrix)}, \tag{6}$$

where A^t stands for the conjugate matrix of A. We point out that this a pointwise property: A varies from one point to another. Thus (4) is the Euclidean line element ds^2 in general curvilinear co-ordinates.

The line element ds^2: As usual in mathematics, one considers an approved theory, reformulates it, and looks for meaningful generalizations. Concepts that had been derived previously, and hence were results, are now axiomatically taken as a starting point to develop an extended theory. So we shall do here, too. Let Ω be a domain in R_n (later on $\Omega = R_n$, and especially $\Omega = R_4$, will be of interest). Given a symmetrical matrix $(g_{kl})_{k,l=1}^n$, i.e., $g_{kl} = g_{lk}$, where $g_{kl}(x)$ are real-valued functions differentiable on Ω up to arbitrary order. Further let $\det(g_{kl}) \neq 0$ on Ω. We call $(g_{kl})_{k,l=1}^n$ the fundamental tensor, which is used to form the line element

$$ds^2 = g_{kl} dx^k dx^l, \quad g_{kl} = g_{kl}(x), \tag{7}$$

where again the summation convention has to be applied. Here $x = (x^1, \ldots, x^n)$ are Cartesian co-ordinates. The decisive alteration as compared with the Euclidean line element is that $(g_{kl})_{k,l=1}^n$ is not necessarily positive definite. For a fixed $x \in \Omega$ it is possible, by analogy with (6), to obtain the normal form

$$A^t(g_{kl})_{k,l=1}^n A = \begin{pmatrix} 1 & & & & 0 \\ & \ddots & & & \\ & & 1 & & \\ & & & -1 & \\ & & & & \ddots \\ 0 & & & & -1 \end{pmatrix}. \tag{8}$$

Here the non-singular matrix $A = A(x)$ depends on x, but not the number of $+1$ and -1 diagonal elements, respectively.

Examples. (1) If $\Omega = R_n$ and if $g_{kl} = \delta_{kl}$, so that $(g_{kl})_{k,l=1}^n$ is the identity matrix, then $ds^2 = dx^k dx^k$ is the Euclidean line element. (2) If $\Omega = R_4$, and if

$$(g_{kl}) = \begin{pmatrix} 1 & & & 0 \\ & 1 & & \\ & & 1 & \\ 0 & & & -1 \end{pmatrix}, \tag{9}$$

then $ds^2 = (dx^1)^2 + (dx^2)^2 + (dx^3)^2 - (dx^4)^2$ is the Lorentz metric, with $\det(g_{kl}) = -1$. Later on this metric will play a decisive role. Sometimes the line element (7) is also called a metric, but this has nothing to do with the metric of a metric space as described in Subsec. 1.3.3.

Transformation of the fundamental tensor: As in the case of the Euclidean line element, besides the Cartesian co-ordinates $x^1, \ldots, x^n$ in Ω we again consider arbitrary curvilinear co-ordinates $x'^1, \ldots, x'^n$, where again the previous conditions shall hold. Conversion of ds^2 in (7) to the curvilinear co-ordinates gives

$$ds^2 = g'_{kl} dx'^k dx'^l \quad \text{where} \quad g'_{kl} = g_{pq} \frac{\partial x^p}{\partial x'^k} \cdot \frac{\partial x^q}{\partial x'^l}. \tag{10}$$

Lemma: *If* (x'^k) *and* (x''^k) *are two curvilinear co-ordinate systems in* Ω, *and if* (g'_{kl}) *and* (g''_{kl}) *are the corresponding fundamental tensors in the sense of* (10), *then*

$$g''_{kl} = g'_{pq} \frac{\partial x'^p}{\partial x''^k} \cdot \frac{\partial x'^q}{\partial x''^l}, \quad \det(g''_{kl}) = \left(\frac{\partial(x'^k)}{\partial(x''^l)}\right)^2 \det(g'_{kl}). \tag{11}$$

Remark 1. Formula (11) is easily deduced from (10). The first formula in (11) is called the transformation law of a covariant tensor of order two. The second formula in (10) is a special case. From (11) it is already observed that the Cartesian co-ordinates no longer play a distinguished role. Still they are primus inter pares, but within the more general theory to be presented in Chap. 29 they will loose this position, too.

Remark 2. From (1) it follows that $g'_{kl} = g'_{lk}$, which means that the symmetry is invariant under co-ordinate transformations.

24.1.3. Tensors

As usual, let Ω be a fixed domain in R_n. For every system (x'^l) of curvilinear co-ordinates in Ω, consider an n^k-tuple

$$T'^{l\ldots m}{}_{p\ldots q}(x'^r), \quad 1 \leq l, \ldots, m, p, \ldots, q \leq n \tag{1}$$

of real-valued functions that are differentiable up to arbitrary order. Here $l, \ldots, m, p, \ldots, q$ are k indices in total, which run from 1 to n independently of one another.

Definition. (1) *is called a tensor field of order k if, for every two systems* (x'^r) *and* (x''^s) *of curvilinear co-ordinates in* Ω,

$$\underbrace{T''^{a\ldots b}}_{\text{contra-variant}}\underbrace{{}_{c\ldots d}}_{\text{co-variant}}(x''^l)=T'^{r\ldots s}{}_{u\ldots v}(x'^w)\,\frac{\partial x''^a}{\partial x'^r}\cdots\frac{\partial x''^b}{\partial x'^s}\,\frac{\partial x'^u}{\partial x''^c}\cdots\frac{\partial x'^v}{\partial x''^d} \tag{2}$$

Remark 1. We have to sum up over doubly occurring indices from 1 to n. The structure of (2) is clear: a superscript on the left-hand side that refers to x'' is also a superscript on the right-hand side (for example a). An analogous statement can be made for subscripts. The summation is always carried out over opposite indices, i.e., a superscript and a subscript each. (2) has to be understood as an identity expressed either in x' or in x'' co-ordinates. The structure of the formula can almost be kept in mind with the aid of the above remarks. The superscripts are called contravariant, the subscripts covariant.

Remark 2. It is essential that every curvilinear co-ordinate system in Ω is assigned a separate n^k-tuple of C^∞-functions. (2) interrelates these separate n^k-tuples, which immediately raises the question whether non-trivial tensor fields exist at all. If one considers a fixed co-ordinate system in Ω (e.g., the Cartesian co-ordinates $x^1, \ldots, x^n$) and an arbitrary n^k-tuple $T^{r\ldots s}{}_{u\ldots v}(x)$ of C^∞-functions defined on Ω, and sets

$$T'^{a\ldots b}{}_{c\ldots d}(x'^l)=T^{r\ldots s}{}_{u\ldots v}(x^m)\,\frac{\partial x'^a}{\partial x^r}\cdots\frac{\partial x'^b}{\partial x^s}\,\frac{\partial x^u}{\partial x'^c}\cdots\frac{\partial x^v}{\partial x'^d},$$

then it is easily verified that in this way a tensor field of order k is obtained (Lemma 24.1.2 is an example for this assertion). Hence there are tensor fields as numberless as the sand of the sea.

Remark 3 (examples). (1) Lemma 24.1.2 now shows that g_{kl} (from now on we shall omit the parentheses) is a covariant tensor field of order two. (2) It will be useful to regard (scalar) functions as tensor fields of order zero, that means $f'(x'^k)=f''(x''^l)$. (3) If $f'(x'^r)$ is a function on Ω, then we set $f'_{,l}(x'^r)=\dfrac{\partial f'}{\partial x'^l}(x'^r)$ for the partial derivatives, and analogously $f''_{,l}=\dfrac{\partial f''}{\partial x''^l}$. This means that the prime „'" indicates the partial derivative with respect to the following indices. This convention also holds for derivatives of higher orders, e.g., for $f''_{,lk}=\dfrac{\partial^2 f''}{\partial x''^l \partial x''^k}$. Now, if $f'(x'^r)=f''(x''^s)$ is a scalar (a tensor field of order zero), then it follows from

$$f''_{,k}=f'_{,l}\,\frac{\partial x'^l}{\partial x''^k}, \tag{3}$$

that $f'_{,l}(x'^r)$ is a covariant tensor field of order one (also called a covariant vector field). Hence the transformation laws for covariant vector fields a'_k and for contravariant vector fields a'^k read, respectively,

$$a''_k=a'_l\,\frac{\partial x'^l}{\partial x''^k}\quad\text{and}\quad a''^k=a'^l\,\frac{\partial x''^k}{\partial x'^l}. \tag{4}$$

Mnemonic remark (or simply, asses' bridge). From the chain rule it follows that

$$\mathrm{d}x''^k=\frac{\partial x''^k}{\partial x'^l}\,\mathrm{d}x'^l. \tag{5}$$

Hence the transformation laws for contravariant and covariant vector fields can be derived by using the chain rule to convert $\mathrm{d}x''^k$ (contravariant, since k is a superscript) into $\mathrm{d}x'^l$ and $f''_{,k}$ (covariant, since k is a subscript) into $f'_{,l}$, which is done according to (3) and (5). The general tensor fields are then built up according to the above principles.

Terminological remark. Sometimes we simply speak of tensors (and vectors) instead of tensor fields (and vector fields). Usually the n^k-tuples $T^{a...b}{}_{c...d}(P)$ at a fixed point $P \in \Omega$ are called tensors, provided that transformation laws of the above form are applicable. But here we exclusively consider tensor fields, so that confusions can only occur if we definitely set our mind on them.

24.1.4. Properties of Tensors

As was just pointed out, here we only consider tensor fields defined on the domain Ω in R_n, but sometimes we choose to simply speak of tensors (and vectors).

Theorem 1. (a) *If $T^{k...l}{}_{u...v}$ and $S^{k...l}{}_{u...v}$ are tensors with the same index pattern (i.e., tensors of the same order which have equal numbers of covariant and contravariant indices, respectively), and if λ and μ are real numbers, then*

$$\lambda T^{k...l}{}_{u...v} + \mu S^{k...l}{}_{u...v} \tag{1}$$

is a tensor of the same type.

(b) *If $T^{k...l}{}_{u...v}$ and $S^{a...b}{}_{c...d}$ are any two tensors (whose index patterns are not necessarily identical), then*

$$T^{k...l}{}_{u...v} S^{a...b}{}_{c...d} = U^{k...la...b}{}_{u...vc...d} \tag{2}$$

is also a tensor.

(c) *(Contraction). If $T^{k...a...l}{}_{u...b...v}$ is a tensor, then $T^{k...a...l}{}_{u...a...v}$ is also a tensor (summation over a).*

Remark 1. A tensor is of type (k_1, k_2) if it has k_1 contravariant and k_2 covariant indices, where $k_1=0, 1, 2, \dots$ and $k_2=0, 1, 2, \dots$ Here $k=k_1+k_2$ is the order of the tensor. Thus in part (a) T and S as well as the tensor in (1) are of type (k_1, k_2). If in part (b) T is of type (k_1, k_2) and S is of type (k_1', k_2'), then U in (2) is of type (k_1+k', k_2+k_2'). If in part (c) $T^{k...a...l}{}_{u...b...v}$ is of type (k_1, k_2), where $k_1 \geqq 1$ and $k_2 \geqq 1$, then $T^{k...a...l}{}_{u...a...v}$ is of type (k_1-1, k_2-1).

Remark 2. If $T^k{}_l$ is a tensor of type (1, 1), then $T^k{}_k$ is a scalar. This is an example for part (c) of the above theorem.

Theorem 2. (a) $\delta^k{}_l = \begin{cases} 1 & \text{for } k=l \\ 0 & \text{for } k \neq l \end{cases}$ *is a tensor of type (1, 1) in all curvilinear co-ordinate systems.*

(b) *If T_k is a covariant vector, then $T_{k'l} - T_{l'k}$ is a tensor of type (0, 2) (curl, or rotation, of T_k).*

(c) *If g_{kl} is the fundamental tensor, and if g^{kl} is determined from the systems of equations $g^{kl} g_{lm} = \delta^k{}_m$, then g^{kl} is a symmetrical tensor of type (2, 0) (the contravariant version of the fundamental tensor).*

Remark 3. If we fix $k=1, \dots, n$ and let m vary from 1 to n, then each of the linear systems of equations $g^{kl} g_{lm} = \delta^k{}_m$ (summation over l) has a unique solution, because $\det(g_{lm}) \neq 0$. The assertions of the theorem can be proved in a rather simple way, but are not trivial.

24.1.5. Metric Geodesics

Metric geodesics are of essential importance in the general theory of relativity: test particles and electromagnetic waves move along metric geodesics in a four-dimensional curved space-time that is pointwise a Lorentz metric. For the moment, however, we confine ourselves to the pointwise Euclidean case we are able to discuss on the basis of the tools we have developed so far. Again let Ω be a domain in R_n, and let the fundamental tensor referred to in Subsec. 24.1.2. be positive definite. This means that pointwise it is possible to transform g_{kl} into the identity matrix, as was done in (24.1.2/6). Then $\mathrm{d}s^2 = g_{kl}\mathrm{d}x^k\mathrm{d}x^l$ is the line element in Ω (where we need not necessarily assume that $x^1, \ldots, x^n$ are Cartesian co-ordinates in Ω). Let $x(t) = (x^1(t), \ldots, x^n(t))$ be a continuously differentiable curve in Ω (Fig. 24.2). By analogy with the Euclidean length measurement described in Subsecs. 9.2.3. and 11.2.3., we set

$$s = L(x_k) = \int_{t_0}^{t} \sqrt{g_{kl}\dot{x}^k\dot{x}^l}\,\mathrm{d}t\,. \tag{1}$$

Since $g_{kl} = g_{kl}(x)$ is positive definite, one has $g_{kl}\dot{x}^k\dot{x}^l \geqq c\dot{x}^k\dot{x}^k$, where $c > 0$. Hence (1) is meaningful. We assume that $\dot{x}^k\dot{x}^k > 0$.

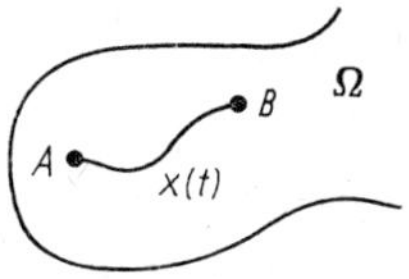

Fig. 24.2

Definition 1. *Curves of minimum length s in the sense of* (1) *which pass through two given points A and B in Ω are called metric geodesics* (Fig. 24.2).

Remark 1. This is a variational problem in the sense of Def. 11.1.3(b), with $L(t, x^k, \dot{x}^k) = \sqrt{g_{kl}\dot{x}^k\dot{x}^l}$; see also Subsec. 11.2.3. and the remarks we made there. s is again called the (arc) length of the curve considered.

Definition 2. *Christoffel's symbols of the first kind are defined as*

$$\{jk, l\} = \frac{1}{2}\,(g_{jl'k} + g_{kl'j} - g_{jk'l})\,, \tag{2}$$

and Christoffel's symbols of the second kind are defined as

$$\left\{{j \atop kl}\right\} = g^{jr}\{kl, r\}\,. \tag{3}$$

Remark 2. Of course the usual summation conventions are applicable to (2) and (3).

Theorem. $x(s) = (x^1(s), \ldots, x^n(s))$ *is a metric geodesic if and only if*

$$\frac{\mathrm{d}^2x^k}{\mathrm{d}s^2} + \left\{{k \atop lm}\right\}\frac{\mathrm{d}x^l}{\mathrm{d}s}\cdot\frac{\mathrm{d}x^m}{\mathrm{d}s} = 0\,, \tag{4}$$

where s is the lenght of curve (arc length) defined by (1).

Remark 3. Formula (4) is obtained from the Euler-Lagrange differential equations (11.1.3/2) if it is taken into account that s is the length of curve, i.e., that $g_{kl}\dfrac{\mathrm{d}x^k}{\mathrm{d}s}\,\dfrac{\mathrm{d}x^l}{\mathrm{d}s} = 1$.

Remark 4. In the (global) Euclidean case the theorem coincides with Theorem 11.2.3. But the decisive advantage of the above considerations is that we obtain a description of metric geodesics which is independent of the curvilinear co-ordinates chosen. If (x^k) and (x'^k) are curvilinear co-ordinates in Ω, and if g_{kl} and g'_{kl} are the corresponding versions of the fundamental tensor, then

$$g_{kl}(x^r)\,\frac{\mathrm{d}x^k}{\mathrm{d}t}\,\frac{\mathrm{d}x^l}{\mathrm{d}t} = g'_{kl}(x'^r)\,\frac{\mathrm{d}x'^k}{\mathrm{d}t}\,\frac{\mathrm{d}x'^l}{\mathrm{d}t}\,.$$

Hence the formulation of the problem is invariant under co-ordinate transformation, and so is the result (4). The equations (4) are of essential importance in the general theory of relativity, but there the fundamental tensor g_{kl} is pointwise a Lorentz metric.

24.2. Classical Field Theory

24.2.1. The Model of Field Theory

In Sec. 12.1. we discussed the subject of modelling in physics. Examples of this conception were the classical point mechanics presented in Sec. 12.2. and the hydrodynamics described in Sec. 16.1. The pretension of field theory goes farther: it is the basis of a good many theories that are described by ordinary and partial differential equations. In spite of this generality, the conception described in Sec. 12.1. can be reified, especially the black box of a "mathematical theory" given in Subsec. 12.1.2. Figure 12.1 now looks as follows (Fig. 24.3). The mathematical theory consists of a C^∞-manifold (in the case considered so far it was a domain Ω in R_n), a fundamental tensor g_{kl}, and systems of ordinary and partial differential equations for tensors and other (invariant) geometric objects. There is an essential difference between absolute and general space-time theories. In absolute space-time theories, Ω and the geometry, in the form of the fundamental tensor g_{kl}, are given. (Examples are the Euclidean space R_{3n} for an n-particle system of classical point mechanics, or the four-dimensional Minkowskian space of special relativity.) Dynamic objects which must be translated from the "physical field" into the "mathematical theory" are then tensors and related geometric entities. Among the arrows  in the above figure, it is only  which is realized here. In general space-time theories the situation is different. Here the geometry, in the form of the fundamental tensor g_{kl} is itself a dynamic object. This means that the geometry is determined only by the given concrete physical problem. This characteristic feature of general relativity will be discussed in greater detail later on. As was already stated in Subsec. 24.1.1., for the moment we confine ourselves to absolute space-time theories.

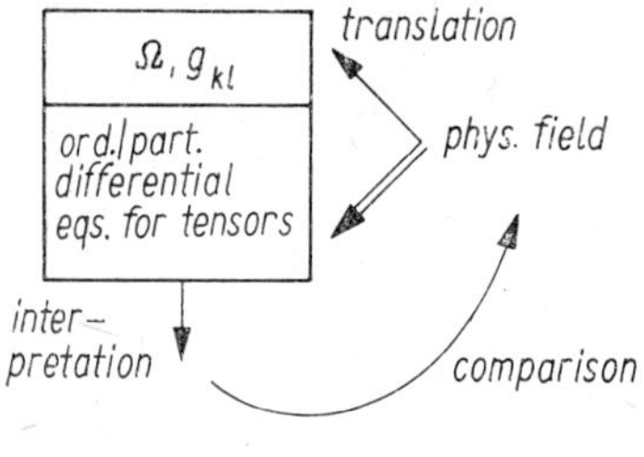

Fig. 24.3

24.2.2. Lagrange Densities

As previously, let Ω be an arbitrary domain in R_n, and let g_{kl} be a fundamental tensor in Ω in the sense of Subsec. 24.1.2. (Fig. 24.4). If x'^k and x''^k are arbitrary curvilinear co-ordinates in Ω, then $x'(\tau) = (x'^1(\tau), \ldots, x'^n(\tau)) = (x''^1(\tau), \ldots, x''^n(\tau)) = x''(\tau)$ is a curve in Ω, where τ is a real parameter. As we did already in Sec. 24.1., from now on we shall tacitly assume that co-ordinate transformations,

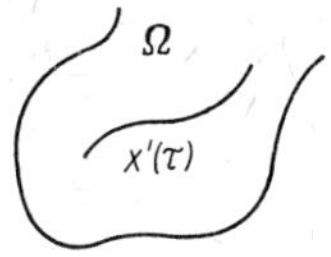

Fig. 24.4

curves etc. are sufficiently smooth, e.g., that they are C^∞-functions. As in Subsec. 24.1.2. we admit only co-ordinate transformations $x''^k = x''^k(x'^l)$ such that $\dfrac{\partial(x''^k)}{\partial(x'^l)} > 0$.

Definition. (a) $L\left(x'^k(\tau), \dfrac{dx'^k}{d\tau}(\tau)\right)$ *is called Lagrange density for curves if, for every two systems of curvilinear co-ordinates* x'^k *and* x''^k,

$$L\left(x'^k(\tau), \frac{dx'^k}{d\tau}(\tau)\right) = L\left(x''^k(\tau), \frac{dx''^k}{d\tau}(\tau)\right). \tag{1}$$

(b) $L(x'^k, \overset{r}{T}{}'^{k\ldots l}{}_{p\ldots q}, \overset{r}{T}{}'^{k\ldots l}{}_{p\ldots q's})$ *is called Lagrange density for the tensors* $\overset{r}{T}{}'^{k\ldots l}{}_{p\ldots q}$ *with* $r = 1, \ldots, R$, *if, for every two systems of curvilinear co-ordinates* x'^k *and* x''^k,

$$L(x'^k, \overset{r}{T}{}'^{k\ldots l}{}_{p\ldots q}, \overset{r}{T}{}'^{k\ldots l}{}_{p\ldots q's}) = \frac{\partial(x''^a)}{\partial(x'^b)} L(x''^k, \overset{r}{T}{}''^{k\ldots l}{}_{p\ldots q}, \overset{r}{T}{}''^{k\ldots l}{}_{p\ldots q's}). \tag{2}$$

Remark 1. As in Def. 11.1.3 we demand that the functions $L = L(u^1, \ldots, u^N)$ are twice continuously differentiable with respect to all their arguments. But it is useful (and sufficient) to require this differentiability not on the total R_n but only on certain domains (which depend on the formulation of the problem). As regards this, confer the examples in Sec. 11.2., e.g., Remark 11.2.3/1. Of course this involves certain restrictions with respect to the curves and tensors admitted.

Remark 2. In part (b) of the definition we consider R tensors, $\overset{r}{T}$, which in general have different index patterns. As previously one has $\overset{r}{T}{}'^{\cdots}{}_{\ldots's} = \dfrac{\partial}{\partial x'^s} \overset{r}{T}{}'^{\cdots}{}_{\ldots}$.

Examples. We consider the Minkowskian space. This means that $\Omega = R_4$, and

$$(g_{kl}(x)) = \begin{pmatrix} 1 & & & 0 \\ & 1 & & \\ & & 1 & \\ 0 & & & -1 \end{pmatrix} \tag{3}$$

is the fundamental tensor in Cartesian co-ordinates, $x = (x^1, x^2, x^3, x^4)$. One has $g = \det g_{kl} = -1$. If x'^k is an arbitrary curvilinear co-ordinate system, then it follows from Lemma 24.1.2 that $g' = \det g'_{kl} < 0$. (According to the transformation law (24.1.2/11), g'_{kl} is known in arbitrary co-ordinate systems x'^k if g_{kl} is given in the co-ordinate system x^k.) If we consider the curve $x(\tau) = (x^1(\tau), x^2(\tau), x^3(\tau), x^4(\tau))$ in the Minkowskian space (with Cartesian co-ordinates), then

$$\left(\frac{ds}{d\tau}\right)^2 = \left(\frac{dx^1}{d\tau}\right)^2 + \left(\frac{dx^2}{d\tau}\right)^2 + \left(\frac{dx^3}{d\tau}\right)^2 - \left(\frac{dx^4}{d\tau}\right)^2.$$

Later on (x^1, x^2, x^3) will be the position (of a particle), and $x^4 = t$ the time. In this sense a curve $x(\tau)$ is called time-like if $\left(\dfrac{ds}{d\tau}\right)^2 < 0$ for all parameter values τ

(which are admitted to competition). $\left(\text{If always } \left(\frac{\mathrm{d}s}{\mathrm{d}\tau}\right)^2 > 0, \text{ then the curve is called space-like.}\right)$ Geometrically this means: if $(\xi^1 - x^1(\tau))^2 + (\xi^2 - x^2(\tau))^2 + (\xi^3 - x^3(\tau))^2 = (\xi^4 - x^4(\tau))^2$ is the right circular cone in R_4 with $x(\tau)$ as a singular point, then the curve $x(\tau)$ is time-like if and only if the tangent $\frac{\mathrm{d}x}{\mathrm{d}\tau} = \left(\frac{\mathrm{d}x^1}{\mathrm{d}\tau}, \frac{\mathrm{d}x^2}{\mathrm{d}\tau}, \frac{\mathrm{d}x^3}{\mathrm{d}\tau}, \frac{\mathrm{d}x^4}{\mathrm{d}\tau}\right)$ always points to the interior of this cone (Fig. 24.5). To give an ex-

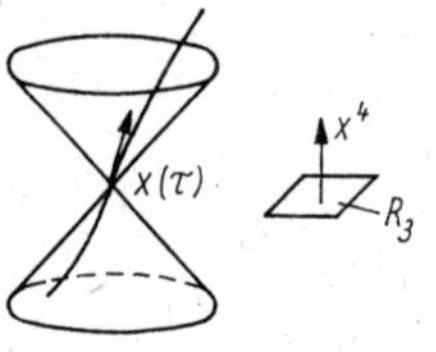

Fig. 24.5

ample of a Lagrange density for curves, we confine ourselves to time-like curves, $x(\tau)$, following the lines of Remark 1. (For the later applications this will be sufficient, since physical processes are described in the Minkowskian space by time-like lines, or lines for which $\left(\frac{\mathrm{d}s}{\mathrm{d}\tau}\right)^2 \leqq 0.$) Now it can be readily verified that

$$L\left(x'^k(\tau), \frac{\mathrm{d}x'^k}{\mathrm{d}\tau}(\tau)\right) = \sqrt{-g'_{kl}\frac{\mathrm{d}x'^k}{\mathrm{d}\tau}\cdot\frac{\mathrm{d}x'^l}{\mathrm{d}\tau}} \tag{4}$$

is a Lagrange density for time-like curves. Here x'^k is an arbitrary curvilinear co-ordinate system in the Minkowskian space. To give an example of a Lagrange density for tensors, we first of all recall Lemma 24.1.2. If x'^k and x''^k are any two curvilinear co-ordinate systems in the Minkowskian space, and if $g' = \det g'_{kl}$ and $g'' = \det g''_{kl}$, then it follows from Lemma 24.1.2 and from $\frac{\partial(x'^k)}{\partial(x''^l)} > 0$ that

$$\sqrt{-g''} = \frac{\partial(x'^k)}{\partial(x''^l)}\sqrt{-g'}\,. \tag{5}$$

Now, if $\tilde{L}$ is a scalar function such that

$$\tilde{L}(x'^k, \overset{r}{T}{}'^{k\ldots l}{}_{p\ldots q}, \overset{r}{T}{}'^{k\ldots l}{}_{p\ldots q's}) = \tilde{L}(x''^k, \overset{r}{T}{}''^{k\ldots l}{}_{p\ldots q}, \overset{r}{T}{}''^{k\ldots l}{}_{p\ldots q's}) \tag{6}$$

for arbitrary co-ordinate systems x'^k and x''^k, then it follows from (5) and (6) that

$$L(x'^k, \overset{r}{T}{}'^{k\ldots l}{}_{p\ldots q}, \overset{r}{T}{}'^{k\ldots l}{}_{p\ldots q's}) = \sqrt{-g'}\,\tilde{L}(x'^k, \overset{r}{T}{}'^{k\ldots l}{}_{p\ldots q}, \overset{r}{T}{}'^{k\ldots l}{}_{p\ldots q's}) \tag{7}$$

is a Lagrange density for the tensors $\overset{r}{T}$. A typical example is

$$L = \sqrt{-g}G^{kl}H_{kl}\,, \tag{8}$$

where G^{kl} is a tensor of type (2, 0) and H_{kl} is a tensor of type (0, 2). (We recall that we have to sum up over doubly occurring indices; $G^{kl}H_{kl}$ is a scalar after the double contraction procedure.)

Remark 3. To avoid misunderstandings, we point out that (1) and (2) are to be understood as identities. With the use of the chain rule and of the transformation law for tensors, one has to convert both sides of (1) and (2) to a fixed co-ordinate system (e.g., to x'^k). Thereafter (1) and (2) must be identities for all (admitted) curves or tensors. The above examples how clearly how this is to be understood.

24.2.3. Lagrangian Formalism

Curves: Let $x(\tau)=(x^1(\tau), \ldots, x^n(\tau))$, $\tau_0 \leqq \tau \leqq \tau_1$, be a curve in the domain Ω in R_n. As usual, let $x^k(\tau)$ be smooth (e.g., twice continuously differentiable) functions. Let $\varphi(\tau)=(\varphi^1(\tau), \ldots, \varphi^n(\tau))$, with $\varphi(\tau_0)=\varphi(\tau_1)=0$, such that $x(\tau)+\varepsilon\varphi(\tau)$ are smooth (twice continuously differentiable) curves in Ω for $0 \leqq \varepsilon \leqq 1$ (Fig. 24.6). If L is a Lagrange density for curves, then we ask for curves $x(\tau)$ such that

$$\int_{\tau_0}^{\tau_1} L\left(x^k(\tau), \frac{\mathrm{d}x^k}{\mathrm{d}\tau}(\tau)\right) \mathrm{d}\tau \tag{1}$$

becomes extremal (or stationary). (1) is independent of the particular choice of the (curvilinear) co-ordinates x^k. By analogy with Subsec. 11.1.3., this formulation of the problem means that, for all $\varphi(\tau)$ with the above properties,

$$\frac{\mathrm{d}}{\mathrm{d}\varepsilon} I(\varepsilon) \big|_{\varepsilon=0} = 0\,, \tag{2}$$

where

$$I(\varepsilon)=\int_{\tau_0}^{\tau_1} L\left(x^k(\tau)+\varepsilon\varphi^k(\tau), \frac{\mathrm{d}}{\mathrm{d}\tau}\left[x^k(\tau)+\varepsilon\varphi^k(\tau)\right]\right) \mathrm{d}\tau\,.$$

Apart from the requirement (24.2.2/1), this is the variational problem with fixed boundary values as per Def. 11.1.3(b).

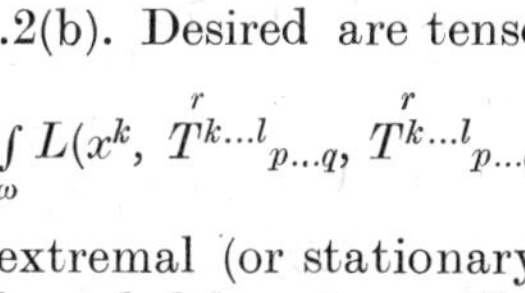

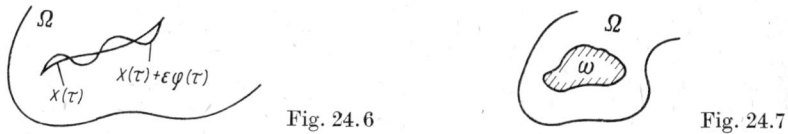

Fig. 24.6

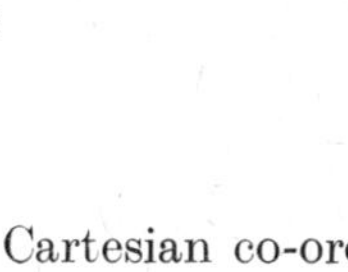

Fig. 24.7

Tensors: Given a domain Ω and a Lagrange density L for tensors in the sense of Def. 24.2.2(b). Desired are tensors $\overset{r}{T}{}^{k\ldots l}{}_{p\ldots q}$ such that

$$\int_\omega L(x^k, \overset{r}{T}{}^{k\ldots l}{}_{p\ldots q}, \overset{r}{T}{}^{k\ldots l}{}_{p\ldots q's})\, \mathrm{d}x \tag{3}$$

becomes extremal (or stationary) (as referred to the Cartesian co-ordinates x^k) for every bounded domain ω with $\bar{\omega} \subset \Omega$ (Fig. 24.7). The formulation of the problem is meaningful (i.e., independent of the particular choice of the co-ordinate systems): if x'^k and x''^k are any two systems of curvilinear co-ordinates in Ω, and if ω' and ω'' are the images of ω obtained by application of $x'^k(x^l)$ and $x''^k(x^l)$, respectively (which are interpreted here as mappings of Ω in R_n), then it follows from Theorem 9.2.2 and from (24.2.2/2) that

$$\int_{\omega''} L(x''^k, \overset{r}{T}''^{\cdots}{}_{\cdots}, \overset{r}{T}''^{\cdots}{}_{\cdots's})\,\mathrm{d}x''$$

$$= \int_{\omega'} L(x'^k, \overset{r}{T}'^{\cdots}{}_{\cdots}, \overset{r}{T}'^{\cdots}{}_{\cdots's})\,\frac{\partial(x'^b)}{\partial(x''^a)}\cdot\frac{\partial(x''^c)}{\partial(x'^d)}\,\mathrm{d}x'$$

$$= \int_{\omega'} L(x'^k, \overset{r}{T}'^{\cdots}{}_{\cdots}, \overset{r}{T}'^{\cdots}{}_{\cdots's})\,\mathrm{d}x'\,.$$

This means that the problem can also be formulated without reference to Cartesian co-ordinates, in a way that is invariant under co-ordinate transformation. The precise formulation of the above requirement now reads as follows. If $0 \leq \varepsilon \leq 1$, and if $\overset{r}{V}{}^{k\ldots l}{}_{p\ldots q}$ are tensors which have the same index pattern as $\overset{r}{T}{}^{k\ldots l}{}_{p\ldots q}$, then let

$$I_\omega(\varepsilon) = \int_\omega L(x^k, \overset{r}{T}{}^{k\ldots l}{}_{p\ldots q} + \varepsilon \overset{r}{V}{}^{k\ldots l}{}_{p\ldots q}, \overset{r}{T}{}^{k\ldots l}{}_{p\ldots q's} + \varepsilon \overset{r}{V}{}^{k\ldots l}{}_{p\ldots q's})\,\mathrm{d}x.$$

Desired are tensors $\overset{r}{T}{}^{k\ldots l}{}_{p\ldots q}$ such that

$$\frac{\mathrm{d}}{\mathrm{d}\varepsilon} I_\omega(\varepsilon)\,|_{\varepsilon=0} = 0 \tag{4}$$

for all bounded domains ω with $\bar{\omega} \subset \Omega$ and all tensors $\overset{r}{V}$ with $\overset{r}{V}{}^{k\ldots l}{}_{p\ldots q} \in D(\omega)$.

Theorem. (**a**) (*Curves*). *The curve $x(\tau)$ is extremal (stationary) in the sense set out above if and only if, for $k = 1, \ldots, n$,*

$$\frac{\partial L}{\partial x^k} - \frac{\mathrm{d}}{\mathrm{d}\tau}\left(\frac{\partial L}{\partial \dot{x}^k}\right) = 0 \tag{5}$$

(*Euler-Lagrange equations*).

(**b**) (*Tensors*). *The tensors (dynamic objects) $\overset{r}{T}{}^{\cdots}{}_{\cdots}$ are extremal (stationary) if and only if, for $r = 1, \ldots, R$,*

$$\frac{\partial L}{\partial \overset{r}{T}{}^{k\ldots l}{}_{p\ldots q}} - \left(\frac{\partial L}{\partial \overset{r}{T}{}^{k\ldots l}{}_{p\ldots q's}}\right)_{'s} = 0\,. \tag{6}$$

Remark. Part (a) is identical with Theorem 11.1.3/1. In Remark 11.1.3/2 it was explained how (5) is to be understood. One obtains a system of n ordinary differential equations. In the same way one has to interpret (6), which now gives a system of partial differential equations. Here the sum is taken over s, while $k, \ldots, l, p, \ldots, q$ vary independently from 1 to n. Further, $r = 1, \ldots, R$. Thus we have reached one of the goals stated in Subsec. 24.1.1.: the derivation of systems of partial differential equations from variational principles. Thus for a concrete physical problem one has to find a Lagrange density L such that (6) is identical with the corresponding system of partial differential equations.

24.3. Examples of Field Theories

24.3.1. Covariant Point Mechanics

As in Chap. 12, we consider a system of n mass points in R_3, with the masses m_l, where $l=1, \ldots, n$. The investigations carried out in Subsec. 12.3.1. can now be described as follows. If $(x^{3r-2}(\tau), x^{3r-1}(\tau), x^{3r}(\tau))$ is the trajectory of the r-th particle in R_3, described in Cartesian co-ordinates, then we set, in the sense of Sec. 24.2.,

$$\Omega = R_{3n+1}, \quad (g_{kl}) = \begin{pmatrix} 1 & & 0 \\ & \ddots & \\ 0 & & 1 \end{pmatrix}. \tag{1}$$

Here $t=x^{3n+1}(\tau)$ is the time. Now the Lagrangian function (= Lagrange density for curves) stated in (12.3.1/1) can easily be rewritten to fit into our present calculus (see below). Here the potential energy $V=V(x^s)$ is a scalar function, which is now required to be independent of t. Then (24.2.3/5) is identical with Theorem 12.3.1. The preceding sections enable us to formulate the same problem in arbitrary curvilinear coordinates. Let again n mass points be given, then we describe the position of the r-th particle by separate curvilinear coordinates in R_3, which are again denoted by $(x^{3r-2}, x^{3r-1}, x^{3r})$. It is easy to imagine problems for which certain curvilinear co-ordinates are much more suitable than Cartesian co-ordinates, which take no account of the specific character of the problem. If, for example, it is known a priori that a mass point will always move on a sphere, then it is natural to use spherical polar co-ordinates with the origin at the centre of the corresponding ball. (1) must now be replaced by

$$\Omega \subset R_{3n+1}, \quad (g_{kl}) = \begin{pmatrix} (G_3) & & 0 \\ & (G_3) & \\ & & \ddots \\ 0 & & & (G_3) \end{pmatrix}. \tag{2}$$

Here (G_3) are 3×3-matrices, where the r-th matrix (corresponding to the r-th particle) takes the form

$$g_{kl} = \frac{m_r}{g} \overset{r}{g}_{\alpha,\beta}(x^{3(r-1)+\gamma}), \quad k=3\,(r-1)+\alpha, \quad l=3(r-1)+\beta\,. \tag{3}$$

Here α, β and γ can assume the values 1, 2, 3, and $r=1, \ldots, n$. Further, $(\overset{r}{g}_{\alpha,\beta})^3_{\alpha,\beta=1}$ are positive definite, symmetrical matrices which belong to three-dimensional line elements in the Euclidean space R_3, as referred to the curvilinear co-ordinates $(x^{3r-2}, x^{3r-1}, x^{3r})$. Examples are Cartesian co-ordinates or polar co-ordinates, where Ω has to be suitably chosen. The $(3n+1)$-th co-ordinate x^{3n+1} in (2) is again the time.

$$L = g_{kl}\,\frac{\mathrm{d}x^k}{\mathrm{d}\tau}\cdot\frac{\mathrm{d}x^l}{\mathrm{d}\tau} - V(x^s) \tag{4}$$

is a Lagrange density for curves, provided that $V(x^s)$ is a scalar function (tensor of order zero).

Theorem. *If V does not explicitly depend on $t=x^{3n+1}$, then the Euler-Lagrange equations* (24.2.3/5) *for the Lagrange density L in* (4) *read*

$$\ddot{x}_l + \begin{Bmatrix} l \\ km \end{Bmatrix} \dot{x}^k \dot{x}^m = -\frac{1}{2}\, g^{kl}\,\frac{\partial V}{\partial x^k} \quad \textit{for} \quad l=1, \ldots, 3n \tag{5}$$

and $\dfrac{\partial^2 t}{\partial \tau^2} = 0.$

Remark 1. One has $\dot{x}^l = \frac{\mathrm{d}x^l}{\mathrm{d}\tau}$. The requirement that V does not explicitly depend on x^{3n+1} means that we consider a conservative system in the sense set out in Subsec. 12.3.3. $\left\{ {l \atop km} \right\}$ are the Christoffel symbols of the second kind as per Def. 24.1.5/2. From $\frac{\mathrm{d}^2 t}{\mathrm{d}\tau^2} = 0$ it follows that we can set $t = \tau$ without loss of generality.

Remark 2. The interpretation of (5) is clear from the way we proceeded so far: with $\tau = t$, $(x^{3r-2}(t), x^{3r-1}(t), x^{3r}(t))$ is the trajectory of the r-th particle, with the time as a parameter. With given initial data $x^k(t_0)$ and $\dot{x}^k(t_0)$, the system of equations (5) has (at least locally) a unique solution. If $V(x^l) \equiv 0$, then it follows from Theorem 12.2.3 that L is constant along a trajectory. The constant is positive. Hence, except for a positive constant, t is identical with the arc length s in Subsec. 24.1.5. For $V(x^k) \equiv 0$, (5) is thus identical with (24.1.5/4).

Consequently, a force-free system, being characterized by $V(x^k) \equiv 0$, *moves along a metric geodesic in* R_{3n}. (6)

We will find this proposition again in the general theory of relativity. It is of special interest (here as well as later on) that this proposition is independent on the particular choice of curvilinear co-ordinates.

Remark 3. According to Newton's axioms,

In an inertial system, a mass point under no forces moves uniformly in a straight line (Lex prima). (7)

(5) shows that there are two kinds of forces that counteract a uniform motion in a straight line, $\ddot{x}^l \equiv 0$: the real force $\frac{1}{2} g^{kl} \frac{\partial V}{\partial x^k}$ and the fictitious force $\left\{ {l \atop km} \right\} \dot{x}^k \dot{x}^m$. The real force is of physical origin (gravitational forces, electromagnetic forces etc.), and its existence has nothing to do with the specific character of the chosen co-ordinates x^k. On the other hand the ficticious force has no physical cause, it is, so to speak, a measure of the deviation of the co-ordinate system chosen from an inertial system. It is however possible to regard (6) as an elegant generalization of (7).

24.3.2. The Maxwell-Lorentz Equations of Electrodynamics

Maxwell's equations are the fundamental equations of electrodynamics. When written in the four-dimensional notation with reference to the Minkowskian space, they are also called Maxwell-Lorentz equations. Here we consider the equations for the free space. In this subsection we shall derive them from a variational principle, in Subsec. 24.3.3. we shall describe interpretations and transformations. Typical problems will be discussed in Sec. 25.3. Later on, within the general theory of relativity, we shall repeatedly discuss the Einstein-Maxwell equations; cf. Subsec. 30.1.3 and Sec. 33.5.

Minkowskian space: As a starting point we consider the Minkowskian space referred to in the examples in Subsec. 24.2.2. So let $\Omega = R_4$, and let the fundamental tensor g_{kl} have the form (24.2.2/3) in Cartesian co-ordinates. If (x^1, x^2, x^3, x^4) are any curvilinear co-ordinates in the Minkowskian space, then the corresponding fundamental tensor is obtained from (24.2.2/3) by well-known conversion rules, cf. (24.1.2/11). From this formula it also follows that $g = \det g_{kl} < 0$, which has also been used in the examples stated in Subsec. 24.2.2.

Theorem. (**a**) *If F^{kl} is a contravariant tensor and if A_k is a covariant vector in the Minkowskian space, then*

$$L=\frac{1}{4}\sqrt{-g}\,F^{kl}g_{kr}g_{ls}F^{rs}-\frac{1}{2}\sqrt{-g}\,F^{kl}\,(A_{k'l}-A_{l'k}) \tag{1}$$

is a Lagrange density in the sense of Def. 24.2.2(b).

(**b**) *The Euler-Lagrange equations as stated in Theorem* 24.2.3(b) *for L with respect to the dynamic objects F^{kl} and A_k read in Cartesian co-ordinates*

$$F_{kl}=A_{k'l}-A_{l'k}\quad \textit{with}\quad F_{kl}=g_{kr}g_{ls}F^{rs} \tag{2}$$

and

$$F^{kl}{}_{,l}=0\,. \tag{3}$$

Remark 1. By Theorem 24.1.4/2(b), $A_{k'l}-A_{l'k}$ is a covariant tensor. Of course the previous convention again has to be applied to (1), (2) and (3): one has to sum up over doubly occurring indices. Then L has the structure of (24.2.2/8), and hence is a Lagrange density for tensors in the Minkowskian space.

Remark 2. As was already pointed out repeatedly, for the moment we are considering absolute space-time theories. Hence the fundamental tensor in L is not a dynamic object, and the tensors $\overset{r}{T}$ in (24.2.3/6) can be identified with F^{kl} and A_k. For arbitrary curvilinear co-ordinate systems in the Minkowskian space one obtains

$$0=\frac{\partial L}{\partial A_k}-\left(\frac{\partial L}{\partial A_{k'l}}\right)_{'l}=\frac{1}{2}\,(\sqrt{-g}\,(F^{kl}-F^{lk}))_{'l}\,, \tag{4}$$

$$0=\frac{\partial L}{\partial F^{kl}}-\left(\frac{\partial L}{\partial F^{kl}{}_{'p}}\right)_{'p}=\frac{2}{4}\sqrt{-g}\,g_{kr}g_{ls}F^{rs}-\frac{1}{2}\sqrt{-g}\,(A_{k'l}-A_{l'k})\,. \tag{5}$$

It is now easily seen that both F_{kl} in (2) and F^{kl} are anti-symmetrical, that is, $F_{kl}=-F_{lk}$ and $F^{kl}=-F^{lk}$. If this is taken into account, then for Cartesian co-ordinates the formulas (2) and (3) follow from (5) and (4), respectively.

Remark 3. The formulas (2) and (3) are the Maxwell-Lorentz equations for the electromagnetic field in the free space. In the next subsection we shall rewrite these formulas into the common, non-relativistic form, which looks much more intricate. As compared with this form, (2), (3) is a very elegant version. But it would be an error to believe that this elegance cannot be surpassed. We refer to Sec. 33.5. See also [64], p. 30, for the various formulations of the Maxwell equations.

Was it a god who wrote these signs?
(Boltzmann on Maxwell's equations)

24.3.3. Interpretation and Transformation of Maxwell's Equations

Interpretation: As regards the modeling described in Subsec. 12.1.2. and its concrete formulation in Subsec. 24.2.1., we have the situation illustrated in Fig. 24.8. The "translation" is accomplished by means of the Lagrange density L in (24.3.2/1). But for the moment it is still unclear how F^{kl} and A_k are connected with physical quantities (we already have done the translation, but just don't know the outcome). In the Minkowskian space we only just consider Cartesian co-ordinates, which we denote by $(x, y, z, t)=(x^1, x^2, x^3, x^4)$. Thus it

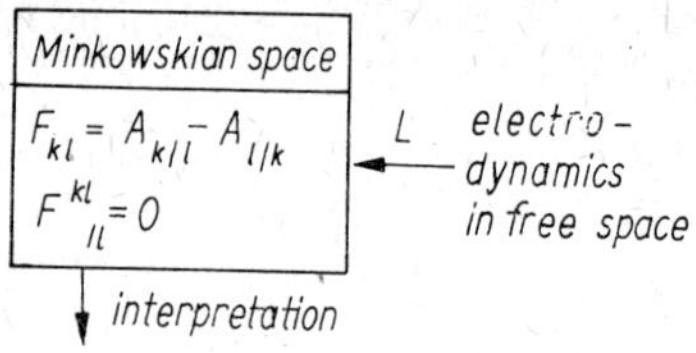

Fig. 24.8

is clear how these co-ordinates are interpreted: $\bar{x} = (x, y, z)$ are the space co-ordinates in the real three-dimensional space, and t is the time. If

$\mathfrak{E}(\bar{x}, t) = (E_x, E_y, E_z)$ is the electric field intensity and
$\mathfrak{B}(\bar{x}, t) = (B_x, B_y, B_z)$ is the magnetic field intensity,

then the tensor F_{kl} is identified with

$$(F_{kl}) = \begin{pmatrix} 0 & -B_z & B_y & -E_x \\ B_z & 0 & -B_x & -E_y \\ -B_y & B_x & 0 & -E_z \\ E_x & E_y & E_z & 0 \end{pmatrix}. \tag{1}$$

Further let

$$(A_k) = (A_x, A_y, A_z, \Phi) = (\mathfrak{A}, \Phi), \tag{2}$$

where $\mathfrak{A} = (A_x, A_y, A_z)$ is the vector potential and Φ is the scalar potential of electrodynamics. All functions depend on $\bar{x} = (x, y, z)$ and t.

Theorem 1. *If F_{kl} and A_k have the meaning stated in* (1) *and* (2), *then the Maxwell equations* (24.3.2/2) *and* (24.3.2/3) *read as follows:*

$$\mathfrak{B} = \operatorname{rot} \mathfrak{A}, \quad \mathfrak{E} = \operatorname{grad} \Phi - \frac{\partial \mathfrak{A}}{\partial t}, \tag{3}$$

$$\operatorname{rot} \mathfrak{B} = \frac{\partial \mathfrak{E}}{\partial t}, \quad \operatorname{div} \mathfrak{E} = 0. \tag{4}$$

Remark 1. Here (3) is the rewritten form of (24.3.2/2) and (4) is the rewritten form of (24.3.2/3). Altogether we have 10 equations, of which 6 are contained in (3), and 4 in (4). We recall the meaning of div, rot, grad:

$$\operatorname{div} \mathfrak{E} = \frac{\partial E_x}{\partial x} + \frac{\partial E_y}{\partial y} + \frac{\partial E_z}{\partial z}, \quad \operatorname{grad} \Phi = \left(\frac{\partial \Phi}{\partial x}, \frac{\partial \Phi}{\partial y}, \frac{\partial \Phi}{\partial z}\right),$$

$$\operatorname{rot} \mathfrak{A} = \left(\frac{\partial A_z}{\partial y} - \frac{\partial A_y}{\partial z}, \frac{\partial A_x}{\partial z} - \frac{\partial A_z}{\partial x}, \frac{\partial A_y}{\partial x} - \frac{\partial A_x}{\partial y}\right).$$

The components of rot $\mathfrak{A}$ are the subdeterminants (the signs being taken into account) of the symbolic matrix

$$\begin{pmatrix} * & * & * \\ \frac{\partial}{\partial x} & \frac{\partial}{\partial y} & \frac{\partial}{\partial z} \\ A_x & A_y & A_z \end{pmatrix}.$$

Finally, $\frac{\partial \mathfrak{A}}{\partial t} = \left(\frac{\partial A_x}{\partial t}, \frac{\partial A_y}{\partial t}, \frac{\partial A_z}{\partial t}\right)$, by analogy with $\frac{\partial \mathfrak{E}}{\partial t}$.

Remark 2. As has been stated, (3) and (4) are the rewritten versions of (24.3.2/2) and (24.3.2/3). Here we have

$$(F^{kl}) = \begin{pmatrix} 0 & -B_z & B_y & E_x \\ B_z & 0 & -B_x & E_y \\ -B_y & B_x & 0 & E_z \\ -E_x & -E_y & -E_z & 0 \end{pmatrix}.$$

Theorem 2. *From* (3) *and* (4) *it follows that*

$$\operatorname{rot} \mathfrak{E} + \frac{\partial \mathfrak{B}}{\partial t} = 0, \quad \operatorname{div} \mathfrak{B} = 0\,, \tag{5}$$

$$\operatorname{rot} \mathfrak{B} - \frac{\partial \mathfrak{E}}{\partial t} = 0, \quad \operatorname{div} \mathfrak{E} = 0\,. \tag{6}$$

Remark 3. This is the common form of Maxwell's equations in the free space in the absence of charges and currents. (5) and (6) can easily be deduced from (3) and (4) (only (5) is new). Conversely, it follows from (5) that $\mathfrak{E}$ and $\mathfrak{B}$ can be represented by (3), provided that $\mathfrak{A}$ and Φ are chosen suitably. (6) is not modified in this form. In the presence of electric charges with the charge density ϱ, and of electric currents with the current density $j = (j_x, j_y, j_z)$, (5) remains unchanged, whereas (6) is replaced by

$$\operatorname{rot} \mathfrak{B} - \frac{\partial \mathfrak{E}}{\partial t} = 4\pi j, \quad \operatorname{div} \mathfrak{E} = 4\pi\varrho\,. \tag{7}$$

One then obtains Maxwell's equations in the free space in the presence of charges and currents. Typical initial value problems for these Maxwell equations will be discussed in Sec. 25.3. Corresponding problems for the Einstein-Maxwell equations will be studied in Sec. 33.5.

Remark 4. Before the Maxwell equations were established, electric and magnetic fields were largely considered separately. The fact that these two fields are coupled so closely, as it is expressed in the equations (5) and (7), may well have been one of the reasons for Boltzmann's comment cited above. The relativistic formulation and the interpretations to be given in Chap. 25 will show even more: what is a purely electric field in one inertial system, may appear as a mixed electric and magnetic field in a different inertial system.

Absolute, true, and mathematical time, of itself, and from its own nature, flows equably without relation to anything external.
Absolute space, in its own nature, without relation to anything external, remains always similar and immovable. (Newton, Principia, 1687, [46])[1])

From this hour on, space as such
and time as such shall recede to
the shadows and only a kind of
union of the two retain significance.
(Minkowski, 1909)

25. Principles of Special Relativity and Electrodynamics

25.1. Lorentz Group and Space-Time[2])

25.1.1. Minkowskian Space and Inertial Systems

The Minkowskian space was introduced in Subsec. 24.2.2. We have $\Omega = R_4$. Further,

$$(\eta_{kl}) = (g_{kl}) = \begin{pmatrix} 1 & & & 0 \\ & 1 & & \\ & & 1 & \\ 0 & & & -1 \end{pmatrix} \qquad (1)$$

is the fundamental tensor in Cartesian co-ordinates, which from now on we shall denote by (x^1, x^2, x^3, x^4) or by (x, y, z, t), as required in the particular case. The latter notation refers to the physical interpretation given in Subsec. 24.3.3: (x, y, z) is the position in the real three-dimensional space, and t is the time. It is painful to dispense with the free choice of arbitrary curvilinear co-ordinate systems in the Minkowskian space, but for the moment it is mathematically and physically necessary. It is our aim to describe the new space-time conception of the theory of relativity. Within special relativity one has to accept constraints that restrict the range of validity of this theory. This is because an unscrupulous handling of this calculus leads to awkward inconsistencies from both the mathematical and the physical point of view, as the twin paradox to be described in Subsec. 25.2.1. will show. These constraints will no longer exist in general relativity. Then the new space-time conception and the possibility of choosing arbitrary curvilinear co-ordinate systems are reconciled with each other. The price to be paid for this is an extension of the calculus presented in Sec. 24.1., especially of its foundations.

Inertial frames: The special theory of relativity refers all its statements to inertial frames. The three-dimensional Euclidean space with the Cartesian

---

[1]) Einstein (1933): "Newton felt by no means comfortable about the concept of absolute space, . . . of absolute rest . . . [and] about the introduction of action at a distance."

[2]) "I regret that it has been necessary for me in this lecture to administer such a large dose of four-dimensional geometry. I do not apologize, because I am really not responsible for the fact that nature in its most fundamental aspect is four-dimensional. Things are what they are . . ." (A. N. Whitehead, The Concept of Nature, 1920).

co-ordinates (x, y, z) (or a subdomain of it) and an associated time measurement t are called an inertial frame if, as referred to it,

every force-free mass point moves uniformly in a straight line. (2)

In our terms this means that the Minkowskian space with the Cartesian co-ordinates (x, y, z, t) and with the above interpretation of position and time is admitted to competition only if (2) is fulfilled. Mathematically this is satisfactory, if only because, in the mathematical part of the considerations to follow, we shall never make use of this definition (but we shall do so in the physical interpretations). From the physical point of view, there arises the question of whether there are inertial frames in nature, and how to find them. If one has a space in which there are no acting forces (i.e., forces due to physical causes), then one can take the trajectories of three mass points moving freely (and not in one plane) to construct a Cartesian co-ordinate system. Then the time scale must be so chosen that the mass points move at a constant velocity. The result is an inertial frame. One can also proceed in a (physically) simpler way: experience shows that a Cartesian co-ordinate system which has its origin fixed at the centre of the sun and is oriented in a rigid relation to the starry sky is an inertial frame, provided that time is measured by a common physical procedure (e.g., by means of an atomic clock).

Remark. The considerations described in Subsec. 24.3.1. now show that with such a conception it is not possible to calculate with arbitrary curvilinear co-ordinate systems.

The ether hypothesis: Let (x, y, z, t) be an inertial frame with the space co-ordinates (x, y, z) and the time t. Consider another frame (x', y', z', t') that moves uniformly in a straight line with respect to (x, y, z) in the three-dimensional space (Fig. 25.1). Time measurement is supposed to be identical in the two frames, that is, $t'=t$. With a suitable normalization one then obtains the Galilean transformation

$$x=x'+v_1t, \quad y=y'+v_2t, \quad z=z'+v_3t, \quad t=t'. \tag{3}$$

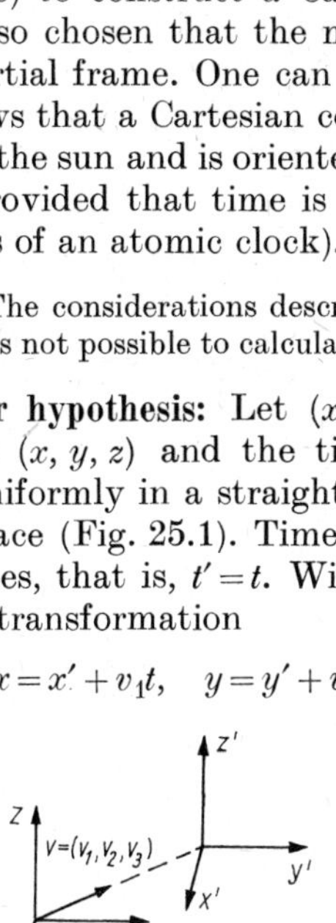

Fig. 25.1

This is a simple geometric conversion. It is immediately seen that (2) also holds for (x', y', z', t'), so that one obtains a new inertial frame. The two frames are fully equivalent, and the laws of classical mechanics are invariant under Galilean transformations. If one applies (3) to equations of electrodynamics, e.g., to (24.3.3/5)–(24.3.3/7), then it is seen that this invariance is lost. The formulas change their form. By the end of the past century and at the beginning of this century, this led to the so-called ether hypothesis. According to this hypothesis there exists a distinguished inertial frame in which both Maxwell's equations and the fundamental equations of mechanics hold in their common form. If one considers a frame of reference that moves uniformly in a straight line with respect to the ether, then one has to convert the formulas of electrodynamics according to (3). The ether was a hypothetical medium in which electromagnetic waves were supposed to propagate. In particular, the velocity of light was assumed to be constant and independent of direction in this ether[1]). For the purpose of this chapter, let us normalize the value of the velocity of light to 1. The constancy of light velocity also opens the possibility of verifying the existence of ether by experiment: if a reference frame moves relative to the ether, e.g., according to (3), then the velocity of light, as referred to this frame, must generally be different from 1. The effect is small, but at that time it was already within the reach of the art of physical experimenting. But the ingenious experiments carried out on this subject altogether yielded negative results: within the limits of measuring accuracy, a deviation of the velocity of light from 1 was not observed.

Constancy of the velocity of light: In 1905, Einstein postulated that:

The velocity of light in empty space is equal with respect to all inertial frames

(=1 in our calculus). Of course this contradicts the classical Newtonian conception of space and time, and especially is inconsistent with (3). Further it was required that the form of Maxwell's equations should remain unchanged under a transformation from one inertial frame to another. It was postulated that all inertial frames are perfectly equivalent and that it is impossible to distinguish one of them by mechanical or electrodynamic experiments of any kind. It is clear that these requirements could only be fulfilled by a new conception of space and time.

25.1.2. World Lines

Before describing the mathematics that underlies special relativity, let us develop the ideas presented at the end of Subsec. 25.1.1. somewhat further. Consider the Minkowskian space with the Cartesian co-ordinates (x, y, z, t), where $\bar{x}=(x, y, z)\in R_3$ is the position in the real three-dimensional space and t is the time (Fig. 25.2). We shall always assume that (x, y, z, t) is an inertial

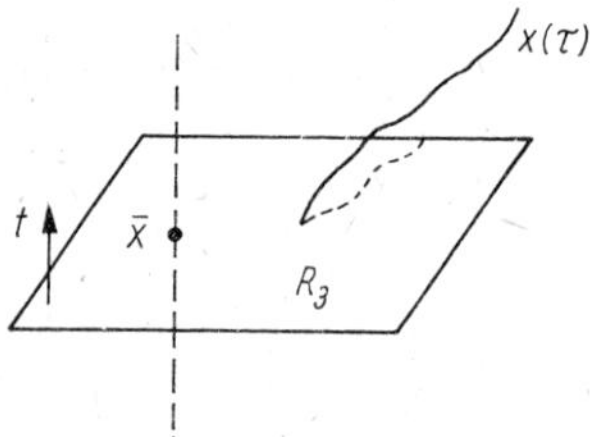

Fig. 25.2

frame. The world line of an observer who is at rest at the position $\bar{x}$ is a straight line which is parallel to the t-axis and passes through $\bar{x}$: the point $(\bar{x}, t)$ indicates that the observer is at the position $\bar{x}$ at the time t. A moving observer has the world line $(x(\tau)=x^1(\tau),\ x^2(\tau),\ x^3(\tau),\ x^4(\tau))$ (solid line in the figure). The trace

[1]) In 1760, Euler said: "Accordingly, light is nothing else but a motion or vibration in the smallest particles of the ether, and this is present everywhere, because of its extraordinary fineness, by virtue of which it penetrates all bodies."

(broken line) in R_3 indicates the position, and $x^4(\tau)$ indicates the time. Let us choose $\tau = t$ as the curve parameter τ. For physically meaningful curves this does not mean any restriction (the time axis is oriented, and the time scale is passed only from $-\infty$ to ∞). If $\bar{u} = (u_1, u_2, u_3)$ is the common, three-dimensional velocity, then one obtains

$$\left(\frac{\mathrm{d}x^k}{\mathrm{d}\tau}\right) = (u_1, u_2, u_3, 1), \quad |\bar{u}| = \sqrt{u_1^2 + u_2^2 + u_3^2} \,. \tag{1}$$

Physical principles: In accordance with the constancy of light velocity, the following is postulated:

If physical information is transported along the world line $x(\tau)$, then $|\bar{u}| \leqq 1$.

In particular this holds for electromagnetic waves (where $|\bar{u}| = 1$) and for the world lines of material particles. If always $|\bar{u}| < 1$, then the curve is called time-like. See also Subsec. 24.2.2. As was already mentioned there, for time-like curves one has

$$\eta_{kl} \frac{\mathrm{d}x^k}{\mathrm{d}\tau} \cdot \frac{\mathrm{d}x^l}{\mathrm{d}\tau} = u_1^2 + u_2^2 + u_3^2 - 1 < 0 \,,$$

and the tangent (1) to $x(\tau)$ always points into the interior of the right circular cone with $x(\tau)$ as a singular point as described there (Fig. 25.3).

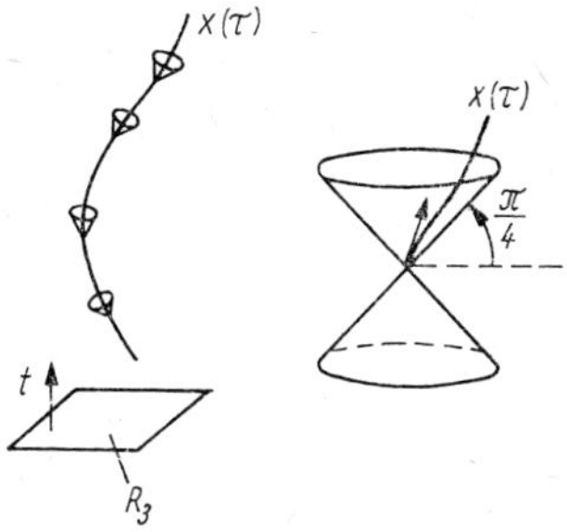

Fig. 25.3

25.1.3. The Lorentz Group

Consider the Minkowskian space with the Cartesian co-ordinates x^k and with (25.1.1/1) as the fundamental tensor. The interpretation as a space-time and the requirement that the co-ordinate system considered shall be an inertial frame are of no interest for the moment. Let $x^k = a^k{}_l x'^l$, with $\det(a^k{}_l) \neq 0$, where, as previously, we have to take the sum over doubly occurring indices (from 1 to 4). With respect to the new co-ordinates x'^l, according to (24.1.2/11)

$$g'_{kl} = \frac{\partial x^r}{\partial x'^k} \cdot \frac{\partial x^s}{\partial x'^l} g_{rs} = a^r{}_k a^s{}_l \eta_{rs} \tag{1}$$

is the associated fundamental tensor.

Definition 1: $x^k = a^k{}_l x'^l$ *is called a (general) Lorentz transformation if* $g'_{kl} = \eta_{kl}$.

Remark 1. This means, what is desired are affine transformations in the Minkowskian space which do not change the fundamental tensor (η_{kl}). Thus it is required that

$$\eta_{kl} = a^r{}_k a^s{}_l \eta_{rs} \,. \tag{2}$$

Theorem 1. *The set of all Lorentz transformations forms a group (the Lorentz group) where the group multiplication is defined as the successive application of the transformations.*

Remark 2. The proof is rather simple. The identical transformation is the unit element. Two properties will be of interest later on. If the determinant is formed on both sides of (2), one obtains

$$-1=(\det a^k{}_l)^2\,(-1), \quad \text{hence} \quad \det a^k{}_l=\pm 1\,. \tag{3}$$

Further one obtains from (2)

$$-1=\eta_{44}=(a^1{}_4)^2+(a^2{}_4)^2+(a^3{}_4)^2-(a^4{}_4)^2, \quad \text{hence} \quad |a^4{}_4|\geqq 1\,. \tag{4}$$

Definition 2. $x^k=a^k{}_l x'^l$ *is called a proper Lorentz transformation if* $g'_{kl}=\eta_{kl}$, $\det a^k{}_l=1$ *and* $a^4{}_4\geqq 1$.

Remark 3. By (3) and (4) the additional requirements (as compared with Def. 1) are meanngful.

Theorem 2. *The set of all proper Lorentz transformations forms a group (the proper Lorentz group).*

Remark 4. So the proper Lorentz group is a subgroup of the Lorentz group.

25.1.4. Special Transformations of the Proper Lorentz Group

The transformation $V(v)$: We look for proper Lorentz transformations of the form

$$V(v)=(a^k{}_l)=\begin{pmatrix}1 & 0 & 0 & 0\\ 0 & 1 & 0 & 0\\ 0 & 0 & \gamma & -v\gamma\\ 0 & 0 & -v\gamma & \gamma\end{pmatrix}. \tag{1}$$

From $\gamma=a^4{}_4\geqq 1$ and $\det V(v)=\gamma^2-v^2\gamma^2=1$ it follows that $\gamma=\dfrac{1}{\sqrt{1-v^2}}$, where $|v|<1$. If we again set $(x^k)=(x, y, z, t)$, and accordingly for (x'^k), then we obtain

$$x'=x, \quad y'=y, \quad z'=\frac{z-vt}{\sqrt{1-v^2}}, \quad t'=\frac{t-vz}{\sqrt{1-v^2}}\,. \tag{2}$$

(The fact that we interchanged the primed and the unprimed co-ordinates as compared with Def. 25.1.3/1 is of no account.) Now it is readily verified that (2) is in fact a proper Lorentz transformation. For we have

$$\begin{aligned}(x')^2+(y')^2+(z')^2-(t')^2&=x^2+y^2+(z'+t')(z'-t')\\ &=x^2+y^2+\frac{1}{1-v^2}(1-v)(z+t)(1+v)(z-t)=x^2+y^2+z^2-t^2\end{aligned}$$

and hence $g'_{kl}=\eta_{kl}$. The transformation inverse to $V(v)$ is $V(-v)$.

Rotations in R_3. Let

$$d=\begin{pmatrix}\alpha_{11} & \alpha_{12} & \alpha_{13}\\ \alpha_{21} & \alpha_{22} & \alpha_{23}\\ \alpha_{31} & \alpha_{32} & \alpha_{33}\end{pmatrix} \quad \text{and} \quad D=\begin{pmatrix}d & 0\\ 0 & 1\end{pmatrix}\begin{matrix}\}\,3\\ \}\,1\end{matrix} \tag{3}$$

where $\alpha_{kr}\alpha_{lr}=\delta_{kl}=\begin{cases}1 \text{ for } k=l\\ 0 \text{ for } k\neq l\end{cases}$ (summation over r from 1 to 3), and $\det d=1$. Then d is a rotation matrix in R_3. Hence in particular one has $(x')^2+(y')^2+(z')^2=x^2+y^2+z^2$. Since $\det d=1$, D is then a proper Lorentz transformation.

The transformation $D^{-1}V(v)D$. Together with D, D^{-1} is also a proper Lorentz transformation, and hence also $D^{-1}V(v)D$. These transformations are of physical interest. We need some structural informations. If $(w_1, w_2, w_3)\in R_3$ is the unit vector transformed into (0, 0, 1) by the rotation d as given in (3), then it is easily seen that

$$D=\left(\begin{array}{ccc|c} & * & & 0\\ & & & 0\\ w_1 & w_2 & w_3 & 0\\ \hline 0 & 0 & 0 & 1\end{array}\right) \quad\text{and}\quad D^{-1}=\left(\begin{array}{ccc|c} & * & w_1 & 0\\ & & w_2 & 0\\ & & w_3 & 0\\ \hline 0 & 0 & 0 & 1\end{array}\right) \tag{4}$$

where $*$ indicates that values of the places in question are of no interest. From (1) we then obtain

$$\begin{aligned} D^{-1}V(v)\,D &= D^{-1}\left(\begin{array}{ccc|c} & & & 0\\ & * & & 0\\ & & & -v\gamma\\ \hline -w_1v\gamma & -w_2v\gamma & -w_3v\gamma & \gamma\end{array}\right)\\ &=\left(\begin{array}{ccc|c} & & & -w_1v\gamma\\ & * & & -w_2v\gamma\\ & & & -w_3v\gamma\\ \hline -w_1v\gamma & -w_2v\gamma & -w_3v\gamma & \gamma\end{array}\right). \end{aligned} \tag{5}$$

25.1.5. Space-Time (Physical Aspects)

Consider any two Cartesian co-ordinate systems (x, y, z) and (x', y', z') in the real three-dimensional space R_3 (Fig. 25.4). Suppose that the observers *0* and *0'* are located at the origins of these systems, and that they do not leave these positions in the course of time. *0* and *0'* agree that they will measure times and lengths by identical physical processes, say, by means of atomic clocks and standardized rigid bodies, or by means of the wavelengths of fixed oscillations.

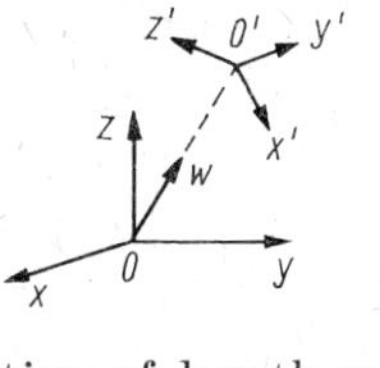

Fig. 25.4

The question of length and time measurement is not so unproblematic as it may appear at a first glance, because for the actual realization one makes use of physical facts which, in their turn, must be described in space and time. For a detailed discussion of this problem the reader is referred to [3] as well as to [26]. Here we suppose that the two observers *0* and *0'* measure lengths and time intervals each by the same method in laboratories which are at rest relative to them. Thus O and O' are provided with quantitative space-times (x, y, z, t) and (x', y', z', t'), respectively, so that it makes sense to speak of velocities. We suppose that O makes the following statement concerning O': O' moves uniformly in a straight line, and the axes of the co-ordinate system (x, y, z) can at any time be transformed into the axes of the co-ordinate system (x', y', z') by applying one and the same time-independent rotation matrix R. Further we suppose that at the time $t=t'=0$ both O and O' are at the same position (that means $x=y=z=x'=y'=z'=0$). Then O' moves at the velocity v in a straight line which passes through $x=y=z=0$ and the direction of which is given by the unit vector $w=(w_1, w_2, w_3)$. Thus at the time t the position of O' is $vtw=t(v_1, v_2, v_3)$. Now the observers O and O' are able to compare their quantitative, physically determined space-times $(x, y, z, t,)$ and (x', y', z', t') by experiment. The Galilean transformation (25.1.1/3) as a conversion formula does not fit the failure of the experiments devised to verify the ether hypothesis (cf. Subsec. 25.1.1.) (here we assume that R is the identity matrix). Rather there arises the problem to transform these two space-times into one another in such a way that they are compatible with the physical postulates stated in Subsecs. 25.1.1. and 25.1.2. and stand experimental tests. The experimental aspect is not our concern. For a description of experiments that support special relativity, the reader is referred to [26]. In the next subsection we shall state by axiom how (x, y, z, t) is transformed into (x', y', z', t'). But before doing so, let us once again recall those properties which are desirable for such a transformation.

I. The systems (x, y, z, t) and (x', y', z', t') are equivalent. If one fixes a procedure that transforms (x, y, z, t) into (x', y', z', t'), then the same procedure must also transform (x', y', z', t') into (x, y, z, t) in case the primed and the unprimed co-ordinates change parts.

II. If (x, y, z, t) is an inertial frame, then (x', y', z', t') is also an inertial frame (and vice versa).

III. The Maxwell equations in their classical form (24.3.3/5)–(24.3.3./7) remain unchanged under a transformation from (x, y, z, t) to (x', y', z', t') (and vice versa).

IV. If $\bar{x}(\tau)$ is the world line of a physical event in (x, y, z, t), then let $\bar{x}'(\tau)$ be the transform of this world line to (x', y', z', t')-co-ordinates. Let the three-dimensional velocities in terms of (25.1.2/1) be denoted by $|\bar{u}|$ and $|\bar{u}'|$. Then we require that $|\bar{u}|<1$ implies $|\bar{u}'|<1$ and that $|\bar{u}|=1$ implies $|\bar{u}'|=1$ (and vice versa).

Remark 1. The statements I, II, and III correspond to the requirements made at the end of Subsec. 25.1.1. The statement IV is a strengthened version of the physical principle stated in Subsec. 25.1.2., which includes the constancy of light velocity as a requirement.

Remark 2. The corrections applied to Newton's mechanics by special relativity are measurable only if the relative velocity v of the two co-ordinate systems (x, y, z, t) and (x', y', z', t') is not too small compared with the velocity of light. Thus it appears resonable to require that, as $|v|\to 0$, the transformation from (x, y, z, t) to (x', y', z', t') approaches the Galilean transformation (25.1.1/3) (where R is the identity matrix).

25.1.6. Space-Time (Mathematical Aspects)

The rotation matrix R, the unit vector $w=(w_1, w_2, w_3)$ and the velocity v shall have the same meaning as in Subsec. 25.1.5. In accordance with the physical principles stated in 25.1.2. we assume that $|v|<1$. ($v<0$ is admitted, so that w can be replaced by $-w$). If one sets

$$\hat{R}=\left(\begin{array}{c|c} R & 0 \\ \hline 0 & 1 \end{array}\right)\begin{matrix}\}\ 3 \\ \}\ 1,\end{matrix} \tag{1}$$

then $\hat{R}$ is a proper Lorentz transformation.

Axiom. *The space-times* (x, y, z, t) *and* (x', y', z', t') *described in Subsec.* 25.1.5. *are transformed into one another by*

$$(x', y', z', t')=\hat{R}D^{-1}\,V(v)\,D(x, y, z, t)\,. \tag{2}$$

Here $V(v)$, D *and* D^{-1} *have the meaning given in* (25.1.4/1) *and* (25.1.4/4).

Remark 1. According to Theorem 25.1.3/2 and to the considerations described in Subsec. 25.1.4., $\hat{R}D^{-1}V(v)D$ is a proper Lorentz transformation.

Remark 2. $\hat{R}$ is unique, whereas D in (25.1.4/4) is not. In order that (2) be meaningful, one has to show that (2) is independent of the particular choice of admissible matrices D. If D_1 and D_2 are any two admissible matrices, then $D_2D_1^{-1}$ transforms the vector (0, 0, 1, 0) into itself and the vector (0, 0, 0, 1) into itself, too, and hence has the form

$$D_2D_1^{-1}=\left(\begin{array}{cc|cc} \multicolumn{2}{c|}{\multirow{2}{*}{$*$}} & 0 & 0 \\ & & 0 & 0 \\ \hline 0 & 0 & 1 & 0 \\ 0 & 0 & 0 & 1 \end{array}\right)=D_3\,.$$

D_3 and D_3^{-1} commute with $V(v)$. Then it follows from $D_2=D_3D_1$ and $D_2^{-1}=D_1^{-1}D_3^{-1}$ that

$$D_2^{-1}V(v)D_2=D_1^{-1}D_3^{-1}V(v)D_3D_1=D_1^{-1}V(v)D_1\,.$$

which proves the desired independence.

Remark 3. Since $T=\hat{R}D^{-1}V(v)D$ is a proper Lorentz transformation, the cones referred to in Subsec. 25.1.2. are mapped onto cones of the same type (Fig. 25.5). Here the upper and the lower half-cones (as referred to the orientation of the t-axis) are again transformed into upper and lower half-cones, respectively. But now the considerations described in Subsec. 25.1.2. immediately show that the requirement IV of Subsec. 25.1.5. is fulfilled. In particular (in the formulation used there) one has $|\bar{u}|=1$ if and only if $|\bar{u}'|=1$: this is the constancy of the velocity of light (normalized to 1).

Remark 4. A motion is rectilinear and uniform if and only if the corresponding world line in the Minkowskian space is a straight line. Since T is linear, straight lines are transformed into straight lines. This shows that requirement II of Subsec. 25.1.5. is also satisfied.

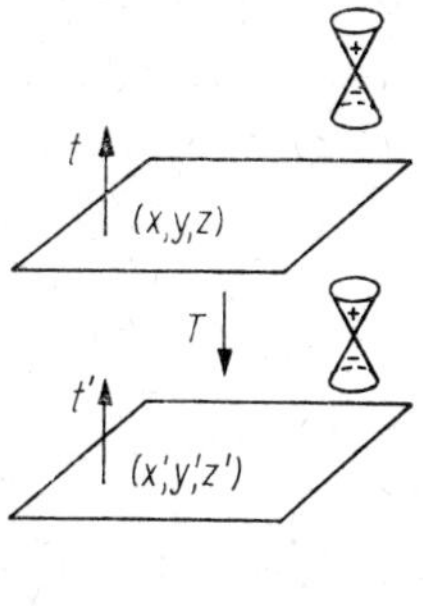

Fig. 25.5

Remark 5. In particular this implies that the observer $0'$ will find that, as referred to his own space-time (x', y', z', t'), 0 moves uniformly in a straight line. $\hat{R}$ and D do not influence the velocity of this motion, so that we may assume that $T = V(v)$ is given by (25.1.4/2). v is obtained from $0 = z' = z - vt$. Since $V(-v)$ is the transformation inverse to $V(v)$, the same argument shows that the velocity of 0, as measured in the space-time (x', y', z', t'), is $-v$. If we now want to apply the axiom by changing the parts of (x, y, z, t) and (x', y', z', t'), then we have to make the following substitutions:

$$(x, y, z, t):\ v,\ \hat{R},\ D$$

$$(x', y', z', t'):\ -v,\ \hat{R}^{-1},\ D\hat{R}^{-1}.$$

Hence requirement I of Subsec. 25.1.5. is fulfilled if

$$(x, y, z, t) = \hat{R}^{-1}\hat{R}D^{-1}V(-v)\, D\hat{R}^{-1}(x', y', z', t') = D^{-1}V(-v)\, D\hat{R}^{-1}(x', y', z', t') .$$

But according to (2) this is true.

Remark 6. Finally, requirement III of Subsec. 25.1.5. is fulfilled, as is shown by the derivation of (24.3.3/5) and (24.3.3/6) from the invariant equations (24.3.2/2) and (24.3.2/3). Thus we have reached our aim: (2) is supported by experiments within the limits of measuring accuracy, and the requirements I–IV of Subsec. 25.1.5. are fulfilled. Finally we recall the (necessary) self-restriction pointed out in Subsec. 25.1.1.: the space-time (x, y, z, t) (and hence, by requirement II if Subsec. 25.1.5., also (x', y', z', t')) is an inertial frame. According to the above considerations, the fundamental equations of electrodynamics are invariant under the proper Lorentz transformations. On the other hand the fundamental equations of mechanics are not invariant under proper Loentz transformations (whereas they are invariant under Galilean transformations). In order to achieve invariance, one thus has to correct the fundamental equations of mechanics (and not those of electrodynamics, as was intended originally, when the Galilean transformation had been taken as a basis).

Remark 7. In the c.g.s. system of units, where $c \sim 3 \cdot 10^5$ km/s is the value of the velocity of light, the last two formulas in (25.1.4/2) read

$$z' = \frac{z - vt}{\sqrt{1 - \frac{v^2}{c^2}}}, \quad t' = \frac{-\frac{vz}{c^2}}{\sqrt{1 - \frac{v^2}{c^2}}}.$$

If $|v|$ is much smaller than c, then these formulas (if interpreted liberally) approach $z' = x - vt$ and $t' = t$. See also Remark 25.1.5/2.

Remark 8. If R is the identity matrix, then from (2) and (25.1.4/5) one obtains the important formula

$$t' = \frac{t - (\bar{v}, \bar{x})}{\sqrt{1 - v^2}}, \quad \text{where} \quad \bar{v} = (v_1, v_2, v_3) = v(w_1, w_2, w_3) \quad \text{and} \quad \bar{x} = (x, y, z) . \tag{3}$$

Here $(\bar{v}, \bar{x}) = v_1 x + v_2 y + v_3 z$ is the common scalar product in R_3.

25.2. Effects of Special Relativity

25.2.1. Time Dilatation and the Twin Paradox

It is not our aim to give a systematic description of the theory of special relativity. We shall confine ourselves to the description of several spectacular effects that illustrate the radical change of the space-time conception established by Einstein as compared with Newton's space-time conception.

Time dilatation: Consider the same situation as described in Subsec. 25.1.5.: two observers 0 and $0'$ with the space-times (x, y, z, t) and (x', y', z', t'), which represent inertial frames (Fig. 25.6). Let $R=E$ be the identity matrix (a simplification which is physically of no account). The observer 0 trips a flash of light each at the times $t=0$ and $t=T>0$, thus marking a time interval of length T in the space-time (x, y, z, t). The world line of 0 is a straight line which passes through $x=y=z=0$ and is parallel to the t-axis (Fig. 25.7). According to (25.1.6/3), the observer $0'$, in his space-time (x', y', z', t'), finds that the two light flashes occur at the times $t'=0$ and $t'=T'=\frac{T}{\sqrt{1-v^2}}$. In particular this gives $T'>T$. In other words, the length of the time interval between two events depends on the choice of the space-time (x', y', z', t'). The time interval is shortest if one chooses a co-ordinate system which is at rest relative to the event.

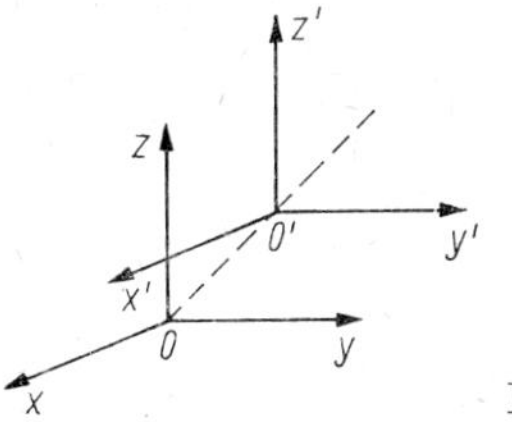

Fig. 25.6

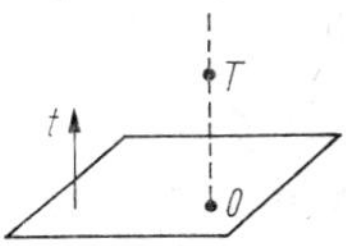

Fig. 25.7

μ-Mesons. An impressive (and experimentally well-supported) example of time dilatation is shown by the μ-mesons. It is known that they are generated at an altitude of 10–30 km above the earth's surface and that (in a rest frame) they have a lifetime of 2.2×10^{-6} s. Although they travel almost at light velocity, according to the classical argumentation they can travel at most a distance of $2.2\times10^{-6}\,\mathrm{s}\times3\times10^5\,\mathrm{km/s}<1$ km. But on the other hand they arrive at the earth's surface. The reason for the above false conclusion is that in a co-ordinate system which is at rest relative to the earth's surface (but just not relative to the μ-meson) a time dilatation occurs, which in the present case amounts to a factor of 10^2 to 10^3.

Twin paradox. There is hardly another effect of the theory of relativity that has been discussed in so great detail and so vehemently as the so-called twin paradox: there is not only a huge crowd of physical and philosophical papers on this subject, but whole books have been written about it. The problem considered is as follows. Suppose there are two twins. One of them stays on the earth, while the other undertakes an extensive excursion to the space. Apart from the starting, turnabout and landing phases (for which the time required will be short compared with the total flight), twin 2 is supposed to move uniformly in a straight line at the velocity v, as referred to a co-ordinate system at the origin of which twin 1 is at rest. After the space flight has been finished, the twins meet again, and each of the two claims to be younger than his twin brother. This collides with the simplest rules of logics, but can be reasoned as follows: let (x, y, z, t) be a space-time in which twin 1 is at rest (Fig. 25.8). Hence his world line is a straight line through 0 that is parallel to the t-axis. In this co-ordinate system the world line of twin 2 is curved. Neglecting the start, turnabout and landing phases, we idealize this curved path into a triangular one, and now we

Fig. 25.8

consider one half of the adventure. If (x', y', z', t') is a corresponding space-time for twin 2, then we have the situation investigated above, which gives a time dilatation of $\frac{T_1}{2} = \frac{T_2}{2\sqrt{1-v^2}}$, which means that $T_1 > T_2$. Here T_1 is the proper time of twin 1 that elapses between departure and return. An analogous statement holds for T_2. The proper time is the time measured by an observer in a laboratory which is at rest relative to himself. It gives the timing for inorganic and organic happenings (that is at rest relative to himself), and hence also governs the biological clock by which he ages. Now one may interchange the parts of the twins, so that one obtains $T_2 > T_1$. Thus one has the paradoxical result that, when the twins meet again, each of the two will be younger than the other. So the above argumentation must be wrong (in case we are confident that the mathematical model developed in Sec. 25.1. is in fact suitable to give a satisfactory, and hence in particular a logically consistent, description of physical reality).

As a first possible objection, one may ask for the legitimacy of the above idealization (i.e. of neglecting the start, landing, and turnabout phases). If one thinks of the space flights as being extended at choice, the start, turnabout, and landing operations always being executed in the same way, then the above idealization appears at least physically plausible.

The error is hidden elsewhere: at most one of the two space-times (x, y, z, t) and (x', y', z', t'), which are rest frames for one of the twins each, can be an inertial frame. If, for instance, (x, y, z, t) is an inertial frame (which can be supposed approximately: if necessary, one settles twin 1 in the centre of the sun, making use of the remarks on inertial systems as given in Subsec. 25.1.1.), then the above considerations which led to $T_1 > T_2$ are admissible within the special theory of relativity: the considerations refer to the inertial system (x, y, z, t). On the other hand, (x', y', z', t') is not a space-time system to which the special theory of relativity can be applied. Hence the parts of the twins cannot be changed in the above argumentation. At this point it turns out that the restriction to inertial systems (x, y, z, t) is necessary in the special theory of relativity. So the contradiction has been eliminated; what remains is a slight uneasiness.

Now, how about the real solution? In fact the twin paradox is an effect of the general rather than the special theory of relativity. If (as supposed above) (x, y, z, t) is an inertial frame, then in general relativity one cannot fail to conclude (without any tricks of argumentation, neglections etc.) that always $T_1 > T_2$ (no matter whether twin 2 departs for a longer or shorter trip, the only thing he must do is to travel). Popularly, it has been known all along: "Resting is rusting".

25.2.2. Lorentz Contraction

In the inertial frame (x, y, z, t) we consider a moving rod (Fig. 25.9). To determine the length of this rod, we consider the trace points of the world lines of the two rod ends in the plane $t = T = \text{const}$. The three-dimensional Euclidean distance of these two trace points is the length of the rod (at the time $t = T$). We now consider two inertial frames (x, y, z, t) and (x', y', z', t'), where (x', y', z') is supposed to move uniformly at the velocity v along the z-axis (Fig. 25.10). Then, by the axiom stated in Subsec. 25.1.6., the two systems transform according to (25.1.4/2). If $(0, 0, 0)$ and $(0, 0, l)$ are the front and the rear end of a rod at rest in the (x, y, z) system, then in the (x', y', z', t') system one obtains

$$z'_a = \frac{0 - vt_a}{\sqrt{1-v^2}}, \quad t'_a = \frac{t_a - 0}{\sqrt{1-v^2}} \quad \text{for the front end, with } t = t_a, \tag{1}$$

$$z'_e = \frac{l - vt_e}{\sqrt{1-v^2}}, \quad t'_e = \frac{t_e - vl}{\sqrt{1-v^2}} \quad \text{for the rear end, with } t = t_e. \tag{2}$$

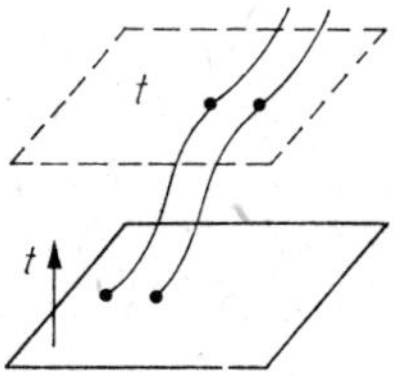

Fig. 25.9

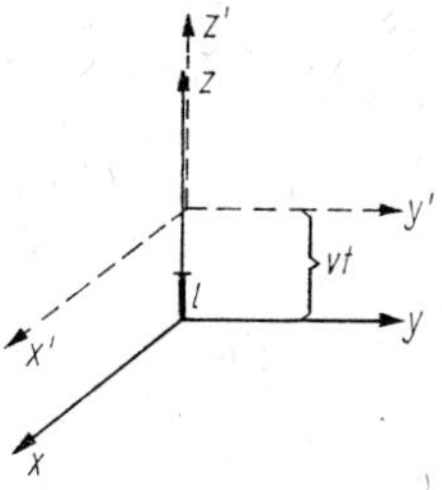

Fig. 25.10

The length of the rod in the (x, y, z, t) system is l. To determine the length of the rod in the (x', y', z', t') system, we have to set $t'_a = t'_e$ and hence $t_a = t_e - vl$. Insertion into (1) and (2) gives

$$l' = z'_e - z'_a = \frac{(1-v^2)l}{\sqrt{1-v^2}} = l\sqrt{1-v^2}.$$

Hence $l' < l$. In a moving inertial frame the rod is shorter than in the rest frame.

25.2.3. The Relativistic Addition Theorem of Velocities

Consider two inertial frames (x, y, z, t) and (x', y', z', t'), where (x', y', z') moves uniformly in a straight line along the z-axis (Fig. 25.11). By (25.1.4/2) one has $x' = x$, $y' = y$,

$$z' = \frac{z - vl}{\sqrt{1-v^2}} \quad \text{and} \quad t' = \frac{t - vz}{\sqrt{1-v^2}}, \tag{1}$$

where v is the relative velocity. Further consider a particle which moves uniformly in a straight line with respect to the (x, y, z, t) system (Fig. 25.12). One has

$$x = u_1 t, \quad y = u_2 t, \quad z = u_3 t, \tag{2}$$

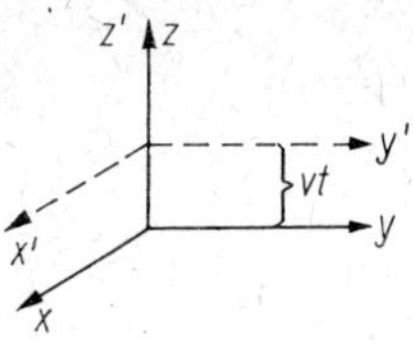

Fig. 25.11

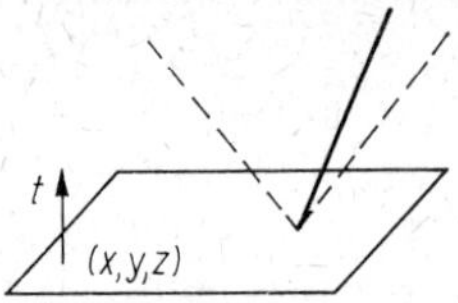

Fig. 25.12

where (u_1, u_2, u_3) is the velocity vector. By inserting (2) into (1) one obtains the following relations for the velocity vector (u_1', u_2', u_3') in the (x', y', z', t') system:

$$u_1' = \frac{x'}{t'} = u_1 \frac{\sqrt{1-v^2}}{1-vu_3}, \quad u_2' = \frac{y'}{t'} = u_2 \frac{\sqrt{1-v^2}}{1-u_3 v}, \quad u_3' = \frac{z'}{t'} = \frac{u_3 - v}{1-vu_3}. \tag{3}$$

If the particle moves at the velocity u in the z-direction, i.e., if $u_1 = u_2 = 0$ and $u_3 = u$, then one obtains the relativistic addition theorem of velocities,

$$u' = \frac{u-v}{1-vu}, \quad (u' = u_3'). \tag{4}$$

The classical counterpart reads $u' = u - v$; thus the factor $(1-vu)^{-1}$ is the relativistic correction. From the formulas (3) and (4) we can once again read off the previous statement that $u_1^2 + u_2^2 + u_3^2 < 1$ implies $u_1'^2 + u_2'^2 + u_3'^2 < 1$ and that $u = 1$ implies $u' = 1$.

25.2.4. The Free Relativistic Particle

In an inertial frame $(x^k) = (x, y, z, t)$, consider a force-free particle that moves uniformly in a straight line. We ask for a Lagrange density for curves which describes this situation. According to (24.2.2/4) we make the setup

$$L = -\alpha \sqrt{-\eta_{kl} \frac{dx^k}{ds} \frac{dx^l}{ds}}. \tag{1}$$

Here α is a real constant (which is unimportant for the time being). Now it follows from (24.2.3/5) that $\frac{d^2x^k}{ds^2} \equiv 0$, provided that s is identified with the geodesic parameter, which means that $-\eta_{kl} \frac{dx^k}{ds} \frac{dx^l}{ds} = 1$. Hence, apart from constants, one has $x^k = u^k s$. From $t = x^4 = u^4 s$ it follows that the motion is rectilinear and uniform, as should be expected.

25.2.5. Proper Time, Mass, and Energy

Proper time: From the normalization $-\eta_{kl} \frac{dx^k}{ds} \frac{dx^l}{ds} = 1$ and from $t = u^4 s$ in Subsec. 25.2.4., it follows for the force-free particle that $1 = -|u|^2 (u^4)^2 + (u^4)^2$, with $|u|^2 = (u_1)^2 + (u_2)^2 + (u_3)^2$ and $u_k = \frac{dx^k}{dt}$. Thus, after choosing a suitable sign one obtains $t = \frac{s}{\sqrt{1-|u|^2}}$. Now it follows from (25.1.6/3) that the arc length s is the proper time of the particle, that means, the time measured in an inertial frame which is at rest relative to the particle. In the general theory of relativity we shall then extend this statement to arbitrary motions (which are not necessarily rectilinear and uniform).

Mass and energy: Let us carry out some other speculative calculations, but without entering into physical reasons (see for instance [11]). For small velocities $|u|$,by substituting t for s one obtains from (25.2.4/1)

$$L=-\alpha'\sqrt{1-|u|^2}=-\alpha'+\frac{\alpha'}{2}|u|^2+\dots \tag{1}$$

The non-relativistic Lagrange density is $L_{\mathrm{cl}}=\frac{m_0}{2}|u|^2$, where m_0 is the mass (rest mass) of the particle. Additive constants are of no importance in Lagrange densities (the Euler-Lagrange equations contain nothing but partial derivatives). Thus a comparison of (1) and L_{cl} suggests choosing $\alpha'=m_0$, so that $L=-m_0\sqrt{1-|u|^2}$. The momentum p_k is obtained from

$$p_k=\frac{\partial L}{\partial u_k}=\frac{m_0 u_k}{\sqrt{1-|u|^2}}=mu_k\,, \tag{2}$$

where $m=\frac{m_0}{\sqrt{1-|u|^2}}$ is the inertial mass. If we calculate the energy by $E=\sum_{k=1}^{3}p_k u_k-L$, then the result is $E=\frac{m_0}{\sqrt{1-|u|^2}}=m$. Thus one obtains the relativistic formulas $m=\frac{m_0}{\sqrt{1-|u|^2}}$, $E=m$, which have to be verified by experiment. If the velocity of light is not normalized to 1, then in the c.g.s. system of units one obtains Einstein's famous mass-energy relation $E=mc^2$.

25.3. The Maxwell Equations

25.3.1. Formulation of the Problems

Maxwell's equations in free space and in the absence of electric charges and currents were described in Theorem 24.3.3/2. We use the notation employed there; in particular we write $\bar{x}=(x, y, z)\in R_3$.

Problem 1 (*initial value problems, classical solutions*). *Given the vector functions $\mathfrak{E}_0(\bar{x})$ and $\mathfrak{B}_0(\bar{x})$ that are three times continuously differentiable on R_3. Desired are vector functions $\mathfrak{E}(\bar{x}, t)$ and $\mathfrak{B}(\bar{x}, t)$ which are twice continuously differentiable on $\overline{R_4^+}$, such that*

$$\operatorname{rot}\mathfrak{E}+\frac{\partial\mathfrak{B}}{\partial t}=0,\quad \operatorname{div}\mathfrak{B}=0 \quad in \quad \overline{R_4^+}\,, \tag{1}$$

$$\operatorname{rot}\mathfrak{B}-\frac{\partial\mathfrak{E}}{\partial t}=0,\quad \operatorname{div}\mathfrak{E}=0 \quad in \quad \overline{R_4^+}\,, \tag{2}$$

and

$$\mathfrak{E}(\bar{x}, 0)=\mathfrak{E}_0(\bar{x}),\quad \mathfrak{B}(\bar{x}, 0)=\mathfrak{B}_0(\bar{x}) \quad in \quad R_3\,. \tag{3}$$

Remark 1. As in Subsec. 24.3.3., $\mathfrak{E}(\bar{x}, t)$ and $\mathfrak{B}(\bar{x}, t)$ each have one x-component, one y-component, and one z-component. The same holds for $\mathfrak{E}_0(\bar{x}) = (E_{0,x}(\bar{x}), E_{0,y}(\bar{x}), E_{0,z}(\bar{x}))$, and accordingly for $\mathfrak{B}_0(\bar{x})$. This is a typical initial value problem, being analogous to the initial value problems for the wave equation in Sec. 19.3. Given are the electric and the magnetic field intensity at the time $t=0$. For physical reasons the above problem should then have a unique solution for $t>0$.

Remark 2. If charges and currents are present, then, as was described in Remark 24.3.3/3, the formulas (1), (2) must be replaced by

$$\operatorname{rot} \mathfrak{E} + \frac{\partial \mathfrak{B}}{\partial t} = 0, \qquad \operatorname{div} \mathfrak{B} = 0\,, \tag{4}$$

$$\operatorname{rot} \mathfrak{B} - \frac{\partial \mathfrak{E}}{\partial t} = 4\pi j, \quad \operatorname{div} \mathfrak{E} = 4\pi \varrho\,. \tag{5}$$

Here ϱ is the charge density and $j=(j_x, j_y, j_z)$ is the current density. The formulation of Problem 1 might immediately be extended to this case. But it is physically reasonable to consider point-like charges, charges on surfaces, currents along lines etc. Then one cannot expect classical solutions of (4), (5) (by analogy to Problem 1). In a way analogous to Sec. 23.2., one will attempt to find distribution solutions. We use the same notations as in Subsec. 23.2.1. Thus in R_4, with $(x, y, z, t) = (\bar{x}, t)$, $E_{0,x} \otimes \delta$ means the tensor product of $E_{0,x} \in D'(R_3)$, with respect to $\bar{x}$, and $\delta \in D'(R_1)$, with respect to t. For abbreviation we set $\mathfrak{E}_0 \otimes \delta = (E_{0,x} \otimes \delta, E_{0,y} \otimes \delta, E_{0,z} \otimes \delta)$ and analogously for $\mathfrak{B}_0 \otimes \delta$. Finally, supp $\mathfrak{E} \subset \overline{R_4^+}$ means that the supports of all components of $\mathfrak{E}$ lie in $\overline{R_4^+}$. Analogous notations are used for other vector distributions.

Problem 2 (*initial value problems, distribution solutions*). *Given* $\mathfrak{E}_0 = (E_{0,x}, E_{0,y}, E_{0,z})$ *and* $\mathfrak{B}_0 = (B_{0,x}, B_{0,y}, B_{0,z})$ *with components in* $D'(R_3)$, $j = (j_x, j_y, j_z)$ *with components in* $D'(R_4)$ *and* supp $j \subset \overline{R_4^+}$, *and* $\varrho \in D'(R_4)$ *with* supp $\varrho \subset \overline{R_4^+}$. *Desired are* $\mathfrak{E} = (E_x, E_y, E_z)$ *and* $\mathfrak{B} = (B_x, B_y, B_z)$, *with components in* $D'(R_4)$ *and* supp $\mathfrak{E} \subset \overline{R_4^+}$, supp $\mathfrak{B} \subset \overline{R_4^+}$, *such that*

$$\operatorname{rot} \mathfrak{E} + \frac{\partial \mathfrak{B}}{\partial t} = \mathfrak{B}_0 \otimes \delta, \qquad \operatorname{div} \mathfrak{B} = 0\,, \tag{6}$$

$$\operatorname{rot} \mathfrak{B} - \frac{\partial \mathfrak{E}}{\partial t} = 4\pi j - \mathfrak{E}_0 \otimes \delta, \quad \operatorname{div} \mathfrak{E} = 4\pi \varrho\,. \tag{7}$$

Remark 3. This is the analogue to Def. 23.2.1(a), where distribution solutions of initial value problems for the wave equation were described. As in Theorem 23.2.1, it can be expected that distribution solutions are classical solutions, provided that all the data are sufficiently smooth.

Theorem. *If* $j=0$ *and* $\varrho=0$, *and if the components of* $\mathfrak{E}_0$ *and* $\mathfrak{B}_0$ *are three times continuously differentiable on* R_3, *then Problem 1 and Problem 2 coincide.*

25.3.2. Initial Value Problems

Theorem 1. *If* $\mathfrak{E}_0(\bar{x})$ *and* $\mathfrak{B}_0(\bar{x})$ *are vector functions three times continuously differentiable on* R_3, *with* $\operatorname{div} \mathfrak{E}_0(\bar{x}) \equiv \operatorname{div} \mathfrak{B}_0(\bar{x}) \equiv 0$, *then Problem 1 has a unique solution. It is given by*

$$\mathfrak{E}(\bar{x}, t) = \frac{1}{4\pi t} \int\limits_{|\bar{y}-\bar{x}|=t} \operatorname{rot} \mathfrak{B}_0(\bar{y})\, ds_{\bar{y}} + \frac{1}{4\pi} \frac{\partial}{\partial t} \left[\frac{1}{t} \int\limits_{|\bar{y}-\bar{x}|=t} \mathfrak{E}_0(\bar{y})\, ds_{\bar{y}} \right],$$

$$\mathfrak{B}(\bar{x}, t) = \frac{1}{4\pi t} \int\limits_{|\bar{y}-\bar{x}|=t} \operatorname{rot} \mathfrak{E}_0(\bar{y})\, ds_{\bar{y}} + \frac{1}{4\pi} \frac{\partial}{\partial t} \left[\frac{1}{t} \int\limits_{|\bar{y}-\bar{x}|=t} \mathfrak{B}_0(\bar{y})\, ds_{\bar{y}} \right].$$

Remark 1. The explicit solutions $\mathfrak{E}(\bar{x}, t)$ and $\mathfrak{B}(\bar{x}, t)$ agree with the solution (19.3.3/2) for the wave equation in three dimensions (provided that u_0 and u_1 are chosen suitably). This also gives an indication of how to treat the Maxwell equations. They are reduced to wave equations for $\mathfrak{E}(\bar{x}, t)$ and $\mathfrak{B}(\bar{x}, t)$. For curved space-times we shall describe this procedure in Sec. 33.5.

Theorem 2. *Problem 2 has a solution if and only if*

$$\operatorname{div} j + \frac{\partial \varrho}{\partial t} = \operatorname{div} \mathfrak{E}_0 \otimes \delta \quad \textit{and} \quad \operatorname{div} \mathfrak{B}_0 = 0 . \tag{1}$$

If this condition is fulfilled, then Problem 2 has a unique solution, which is given by

$$\mathfrak{E} = G * \left(-\operatorname{grad} \varrho - \frac{\partial j}{\partial t} + (\operatorname{rot} \mathfrak{B}_0) \otimes \delta + \mathfrak{E}_0 \otimes \delta' \right), \tag{2}$$

$$\mathfrak{B} = G * (\operatorname{rot} j - (\operatorname{rot} \mathfrak{E}_0) \otimes \delta + \mathfrak{B}_0 \otimes \delta') . \tag{3}$$

Remark 2. According to our previous convention, div, grad and rot always refer to x, y, z, but not to t. The above theorem is proved by reducing it to Theorem 23.2.2; in particular G is the fundamental solution of the three-dimensional wave equation as given in Theorem 23.1.4. If all the distributions concerned are regular, then $\mathfrak{E}$ and $\mathfrak{B}$ can be represented explicitly in the sense of (23.2.2/2). The above theorem was proved by Leopold [38].

26. Self-Adjoint Operators in the Hilbert Space

26.1. Unbounded Operators

26.1.1. Closed Operators

In Sec. 17.3. we developed the geometry of the Hilbert space. In particular we recall that a Hilbert space in the sense of our terminology is always complex and separable. Thus Hilbert spaces are special separable Banach spaces. Linear operators (also called mappings) were described in Def. 20.2.1. We shall now use the same notations as introduced there. Chaps. 20 and 21 contain a description of the theory of bounded (or continuous) operators in Banach spaces and Hilbert spaces. But it turns out that many operators of mathematical and physical interest are not bounded. Here the so-called self-adjoint operators in the Hilbert space are in the centre of interest. In the present chapter we shall develop the theory of these operators. In this chapter we consider linear operators A acting in a Hilbert space H, which means that the (linear) domain of definition $D(A)$ as well as the range $R(A)$ of A, in the sense set out in Subsecs. 20.2.1. and 20.2.3., lie in H.

Definition. *Let H be a Hilbert space, and let A be a linear operator with $D(A)\subset H$ and $R(A)\subset H$.*

(**a**) *For $x\in D(A)$ and $y\in D(A)$ we define $[x, y]=(x, y)+(Ax, Ay)$.*

(**b**) *A is said to be closed if the limit element x of every convergent sequence $\{x_k\}_{k=1}^{\infty}\subset D(A)$ with $Ax_n\to y$ belongs to $D(A)$, and if $Ax=y$.*

Remark 1. It is immediately verified that $[x, y]$ is a scalar product on $D(A)$. Hence it is possible to introduce a norm on $D(A)$ by setting $\|x\|_A=\sqrt{[x, x]}$. The question is whether $D(A)$ is complete with respect to this norm.

Remark 2. Thus in part (b) of the definition one has to consider all convergent sequences $\{x_k\}$ in $D(A)$ whose images $\{Ax_k\}$ are also convergent. This is a typical construction for not bounded operators. For, if A is bounded (in the sense of Def. 20.2.1) then $x_n\to x$ immediately implies that $Ax_n\to Ax$. Hence, in particular, every bounded operator A with $D(A)=H$ and $R(A)\subset H$ is closed. Thus it is especially not bounded operators that are of interest, as is also shown by the following theorem.

Theorem. (**a**) *If A is a closed operator with $D(A)=H$, then A is bounded.*

(**b**) *A is closed if and only if $D(A)$ is closed with respect to the norm $\|x\|_A$.*

Remark 3. The proof of part (a) is not so simple as it would be expected at a first glance. It is the converse of the statement made in Remark 2.

Remark 4. Part (b) says $D(A)$, with the norm $\|x\|_A$, is a Banach space if A is closed. If this Banach space is separable, then we even have a Hilbert space, because $\|x\|_A$ has been derived from a scalar product.

26.1.2. Closable Operators

Definition. *The linear operator A is said to be closable if there exists a closed operator B, in the sense of Def. 26.1.1, which is an extension of A.*

Remark 1. All operators in this chapter are linear and act in a Hilbert space H. An operator B is called extension of the operator A if $D(B)\supset D(A)$ and $Bx=Ax$ for $x\in D(A)$. For this we also write $B\supset A$ or $A\subset B$.

Theorem. *If A is closable, then there exists a unique minimal closed extension of A. It is denoted by $\bar{A}$, and its domain of definition, $D(\bar{A})$, is the completion of $D(A)$ with respect to the norm $\|\cdot\|_A$, as referred to in Remark 26.1.1/1.*

Remark 2. $\bar{A}$ is called closure of A. Hence the assertion is that $B\supset\bar{A}$ if B is a closed extension of A. The last part of the theorem says that $D(\bar{A})$ consists of all elements $x\in H$ for which there is a fundamental sequence (Cauchy sequence) $\{x_k\}_{k=1}^{\infty}$ in $D(A)$ (with respect to the norm $\|\cdot\|_A$) such that $x_k\to x$ in H as $k\to\infty$. The statement that $\{x_k\}_{k=1}^{\infty}$ is a fundamental sequence in $D(A)$ with respect to the norm $\|\cdot\|_A$ is equivalent with the statement that $\{x_k\}_{k=1}^{\infty}$ and $\{Ax_k\}_{k=1}^{\infty}$ are fundamental sequences in H. Hence it follows that $Ax_k\to\bar{A}x$ in addition to $x_k\to x$.

26.1.3. Adjoint Operators

Definition. *If A is a linear operator whose domain of definition $D(A)$ is dense in H, then we set*

$$D(A^*)=\{y \mid y\in H,\ \exists y^*\in H,\ \text{with } (Ax, y)=(x, y^*) \text{ for all } x\in D(A)\}\,,$$
$$A^*y=y^* \quad \text{for} \quad y\in D(A^*)\,.$$

Remark 1. y^* is uniquely determined. From $(Ax, y)=(x, y^*)=(x, y^{**})$ for all $x\in D(A)$ it follows that $(x, y^*-y^{**})=0$ for all x in the dense set $D(A)$. But then $y^*=y^{**}$. Hence $A^*y=y^*$ is meaningful. It is immediately seen that $D(A^*)$ is a linear set in H and that A^* is a linear operator in H. A^* is called adjoint operator of A. This is obviously a generalization of Def. 21.1.4.

Theorem. *Let A be a linear operator whose domain of definition $D(A)$ is dense in H.*
(a) *The adjoint operator A^* is closed.*
(b) *If $B\supset A$, then $A^*\supset B^*$.*
(c) *If A is closable, then $(\bar{A})^*=A^*$.*

Remark 2. Proposition (b) is meaningful, because if $D(A)$ is dense in H then $D(B)$ is also dense in H.

26.1.4. Symmetric and Self-Adjoint Operators

Definition 1. *Let A be a linear operator whose domain of definition $D(A)$ is dense in the Hilbert space H.*
(a) *A is called symmetric if $(Ax, y)=(x, Ay)$ for all $x\in D(A)$ and $y\in D(A)$.*
(b) *A is called self-adjoint if $A=A^*$.*

Remark 1. If A is self-adjoint, then $(Ax, y)=(x, A^*y)=(x, Ay)$ for all $x\in D(A)$ and $y\in D(A)$. Hence self-adjoint operators are symmetric.

Theorem. (a) *If A is symmetric, then A is closable, and $\bar{A}$ is also symmetric.*
(b) *If B is a symmetric extension of A, then $B\subset A^*$.*
(c) *A linear operator A with a dense domain of definition $D(A)$ is symmetric if and only if (Ax, x) is real for all $x\in D(A)$.*

Remark 2. Part (b) says that every symmetric extension B of the operator A (which is necessarily also symmetric) is a restriction of A^*. In particular, every self-adjoint operator has no proper symmetric extensions.

Remark 3. The proof of (c) is largely based on the fact that H is a complex space.

Definition 2. *A symmetric operator A is called essentially self-adjoint if its closure $\bar{A}$ is self-adjoint.*

Remark 4. By part (a) of the theorem the definition is meaningful. In the later applications to quantum mechanics we shall be interested in self-adjoint operators. On the other hand, by Remark 26.1.2/2 the operator $\bar{A}$ is known if A is known. Thus for physical applications it will suffice to know that A is essentially self-adjoint.

26.1.5. Criteria for the Self-Adjointness of Operators

The self-adjoint and the essentially self-adjoint operators take a central position in the theory of this chapter as well as in the following applications. The criteria to be described now will already show what remarkable properties are exhibited by self-adjoint operators.

Theorem 1. *Let A be a symmetric operator.*
(a) *If $D(A)=H$, then A is self-adjoint and bounded.*

(b) *A is self-adjoint if there exists a complex number λ such that $R(A-\lambda E)=R(A-\bar{\lambda}E)=H$.*

(c) *A is self-adjoint if and only if $R(A-\lambda E)=H$ for every complex number λ with* $\operatorname{Im}\lambda\neq 0$.

Remark 1. Part (a) is a simple implication of $A=A^*$ and Theorem 26.1.1(a).

Remark 2. Part (c) shows that for self-adjoint operators A the behaviour of $A-\lambda E$ will be of special interest if λ is real. As usual, E is the identity operator. Part (c) follows from the inequality

$$\|Ax-\lambda x\|\geqq|\operatorname{Im}\lambda|\,\|x\| \quad \text{for all} \quad x\in D(A)\,.$$

Remark 3. As previously, we shall also here consider the family of operators $A-\lambda E$ rather than an individual operator A. In this connection the following statement is of interest. If λ is a complex number and if A is a symmetric operator, then $(A-\lambda E)^*=A^*-\bar{\lambda}E$, where $D(A-\lambda E)=D(A)$. Further, as in Theorem 21.2.2, one has

$$H=\overline{R(A-\lambda E)}\oplus N(A^*-\bar{\lambda}E)\,,$$

where the symbols have the same meaning as in Subsecs. 21.2.1. and 21.2.2.

Theorem 2. *Let A be a self-adjoint, and B a symmetric operator, with $D(B)\supset D(A)$. If there exist two real numbers c and δ such that $c\geqq 0$, $0\leqq\delta<1$, and if*

$$\|Bx\|\leqq\delta\|Ax\|+c\,\|x\| \qquad \forall x\in D(A)\,,$$

then $A+B$, with $D(A+B)=D(A)$, is self-adjoint.

Remark 4. This theorem is of interest especially in physical applications. It says that the self-adjointness of an operator A is stable under the action of small symmetric perturbations B.

26.2. The Spectrum of Self-Adjoint Operators

26.2.1. The Spectra $\tilde{D}_A$ and $\tilde{C}_A$

The resolvent set and the spectrum of a bounded operator were considered in Subsec. 20.2.3. Let us now extend these investigations to unbounded operators in the Hilbert space („unbounded" means „not necessarily bounded"; thus in particular a bounded operator is a special unbounded operator: this is mathematical scientese). As everywhere in this chapter, all operators A are linear, the domain of definition $D(A)$ and the range $R(A)$ lie in one and the same Hilbert space H. E denotes the identity operator. $L(H)$ has the same meaning as in Subsec. 20.2.3. Finally, C_1 is the complex plane.

Definition 1. *If A is a linear operator whose domain of definition $D(A)$ is dense in the Hilbert space H, then*

$$M_A=\{\lambda\mid\lambda\in C_1,\ \exists\ (A-\lambda E)^{-1} \quad \text{and} \quad (A-\lambda E)^{-1}\in L(H)\}$$

is called the resolvent set (of A), and $S_A=C_1\backslash M_A$ is called the spectrum (of A). Further we define

$$\tilde{D}_A=\{\lambda\mid\exists x\in D(A) \quad \text{with} \quad \|x\|=1 \text{ and } Ax=\lambda x\}\,,$$

and $\tilde{C}_A=\{\lambda\mid\exists\ (A-\lambda E)^{-1}$, but does not belong to $L(H)\}$.

Remark 1. The first part is largely a repetition of Def. 20.2.3. As previously, $(A-\lambda E)^{-1}$ denotes the inverse operator of $A-\lambda E$, which exists if and only if $A-\lambda E$ provides a one to one mapping of $D(A)=D(A-\lambda E)$ onto $R(A-\lambda E)$. It is easily seen that all possibilities are exhausted by M_A, $\tilde{D}_A$ and $\tilde{C}_A$. In particular one has $S_A=\tilde{D}_A\cup\tilde{C}_A$. As previously, $\tilde{D}_A$ is the set of all eigenvalues of the operator A. If $\lambda\in\tilde{D}_A$ is an eigenvalue, then $x\in D(A)$, with $x\neq 0$ and $Ax=\lambda x$, is called the (corresponding) eigenelement (eigenvector). One has $\tilde{D}_A\cap\tilde{C}_A=\emptyset$.

Remark 2. Thus the spectrum S_A is subdivided into the eigenvalue spectrum $\tilde{D}_A$ and the residual spectrum $\tilde{C}_A$. This suggests itself, but is not convenient for the following considerations. As in Def. 21.2.1 the dimension of $N(A-\lambda E)$ is called the multiplicity of the eigenvalue $\lambda\in\tilde{D}_A$. It is our aim to slightly modify the spectra $\tilde{D}_A$ and $\tilde{C}_A$. Roughly speaking, we want to take the eigenvalues of infinite multiplicities out of $\tilde{D}_A$ and assign them to $\tilde{C}_A$. For a precise formulation we will need the concept of a Weyl sequence.

Definition 2. *If A is a linear operator whose domain of definition $D(A)$ is dense in the Hilbert space H, then $\{x_k\}_{k=1}^{\infty}\subset D(A)$ is called Weyl sequence for the point λ if $\{x_k\}_{k=1}^{\infty}$ is bounded and not precompact, and if $Ax_k-\lambda x_k\to 0$ as $k\to\infty$.*

Remark 3. Precompact sets were described in Def. 20.1.2/1(c). The prototype of a bounded non-precompact set in the Hilbert space is a orthonormal sequence $\{x_k\}_{k=1}^{\infty}$. The operator $A-\lambda E$ transforms "poor (Weyl) sequences" (from the convergence point of view) into "good sequences". One can at least surmise that points λ for which there exists a Weyl sequence are of special interest.

Theorem. *Let A be a self-adjoint operator.*

(a) *S_A is a subset of the real numbers, and*

$$\|(A-\lambda E)^{-1}\|\leq\frac{1}{|\mathrm{Im}\,\lambda|}\quad \textit{for}\quad \mathrm{Im}\,\lambda\neq 0\,.$$

(b) *λ belongs to $\tilde{D}_A$ if and only if $\overline{R(A-\lambda E)}$ is a proper subspace of H.*

(c) *λ belongs to $\tilde{C}_A$ if and only if $\overline{R(A-\lambda E)}=H$ and $R(A-\lambda E)\neq H$.*

(d) *For every point $\lambda\in\tilde{C}_A$ there exists a Weyl sequence.*

(e) *For every eigenvalue of infinite multiplicity there exists a Weyl sequence.*

Remark 4. Part (a) is an implication of Theorem 26.1.5/1 and Remark 26.1.5/2. Likewise, it is easy to see that proposition (e) is true: every orthonormal sequence $\{x_k\}_{k=1}^{\infty}\subset N(A-\lambda E)$ is a Weyl sequence for the point λ.

26.2.2. The Spectra D_A and C_A

By means of Weyl sequences it is now possible to repartition the spectrum of self-adjoint operators.

Definition. *If A is a self-adjoint operator, then*

$$D_A=\{\lambda \mid \lambda \textit{ is an eigenvalue of finite multiplicity}\}$$

is called the discrete spectrum (point spectrum), and

$$C_A=\{\lambda \mid \exists \textit{ a Weyl sequence for } \lambda\}$$

is called the continuous spectrum of A.

Remark 1. By Theorem 26.2.1 we have $\tilde{C}_A\subset C_A$ and $D_A\subset\tilde{D}_A$; further the eigenvalues of infinite multiplicities belong to C_A. Hence it is clear that $\tilde{C}_A\cup\tilde{D}_A=S_A\subset C_A\cup D_A$. The arising problem of whether C_A can also contain points of the resolvent set is answered negatively by the following theorem.

Theorem. *Let A be a self-adjoint operator.*

(**a**) $S_A = C_A \cup D_A$.

(**b**) *If* $\lambda_1 \in D_A$, $\lambda_2 \in D_A$, $Ax_1 = \lambda_1 x_1$, $Ax_2 = \lambda_2 x_2$, *and if* $\lambda_1 \neq \lambda_2$, *then* $(x_1, x_2) = 0$.

Remark 2. This is the desired better partition of the spectrum S_A (better for our purposes). Part (b) follows from

$$\lambda_1(x_1, x_2) = (Ax_1, x_2) = (x_1, Ax_2) = \lambda_2(x_1, x_2) .$$

Hence eigenelements corresponding to different eigenvalues are always orthogonal.

26.2.3. Compact Self-Adjoint Operators

The theory of Riesz and Schauder presented in Sec. 21.2. shows that compact operators are of special importance. On the other hand it is shown by the preceding investigations that self-adjoint operators also play a special role.

Thus it is clear that self-adjoint compact operators will play quite an extraordinary role. $L(H)$ shall have the same meaning as in Subsec. 20.2.3.

Theorem 1. *If $A \in L(H)$ is self-adjoint, then $S_A \subset [-\|A\|, \|A\|]$, where at least one of the two points $\|A\|$ and $-\|A\|$ belongs to S_A.*

Remark 1. $S_A \subset [-\|A\|, \|A\|]$ follows immediately from Theorem 26.2.1(a) and Theorem 20.2.3. What is new in this theorem is the statement that either $\|A\|$ or $-\|A\|$ belong to the spectrum.

Theorem 2. *Let $A \in L(H)$ be self-adjoint and compact.*

(**a**) $S_A \subset [-\|A\|, \|A\|]$ *and* $C_A = \{0\}$.

(**b**) *D_A consists of an at most countably infinite number of different eigenvalues, which can accumulate only at the point 0. Every eigenvalue that is different from 0 is of finite multiplicity.*

(**c**) *There exists an orthonormal system $\{x_k\} \subset H$ such that $Ax_k = \lambda_k x_k$ and*

$$Ax = \sum \lambda_k(x, x_k)\, x_k \quad \textit{for every} \quad x \in H . \tag{1}$$

Remark 2. In the theorem it has been assumed that H is infinite-dimensional (otherwise $C_A = \emptyset$). The eigenvalues can be ordered by their absolute values, taking into account their multiplicities: $|\lambda_1| \geqq |\lambda_2| \geqq |\lambda_3| \geqq \ldots$, where $\lambda_k \to 0$ (in case there are infinitely many non-zero eigenvalues; Fig. 26.1). For a fixed eigenvalue λ, the characteristic manifold $N(A - \lambda E)$ can be spanned by an orthonormal system. The class of all these systems then forms the system $\{x_k\}$ in part (c). (By Theorem 26.2.2(b), eigenelements corresponding to different eigenvalues are always orthogonal). The sum in (1) needs only to be taken over those k for which $\lambda_k \neq 0$.

Theorem 3. *A self-adjoint operator $A \in L(H)$ is compact if and only if $C_A = \{0\}$.*

Remark 3. It has again been assumed that H is infinite-dimensional.

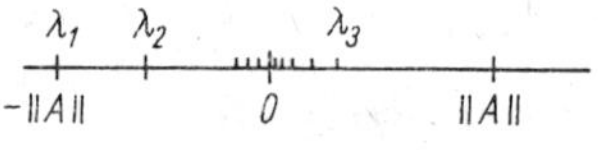

Fig. 26.1

26.3. Spectral Families

26.3.1. Definitions

Definition. *A family* $\{E_\lambda\}_{\lambda \in R_1}$ *of projectors in the Hilbert space* H *is called a spectral family if it has the following properties:*

(a) *For all* $x \in H$, $\lim_{\lambda \to -\infty} E_\lambda x = 0$ *and* $\lim_{\lambda \to \infty} E_\lambda x = x$.

(b) *For all* $x \in H$ *and all* $\mu \in R_1$, $\lim_{\lambda \uparrow \mu} E_\lambda x = E_\mu x$.

(c) *For all* $\lambda \in R_1$ *and* $\mu \in R_1$, $E_\lambda E_\mu = E_{\min(\lambda,\mu)}$.

Remark 1. Projectors (or projection operators) were described in Subsec. 21.1.5. The convergences in (a) and (b) are of course to be understood in terms of the norm in H. As previously, $\lambda \uparrow \mu$ means that $\lambda \to \mu$ and $\lambda < \mu$. By analogy we write $\lambda \downarrow \mu$ for $\lambda \to \mu$ and $\lambda > \mu$. From (c) it follows that $E_\lambda E_\mu = E_\mu E_\lambda$. Further $E_\lambda^2 = E_\lambda$, which, by Theorem 21.1.5, is a necessary property of projectors. The same theorem shows that projectors are self-adjoint.

Lemma. *For all* $x \in H$ *and all* $\mu \in R_1$, $\lim_{\lambda \downarrow \mu} E_\lambda x = E_{\mu+0} x$ *exists. Here* $E_{\mu+0}$ *is a projector.*

Remark 2. According to Remark 21.1.5/2, $H^\lambda = \{x \mid x \in H,\ E_\lambda x = x\}$ is the projection space corresponding to E_λ. By analogy we define $H^{\lambda+0} = \{x \mid x \in H,\ E_{\lambda+0} x = x\}$.

Remark 3. The properties (a) and (c) in the above definition are essential, while (b) and the lemma are normalizations.

26.3.2. Properties

All notations to be used shall have the same meaning as above.

Theorem. (a) *For* $-\infty < \lambda \leq \mu < \infty$, $H^\lambda \subset H^\mu \subset H^{\mu+0}$ *and* $\overline{\bigcup_\lambda H^\lambda} = H$. *Further* $E_\mu - E_\lambda$ *is the projector with the projection space* $H^\mu \ominus H^\lambda$, *and* $E_{\mu+0} - E_\lambda$ *is the projector with the projection space* $H^{\mu+0} \ominus H^\lambda$.

(b) *For* $-\infty < \lambda < \mu < \infty$, $H^{\lambda+0} \subset H^\mu$, *and* $E_\mu - E_{\lambda+0}$ *is the projector with the projection space* $H^\mu \ominus H^{\lambda+0}$.

Remark 1. $H^\mu \ominus H^\lambda$ is the space which complements H^λ orthogonally to H^μ (Fig. 26.2) Hence, in terms of Subsec. 17.3.5. one has $H^\mu = H^\lambda \oplus (H_\mu \ominus H^\lambda)$. An analogous relation holds for $H^{\mu+0} \ominus H^\lambda$.

Remark 2. Thus H^λ is a monotonically increasing sequence of Hilbert spaces which exhaust the total Hilbert space H as $\lambda \to \infty$: $\bigcup_\lambda H^\lambda$ is dense in H.

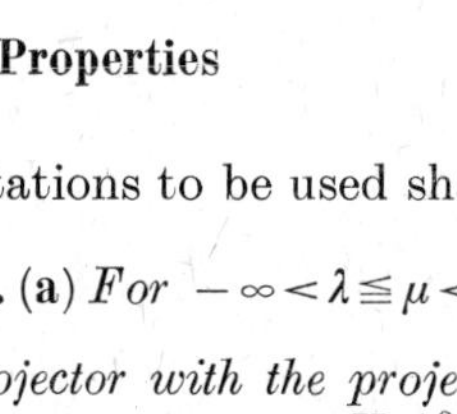

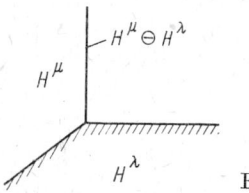

Fig. 26.2

26.4. Spectral Operators

26.4.1. Riemann-Stieltjes Integrals for Functions

Before considering Riemann-Stieltjes integrals for spectral families in the following subsections, it is useful to study corresponding integrals for functions. If $h(t)$ is a real-valued increasing (non-decreasing) left-continuous function on R_1 (Fig. 26.3), then $h(t)$ generates a Borel measure $\mathrm{d}h(t)$ in the sense of Subsec. 13.2.4. If $\varphi(t)$ is a real-valued continuous function on R_1, and if $-\infty < a \leq b < \infty$, then according to Chap. 14 we can form the integral $\int_{\{a,b\}} \varphi(t)\,\mathrm{d}h(t)$. Here $\{a, b\}$ means (a, b) or $[a, b)$ or $(a, b]$ or $[a, b]$ ($\{a, b\} = \{a\}$ is also admitted). If $\varphi(t)$ is complex-valued, then of course we set

$$\int \varphi(t)\,\mathrm{d}h(t) = \int \mathrm{Re}\,\varphi(t)\,\mathrm{d}h(t) + \mathrm{i}\int \mathrm{Im}\,\varphi(t)\,\mathrm{d}h(t)\,.$$

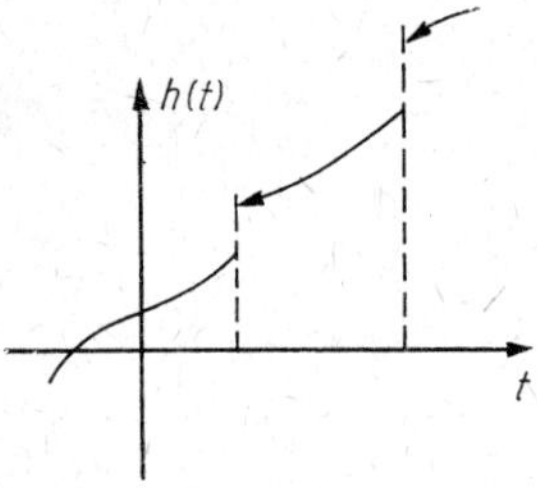

Fig. 26.3

But this special case of the general integration theory can also be treated in terms of the Riemann integral discussed in Sec. 3.2. Let $a = a_0 < a_1 < \ldots < a_n < a_{n+1} = b$ be the partition of (3.2.1/1), and let $d(Z) = \max_{k=0,\ldots,n} |a_{k+1} - a_k|$. If $t_k \in [a_k, a_{k+1}]$, then

$$\int_{[a,b)} \varphi(t)\,\mathrm{d}h(t) = \lim_{d(Z)\to 0} \sum_{k=0}^{n} \varphi(t_k)\,(h(a_{k+1}) - h(a_k))\,. \tag{1}$$

If $\{a, b\} = (a, b)$, then one has to replace $h(a_0)$ in (1) by $h(a_0+0)$. If $\{a, b\} = (a, b]$, then in (1) we have to replace $h(a_0)$ by $h(a_0+0)$ and $h(a_{n+1})$ by $h(a_{n+1}+0)$. If $\{a, b\} = [a, b]$, then in (1) we have to replace $h(a_{n+1})$ by $h(a_{n+1}+0)$. (1) and the modifications indicated are the analogue to Lemma 3.2.1/2. The formulation also makes it clear that (1) is independent of the particular choice of the t_k. If $\int_{R_1} \varphi(t)\,\mathrm{d}h(t)$ exists (in the sense of the integration theory described in Chap. 14) then it is clear that

$$\int_{R_1} \varphi(t)\,\mathrm{d}h(t) = \lim_{N\to\infty} \int_{[-N,N]} \varphi(t)\,\mathrm{d}h(t)\,. \tag{2}$$

Complex distribution functions: If $h(t) = h_1(t) - h_2(t) + \mathrm{i}h_3(t) - \mathrm{i}h_4(t)$, where the $h_k(t)$ are real-valued increasing left-continuous functions, then one sets

$$\int_{R_1} \varphi(t)\,\mathrm{d}h(t) = \int_{R_1} \varphi(t)\,\mathrm{d}h_1(t) - \int_{R_1} \varphi(t)\,\mathrm{d}h_2(t) + \mathrm{i}\int_{R_1} \varphi(t)\,\mathrm{d}h_3(t) - \mathrm{i}\int_{R_1} \varphi(t)\,\mathrm{d}h_4(t)\,. \tag{3}$$

Here $\varphi(t)$ is a complex-valued continuous function on R_1 for which all integrals on the right-hand side of (3) exist. It is clear that $\int_{R_1} \varphi(t)\,\mathrm{d}h(t)$ has the usual properties of integrals.

Examples. Let $\{E_t\}_{t\in R_1}$ be a spectral family. Then $h(t)=\|E_t x\|^2$, with $x\in H$, is a real-valued increasing left-continuous function. If $x\in H$ and $y\in H$, then $h(t)=(E_t x, y)$ is a complex distribution function in the above sense. This follows from

$$(E_t x, y)=\left\| E_t \frac{x+y}{2}\right\|^2-\left\| E_t \frac{x-y}{2}\right\|^2+\mathrm{i}\left\| E_t \frac{x+\mathrm{i}y}{2}\right\|^2-\mathrm{i}\left\| E_t \frac{x-\mathrm{i}y}{2}\right\|^2 .$$

Thus we can consider Riemann-Stieltjes integrals of the form

$$\int_{R_1} \varphi(t)\, \mathrm{d}\,\|E_t x\|^2 \quad \text{and} \quad \int_{R_1} \varphi(t)\, \mathrm{d}(E_t x, y) .$$

26.4.2. Riemann-Stieltjes Integrals for Spectral Families on Finite Intervals

Our aim is to replace the function $h(t)$ referred to in the preceding subsection by the function $E_t x$, whose range is a Hilbert space. Here $\{E_t\}_{t\in R_1}$ is a spectral family, and $x\in H$. The symbols to be used shall have the same meaning as in the preceding subsection, i.e. $-\infty<a\leqq b<\infty$, the partition $a=a_0<a_1<\ldots<a_{n+1}=b$, the points t_k, and $d(Z)$.

Theorem. *Let $\{E_t\}_{t\in R_1}$ be a spectral family in the Hilbert space H. If $\varphi(t)$ is a complex-valued continuous function on R_1, then, for every $x\in H$,*

$$\lim_{d(Z)\to 0} \sum_{k=0}^{n} \varphi(t_k)\,(E_{a_{k+1}}x-E_{a_k}x)=\int_{[a,b)} \varphi(t)\,\mathrm{d}E_t x \tag{1}$$

exists as a limit in H. The limit element belongs to the projection space of E_b-E_a, and

$$\Big\|\int_{[a,b)} \varphi(t)\,\mathrm{d}E_t x\Big\|^2=\int_{[a,b)} |\varphi(t)|^2\,\mathrm{d}\,\|E_t x\|^2 . \tag{2}$$

If $x_1\in H$ and $x_2\in H$, and if ϱ_1 and ϱ_2 are complex numbers, then

$$\int_{[a,b)} \varphi(t)\,\mathrm{d}E_t\,(\varrho_1 x_1+\varrho_2 x_2)=\varrho_1\int_{[a,b)} \varphi(t)\,\mathrm{d}E_t x_1+\varrho_2\int_{[a,b)} \varphi(t)\,\mathrm{d}E_t x_2 . \tag{3}$$

Remark 1. It is clear how one has to understand (1): the left-hand side of (1) exists as a limit in H, which is independent of the particular choice of the partitions and independent of the points t_k. The right-hand side of (1) is then the definition of a Riemann-Stieltjes integral whose values are elements of a Hilbert space, by analogy with (26.4.1/1). The fact that this integral belongs to the projection space of E_b-E_a is verified rather easily. The right-hand side of (2) is an ordinary Riemann-Stieltjes integral for functions in the sense of Subsec. 26.4.1.

Remark 2. By analogy with Subsec. 26.4.1. one may introduce corresponding integrals $\int_{\{a,b\}} \varphi(t)\,\mathrm{d}E_t x$ for the other interval types. If one replaces $E_{a_0}x$ in (1) by $E_{a_0+0}x$, then one obtains $\int_{(a,b)} \varphi(t)\,\mathrm{d}E_t x$. Further,

$$\int_{(a,b]} \varphi(t)\,\mathrm{d}E_t x=\int_{(a,b)} \varphi(t)\,\mathrm{d}E_t x+\varphi(b)\,(E_{b+0}-E_b)x ,$$

$$\int_{[a,b]} \varphi(t)\,\mathrm{d}E_t x=\int_{[a,b)} \varphi(t)\,\mathrm{d}E_t x+\varphi(b)\,(E_{b+0}-E_b)x .$$

(2) and (3) remain true if $[a, b)$ is replaced by $\{a, b\}$. Further the integrals belong to the corresponding projection spaces, e.g.,

$$\int_{(a,b)} \varphi(t)\,\mathrm{d}E_t x\in H^b\ominus H^{a+0} ,$$

$$\int_{\{a\}} \varphi(t)\,\mathrm{d}E_t x=\varphi(a)\,(E_{a+0}-E_a)x\in H^{a+0}\ominus H^a .$$

26.4.3. Riemann-Stieltjes Integrals for Spectral Families on R_1

By analogy with (26.4.1/2) let us extend the considerations of the preceding subsection to R_1. As usual, H is a Hilbert space.

Theorem. *Let $\{E_t\}_{t\in R_1}$ be a spectral family, let $\varphi(t)$ be a complex-valued continuous function on R_1, and let*

$$M=\{y \mid y\in H, \int_{R_1} |\varphi(t)|^2 \,\mathrm{d}\, \|E_t y\|^2 < \infty\} . \tag{1}$$

(a) *M is a linear set dense in H.*

(b) $$\lim_{N\to\infty} \int_{(-N,N)} \varphi(t)\,\mathrm{d}E_t x = \int_{R_1} \varphi(t)\,\mathrm{d}E_t x \tag{2}$$

exists if and only if $x\in M$.

(c) *For $x\in M$,*

$$\| \int_{R_1} \varphi(t)\,\mathrm{d}E_t x\|^2 = \int_{R_1} |\varphi(t)|^2 \,\mathrm{d}\, \|E_t x\|^2 . \tag{3}$$

Remark. In (2) it is possible to replace $\lim\limits_{N\to\infty} \int\limits_{(-N,N)}$ by $\lim\limits_{\substack{\alpha\to\infty\\ \beta\to-\infty}} \int\limits_{(\beta,\alpha)}$. It is clear that the right-hand side of (2) is the definition of the corresponding integral, provided that the left-hand side of (2) exists.

26.4.4. Spectral Operators

Definition. *If $\{E_t\}_{t\in R_1}$ is a spectral family, and if $\varphi(t)$ is a complex-valued continuous function on R_1, then A, given by*

$$Ax = \int_{R_1} \varphi(t)\,\mathrm{d}E_t x$$

with the domain of definition

$$D(A)=\{x \mid \int_{R_1} |\varphi(t)|^2 \,\mathrm{d}\, \|E_t x\|^2 < \infty\}$$

is a spectral operator.

Remark 1. From Theorem 26.4.3 and from (26.4.2/3) it is easily concluded that A is a linear operator whose domain of definition $D(A)$ is dense in H. This raises two questions: (1) What are the properties of spectral operators? (2) How general is this concept, which operators can be represented as spectral operators?

Theorem 1. *Let A be a spectral operator with respect to the spectral family $\{E_t\}_{t\in R_1}$ and the complex-valued continuous function $\varphi(t)$.*

(a) *A is a closed operator with a dense domain of definition $D(A)$.*

(b) *The adjoint operator A^* is also a spectral operator, and*

$$A^*x = \int_{R_1} \overline{\varphi(t)}\,\mathrm{d}E_t x \quad \text{and} \quad D(A^*)=D(A) . \tag{1}$$

(c) *For $x\in D(A)$ and $y\in H$,*

$$(Ax, y) = \int_{R_1} \varphi(t)\,\mathrm{d}(E_t x, y) . \tag{2}$$

(d) *A is self-adjoint if and only if $\varphi(t)$ is real-valued.*

Remark 2. Since $D(A)$ is dense in the Hilbert space H, by Subsec. 26.1.3. it is possible to form the adjoint operator A^*. Then (d) follows from (1). By Subsec. 26.4.1. formula (2) is meaningful, where the formulation in part (c) is meant to include the convergence of the integral in (2).

Theorem 2. *If $\{E_t\}_{t\in R_1}$ is a spectral family, and if $\varphi(t)$ and $\psi(t)$ are any two continuous functions on R_1, then let*

$$Ax = \int_{R_1} t \mathrm{d}E_t x \,, \qquad x \in D(A) \,, \tag{3}$$

$$\varphi(A)\, x = \int_{R_1} \varphi(t)\, \mathrm{d}E_t x, \qquad x \in D(\varphi(A)) \,, \tag{4}$$

with an analogous formula assumed for $\psi(A)$. Then

$$\overline{\varphi(A) \cdot \psi(A)} = \overline{\psi(A) \cdot \varphi(A)} = (\varphi\psi)\,(A) \,. \tag{5}$$

Further $\varphi(A) = A^n$ if $\varphi(t) = t^n$, $n = 1, 2, 3, \ldots$

Remark 3. By Theorem 1, A is self-adjoint. (4) then says that we can form functions of self-adjoint spectral operators. (5) shows that operator functions can be handled in calculations like ordinary functions. Here the domain of definition,

$$D(\varphi(A)\, \psi(A)) = \{x \mid x \in D(\psi(A)), \quad \psi(A)x \in D(\varphi(A))\}$$

is the set of all elements $x \in H$ on which $\varphi(A)$ $(\psi(A)\, x)$ can be carried out. Then $\varphi(A)\, \psi(A)$ is a closable operator, and $\overline{\varphi(A)\psi(A)}$ is the closure. The last assertion of the theorem shows that this construction is compatible with the formation of the product, $A^2 = AA$, $A^n = AA^{n-1}$.

26.4.5. The Fundamental Theorem of Spectral Theory

Theorem. *Every self-adjoint operator A in the Hilbert space H, with the domain of definition $D(A)$, has a unique spectral family $\{E_t\}_{t\in R_1}$ such that A and $D(A)$ can be represented as*

$$Ax = \int_{R_1} t\, \mathrm{d}E_t x, \quad D(A) = \{x \mid x \in H, \ \int_{R_1} t^2\, \mathrm{d}\|E_t x\|^2 < \infty\} \,.$$

Remark 1. This is one of the most fundamental and most beautiful theorems of operator theory. It shows that (26.4.4/3) does not describe a narrow subset of the self-adjoint operators (as might be guessed at a first glance), but rather represents the most general self-adjoint operator. Thus in particular we can form operator functions $\varphi(A)$ of arbitrary self-adjoint operators A. $\sin A$, $|A|$, or $\sqrt{|A|}$ are meaningful, because $\sin t$, $|t|$, and $\sqrt{|t|}$ are admissible functions.

Remark 2. The proofs for the theorems in Subsec. 26.4.4. are not trivial, but yet comparatively simple. On the other hand a complete proof of the above theorem is highly intricate.

26.4.6. The Spectrum of Self-Adjoint Operators

In Subsec. 26.2.2. we considered the spectra D_A, C_A and S_A of a self-adjoint operator A. Now we think of A as being represented in the form of Theorem 26.4.5. The spaces H^λ and $H^{\lambda+0}$ shall have the same meaning as in Subsec. 26.3.1. with respect to the spectral family $\{E_t\}_{t\in R_1}$, which, by Theorem 26.4.5, is uniquely determined by A.

Theorem. (a) *For every real number λ, $N(A-\lambda E)=H^{0+u}\ominus H^{\lambda}$.*

(b) *The real number λ belongs to C_A if and only if $H^{\lambda+\varepsilon}\ominus H^{\lambda-\varepsilon}$ is infinite-dimensional for every number $\varepsilon>0$.*

Remark. If λ is an eigenvalue, then $H^{\lambda+0}\ominus H^{\lambda}$ is the corresponding characteristic manifold, i.e., the set of all eigenelements for the eigenvalue λ and the zero element. $\lambda\in D_A$ is true if and only if $H^{\lambda+0}\ominus H^{\lambda}$ is finite-dimensional but consists not only of the zero element. The theorem thus allows a remarkable characterization of the spectra D_A and C_A by properties of the corresponding spectral family in the sense of Theorem 26.4.5. This shows the close relation between spectrum and spectral family.

26.4.7. Operators with a Pure Point Spectrum

The spectra D_A and C_A for a self-adjoint operator A shall have the same meaning as in Subsec. 26.2.2.

Definition. *A self-adjoint operator A is called operator with a pure point spectrum if $C_A=\emptyset$.*

Remark 1. Thus, by Theorem 26.2.2, the spectrum $S_A=D_A$ consists only of eigenvalues of finite multiplicities. Such operators play an important role in physical applications.

Theorem 1. *An operator with a pure point spectrum is not bounded.*

Remark 2. Hence a bounded self-adjoint operator has a nonvoid continuous spectrum C_A. As previously, here we have again tacitly assumed that H is infinite-dimensional.

Theorem 2. *Let A be an operator with a pure point spectrum. Then A has a countably infinite number of different eigenvalues of finite multiplicities, which cannot accumulate in finiteness. Further there exists a complete orthonormal system $\{x_k\}_{k=1}^{\infty}\subset\subset D(A)$, such that $Ax_k=\lambda_k x_k$,*

$$D(A)=\left\{x \mid x\in H,\ \sum_{k=1}^{\infty}\lambda_k^2\,|(x,x_k)|^2<\infty\right\}, \tag{1}$$

$$Ax=\sum_{k=1}^{\infty}\lambda_k(x,x_k)x_k \quad \textit{for} \quad x\in D(A). \tag{2}$$

Remark 3. This is the analogue to Theorem 26.2.3/2. As we did there, we have assumed that H is infinite-dimensional. If the eigenvalues are ordered, taking into account their multiplicities, then one obtains $0\leq|\lambda_1|\leq|\lambda_2|\leq\ldots$, with $|\lambda_k|\to\infty$ as $k\to\infty$. In every characteristic manifold one can freely choose an orthonormal system. The set of all these systems then forms the system $\{x_k\}$ referred to in the theorem. (By Theorem 26.2.2(b) the eigenelements corresponding to different eigenvalues are always orthogonal).

Remark 4. The representation (1), (2) is the concrete formulation of Theorem 26.4.5, representing the infinite-dimensional variant of the well-known transformation to principal axes of surfaces of second order in analytical geometry. Theorem 26.4.5 then is the continuous analogue.

27. Differential Operators and Orthogonal Functions

27.1. Classical Orthogonal Functions

27.1.1. Introductory Remark

The spectral theory presented in Chap. 26, and especially Theorem 26.4.7/2 and Remark 26.4.7/3, can be used effectively in studying ordinary and partial differential operators. The method can be outlined as follows. One considers a differential expression, e.g., $Ay=-y''$ in the interval (a,b), where $-\infty<a<b<\infty$. Now one looks for suitable domains of definition $D(A)$, such that A becomes essentially self-adjoint (cf. Def. 26.1.4/2) in the Hilbert space $L_2(a,b)$. Then the closure $\bar{A}$ is self-adjoint, and the theory of Chap. 26 is applicable to $\bar{A}$. What is aimed at is that $\bar{A}$ is an operator with a pure point spectrum in the sense of Def. 26.4.7. Then it is possible to make use of Theorem 26.4.7/2 and Remark 26.4.7/3. We attempt an explicit determination of the eigenvalues and eigenelements (eigenvectors) of $\bar{A}$. Then it follows from Subsec. 26.4.7. that the corresponding orthonormal system of eigenelements is complete in the Hilbert space concerned. Subsequently it is possible to expand arbitrary elements of the Hilbert space considered in terms of such systems. Thus one has a situation as described in Chap. 18 for n-dimensional trigonometric series and Legendre polynomials. The usefulness of such orthogonal expansions can be evaluated from, say, Sec. 19.5., where separation setups for partial differential equations have been treated on this basis. For us, however, the considerations to be presented in this chapter are also intended as a preparation for the following chapter on quantum mechanics.

27.1.2. Trigonometric Functions

In this subsection we take as a basis the (complex) Hilbert space $H=L_2(\Omega)$, with $\Omega=(a,b)$, $-\infty<a<b<\infty$, in the sense of Subsec. 17.3.2. Instead of $L_2((a,b))$ we use the simpler notation $L_2(a,b)$. By $C^\infty([a,b])$ we denote the class of all complex-valued functions differentiable up to arbitrary order on (a,b), all derivatives of which can be continued continuously to $[a,b]$.

Theorem. (a) *The operator* $(A_Df)(x)=-f''(x)$ *with the domain of definition*

$$D(A_D)=\{f \mid f\in C^\infty([a,b]),\ f(a)=f(b)=0\}$$

is essentially self-adjoint in $L_2(a,b)$. $\bar{A}_D$ *is an operator with a pure point spectrum. The eigenvalues are* $\lambda_k=\dfrac{\pi^2k^2}{(b-a)^2}$, *with* $k=1,2,\ldots$ *These eigenvalues are simple ones, and* $f_k(x)=\sqrt{\dfrac{2}{b-a}}\sin\dfrac{x-a}{b-a}\pi k$ *are the corresponding orthonormal eigenfunctions.*

(b) *The operator* $(A_Nf)(x)=-f''(x)$ *with the domain of definition*

$$D(A_N)=\{f \mid f\in C^\infty([a,b]),\ f'(a)=f'(b)=0\}$$

is essentially self-adjoint in $L_2(a, b)$. $\bar{A}_N$ *is an operator with a pure point spectrum. The eigenvalues are* $\lambda_k = \frac{\pi^2 k^2}{(b-a)^2}$, *with* $k = 0, 1, 2, \ldots$ *These eigenvalues are simple ones, and* $f_0(x) = \frac{1}{\sqrt{b-a}}$ *as well as* $f_l(x) = \sqrt{\frac{2}{b-a}} \cos \frac{x-a}{b-a} \pi l$, *with* $l = 1, 2, \ldots$, *are the corresponding orthonormal eigenfunctions.*

Remark. As was already mentioned in Subsec. 27.1.1., it follows from this theorem as well as from the considerations in Subsec. 26.4.7. that both

$$\left\{\sqrt{\frac{2}{b-a}} \sin \frac{x-a}{b-a} \pi k\right\}_{k=1}^{\infty} \quad \text{and} \quad \left\{\frac{1}{\sqrt{b-a}}, \sqrt{\frac{2}{b-a}} \cos \frac{x-a}{b-a} \pi k\right\}_{k=1}^{\infty}$$

are complete orthonormal systems in $L_2(a, b)$. On the basis of the spectral theory of self-adjoint operators, for the one-dimensional case we have thus obtained the same result as in Theorem 18.1.4 and in Remark 18.1.4/1. These considerations can also be extended to the n-dimensional case.

27.1.3. Hermite's Functions

Let $C_0^\infty(R_1) = D(R_1)$ be the class of all complex functions with bounded support that are differentiable up to arbitrary order on R_1; cf. Def. 14.6.4/2 and Subsec. 22.1.2.

Theorem. *The Hermitian differential operator* $(Af)(x) = -f''(x) + x^2 f(x)$, *with the domain of definition* $D(A) = C_0^\infty(R_1)$, *is essentially self-adjoint in* $L_2(R_1)$. $\bar{A}$ *is an operator with a pure point spectrum. The eigenvalues are* $\lambda_k = 2k+1$, *with* $k = 0, 1, 2, \ldots$ *These eigenvalues are simple ones, and with suitably chosen positive constants* c_k,

$$H_k(x) = c_k e^{\frac{x^2}{2}} \frac{d^k}{dx^k} (e^{-x^2})$$

are the corresponding orthonormal eigenfunctions.

Remark 1. The first part of the theorem can be essentially generalized: if $p(x)$ is a real-valued continuous function on R_1, such that $p(x) \to \infty$ as $|x| \to \infty$, then

$$(Af)(x) = -f''(x) + p(x)\, f(x), \quad D(A) = C_0^\infty(R_1)\,,$$

is essentially self-adjoint in $L_2(R_1)$. Further, $\bar{A}$ is an operator with a pure point spectrum in this case, too.

Remark 2. The considerations in Subsec. 26.4.7. show that $\{H_k(x)\}_{k=0}^{\infty}$ is a complete orthonormal system in $L_2(R_1)$. Thus it is possible to represent every function $f(x) \in L_2(R_1)$ as

$$f(x) = \sum_{k=0}^{\infty} (f, H_k)_{L_2(R_1)}\, H_k(x)\,.$$

If λ is not an eigenvalue, then the (generalized) Hermite differential equation $\bar{A}g - \lambda g = f$ has the unique solution

$$g = \sum_{k=0}^{\infty} \frac{(f, H_k)_{L_2(R_1)}}{2k+1-\lambda} H_k(x)\,.$$

Remark 3. It is immediately seen that

$$h_k(x) = \frac{1}{c_n} e^{\frac{x^2}{2}} H_k(x) = e^{x^2} \frac{d^k}{dx^k} (e^{-x^2}),$$

where $k = 0, 1, 2, \ldots$, is a polynomial of degree k. These are the well-known Hermite polynomials.

27.1.4. Legendre's Functions

In Subsec. 18.2.2. we already investigated Legendre polynomials. On the basis of the spectral theory of self-adjoint operators we shall now essentially extend the considerations described there.

Theorem 1. *The Legendre differential operator* $(Af)(x) = -((1-x^2) f'(x))'$, *with the domain of definition* $D(A) = C^\infty([-1, 1])$, *is essentially self-adjoint in* $L_2(-1, 1)$. $\bar{A}$ *is an operator with a pure point spectrum. The eigenvalues are* $\lambda_k = k(k+1)$, *with* $k = 0, 1, 2, \ldots$ *These eigenvalues are simple ones, and the Legendre polynomials*

$$L_k(x) = \sqrt{\frac{2k+1}{2}}\, 2^{-k} \frac{1}{k!} \frac{d^k}{dx^k} [(1-x^2)^k] \tag{1}$$

are the corresponding orthonormal eigenfunctions.

Remark 1. The notations $C^\infty([-1, 1])$ and $L_2(-1, 1)$ have the same meaning as in Subsec. 27.1.2. Apart from normalizing factors, (1) agrees with (18.2.2/1). Again it follows from Subsec. 26.4.7. that $\{L_k(x)\}_{k=0}^\infty$ is complete in $L_2(-1, 1)$. Thus an arbitrary function $f(x) \in L_2(-1, 1)$ can be expanded in terms of Legendre polynomials, as was stated in Remark 18.2.2/2.

Theorem 2. *For* $m = 0, 1, 2, \ldots$, *the differential operator*

$$(A_m f)(x) = -((1-x^2) f'(x))' + \frac{m^2}{1-x^2} f(x), \quad \textit{with}$$

$$D(A_m) = \{f \mid f(x) = (1-x^2)^{\frac{m}{2}} P(x),\ P(x)\ \textit{polynomial}\}$$

as a domain of definition, is essentially self-adjoint in $L_2(-1, 1)$. $\bar{A}_m$ *is an operator with a pure point spectrum. The eigenvalues are* $\lambda_k = k(k+1)$, *with* $k = m, m+1, m+2, \ldots$ *These eigenvalues are simple ones, and*

$$L_k^m(x) = \sqrt{\frac{(k-m)!}{(k+m)!}} (1-x^2)^{\frac{m}{2}} \frac{d^m}{dx^m} L_k(x)$$

are the corresponding orthonormal eigenfunctions.

Remark 2. For $m = 0$ one largely obtains Theorem 1. As usual we have set $0! = 1$ (in (1) too). $L_k^m(x)$ are called associated Legendre functions. Again the completeness of the system $\{L_k^m(x)\}_{k=m}^\infty$ in $L_2(-1, 1)$ is deduced from Subsec. 26.4.7.

27.1.5. Laguerre's Functions

By $C^\infty([0, \infty))$ we denote the class of all complex-valued functions $f(x)$ differentiable up to arbitrary order on $(0, \infty)$, which vanish for large values of x and all of whose derivatives can be continued continuously to the point 0.

Theorem. *For $\alpha > -1$, the Laguerre differential operator*

$$(A_\alpha f)(x) = -4(xf'(x))' + \left(x + \frac{\alpha^2}{x}\right) f(x), \quad \textit{with}$$

$$D(A_\alpha) = \{f \mid x^{-\frac{\alpha}{2}} f(x) \in C^\infty([0, \infty))\}$$

as a domain of definition, is essentially self-adjoint in the Hilbert space $L_2(0, \infty)$. $\bar{A}_\alpha$ is an operator with a pure point spectrum. The eigenvalues are $\lambda_k = 2(2k+1+\alpha)$, with $k = 0, 1, 2, \ldots$ These eigenvalues are simple ones, and

$$L_{\alpha,k}(x) = c_{k,\alpha} e^{\frac{x}{2}} x^{-\frac{\alpha}{2}} \frac{\mathrm{d}^k}{\mathrm{d}x^k} (e^{-x} x^{n+\alpha})$$

with a suitable choice of the positive constants $c_{k,\alpha}$, are the corresponding orthonormal eigenfunctions.

Remark 1. $L_{\alpha,k}(x)$ are the Laguerre functions. Again it follows from Subsec. 26.4.7. that $\{L_{\alpha,k}(x)\}_{k=0}^\infty$ is a complete orthonormal system in $L_2(0, \infty)$. In particular it is possible to expand every function $f(x) \in L_2(0, \infty)$ in terms of Laguerre functions, that means,

$$f(x) = \sum_{k=0}^\infty (f, L_{\alpha,k})_{L_2(0,\infty)} L_{\alpha,k}(x) .$$

Remark 2. It is immediately seen that

$$l_{\alpha,k}(x) = \frac{1}{c_{k,\alpha}} x^{-\frac{\alpha}{2}} e^{\frac{x}{2}} L_{\alpha,k}(x) = e^x x^{-\alpha} \frac{\mathrm{d}^k}{\mathrm{d}x^k} (e^{-x} x^{n+\alpha})$$

are polynomials, which are called Laguerre polynomials.

Remark 3. The above theorem was established by U. Grünewald (dissertation, Jena, 1975).

27.2. Surface Harmonics

27.2.1. Beltrami's Differential Operator

The Laplace operator $\Delta = \sum_{k=1}^n \frac{\partial^2}{\partial x_k^2}$ in R_n and in domains $\Omega \subset R_n$ was investigated in detail in Chaps. 19 and 23. Later on, in Chap. 33, we shall consider the Laplacian and the wave operator in general curved spaces. At present we are interested in the analogue to Δ on the unit sphere in R_n, i.e., on $\omega_n = \{x \mid |x| = 1\}$ (Fig. 27.1). Since we do not intend to use the (rather complicated) techniques to be introduced later on, we content ourselves with a specific construction. Let ϑ be the general point on ω_n, and let $C^\infty(\omega_n)$ be the class of all complex-valued func-

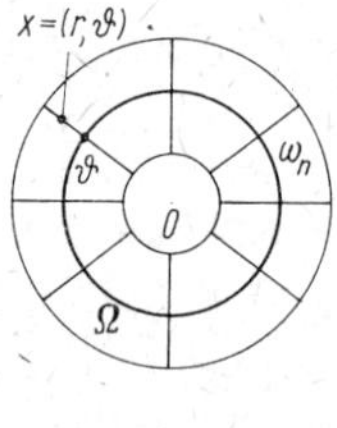

Fig. 27.1

tions that are differentiable on ω_n up to arbitrary order. If ds is the surface element on ω_n in the sense of Subsec. 9.2.5., then

$$(f, g)_{L_2(\omega_n)} = \int_{\omega_n} f(\vartheta)\, \overline{g(\vartheta)}\, \mathrm{d}s, \quad f \in C^\infty(\omega_n), \quad f \in C^\infty(\omega_n)\,, \tag{1}$$

is a scalar product. Let $L_2(\omega_n)$ be the Hilbert space obtained by completion of $C^\infty(\omega_n)$ with respect to the norm corresponding to (1). This space can also be described as follows. Let $r = |x|$ be the distance of $x \in R_n$ from the point 0. Then let $\Omega = \left\{ x \,\middle|\, \frac{1}{2} < x < 2 \right\}$. Now it is possible to identify $x \in \Omega$ uniquely by r and $\vartheta \in \omega_n$, $x = (r, \vartheta)$. We now continue $f \in C^\infty(\omega_n)$ continuously to Ω by setting $f_\Omega(x) = f(\vartheta)$ (this means that $f_\Omega(x)$ is constant along the radii from the origin). Now it is possible to regard $L_2(\omega_n)$ as a completion of $\{f_\Omega(x)$ with $f(\vartheta) \in C^\infty(\omega_n)\}$ in $L_2(\Omega)$, i.e., as a subspace of $L_2(\Omega)$. Such a point of view (though possibly not quite satisfactory from the logical aspect) will suffice for our purposes.

Definition. $(Bf)(\vartheta) = -r^2 \Delta f_\Omega(x)$, *with* $f \in C^\infty(\omega_n)$, *is called Beltrami's (second) differential operator*, $n \geqq 2$.

Remark 1. If $f \in C^\infty(\omega_n)$, then $f_\Omega(x)$ with $x \in \Omega$ has the above meaning. Then

$$-r^2 \Delta f_\Omega(x) = -(x_1^2 + \dots + x_n^2) \sum_{k=1}^{n} \frac{\partial^2 f_\Omega}{\partial x_k^2}(x)$$

is a function differentiable up to arbitrary order on Ω. If we again set $x = (r, \vartheta)$, then it turns out that this function does not depend on r, i.e., that it is constant along the radii from the origin. According to the above identification one then has $-r^2 \Delta f_\Omega(x) \in C^\infty(\omega_n)$. It is in this sense that the above definition has to be understood. One may ask for an intrinsic description of B, which does not make use of the continuation from ω_n to Ω. Systematic considerations of this kind are described in Chap. 33. On the other hand it is possible to describe B explicitly by means of n-dimensional polar co-ordinates; cf. [66], p. 418. In the following remark we confine ourselves to the cases $n = 2$ and $n = 3$.

Remark 2. In the case $n = 2$ we identify the points $\vartheta \in \omega_2$ in the usual way by the angle ϑ (Fig. 27.2). Here $0 \leqq \vartheta < 2\pi$. With respect to the polar co-ordinates (r, ϑ) one obtains

$$\Delta u(x) = \frac{1}{r} \frac{\partial}{\partial r}\left(r \frac{\partial u}{\partial r} \right) + \frac{1}{r^2} \frac{\partial^2 u}{\partial \vartheta^2}\,.$$

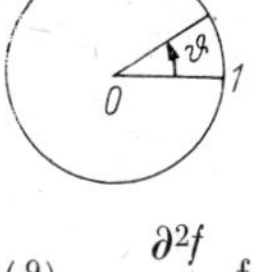

Fig. 27.2

Hence $(Bf)(\vartheta) = -\dfrac{\partial^2 f}{\partial \vartheta^2}$ for $n = 2$ and $f \in C^\infty(\omega_2)$. In the case $n = 3$ we use the polar co-ordinates defined in Subsec. 12.4.1., which gives

$$\Delta u(x) = \frac{1}{r^2} \frac{\partial}{\partial r}\left(r^2 \frac{\partial u}{\partial r} \right) + \frac{1}{r^2 \sin \vartheta} \frac{\partial}{\partial \vartheta}\left(\sin \vartheta \frac{\partial u}{\partial \vartheta} \right) + \frac{1}{r^2 \sin^2 \vartheta} \frac{\partial^2 u}{\partial \varphi^2}\,.$$

Hence

$$(Bf)(\vartheta, \varphi) = -\frac{1}{\sin \vartheta} \frac{\partial}{\partial \vartheta}\left(\sin \vartheta \frac{\partial f}{\partial \vartheta} \right) - \frac{1}{\sin^2 \vartheta} \frac{\partial^2 f}{\partial \varphi^2} \tag{2}$$

for $n = 3$ and $f \in C^\infty(\omega_3)$. The disadvantage of (2) is that the polar co-ordinate system (r, ϑ, φ) becomes singular for $\vartheta = 0$ and $\vartheta = \pi$.

27.2.2. Surface Harmonics as Eigenfunctions

In Remark 23.1.2/2 we described harmonic polynomials $P(x)$ in R_n. A polynomial is called homogeneous if $P(x) = \sum_{|\alpha|=m} b_\alpha x^\alpha$, where, as previously, $x^\alpha = x_1^{\alpha_1} \dots x_n^{\alpha_n}$. Such a homogeneous polynomial can be written as $P(x) = r^m S(\vartheta)$, where $S(\vartheta) \in C^\infty(\omega_n)$. We shall retain all the notations of Subsec. 27.2.1.

Definition. *If* $k = 0, 1, 2, \dots$, *then* $S(\vartheta) \in C^\infty(\omega_n)$ *is called surface harmonic of degree* k *if* $r^k S(\vartheta) = P(x)$ *is a homogeneous harmonic polynomial in* R_n.

Remark 1. $S(\vartheta) \equiv 1$ is a surface harmonic of degree zero, $S(\vartheta) = \frac{x_k}{r}$ is one of degree one, $S(\vartheta) = \frac{x_1^2 - x_2^2}{r^2}$ is one of degree two, etc.

Theorem. *Beltrami's differential operator* B *as per Def. 27.2.1 is essentially self-adjoint in the Hilbert space* $L_2(\omega_n)$. $\overline{B}$ *is an operator with a pure point spectrum. The different eigenvalues are* $\lambda_k = k(k+n-2)$, *with* $k = 0, 1, 2, \dots$ *The multiplicity of the eigenvalue* λ_k *is* $V_n(k) = \binom{k+n-1}{n-1} - \binom{k+n-3}{n-1}$. *The corresponding characteristic manifold consists of the surface harmonics of degree* k.

Remark 2. Here $\binom{m}{k} = 0$ if $m < k$. From the theorem it follows that there are $V_n(k)$ linearly independent homogeneous harmonic polynomials of degree k in R_n. Further, the characteristic manifold corresponding to the eigenvalue λ_k can be spanned by $V_n(k)$ orthonormal surface harmonics $S_r^{(k)}(\vartheta)$. Since eigenfunctions corresponding to different eigenvalues are always orthogonal, it then follows from Subsec. 26.4.7. that $\{S_r^{(k)}\}_{\substack{k=0,1,2,\dots\\ r=1,\dots,V_n(k)}}$ is a complete orthonormal system in the space $L_2(\omega_n)$. An arbitrary function $f(\vartheta) \in L_2(\omega_n)$ can thus be expanded in $L_2(\omega_n)$ in terms of surface harmonics,

$$f(\vartheta) = \sum_{k=0}^{\infty} \sum_{r=1}^{V_n(k)} (f, S_r^{(k)})_{L_2(\omega_n)} S_r^{(k)}(\vartheta).$$

Remark 3. Of special interest are the cases $n = 2$ and $n = 3$. One has $V_2(0) = 1$ and $V_2(k) = 2$ for $k = 1, 2, 3, \dots$ Further, $V_3(k) = 2k+1$. The case $n = 3$ will be dealt with in greater detail in the next subsection. For $n = 2$ it follows from Remark 27.2.1/2 that $(Bf)(\vartheta) = -f''(\vartheta)$. In this case the eigenvalues are $\lambda_k = k^2$, where $k = 0, 1, 2, \dots$ The corresponding orthononormal eigenfunctions are $\frac{1}{\sqrt{2\pi}} e^{\pm ik\vartheta}$ (for $k = 1, 2, 3, \dots$ one has two functions each, and one function for $k = 0$). Hence

$$\left\{ \frac{1}{\sqrt{2\pi}}, \frac{1}{\sqrt{2\pi}} e^{ik\vartheta}, \frac{1}{\sqrt{2\pi}} e^{-ik\vartheta} \right\}_{k=1}^{\infty}$$

is a complete orthonormal system in the Hilbert space $L_2(\omega_2) = L_2(0, 2\pi)$. This agrees with Remark 18.1.1/4 and Theorem 18.1.3. Thus, in the case $n = 2$, $r^k e^{\pm ik\vartheta} = (x_1 \pm ix_2)^k$ are harmonic polynomials, provided that the representation in polar co-ordinates, $x_1 = r \cos \vartheta$, $x_2 = r \sin \vartheta$, is used.

27.2.3. Three-Dimensional Surface Harmonics

We now discuss the case $n=3$. As in Remark 27.2.1/2, let (r, ϑ, φ) be the polar co-ordinates referred to in Subsec. 12.4.1., $0<r<\infty$, $0<\vartheta<\pi$, $0\leqq\varphi<2\pi$. The Beltrami operator B now has the same meaning as in (27.2.1/1). Finally, we recall the associated Legendre functions $L_k^m(x)$ referred to in Subsec. 27.1.4.

Theorem. $\{S_m^{(k)}(\vartheta, \varphi)\}_{\substack{k=0,1,2,\ldots\\ m=-k,-k+1,\ldots,0,\ldots k-1,k}}$, given by

$$S_m^{(k)}(\vartheta, \varphi)=\frac{1}{\sqrt{2\pi}}\, L_k^{|m|}(\cos\vartheta)\, e^{im\varphi}\,, \tag{1}$$

is a complete orthonormal system of three-dimensional surface harmonics.

Remark 1. Here $\{S_m^{(k)}(\vartheta, \varphi)\}_{m=-k,\ldots,0,\ldots,k}$ is an orthonormal system of $2k+1$ three-dimensional surface harmonics of degree k. Every other surface harmonic of degree k can be represented as a linear combination of these $2k+1$ surface harmonics of degree k. Then it follows from Theorem 27.2.2 that these surface harmonics are the eigenfunctions of the three-dimensional Beltrami operator in (27.2.1/1).

Remark 2. As was already pointed out repeatedly, every function $f(\vartheta, \varphi)\in L_2(\omega_3)$ can be expanded in terms of the system $\{S_m^{(k)}\}$. Further it can be observed from the above considerations that $r^k S_m^{(k)}(\vartheta, \varphi)$ are homogeneous harmonic polynomials of degree k in R_3.

... it is this as well as the numerous other applications from which mathematics derives whatever reputation it enjoys in the wider public.
(Hilbert, Naturerkennen und Logik, 1930)

28. Principles of Quantum Mechanics

28.1. Axiomatics of Quantum Mechanics

28.1.1. The Hilbert Space Model

In Sec. 12.1. we discussed mathematical models of physical theories. In the case of quantum mechanics, the schematic representation of Subsec. 12.1.2. can be reified as follows, where the numbers indicate the sections in which the subjects in question are dealt with (Fig. 28.1). In this section we shall consider the corresponding mathematical theory, referring to the spectral theory of self-adjoint operators of Chap. 26. As in this theory, we also here assume H to be always a (complex separable) infinite-dimensional Hilbert space.

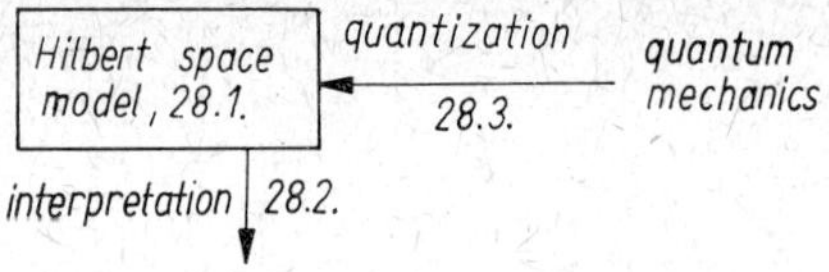

Fig. 28.1

Definition. *A quantum mechanical system $\mathfrak{S}=\{\mathcal{H}, \Psi\}$ consists of a self-adjoint operator $\mathcal{H}$ in H and a one-dimensional subspace Ψ in H. $\mathcal{H}$ is called the Hamiltonian operator of $\mathfrak{S}$, and Ψ the state of $\mathfrak{S}$.*

Remark 1. What we are considering now is the definition of mathematical concepts which, for the time being, have no physical content. The wording, however, already indicates later physical interpretations.

Remark 2. A one-dimensional subspace Ψ of H consists of all elements $\lambda\psi$, where $\psi \in H$, with $\|\psi\|=1$, is a fixed element and λ is an arbitrary complex number. Here ψ is unique except for a complex factor of absolute value 1. In this sense, from now on we shall also call $\mathfrak{S}=\{\mathcal{H}, \psi\}$, with $\|\psi\|=1$, a quantum mechanical system, and ψ the corresponding state. Thus ψ may also be replaced by $\psi e^{i\varrho}$, where ϱ is a real number. Consequently, all the concepts to be defined must be invariant under the substitution $\psi \to \psi e^{i\varrho}$.

Now I know for certain that the quantum of action has a much more fundamental significance than I originally suspected.
(M. Planck, at the end of his life)

28.1.2. The Dynamics of Quantum Mechanical Systems

We consider a quantum mechanical system that depends on time (more precisely, on a real parameter that we shall call time), $\mathfrak{S}(t)=\{\mathcal{H}, \psi(t)\}$. Here the Hamiltonian operator $\mathcal{H}(t)=\mathcal{H}$ is assumed to be independent of time, so that it is only the state $\psi(t)$ which is variable. In the sense of Remark 28.1.1/2 one has $\|\psi(t)\|=1$. We shall now establish by axiom what $\psi(t)$ will be for $t>0$ if $\psi(0)$ and $\mathcal{H}$ are known. Here $\hbar=\frac{h}{2\pi}$, h being Planck's quantum of action, $h=6.62 \cdot 10^{-27}$ erg · sec.[1])

Axiom. *The quantum mechanical system $\mathfrak{S}=\{\mathcal{H}, \psi\}$ at the time $t=0$ transforms into the quantum mechanical system $\mathfrak{S}(t)=\{\mathcal{H}, \psi(t)\}$, with $\psi(t)=e^{-i\frac{t}{\hbar}\mathcal{H}}\psi$ at the time $t \geqq 0$.*

Remark 1. Here $e^{-i\frac{t}{\hbar}\mathcal{H}}$ has the meaning indicated in Theorem 26.4.5 and Theorem 26.4.4/2, with $A=\mathcal{H}$ and $\varphi(t)=e^{-i\frac{t}{\hbar}}$. According to Subsec. 26.4.4. and Theorem 26.4.3, $e^{-i\frac{t}{\hbar}\mathcal{H}}$ is a unitary operator in the Hilbert space H. Hence one has $\|\psi(t)\|=\|\psi\|=1$. In view of Remark 28.1.1/2, the axiom is meaningful. It describes the dynamics of the system $\mathfrak{S}$: the variation of the state $\psi(t)$ when the Hamiltonian operator $\mathcal{H}$ is constant.

---

[1]) E. Segrè writes in [57, pp. 61, 87]: "It often seems as if physics follows a predestined line and that great scientists have simply accelerated progress. If one scientist had not been there, another would soon have taken his place and found the same thing. One important exception to this is the discovery of the quantum of action . . . The whole tone of Planck's first, and even later, work gives the impression that quantization was little more to him than a calculational device."

Remark 2 (Schrödinger's equation). By analogy with R_1 and C_1, it is possible to introduce a concept of derivative in H. $\{\psi(t)\}_{t\geq 0} \subset H$ is called differentiable if, for $t \geq 0$,

$$\lim_{\delta \to 0} \frac{\psi(t+\delta)-\psi(t)}{\delta} = \psi'(t) \quad \text{in} \quad H$$

exists. (For $t=0$, take the limit on the right, $\delta \downarrow 0$.) If $\psi(t)$ has the meaning set out in the axiom, and if $\psi \in D(\mathcal{H})$, then it can be shown that $\psi(t)$ is differentiable and solves the abstract Schrödinger equation

$$\psi'(t) = -\frac{\mathrm{i}}{\hbar} \mathcal{H}\psi(t), \quad \psi(0) = \psi \in D(\mathcal{H}). \tag{1}$$

In particular one always has $\psi(t) \in D(\mathcal{H})$. The following converse can also be shown to be true: if $\psi \in D(\mathcal{H})$, then the differential equation (1) has a unique solution $\psi(t)$, with $\psi(0)=\psi$. For $\psi \in D(\mathcal{H})$ the axiom and (1) are equivalent. The axiom is more general, because it applies to arbitrary $\psi \in H$, whereas (1) is more elegant. As the states of physical interest, which we shall consider later on, belong to $D(\mathcal{H})$, there is no substantial difference between the abstract Schrödinger equation (1) and the axiom.

28.1.3. Stationary States

Definition. *A quantum mechanical system* $\mathfrak{S} = \{\mathcal{H}, \psi\}$ *is called stationary (or stable) if* $\mathfrak{S}(t) = \mathfrak{S}$ *for* $t \geq 0$ *in the sense of Axiom* 28.1.2.

Remark 1. According to Axiom 28.1.2 and Remark 28.1.1/2, $\mathfrak{S}$ is stationary if there exists a real-valued function $\varrho(t)$ such that $\psi(t) = e^{\mathrm{i}\varrho(t)}\psi$. This means that the system always remains in the same state. Such stationary states are of special interest. They serve, for example, as a basis for the quantum mechanical explanation of the existence of atoms and molecules.

Theorem. *A state* ψ *of a quantum mechanical system* $\mathfrak{S} = \{\mathcal{H}, \psi\}$ *is stationary if and only if* ψ *is an eigenelement of the Hamiltonian operator* $\mathcal{H}$.

Remark 2 (energy level). Hence ψ is stationary if and only if there exists a real number E such that $\mathcal{H}\psi = E\psi$. E is the corresponding eigenvalue, called energy level.

Remark 3. The proof of the theorem is easy in one direction. From $\mathcal{H}\psi = E\psi$ it follows that $(E_{E+0} - E_E)\,\psi = \psi$, and hence

$$\psi(t) = e^{-\mathrm{i}\frac{t}{\hbar}\mathcal{H}}\psi = e^{-\mathrm{i}\frac{t}{\hbar}E}\psi$$

according to Theorem 26.4.4/2. Proving the converse is somewhat more difficult.

> ... the opposite of a shallow truth is untrue, but the opposite of a deep truth is also true ...
> (N. Bohr)
> ... what greater thing can Man give to Man than truth?
> (F. Schiller in his inaugural address as a professor of history, Jena, 1789)

28.2. Interpretations

28.2.1. Bohr's Postulate

The interpretations in the sense of the scheme of Subsec. 28.1.1. for the Hilbert space model can in effect be summarized into two rules of interpretation: Bohr's postulate and the statistical interpretation to be described in Subsec. 28.2.2. Let us now consider the following situation: given a quantum mechanical system with the Hamiltonian operator $\mathcal{H}$, which is in a stationary state ψ_1, that means $\mathcal{H}\psi_1 = E_1\psi_1$. The system can remain in this state for an arbitrary period of time. Later on we shall identify the hitherto considered (mathematical) quantum mechanical systems with actual (physical) ones. Such systems can jump spontaneously from one stationary state ψ_1 to another stationary state ψ_2, e.g., by external influences such as electromagnetic radiation, where the Hamiltonian operator $\mathcal{H}$ remains unchanged. Hence one has $\mathcal{H}\psi_2 = E_2\psi_2$. Physically, this is also interpreted by saying that the system characterized by the Hamiltonian operator $\mathcal{H}$ jumps from the stationary energy level E_1 to the stationary energy level E_2. When this happens, electromagnetic radiation is absorbed or emitted. Before and after this transition the system is in stationary states, i.e., in states that are invariable in time.

Bohr's postulate (1*st rule of interpretation*). *If $\mathcal{H}$ is the Hamiltonian operator of a quantum mechanical system that jumps spontaneously from the stationary state ψ_1, with $\mathcal{H}\psi_1 = E_1\psi_1$, into the stationary state ψ_2, with $\mathcal{H}\psi_2 = E_2\psi_2$, then the transition involves absorption or emission of electromagnetic radiation of the frequency* $\nu = \frac{1}{\mathrm{h}}|E_1 - E_2|$.

Remark 1. Here h is again Planck's quantum of action, as referred to in Subsec. 28.1.2. As the Hilbert space H is separable, $\mathcal{H}$ has an at most countably infinite number of eigenvalues. Hence there is an at most countably infinite number of sharp frequencies in the sense of the postulate. This is the basis for the explanation of the absorption and emission spectra of atoms and molecules.

Ground state. *A stationary quantum mechanical system $\{\mathcal{H}, \psi\}$ tends to assume a state of minimum energy.*

Remark 2. Hence, the quantities of interest are the lowest eigenvalue E of the Hamiltonian operator $\mathcal{H}$ and the corresponding eigenelements ψ. If the stationary system is undisturbed, then it is in such a state ψ. Electromagnetic radiation of sharp frequencies ν, in the sense of the postulate, excites the system so that it jumps into a state of higher energy, absorbing electromagnetic radiation of suitable frequencies. Later on the system spontaneously returns to lower energy levels, emitting electromagnetic radiation of corresponding frequencies.

Remark 3. The above considerations presuppose that $\mathcal{H}$ has eigenvalues and that there exists a lowest eigenvalue. As we shall see later on, many actual quantum mechanical operators have this property.

> God does not play at dice
> (Einstein, who disagreed with the statistical interpretation of quantum mechanics)
>
> God casts the die, not the dice
> (enhanced literary variant)

28.2.2. Statistical Interpretation of Quantum Mechanics

We shall now extend the set of problems to be considered by admitting "questions" to be put to a given quantum mechanical system $\mathfrak{S} = \{\mathcal{H}, \psi\}$. Such a question shall be connected with a measuring instruction that yields a real number as a measured value. For example, if the (actual physical) quantum mechanical system of the hydrogen atom is given in a certain state, one may ask for, say, the position or momentum of the electron or for the distance between the atomic nucleus and the electron (naturally such questions already involve some sort of a model conception of the hydrogen atom). The problem is to fit such questions into the above theory. To this end, every (meaningful) question is assigned a self-adjoint operator A, which is also called observable in this connection. The method of this assignment will be described later on. Let $\{E_\lambda\}_{\lambda \in R_1}$ be the spectral family of A in the sense of Theorem 26.4.5. By analogy with Subsec. 26.4.2. we set

$$\begin{aligned}
E_B &= E_b - E_a && \text{for } B=[a,b), && -\infty < a < b \leqq \infty,\\
E_B &= E_b - E_{a+0} && \text{for } B=(a,b), && -\infty \leqq a < b \leqq \infty,\\
E_B &= E_{b+0} - E_a && \text{for } B=[a,b], && -\infty < a \leqq b < \infty,\\
E_B &= E_{b+0} - E_{a+0} && \text{for } B=(a,b], && -\infty \leqq a < b < \infty.
\end{aligned}$$

where $E_{-\infty+0}=0$ and $E_\infty = E$.

Statistical interpretation (*2nd rule of interpretation*). *The probability that the measurement corresponding to the observable A yields a value that lies in the interval B is $\|E_B\psi\|^2 = w(\psi, A, B)$.*

Remark 1. If $B=(-\infty, \lambda)$, then $E_B = E_\lambda$. From Subsec. 26.4.1. it follows that $h(\lambda) = \|E_\lambda \psi\|^2$, with $\|\psi\|=1$, is a distribution function. One has $h(\infty)=1$ and

$$\|(E_b - E_a)\,\psi\|^2 = \|E_b\psi\|^2 - \|E_a\psi\|^2 = h(b) - h(a)$$

for $b > a$. Now it follows from Subsecs. 13.3.4. and 13.3.5. that $w(\psi, A, B)$ can be uniquely continued to become a probability measure on the Borel sets of R_1. This justifies the notation.

Remark 2. Thus the question of classical physics "What value will be obtained from the measurement?" must be replaced in quantum mechanics by the question "What is the probability for the value to fall into a certain set?". Einstein could never prevail on himself to accept this interpretation, even though he conceded that it is logically consistent and very successful. Rather he believed in deeper deterministic relations, the coarsening of which leads to statements of the above form.

Remark 3. If ψ is an eigenelement of A corresponding to the eigenvalue a, then it follows from Theorem 26.4.6. that $\psi = (E_{a+0} - E_a)\psi$. With $B=\{a\}$ one then has $w(\psi, A, \{a\}) = \|\psi\|^2 = 1$. Thus in the measurement one obtains the value a with probability 1.

28.2.3. Heisenberg's Uncertainty Principle

Definition. *Let $\mathfrak{S} = \{\mathcal{H}, \psi\}$ be a quantum mechanical system, and let A be an observable. If $\varphi \in D(A)$, then $(A\varphi, \varphi)$ is called the mean value, and $\delta(\varphi, A) = \|A\varphi - (A\varphi, \varphi)\varphi\|$ the variance, of the measurement corresponding to A.*

Remark 1. It follows from Subsec. 28.2.2. that $h(\lambda) = \|E_\lambda\varphi\|^2 = w(\varphi, A, (-\infty, \lambda))$ is the distribution function which generates the probability measure w on R_1. Here $\{E_\lambda\}_{\lambda\in R_1}$ is the spectral family that belongs to A in the sense of Theorem 26.4.5. On the basis of the considerations described in Subsec. 26.4.4. it is not very difficult to show that

$$(A\varphi, \varphi) = \int_{R_1} \lambda\, dh(\lambda), \quad \delta^2(\varphi, A) = \int_{R_1} |\lambda - (A\varphi, \varphi)|^2\, dh(\lambda)\,. \tag{1}$$

But these are the well-known definitions of probability theory. The magnitude of $\delta(\varphi, A)$ indicates the accuracy with which the mean value $(A\varphi, \varphi)$ is assumed.

Remark 2. Let A be an operator with a pure point spectrum as defined in Subsec. 26.4.7. Let $\{\lambda_k\}_{k=1}^{\infty}$ denote the eigenvalues. Then the total mass of the probability measure $w(\varphi, A, B)$ (where B now is a Borel set in R_1) concentrates at the points λ_k, i.e., $w(\varphi, A, \{\lambda_k\}_{k=1}^{\infty})=1$. Thus one of the points λ_k is assumed with probability 1 in a measurement. If $\varphi=\sum_{k=1}^{\infty} c_k\varphi_k$, where $\{\varphi_k\}_{k=1}^{\infty}$ is a complete orthonormal system of eigenelements of the operator A, i.e., $A\varphi_k=\lambda_k\varphi_k$, then (1) reduces to

$$(A\varphi, \varphi)=\sum_{k=1}^{\infty} \lambda_k w(\varphi, A, \{\lambda_k\})=\sum_{k=1}^{\infty} \lambda_k\, |c_k|^2\,,$$

$$\delta^2(\varphi, A)=\sum_{k=1}^{\infty} |\lambda_k-(A\varphi, \varphi)|^2\, w(\varphi, A, \{\lambda_k\})=\sum_{k=1}^{\infty} |\lambda_k-(A\varphi, \varphi)|^2\, |c_k|^2\,.$$

Here $\sum_{k=1}^{\infty} |c_k|^2=1$.

Remark 3. A measurement is called sharp if $\delta(\varphi, A)=0$. From the definition of the variance and from (1) it is easy to observe the following properties: 1. A measurement is sharp if and only if φ is an eigenelement of A. If μ is the corresponding eigenvalue, that means if $A\varphi=\mu\varphi$, then $(A\varphi, \varphi)=\mu$. 2. A measurement is sharp if and only if the mean value $(A\varphi, \varphi)$ is assumed with probability 1. If $A=\mathcal{H}$ is the Hamiltonian operator of the system, then it follows that the energy of the system can be measured sharply if and only if φ is a stationary state (cf. Subsec. 28.2.1.).

Theorem. *Let $\mathfrak{S}=\{\mathcal{H}, \varphi\}$ be a quantum mechanical system. Let A and B be observables. If $\varphi\in D(A)\cap D(B)$, $A\varphi\in D(B)$, and $B\varphi\in D(A)$, then*

$$\delta(\varphi, A)\cdot\delta(\varphi, B)\geqq\frac{1}{2}\,|(BA\varphi-AB\varphi, \varphi)|\,. \tag{2}$$

Remark 4. Heisenberg's uncertainty principle is deduced from this as a special case: if $(BA-AB)\,\varphi=\frac{\hbar}{i}\,\varphi$, then

$$\delta(\varphi, A)\;\delta(\varphi, B)\geqq\frac{\hbar}{2}\,. \tag{3}$$

If one substitutes ϱ for $\frac{\hbar}{i}$, then (3) holds with $\frac{\varrho}{2}$ instead of $\frac{\hbar}{2}$. Formula (3) means that a relatively sharp measurement with respect to A (i.e., a small value of $\delta(\varphi, A)$) involves a relatively unsharp measurement with respect to B (i.e., a high value of $\delta(\varphi, B)$). Moreover, Remark 3 shows that there are no eigenelements of A or B with the additional properties mentioned in the above theorem.

Conceptions without experience are void;
experience without conceptions is blind
(A. Einstein)

28.3. Quantization

28.3.1. The Quantization Rule

To complete the scheme shown in Subsec. 28.1.1., we still have to answer the question of how an actual physical problem of quantum mechanics is translated into our calculus. For that purpose it is decisive to obtain the corresponding

Hamiltonian operator $\mathcal{H}$. Its spectral properties then determine the stationary states of associated quantum mechanical systems as well as the absorption and emission spectra of electromagnetic radiation.

Quantization rule. *Consider an n-particle system of mechanics and electrodynamics with the classical Hamiltonian function* $f(x_1, \ldots, x_{3n}, p_1, \ldots, p_{3n})$, *where* $x_1, \ldots, x_{3n}$ *are Cartesian space co-ordinates and* $p_1, \ldots, p_{3n}$ *are momentum co-ordinates. The Hamiltonian operator* $\mathcal{H}$ *of the corresponding quantum mechanical system is then obtained from the differential expression* $\mathcal{H}\varphi = f\left(x_k, \frac{\hbar}{\mathrm{i}} \frac{\partial}{\partial x_k}\right) \varphi$.

Remark 1. The procedure goes as follows. At first the classical Hamiltonian function $f(x_k, p_k)$ (with $k=1, \ldots, 3n$) is formed according to the rules of point mechanics and electrodynamics. Here $(x_{3k-2}, x_{3k-1}, x_{3k})$, $k=1, \ldots, n$, is the position of the k-th particle, and $(p_{3k-2}, p_{3k-1}, p_{3k})$ is its momentum. Then x_k is replaced by the multiplication operator x_k, and p_k by the differential operator $\frac{\hbar}{\mathrm{i}} \frac{\partial}{\partial x_k}$, where $\hbar$ has the meaning indicated in Subsec. 28.1.2. This gives the differential expression $\mathcal{H}\varphi = f\left(x_k, \frac{\hbar}{\mathrm{i}} \frac{\partial}{\partial x_k}\right) \varphi$. What remains is to choose a suitable Hilbert space H, which will in many cases be $H = L_2(R_{3n})$, and to fix a domain of definition $D(\mathcal{H})$. To get connection with the theory, $\mathcal{H}$ must be self-adjoint. But it would even suffice for $\mathcal{H}$ to be essentially self-adjoint. In this case we could choose $\overline{\mathcal{H}}$ as the Hamiltonian operator.

Remark 2. The quantization rule is not free of any doubt. On the one hand it necessitates several complementary additions, such as the choice of the Hilbert space H and of the domain of definition $D(\mathcal{H})$. On the other hand, however, even the formation of the differential expression $f\left(x_k, \frac{\hbar}{\mathrm{i}} \frac{\partial}{\partial x_k}\right)$ is not in all cases unambiguous. If one considers, for example, the term $x_1 p_1$, then of course $x_1 p_1 = p_1 x_1$. The quantization of these two expressions yields $x_1 \frac{\hbar}{\mathrm{i}} \frac{\partial}{\partial x_1} \varphi$ and

$$\frac{\hbar}{\mathrm{i}} \frac{\partial}{\partial x_1} (x_1 \varphi) = \frac{\hbar}{\mathrm{i}} x_1 \frac{\partial \varphi}{\partial x_1} + \frac{\hbar}{\mathrm{i}} \varphi ,$$

which terms are not equal. This is because the translation is done from the commutative number domain of real numbers to the non-commutative domain of operators. The physicists behave very wise in such situations, applying the quantization to $\frac{1}{2}(x_1 p_1 + p_1 x_1)$.

Remark 3. We have formulated the quantization rule for particles in R_3. For one- or two-dimensional problems, of course, a corresponding rule holds. What is decisive is always the substitution $x_k \to x_k$ and $p_k \to \frac{\hbar}{\mathrm{i}} \frac{\partial}{\partial x_k}$.

28.3.2. Examples of Quantization

One-dimensional Motion of a Free Particle: Consider a particle of mass m that is free to move in R_1 without external forces acting upon it. The classical Hamiltonian function is $f(x, p) = \frac{p^2}{2m}$. Quantization gives

$$\mathcal{H}\varphi = f\left(x, \frac{\hbar}{\mathrm{i}} \frac{\mathrm{d}}{\mathrm{d}x}\right) \varphi = \frac{1}{2m} \left(\frac{\hbar}{\mathrm{i}} \frac{\mathrm{d}}{\mathrm{d}x}\right)^2 \varphi = -\frac{\hbar^2}{2m} \frac{\mathrm{d}^2 \varphi}{\mathrm{d}x^2}. \tag{1}$$

It suggests itself to choose $H = L_2(R_1)$ as the Hilbert space and $D(\mathcal{H}) = C_0^\infty(R_1)$ as the domain of definition. Here, as previously, $C_0^\infty(R_1) = D(R_1)$ is the class of all complex-valued functions with bounded support which are differentiable on R_1 up to arbitrary order. Thus one observes that in (1) the derivation is handled quite formally.

Harmonic Oscillator: The classical harmonic oscillator has been dealt with in Subsec. 12.3.4. The Hamiltonian function is $f(x, p) = \frac{p^2}{2m} + \frac{k}{2} x^2$. Quantization gives

$$\mathcal{H}\varphi = -\frac{\hbar^2}{2m} \frac{d^2\varphi}{dx^2} + \frac{k}{2} x^2 \varphi . \tag{2}$$

Again we choose $H = L_2(R_1)$ to be the Hilbert space, and $D(\mathcal{H}) = C_0^\infty(R_1)$ to be the domain of definition.

Atoms: Consider an atom whose nucleus is fixed at the origin of the co-ordinate system, and let it have n electrons. The position and the momentum co-ordinates of the k-th electron are $(x_{3k-2}, x_{3k-1}, x_{3k})$ and $(p_{3k-2}, p_{3k-1}, p_{3k})$, respectively. The Hamiltonian function is

$$f(x_1, \ldots, p_{3n}) = \frac{1}{2m} \sum_{k=1}^{3n} p_k^2 - Z\varepsilon^2 \sum_{k=1}^{n} \frac{1}{r_k} + \varepsilon^2 \sum_{l>k} \frac{1}{r_{l,k}} . \tag{3}$$

Here m and ε denote, respectively, the mass and the charge of the electron, and Z is the atomic charge. $Z = n$ means neutral atoms, whereas $Z \neq n$ means charged ions. Quantization gives

$$\mathcal{H}_{\text{atom}}\, \varphi = -\frac{\hbar^2}{2m} \Delta\varphi + \left(-Z\varepsilon^2 \sum_{k=1}^{n} \frac{1}{r_k} + \varepsilon^2 \sum_{l>k} \frac{1}{r_{l,k}} \right) \varphi . \tag{4}$$

Here $\Delta\varphi = \sum_{r=1}^{3n} \frac{\partial^2 \varphi}{\partial x_r^2}$. We choose $H = L_2(R_{3n})$ as the Hilbert space and $D(\mathcal{H}_{\text{atom}}) = C_0^\infty(R_{3n})$ as the domain of definition. We recall that $C_0^\infty(R_{3n}) = D(R_{3n})$ is the class of all complex-valued functions with bounded support which are differentiable on R_{3n} up to arbitrary order. Modification of (4) in an obvious way yields a corresponding differential expression for molecules.

Hydrogen Atom in an Electromagnetic Field: Consider the neutral hydrogen atom in an external electromagnetic field that is constant in time, having the electric field intensity $\mathfrak{E}(x)$ and the magnetic field intensity $\mathfrak{B}(x)$. Here $x = (x_1, x_2, x_3) \in R_3$. According to (24.3.3/3), let $\mathfrak{B}(x) = \operatorname{rot} \mathfrak{A}(x)$ and $\mathfrak{E}(x) = \operatorname{grad} \Phi(x)$, where $\mathfrak{A}(x) = (A_1(x), A_2(x), A_3(x))$ is the vector potential and $\Phi(x)$ is the scalar potential. Without loss of generality one can assume that $\operatorname{div} \mathfrak{A}(x) \equiv 0$. Now one has to set $Z = n = 1$ in (1) and, moreover, to take into account the electromagnetic field. According to the rules of electrodynamics one obtains

$$f(x_1, x_2, x_3, p_1, p_2, p_3) = \frac{1}{2m} \sum_{k=1}^{3} \left(p_k - \frac{\varepsilon}{c} A_k(x) \right)^2 - \frac{\varepsilon^2}{r} - \varepsilon\Phi(x) \tag{5}$$

as the classical Hamiltonian function. Here it has again been assumed that the atomic nucleus is fixed at the origin. (x_1, x_2, x_3) is the position, and (p_1, p_2, p_3) is the momentum of the electron, $r = \sqrt{x_1^2 + x_2^2 + x_3^2}$. Further, m is the mass, ε is the charge of the electron, and c is the velocity of light. By applying the quantization rule, taking into account that div $\mathfrak{A}(x) \equiv 0$, one obtains

$$\mathcal{H}_{\mathfrak{A},V}\varphi = -\frac{\hbar^2}{2m}\Delta\varphi - \frac{\hbar\varepsilon}{mci}\sum_{k=1}^{3} A_k(x)\frac{\partial\varphi}{\partial x_k} + \left(\frac{\varepsilon^2}{2mc^2}|\mathfrak{A}(x)|^2 - \frac{\varepsilon^2}{r} - \varepsilon\Phi(x)\right)\varphi\,, \tag{6}$$

where $|\mathfrak{A}(x)|^2 = A_1^2(x) + A_2^2(x) + A_3^2(x)$, and the delta operator is defined as $\Delta = \frac{\partial^2}{\partial x_1^2} + \frac{\partial^2}{\partial x_2^2} + \frac{\partial^2}{\partial x_3^2}$. We choose $H = L_2(R_3)$ as the Hilbert space and $D(\mathcal{H}_{\mathfrak{A},V}) = C_0^\infty(R_3)$ as the domain of definition.

Hydrogen Atom in a Weak, Constant Magnetic Field: The following case is of special interest:

$$\Phi(x) \equiv 0, \quad A_1(x) = -\frac{A}{2}x_2, \quad A_2(x) = \frac{A}{2}x_1, \quad A_3(x) \equiv 0\,,$$

where A is a constant. One obtains $\mathfrak{E}(x) \equiv 0$, div $\mathfrak{A}(x) \equiv 0$, and $\mathfrak{B}(x) = (0, 0, A)$. Consequently, the field in question is a magnetic field that is constant in space and time, its direction being that of the x_3-axis. If $|A|$ is small, one has a weak magnetic field. In this case it is deemed justified to neglect the annoying term $\frac{\varepsilon^2}{2mc^2}|\mathfrak{A}(x)|^2$. This leads to the following set-up for the hydrogen atom in a constant, weak magnetic field $\mathfrak{B}(x) = (0, 0, A)$:

$$\mathcal{H}_{\text{Zee}}\varphi = -\frac{\hbar^2}{2m}\Delta\varphi - \frac{\hbar\varepsilon}{2mci}A\left(x_1\frac{\partial\varphi}{\partial x_2} - x_2\frac{\partial\varphi}{\partial x_1}\right) - \frac{\varepsilon^2}{r}\varphi, \tag{7}$$ [1]

where $H = L_2(R_3)$ is again the associated Hilbert space, and $D(\mathcal{H}_{\text{Zee}}) = C_0^\infty(R_3)$ is the domain of definition.

Remark. For a number of important physical examples we have described differential expressions $\mathcal{H}\varphi$ obtained by the quantization rule of Subsec. 28.3.1. In all cases the indicated domains of definition $D(\mathcal{H})$ and Hilbert spaces H guarantee that $\mathcal{H}$ is essentially self-adjoint (where in (5) it is necessary to impose physically reasonable additional conditions upon $A_k(x)$ and $\Phi(x)$). Hence $\overline{\mathcal{H}}$ can be used as the Hamiltonian operator of the respective problems. Moreover, the considerations show that, although the quantization rule provides decisive information as to the structure of the differential operators to be used, suitable modifications are applied on a case by case basis. An example of this is $\mathcal{H}_{\text{Zee}}$. Later on we shall construct new Hamiltonian operators, where we shall depart even farther from the quantization rule.

28.4. Single-Particle Problems

28.4.1. One-Dimensional Motion of a Free Particle

Theorem 1. *The operator* $\mathcal{H}\varphi = -\frac{\hbar^2}{2m}\frac{d^2\varphi}{dx^2}$, *with* $D(\mathcal{H}) = C_0^\infty(R_1)$, *is essentially self-adjoint in the Hilbert space* $H = L_2(R_1)$. *One has* $C_{\overline{\mathcal{H}}} = [0, \infty)$ *and* $\tilde{D}_{\overline{\mathcal{H}}} = \emptyset$.

Definition. $\overline{\mathcal{H}}$ *of Theorem 1 is the Hamiltonian operator of the quantum mechanical system of a free particle of mass* m *that moves in* R_1.

---

[1]) This operator is used to explain the so-called Zeeman effect (cf. Subsec. 28.5.2.).

Remark 1. The definition is suggested by Theorem 1 and (28.3.2/1). $C_{\overline{\mathcal{H}}}$ and $\tilde{D}_{\overline{\mathcal{H}}}$ have the meaning indicated in Subsecs. 26.2.1. and 26.2.2. Theorem 28.1.3 then shows that the free particle moving in one dimension has no stationary state, i.e., it is not stable. According to Remark 28.2.3/3 this implies that there is no state of this system in which its energy can be measured sharply.

Remark 2. Now let $\mathfrak{S}=\{\overline{\mathcal{H}}, \psi\}$, be the quantum mechanical system of the free particle moving in one dimension, which is in the state $\psi \in L_2(R_1)$. In the sense of Subsec. 28.2.2. we ask: "What is the position of the particle?" We have to assign an observable A to this question. It suggests itself to resort again to the quantization rule of Subsec. 28.3.1., so that the question "Which point $x \in R_1$ is the position of the particle?" is assigned the observable A,

$$A\varphi = x\varphi(x), \quad D(A)=\{\varphi \mid \varphi(x)\in L_2(R_1),\, x\varphi(x)\in L_2(R_1)\}\,. \tag{1}$$

Theorem 2. *Let* $-\infty \leqq a < b \leqq \infty$, *and let* ψ *be the state of the free particle moving in one dimension. Then* $\int\limits_a^b |\psi(x)|^2 \,\mathrm{d}x$ *is the probability that the position of the particle is in the interval* (a, b).

Remark 3. The theorem is a consequence of the statistical interpretation referred to in Subsec. 28.2.2. It is relatively easy to show that A of (1) is a self-adjoint operator in $L_2(R_1)$. The spectral family $\{E_\lambda\}_{\lambda \in R_1}$ corresponding to A in the sense of Subsec. 26.4.5. is

$$(E_\lambda\varphi)(x) = \chi(-\infty, \lambda)(x)\, \varphi(x) \quad \text{for} \quad \varphi \in L_2(R_1)\,,$$

where $\chi(-\infty, \lambda)(x)$ is the characteristic function of $(-\infty, \lambda)$, i.e. $\chi(-\infty, \lambda)(x)=1$ for $x<\lambda$ and $\chi(-\infty, \lambda)(x)=0$ for $x \geqq \lambda$. Hence the proposition of Theorem 2 is easily deduced.

28.4.2. The Harmonic Oscillator

Theorem. *The operator* $\mathcal{H}\varphi = -\dfrac{\hbar^2}{2m}\dfrac{d^2\varphi}{dx^2} + \dfrac{k}{2}x^2\varphi$, *with* $D(\mathcal{H}) = C_0^\infty(R_1)$, *is essentially self-adjoint in the Hilbert space* $L_2(R_1)$. $\overline{\mathcal{H}}$ *is an operator with a pure point spectrum. The eigenvalues are* $E_l = \hbar\sqrt{\dfrac{k}{m}}\left(l+\dfrac{1}{2}\right)$, $l=0, 1, 2, \ldots$ *These eigenvalues are simple, and with a suitable choice of the positive constants* c_l *in Theorem* 27.1.3. *one obtains* $H_l((km)^{\frac{1}{4}}\hbar^{-\frac{1}{2}}x)$ *as the corresponding orthonormal eigenfunctions.*

Remark 1. Apart from the physical constants this is a repetition of Theorem 27.1.3. In particular, $H_l(x)$ are the Hermite functions of Subsec. 27.1.3.

Definition. $\overline{\mathcal{H}}$ *of the above theorem is the Hamiltonian operator of the quantum mechanical system of the harmonic oscillator.*

Remark 2. The definition is suggested by (28.3.2/2) and the above theorem. For the classical harmonic oscillator, m and k have the meaning indicated in Subsec. 12.3.4. Hence the quantum mechanical harmonic oscillator has a countably infinite number of stationary states $H_l((km)^{\frac{1}{4}}\hbar^{-\frac{1}{2}}x)$. The ground state is $H_0((km)^{\frac{1}{4}}\hbar^{-\frac{1}{2}}x) = c_0 e^{-\frac{1}{2}(km)^{\frac{1}{2}}\hbar^{-1}x^2}$. The corresponding energy levels are E_l[1]).

1) A detailed description of mathematical aspects of single-particle problems in quantum mechanics is given in [54].

28.4.3. The Relativistic Free Particle in $\boldsymbol{R}_3$

By analogy with Subsec. 28.4.1., it is possible to deal with a free particle in R_3. Then $\mathcal{H}$ has the form $\mathcal{H}\varphi = -\frac{\hbar^2}{2m}\Delta\varphi$, where Δ is the three-dimensional Laplacian operator (delta operator). The dynamics of the corresponding quantum mechanical system is then described by the Schrödinger equation $\frac{\partial\varphi}{\partial t} = \frac{\mathrm{i}\hbar}{2m}\Delta\varphi$ of (28.1.2/1), with $\varphi = \varphi(x, t)$, $x = (x_1, x_2, x_3)$. This equation clearly shows that our approach chosen so far has been non-relativistic (the Hamiltonian function, too, in the quantization rule of Subsec. 28.3.1. is based on non-relativistic mechanics and electrodynamics). The Lorentz transformations of the special theory of relativity described in Chap. 25 mix space and time. Thus it is clear that a Schrödinger equation devised to describe a relativistic quantum mechanical particle must contain derivatives of equal order with respect to space and time. If one keeps the approved form (28.1.2/1) of the abstract Schrödinger equation, then one has to find for $\mathcal{H}$ a differential expression of first order with respect to the space variables. On the other hand, $-\Delta$ (multiplied by suitable constants) has proven successful as the Hamiltonian operator for the free non-relativistic particle. Combination of both ideas suggests to use operators of the form $\sqrt{-\Delta}$. Within the spectral theory of Chap. 26, $\sqrt{-\Delta}$ can be given a precise meaning. Unfortunately the result is not a differential operator of first order. Dirac has shown the way out of this dilemma by the following construction. Instead of using $L_2(R_3)$ as a base space, consider the Hilbert space

$$L_2^2(R_3) = \{f \mid f = \{f_1, f_2, f_3, f_4\},\ f_k \in L_2(R_3)\} \tag{1}$$

with the scalar product $(f, g)_{L_2^2(R_3)} = \sum_{k=1}^{4} (f_k, g_k)_{L_2(R_3)}$, $g = \{g_1, g_2, g_3, g_4\}$. (The superscript 2 in $L_2^2(R_3)$ is the exponent of $2^2 = 4$; later on we shall investigate spaces of type $L_2^n(R_k)$ with 2^n components.) What is desired is a differential expression $\mathcal{H}$ of first order in the space variables, the square of which, $\mathcal{H}^2$, coincides with $(-\Delta, -\Delta, -\Delta, -\Delta)$, except for constant factors. Such a result would fully take account of the above heuristic remarks. The following considerations show that these wishes are realizable.

Matrices: *If* $\sigma = (\sigma_{k,l})_{k,l=1}^N$ is an N-rowed square matrix with complex elements, then $\sigma^* = (\sigma_{k,l}^*)_{k,l=1}^N$, with $\sigma_{k,l}^* = \bar{\sigma}_{l,k}$, is called the adjoint matrix. σ is called self-adjoint if $\sigma = \sigma^*$. If one regards matrices as mappings in the complex space C_N, then these definitions coincide with the customary concepts of the theory of operators. Square N-rowed matrices $\sigma_1, \ldots, \sigma_M$ form an anticommutative normalized system if $\sigma_k\sigma_l = -\sigma_l\sigma_k$ for $l \neq k$ and if σ_k^2 is the identity matrix, where $l = 1, \ldots, M$ and $k = 1, \ldots, M$. Here we consider two-rowed and four-rowed matrices. The four-rowed matrices will also be represented as $\begin{pmatrix}\alpha_{11} & \alpha_{12}\\ \alpha_{21} & \alpha_{22}\end{pmatrix}$, where the α_{kl}'s are two-rowed matrices.

Lemma. (a) *The self-adjoint two-rowed square matrices*

$$\sigma_1 = \begin{pmatrix}0 & 1\\ 1 & 0\end{pmatrix},\quad \sigma_2 = \begin{pmatrix}0 & -\mathrm{i}\\ \mathrm{i} & 0\end{pmatrix},\quad \sigma_3 = \begin{pmatrix}1 & 0\\ 0 & -1\end{pmatrix} \tag{2}$$

form an anticommutative normalized system.

(b) *The self-adjoint four-rowed square matrices*

$$\alpha_k = \begin{pmatrix} \bar{0} & -\mathrm{i}\sigma_k \\ \mathrm{i}\sigma_k & \bar{0} \end{pmatrix}, \quad k=1, 2, 3, \quad \alpha_4 = \begin{pmatrix} \bar{1} & \bar{0} \\ \bar{0} & -\bar{1} \end{pmatrix}, \quad \alpha_5 = \begin{pmatrix} \bar{0} & \bar{1} \\ \bar{1} & \bar{0} \end{pmatrix} \tag{3}$$

form an anticommutative normalized system. Here $\sigma_1, \sigma_2, \sigma_3$ *are the matrices of* (2), *and* $\bar{1} = \begin{pmatrix} 1 & 0 \\ 0 & 1 \end{pmatrix}$ *and* $\bar{0} = \begin{pmatrix} 0 & 0 \\ 0 & 0 \end{pmatrix}$.

Remark 1. σ_1, σ_2, and σ_3 are the Pauli spin matrices, $\alpha_1, \ldots, \alpha_5$ are the Dirac spin matrices.

The Operator $\mathcal{H}^{\text{Dirac}}$: With α_k being the Dirac spin matrices, we set

$$\mathcal{H}^{\text{Dirac}}\varphi = \frac{c\hbar}{\mathrm{i}} \sum_{k=1}^{3} \alpha_k \frac{\partial\varphi}{\partial x_k} + mc^2\alpha_4\varphi\,. \tag{4}$$

Here $\hbar$ has the meaning indicated in Subsec. 28.1.2, c is the velocity of light, and m is the mass of the particle considered. Further, $\varphi = \{\varphi_1(x), \varphi_2(x), \varphi_3(x), \varphi_4(x)\}$, with $\varphi_k(x) \in C_0^\infty(R_3)$ and $x \in R_3$. Finally, $\alpha_4\varphi$ and $\alpha_k \frac{\partial\varphi}{\partial x_k}$, with $\frac{\partial\varphi}{\partial x_k} = \left\{\frac{\partial\varphi_1}{\partial x_k}, \frac{\partial\varphi_2}{\partial x_k}, \frac{\partial\varphi_3}{\partial x_k}, \frac{\partial\varphi_4}{\partial x_k}\right\}$, have the usual meaning of the multiplication of a matrix by a vector, where φ and $\frac{\partial\varphi}{\partial x_k}$ are regarded as column vectors (column matrices). As desired, $\mathcal{H}^{\text{Dirac}}$ is a differential expression of first order. Using the above lemma it can be shown that

$$(\mathcal{H}^{\text{Dirac}})^2\,\varphi = \{B\varphi_1, B\varphi_2, B\varphi_3, B\varphi_4\}$$

where $B\psi = m^2c^4\psi - c^2\hbar^2\Delta\psi$. This yields us the connection to the above heuristic considerations, so that we can attempt to use $\mathcal{H}^{\text{Dirac}}$ for the quantum mechanical description of the relativistic free particle in R_3.

Theorem. *The operator* $\mathcal{H}^{\text{Dirac}}$ *of* (4), *with the domain of definition*

$$D(\mathcal{H}^{\text{Dirac}}) = \{\varphi \mid \varphi = \{\varphi_1, \varphi_2, \varphi_3, \varphi_4\},\ \varphi_k \in C_0^\infty(R_3)\}$$

is essentially self-adjoint in the Hilbert space $L_2^2(R_3)$. *One has*

$$\tilde{D}_{\overline{\mathcal{H}}^{\text{Dirac}}} = \emptyset \quad \text{and} \quad C_{\overline{\mathcal{H}}^{\text{Dirac}}} = (-\infty, -mc^2] \cup [mc^2, \infty)\,. \tag{5}$$

Definition. $\overline{\mathcal{H}}^{\text{Dirac}}$ *is the Hamiltonian operator of the quantum mechanical system of a relativistic free particle in* R_3.

Remark 2. The above theorem and the preparatory considerations suggest this definition. It can be shown that the corresponding abstract Schrödinger equation (28.1.2/1) is in fact tailored for the requirements of the special theory of relativity as referred to in Chap. 25. Again it follows from Subsec. 28.1.3. that the relativistic free particle, too, has no stationary state.

Remark 3. Before the introduction of $\mathcal{H}^{\text{Dirac}}$ in 1928, physicists were accustomed to the fact that all quantum mechanical operators $\mathcal{H}$ were bounded from below, i.e., $S \subset [\mu, \infty)$, where μ is a suitable real number. To this there corresponded the semi-boundedness of the possible energy levels, which was considered reasonable from the physical point of view. The unexpected and unintended effect (5) decisively contradicted this conception. It induced Dirac to predict the positron, a particle that has a mass equal to that of the electron, but the charge $+|\varepsilon|$. Later on the positron has been detected by experiment. This leads to a conclusion well known among mathematicians and physicists:

"*Formulae are wiser than men*".

28.5. The Hydrogen Atom

28.5.1. The Hydrogen Atom without Spin

Let $S_M^{(L)}(\vartheta, \varphi)$ again denote the three-dimensional spherical harmonics of Theorem 27.2.3, and let $L_{\alpha,k}(r)$ denote the Laguerre functions of Theorem 27.1.5.

Theorem. *The operator* $\mathcal{H}_H$,

$$\mathcal{H}_H\varphi = -\frac{\hbar^2}{2m}\,\Delta\varphi - \frac{\varepsilon^2}{r}\,\varphi \quad \textit{with} \quad D(\mathcal{H}_H) = C_0^\infty(R_3) \tag{1}$$

is essentially self-adjoint in the Hilbert space $L_2(R_3)$. *One has*

$$C_{\overline{\mathcal{H}}_H} = [0, \infty) \quad \textit{and} \quad \tilde{D}_{\overline{\mathcal{H}}_H} = D_{\overline{\mathcal{H}}_H} = \left\{-\frac{m\varepsilon^4}{2\hbar^2 N^2}\right\}_{N=1}^\infty. \tag{2}$$

The eigenraum associated with the eigenvalue $-\dfrac{m\varepsilon^4}{2\hbar^2 N^2}$ *has the dimension* N^2 *(multiplicity of the eigenvalue), being spanned by the orthogonal eigenfunctions*

$$\{f_{N,L,M}(x)\}_{\substack{L=0,\dots,N-1\\ M=-L,\dots,0,\dots,L}} \tag{3}$$

where $f_{N,L,M}(x) = r^{-\frac{1}{2}} L_{2L+1,N-L-1}\left(\dfrac{2m\varepsilon^2}{\hbar^2 N}\, r\right) S_M^{(L)}(\vartheta, \varphi)$.

Remark 1. $\mathcal{H}_H$ is the operator $\mathcal{H}_{\text{atom}}$ of (28.3.2/4), with $Z=n=1$, or the operator $\mathcal{H}_{\mathfrak{A},V}$ of (28.3.2/6), in the absence of electromagnetic fields. All the above quantities have the meaning indicated there. $\mathcal{H}_H$ corresponds to the neutral hydrogen atom. Thus the theorem suggests the following definition.

Definition. $\overline{\mathcal{H}}_H$ *is the Hamiltonian operator of the quantum mechanical system of the hydrogen atom, with its nucleus being fixed at the origin.*

Remark 2. Later on we shall describe other Hamiltonian operators for the hydrogen atom. In all cases the object will be the neutral hydrogen atom with the nucleus fixed at the origin. $\overline{\mathcal{H}}_H$ corresponds to the non-relativistic hydrogen atom without spin. Later on we shall investigate the non-relativistic hydrogen atom with spin (Subsec. 28.5.3.) and the relativistic hydrogen atom (Subsec. 28.5.4.). The multitude of physically meaningful Hamiltonian operators for the hydrogen atom again illustrates that the quantization rule of Subsec. 28.3.1. is only one tool (though a very useful one) that must be complemented by additional considerations which are in part extensive.

Ground state: From Subsec. 28.1.3. it follows that the hydrogen atom has infinitely many stationary states, where $E_N = -\dfrac{m\varepsilon^4}{2\hbar^2 N^2}$ are the corresponding energy levels (Fig. 28.3). According to Subsec. 28.2.1., the ground state is the state of minimum energy, i.e.,

$$\psi(x) = c_1 r^{-\frac{1}{2}} L_{1,0}\left(\frac{2m\varepsilon^2}{\hbar^2}\, r\right) S_0^{(0)}(\vartheta, \varphi) = c_2 e^{-\frac{m\varepsilon^2}{\hbar^2} r} \quad \text{where} \quad E_1 = -\frac{m\varepsilon^4}{2\hbar^2}. \tag{4}$$

Here c_1 and c_2 must be so chosen that $\|\psi\|_{L_2(R_1)} = 1$.

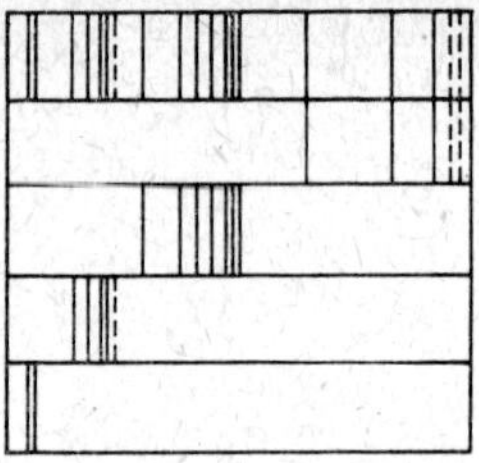

Fig. 28.2

Spectrum of hydrogen: Now the absorption and the emission spectrum of the hydrogen atom can be calculated on the basis of Bohr's postulate stated in Subsec. 28.2.1. The spectrum consists of sharp frequencies. With

$$\mathrm{R} = \frac{2\pi^2 m\varepsilon^4}{\mathrm{h}^3} \quad \text{(Rydberg-constant)}$$

one obtains the following possible frequencies:

$$\mathrm{R}\left(\frac{1}{N_1^2} - \frac{1}{N_2^2}\right), \quad N_2 > N_1 \geqq 1 . \tag{5}$$

It is customary (and contingent upon historical usage) to combine these possible frequencies into series: $N_1 = 1$ with $N_2 = 2, 3, \ldots$ is the Lyman series, $N_1 = 2$ with $N_2 = 3, 4, \ldots$ is the Balmer series etc. Fig. 28.2 is a qualitative representation of these series. The frequencies of the Balmer series fall into the range of visible light and can be readily observed by optical means. The experimental values are in excellent agreement with these theoretical values. Only high-precision measurements reveal deviations, which shall be dealt with later on in the relativistic theory.

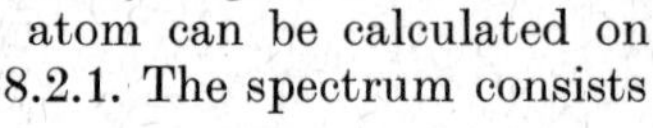

Fig. 28.3

Bohr radius: Bohr stated the radius of the hydrogen atom to be

$$a = \frac{\hbar^2}{m\varepsilon^2} \sim 0{,}53 \cdot 10^{-8}\ \mathrm{cm} .$$

This raises the question whether this number can be found in the above calculus. To this end we put the "question", in the sense of Subsec. 28.2.2.: what is the distance between the electron and the atomic nucleus? As the nucleus is fixed at the origin, this is the question for the distance r of the electron from the origin. According to the quantization rule of Subsec. 28.3.1. (and by analogy with (28.4.1/1)) we assign to this question the self-adjoint operator A,

$$(A\varphi)(x) = |x|\, f(x), \quad D(A) = \{\varphi \mid \varphi(x) \in L_2(R_3),\ |x|\ \varphi(x) \in L_2(R_3)\} ,$$

in the space $L_2(R_3)$. If the hydrogen atom is in the ground state $\psi(x)$ of (4), one obtains the following result: the probability that the distance of the electron from the origin lies between $\varkappa$ and λ, where $0 \leqq \varkappa \leqq \lambda \leqq \infty$, is

$$\|(E_\lambda - E_\varkappa)\, \psi\|^2_{L_2(R_3)} = c \int_\varkappa^\lambda r^2 \mathrm{e}^{-\frac{2}{a} r}\, \mathrm{d}r .$$

Here $\{E_t\}_{t \in R_1}$ is the spectral family of A. Hence the "most probable" value is the maximum of the probability density $\varrho(r) = cr^2 e^{-\frac{2}{a}r}$. But this is $r = a$ (see Fig. 28.4). In this way we also find the Bohr radius in the theory described here.

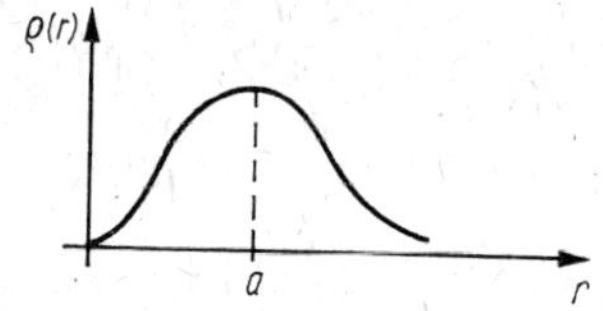

Fig. 28.4

Quantum numbers: $N = 1, 2, \ldots$ in (3) is called principal quantum number, $L = 0, \ldots, N-1$ is called secondary quantum number, and $M = -L, \ldots, L$ is the magnetic quantum number. Later on we shall return to the meaning of these numbers.

28.5.2. The Zeeman Effect

The symbols have the same meaning as in the preceding subsection, especially the functions $f_{N,L,M}(x)$ and the Rydberg constant R. Further let

$$\mu_B = \frac{|\varepsilon|\, \hbar}{2mc}$$

be the Bohr magneton.

As usual, m and ε are the mass and the charge, respectively, of the electron, h is Planck's quantum of action, $\hbar = \frac{h}{2\pi}$, and c is the velocity of light.

Theorem. *For every real number A, the operator $\mathcal{H}_{\text{Zee}}$ of* (28.3.2/7), *with $D(\mathcal{H}_{\text{Zee}}) = C_0^\infty(R_3)$, is essentially self-adjoint in the Hilbert space $L_2(R_3)$. One has*

$$\tilde{D}_{\overline{\mathcal{H}}_{\text{Zee}}} = D_{\overline{\mathcal{H}}_{\text{Zee}}} = \left\{ -\frac{\text{Rh}}{N^2} - \mu_B A M \right\}_{\substack{N=1,2,\ldots \\ M=-N+1,\ldots,N-1}}. \tag{1}$$

The eigenraum associated with the eigenvalue $-\frac{\text{Rh}}{N^2} - \mu_B A M$ *has the dimension* $N - |M|$ *(multiplicity of the eigenvalue), and is spanned by the orthogonal eigenfunctions* $\{f_{N,L,M}(x)\}_{L=|M|,\ldots,N-1}$.

Remark 1. The theorem and the considerations described in Subsec. 28.3.2. suggest the following definition.

Definition. $\overline{\mathcal{H}}_{\text{Zee}}$ *is the Hamiltonian operator of the quantum mechanical system of a hydrogen atom, with its nucleus being fixed at the origin, in a constant, weak magnetic field* $\mathfrak{B}(x) = (0, 0, A)$.

Remark 2. It follows from (1) that the eigenvalues $-\frac{\text{Rh}}{N^2}$ of the hydrogen atom without a magnetic field split up into $2N-1$ equidistant eigenvalues when a weak, constant magnetic field is switched on. Here the sum of the multiplicities of the splitted eigenvalues of $\overline{\mathcal{H}}_{\text{Zee}}$ for a fixed N is equal to N^2, i.e., equal to the multiplicity of the eigenvalue $-\frac{\mathrm{Rh}}{N^2}$ of $\overline{\mathcal{H}}_H$ (the eigenraum of $-\frac{\mathrm{Rh}}{N^2}$ for $\overline{\mathcal{H}}_H$ is decomposed; Fig. 28.5). With a suitable choice of A it may well happen that

$$-\frac{\mathrm{Rh}}{N_1^2}-\mu_B A M_1=-\frac{\mathrm{Rh}}{N_2^2}-\mu_B A M_2$$

for $N_1 \neq N_2$. These eigenvalues then coincide "accidentally", and the theorem has to be modified in an obvious way.

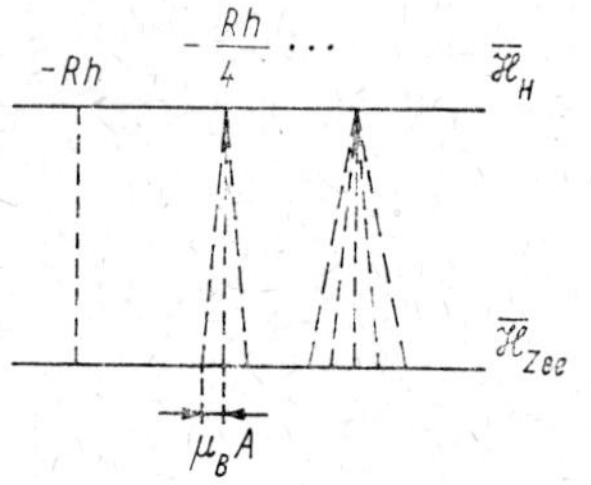

Fig. 28.5

Spectrum and selection rules. By analogy with (28.5.1/5), now the values

$$\mathrm{R}\left(\frac{1}{N_1^2}-\frac{1}{N_2^2}\right)+\frac{\mu_B}{\mathrm{h}} A(M_1-M_2)>0 \tag{2}$$

are possible frequencies of absorbed or emitted electromagnetic waves. Here $N_1=1,2,\ \ldots$ and $N_2=1,2,\ \ldots,$ while $M_1=-N_1+1,\ \ldots,\ N_1-1$ and $M_2=-N_2+1, \ldots, N_2-1$. However, it turns out that only a few of these possible frequencies are in fact realized. This is due to the selection rules. One may put the question of what is the probability for the system to jump from the (normalized) state $f_{N_1,L_1,M_1}(x)$ into the (normalized) state $f_{N_2,L_2,M_2}(x)$. We shall not discuss this theory here. It yields the selection rules (for dipole radiation): the transition from $f_{N_1,L_1,M_1}(x)$ to $f_{N_2,L_2,M_2}(x)$ has a positive probability only if $|L_1-L_2|=1$ and $|M_1-M_2|\leq 1$. With these qualifications, the frequencies (2) are in good agreement with experimental findings.

Quantum numbers. It is now clear why M was called "magnetic quantum number": it is responsible for the splitting of the spectrum upon switching on a constant magnetic field.

28.5.3. The Hydrogen Atom with Spin

Anomalous Zeeman effect. The frequencies (28.5.2/2) (normal Zeeman effect) have been verified by experiment (taking into account the selection rules). There are, however, additional splittings which resemble those of (28.5.2/2) but are not covered by this formula: this is the anomalous Zeeman effect. Again one may attempt to explain these frequencies in the sense of Bohr's postulate. What is desired is a Hamiltonian operator fitting the problem, which has suitable eigenvalues and deviates as little as possible (and in a physically plausible manner) from the proven operators $\overline{\mathcal{H}}_H$ and $\overline{\mathcal{H}}_{\mathrm{Zee}}$. The experimental data of the anomalous Zeeman effect suggest doubling the spectrum of $\overline{\mathcal{H}}_H$ (more precisely:

the dimension of the eigenraum associated with the eigenvalue $-\frac{\mathrm{Rh}}{N^2}$ should be $2N^2$ instead of N^2), which then splits up when a weak, constant magnetic field $\mathfrak{B}(x)=(0, 0, A)$ is applied. Quantitatively, the eigenvalue $-\mathrm{Rh}$ (which is now twofold, having been simple before) should split up into $-\mathrm{Rh}+\mu_B A$ and $-\mathrm{Rh}-\mu_B A$. The other eigenvalues should behave correspondingly. Mathematically, such a repair is easy to accomplish. By analogy with (28.4.3/1), we consider the Hilbert space

$$L_2^1(R_3)=\{f \mid f=\{f_1, f_2\}, f_k \in L_2(R_3)\} \tag{1}$$

with the scalar product

$$(f, g)_{L_2^1(R_3)}=(f_1, g_1)_{L_2(R_3)}+(f_2, g_2)_{L_2(R_3)}, \quad g=\{g_1, g_2\}.$$

The spectrum of the operator $\overline{\mathcal{H}}_H$ is doubled by considering the self-adjoint operator $\{\overline{\mathcal{H}}_H\varphi_1, \overline{\mathcal{H}}_H\varphi_2\}$ in $L_2^1(R_3)$. An analogous approach is chosen for $\overline{\mathcal{H}}_{\text{Zee}}$. The desired (and necessary) correction corresponding to the anomalous Zeeman effect can now be attained as follows. All the symbols used have the meaning indicated previously. Derivatives $\Delta\varphi$ and $\frac{\partial\varphi}{\partial x_1}$, with $\varphi=\{\varphi_1, \varphi_2\}$, are defined as in Subsec. 28.4.3.

Theorem. *The operator* $\mathcal{H}_{\text{Zee}}^{\text{spin}}$,

$$\mathcal{H}_{\text{Zee}}^{\text{spin}}\varphi=-\frac{\hbar^2}{2m}\Delta\varphi-\frac{\varepsilon^2}{r}\varphi+\frac{\mu_B}{\mathrm{i}}A\left(x_1\frac{\partial\varphi}{\partial x_2}-x_2\frac{\partial\varphi}{\partial x_1}\right)+\mu_B A\{\varphi_1, -\varphi_2\} \tag{2}$$

with $D(\mathcal{H}_{\text{Zee}}^{\text{spin}})=\{\varphi \mid \varphi=\{\varphi_1, \varphi_2\}, \varphi_k \in C_0^\infty(R_3)\}$ *is essentially self-adjoint in the Hilbert space* $L_2^1(R_3)$. *One has*

$$\tilde{D}_{\overline{\mathcal{H}}_{\text{Zee}}^{\text{spin}}}=D_{\overline{\mathcal{H}}_{\text{Zee}}^{\text{spin}}}=\left\{-\frac{\mathrm{Rh}}{N^2}-\mu_B A M \pm \mu_B A\right\}_{\substack{N=1,2,\ldots \\ M=-N+1,\ldots,N-1}} \tag{3}$$

Remark 1. The modification made to the doubled operator $\mathcal{H}_{\text{Zee}}$ of Subsec. 28.5.2. is the "spin term" $S\varphi=\mu_B A\{\varphi_1, -\varphi_2\}$. This operator is very simple. If shifts the spectrum of $\overline{\mathcal{H}}_{\text{Zee}}$ by $\mu_B A$ to the right in one component and to the left in the other. This is just the desired effect. Moreover it is observed that the eigenraums associated with the respective eigenvalues are spanned by the orthogonal elements

$$f_{N,L,M}(x)\uparrow=\{f_{N,L,M}(x), 0\}, \quad f_{N,L,M}(x)\downarrow=\{0, f_{N,L,M}(x)\}. \tag{4}$$

N, L, M have the previous meaning. (We dispense with the elementary but somewhat lengthy procedure of clarifying the correspondences between the respective eigenvalues and eigenfunctions.) The preceding considerations as well as the requirements described above now suggest the following definition.

Definition. $\overline{\mathcal{H}}_{\text{Zee}}^{\text{spin}}$ *is the Hamiltonian operator of the quantum mechanical system of a hydrogen atom, with its nucleus fixed at the origin, in a constant, weak magnetic field* $\mathfrak{B}(x)=(0, 0, A)$, *where the effect of spin has been taken into account.*

Remark 2. In the physical interpretation, an additional degree of freedom of the electron, called spin, is introduced to explain the anomalous Zeeman effect. The spin has only two possible states (orientations), usually denoted by ↑ and ↓. They correspond to the elements (4) in our calculus. The spin is interpreted as an intrinsic magnetic moment of the electron that can assume the values μ_B and $-\mu_B$, responding accordingly to an external magnetic field.

Spectrum. By analogy with (28.5.1/5) and (28.5.2/2), now

$$R\left(\frac{1}{N_1^2}-\frac{1}{N_2^2}\right)+\frac{\mu_B}{\mathrm{h}}A(M_1-M_2)>0 \quad \text{and}$$

$$R\left(\frac{1}{N_1^2}-\frac{1}{N_2^2}\right)+\frac{\mu_B}{\mathrm{h}}A(M_1-M_2\pm 2)>0$$

are the possible frequencies of absorbed or emitted electromagnetic waves. Again $N_1=1, 2, \ldots$ and $N_2=1, 2, \ldots$, while $M_1=-N_1+1, \ldots, N_1-1$ and $M_2=-N_2+1, \ldots, N_2-1$. These possibilities are again restricted by the selection rules, but give a good description of the normal and the anomalous Zeeman effect otherwise.

Perturbations. Generally the eigenvalues of $\overline{\mathcal{H}}_H$, $\overline{\mathcal{H}}_{\text{Zee}}$, and $\overline{\mathcal{H}}^{\text{spin}}_{\text{Zee}}$ are degenerate (i.e., not simple). According to our calculus described so far, every normalized eigenelement of an eigenvalue is a stationary state. But physical reality can only be attributed to such stationary states that undergo only slight variations as a result of small perturbations of the corresponding Hamiltonian operator. If under such a (physically reasonable) perturbation a degenerate eigenvalue is split up into a multitude of simple eigenvalues, then physical relevance can be ascribed only to those eigenelements of the unperturbed Hamiltonian operator which result from the eigenelements of the perturbed operator as the perturbation approaches zero. What is then left of the originally K-dimensional eigenraum are only K orthonormal eigenelements of physical relevance. If $\mathcal{H}^{\text{spin}}_{\text{Zee}}$ is modified by

$$\mathcal{H}^{\text{spin}}_{\text{Zee}}\varphi+h(r)\varphi+\delta A\{\varphi_1-\varphi_2\},$$

then $h(r)$ can be regarded as a (physically reasonable) perturbation of $\dfrac{\varepsilon^2}{r}$, and δ as a perturbation of μ_B. With a suitable choice of $h(r)$ and δ, the K-degenerate eigenvalues of (3) then split up into K simple eigenvalues. Further it turns out that the orthonormal eigenfunctions of physical relevance agree with the system (4).

Quantum numbers. The result is as follows: the system (4) describes the physically meaningful stationary states of the hydrogen atom (with the nucleus fixed at the origin) in the constant, weak magnetic field $\mathfrak{B}(x)=(0, 0, A)$, where the spin has been taken into account. Here the following notation has been used for the quantum numbers:

principal quantum number $N=1, 2, \ldots,$
secondary quantum number $L=0, 1, \ldots, N-1,$
magnetic quantum number $M=-L, \ldots, 0, \ldots, L,$
spin quantum number $S=\uparrow, \downarrow.$

28.5.4. The Relativistic Hydrogen Atom

The operator $\mathcal{H}^{\text{Dirac}}$ has the same meaning as in Subsec. 28.4.3. Further let $\alpha = \frac{\varepsilon^2}{\hbar c} \sim \frac{1}{137}$ be the Sommerfeld fine structure constant. It is physically dimensionless. $\left(\right.$Originally it had been assumed that α is exactly equal to $\frac{1}{137}$. This is the reason why it is written $\alpha \sim \frac{1}{137}$ even today.$\left.\right)$

Theorem. *The operator* $\mathcal{H}_{\text{H}}^{\text{Dirac}}$,

$$\mathcal{H}_{\text{H}}^{\text{Dirac}}\varphi = \mathcal{H}^{\text{Dirac}}\varphi - \frac{\varepsilon^2}{r}\varphi, \quad D(\mathcal{H}_{\text{H}}^{\text{Dirac}}) = D(\mathcal{H}^{\text{Dirac}}), \tag{1}$$

is essentially self-adjoint in the Hilbert space $L_2^2(R_3)$. *One has*

$$C_{\overline{\mathcal{H}}_{\text{H}}^{\text{Dirac}}} = (-\infty, -mc^2] \cup [mc^2, \infty) \tag{2}$$

and

$$\tilde{D}_{\overline{\mathcal{H}}_{\text{H}}^{\text{Dirac}}} = D_{\overline{\mathcal{H}}_{\text{H}}^{\text{Dirac}}} = \{E_{n,k}\}_{\substack{n=0,1,2,\ldots\\ k=1,2,3,\ldots}}$$

where

$$E_{n,k} = mc^2\left(1 + \frac{\alpha^2}{(n + \sqrt{k^2 - \alpha^2})^2}\right)^{-\frac{1}{2}}. \tag{3}$$

Remark 1. The proof exhibits a peculiarity: it uses the fact that $\alpha < \frac{1}{2}$. That is, the knowledge of the numerical values of physical constants is required. The point spectrum $D_{\overline{\mathcal{H}}_{\text{H}}^{\text{Dirac}}}$ accumulates from the left at the point mc^2 (Fig. 28.6).

Remark 2. The operator $\mathcal{H}_{\text{H}}^{\text{Dirac}}$ is derived from the operator $\mathcal{H}^{\text{Dirac}}$ of the relativistic free particle in R_3 in the same way as the operator $\mathcal{H}_{\text{H}}$ of Subsec. 28.5.1. is derived from the operator $-\frac{\hbar^2}{2m}\Delta$ of the non-relativistic free particle in R_3. The above theorem then suggests the following definition.

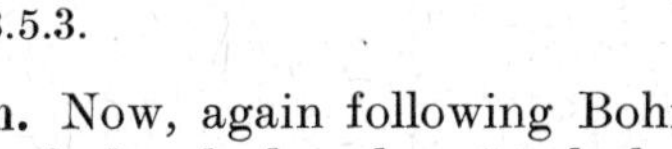

Fig. 28.6

Definition. $\overline{\mathcal{H}}_{\text{H}}^{\text{Dirac}}$ *is the Hamiltonian operator of the quantum mechanical system of the relativistic hydrogen atom with its nucleus fixed at the origin.*

Remark 3. If one considers the relativistic hydrogen atom in a constant, weak magnetic field, it turns out that the above formulation automatically covers the spin in the sense of Subsec. 28.5.3.

Spectrum. Now, again following Bohr's postulate of Subsec. 28.2.1., the frequencies of absorbed and emitted electromagnetic waves can be stated. It is of interest to compare them with the frequencies (28.5.1/5) of the non-relativistic hydrogen atom. Expansion of (3) with respect to α gives

$$E_{n,k} - mc^2 = -\frac{m\varepsilon^4}{2\hbar^2(n+k)^2}\left[1 + \frac{\alpha^2}{k(n+k)} - \frac{3}{4}\frac{\alpha^2}{(n+k)^2} + \ldots\right], \tag{4}$$

where the remainder contains terms with α^4, α^6, ... Since nothing but the differences of $E_{n,k}$ are of interest, the term mc^2 has no effect. Setting $N=n+k$ and comparing this with (28.5.1/2), one observes that the relativistic correction is less than $\alpha^2 10^2$ %, i.e., less than 10^{-2} %. A minute difference, which can, however, be detected by high-precision measurements. It is also seen that the non-relativistic eigenvalue $-\frac{m\varepsilon^4}{2\hbar^2(n+k)^2}$ splits up into different relativistic eigenvalues that are very close to one another. A corresponding splitting then occurs with the frequencies in the sense of Bohr's postulate: this is the fine structure of the hydrogen atom. These minute splittings are in good agreement with experimental findings.

28.6. Atoms and the Periodic System of the Elements

28.6.1. Atoms without Spin

Theorem. *The operator* $\mathcal{H}_{\text{atom}}$ *of* (28.3.2/4), *with* $D(\mathcal{H}_{\text{atom}})=C_0^\infty(R_{3n})$, *is essentially self-adjoint in the Hilbert space* $L_2(R_{3n})$.

Remark 1. From the considerations in Subsec. 28.3.2. it follows that $\mathcal{H}_{\text{atom}}$ might be suitable to describe the ionized atom whose nucleus is fixed at the origin. The theorem suggests the following definition.

Definition. $\overline{\mathcal{H}}_{\text{atom}}$ *is the Hamiltonian operator of the quantum mechanical system of an atom (without spin and without Pauli's principle being considered) with the atomic charge Z and n electrons, the atomic nucleus being fixed at the origin.*

Spectrum: In contrast to the hydrogen atom, the spectrum of $\overline{\mathcal{H}}_{\text{atom}}$ cannot in general be determined explicitly. There is, however, a lot of qualitative information. For neutral atoms, i.e., $Z=n$, the following is known. There is a number μ such that $C_{\overline{\mathcal{H}}_{\text{atom}}}=[\mu,\infty)$; on the left of μ there exist infinitely many eigenvalues of finite multiplicities, which accumulate at μ (Fig. 28.7). Further, in $[\mu,\infty)$ there exist infinitely many eigenvalues, too, with an infinite number of accumulation points. The physically interesting part of the spectrum, which lies in $(-\infty,\mu]$, is hydrogen-like.

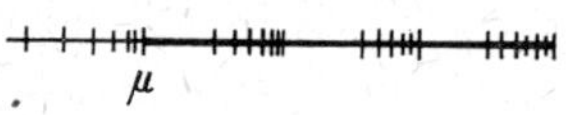

Fig. 28.7

Remark 2. The investigations on the hydrogen atom show that $\overline{\mathcal{H}}_{\text{atom}}$ cannot yet be the last word. The spin, which we had found an essential new aspect in the case of the hydrogen atom, has not been taken into account. Moreover we must try to introduce the so-called Pauli principle into our axiomatics, which will lead to a deeper understanding of the periodic system of the elements.

28.6.2. The Space $L_{2,A}^n(R_{3n})$

We shall now develop the necessary mathematical tools for our further considerations.

The space $L_2^n(R_{3n})$. Let $\bar{x}_k=(x_{3k-2}, x_{3k-1}, x_{3k})\in R_3$, with $k=1,\ldots,n$. By analogy with (28.4.3/1), let $L_2^n(R_{3n})$ be the Hilbert space

$$\{f \mid f = \{f_0(\bar{x}_1, \ldots, \bar{x}_n), f_1(\bar{x}_1, \ldots, \bar{x}_n), \ldots, f_{2^n-1}(\bar{x}_1, \ldots, \bar{x}_n)\}, f_k \in L_2(R_{3n})\}$$

with the scalar product

$$(f, g)_{L_2^n(R_{3n})} = \sum_{k=0}^{2^n-1} (f_k, g_k)_{L_2(R_{3n})},$$

where $g = \{g_0, \ldots, g_{2^n-1}\} \in L_2^n(R_{3n})$.

The group $\mathfrak{S}_n$. As usual, $\mathfrak{S}_n$ is the group of the $n!$ different permutations of the natural numbers from 1 to n, i.e.,

$$q(1, \ldots, n) = (q_1, \ldots, q_n) \quad \text{and} \quad q(p_1, \ldots, p_n) = (q_{p_1}, \ldots, q_{p_n}) .$$

The group multiplication is defined as successive application; q^{-1}, with $q^{-1}(q_1, \ldots, q_n) = (1, \ldots, n)$, is the transformation inverse to q, the identical transformation is the identity element. Finally, let the sign $(-1)^q$ be determined by

$$\prod_{k>l}^{1,\ldots,n} (y_k - y_l) = (-1)^q \prod_{k>l} (y_{q_k} - y_{q_l}) .$$

If $(-1)^q = 1$, then q is called an even permutation, and an odd one if $(-1)^q = -1$.

The operation qf. Every integer j, $0 \leq j \leq 2^n - 1$, has a unique representation as an n-digit binary number, $j = \alpha_1 \alpha_2 \ldots \alpha_n$, where α_r is either 0 or 1. If $q \in \mathfrak{S}_n$, then we set

$$q[j] = \alpha_{q_1} \ldots \alpha_{q_n} ,$$

if $j = \alpha_1 \ldots \alpha_n$ in the binary representation. If $f \in L_2^n(R_{3n})$, then

$$qf = \{f_0(\bar{x}_{q_1}, \ldots, \bar{x}_{q_n}), \ldots, f_j(\bar{x}_{q_1}, \ldots, \bar{x}_{q_n}), \ldots, f_{2^n-1}(\bar{x}_{q_1}, \ldots, \bar{x}_{q_n})\} \tag{1}$$

generates a unitary operator in $L_2^n(R_{3n})$ where f_j stands on the place $q[j]$. In (1) it is not only the co-ordinates $\bar{x}_1, \ldots, \bar{x}_n$ that are permuted, but also the places of the elements f_j. One has $q[0] = 0$ and $q[2^n - 1] = 2^n - 1$.

Definition.

$$L_{2,A}^n(R_{3n}) = \{f \mid f \in L_2^n(R_{3n}) \quad \textit{where} \quad qf = (-1)^q f \quad \textit{for all} \quad q \in \mathfrak{S}_n\} .$$

Remark 1. Thus $L_{2,A}^n(R_{3n})$ is defined to contain all those elements of $L_2^n(R_{3n})$ which are antisymmetric under the transformation $f \to qf$. Correspondingly, it is possible to define a space $L_{2,S}^n(R_{3n})$ that consists of all elements of $L_2^n(R_{3n})$ with $qf = f$ for all $q \in \mathfrak{S}_n$. This space is of physical interest, too (like the space $L_{2,A}^n(R_{3n})$).

Theorem. (**a**) *$L_{2,A}^n(R_{3n})$ is an infinite-dimensional closed subspace of $L_2^n(R_{3n})$ (and hence a Hilbert space). The corresponding projection operator P is*

$$Pf = \frac{1}{n!} \sum_{q \in \mathfrak{S}_n} (-1)^q qf \quad \textit{for} \quad f \in L_2^n(R_{3n}) .$$

(**b**) *If $h_1(\bar{y}), \ldots, h_k(\bar{y})$, with $\bar{y} = (y_1, y_2, y_3)$, is an orthonormal system in $L_2(R_3)$, and if $f = \{0, \ldots, 0, \underbrace{h_{l_1}(\bar{x}_1) \ldots h_{l_n}(\bar{x}_n)}_{j\text{-th place}}, 0, \ldots, 0\}$, then $Pf = 0$ if and only if there exist at least two indices r and s, with $1 \leq r < s \leq n$, such that $h_{l_r}(\bar{y}) = h_{l_s}(\bar{y})$ and $\alpha_r = \alpha_s$. Here $j = \alpha_1 \ldots \alpha_n$ is the binary representation of j.*

Remark 2. Part (b) looks relatively harmless. The proof, however, is not so simple. Hidden behind this formulation is the Pauli principle. Elements f of the above form with $Pf \neq 0$ will be of interest later on. For a given k it will then be a combinatorial problem to insert suitable products of the functions $h_r(\bar{y})$ at suitable places.

28.6.3. Atoms with Spin

Theorem. *The operator* $\mathcal{H}_{\text{atom}}^{\text{Pauli}}$,

$$\mathcal{H}_{\text{atom}}^{\text{Pauli}}\, \varphi = -\frac{\hbar^2}{2m}\, \Delta\varphi + \left(-Z\varepsilon^2 \sum_{k=1}^{n} \frac{1}{r_k} + \varepsilon^2 \sum_{l>k} \frac{1}{r_{l,k}} \right) \varphi \tag{1}$$

with

$$D(\mathcal{H}_{\text{atom}}^{\text{Pauli}}) = \{\varphi \mid \varphi = \{\varphi_0, \ldots, \varphi_{2^n-1}\} \in L_{2,\mathrm{A}}^n(R_{3n}),\ \varphi_k \in C_0^\infty(R_{3n})\} \tag{2}$$

is essentially self-adjoint in the Hilbert space $L_{2,A}^n(R_{3n})$.

Spin and Pauli principle (*first version*). The transition from $\mathcal{H}_{\text{atom}}$ of (28.3.2/4) and Theorem 28.6.1 to $\mathcal{H}_{\text{atom}}^{\text{Pauli}}$ can be described as consisting of two steps. First we substitute $L_2^n(R_{3n})$ for $L_2(R_{3n})$. Physically this corresponds to taking into account the spin. According to Subsec. 28.5.3. the spin is interpreted as a property of the electron, where the spin quantum number S has the possibilities $\uparrow$ and $\downarrow$. In this interpretation, n electrons permit 2^n spin configurations. Setting $\uparrow = 0$ and $\downarrow = 1$, one may interpret a spin configuration $\uparrow\downarrow\downarrow\downarrow \ldots \uparrow = j$ as a binary representation of j and assign the place j in $\{\varphi_0, \ldots, \varphi_{2^n-1}\}$ to this configuration. Thus each of the 2^n components $L_2(R_{3n})$ of $L_2^n(R_{3n})$ represents a certain spin configuration of the n electrons. To the transition from $L_2(R_{3n})$ to $L_2^n(R_{3n})$ thus there corresponds the transition from atoms without spin to atoms with spin. However, it turns out that $L_2^n(R_{3n})$ is too large as a base space: many of its eigenfunctions do not allow a physical interpretation and contradict the Pauli principle. Roughly speaking, this principle says that, if $L_2^n(R_{3n})$ instead of $L_2(R_{3n})$ is taken as the base space, only those eigenelements of $\mathcal{H}_{\text{atom}}$ which belong to $L_{2,A}^n(R_{3n})$ are of physical relevance. The theorem is adapted to this situation. The fact that (1) applies to $L_{2,A}^n(R_{3n})$ follows from the commutativity of $\mathcal{H}_{\text{atom}}^{\text{Pauli}}$ and P of Theorem 28.6.2, which is formally written as $\mathcal{H}_{\text{atom}}^{\text{Pauli}} P = P\mathcal{H}_{\text{atom}}^{\text{Pauli}}$. If one uses the above interpretation of the place $j = \downarrow\downarrow\uparrow \ldots \downarrow$ as a characterization of a certain spin configuration, then it is now possible to give a physical interpretation of the operation qf of (28.6.2/1). $q \in \mathfrak{S}_n$ permutes the places $\bar{x}_1, \ldots, \bar{x}_n$ of the n electrons, $\bar{x}_{q_1} \to \bar{x}_1, \ldots, \bar{x}_{q_n} \to \bar{x}_n$. If the n electrons have a spin configuration $\uparrow\downarrow \ldots \uparrow = j$, then under this permutation the places of the spin configuration are carried over without being changed; the function on the place $j = \alpha_1 \ldots \alpha_n$ is then shifted (under the permutation $\bar{x}_{q_k} \to \bar{x}_k$) to the place $q[j] = \alpha_{q_1} \ldots \alpha_{q_n}$. Now the above, rough first version of the Pauli principle can be reformulated as follows: $\mathcal{H}_{\text{atom}}^{\text{Pauli}}$ admits only those stationary states φ which are antisymmetric under a permutation of the n electrons with preservation of the spin configuration, that is, $q\varphi = (-1)^q \varphi$. In the next section we shall make this more precise. The theorem as well as the statements made above now suggest the following definition.

Definition. $\mathcal{H}_{\text{atom}}^{\text{Pauli}}$ *is the Hamiltonian operator of the quantum mechanical system of an atom (where the spin as well as the Pauli principle have been taken into* *account) with atomic charge Z and n electrons, the atomic nucleus of which is fixed at the origin.*

Spectrum. For neutral atoms, i.e., $Z=n$, the interesting part of the spectrum of $\overline{\mathcal{H}}^{\text{Pauli}}_{\text{atom}}$ is again hydrogen-like: there exists a real number ν such that $C_{\overline{\mathcal{H}}^{\text{Pauli}}_{\text{atom}}} = [\nu, \infty)$, and on the left of ν there exist infinitely many eigenvalues of finite multiplicities, which accumulate at ν (Fig. 28.8). However, other than in the case of the hydrogen atom there are also eigenvalues in $[\nu, \infty)$ (for $n \geqq 2$). In particular there exist infinitely many stationary states (what a luck, for otherwise the atoms would not know why they are immortal). There is a lowest eigenvalue, and hence there are also ground states.

Fig. 28.8

28.6.4. The Pauli Principle

The entities we are interested to know are the lowest eigenvalue of $\overline{\mathcal{H}}^{\text{Pauli}}_{\text{atom}}$ as well as the associated eigenraum. Unfortunately there is no chance of an explicit analytical determination. Following the best physical custom, it is attempted to replace $\overline{\mathcal{H}}^{\text{Pauli}}_{\text{atom}}$ by a simpler operator, hoping that the simpler operator will approximately show the same (qualitative and quantitative) behaviour as $\overline{\mathcal{H}}^{\text{Pauli}}_{\text{atom}}$. What is annoying in (28.6.3/1) are the electron-electron interaction terms $\varepsilon^2 \sum_{l>k} \frac{1}{r_{l,k}} \varphi$. One tries to replace this term by $\sum_{k=1}^{n} V(r_k)\, \varphi$. A suitable choice of V will also be effective in the sense set out above. The operator modified in this way is much simpler, then n electrons are decoupled, and the operator is in effect separated into n operators of the hydrogen type. If the approximate quantitative correctness of the description is dispensed with, one may set $V(r) \equiv 0$. This then gives the operator $\mathring{\mathcal{H}}^{\text{Pauli}}_{\text{atom}}$,

$$\mathring{\mathcal{H}}^{\text{Pauli}}_{\text{atom}}\, \varphi = -\frac{\hbar^2}{2m} \Delta\varphi - Z\varepsilon^2 \sum_{r=1}^{n} \frac{1}{r_k} \varphi\,, \tag{1}$$

$$D(\mathring{\mathcal{H}}^{\text{Pauli}}_{\text{atom}}) = D(\overline{\mathcal{H}}^{\text{Pauli}}_{\text{atom}})\,,$$

which is again considered in the Hilbert space $L^n_{2,A}(R_{3n})$. It is hoped (not in vain, as the success shows) that $\mathring{\mathcal{H}}^{\text{Pauli}}_{\text{atom}}$ gives a qualitatively correct description of the spectrum of $\overline{\mathcal{H}}^{\text{Pauli}}_{\text{atom}}$, but at least of the lowest eigenvalue and of the structure of the associated eigenraum. Apart from Z, $\mathring{\mathcal{H}}^{\text{Pauli}}_{\text{atom}}$ is separated into n Hamiltonian operators of the hydrogen type according to (28.5.1/1) in each of the 2^n components of $\varphi = \{\varphi_0, \ldots, \varphi_{2^n-1}\}$. If in (28.5.1/1) ε^2 is replaced by $Z\varepsilon^2$, then the corresponding operator has the eigenvalues $-\frac{\mathrm{R}hZ^2}{N^2}$, $N = 1, 2, \ldots$, where R is again the Rydberg constant. Correspondingly, one has to replace ε^2 by $Z\varepsilon^2$ in (28.5.1/3). We shall, however, keep the notation $f_{N,L,M}(x)$ for the function modified in this manner. Now the eigenvalues and eigenfunctions of $\mathring{\mathcal{H}}^{\text{Pauli}}_{\text{atom}}$ can be constructed of $-\frac{\mathrm{R}hZ^2}{N^2}$ and $f_{N,L,M}(x)$, where it must, however, be taken into consideration that the base space is $L^n_{2,A}(R_{3n})$ (and not $L^n_2(R_{3n})$).

Theorem. *The operator* $\mathring{\mathcal{H}}^{\text{Pauli}}_{\text{atom}}$ *stated in* (1), *with the domain of definition* $D(\mathring{\mathcal{H}}^{\text{Pauli}}_{\text{atom}}) = D(\mathcal{H}^{\text{Pauli}}_{\text{atom}})$ *according to* (28.6.3/2), *is essentially self-adjoint in the Hilbert space* $L^n_{2,A}(R_{3n})$. *If* $N(n)$ *is a non-negative integer such that*

$$2\sum_{N=0}^{N(n)} N^2 \leqq n < 2\sum_{N=0}^{N(n)+1} N^2, \quad r(n) = n - 2\sum_{N=0}^{N(n)} N^2\,, \tag{2}$$

then $E = -2\text{Rh}Z^2N(n) - \text{Rh}Z^2\dfrac{r(n)}{(N(n)+1)^2}$ *is the lowest eigenvalue of the closure of* $\mathring{\mathcal{H}}^{\text{Pauli}}_{\text{atom}}$. *The associated eigenraum is spanned by*

$$P\{0, \ldots, 0, \underset{j\text{-th place}}{\prod_{t=1}^{n} f_{N_t,L_t,M_t}(\bar{x}_t)}, 0, \ldots, 0\}\,. \tag{3}$$

Here the quadruples $(N_t, L_t, M_t, \alpha_t)$, $t = 1, \ldots, n$, *are pairwise different, where* $j = \alpha_1 \ldots \alpha_n$ *is the binary representation of* $j = 0, \ldots, 2^n - 1$. *Further,* $N_t \leqq N(n) + 1$, *where exactly* $r(n)$ *of the numbers* N_t *are equal to* $N(n) + 1$, *and* $L_t = 0, \ldots, N_t - 1$ *as well as* $M = -L_t, \ldots, 0, \ldots, L_t$.

Remark 1. If (1) would be considered in $L^n_2(R_{3n})$, then $-\text{Rh}Z^2n$ would be the lowest eigenvalue, and the associated eigenraum would then be spanned by the elements $\left\{0, \ldots, 0, \prod_{t=1}^{n} f_{1,0,0}(\bar{x}_t), 0, \ldots, 0\right\}$. Due to the restriction to $L^n_{2,A}(R_{3n})$ the lowest eigenvalue E increases substantially. The decisive effect is due to (3) with the indicated qualifications. This is a relatively simple reformulation of Theorem 28.6.2(b). In particular it is seen that the elements of (3) are not zero.

Definition. *The ground state of an atom with the atomic charge Z and the nucleus fixed at the origin is qualitatively described by the normalized element* (3), *where the quadruples* $(N_t, L_t, M_t, \alpha_t)$ *satisfy the above conditions and* $\alpha_1 \ldots \alpha_n = j$ *characterizes the spin configuration of the n electrons.*

Remark 2. In Subsec. 28.5.3. we had found by perturbation considerations that for a fixed N physical relevance can be attributed to the functions $f_{N,L,M}(x)$ alone (and not to their linear combinations). Now we also stick to this statement, as is shown by (3) and the above definition. The magnetic quantum number would, however, become important not until magnetic fields are applied.

Pauli principle: According to Theorem 28.6.2(b), (3) is different from zero (and hence, after being normalized, describes a stationary state) if and only if the n quadruples $(N_t, L_t, M_t, \alpha_t)$ are pairwise different. This is the usual wording of the Pauli principle. If the additional conditions stated at the end of the above theorem are satisfied, then a state of minimum energy is realized, i.e. the system is in the ground state. Higher energy levels represent excited states. According to Bohr's postulate, the jump transitions between these states yield possible frequencies of absorbed or emitted electromagnetic waves. For $n = 1$ (hydrogen atom) the Pauli principle is not effective; in this case one has the system (28.5.3/4).

Aufbauprinciple: The following interpretation of (3) is customary in physics. Each of the n electrons is characterized by a quadruple of 4 quantum numbers $(N_t, L_t, M_t, \alpha_t)$, where N, L, M and $\alpha = S$ have the same meaning as at the end of Subsec. 28.5.3. The state of the atom is then built up from the "states" of the electrons in the sense of formula (3). To what extent such an aufbauprinciple, which neglects electron-electron interaction, yields a satisfactory description, must be shown by comparison with the experiment.

28.6.5. Periodic System of the Elements

Definition. *Two (neutral) atoms with the atomic charges* n_1 *and* n_2 *(number of electrons) are chemically related if* $r(n_1)=r(n_2)$ *in* (28.6.4/2).

Periodic system. According to this definition, the system of chemical elements exhibits periods of length $2N^2$, where $N=1, 2, 3, \ldots$. The first three periods then include 2, 8, and 18 elements. The first period includes the elements H (hydrogen) and He (helium). According to Def. 28.6.4, functions $f_{N,L,M}(\bar{x})$ with $N=1$ will suffice for the construction of the ground state. If $n=3$ (lithium, Li), then a function $f_{N,L,M}(\bar{x})$ with $N=2$ is required in addition. This period covers the atomic charges up to $n=10$ (neon, Ne). The next period starts from $n=11$, that is sodium, Na. Chemically related elements are H, Li, Na with $r(n)=1$, and He, Ne with $r(n)=0$.

Shell model. The aufbauprinciple of Subsec. 28.6.4 now leads to the shell model. The individual shells are determined by the principal quantum number N, and the shell N accommodates a maximum of $2N^2$ electrons. The atom provides for as favourable an energetic arrangement of its n electrons as possible. To this tendency there corresponds a progressive filling-up of the shells from the interior to the exterior ones. Then the outermost, that is, the $(N(n)+1)$-st shell is occupied by $r(n)$ electrons that are responsible for the chemical behaviour of the atom.

Perturbations. The qualitative approximation according to Def. 28.6.4 becomes increasingly questionable as n increases. Moreover, following the lines of the perturbation considerations of Subsec. 28.5.3., one would expect that, in addition to N, the secondary quantum number L will also influence the magnitude of the eigenvalue. On the other hand, the magnetic quantum number M has no effect. In a comparison of the above considerations with experimental data, the first deviation is found to occur for $n=19$. For larger values of n these deviations increase more and more. What is more important than these deviations, however, is the remarkable fact that it is possible to give an explanation of the periodic system of elements on the basis of quantum mechanical reasoning.

29. Geometry on Manifolds I (Tensors)

29.1. Manifolds

29.1.1. The Paracompact Hausdorff Space

Topology. A system $\mathfrak{B}$ of subsets of a given set M is called a topology in M if $\mathfrak{B}$ has the following three properties.
1. $\emptyset \in \mathfrak{B}$ and $M \in \mathfrak{B}$ (as usual, the void set is considered to be a subset of every set).
2. Finite intersections of sets in $\mathfrak{B}$ belong to $\mathfrak{B}$.
3. Finite or infinite unions of sets in $\mathfrak{B}$ belong to $\mathfrak{B}$.

The system $\mathfrak{B}$ with the above properties is also called a system of open sets. $B \in \mathfrak{B}$ is called a neighbourhood of a point x of M if $x \in B$.

Hausdorff space. A set M with a topology $\mathfrak{B}$ is called a Hausdorff space if for every pair of different points $x \in M$ and $y \in M$ there exist sets $B_x \in \mathfrak{B}$ and $B_y \in \mathfrak{B}$ such that $B_x \cap B_y = \emptyset$ (Fig. 29.1). That is, $\mathfrak{B}$ contains enough elements to be used for a separation of the points of M.

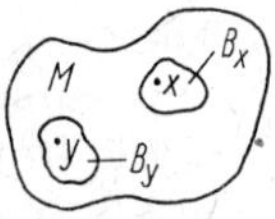

Fig. 29.1

Compact sets. A set C in a Hausdorff space M with the topology $\mathfrak{B}$ is called compact if in every covering of C by a (not necessarily countable) number of open sets it is possible to find a finite number of sets that suffice to cover C. In other words, $\bigcup_{i \in I} B_i \supset C$ with $B_i \in \mathfrak{B}$ implies that there exist sets B_{i_k} such that $\bigcup_{k=1}^{N} B_{i_k} \supset C$.

Closure. A is called a closed set in the Hausdorff space M with the topology $\mathfrak{B}$ if there exists a set $B \in \mathfrak{B}$ such that $A = M \setminus B$. Let $\mathfrak{A}$ be the system of closed sets in M. If C is an arbitrary set in M, then $\bar{C} = \cap D$, where the intersection is taken over all $D \in \mathfrak{A}$ with $D \supset C$. $\bar{C}$ is a closed set, called the closure of C.

Definition. *Let M be a Hausdorff space with the topology $\mathfrak{B}$.*

(**a**) *A covering $\{U_i\}_{i \in I} \subset \mathfrak{B}$ of M, that is $\bigcup_{i \in I} U_i = M$, is called a refinement of a covering $\{V_k\}_{k \in K} \subset \mathfrak{B}$ of M if for every $i \in I$ there exists a $k = k(i) \in K$ such that $U_i \subset V_k$.*

(**b**) *A covering $\{U_i\}_{i \in I} \subset \mathfrak{B}$ of M is called locally finite if for every $x \in M$ there exists a neighbourhood V_x such that $U_i \cap V_x$ is nonvoid for only finitely many parameter values $i \in I$.*

(**c**) *M is called paracompact if for every covering $\{V_k\}_{k \in K} \subset \mathfrak{B}$ of M there exists a locally finite covering $\{U_i\}_{i \in I} \subset \mathfrak{B}$ that is a refinement of $\{V_k\}_{k \in K}$.*

(**d**) *M is called locally compact if for every $x \in M$ there exists a $B \in \mathfrak{B}$ such that $x \in B$ and $\bar{B}$ is compact.*

Remark 1. The R_n is a paracompact Hausdorff space, provided that the usual open sets are chosen as a topology. This is also true for arbitrary sets in R_n that are open in the usual sense.

Remark 2. From now on, the term "covering" shall always be understood to mean a "covering by open sets". Further, from now on we shall exclusively use finite or countably infinite coverings. It can be shown that this does not involve a loss of generality, provided that the Hausdorff space is connected. Here, the set $B \in \mathfrak{B}$ (and especially also M) is said to be connected if B cannot be represented as a union $B = B_1 \cup B_2$ of two nonvoid open disjoint sets.

29.1.2. C^∞-Manifolds

From now on we shall always assume that $(M, \mathfrak{B})$ is a paracompact connected Hausdorff space with the topology $\mathfrak{B}$.

Induced topology. If $B \in \mathfrak{B}$ is connected, then $(B, \mathfrak{B}_B)$, with $\mathfrak{B}_B = \{C \mid C \in \mathfrak{B}, C \subset B\}$ is again a paracompact connected Hausdorff space. $\mathfrak{B}_B$ is called the topology induced by $\mathfrak{B}$.

Topological homeomorphism. Let $(M_1, \mathfrak{B}_1)$ and $(M_2, \mathfrak{B}_2)$ be two Hausdorff spaces. A topological homeomorphism is a one-to-one mapping F of M_1 onto M_2, such that $F(\mathfrak{B}_1) = \mathfrak{B}_2$. (That is, the system of open sets $\mathfrak{B}_1$ is mapped one-to-one onto the system of open sets $\mathfrak{B}_2$.)

Dimension. A Hausdorff space $(M, \mathfrak{B})$ has the dimension n if for every $x \in M$ there exists a (connected) neighbourhood V_x that can be mapped homeomorphically (with the use of the induced topology) onto a (connected) open set in R_n (Fig. 29.2). (Open sets in R_n shall always have the usual meaning.)

Atlas. Let $M = \bigcup_{k=1}^{\infty} V_k$ be a locally finite covering of the n-dimensional Hausdorff space $(M, \mathfrak{B})$. If F_k is a homeomorphic mapping of V_k onto an open set in R_n, then $\{V_k; F_k\}_{k=1}^{\infty}$ is called atlas. M is then described by local co-ordinates (or local charts).

Equivalence classes of atlases (Fig. 29.3). If $\{V_k; F_k\}$ is an atlas of $(M, \mathfrak{B})$, then let $F_k(V_k) = U_k \subset R_n$. If $V_k \cap V_l \neq \emptyset$, then $F_l F_k^{-1}$ is a one-to-one mapping of $F_k(V_k \cap V_l)$ onto $F_l(V_k \cap V_l)$. From now on we shall always require that all of these mappings be differentiable up to arbitrary order, the term "atlas" shall be understood to imply this property. Two atlases $\{V_k; F_k\}$ and $\{W_l; G_l\}$ are

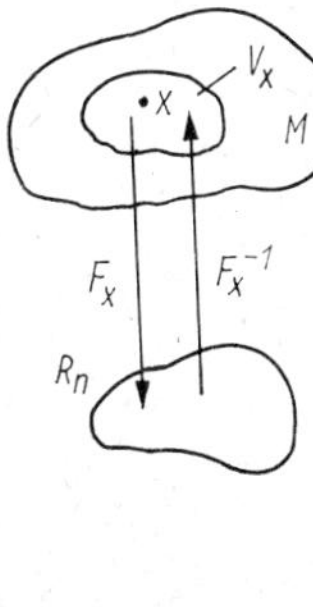

Fig. 29.2

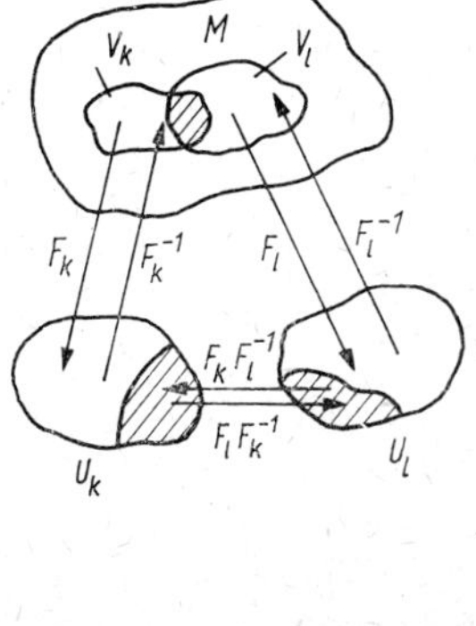

Fig. 29.3

called compatible if $\{V_k, W_l; F_k, G_l\}_{k,l=1}^{\infty}$ is an atlas (this is a demand made upon the above properties of differentiability). The term "compatible" is an equivalence relation, and hence it is possible to form equivalence classes of compatible atlases in the class of all atlases of a given n-dimensional Hausdorff space.

Orientation. An atlas $\{V_k; F_k\}$ is called orientable if the Jacobian determinants of all the above mappings $F_l F_k^{-1}$ of $F_k(V_k \cap V_l)$ onto $F_l(V_k \cap V_l)$ are positive.

Definition. *An n-dimensional orientable C^∞-manifold consists of an n-dimensional paracompact connected Hausdorff space and an equivalence class of atlases that contains at least one orientable atlas.*

Remark 1. The requirement that the space be connected is not necessary, but convenient.

Remark 2. Examples of orientable C^∞-manifolds are connected open domains in R_n, but also smooth surfaces like those dealt with in Subsec. 8.1.5. An example of a non-orientable C^∞-manifold is the well-known Möbius strip. From now on we shall only consider orientable manifolds and orientable atlases, so that we need not mention this explicitly.

Remark 3. Here we restrict ourselves to C^∞-manifolds. In the same way it would be possible to investigate C^k-manifolds and analytical manifolds.

Diffeomorphic mappings (Fig. 29.4). Let $(M_1, \mathfrak{B}_1)$ and $(M_2, \mathfrak{B}_2)$ be two n-dimensional orientable C^∞-manifolds. Let $\{V_{k,1}; F_{k,1}\}$ and $\{V_{k,2}; F_{k,2}\}$ be two associated atlases. A topological homeomorphism (homeomorphic mapping) Φ of M_1 onto M_2 is called diffeomorphism if for all k and l, $F_{l,2}\Phi F_{k,1}^{-1}$ is a C^∞-mapping of

$$F_{k,1}\,(V_{k,1} \cap \Phi^{-1} V_{l,2}) \quad \text{onto} \quad F_{l,2}\,(\Phi V_{k,1} \cap V_{l,2})$$

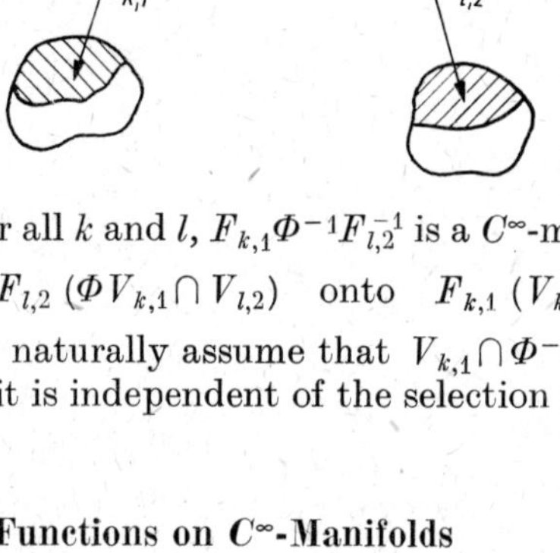

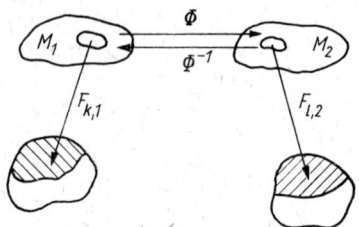

Fig. 29.4

and if, for all k and l, $F_{k,1}\Phi^{-1}F_{l,2}^{-1}$ is a C^∞-mapping of

$$F_{l,2}\,(\Phi V_{k,1} \cap V_{l,2}) \quad \text{onto} \quad F_{k,1}\,(V_{k,1} \cap \Phi^{-1} V_{l,2})\,.$$

Here we naturally assume that $V_{k,1} \cap \Phi^{-1} V_{l,2} \neq \emptyset$. The definition is meaningful because it is independent of the selection of atlases.

29.1.3. Functions on C^∞-Manifolds

If f is a real function on an n-dimensional C^∞-manifold M (that is, a mapping that assigns to every point of M a real number) then it is possible to transfer f to R_n: if $\{V_k; F_k\}_{k=1}^{\infty}$ is an atlas, then let $f_k(x) = f(F_k^{-1}x)$ with $x \in U_k$. The class of all functions f for which the functions $f_k(x)$ are differentiable on U_k up to arbitrary order, where $k = 1, 2, \ldots,$ is denoted by $C^\infty(M)$. This definition does not depend on the choice of the atlas. Let $C_0^\infty(M)$ be the class of the functions $\varphi \in C^\infty(M)$ with a compact support, supp φ, where supp φ is the closure of $\{p \mid p \in M, \varphi(p) \neq 0\}$ (cf. Def. 14.6.4/2).

Lemma (*partition of unity*). *If M is an n-dimensional C^∞-manifold, and if $\{V_k; F_k\}_{k=1}^\infty$ is an associated atlas, then there exist functions $\varphi_k \in C_0^\infty(M)$, $0 \leq \varphi_k \leq 1$, such that $\sum_{k=1}^{\infty} \varphi_k(p) = 1$ for all $p \in M$ and* supp $\varphi_k \subset V_l$ *for $k = 1, 2, \ldots$ for a suitable choice of $l = l(k)$.*

Remark. This lemma can be reduced to Lemma 22.1.4.

29.2. Geometric Objects

29.2.1. Fibre Bundles

Over every point μ of the n-dimensional C^∞-manifold M we erect the fibre $N = R_k$, where R_k is again the k-dimensional Euclidean space. (For the time being, this specialization of N is irrelevant, and so is the assumption that one and the same fibre is erected over every point $\mu \in M$.) Thus we consider the fibre bundle $M \times N = \{(\mu, \nu) \mid \mu \in M, \nu \in N\}$, where (in contrast to the literature) we dispense with introducing a topology on $M \times N$. We also write N_μ in order to indicate that the fibre N over $\mu \in M$ is meant.

Definition. *A geometric object is a section of a fibre bundle $M \times N$, which means one-to-one correspondence $\mu \in M \to \nu(\mu) \in N_\mu$* (Fig. 29.5).

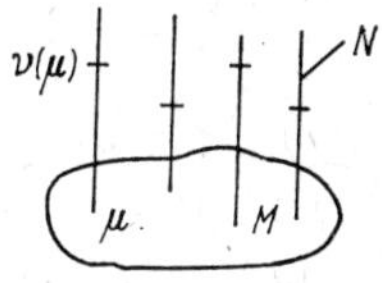

Fig. 29.5

Transitivity. Let $\{V_k; F_k\}_{k=1}^\infty$ be an atlas of the n-dimensional C^∞-manifold M, and let $U_k = F_k(V_k)$ be the image of V_k in R_n. Then we extend the mapping F_k from V_k to $V_k \times N$:

$$F_k(\mu, \nu) = (F_k\mu, F_k^\mu \nu) \in U_k \times N\,, \tag{1}$$

where $F_k^\mu \nu$ is a one-to-one mapping of the fibre N over $\mu \in M$ onto the fibre N over $F_k\mu$. In particular F_k^{-1} exists as a mapping of $U_k \times N$ onto $V_k \times N$. Then $F_l F_k^{-1} = F_{l,k}$ is a one-to-one mapping of $F_k(V_k \cap V_l) \times N$ onto $F_l(V_k \cap V_l) \times N$ with

$$F_{l,k}(F_k\mu, F_k^\mu \nu) = (F_l\mu, F_l^\mu \nu)\,.$$

If $V_k \cap V_l \cap V_i \neq \emptyset$ and $G = F_i(V_k \cap V_l \cap V_i)$, then this immediately implies the transitivity relation

$$F_{l,k} F_{k,i}(\varkappa, \nu) = F_{l,i}(\varkappa, \nu) \quad \text{for} \quad (\varkappa, \nu) \in G \times N\,. \tag{2}$$

Local geometric object. If M is an n-dimensional C^∞-manifold with the atlas $\{V_k; F_k\}$, then let us consider the sets $U_k \times N$. We extend the C^∞-mappings $F_l F_k^{-1}$ from $F_k(V_k \cap V_l)$ onto $F_l(V_k \cap V_l)$ to $F_k(V_k \cap V_l) \times N$ in a manner analogous to (1): the fibre N over $F_k\mu$ is mapped one-to-one onto the fibre N over $F_l\mu$, $\mu \in V_k \cap V_l$. If the mapping extended in this manner is again denoted by $F_{l,k}$, and if (2) is satisfied, then $\{U_k \times N\}_{k=1}^{\infty}$ is called a local fibre bundle. A section of this local fibre bundle is called a local geometric object.

Theorem. *A geometric object is transformed by* (1) *into a local geometric object. Every local geometic object can be obtained in this way from a geometric object.*

Remark 1. The geometric object which generates a given local geometric object according to (1) is of course not unique, because the one-to-one mappings of the fibres N onto one another are largely arbitrary. It is also decisive that we have not topologized $M \times N$ and $U_k \times N$. For the above theorem is not generally true if topologies are involved, where, for the moment, we do not explain precisely what this means. For us it is of importance that the theorem enables us to identify geometric and local geometric objects. From now on we can restrict ourselves to local geometric objects given by explicit transformation rules $F_{l,k}$ that satisfy (2).

Remark 2. In Chaps. 29 and 32 we shall give a description of the geometry on manifolds which is sufficient for our purposes. More detailed presentations that are addressed especially to readers who are interested in physical applications can be found in [1, 9]. We also refer to the standard book [33] as well as to [12, 60].

29.2.2. Tensor Densities

As usual, let M be an n-dimensional C^∞-manifold. If $\{V_k; F_k\}$ is an associated atlas, then let again $U_k = F_k(V_k)$ be a domain in R_n. We now consider arbitrary U_k and U_l, $l \neq k$, denoting the Cartesian co-ordinates by x^a in U_k and by x'^a in U_l (Fig. 29.6). Then we write $F_l F_k^{-1}$ as $x'^a(x^b)$. As in Subsec. 24.1.2., $\dfrac{\partial(x'^a)}{\partial(x^b)}$ (>0) is the associated Jacobian determinant. Further we again use the summation convention referred to in Subsec. 24.1.2.

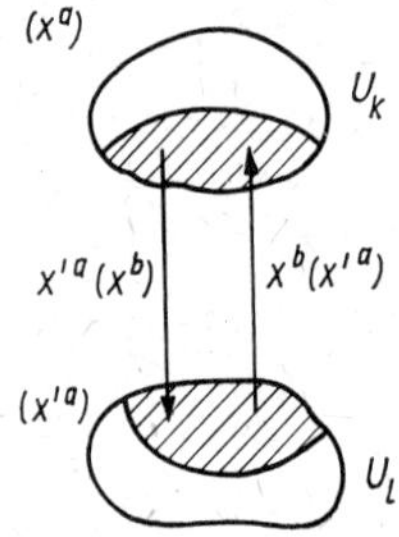

Fig. 29.6

Theorem. *If w is a real number, and if R_{n^K}, $K = 0, 1, 2, \ldots$, is the above fibre N, then $T^{a\ldots b}{}_{c\ldots d}$ (where the K parameters $a, \ldots, b, c, \ldots, d$ run from 1 to n each) is a geometric object if*

$$T'^{a\ldots b}{}_{c\ldots d}(x') = \left(\frac{\partial(x^p)}{\partial(x'^q)}\right)^w \frac{\partial x'^a}{\partial x^r} \cdots \frac{\partial x'^b}{\partial x^s}\frac{\partial x^u}{\partial x'^c} \cdots \frac{\partial x^v}{\partial x'^d} T^{r\ldots s}{}_{u\ldots v}(x) \tag{1}$$

holds for any pair of local charts U_k and U_l of the above form which belong to arbitrary admissible atlases (tensor density fields of rank K, with the weight w and with K_1 contravariant components $a, \ldots, b$ and K_2 covariant components $c, \ldots, d$, where $K_1 + K_2 = K$).

Remark 1. According to Theorem 29.2.1 we need not distinguish any longer between geometric and local geometric objects. In this sense we suppose that an n^K-tuple $T^{a\ldots b}{}_{c\ldots d}(x)$ of real numbers be given at every point $x \in U_k$. If $x' \in U_{l'}$, then, for the purpose of distinction, we denote this n^K-tuple by $T'^{a\ldots b}{}_{c\ldots d}$. Then the theorem reads as follows: if (1) holds, then the transitivity condition (29.2.1/2) is fulfilled, and we obtain a local geometric object that can be transformed into a geometric object according to Theorem 29.2.1. The transitivity condition is in effect a consequence of the chain rule.

Remark 2. If $w=0$, then we speak of tensor fields, as we did in Subsec. 24.1.3.

Terminology. From now on we shall use the term "tensor" instead of "tensor field", and the term "tensor density" instead of "tensor density field"; cf. Subsec. 24.1.3.

Remark 3. In the sense of (1), we shall call the scalar density $T'(x') = \left(\frac{\partial(x^p)}{\partial(x'^q)}\right)^w T(x)$ a tensor density of rank zero (and of weight w). It is a geometric object with $N = R_1$. For $w=0$ one obtains functions. If $K=1$, that means $N = R_n$, then T^a, satisfying (1) with $w=0$, is called a contravariant vector, and T_a, satisfying (1) with $w=0$, is called a covariant vector; cf. Subsec. 24.1.3., especially the mnemotechnical remark stated there.

29.3. Tensor Analysis

29.3.1. Fundamental Operations for Tensor Densities

We are now going to generalize Subsec. 24.1.4. A tensor density is said to be of the type $(k, l; w)$ if it has k contravariant and l covariant components and is of the weight w. If $w=0$, then we write (k, l) instead of $(k, l; 0)$. Let $\mathfrak{S}_k$ be the group of permutations referred to in Subsec. 28.6.2., where $(-1)^p$, $p \in \mathfrak{S}_k$, has the meaning indicated there.

Theorem 1. (a) *If $T^{a\ldots b}{}_{c\ldots d}$ and $S^{a\ldots b}{}_{c\ldots d}$ are tensor densities of the type $(k, l; w)$, and if λ and μ are real numbers, then $\lambda T^{a\ldots b}{}_{c\ldots d} + \mu S^{a\ldots b}{}_{c\ldots d}$ is a tensor density of the same type.*

(b) *If $T^{a\ldots b}{}_{c\ldots d}$ is a tensor density of the type $(k_1, l_1; w_1)$, and if $S^{u\ldots v}{}_{r\ldots s}$ is a tensor density of the type $(k_2, l_2; w_2)$, then $T^{a\ldots b}{}_{c\ldots d} S^{u\ldots v}{}_{r\ldots s}$ is a tensor density of the type $(k_1+k_2, l_1+l_2; w_1+w_2)$.*

(c) *If $T^{a\ldots b}{}_{c\ldots d}$ is a tensor density of the type $(k, l; w)$, then*

$$T^{(a\ldots b)}{}_{c\ldots d} = \frac{1}{k!} \sum_{p \in \mathfrak{S}_k} T^{p(a\ldots b)}{}_{c\ldots d} \tag{1}$$

and

$$T^{[a\ldots b]}{}_{c\ldots d} = \frac{1}{k!} \sum_{p \in \mathfrak{S}_k} (-1)^p T^{p(a\ldots b)}{}_{c\ldots d} \tag{2}$$

are tensor densities of the same type. An analogous proposition is true for the covariant components.

(d) (*Contraction*). *If $T^{r\ldots a\ldots s}{}_{u\ldots b\ldots v}$ is a tensor density of the type $(k, l; w)$, then $T^{r\ldots a\ldots s}{}_{u\ldots a\ldots v}$ (summation over a) is a tensor density of the type $(k-1, l-1; w)$.*

Remark 1. Remarks 1 and 2 of Subsec. 24.1.4. hold analogously for tensor densities, too.

Remark 2. A tensor density $T^{a\ldots b}{}_{c\ldots d}$ is called symmetrical if $T^{a\ldots b}{}_{c\ldots d}=T^{(a\ldots b)}{}_{c\ldots d}$, and antisymmetrical if $T^{[a\ldots b]}{}_{c\ldots d}=T^{a\ldots b}{}_{c\ldots d}$. By way of example, T^{ab} is symmetrical if $T^{ab}=T^{ba}$, and antisymmetrical if $T^{ab}=-T^{ba}$.

Theorem 2. (a) $\delta^a{}_b$, *with* $\delta^a{}_b=1$ *for* $a=b$ *and* $\delta^a{}_b=0$ *for* $a\neq b$, *is a tensor of the type* (1, 1).

(b) $\varepsilon^{a\ldots b}$ (*n indices*), *with* $\varepsilon^{a\ldots b}=(-1)^p$ *if* $(a, \ldots, b)=p(1, \ldots, n)$ *is a permutation of the numbers* $1, \ldots, n$, *and* $\varepsilon^{a\ldots b}=0$ *otherwise, is a tensor density of the type* $(n, 0;\ 1)$ (*Levi-Civita tensor density*).

(c) *If* T_{ab} *is a tensor density of the type* $(0, 2;\ w)$ *with* $\det T_{ab}>0$, *then* $(\det T_{ab})^\varkappa$ *is a scalar density of weight* $\varkappa(nw+2)$. *Here* $\varkappa$ *is a real number.*

Remark 3. $\det T_{ab}$ is the determinant of $(T_{ab})_{a,b=1}^n$. In the sense of (29.2.2/1) one has

$$(\det T_{ab})'=\left(\frac{\partial(x^p)}{\partial(x'^q)}\right)^{nw+2}\det T_{ab}\,. \tag{3}$$

Since $\dfrac{\partial(x^p)}{\partial(x'^q)}>0$, the requirement that $\det T_{ab}>0$ is meaningful and independent of the choice of the local co-ordinates.

29.3.2. Differential Operations

If $\varphi\in C^\infty(M)$, then (keeping the notation) φ can be regarded as C^∞-function on the local charts $U_k=F_k(V_k)$. All the symbols used have the previous meaning. As in Subsec. 24.1.3., let $\varphi_{,k}=\dfrac{\partial\varphi}{\partial x^k}$, $\varphi_{,kl}=\dfrac{\partial^2\varphi}{\partial x^k\partial x^l}$ etc. It is now easily verified that $\varphi_{,k}$ is a covariant vector. On the other hand, $\varphi_{,kl}$ is (generally) not a covariant tensor.

Theorem. (a) *If* T_k *is a covariant vector, then* $T_{k,l}-T_{l,k}$ *is a tensor of the type* (0, 2) (*curl, or rotor, of* T_k).

(b) *If* T^k *is a tensor density of the type* $(1, 0;\ 1)$, *then* $T^k{}_{,k}$ *is a scalar density of the weight* $w=1$ (*divergence of* T^k).

(c) *If* T^{kl} *is an antisymmetrical tensor density of the type* $(2, 0;\ 1)$ (*that is, if* $T^{kl}=-T^{lk}$), *then* $T^{kl}{}_{,k}$ *is a tensor density of the type* $(1, 0;\ 1)$.

Remark. In the above theorem as well as in our further considerations we always assume that the components of vectors, tensors, tensor densities etc. are differentiable up to arbitrary order (with respect to the local co-ordinates). As M is a C^∞-manifold, this assumption is reasonable.

29.3.3. Integrals on Manifolds

If $\{V_k;\ F_k\}$ is an atlas for the n-dimensional C^∞-manifold M, then let $\{\varphi_k\}_{k=1}^\infty$ be a corresponding partition of unity in the sense of Lemma 29.1.3. As previously, let $U_k=F_k(V_k)$. Further suppose that the support of a tensor density $T^{a\ldots b}{}_{c\ldots d}$ on M be the closure $\overline{C}$ of the set C, where C consists of all those points $m\in M$ for which at least one component $T^{a\ldots b}{}_{c\ldots d}(F_k m)$ is different from zero (where it is assumed that $m\in V_k$). Since tensor densities are linear geometric objects, this definition is independent of the choice of the atlas.

Definition. *If f is a continuous scalar density of weight 1 with a compact support on M, then*

$$\int_M f\, \mathrm{d}m = \sum_{k=1}^{\infty} \int_{U_k} (f \cdot \varphi_k)(F_k^{-1}x)\, \mathrm{d}x\,. \tag{1}$$

Remark 1. As f has a compact support, it can be expected that only finitely many integrals on the right-hand side of (1) are different from zero. Further, (1) is independent of the choice of the atlases and the functions φ_k. This is in effect an implication of Theorem 9.2.2, if one takes into consideration that the Jacobian determinants of the mappings considered here are always positive. It is also observed that the weight $w=1$ is necessary.

Theorem. *If T^k is a (continuously differentiable) tensor density of the type* (1, 0; 1) *with a compact support, then* $\int_M T^k{}_{,k}\, \mathrm{d}m = 0$.

Remark 2. Theorem 29.3.2 (b) shows that the above definition is applicable to $f = T^k{}_{,k}$. The theorem is in effect an implication of Theorem 9.3.1/2.

29.4. Affine Spaces

29.4.1. Affine Transformations

The symbols in the following theorem have the same meaning as in Subsec. 29.2.2.

Theorem. *If R_{n^3} is the fibre N, then Γ^a_{bc} (where the indices a, b, c run from 1 to n) is a (nonlinear) geometric object if*

$$\Gamma'^a_{bc}(x') = \frac{\partial x'^a}{\partial x^r} \frac{\partial x^s}{\partial x'^b} \frac{\partial x^t}{\partial x'^c} \Gamma^r_{st}(x) + \frac{\partial x'^a}{\partial x^r} \frac{\partial^2 x^r}{\partial x'^b x'^c} \tag{1}$$

holds for every pair U_k and U_l of local charts (affine transformation).

Remark 1. For the notations, see also Remark 29.2.2/1. To prove the above theorem, one has to check the transitivity relation (29.2.1/2).

Remark 2. It is immediately seen that $\Gamma^a_{bc} - \Gamma^a_{cb}$ is a tensor (torsion tensor.) An affine transformation is called symmetrical if $\Gamma^a_{bc} = \Gamma^a_{cb}$.

Definition. *An affine space consists of an (n-dimensional orientable) C^∞-manifold and a symmetrical affine transformation.*

Remark 3. We thus restrict ourselves to "torsion-free" affine spaces.

29.4.2. Normal Co-ordinates

Theorem. *If M is an affine space and if $m \in M$, then there exist an atlas $\{V_k; F_k\}_{k=1}^{\infty}$ in the equivalence class of the associated atlases and a number l such that $m \in V_l$ and*

$$\Gamma^a_{bc}(x_0) = 0 \quad \text{for} \quad 1 \leq a, b, c \leq n \quad \text{and} \quad x_0 = F_l m\,. \tag{1}$$

Remark. The Cartesian co-ordinates x in the local chart $U_l = F_l(V_l)$ are called normal co-ordinates with respect to x_0. The existence of such normal co-ordinates is proved by utilizing the nonlinearity in (29.4.1/1). The theorem looks harmless, but the normal co-ordinates are an extremely efficient tool in affine spaces.

29.4.3. Covariant Differentiation

If φ is a function, then $\varphi_{,k}$ is a covariant vector. Higher-order derivatives of φ, however, in general do not yield geometric objects. This is also true for arbitrary (including first-order) derivatives of tensor densities. Therefore we have to look out for a substitute which on the one hand exhibits typical properties of partial derivatives, while on the other hand it transforms tensor densities again into tensor densities. This is the covariant differentiation, which is denoted by a semicolon ";".

Theorem. *If $T^{a\ldots b}{}_{c\ldots d}$ is a tensor density of the type $(k, l; w)$ in an affine space, then*

$$T^{a\ldots b}{}_{c\ldots d;e} = T^{a\ldots b}{}_{c\ldots d,e} + \Gamma^a_{re} T^{r\ldots b}{}_{c\ldots d} + \ldots + \Gamma^b_{re} T^{a\ldots r}{}_{c\ldots d} - \Gamma^r_{ce} T^{a\ldots b}{}_{r\ldots d} - \ldots - \Gamma^r_{de} T^{a\ldots b}{}_{c\ldots r} - w\Gamma^r_{re} T^{a\ldots b}{}_{c\ldots d} \tag{1}$$

is a tensor density of the type $(k, l+1; w)$.

Remark 1. Here (1) is again understood as a representation in local co-ordinates; the components of $T^{a\ldots b}{}_{c\ldots d}$ are assumed to be differentiable up to arbitrary order and, as usual, one has to take the sum over doubly occurring indices.

Remark 2. If φ is a function, then we also write $\varphi_{,k} = \varphi_{;k}$. Examples of (1) are

$$T^a{}_{;e} = T^a{}_{,e} + \Gamma^a_{re} T^r$$

for a contravariant, and

$$T_{a;e} = T_{a,e} - \Gamma^r_{ae} T_r$$

for a covariant vector.

Remark 3. In normal co-ordinates one has

$$T^{a\ldots b}{}_{b\ldots d;e}(x_0) = T^{a\ldots b}{}_{c\ldots d,e}(x_0)\,.$$

Hence in particular it follows that the usual calculating rules for partial derivatives, especially the Leibniz theorem, are also applicable to covariant derivatives.

Remark 4. From Remark 3 it follows that $\delta^a{}_{b;e} = 0$.

29.4.4. Translation

Let M be an affine space, and let U be a local chart of M in R_n. Consider a smooth (once continuously differentiable) curve γ in U, which is assumed to be given by $x^k(\tau)$, where $0 \leqq \tau \leqq 1$, and $k = 1, \ldots, n$ (Fig. 29.7). The tangent $\dfrac{\mathrm{d}x^k}{\mathrm{d}\tau}$ is transformed like a contravariant vector under a co-ordinate transformation $x' = x'(x)$: $\dfrac{\mathrm{d}x'^k}{\mathrm{d}\tau} = \dfrac{\partial x'^k}{\partial x^l} \cdot \dfrac{\mathrm{d}x^l}{\mathrm{d}\tau}$.

Definition. (a) *A contravariant vector T^k is called parallel along γ if*

$$T^k{}_{;l} \frac{\mathrm{d}x^l}{\mathrm{d}\tau} = 0 \quad \textit{for} \quad k = 1, \ldots, n \quad \textit{and} \quad 0 \leqq \tau \leqq 1\,. \tag{1}$$

(b) *A covariant vector T_k is called parallel along γ if*

$$T_{k;l} \frac{\mathrm{d}x^l}{\mathrm{d}\tau} = 0 \quad \textit{for} \quad k = 1, \ldots, n \quad \textit{and} \quad 0 \leqq \tau \leqq 1\,. \tag{2}$$

Fig. 29.7

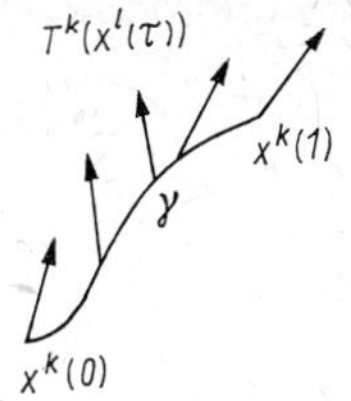

Fig. 29.8

Remark 1. Let us recall that the term "vector" always means "vector field", where the components are assumed to be differentiable up to arbitrary order. As $\frac{dx^k}{d\tau}$ is transformed like a contravariant vector, the definitions are independent of the choice of the charts. In particular the definition can immediately be extended to curves γ that do not lie in one chart.

Remark 2. (1) and (2) are systems of ordinary differential equations of first order for T^k. (1) yields

$$\frac{dT^k}{d\tau}=T^k{}_{,l}\frac{dx^l}{d\tau}=-\Gamma^k_{rl}\cdot T^r\frac{dx^l}{d\tau}, \tag{3}$$

where one has to insert $x^k=x^k(\tau)$. If $T^k(x^l(0))$ is given, then from (3) one obtains a unique solution $T^k(x^l(\tau))$: the vector $T^k(x^l(0))$ undergoes a unique parallel shift along γ (Fig. 29.8). An analogous result is obtained for (2).

Teleparallelism. M is called a space with teleparallelism if this parallel shift is path-independent. The conventional translation in R_n exemplifies this. Generally, however, the parallel shift depends on the path γ.

29.4.5. Affine Geodesics

Definition. *A smooth curve* $\gamma\sim x^k(\tau)$ (Fig. 29.9) *in the affine space* M *is called an affine geodesic if*

$$\frac{d^2x^k}{d\tau^2}+\Gamma^k_{st}\frac{dx^s}{d\tau}\cdot\frac{dx^t}{d\tau}=0 \tag{1}$$

for $k=1, \ldots, n$ *and for a suitable parameter* τ.

Remark 1. If T^k is a contravariant vector that coincides with $\frac{dx^k}{d\tau}$ along γ (there will always exist such vectors), then (1) and

$$T^k{}_{;t}\frac{dx^t}{d\tau}=0 \quad \text{along} \quad \gamma \tag{2}$$

are equivalent. This follows from (29.4.4/3). Roughly speaking, this means that γ is an affine geodesic if and only if the covariant tangent vector $\frac{dx^k}{d\tau}$ is parallel along γ.

Fig. 29.9

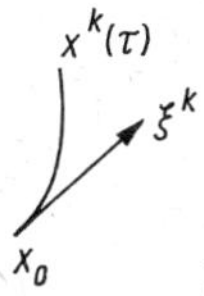

Fig. 29.10

Remark 2. (1) is a system of ordinary differential equations of second order. If U is a local chart, and if $x_0 \in U$ and $\xi^k \in R_n$, then locally there exists exactly one affine geodesic $x(\tau) = (x^k(\tau))$ such that $x(0) = x_0$ and $\frac{dx^k}{d\tau}(0) = \xi^k$ (Fig. 29.10).

Remark 3. If $\sigma = \sigma(\tau)$ is a strictly monotonic, differentiable function, then γ can also be parametrized by σ. Then (1) reads

$$\frac{d^2x^k}{d\sigma^2} + \Gamma^k_{st} \frac{dx^s}{d\sigma} \cdot \frac{dx^t}{d\sigma} = -\frac{\sigma''}{\sigma'^2} \frac{dx^k}{d\sigma}.$$

If one wants to have the form (1) again, with σ instead of τ, then this requires that $\sigma'' = 0$. Such parameters are called affine or geodesic.

Theorem. *σ is a geodesic parameter for the affine geodesic γ if and only if $\sigma = c\tau + d$. Here τ is the parameter of* (1), *c and d are real numbers, $c \neq 0$.*

29.4.6. Riemann's Tensor

Theorem. *If M is an affine space, then*

$$B^a{}_{bcd} = \Gamma^a_{bc'd} - \Gamma^a_{bd'c} + \Gamma^a_{dr}\Gamma^r_{bc} - \Gamma^a_{cr}\Gamma^r_{bd} \tag{1}$$

is a tensor of the type (1, 3) (*Riemann's tensor*). *If T^a is an arbitrary contravariant vector, then*

$$T^a{}_{;cd} - T^a{}_{;dc} = B^a{}_{bcd} T^b. \tag{2}$$

Remark 1. If $T^a{}_{;c} = S^a{}_c$, then $T^a{}_{;cd} = S^a{}_{c;d}$ means twice repeated covariant differentiation.

Remark 2. (2) is the starting point, whereas formula (1) gives an explicit expression, which is, however, not so very interesting. As is well known, $\frac{\partial^2}{\partial x^c \partial x^d} = \frac{\partial^2}{\partial x^d \partial x^c}$. Hence $B^a{}_{bcd}$ in (2) is a measure of the degree to which the covariant differentiation differs from the partial one. An affine space M is called locally Euclidean if $\Gamma^a{}_{bc} \equiv 0$ is achievable with a suitable choice of an atlas. Then $B^a{}_{bcd} = 0$, the covariant derivative and the partial one coincide. In the general case $B^a{}_{bcd}$ expresses the degree to which M differs from a local Euclidean space, i.e., the "curvature" of M. One might attempt to replace T^a in (2) by a scalar function f. However, it is easily seen that $f_{;ab} = f_{;ba}$. Hence a deviation from the familiar commutative law is encountered for the first time for vectors.

Remark 3. In normal co-ordinates for x_0, (1) reduces to

$$B^a{}_{bcd}(x_0) = \Gamma^a{}_{bc'd}(x_0) - \Gamma^a{}_{bd'c}(x_0). \tag{3}$$

Symmetries: Using (3), one shows that

$$\begin{aligned} &B^a{}_{bcd} = -B^a{}_{bdc}, \\ &B^a{}_{[bcd]} = 0 = B^a{}_{bcd} + B^a{}_{cdb} + B^a{}_{dbc}, \\ &B^a{}_{b[cd;e]} = 0 = B^a{}_{bcd;e} + B^a{}_{bde;c} + B^a{}_{bec;d}. \end{aligned} \tag{4}$$

(4) is called *Bianchi identity.*

29.4.7. Flat Affine Spaces

Definition. *Let M be an affine space.*

(**a**) *M is called flat if $B^a{}_{bcd} \equiv 0$.*

(**b**) *M is called locally Euclidean if in the equivalence class of the associated atlases there exists an atlas $\{V_k; F_k\}_{k=1}^{\infty}$ such that $\Gamma^a{}_{bc}(x) \equiv 0$ for $x \in U_k = F_k(V_k)$.*

(**c**) *A contravariant vector T^k is called constant if $T^k{}_{;l} = 0$ for $k = 1, \ldots, n$ and $l = 1, \ldots, n$.*

Remark. The terms (a) and (b) express additional properties of M that make M closer resemble the Euclidean space R_n. Let us recall the concept of teleparallelism mentioned in Subsec. 29.4.4. With $\Gamma^a{}_{bc} \equiv 0$, R_n is flat, locally Euclidean, and exhibits teleparallelism. Moreover there exist nontrivial constant vectors.

Theorem. *If M is an affine space, then the following four propositions are equivalent.* **1.** *M is flat;* **2.** *M is locally Euclidean;* **3.** *M exhibits teleparallelism;* **4.** *If U is a local chart, and if $x_0 \in U$ and $\xi^k \in R_n$, then there exists exactly one constant contravariant vector T^k such that $T^k(x_0) = \xi^k$.*

29.5. Metric Spaces

29.5.1. Fundamental Tensor

By a C^∞-manifold we shall always understand an n-dimensional orientable connected manifold (cf. Def. 29.1.2) without continually referring to these additional properties. On such a manifold M, let us consider a tensor g_{kl} of the type (0, 2) with the properties

$$g_{kl} = g_{lk} \quad \text{(symmetry) and} \quad \det g_{kl} \neq 0\,. \tag{1}$$

Suppose that the components of g_{kl} are differentiable up to arbitrary order. $\det g_{kl}$ is the determinant of $(g_{kl})_{k,l=1}^n$, which is a scalar density according to (29.3.1/3). Hence it follows that (1) (including the symmetry) is independent of the choice of local charts, and hence meaningful. If $m \in M$, then it is possible to find an admissible atlas $\{V_k; F_k\}_{k=1}^{\infty}$, with $m \in V_r$, such that g_{kl} (being interpreted as a symmetrical matrix) assumes the normal form

$$(g_{kl}(x_0)) = \begin{pmatrix} 1 & & & & & 0 \\ & \ddots & & & & \\ & & 1 & & & \\ & & & -1 & & \\ & & & & \ddots & \\ 0 & & & & & -1 \end{pmatrix} \begin{matrix} \left.\vphantom{\begin{matrix}1\\ \ddots\\ 1\end{matrix}}\right\} s_+ \\ \left.\vphantom{\begin{matrix}-1\\ \ddots\\ -1\end{matrix}}\right\} s_- \end{matrix} \tag{2}$$

at the point $x_0 = F_r m$. This is a proposition of analytical geometry. In this case s_+ and s_- (the numbers of the 1's and -1's) are invariant: every admissible transformation that leads to (2) yields the same numbers. For reasons of continuity it follows that s_+ and s_- are also independent of x_0, and hence represent characteristic numbers on M. The term $s = s_+ - s_- = 2s_+ - n$ is called signature. Two cases are of special interest:

$s = n$: positive definite metric ,

$s = n - 2$: Lorentz metric ,

where g_{kl} is also called metric on M. If g_{kl} is a Lorentz metric, let us consider the zero manifold of

$$g_{kl}(x_0)\,\xi^k\xi^l = \sum_{r=1}^{n-1} (\xi^r)^2 - (\xi^n)^2 \quad \text{in} \quad R_n\,.$$

It is a circular cone.

Conoid (Fig. 29.11): A conoid is obtained from the cone $\sum_{r=1}^{n-1} (\xi^r)^2 = (\xi^n)^2$ by a one-to-one mapping $\eta = \eta(\xi)$ that is defined and differentiable up to arbitrary order in a neighbourhood of $0 \in R_n$. $\eta(0)$ is called the origin of the conoid.

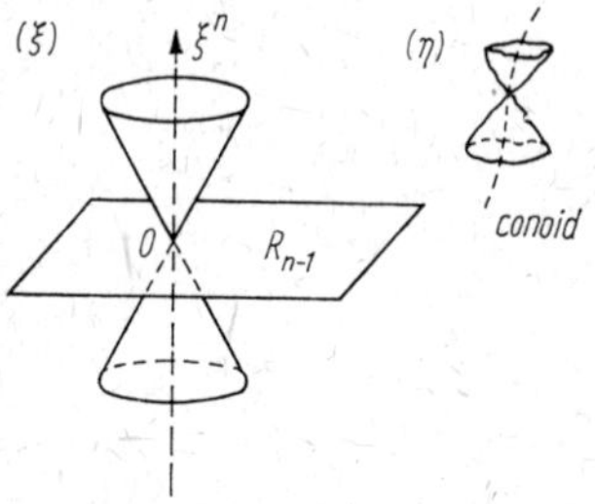

Fig. 29.11

Congruence of Lines (Fig. 29.12): If M is a C^∞-manifold with the Lorentz metric, then there exists a local conoid at every point $x \in U_k$ of the local charts $U_k = F_k(V_k)$. The directions of the local axis of symmetry at the origin of the conoid generate on M a congruence of lines that is free from singularities and covers M simply and without gaps.

Theorem. *Let M be a C^∞-manifold.*

(a) *In any case there exists a positive definite metric on M.*

(b) *There exists a Lorentz metric on M if and only if there exists a (simple, gapless) congruence of lines that is free from singularities.*

(c) *If M is not compact, then there exists a Lorentz metric.*

Remark 1. According to (b) there is no Lorentz metric on a sphere in R_3, whereas there exists a Lorentz metric on a torus (see Fig. 29.13).

Remark 2. g_{kl} with the property (1) is called fundamental tensor. Here g_{kl} is thought to be fixed on M. Confer Subsec. 24.1.2. for reference.

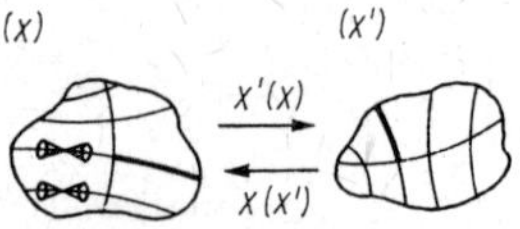

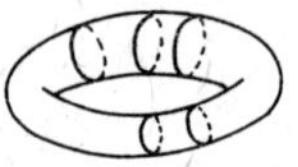

Fig. 29.13

29.5.2. Index Shifting

Let M be a C^∞-manifold with a given fundamental tensor g_{kl}.

Lemma. *If g^{kl} is determined from $g^{kl}g_{lm} = \delta^k{}_m$, then g^{kl} is a tensor of the type* (2, 0).

Remark. Confer Theorem 2(c) as well as Remark 3 in Subsec. 24.1.4.

Index shifting. If $T^{ab...c}{}_{de...f}$ is a tensor density, then we consider the tensor densities

$$T^{b...c}{}_{ade...f} = g_{ar} T^{rb...c}{}_{de...f}, \qquad T^{ab...cd}{}_{e...f} = g^{dr} T^{ab...c}{}_{re...f}$$

to be not essentially different from $T^{a...}{}_{...f}$. By means of the fixed tensors g_{ab} and g^{ab} it is thus possible to raise and to lower indices. Then there are covariant and contravariant versions of tensor densities etc. These procedures of index shifting must always be stated explicitly in each particular case, including the points where the indices have to be placed, in order to avoid misunderstandings.

29.5.3. Characteristic Surfaces

Definition. *If M is a C^∞-manifold with a Lorentz metric g^{kl}, and if $\omega(x)$ is a C^∞-function in M, then $\Omega = \{x \mid \omega(x) = 0\}$, with $\sum_{r=1}^{n} \omega_{,r}^2(x) > 0$ for $x \in \Omega$, is called a characteristic surface if $g^{kl}(x)\,\omega_{,k}(x)\,\omega_{,l}(x) = 0$ for $x \in \Omega$.*

Remark 1. This is a definition in local charts. Since $\omega_{,k}$ is transformed like a covariant vector, it is independent of the choice of the local charts and can in particular be transferred from one local chart to another. Hence the above definition is a global one.

Theorem 1. *If U is a local chart, then for every point $x_0 \in U$ there exists exactly one conoid with origin x_0 that is a characteristic surface (characteristic conoid).*

Remark 2. x_0 does not belong to this characteristic surface, because the latter has a singularity at this point.

Remark 3. If $M = R_n$, where g_{kl} has the form (29.5.1/2) with $s_- = 1$, then the cones

$$0 = \omega(x + x_0) = \sum_{k=1}^{n-1} (x^k)^2 - (x^n)^2$$

are the desired characteristic conoids. For $n = 4$ (Minkowskian space) one obtains the light cones of the special theory of relativity.

Rays. If $x_1 \neq x_0$ is a point on the characteristic conoid for x_0, then the characteristic conoids for x_0 and x_1 touch each other along a C^∞-curve $x^k(\tau)$ (Fig. 29.14). These curves are called rays. Thus one obtains a distinguished congruence of lines on the characteristic conoid. In the case of the cone referred to in Remark 3, the curves in question are straight lines through the origin of the cone.

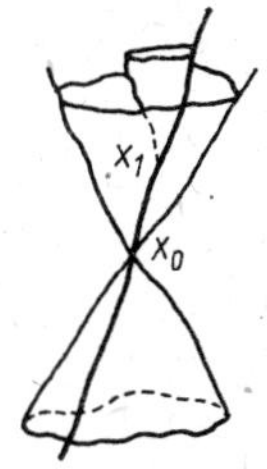

Fig. 29.14

Theorem 2. *If $x^k(\tau)$ is a ray, then $g_{kl} \dfrac{dx^k}{d\tau} \dfrac{dx^l}{d\tau} = 0.$*

Remark 4. As $\dfrac{dx^k}{d\tau}$ is transformed like a contravariant vector, the above proposition is meaningful (independent of the co-ordinate system).

29.5.4. Metric Geodesics

If M is a C^∞-manifold with a given fundamental tensor g_{kl}, then the Christoffel symbols $\{jk, l\}$ and $\begin{Bmatrix} j \\ kl \end{Bmatrix}$ have the same meaning as in the formulas (2) and (3) of Subsec. 24.1.5.

Definition 1. *If M is an n-dimensional C^∞-manifold with a Lorentz metric, then every C^∞-curve $x^a(s)$ that solves the system of differential equations*

$$\frac{\mathrm{d}^2 x^a}{\mathrm{d}s^2} + \begin{Bmatrix} a \\ bc \end{Bmatrix} \frac{\mathrm{d}x^b}{\mathrm{d}s} \frac{\mathrm{d}x^c}{\mathrm{d}s} = 0, \quad a = 1, \dots, n, \tag{1}$$

is called a metric geodesic (line).

Remark 1. This definition is an extension of Theorem 24.1.5 to manifolds with a Lorentz metric. Of course it is also possible to generalize the definition to manifolds with arbitrary fundamental tensors, but it is of no interest for us to do so.

Remark 2. If U is a local chart, and if $x_0 \in U$ and $\xi^k \in R_n$, then locally there exists exactly one metric geodesic $x(s) \sim x^k(s)$ such that $x(0) = x_0$ and $\dfrac{\mathrm{d}x^k}{\mathrm{d}s}(0) = \xi^k$.

Lemma. *If $x^k(s)$ is a metric geodesic, then there exists a real number c such that*

$$g_{kl}(x^r(x)) \frac{\mathrm{d}x^k}{\mathrm{d}s} \frac{\mathrm{d}x^l}{\mathrm{d}s} \equiv c. \tag{2}$$

Definition 2. *A metric geodesic is called time-like if $c < 0$, null geodesic if $c = 0$, and space-like (geodesic) if $c > 0$ in (2).*

Remark 3. If $x \in U$ (chart), then a contravariant vector ξ^k is called time-like (at the point x) if $g_{kl}(x)\xi^k\xi^l < 0$, null if $g_{kl}\xi^k\xi^l = 0$, and space-like if $g_{kl}(x)\xi^k\xi^l > 0$ (Fig. 29.15). Hence a metric geodesic is time-like if its tangent vector $\dfrac{\mathrm{d}x^k}{\mathrm{d}\tau}$ is time-like. Analogous statements are true for the null and space-like properties. Time-like vectors point to the interior of the characteristic conoid.

Remark 4. The time-like metric geodesics (and, with a suitable modification, the space-like ones too) can be derived from a variational principle, as was done in Subsec. 24.1.5. If $x(\tau) \sim x^k(\tau)$ is a time-like curve, i.e., if $\dfrac{\mathrm{d}x^k}{\mathrm{d}\tau}$ is always time-like, then let

$$L_{x_0 x_1} = \int_{\tau_0}^{\tau_1} \sqrt{-g_{kl}(x(\tau)) \frac{\mathrm{d}x^k}{\mathrm{d}\tau} \frac{\mathrm{d}x^l}{\mathrm{d}\tau}}\, \mathrm{d}\tau \tag{3}$$

be the arc length of $x(\tau)$ (Fig. 29.16). Further let $x_0 = x(\tau_0)$, $x_1 = x(\tau_1)$, and $\tau_0 < \tau_1$. It is easily seen that (3) is invariant under co-ordinate transformations $x'^k = x'^k(x^l)$ and under parame-

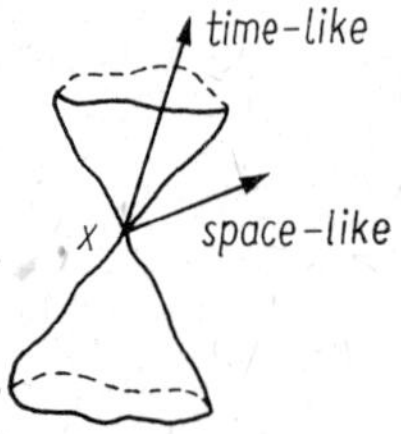

Fig. 29.15

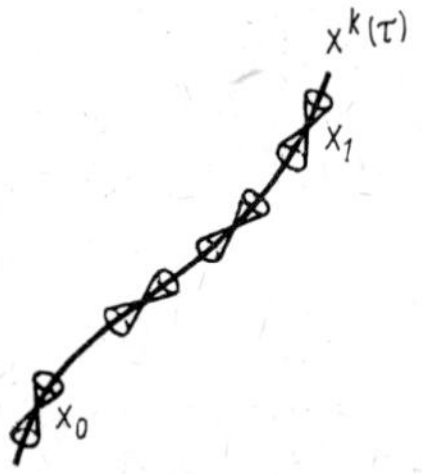

Fig. 29.16

ter transformations $\tau=\tau(s)$. s can be so chosen that (2) holds with $c=-1$. Then $L_{x_0x_1}=|s_0-s_1|$, i.e., s is the arc length. It is unique except for $s\to s+s_0$ and $s\to -s+s_0$; s_0 real. As in Subsec. 24.1.5., one can now look for curves $x^k(\tau)$ of extremal arc length between two points. This is a variational problem, and as a result one obtains (by analogy with Theorem 24.1.5) time-like metric geodesics.

Remark 5. Metric geodesics are identical with the solutions of the following system of differentiale equations (canonical form):

$$\frac{dx^k}{ds}=\frac{\partial H}{\partial \xi_k}, \quad \frac{d\xi_k}{ds}=-\frac{\partial H}{\partial x^k}, \quad H=\frac{1}{2}g^{kl}(x)\,\xi_k\xi_l. \tag{4}$$

Theorem. *The null geodesics coincide with the rays of Subsec.* 29.5.3.

29.5.5. Geodesically Convex Domains

Definition. *If M is a C^∞-manifold with a Lorentz metric, then an open set D in M is called geodesically convex (or extremal-convex) if any two points of D can be connected in D by exactly one metric geodesic.*

Remark 1. This is the analogue to the concept of convex domains in R_n.

Theorem. *Every point of M has a geodesically convex neighbourhood.*

Remark 2. The theorem is plausible but not trivial.

29.5.6. Metric Spaces

Lemma 1. *If M is a C^∞-manifold with the fundamental tensor g_{kl}, then $\Gamma^a_{bc}=\left\{ {a \atop bc} \right\}$ is an affine transformation in the sense of Theorem* 29.4.1.

Remark 1. $\left\{ {a \atop bc} \right\}$ is the Christoffel symbol of (29.1.5/3). A comparison of the affine geodesics of (29.4.5/1) and the metric geodesics of (29.5.4/1) now suggests the following definition.

Definition 1. *A metric space is an (n-dimensional orientable) C^∞-manifold with a fundamental tensor g_{kl} and the (symmetrical) affine transformation $\Gamma^a_{bc}=\left\{ {a \atop bc} \right\}$.*

Remark 2. Consequently, metric spaces are special affine spaces (this definition differs from the usual terminology). All the considerations on affine spaces of Sec. 29.4. can also be applied to metric spaces, especially the concept of the covariant derivative. Since affine and metric geodesics are identical, from now on we shall only speak of geodesics. The term "metric space" is customary, but should not be confused with the metric space of Subsec. 1.3.3. A Lorentz-metric space is a metric space with a Lorentz metric.

Theorem 1. $g_{ab;c}=g^{ab}{}_{;c}=\delta^a{}_{b;c}=0.$

Remark 3. This means that the fundamental tensor behaves like a constant under covariant differentiation.

Lemma 2 (*normal co-ordinates*). *If M is a Lorentz-metric space, and if $m\in M$, then there exist an atlas $\{V_k;\,F_k\}$ in the equivalence class of the associated atlases and a* *number l such that $m \in V_l$ and*

$$g_{ab}(x_0) = \begin{pmatrix} 1 \cdot & & 0 \\ & \cdot & \\ 0 & 1 & -1 \end{pmatrix}, \quad g_{ab'c}(x_0) = 0, \quad x_0 = F_l m .$$

Remark 4. It is easily seen that then $\left\{ {a \atop bc} \right\} (x_0) = 0$. This connects our considerations to the normal co-ordinates of Subsec. 29.4.2.

Definition 2. *A Lorentz-metric space is called metric-flat if, in a suitable atlas,* $g_{ab}(x) \equiv \begin{pmatrix} 1 \cdot & & 0 \\ & \cdot & \\ 0 & 1 & -1 \end{pmatrix}$.

Theorem 2. *A Lorentz-metric space is metric-flat if and only if it is flat.*

Remark 5. This is a supplement to Subsec. 29.4.7.

29.5.7. Riemann's Tensor and Related Tensors

In a metric space we write $R^a{}_{bcd} = B^a{}_{bcd}$ for the Riemann tensor of Subsec. 29.4.6. Further let

$$R_{abcd} = g_{ar} R^r{}_{bcd} .$$

Theorem 1.

$$R_{abcd} = -R_{abdc} = -R_{bacd} = R_{cdab} ,$$
$$R_{a[bcd]} = 0 = R_{abcd} + R_{acdb} + R_{adbc} ,$$
$$R_{ab[cd;e]} = 0 = R_{abcd;e} + R_{abde;c} + R_{abec;d} \quad \textit{(Bianchi's identity)} .$$

Remark 1. The last two formulas are derived from Subsec. 29.4.6.

Remark 2. In view of the symmetry relations, the number of independent components of R_{abcd} reduces to $\frac{1}{12} n^2 (n^2-1)$. For $n=2$, all the components of R_{abcd} can be expressed by R_{1212}.

Definition. *Ricci's curvature tensor is defined as* $R_{ab} = R^r{}_{arb}$ *or* $R^a{}_b = g^{ar} R_{br}$. *The curvature scalar is defined as* $R = R^a{}_a$. *Einstein's tensor is defined as* $G_{ab} = R_{ab} - \frac{1}{2} R g_{ab}$ *or* $G^a{}_b = g^{ar} G_{br}$.

Theorem 2. *One has* $R_{ab} = R_{ba}$ *(and hence also* $G_{ab} = G_{ba}$*). Moreover,* $G^a{}_{b;a} = 0$.

Remark 3. $G^a{}_{b;a} = 0$ is a fundamental equation. It forms the basis of the general theory of relativity.

Remark 4. The Riemann tensor and the Ricci tensor have the forms

$$R_{abcd} = \frac{1}{2} (g_{ac'bd} + g_{bd'ac} - g_{ad'bc} - g_{bc'ad})$$
$$+ g^{rs}(\{ac, r\} \{bd, s\} - \{ad, r\} \{bc, s\}) ,$$

$$R_{ab} = \left\{ {r \atop ar} \right\}_{\prime b} - \left\{ {r \atop ab} \right\}_{\prime r} + \left\{ {r \atop sa} \right\} \left\{ {s \atop rb} \right\} - \left\{ {r \atop ab} \right\} \left\{ {s \atop rs} \right\} . \tag{1}$$

Hence R_{abcd}, R_{ab}, R, and G_{ab} are differential expressions of second order, which are linear in the derivatives of second order, but not in g_{ab} and its first derivatives.

Einstein on the concepts of space and time: "which the physicists, under the stress of facts, had to get down from the Olympus of a-priori in order to repair them and to restore them to a usable condition."

30. General Theory of Relativity I (Fundamental Equations)

30.1. Variational Principles

30.1.1. Lagrangian Formalism

How to derive field equations from variational principles has been described in detail in Sec. 24.2. Therefore we restrict ourselves to a few remarks which link those previous considerations up to Chap. 29. Given a C^∞-manifold M which shall be transformed by a fundamental metric tensor g_{kl} to give a metric space in the sense of Def. 29.5.6/1. In contrast to the absolute space-time theories of Subsec. 24.2.1., however, g_{kl} is not given but rather a dynamic object. But g_{kl} is required to be always a 4-dimensional Lorentz metric in the sense of Subsec. 29.5.1. ($n=4$ in the notation used there). Hence it is general space-time theories which are considered. The investigations described in Sec. 24.2. were local ones. Consequently the change from the domain Ω fixed there to an arbitrary C^∞-manifold M does not raise any difficulties. Lagrange densities for curves are defined in the same way as in Def. 24.2.2(a). For time-like curves, (24.2.2/4) represents an example we have already made use of in Remark 29.5.4/4. Def. 24.2.2(b) is extended in an obvious way as follows. A Lagrange density for geometric objects is a scalar density of weight $w=1$ (cf. Subsec. 29.2.2.), having the form

$$L=L(x^k, \overset{r}{T}{}^{k\ldots l}{}_{p\ldots q}, \overset{r}{T}{}^{k\ldots l}{}_{p\ldots q,s}),$$

where $\overset{r}{T}{}^{k\ldots}{}_{q\ldots}$ are the components of geometric objects (tensor densities and affine transformations being sufficient), $r=1, \ldots, N$. The requirement that L be a scalar density of weight $w=1$ is identical with (24.2.2/2). If g_{kl} is a Lorentz metric, then, according to Theorem 29.3.1/2,

$$L=\sqrt{-g} \quad \textit{with} \quad g=\det g_{kl}$$

is a Lagrange density; cf. (24.2.2/5). Then (24.2.2/8) is also an example of a Lagrange density, G^{kl} and H_{kl} being tensors as defined there. The Lagrangian formalism can be taken over from 24.2.3. without any change. In particular the Euler-Lagrange equations (24.2.3/6) are applicable.

30.1.2. Einstein's Equations

Given a 4-dimensional manifold M, a Lorentz metric g_{kl}, and an affine transformation Γ^k_{lm} on M. Let Λ be a real number, then

$$L=\sqrt{-g}\,(R-2\Lambda) \quad \text{with} \quad R=g^{kl}(\Gamma^a_{ka,l}-\Gamma^a_{kl,a}+\Gamma^a_{kb}\Gamma^b_{la}-\Gamma^a_{kl}\Gamma^b_{ab})$$

is a Lagrange density with the dynamic objects g^{kl} and Γ^a_{bc}: from Theorem 29.4.6 and the usual contraction procedures it follows that R is a scalar.

Theorem. *The Euler-Lagrange equations* (24.2.3/6) *read*

$$\Gamma^a_{bc} = \begin{Bmatrix} a \\ bc \end{Bmatrix}, \quad G^{kl} + \Lambda g^{kl} = 0 . \tag{1}$$

Remark 1. $\begin{Bmatrix} a \\ bc \end{Bmatrix}$ is the Christoffel symbol used in Subsec. 24.1.5. and G^{kl} denotes the Einstein tensor as defined by Def. 29.5.7. This skilful decoupling of the Lorentz metric g_{kl} and the affine transformation Γ^a_{bc} is called Palatini's method. Naturally it is hoped that the "correct" affine transformation Γ^a_{bc} is then established automatically, so that M is a metric space in the sense of Def. 29.5.6/1. Then, according to Subsec. 29.5.7., R is the usual curvature scalar.

Remark 2. The second set of equations in (1), i.e.,

$$R^{kl} - \frac{1}{2} Rg^{kl} + \Lambda g^{kl} = 0 \tag{2}$$

are the famous Einstein field equations for the vacuum. With $\Lambda = 0$, these equations (including the energy-momentum tensor referred to in Remark 3) were set up by Einstein and Hilbert almost simultaneously in 1915. The elegant derivation from a variational principle is due to Hilbert. Later on Einstein added the term Λg^{kl} with the cosmological constant Λ in order to avoid (as he believed) undesirable cosmological consequences. He believed in a static (time-invariable) universe (and so did, presumably, all the physicists of those days). Formula (2) with $\Lambda = 0$ did not lead to this result. After Hubbles' discoveries, which suggested an expanding universe, Einstein regarded the introduction of the cosmological term Λg^{kl} as a serious error.

Remark 3. In the general case, Einstein's equations read

$$R^{kl} - \frac{1}{2} Rg^{kl} + \Lambda g^{kl} = \varkappa T^{kl} , \tag{3}$$

where T^{kl} denotes the energy-momentum tensor. It includes all external influences such as masses, electromagnetic fields etc. $\varkappa$ is a coupling constant. (29.5.7./1) shows that (3) is a set of 10 partial second-order differential equations. It contains the second derivatives of g_{kl} linearly, in contrast to g_{kl} and their first derivatives. (3) also shows that the geometry (given by g^{kl}) and matter (given by T^{kl}) are closely coupled: Mach's principle.

30.1.3. Einstein-Maxwell Field Equations

The question arises whether the general equations (30.1.2/3) can also be deduced from a variational principle. In special cases this is possible. For an electromagnetic field, the Maxwell-Lorentz equations had been derived from a variational principle in Subsec. 24.3.2. The function given in (24.3.2/1) is also a Lagrange density in the sense of Subsec. 30.1.1. (cf. Theorem 29.3.2 (a)). Thus, keeping the notation of Subsecs. 24.3.2. and 30.1.2., it suggests itself to combine the Lagrange densities used there,

$$L = \sqrt{-g} \left\{ -\frac{R}{2} + \Lambda - \varkappa g^{kl} g^{rs} \left[\frac{1}{2} F_{kr} F_{ls} - F_{kr} (A_{l's} - A_{s'l}) \right] \right\}$$

with

$$R = g^{kl} (\Gamma^a_{ka'l} - \Gamma^a_{kl'a} + \Gamma^a_{kb} \Gamma^b_{la} - \Gamma^a_{kl} \Gamma^b_{ab}) .$$

Again L is a Lagrange density. The dynamic objects are g^{kl}, Γ^a_{bc}, F_{kl}, and A_k.

Theorem. *The Lagrange equations* (24.2.3/6) *read as follows:*

$$F_{kl}=A_{k\prime l}-A_{l\prime k}, \quad (\sqrt{-g}\, F^{kl})_{\prime l}=0, \tag{1}$$

$$\Gamma^a_{bc}=\begin{Bmatrix} a \\ bc \end{Bmatrix}, \quad R^{kl}-\frac{1}{2}Rg^{kl}+\Lambda g^{kl}=\varkappa T^{kl} \tag{2}$$

where

$$T^{kl}=F^{kr}F^{l}{}_{r}-\frac{1}{2}g^{kl}F^{rs}F_{rs}\,. \tag{3}$$

Remark 1. The first set of equations in (2) again represent the "correct" affine transformations so that M again becomes a Lorentz-metric space. The equations (1) are the Maxwell equations of Subsec. 24.3.2. in curved space-time coordinates. (3) is the usual energy tensor for the Maxwell field, if the interpretation of (24.3.3/1) is used here, too. Then the Einstein-Maxwell equations are made up of the second sets of equations in (1) and (2): this is a coupled set of 14 partial differential equations for g^{kl} and F^{kl}, where F_{kl} has the structure of the first equation in (1). In particular it is seen that the Lorentz metric g_{kl} of the space is influenced by the eletrocmagnetic field.

Remark 2. As F^{kl} is antisymmetric and $\sqrt{-g}$ is a scalar density of weight $w=1$, it follows from Theorem 29.3.2 that

$$F^{kl}{}_{;l}=\frac{1}{\sqrt{-g}}(\sqrt{-g}\,F^{kl})_{\prime l}\,. \tag{4}$$

Observe that the contravariant vectors on both sides of (4) coincide in normal co-ordinates (cf. Lemma 29.5.6/2). Hence, the second set of equations in (1) can also be written as $F^{kl}{}_{;l}=0$. These are the Maxwell equations in the vacuum (without electric charges and currents). If charges and currents are present, then the current density j and the charge density ϱ referred to in Remark 24.3.3/3 are combined into the four-vector of current, σ^k. The Maxwell equations then read as follows:

$$F^{kl}{}_{;l}=\sigma^k\,. \tag{5}$$

Remark 3. If it is assumed that F^{kl} influences the Lorentz metric only insignificantly, then the Einstein-Maxwell equations can be decoupled: in this case (5) and the first equation of (1) are considered for a given Lorentz metric g_{kl} (and a given four-vector of current).

Newton, you found the only way which, in your age, was just about possible for a man of highest thought and creative power. (A. Einstein)

Einstein during a lecture in Pasadena, 1930.

30.1.4. Some Remarks Einstein Made on Relativity and Quantum Theory

In ref. [43], § 17.7, one may find quotations from fundamental papers by Einstein and Hilbert, which illustrate the creation of the general theory of relativity. Some of Einstein's most important and most famous papers on this subject are included in [84]. In a very impressive way these papers illustrate the final phase of the struggle for a definitive formulation of the fundamental equations of general relativity, i.e., of Eqs. (30.1.2/3) (with $\Lambda = 0$). Here we quote some passages. (We have changed part of the notation to fit it to that used in this book.) In "On General Relativity" (Prussian Academy of Sciences, Proceedings, Nov. 11th, 1915) Einstein wrote:

> "Like special relativity is founded on the postulate that its equations shall be covariant under linear orthogonal transformations, the theory to be described here rests on the postulate of the covariance of all sets of equations under transformations with the substitution determinant 1. Hardly anyone who has really comprehended this theory will be able to resist its fascination; it means a true triumph of the method of the general differential calculus established by Gauss, Riemann, Christoffel, Ricci, and Levi-Civita . . . On the basis of the statements made so far, it suggests itself to set up the field equations of gravitation in the form
>
> $$R^0_{\mu\nu} = \varkappa T_{\mu\nu} ,$$
>
> because we already know that these equations are covariant under any transformations with determinant 1."

(Here $R^0_{\mu\nu}$ means part of the Ricci tensor $R_{\mu\nu}$). In "On General Relativity (Supplement)" (Prussian Academy of Sciences, Berlin, Proceedings, Nov. 18th, 1915) Einstein refers to the preceding paper as follows:

> "In a recently published investigation I have shown how a theory of gravitational fields can be founded on Riemann's covariant theory of multi-dimensional manifolds. Here it shall now be shown that an even stricter logical construction of the theory can be achieved by introducing an additional hypothe-

sis, indeed a bold one, on the structure of matter . . . In what follows we assume that the condition $T^{\mu}_{\ \mu}=0$ is in fact generally satisfied . . . If we now stipulate the field equations of gravitation to read

$$R_{\mu\nu}=\varkappa T_{\mu\nu}\,,$$

we have thus obtained generally covariant field equations."

At last, in "The Field Equations of Gravitation" (Prussian Academy of Sciences, Berlin, Proceedings, Dec. 12th, 1915) Einstein reaches the final goal of his efforts for setting up the general field equations:

"In two recently published communications I have shown how to obtain field equations of gravitation which meet the postulate of general relativity, i.e. which, in their general form, are covariant with respect to arbitrary substitutions of the space-time variables. Here the course of development was as follows: initially I found equations which included Newton's theory as an approximation and were covariant with respect to any substitutions with determinant 1. Then I found out that, if the scalar of the energy tensor of "matter" vanishes, there are generally covariant equations corresponding to those mentioned above. The co-ordinate system then had to be specialized by the simple rule that $\sqrt{-g}$ must become 1, which leads to an eminent simplification of the equations of the theory. Here, however, as mentioned above, it was necessary to introduce the hypothesis that the scalar of the energy tensor of matter vanishes. But lately I discovered that it is possible to do without that hypothesis on the energy tensor of matter if the energy tensor of matter is inserted in the field equations in a way slightly different from that stated in my previous communications . . . We set

$$R_{im}=\varkappa\left(T_{im}-\frac{1}{2}\,g_{im}T\right), \tag{1}$$

where $g^{\varrho\sigma}T_{\varrho\sigma}=T^{\sigma}_{\ \sigma}=T$ ... This at last completes the general theory of relativity as a logical system. The postulate of relativity in its most general form, which makes the space-time co-ordinates physically insignificant parameters, leads, of compelling necessity, to a well-determined theory of gravitation, which explains the advance of perihelion of the Mercury."

It is not difficult to see that (1) is identical with (30.1.2/3) for $\Lambda=0$. In "Geometry and Experience" (Prussian Academy of Sciences, Berlin, Proceedings, Feb. 3rd, 1921), Einstein comments on the role of mathematics as follows.

"Above all other sciences, mathematics is held in high estimation for one reason: its theorems are absolutely certain and indisputable, whereas those of all other sciences are, to some degree, disputed and always in danger of being invalidated by recently discovered facts ... But that high estimation of mathematics, on the other hand, rests on the fact that it is also mathematics that provides the exact natural sciences with a measure of certainty which they could not achieve without mathematics . . . As far as the theorems of mathematics refer to reality they are not certain, and as far as they are certain they do not refer to reality . . . The reason why I put a special significance on this conception of geometry described above is that without this conception it would have been impossible for me to establish the theory of relativity. For without it the following consideration would have been impossible: in a frame of reference which rotates with respect to an inertial frame, owing to Lorentz contraction the laws of placement for rigid bodies do not correspond to the rules of Euclidean geometry; consequently, if non-inertial frames are allowed as equivalent frames Euclidean geometry must be abandoned."

The following quotations from Einstein have been taken from [13], and were extracted from Einstein's personal letters.

"The actual object of my research was always the simplification and unification of the physical theoretical system. I achieved this object satisfactorily for the macroscopic phenomena, but not for the phenomena of quanta and the atomistic structure. I hold that the modern quantum theory, too, in spite of considerable successes, is still far from a satisfactory solution of the latter set of problems." (1932)

"Coming from the sceptical empiricism of, say, Mach's school, by the problem of gravitation I was made a believing rationalist, i.e., one who seeks the only reliable source of truth in mathematical simplicity." (1938)

"Belief in the conceivability of reality by something logically simple and coherent ... It seems hard to get wise to the Lord's game. But that he plays at dice or makes use of 'telepathic' means (as is suggested to him by the present quantum theory) I cannot believe for a minute. (1942)

"The more one hunts after the quanta, the better they hide." (1924)

In a talk with E. Straus (Einstein's assistant from 1944 to 1948), Einstein said:

"What really intrigues me to know is whether God could have made the world differently; that means whether the requirement for logical simplicity leaves any freedom at all."

Finally we mention the following version of the last quotation:

"I want to know how God created this world. I am not interested in this or that phenomenon, in the spectrum of this or that element. I want to know *His* thoughts, the rest are details".

Remark 1. Hilbert found the general field equations (1) (to which (30.1.2/3) corresponds with $\Lambda=0$) almost simultaneously with Einstein in November, 1915. The deduction of (30.1.2/3) from a variational principle, as was indicated in Subsec. 30.1.2., corresponds to Hilbert's approach. Hilbert always pointed out that the priority is due to Einstein, and that his investigation founded on Einstein's earlier work. Nevertheless it is strange that Hilbert's important feat is always referred to only on passing (as was done also here indeed).

Remark 2. "If Raphael, returned to life, were to paint a modern *School of Athens* of physicists, I think he would include Einstein with Galileo, Newton, and Maxwell pointing toward the heavens, while Faraday and Rutherford would point toward the earth" (E. Segrè, [57, p. 78]).

30.2. The Energy-Momentum Tensor

30.2.1. Killing Vectors and Laws of Conservation

The object of Sec. 30.2. is to investigate the tensor T^{kl} and the coupling constand $\varkappa$ from (30.1.2/3) more closely. Some preliminary considerations are useful for that purpose. From (30.1.2/3) and Theorem 29.5.7/2 it follows that

$$T^{kl}=T^{lk} \quad \text{and} \quad T^{kl}{}_{;l}=0 \tag{1}$$

are necessary conditions for the energy-momentum tensor.

Definition. *The contravariant vector K^l is called a Killing vector if $K^r{}_{;s}+K^s{}_{;r}=0$ holds for all r and s.*

Remark 1. As was stipulated previously, we speak of vectors rather than (more exactly) of vector fields.

Theorem (*Law of conservation*). *If the conditions* (1) *are satisfied and K^r is a Killing vector, then*

$$(\sqrt{-g}P^k)_{,k}=0 \quad \textit{where} \quad P^k=T^{kl}K_l\,.$$

Remark 2. $P^k{}_{;k}=0$ can be verified immediately. Then the proposition follows from Theorem 29.3.2(b) and a consideration using local co-ordinates. This differential law of conservation can easily be transformed into an integral one by utilizing the Gaussian theorem 9.3.1/2.

Remark 3. Non-trivial Killing vectors thus lead to (differential and integral) laws of conservation. The existence of non-trivial Killing vectors, however, is always connected with symmetrics in the given metric space. An arbitrary Lorentz-metric space does not have any non-trivial Killing vectors. There are at most 10 linearly independent Killing vectors (at most $\frac{n(n+1)}{2}$ vectors in an n-dimensional metric space). The maximum number 10 is assumed if and only if the space has a constant curvature, i.e., if

$$R_{abcd}=\frac{R}{12}(g_{ac}g_{bd}-g_{bc}g_{ad})\,.$$

Remark 4. In the Minkowskian space, the 10 linearly independent Killing vectors are given by

$$\xi_a=c_a+\varepsilon_{ab}x^b, \quad \varepsilon_{ab}=-\varepsilon_{ba}\,,$$

where c_a and ε_{ab} are constants.

30.2.2. The Covariance Principle

If at some point of a 4-dimensional Lorentz-metric space normal co-ordinates in the sense of Lemma 2 and Remark 4 in Subsec. 29.5.6. are chosen, then at that point covariant derivatives become partial ones, and in a neighbourhood of this point one has approximately the same conditions as in the Minkowskian space. Thus it suggests itself to regard normal coordinates as approximately representing local inertial frames in the sense of the special theory of relativity; cf. Subsec. 25.1.1. This suggests the following approach to the derivation of physical laws within the general theory of relativity.

Comma-semicolon rule (*covariance principle*). *Find the special-relativistic formulation of those physical laws which have nothing to do with gravitation, and then substitute covariant derivatives for partial ones (i.e., semicolons for commas).*

Remark 1. The Maxwell equations $F^{kl}{}_{,l}=0$ of (24.3.2/3) and $F^{kl}{}_{;l}=0$ of Remark 30.1.3/2 represent one example.

Remark 2. In the comma-semicolon rule, the same difficulties occur as in the quantization rule of Subsec. 28.3.1.: the translation is done from a commutative range to a non-commutative one. In the case of the quantization rule, (commutative) real numbers are transformed into (non-commutative) operators, while in the case of the comma-semicolon rule (commutative) partial derivatives are transformed into (non-commutative) covariant ones. Thus in differential expressions of second and higher orders one has a great number of variants. To choose the "correct" one is a matter of physical intuition: it is not possible indeed to develop a higher theory from a lower one in a purely deductive way.

Remark 3. The above considerations and Remark 30.2.1/4 show that the differential laws of conservation of the special relativity theory hold approximately in the general theory of relativity. The basis of this was the existence of Killing vectors and the requirement that $T^{kl}{}_{,l}=0$ in the Minkowskian space. In the sense of the rule stated above, $T^{kl}{}_{,l}=0$ is translated into $T^{kl}{}_{;l}=0$. Thus a physical motive for (30.2.1/1) has been found: it is desired that the differential laws of conservation of the special theory of relativity may hold approximately in the general theory of relativity.

30.2.3. Energy-Momentum Tensor for Ideal Liquids

Naturally, the covariance principle will also be extended to the energy-momentum tensor T^{kl}, where $\varkappa$ in (30.1.2/3) is a coupling constant to be determined later. It appears at least plausible to treat world lines of particles (e.g., galaxies) in a curved four-dimensional space-time in the sense of the covariance principle according to the classical laws of ideal liquids, and to establish the corresponding energy tensor. Let V^k be the contravariant representation of the tangent vector of the time-like world line of a flowing particle (Fig. 30.1), then we normalize $V^k V_k = -1$. The energy tensor then reads

$$T^{kl} = (\mu + p) V^k V^l + p g^{kl} , \tag{1}$$

Fig. 30.1

where the energy density μ and the pressure p are scalar functions (μ includes the mass density ϱ). As was already indicated, this all happens in a 4-dimensional Lorentz-metric space with g_{kl} as a fundamental tensor. From (30.1.2/3) and $T^{kl}{}_{;l}=0$ one obtains

$$R^{kl} - \frac{1}{2} R g^{kl} + \Lambda g^{kl} = \varkappa T^{kl} , \tag{2}$$

$$\mu_{;l} V^l + (\mu + p) V^l{}_{;l} = 0, \qquad V_l V^l = -1 , \tag{3}$$

$$(\mu + p) V^k{}_{;l} V^l + (g^{kl} + V^k V^l) p_{;l} = 0 . \tag{4}$$

This fundamental system underlies many local and cosmological considerations.

30.2.4. Comparison with Newton's Theory of Gravitation

The tensor T^{kl} in (30.1.2/3) is established on the basis of classical energy-momentum tensors according to the covariance principle. What remains is the determination of $\varkappa$. To this end, consider the Newtonian limiting case: for a given mass density ϱ in R_3, $T_{44} = \varrho c^2$, in the CGS-system, is the dominant component in the energy tensor (c is the velocity of light), where we suppose that there exist no electromagnetic fields etc. which make contributions to the energy tensor. Let the Lorentz metric be $g_{kl}=\eta_{kl}+f_{kl}$, where η_{kl} is the metric in the Minkowskian space (cf. (25.1.1/1)) and f_{kl} is a small perturbation. If this set-up is introduced in (30.2.3/2) for $\Lambda=0$, then an approximate calculation gives $\Delta f_{44}=-\varkappa\varrho c^2$, where Δ is the (three-dimensional) Laplace operator. The Newtonian gravitational potential Φ is known to be determined from $\Delta\Phi=4\pi G\varrho$, where G is the Newtonian constant of gravitation. It turns out that one has to set $\Phi=-\frac{c^2}{2}f_{44}$. Then a comparison of the two potential equations yields Einstein's constant

$$\varkappa=\frac{8\pi G}{c^4}=2\cdot 10^{-48}\,g^{-1}\,cm^{-1}\,s^2\,,$$

just a tiny value! For details refer to any serious textbook of General Relativity, e.g. [61], pp. 89–91.

30.3. Equations of Motion

30.3.1. Test Particles and Electromagnetic Waves

Newton's theory of gravitation or Maxwell's field equations still give no information of the equations of motion of particles in the corresponding fields. These equations must be postulated separately. See for instance the considerations on inertial frames as described in (25.1.1/2). Originally it was held that Einstein's field equations must (and can) also be complemented by independent equations of motion for test particles. Test particles are particles with a positive mass m which move under no forces in a given 4-dimensional Lorentz-metric space without influencing the latter. The stipulation was as follows: a test particle moves along a time-like geodesic line. Later on it was realized that, to some degree of necessity, this requirement follows from Einstein's equations. The argumentation, being physically plausible, though somewhat macabre from the mathematical point of view, may go along the following lines. Let $x^k(s)$ be the path of the particle, then one sets $V^k=\frac{dx^k}{ds}$ for $x^k=x^k(s)$ (point of the trajectory), and $V^k=0$ otherwise. (This is a contravariant vector distribution, as will be dealt with in Subsec. 32.3.2.) Here s is the curve length in the sense of Subsec. 29.5.4., so that $V^lV_l=-1$ is satisfied for $x^k=x^k(s)$. For $p=0$ as well as $\mu=\varrho=m$ one then obtains the energy-momentum tensor (30.2.3/1) in the sense of the covariance principle. Then (30.2.3/4) gives

$$\left(\frac{dx^k}{ds}\right)_{;l}\frac{dx^l}{ds}=0,\quad k=1,2,3,4\,. \tag{1}$$

According to Remark 29.4.5/1 these are (time-like) geodesic lines.

Axiom. (**a**) *Test particles move along a time-like geodesic line.*

(**b**) *Electromagnetic waves propagate along rays.*

(**c**) *World lines which describe physical processes including a transport of information are time-like or null-like at every point.*

Remark 1. Part (a) justifies the above considerations. Rays (=null geodesics) were dealt with in 29.5.3. and 29.5.4. Part (c) means that $g_{kl}\frac{\mathrm{d}x^k}{\mathrm{d}\tau}\frac{\mathrm{d}x^l}{\mathrm{d}\tau}\leqq 0$ is valid at every point $x^k=x^k(\tau)$ of the curve. Of course this all happens in a 4-dimensional Lorentz-metric space.

Remark 2. Part (b) must be interpreted in a large-minded way. Consider a flash of light at the point P in the Minkowskian space; it will propagate along straight lines (=rays) on the associated light cone towards infinity. Here the propagation of light is described by the Maxwell equations, cf. Subsec. 25.3.2. These solutions show that the Huyghenian property is satisfied, cf. Subsec. 19.3.4. In curved space-times the Huyghenian property is generally lost (cf. Chap. 33, where we shall refer to such problems in detail). A flash of light generated at the point P of a curved space-time will generally be recorded at all world points in the interior and on the surface of the associated characteristic conoid. Part (b) of the axiom can thus be understood to imply that a light ray propagates along a zigzag path whose sections are rays. The (mathematical) cause of this scattering effect lies in the curvature of space (Fig. 30.2.).

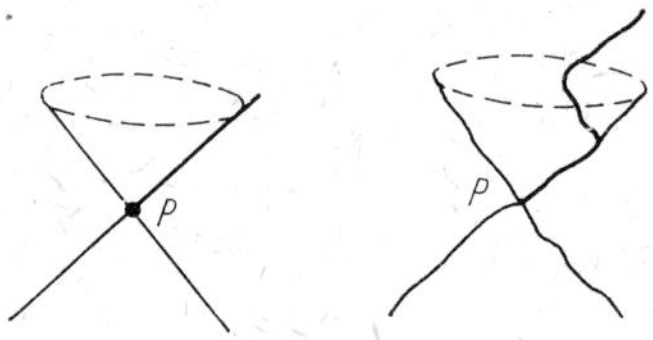

Fig. 30.2

Remark 3. Part (a) of the axiom can (almost) be derived from Einstein's field equations (for, speaking more strictly, from $T^{kl}{}_{;l}=0$). The question arises whether this startling trick will also work for other equations which, at a first glance, would be regarded as independent. If $T^{kl}{}_{;l}=0$ is applied to the energy tensor T^{kl} for the Maxwell field (30.1.3/3), then one obtains the Maxwell equations for the vacuum, $F^{kl}{}_{;l}=0$, as described in 30.1.3., as a generic result (i.e., in the general case). For details (in particular with respect to the "generic" character) refer to [43], pp. 471–473. Consequently, the Maxwell equations for the vacuum can also (almost) be derived from Einstein's equations.

Resting is rusting.
(proverb)

30.3.2. Proper Time and Twin Paradox

Axiom. *The proper time of an observer is equal to the curve length of his time-like world line.*

Remark 1. Thus what is considered is an observer (i.e., physical events) with the time-like world line $x^k(s)$, where s is the curve length as defined in Remark 29.5.4/4. Differences in s, that means time intervals, are uniquely determined by this method. Physically, one has the same situation as at the beginning of Subsec. 25.1.5.: an observer determines his proper time by a fixed physical procedure (e.g., by an atomic clock which is at rest with respect to him, thus always being at the same location as the observer himself). This fixes the timing which governs all the inorganic, organic and biological events which happen at rest with respect to him (including the aging of the observer).

Remark 2. The general theory of relativity defines: time = path, while a proverb claims that time = money. The conclusion is that path = money. This is paradoxical, for the bulk of the money is known to be earned by sitting.

Twin paradox. Now the twin paradox of Subsec. 25.2.1. finds a simple solution. Two world lines extending from P to Q along the paths 1 and 2, respectively (Fig. 30.3), generally have different curve lengths. Hence the corresponding proper-time differences are different, too. A pair of twins who are the same age at P may thus be of different ages when they meet again at Q. Let twin 1 be a test particle: it follows the free interplay of forces, and hence moves along a time-like geodesic line of length s_1 from P to Q. Twin 2 offers resistance, e.g., in the form of extended space trips; let its path from P to Q have the length s_2. In view of Remark 29.5.4/4, s_1 is extremal. It is easily verified that s_1 is maximal. Hence $s_2 < s_1$ (all the considerations are local). Consequently the dynamic twin 2 is younger than its inertial brother 1. The popular voice has known that all along: resting is rusting.

Fig. 30.3

30.4. Schwarzschild's Solution

30.4.1. The Birkhoff Theorem

In R_3 with the Cartesian co-ordinates x^1, x^2, x^3, and $r = \sqrt{(x^1)^2 + (x^2)^2 + (x^3)^2}$, let $\omega = \{(x^1, x^2, x^3) \mid r_i < r < r_a\}$. Here $0 < r_i < r_a \leqq \infty$. A Lorentz metric $g_{kl}(x^a)$ in the 4-dimensional space-time $\omega \times (-\infty, \infty)$ is called rotation-symmetric if $g_{kl}(x^a)$ depends on r and $x^4 = t$ alone.

Theorem (*Birkhoff's theorem*). *After choosing suitable co-ordinates in* $\omega \times (-\infty, \infty)$ *any rotation-symmetric solution* g_{kl} *of the Einstein equations in the vacuum,*

$$R^{kl} - \frac{R}{2} g^{kl} = 0, \tag{1}$$

can be represented as

$$ds^2 = (dx^1)^2 + (dx^2)^2 + (dx^3)^2 + \left[\left(1 - \frac{2M}{r}\right)^{-1} - 1\right] dr^2 - \left(1 - \frac{2M}{r}\right) dt^2 \tag{2}$$

where M is an arbitrary positive constant.

Remark 1. We use the common notation $ds^2 = g_{kl} dx^k dx^l$. It may be supposed that (1) is written in x^1, x^2, x^3, and $x^4 = t$. The manifold $\omega \times (-\infty, \infty)$ then consists of one chart provided with suitable coordinates $x^1, \ldots, x^4$. To make g_{kl} a Lorentz metric, we require that $2M < r_i < r_a \leqq \infty$ (Schwarzschild's exterior solution), cf. Fig. 30.4. Then $2M$ has the dimension of a length, and is called the Schwarzschild radius. What is remarkable is the fact that g_{kl} depends on r alone, but not on t. Thus one obtains (although this has not been required) a stationary solution of (1) (i.e., a solution independent of t).

Remark 2. If the polar coordinates r, ϑ, φ of Subsec. 12.4.1. are used, then (2) reads as follows:

$$ds^2 = \left(1 - \frac{2M}{r}\right)^{-1} dr^2 + r^2 d\vartheta^2 + r^2 \sin^2 \vartheta \, d\varphi^2 - \left(1 - \frac{2M}{r}\right) dt^2, \tag{3}$$

(cf. (12.4.1/2)). This is the usual form. It is simpler, but pretends singularities as $\vartheta \to 0$ and $\vartheta \to \pi$. Consequently a representation in polar co-ordinates requires several charts. Then (3) is a local representation for a suitable choice of the x^3-axis.

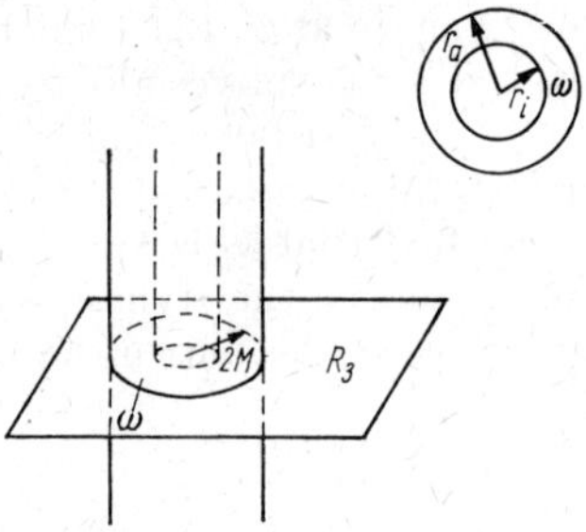

Fig. 30.4

Interpretation. As $r \to \infty$, ds^2 in (2) (or (3)) tends towards the metric of the Minkowskian space, to which we had attributed physical reality. Here x^1, x^2, x^3 were space co-ordinates and $x^4 = t$ was the time co-ordinate. Therefore it suggests itself to keep this interpretation in the empty Einstein space,

$$\{(x^1, x^2, x^3) \mid 2M < r < \infty\} \times (-\infty, \infty)$$

with the Lorentz metric (2). Then a comparison with Newton's mechanics as described in Subsec. 30.2.4. gives

$$M = m \frac{\mathrm{G}}{\mathrm{c}^2} \quad \text{in the CGS-system .} \tag{4}$$

Here m is the mass of a particle at the origin, G is the Newtonian constant of gravitation, and c is the velocity of light. Hence the interpretation is as follows: a particle of mass m located at the origin of R_3 generates a space-time in $\{(x^1, x^2, x^3) \mid 2M < r < \infty\} \times (-\infty, \infty)$, the metric of the space-time being given by (2). (x^1, x^2, x^3) are space co-ordinates and x^4 is a time co-ordinate (which must not, however, be confused with the proper time of particles moving within this space-time). For the earth one obtains $2M = 8.8$ mm, and for the sun, $2M = 3$ km. The Schwarzschild radius becomes interesting only for high-mass, but relatively small stars.

30.4.2. Eddington's Form of the Schwarzschild Solution

Let $\tilde{\mathfrak{M}}$ be a metric space in the sense of Def. 29.5.6/1, with the metric $\tilde{g}_{kl}$. If $\mathfrak{M}$ is an open connected proper subset of $\tilde{\mathfrak{M}}$, then according to 29.1.2. one may establish the induced topology on $\mathfrak{M}$ (Fig. 30.5). By a suitable restriction of the associated equivalence class of atlases in the sense of Def. 29.1.2 and by $g_{kl}(P) = \tilde{g}_{kl}(P)$ for $P \in \mathfrak{M}$, $\mathfrak{M}$ is made a metric space. The equivalence class of permissible atlases for $\mathfrak{M}$, however, is much larger than the set of those atlases which result from the restriction of permissible atlases on $\tilde{\mathfrak{M}}$ to $\mathfrak{M}$. In particular one may imagine permissible atlases which become singular as the border $\partial\mathfrak{M} = \overline{\mathfrak{M}} \setminus \mathfrak{M}$ of

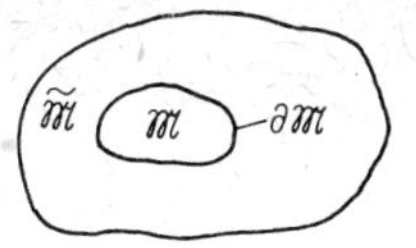

Fig. 30.5

$\mathfrak{M}$ is approached. Hence by suitable atlases (local charts) the origin of $\mathfrak{M}$ from $\tilde{\mathfrak{M}}$ can be completely obliterated. Now it is of interest to invert this questioning: given a metric space $\mathfrak{M}$ with the metric g_{kl}. The question is whether, in the sense stated above, $\mathfrak{M}$ can be embedded in a larger metric space $\tilde{\mathfrak{M}}$. The considerations show that one must be very cautious in judging this problem. An unsuitably chosen atlas in $\mathfrak{M}$ might pretend singularities which in reality do not exist. This is exemplified by the following investigations.

Eddington's metric. We consider the Schwarzschild metric (30.4.1/2) in $\mathfrak{M} = \{(x^1, x^2, x^3) \mid 2M < r < \infty\} \times (-\infty, \infty)$. (We write $\mathfrak{M}$ instead of M in order to avoid confusion with the Schwarzschild radius.) From now on we shall always use the form (30.4.1/3), where it is clear how the apparent singularities occurring as $\vartheta \to 0$ and $\vartheta \to \pi$ must be eliminated. Introducing new local coordinates in $\mathfrak{M}$ by

$$x^k = x'^k \quad \text{for} \quad k = 1, 2, 3 \quad \text{and} \quad x^4 = x'^4 \pm 2M \ln\left(\frac{r'}{2M} - 1\right),$$

one obtains $ds^2 = g'_{kl} dx'^k dx'^l$, where

$$g'_{kl} = \begin{pmatrix} 1 + \frac{2M}{r} & 0 & 0 & \pm\frac{2M}{r} \\ 0 & r^2 & 0 & 0 \\ 0 & 0 & r^2 \sin^2\vartheta & 0 \\ \pm\frac{2M}{r} & 0 & 0 & -\left(1 - \frac{2M}{r}\right) \end{pmatrix}. \tag{1}$$

If $x^4 = x'^4 + \ldots$, then $g'_{14} = +\frac{2M}{r}$, and if $x^4 = x'^4 - \ldots$, then $g'_{14} = -\frac{2M}{r}$. The apparent singularities which occur as $\vartheta \to 0$ and $\vartheta \to \pi$ can again be eliminated by changing to the Cartesian co-ordinates x^k. It is now easily seen that the Eddington metric can be extended from $\mathfrak{M}$ to

$$\tilde{\mathfrak{M}} = \{(x^1, x^2, x^3) \mid 0 < r < \infty\} \times (-\infty, \infty)$$

($r = 2M$ is not a singularity), and that $\tilde{g}_{kl} = g'_{kl}$ is a solution in $\tilde{\mathfrak{M}}$ of Einstein's equation for the vacuum $R'^{kl} - \frac{R'}{2} g'^{kl} = 0$.

Interpretation. In the transformation $x'^k = x'^k(x^l)$, it is only the time co-ordinate t which is transformed into t'. The space co-ordinates x^1, x^2, x^3 remain unchanged. Hence we can take over the interpretation from Subsec. 30.4.1. and extend it from $\mathfrak{M}$ to $\tilde{\mathfrak{M}}$.

30.5. The Classical Effects of the General Theory of Relativity[1])

30.5.1. Planetary Motion

In Sec. 12.4. we had dealt with the planetary motion as described by Newton's theory of gravitation. We shall now investigate the same problem using Einstein's theory. The aspects we are interested in are the differences from Newton's theory which occur in this case.

The model. Let the sun be fixed at the origin of the R_3. According to the interpretation given in Subsec. 30.4.1. it generates (in the otherwise empty space) a space-time metric in $\mathfrak{M} = \{(x^1, x^2, x^3) \mid 2M < r < \infty\}$, with the Schwarzschild metric (2) or (3) stated in Subsec. 30.4.1. Here $2M = 3$ km. Planets are treated as test particles which do not influence this geometry. According to Axiom 30.3.1(a), planets thus move along time-like geodesic lines in $\mathfrak{M}$ (with the Schwarzschild metric). Here x^1, x^2, x^3 are interpreted as space co-ordinates of the real 3-dimensional space. Using the curve length s (proper time), one has the following set of differential equations to be solved,

$$\frac{\mathrm{d}^2 x^k}{\mathrm{d}s^2} + \begin{Bmatrix} k \\ lm \end{Bmatrix} \frac{\mathrm{d}x^l}{\mathrm{d}s}\frac{\mathrm{d}x^m}{\mathrm{d}s} = 0, \quad g_{kl}\frac{\mathrm{d}x^k}{\mathrm{d}s}\frac{\mathrm{d}x^l}{\mathrm{d}s} = -1\,, \tag{1}$$

If again polar coordinates r, ϑ, φ, and t are introduced as functions of s, then the following set of equations follows from (1) (without the last equation) and (30.4.1/3).

$$\frac{\mathrm{d}}{\mathrm{d}s}\left(R\frac{\mathrm{d}t}{\mathrm{d}s}\right) = 0 \quad \text{with} \quad R(r) = 1 - \frac{2M}{r}\,, \tag{2}$$

$$\frac{\mathrm{d}}{\mathrm{d}s}\left(R^{-1}\frac{\mathrm{d}r}{\mathrm{d}s}\right) + \frac{1}{2}\frac{\mathrm{d}R}{\mathrm{d}r}\left(\frac{\mathrm{d}t}{\mathrm{d}s}\right)^2 + \frac{1}{2}\frac{\mathrm{d}R}{\mathrm{d}r}\left(\frac{\mathrm{d}r}{\mathrm{d}s}\right)^2 - r\left(\frac{\mathrm{d}\vartheta}{\mathrm{d}s}\right)^2 - (r\sin^2\vartheta)\left(\frac{\mathrm{d}\varphi}{\mathrm{d}s}\right)^2 = 0\,, \tag{3}$$

$$\frac{\mathrm{d}}{\mathrm{d}s}\left(r^2\frac{\mathrm{d}\vartheta}{\mathrm{d}s}\right) - r^2\sin\vartheta\cos\vartheta\left(\frac{\mathrm{d}\varphi}{\mathrm{d}s}\right)^2 = 0\,, \tag{4}$$

$$\frac{\mathrm{d}}{\mathrm{d}s}\left(r^2\sin^2\vartheta\,\frac{\mathrm{d}\varphi}{\mathrm{d}s}\right) = 0\,. \tag{5}$$

If the initial values (12.4.1/6) as well as $t(0)$ and $\dot{t}(0)$ are given, Eqs. (2)–(5) have a unique solution (at least locally). Here $\dot{\vartheta} = \frac{\mathrm{d}\vartheta}{\mathrm{d}s}$ etc. This is the analogue to Eqs. (3)–(6) of Subsec. 12.4.1.

Plane orbits and law of areas. From (4), the initial conditions $\vartheta(0) = \frac{\pi}{2}$ and

[1]) We quote few sentences from [53, p. 9]: "Special relativity and special relativistic quantum theory have been checked literally billions of times. But for many years only small and poorly measured effects within the solar system indicated that general relativity gave better answers than combining its special relativistic and Newtonian limits ad hoc. Today, more accurate measurements within the solar system, the tentative success of general relativistic models for white dwarf stars and pulsars, the possible discovery of black holes and of the gravitational radiation predicted by general relativity, and the tentative success of general relativistic cosmology have given general relativity a somewhat firmer empirical foundation."

$\frac{d\vartheta}{ds}(0)=0$ as well as from the uniqueness of the solution, one finds that $\vartheta\equiv\frac{\pi}{2}$. Then from (5) it follows that $r^2\frac{d\varphi}{ds}\equiv h$. In other words, Theorem 12.4.2/2 is also true within Einstein's theory.

Advance of perihelion. As in Subsec. 12.4.1., $s=0$ shall be chosen to define that point of the planetary orbit which is next to the sun (perihelion). As in Subsec. 12.4.3. we use φ instead of s as a parameter, setting $r(s)=\frac{1}{\varrho(\varphi)}$. Then (using the last equation in (1)) one obtains

$$\frac{d^2\varrho}{d\varphi^2}+\varrho=\frac{M}{h^2}+3M\varrho^2\,. \tag{6}$$

Apart from the term $3M\varrho^2$, this is just the differential equation for $\varrho(\varphi)$ as given in Subsec. 12.4.3., which had the solution (12.4.3/1),

$$\frac{M}{h^2}\,(1+e\cos\varphi),\quad 0<e<1\,. \tag{7}$$

This was an ellipse with the perihelion given by $s=0$ and $\varphi(0)=0$. For the Mercury, the planet next to the sun, one has

$$3M\varrho^2/\frac{M}{h^2}\sim 7{,}7\cdot 10^{-8}\,.$$

For the other planets this dimensionless number is even smaller. Thus the relativistic correction $3M\varrho^2$ in (6) turns out to be downright tiny. It suggests itself to look for an approximate solution of (6) which has the structure of (7). To this end, $\cos\varphi$ in (7) is replaced by $\cos\varepsilon\varphi$, where ε shall be a number close to 1. The best value of ε yields the approximate solution

$$\varrho(\varphi)\sim\frac{M}{h^2}\left[1+e\cos\left(1-3\,\frac{M^2}{h^2}\right)\varphi\right] \tag{8}$$

of (7). As stipulated above, $\varphi=0$ is a perihelion. The next perihelion is passed when $\varphi\left(1-\frac{3M^2}{h^2}\right)=2\pi$, i.e., for $\varphi\sim 2\pi\left(1+\frac{3M^2}{h^2}\right)$. Consequently a perihelion motion of $\Delta=6\pi\,\frac{M^2}{h^2}$ results. Hence the Kepler-Newton ellipse becomes a (slowly rotating) rosette (Fig. 30.6). The corresponding time-like geodesic line in $\mathfrak{M}$ is a spiral in the t-direction, whose projection yields the rosette. The table below compares these theoretical data with experimental ones. The figures indicate the advance of perihelion in angular second per century. (Mutual perturbing effects of the planets have already been eliminated here.)

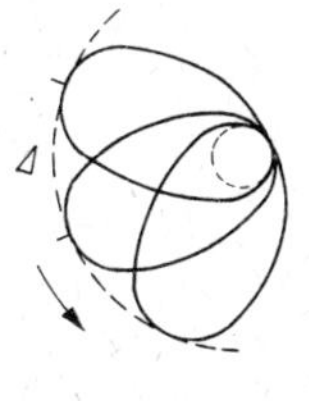

Fig. 30.6

	Theory	Observation
Mercury	43.03	43.11 ± 0.45
Venus	8.6	8.4 ± 4.8
Earth	3.8	5.0 ± 1.2
Mars	1.35	

30.5.2. Deflection of Light

The problem to be discussed now is the behaviour of electromagnetic waves in the vicinity of heavy masses.

Model. Again we choose the same space-time $\mathfrak{M}$ as in Subsec. 30.5.1. with the Schwarzschild metric (2) or (3) given in Subsec. 30.4.1. Again let $2M = 3$ km (that means the heavy mass is the sun). According to Axiom 30.1.3(b) electromagnetic waves (with possible scattering effects being neglected) propagate along rays (null-geodesics). x^1, x^2, x^3 are again interpreted as space co-ordinates of a real 3-dimensional space. Then one has to consider the set of differential equations (30.5.1/1), where in the last equation 0 must be substituted for -1. A transformation to polar co-ordinates leaves the equations (2)–(5) of Subsec. 30.5.1. unchanged.

Deflection of light. With the initial conditions $\vartheta(0) = \frac{\pi}{2}$ and $\frac{\mathrm{d}\vartheta}{\mathrm{d}s}(0) = 0$, one again obtains $\vartheta(s) \equiv \frac{\pi}{2}$. This means that rays of light likewise propagate within a fixed plane, which is the x^1,x^2-plane in the present case. With the use of $g_{kl} \frac{\mathrm{d}x^l}{\mathrm{d}s} \frac{\mathrm{d}x^k}{\mathrm{d}s} = 0$, the analogue to (30.5.1/6) now reads as follows:

$$\frac{\mathrm{d}^2\varrho}{\mathrm{d}\varphi^2} + \varrho = 3M\varrho^2 , \tag{1}$$

where $\varrho = \frac{1}{r}$ and φ have the same meaning as in 30.5.1. The approximate equation $\frac{\mathrm{d}^2\tilde{\varrho}}{\mathrm{d}\varphi^2} + \tilde{\varrho} = 0$ has the solution $\frac{1}{\tilde{r}} = \tilde{\varrho} = \frac{\cos\varphi}{R}$; and hence $x^1 = \tilde{r}\cos\varphi \equiv R$. Insertion of this zero-order approximation in the right-hand side of (1) gives the approximate solution

$$x = R - \frac{M}{R} \frac{x^2 + 2y^2}{\sqrt{x^2 + y^2}}, \quad x = x^1, \quad y = x^2 ,$$

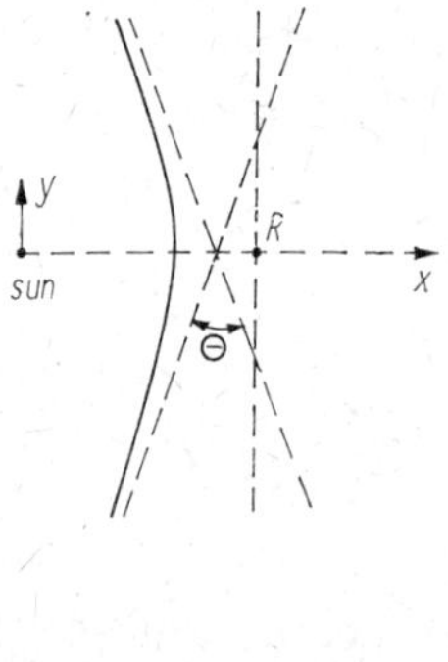

Fig. 30.7

with the asymptotes $x \sim R - \frac{2M}{R}|y|$. Let R be the radius of the sun, then the angle between the two asymptotes is $\Theta = 1.75''$ (angular seconds), cf. Fig. 30.7. Consequently, light is deflected by heavy masses. This effect can be verified experimentally by comparing night photographs of a celestial zone with corresponding photographs taken during a total solar eclipse (Fig. 30.8). The measured values lie between $1.28''$ and $2.71''$. The covering of strong radio sources in the universe by the sun yields a quantitatively better confirmation of this effect (with a deviation of less that 10 %).

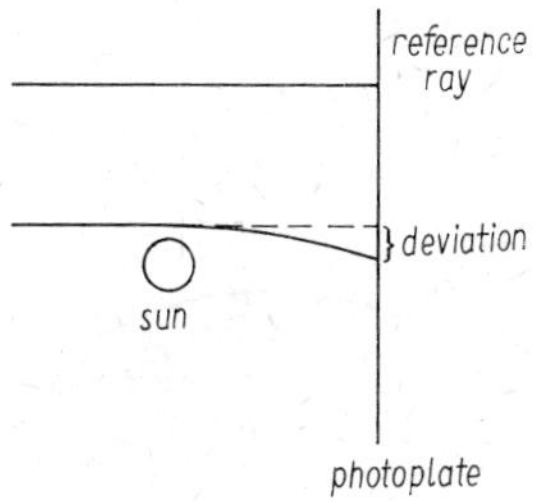

Fig. 30.8

30.5.3. Red Shift in the Gravitational Field

Frequency shifts. Consider a 4-dimensional Lorentz-metric space with the stationary metric

$$ds^2 = g_{\alpha\beta} dx^\alpha dx^\beta + g_{44}(dt)^2, \quad \alpha = 1, 2, 3 \quad \text{and} \quad \beta = 1, 2, 3,$$

where the sum has to be taken over doubly occurring indices (from 1 to 3). "Stationary" means that g_{kl} depends on $x = (x^1, x^2, x^3)$, but not on $t = x^4$. Then from Subsec. 29.5.3. one may conclude that, if $\omega(x, t) = 0$ is a characteristic surface, this is also true for $\omega(x, t+c) = 0$ for any real number c: invariance of translation in the t-direction. But then characteristic conoids, and hence also rays (cf. Subsec. 29.5.3.) are likewise translation-invariant in the t-direction. According to Axiom 30.3.1(b), rays are world lines of electromagnetic waves. $x = (x^1, x^2, x^3)$ is interpreted as a position in a real 3-dimensional space, and t as a co-ordinate time. Hence the world line of a non-moving observer x_0 is a straight line passing through x_0 in the direction parallel to the t-axis (Fig. 30.9). Then for the proper time T_0 of x_0 one has $dT_0^2 = -g_{44}(x_0)dt^2$. Consequently, $\Delta T_0 = \sqrt{-g_{44}(x_0)}\, \Delta t$. An analogous result is obtained for x_1. Let the observer x_0 trip an emission of light with N cycles in the co-ordinate time interval Δt. The observer x_1 also records this light as N cycles in Δt. Then the frequency ν_0 at x_0, which must of course be referred to the proper time, is given by

$$\nu_0 = \frac{N}{\Delta T_0} = \frac{N}{\sqrt{-g_{44}(x_0)}\, \Delta t},$$

and an analogous relation holds for ν_1 at x_1. This gives

$$\nu_1 = \nu_0 \sqrt{\frac{g_{44}(x_0)}{g_{44}(x_1)}}. \tag{1}$$

In view of $|g_{44}(x_1)| > |g_{44}(x_0)|$, it follows that a red shift of the spectrum, i.e., a frequency reduction, occurs.

Schwarzschild metric. The above conditions are satisfied for the Schwarzschild metric (30.4.1/3), with $g_{44}(x) = -\left(1 - \frac{2M}{r}\right)$. Considering that in real cases $\frac{M}{r}$ is a very small value, one obtains (approximately)

$$\frac{\Delta\nu}{\nu_0} \doteq M\left(\frac{1}{r_0} - \frac{1}{r_1}\right) \quad \text{with} \quad 2M < r_0 < r_1 < \infty \quad \text{and} \quad \Delta\nu = \nu_0 - \nu_1$$

as a red shift in the Schwarzschild metric.

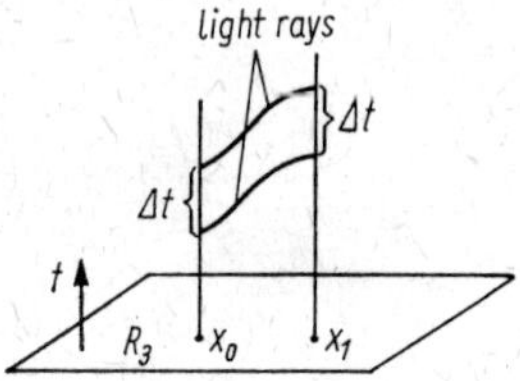

Fig. 30.9

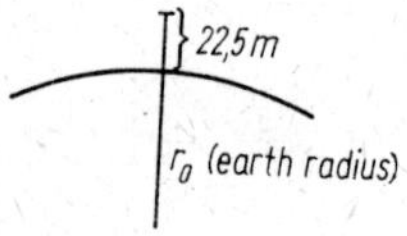

Fig. 30.10

Test (Fig. 30.10). Choosing the radius of the earth to be r_0, and setting $r_1 = r_0 + 22.5$ m, one obtains $\frac{\Delta\nu}{\nu_0} = 2.5 \cdot 10^{-15}$. It is past belief, but yet this effect can be verified by experiment. The result,

$$\frac{\Delta\nu}{\nu_0} = 2.5 \cdot 10^{-15}\,(1.05 \pm 0.10)$$

at the same time yields one of the best quantitative confirmations of the general theory of relativity.

> Now, my suspicion is that the universe
> is not only queerer than we suppose,
> but queerer than we *can* suppose.
> (J. B. S. Haldane, "Possible Worlds")

31. General Theory of Relativity II (Singularities, Black Holes, Cosmology)

31.1. Singular Manifolds

31.1.1. Criteria

The modern theory of relativity is characterized by deep-going studies on singularities in curved space-times. In Secs. 31.1. and 31.2. we shall try to give an impression of these studies. Here we restrict ourselves to a few (as we hope) typical concepts. For details, necessary more precise definitions and a great number of other fundamental concepts we refer to [24]. This book has also been taken as a basis for the following presentations. Further we refer to Chap. 33, which can be considered as a continuation of the present considerations. A description of the present state of the theory of relativity has been given in [81]. As in Chap. 30, M is always assumed to be a 4-dimensional Lorentz-metric space: a (curved) space-time. We ask for mathematical and physical criteria according to which such a space-time can be evaluated.

Definition 1. *A space-time M is called time-like complete if every time-like geodesic and every null geodesic is part of a corresponding geodesic whose geodesic parameter runs through all real numbers. M is called singular if M is not time-like complete.*

Remark 1. In view of Theorem 29.4.5, the definition is meaningful. Consequently, the question is whether every time-like geodesic and every null geodesic can be so extended that its geodesic parameter s runs from $-\infty$ to $+\infty$. Such a question appears very meaningful, because s (after being normalized) is the proper time of test particles. If M is complete, then a test particle can thus travel over an infinitely long time without anything happening.

Remark 2. There are several other (in part more restrictive) concepts of completeness, which have been developed by Penrose, Hawking and Ellis. We refer to [24].

Remark 3. The Minkowskian space is time-like complete. Here the geodesics are identical with the straight lines. However, if one removes a point or a closed set ω from the Minkowskian space, then the remainder is a singular space-time (Fig. 31.1).

Fig. 31.1

Problem. The last example illustrates the problem encountered: can a given singular space-time M be diffeomorphically embedded into a larger, time-like complete space-time $\tilde{M}$? Speaking more precisely, what is desired is a time-like complete space-time $\tilde{M}$ and an open set Ω in $\tilde{M}$ with the following property: if, as described at the beginning of Subsec. 30.4.2., Ω is transformed into a space-time, then there is a diffeomorphism of M onto Ω in the sense of Subsec. 29.1.2. Generally this is not possible. This compels introducing new concepts.

Definition 2. *A space-time M is called locally inextendible if there is no open set ω of M with the following properties:*
1. *$\bar{\omega}$ (the closure of ω in M) is not compact,*
2. *ω can be diffeomorphically embedded into a space-time $\tilde{M}$ in such a way that $\bar{\tilde{\omega}}$ (the closure of the image $\tilde{\omega}$ of ω in $\tilde{M}$) is compact.*

M is called locally extendible otherwise.

Remark 4. Compact sets have been described in Subsec. 29.1.1. Roughly speaking, the definition is as follows: since $\bar{\omega}$ is not compact in M, ω borders on the boundary of M along the bold solid line, cf. Fig. 31.2 (this is not a precise formulation). If ω can be extended beyond this boundary into $\widetilde{M}$, then $\bar{\omega}$ is compact in $\widetilde{M}$. Then all the boundaries of ω lie in the interior of $\widetilde{M}$. The definition is in effect the precise formulation of this illustrative description.

Remark 5. If, for example, $g(x)=\det g_{kl}(x)\to 0$ as $x\in\omega$ tends towards the bold line segment in Fig. 31.2, then this shows that an extension beyond this part of the boundary is not possible.

Fig. 31.2

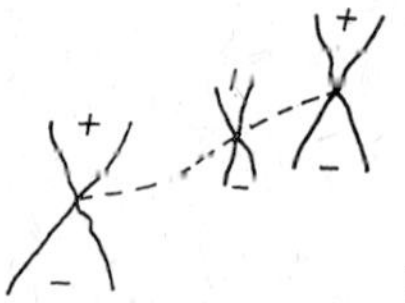

Fig. 31.3

Time-orientable space-times. As the case may be, we shall place additional, physically reasonable conditions on a space-time as well as on the energy-momentum tensors T^{kl}. Thus the condition that there shall be no time-like closed lines is physically reasonable (chronology condition). This simply means that a journey into one's own past shall be impossible. An essential concept is that of time orientability: a space-time M is called time-orientable if the local light cones (local characteristic conoids) can be subdivided in M in a consistent and continuous manner into a "future cone" + and a "past cone" − (Fig. 31.3). Locally (within one chart) this is always practicable (see also Subsec. 33.2.2.), whereas globally, i.e., on the whole space-time M, it is an essential additional requirement, which is physically highly reasonable.

Requirements imposed on T^{kl}. The energy-momentum tensor T^{kl} and the metric g_{kl} of the space-time M satisfy Einstein's equations (30.1.2/3). The following (mutually independent) requirements on T^{kl} are physically reasonable.
(*a*) (*Weak energy condition*): For all time-like vectors V_k, $T^{kl}V_kV_l\geqq 0$.
(*b*) (*Time-like convergence condition*): For all time-like vectors V_k, $R^{kl}V_kV_l\geqq 0$.
(*c*) (*Generic condition*): For every time-like or null geodesic, the tensor $V_{[k}R_{l]pq[r}V_{s]}V^pV^q$ is different from zero at one point at least of this geodesic. V_k is the tangent vector to the geodesic.

Remark 6. In each particular case we shall exactly indicate what kind of additional conditions are imposed on a space-time.

31.1.2. The Schwarzschild-Eddington-Kruskal Metric

Let us apply the considerations of Subsec. 31.1.1. to the Schwarzschild metric (30.4.1/3) in $\mathfrak{M}$ and to the Schwarzschild-Eddington metric (30.4.2/1) in $\widetilde{\mathfrak{M}}$. $\mathfrak{M}$ and $\widetilde{\mathfrak{M}}$ are the space-times of Subsec. 30.4.2. If the domain ω has the shape indicated in Fig. 31.4, then $\bar{\omega}$ is not compact in $\mathfrak{M}$, but it is compact in $\widetilde{\mathfrak{M}}$. Hence $\mathfrak{M}$ is locally extendible, and hence it is not complete. $\widetilde{\mathfrak{M}}$ is the extension of $\mathfrak{M}$. The curvature scalar in $\widetilde{\mathfrak{M}}$ is $R = \frac{48\,M^2}{r^6}$, which becomes singular as $r \to 0$. On the other hand, the proper time (measured in the CGS-system) of a time-like line that starts from a point $0 < r < 2M$ (the counting of the proper time is also started at this point) and moves in the positive t-direction is at most $1.54 \cdot 10^{-5}\,\frac{M}{M_\odot}$ sec. ($M_\odot = 1.5$ km is the Schwarzschild radius of the sun). From these two facts it follows that $\widetilde{\mathfrak{M}}$ is not time-like complete but cannot be extended beyond $r = 0$. Nevertheless it turns out that $\widetilde{\mathfrak{M}}$ is locally extendible.

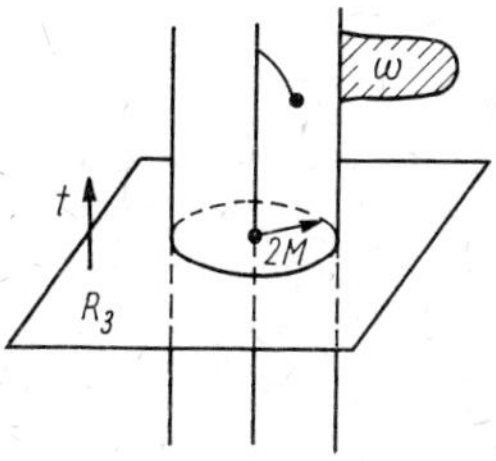

Fig. 31.4

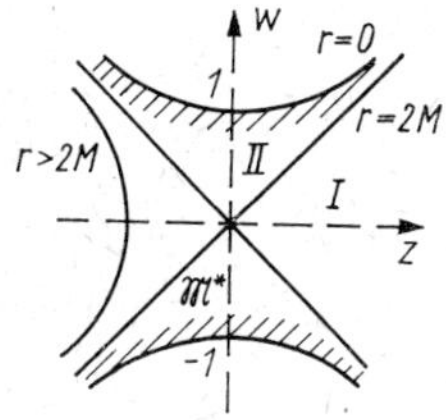

Fig. 31.5

Kruskal metric. The Kruskal metric has the form

$$ds^2 = \frac{32M^3}{r}\,e^{-\frac{r}{2M}}\,(dz^2 - dw^2) + r^2(d\vartheta^2 + \sin^2\vartheta\, d\varphi^2). \tag{1}$$

It is obtained from the Schwarzschild metric (30.4.1/3) by substituting $z = z(r, t)$, $w = w(r, t)$, while φ and ϑ are left unchanged, where

$$z^2 - w^2 = \frac{1}{2M}\,(r - 2M)\,e^{\frac{r}{2M}}, \qquad \frac{w}{z} = \frac{e^{\frac{t}{4M}} - e^{-\frac{t}{4M}}}{e^{\frac{t}{4M}} + e^{-\frac{t}{4M}}}. \tag{2}$$

This also explains the terms with $r = r(z, w)$ as well as with φ and ϑ in (1). $r =$ const are hyperbolas in a z,w-plane (Fig. 31.5), where the uninteresting ϑ-φ-dependence is omitted. Kruskal's space-time $\mathfrak{M}^*$ then lies between the two branches of the hyperbola $w^2 = z^2 + 1$. Here the Schwarzschild space-time $\mathfrak{M}$ becomes the region I, i.e., $\{(w, z) \mid z > 0,\ |w| < z\}$, while Eddington's space-time $\widetilde{\mathfrak{M}}$ becomes the region I + II, i.e., $\{(w, z) \mid (w, z) \in \mathfrak{M}^*,\ w + z > 0\}$. (Here it is always necessary to add ϑ and φ, so that 4-dimensional manifolds are obtained.) Consequently $\mathfrak{M}^*$ is an extension of $\widetilde{\mathfrak{M}}$. It is not complete (for the above considerations for $r \to 0$ concerning the Eddington metric are still applicable), but it is locally inextendible.

Light cone. The straight lines $z = \pm w +$ const are null geodesics in $\mathfrak{M}^*$. This follows from (1) with $\vartheta \equiv$ const and $\varphi \equiv$ const. $\mathfrak{M}$ is time-orientable. Let the direction of increasing w-values be the future direction. To this there corresponds the future cone drawn in I. The future direction is continued continuously to $\mathfrak{M}^*$. This leads to the following interesting situation (Fig. 31.6): information can be transmitted from I (that is $\mathfrak{M}$) to II (that is $\widetilde{\mathfrak{M}}\backslash\mathfrak{M}$) on time-like or null world lines. On the other hand, no information will pass from II to I. In our terminology to be introduced later on, II is a black hole.

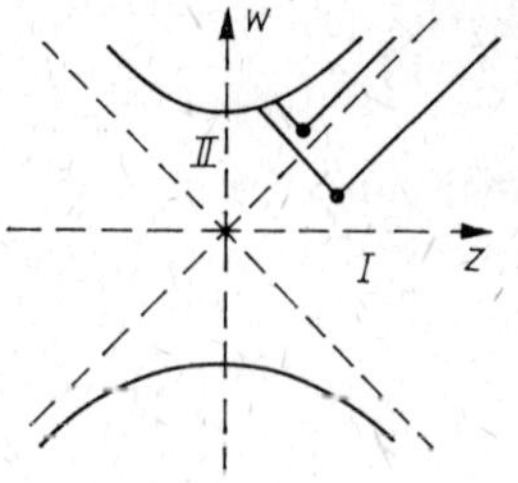

Fig. 31.6

31.1.3. Closed Trapped Surfaces

Closed trapped surfaces are of fundamental importance in the singularity theory of space-times developed by Penrose, Hawking, Ellis and other workers.

Definition. *Let M be a time-orientable space-time. Let F be a compact space-like two-dimensional surface without boundary in M. Then F is called a future-directed closed trapped surface if each of the future-directed families of rays orthogonal to F converges (past-directed closed trapped surfaces are defined analogously).*

Remark 1. Time-orientable space-times have been defined in Subsec. 31.1.1. A surface F (which is naturally regarded as a C^∞-surface) is called space-like if the tangent vectors $V^k = \frac{\mathrm{d}x^k}{\mathrm{d}\tau}$ of every C^∞-curve $x^k(\tau)$ that extends within F are space-like, i.e., if $g_{kl}V^kV^l > 0$. If the local future light cones are erected at every point of F, then the enveloping surfaces obtained are two characteristic surfaces (Fig. 31.7). For details cf. Subsecs. 33.1.2. and 33.1.3. There it is also set out that these two characteristic surfaces are spanned by distinguished rays that are orthogonal to F (cf. (1) and (2) in Subsec. 33.1.3.). These are the two orthogonal future-directed families of rays mentioned in the definition. What is meant by the convergence of such a system of rays is intuitively clear (cf. the concept of caustic in Subsec. 33.1.3.). A precise formulation is, however, rather complicated; cf. [24], p. 262 and p. 101. A first version might read as follows (Fig. 31.8). Consider an (infinitesimal) two-dimensional surface element ΔF in F. The rays passing through ΔF (where we fix one of the two systems) are described by (33.1.3/2). If $s > 0$ is fixed (where $s = 0$ corresponds to the surface F), then one obtains a surface element $(\Delta F)(s)$. The 4-dimensional Lorentz metric g_{kl} induces a (locally Euclidean) 2-dimensional metric on ΔF and on $(\Delta F)(s)$ By means of this metric it is possible to form the surface measure $|\Delta F(s)|$. Convergence of rays now means that $\frac{\mathrm{d}}{\mathrm{d}s}|\Delta F(s)| < 0$. (Although it is not immediately clear whether such a definition of convergence is in agreement with that given in [24], pp. 262/267, it gives an at least qualitatively correct picture.)

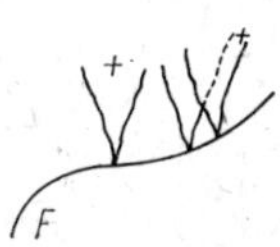

Fig. 31.7

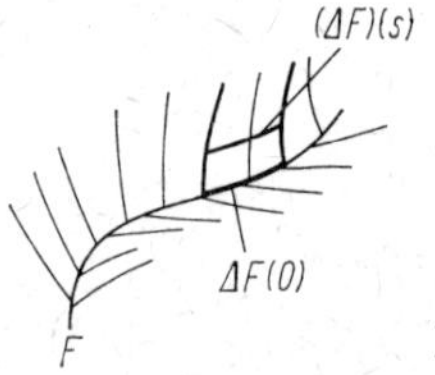

Fig. 31.8

Remark 2. If F is the surface of a 3-dimensional ball in the plane $x^4=0$ in the Minkowskian space, then the two characteristic surfaces mentioned above are cones (Fig. 31.9). The associated rays (straight lines in this case) converge for the "interior" cone and diverge for the "exterior" one. In the Minkowskian space there are no closed trapped surfaces. The corresponding picture in the Eddington metric (30.4.2/1) with "+" looks different. If one chooses for F the surface of a 3-dimensional ball with centre 0 and radius $R<2M$ (Fig. 31.10), then the discussion in Subsec. 31.2.1. shows that now both of the characteristic surfaces mentioned above extend "inwardly". Both families of rays converge. F is a closed trapped surface.

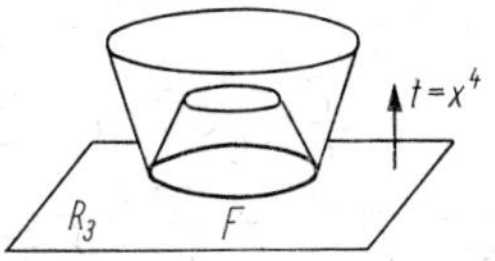

Fig. 31.9

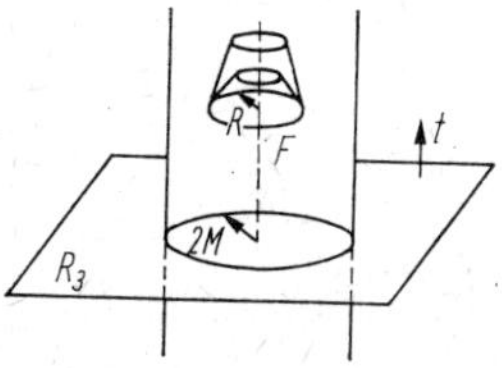

Fig. 31.10

Remark 3. The last example demonstrates the physical meaning of the concept introduced above. If light is emitted on F, this light will propagate into the interior of the ball. The light is trapped (trapped surface).

31.1.4. Singularities

Until as late as the sixties, the following opinion was widespread among workers on relativity theory: the singularities of the known explicit solutions of Einstein's equations are due to the high symmetries of the corresponding metrics (the Schwarzschild-Eddington metric turning King's evidence). It was believed (and hoped) that general (non-symmetric, realistic) solutions of Einstein's equations would avoid such singularities. The clarification of this question grew more and more urgent. On the one hand the cosmological consequences of such singularities were known for a long time: Friedman's models, big bang (cf. Sec. 31.3.). On the other hand the results of modern astrophysics suggested the existence of stars with an incredibly high density of mass: white-dwarf stars, pulsars, neutron stars (cf. Subsec. 31.2.2.). Spherical stars whose mass density is so high that their radius is smaller than the Schwarzschild radius appeared, and still appear, quite real from the physical point of view (black holes). For this the Eddington-Schwarzschild metric provides an impressive theoretical description (cf. Sec. 31.2.). Thus the question whether singularities are locally and globally unavoidable for real space-times is not only of mathematical but also of physical interest. Since 1965 (Penrose) these problems have been systematically studied within a qualitative theory. Roughly speaking, the result (which is positive or negative depending on the respective point of view) reads as follows. Singularities in real space-times are not the exception (due to symmetries) but the rule. Reference [24] (especially Chapter 8) presents a systematic description of these results, which are due to Penrose, Geroch, Hawking, and Ellis. Here we quote a typical theorem that shall give an impression of this theory.

Theorem. (*Hawking, Penrose, 1970*). *Let M be a time-orientable space-time without time-like closed lines. If T^{kl} is the energy-momentum tensor in the sense of Einstein's equations* (30.1.2/3), *then let the demands* (b) (*time-like convergence condition*) *and* (c) (*generic condition*) *made upon T^{kl} at the end of Subsec.* 31.1.1. *be satisfied. If M has a future-directed closed trapped surface, then M is singular* (*in the sense of Def.* 31.1.1/1).

Remark. It is no longer the question of explicit solutions of (30.1.2/3) that is considered here. The propositions are purely qualitative. The fact that the existence of closed trapped surfaces leads to singularities appears physically plausible. This can be realized from the example of the Schwarzschild-Eddington metric. Since the kind of time orientation is irrelevant, the closed trapped surface referred to in the theorem may also be past-directed.

31.1.5. Black Holes

A precise definition of a black hole within Penrose's qualitative theory is rather complicated (cf. [24]). Here we content ourselves with a somewhat poorly defined presentation, which is however hoped to correctly describe the essence of this fundamental concept. Consider a time-orientable manifold M. Without loss of generality we assume that M can be represented by a single chart: let M be an open subset of $R_4 = R_3 \times (-\infty, \infty)$. (Otherwise the following investigations have to be interpreted locally.) The future direction shall be that of increasing t-values ($t = x^4$). In R_4, let S be an open set consisting of a finite number of tubular domains as indicated in Fig. 31.11. In particular, let, for every $t \in R_1$, $\overline{S \cap \{t = \text{const}\}}$ be a 3-dimensional compact set consisting of finitely many connected components. Let $B(t)$ be such a connected component, and let $\partial B(t)$ be its 2-dimensional boundary, i.e., $\partial B(t) = \overline{B(t)} \setminus B(t)$ (in R_3, with t being fixed). The volume is denoted by $|\partial B(t)|$. Now let M be the union of $R \setminus S$ and an open set S_0 with $S_0 \subset S$, that is, the exterior of the tubular domains and parts of the tubular domains. The sets 4 or 5, but also 1–3 or 2–3 (Fig. 31.11) are called branches of S.

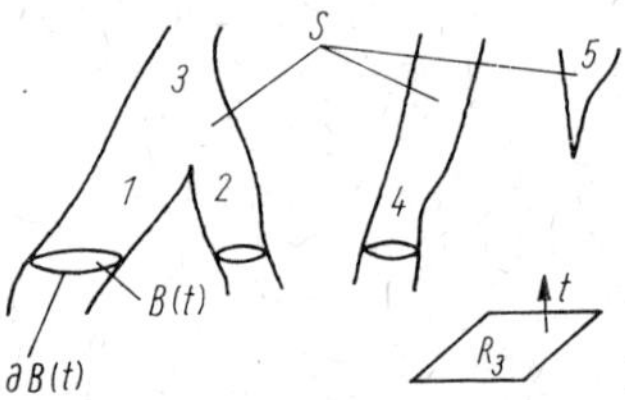

Fig. 31.11

Fig. 31.12

Black hole: *A branch A of S is called black hole if, for every point $P \in A \cap M$, the future-directed light cone* (*characteristic conoid*) *with origin P is also included in $A \cap M$* (Fig. 31.12).

White hole: *A branch A of S is called white hole if, for every point $P \in A \cap M$, the past-directed light cone with origin P is also included in $A \cap M$.*

Remark 1. The precise definition looks more complicated. It consists of a constructive description of $B(t)$. The question for the final stage of a black hole is of special interest. What will happen to $B(t)$ as $t \to \infty$ (stationary black hole)?

Conjecture. *The final stage of a black hole is described by a Kerr-Newman metric.*

Remark 2. The Kerr-Newman metric contains 3 parameters, which correspond to the mass, the angular momentum, and the charge of the black hole. Special cases are the Eddington metric (non-moving neutral black hole; angular momentum and charge are zero) and the Kerr metric (rotating neutral black hole; charge is zero). These cases will be discussed in detail in Sec. 31.2. It appears that the above conjecture is not yet finally proved. But there are numerous theorems that point to this direction (naturally on the basis of a precise definition of a black hole); cf. [24], p. 331, and [43], p. 876. The surprising fact is that a stationary black hole does away with all physical particularities and reduces to three numerical measures (mass, angular momentum, charge). The pleasurable fact is that the associated space-time is explicitly known (since 1965).

Remark 3. The physical meaning of the definition of a black hole is clear. Information can enter a black hole $A \cap M$ from outside, that is from $R_4 \backslash S$, along time-like or null world lines, but no information can pass from $A \cap M$ to $R_4 \backslash S$. We shall study this effect in detail later on, see also the considerations on the Kruskal metric set out in Subsec. 31.1.2.

Remark 4. The above conjecture suggests that black holes cannot disappear, whereas they can arise in the course of time, or also collide with each other (5 or 1–3 with 2–3 in Fig. 31.11). The following assertion can be proved on the basis of Penrose's theory.

Theorem (*Hawking*, 1971). *If two black holes $B_1(t')$ and $B_2(t')$ collide, then a new black hole $B_3(t'')$ emerges, and*

$$|\partial B_3(t'')| > |\partial B_1(t')| + |\partial B_2(t')| . \tag{1}$$

Remark 5. t' is a time long before the collision (then $B_1(t')$ and $B_2(t')$ are almost stationary), and t'' is a time long after the collision (then $B_3(t'')$ is almost stationary).

Remark 6. There arises the question whether black holes are physically real. Black holes are, by definition, invisible. But they influence stars and gas clouds in their close neighbourhood, causing X-ray radiation and the like. On this basis, astrophysics has spotted two candidates for black holes, of which Cygnus X1 in the Swan constellation is the better known one. The satellites "Uhuru" (that means "freedom" in Swahili), 1970, and "Copernicus", 1973, for the measurement of X-radiation, have supported the suspicion; cf. [20], p. 134. We also refer to [63] as well as to [58], p. 264, and especially to [45], pp. 85–87. It is rumoured that there are workers on relativity theory who assess the situation as follows:

There are holes either in the heavens
or in the General Theory of Relativity.

31.2. Theory of Black Holes, Evolution of Stars

31.2.1. The Eddington Metric

As was mentioned in Remark 31.1.5/2, a stationary (that means, invariable in time) non-moving neutral black hole is described by the Eddington metric (30.4.2/1). The term "non-moving" here refers to the starry sky, which is regarded as an inertial frame. The space-time $\mathfrak{M}$ of Subsec. 30.4.2. is time-orientable. We choose the future direction to be that of increasing t-values ($t = x^4$). To obtain a black hole, we then have to choose the "+" sign in (30.4.2/1), i.e.,

$$ds^2 = \left(1 + \frac{2M}{r}\right) dr^2 + 4\, \frac{M}{r}\, dr dt - \left(1 - \frac{2M}{r}\right) dt^2 + r^2 d\vartheta^2 + r^2 \sin^2 \vartheta\, d\varphi^2. \tag{1}$$

Using the notation of Subsec. 31.1.5., one has $S=\{(x^k) \mid 0 \leq r < 2M, t \in R_1\}$ and $S_0=\{(x^k) \mid 0<r<2M,\ t \in R_1\}$. To show that S_0 is a black hole, we restrict ourselves to r vs. t diagrams (i.e. to lines with $\varphi \equiv$ const and $\vartheta \equiv$ const). The rays $ds^2=0$ are then described by the differential equations $dr+dt=0$ and $dt=dr\left(\dfrac{2}{1-\dfrac{2M}{r}}-1\right)$, which are shown in Figs. 31.13 and 31.14. Then the future-directed light cones have the shape indcated in Fig. 31.15. It is seen that S_0 is a black hole. By choosing the "$-$" sign in (30.4.2/1) we would have obtained a white hole.

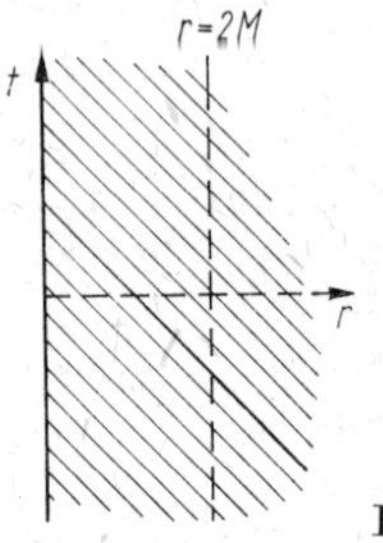

Fig. 31.13

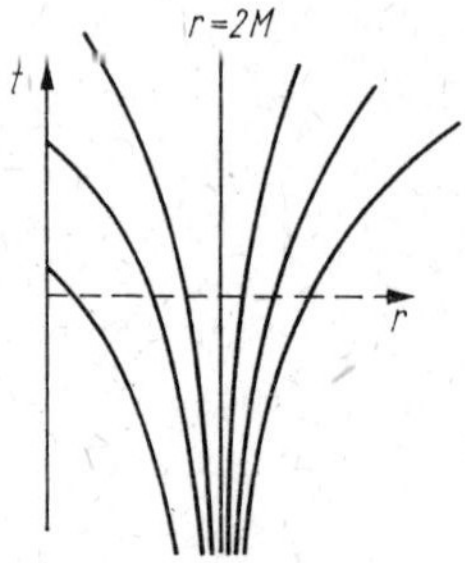

Fig. 31.14

Interpretation. As we stressed on several previous occasions we attribute physical reality to the Eddington metric (= Schwarzschild-Eddington metric) as describing the gravitational field of a heavy mass (cf. the interpretations given

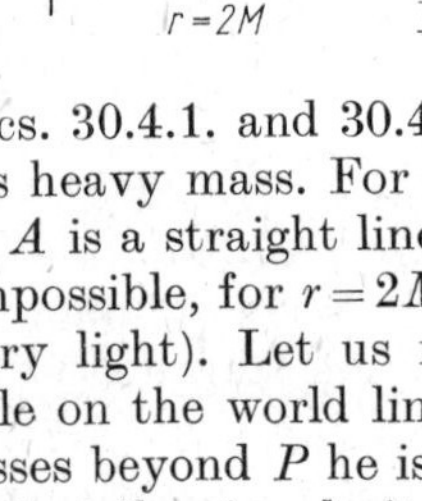

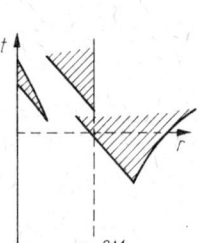

Fig. 31.15

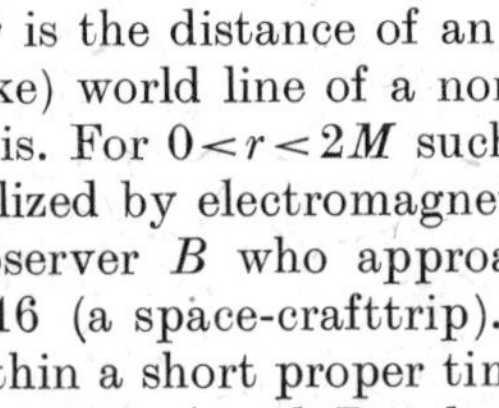

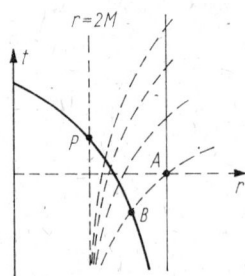

Fig. 31.16

in Subsecs. 30.4.1. and 30.4.2.). In particular, r is the distance of an observer from this heavy mass. For $r>2M$, the (time-like) world line of a non-moving observer A is a straight line parallel to the t-axis. For $0<r<2M$ such a world line is impossible, for $r=2M$ it can only be realized by electromagnetic waves (stationary light). Let us now consider an observer B who approaches the black hole on the world line drawn in Fig. 31.16 (a space-crafttrip). As soon as he passes beyond P he is hopelessly lost. Within a short proper time he will crash against the singularity $r=0$ (cf. Subsec. 31.1.2.). A and B judge all what happens by referring it to their respective proper times (i.e., to the arc lengths s of their world lines). For A (with $r \equiv$ const, $\vartheta \equiv$ const, $\varphi \equiv$ const) the proper time is proportional to t. The non-moving observer A, when looking at his travelling colleague B, finds that the latter slowly approaches the black hole but (as referred to A's proper time) will never arrive there. On the other hand, B finds himself, regretfully, bound to disappear in the black hole very soon (within a finite period of B's proper time).

31.2.2. Stars

It is believed that nowadays the general life history of a star is known, although a few details still have to be clarified. Extensive calculations have been carried out for that purpose. They have been based on thermodynamics, nuclear, and other processes. The models of stars obtained in this way are in good agreement with the star types observed. According to these studies, the life history of a star is as follows.

Birth. Under the influence of gravitation, a cold gas cloud contracts. It is heated up. If the mass m of the gas cloud is less than $10^{-2}\, m_\odot$ (where $m_\odot$ is the mass of the sun), then the cloud will not reach the temperature necessary to initiate the process H $\to$ He (conversion of hydrogen to helium accompanied by a liberation of energy). The gas cloud will then disintegrate.

Lifetime. If $m > 10^{-2} m_\odot$, then (after a transitional phase) a stable radiating star emerges. If the star has a very great mass, then there may be other nuclear processes occurring in parallel to or after H $\to$ He (the star burns its ash). This stable phase will last about $2 \times 10^{10} \left(\frac{m_\odot}{m}\right)^2$ years, and comes to an end when the nuclear fuel is exhausted. The formula shows that stars of great mass will die early.

Death. If $m < 1.2\, m_\odot$ (*Chandrasekhar limit*), the star will end as a white dwarf (and later on presumably as a black one). Such stars have radii from 3×10^3 to 2×10^4 km, a density of mass, μ, of about $10^6\, \frac{\mathrm{g}}{\mathrm{cm}^3}$, and consist of bare nuclei and free electrons (electron gas). If $m > 1.2\, m_\odot$, the star will throw off most of its mass into the universe, giving rise to the astronomical spectacle of a supernova. What is left turns into a neutron star. Stars of this type have radii that range between 6 and 100 km, a mass density μ of about 10^{13} to $10^{15}\, \frac{\mathrm{g}}{\mathrm{cm}^3}$, and consist of neutrons (neutron gas). It is believed that pulsars are neutron stars. The best known example of a supernova and an associated pulsar is the Crab nebula (this supernova was observed and described by the Chinese in 1054).

Remark 1. The above picture is believed to be qualitatively and (apart from common factors) quantitatively correct. It is supported by numerous theoretical considerations and experimental data (especially of radioastronomy). We refer to [43], §§ 24.2 and 24.3, [56], Chap. 1, and to [20, 58, 67]. For us those contributions are of interest which can be made by the theory of relativity to this question. ·

Critical quantities. Let us again consider the Eddington metric. A contracting (spherical) star becomes a black hole if its radius r decreases below its Schwarzschild radius $r_c = 2M$ (Fig. 31.17). In the CGS-system, according to (30.4.1/4) one obtains

$$r \leqq 2\, \frac{\mathrm{G}}{\mathrm{c}^2} \cdot \frac{4}{3}\, r^3 \pi \mu\,,$$

where μ is the mass density. If in this formula r is replaced by the mass $m = \frac{4}{3}\pi r^3\mu$, then one obtains a critical mass m_c. This gives

$$r \geqq r_c \sim \frac{10^{13}}{\sqrt{\mu}}\ \mathrm{cm} \quad \text{and} \quad m \geqq m_c \sim 4 \cdot \frac{10^{39}}{\sqrt{\mu}}\ \mathrm{g}, \tag{1}$$

where μ has to be inserted in $\frac{\mathrm{g}}{\mathrm{cm}^3}$. Some typical numerical values have been entered in the diagrams of Figs. 31.18 and 31.19. Above the critical curves, the star is lost beyond its Schawrzschild radius, becoming a black hole.

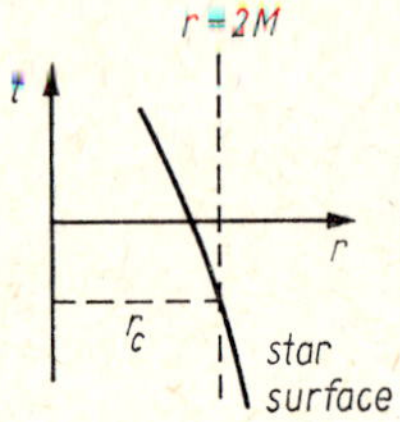

Fig. 31.17

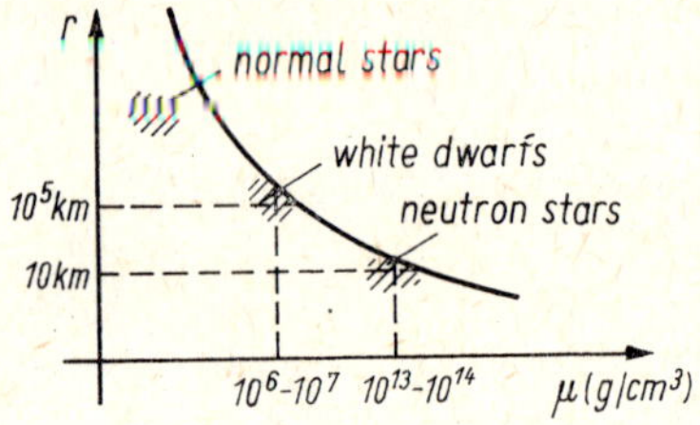

Fig. 31.18

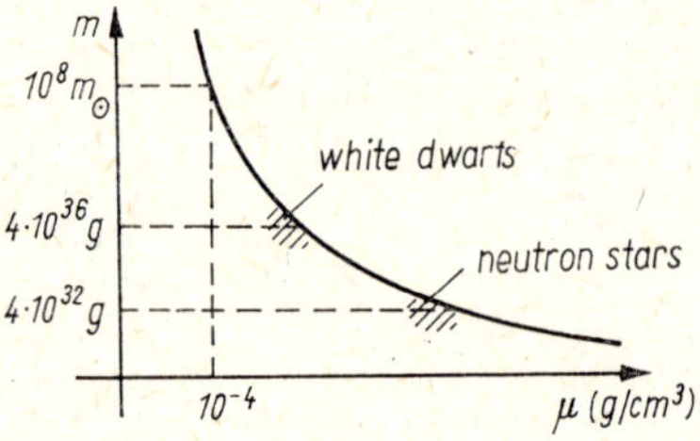

Fig. 31.19

Remark 2. If $\mu \sim 10^{-4} \frac{g}{\mathrm{cm}^3}$ (density of atmosphere), then the critical mass is $m_c \sim 10^8\, m_\odot$. For such a low mass density the nuclear conversion H$\to$He is not possible. So the existence of huge cold masses forming a black hole is at least theoretically conceivable.

Remark 3. Reference figures for the sun, $\odot$, and the earth, E, are as follows: $r_\odot = 7 \times 10^5$ km, $\mu_\odot = 1.4 \frac{\mathrm{g}}{\mathrm{cm}^3}$, $m_\odot = 2 \times 10^{33}$ g, $r_E = 6 \times 10^3$ km, and $\mu_E = 5.3 \frac{\mathrm{g}}{\mathrm{cm}^3}$.

Oh, Be A Fine Girl, Kiss Me Right Now, Sweet!
(Crib used to learn the spectral classes)

31.2.3. The Hertzsprung-Russell Diagram and the Celestial Scale

Books on modern astrophysics and radio astronomy are more thrilling to read than many a thriller [20, 28, 56, 58, 67], for example in comparison with [10]. In the preceding subsection we indicated some quantitative data of stars. Other information (especially about distances in the universe) is of interest for the later cosmological cosiderations. Therefore let us make some remarks on distance measurement.

1st step (*of the celestial scale*). If a star that is not so very far away is observed from different points of the earth's orbit (Fig. 31.20), then its position exhibits slight variations relative to the other fixed stars (angle α). As the diameter of the earth's orbit is known, the distance of the star from the earth can be calculated from these observations. For many stars this effect is too small to be measurable.

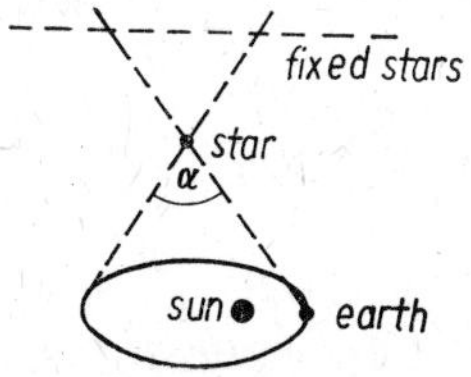

Fig. 31.20

The Hertzsprung-Russell diagram (Fig. 31.21). On the other hand, however, the number of those stars whose distances can be determined by this geometric method is so large that it is possible to find a remarkable relationship between the spectral classes of the stars and their absolute brightness (multiple of the magnitude of the sun in the diagram). The spectral classes (set of all characteristic lines in the spectrum of light that are assigned to certain chemical elements) correspond to surface temperatures. The results is the Hertzsprung-Russell diagram, with a principle line and several secondary lines.

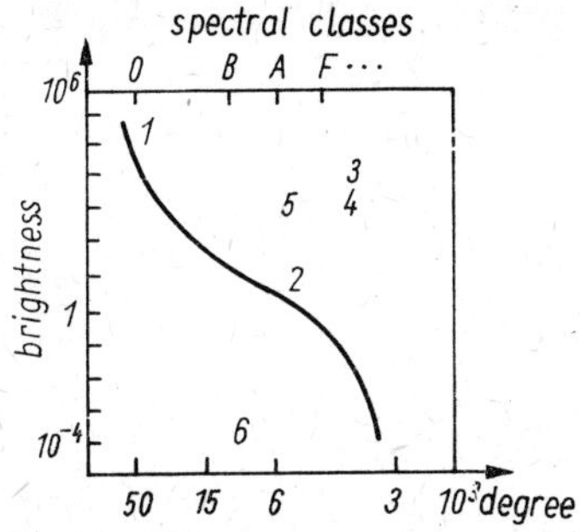

Fig. 31.21
1 extremely hot blue stars; *2* sun-type stars (G stars); *3*, *4* red giants; *5* variable stars; *6* white dwarfs

Variable stars. The geometric method yields, however, yet another remarkable result. There are stars whose stellar brightness shows periodic fluctuations (the period being a few days). For some of the variable stars (a sufficient number of them) the geometric method enables us to measure the distance, and hence

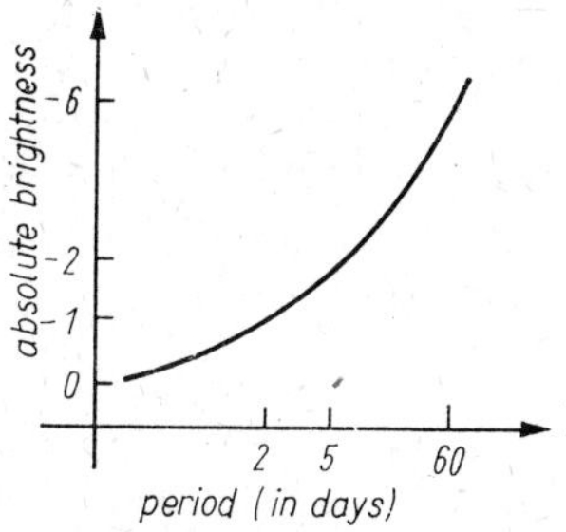

Fig. 31.22

to determine the absolute stellar brightness. It turns out that the period increases in the same measure as the brightness of the star (Fig. 31.22).

2nd step. It is now supposed that the Hertzsprung-Russell diagram as well as the law of brightness periods of the variable stars are applicable to all stars (not only to those whose distances can be determined geometrically). But this makes it possible to determine the distances of all stars: by determining the spectral class one obtains the absolute brightness, and from a comparison with the observed brightness one calculates the distance.

3rd step. In the late twenties, Hubble discovered variable stars in several galaxies. By means of the method of the second step he was able to determine the distances of those stars (and hence those of the galaxies). On the other hand the light of (most of) the galaxies exhibits a red shift of the spectral lines. If this red shift (decrease in frequency) is interpreted as a Doppler effect, one obtains an escape speed of these galaxies from our Milky Way which is proportional to their distances (cf. 31.3.1). Again it is supposed that this law is universally valid. This makes it possible to determine the distance of far galaxies from the Milky Way by measuring the red shift.

Star evolution. The course of life of stars can be traced in the Hertzsprung-Russell diagram (Fig. 31.23; according to [86]). The first diagram shows the evolution of a star of the type of our sun: origin 1, principal line 2, red giant 3, ejection of matter 4, white dwarf 5. The second diagram characterizes the evolution of a massive star ($m > 3m_\odot$): origin 1, principle line 2, red giant and supernova 3, neutron star 4.

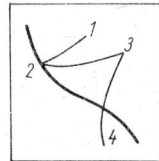

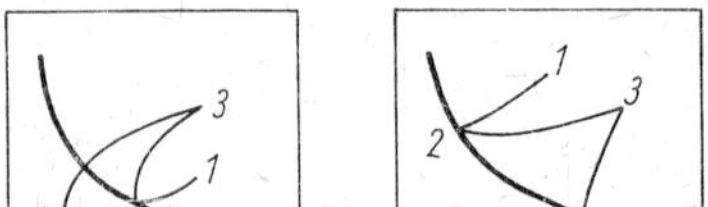

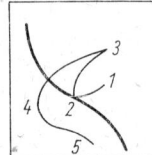

Fig. 31.23

31.2.4. The Kerr Metric

As mentioned in Remark 31.1.5/2, a stationary (i.e. invariable in time) rotating neutral black hole is described by a Kerr metric. Here the term "rotating" refers to the starry sky, which is considered to be an inertial frame. The angular momentum, too, is measured by a non-moving observer far away from the black hole. The Kerr metric is a solution of Einstein's equations in empty space, $R^{kl} - \frac{1}{2} R g^{kl} = 0$, having the form

$$ds^2 = \varrho^2 \left(\frac{dr^2}{\Delta} + d\vartheta^2\right) + (r^2 + a^2)\sin^2\vartheta d\varphi^2 - dt^2 + \frac{2Mr}{\varrho^2}(a \sin^2\vartheta d\varphi - dt)^2, \quad (1)$$

$$\varrho^2 = \varrho^2(r, \Theta) = r^2 + a^2 \cos^2\vartheta, \quad \Delta = \Delta(r) = r^2 - 2Mr + a^2 . \quad (2)$$

Here M and a are constants, $0 < r < \infty$, $0 < \vartheta < \pi$, and $0 \leqq \varphi < 2\pi$. For $a = 0$ one obtains the Schwarzschild metric (30.4.1/3). This suggests setting $M = m\frac{G}{c^2}$, as was done previously in (30.4.1/4), where m is the mass of the black hole. Further it turns out that Ma is proportional to the angular momentum, the body rotating about the axis $\vartheta = 0$ or $\vartheta = \pi$. From (2) it follows that (1) becomes singular for $\Delta(r) = 0$. This gives

$$r_+ = M + \sqrt{M^2 - a^2} \quad \text{and} \quad r_- = M - \sqrt{M^2 - a^2}\,. \tag{3}$$

The case $M < a$ (Fig. 31.24). For $M < a$ there occur no singularities $\Delta(r) = 0$. By a transformation of r, Θ, φ, t into x, y, z, $\bar{t}$ it is possible to obtain the form

$$\begin{aligned} ds^2 &= dx^2 + dy^2 + dz^2 - d\bar{t}^2 \\ &\quad + \frac{2Mr^3}{r^4 + a^2 z^2} \left(\frac{r(x\,dx + y\,dy) - a(x\,dy - y\,dx)}{r^2 + a^2} + \frac{z\,dz}{r}\, d\bar{t} \right)^2 . \end{aligned} \tag{4}$$

Here x, y, z and r are related by $\frac{x^2 + y^2}{a^2 + r^2} + \frac{z^2}{r^2} = 1$. One thus obtains confocal ellipsoids in R_3, which contract to the circular disk $x^2 + y^2 = a^2$ in the plane $z = 0$ as $r \to 0$. (4) tends towards the Minkowski metric as $r \to \infty$. This suggests to interpret (x, y, z) as real space co-ordinates. (4) is the analogue to (30.4.1/2) and describes the gravitational field of a heavy mass rotating about the z-axis (this mass is not a black hole).

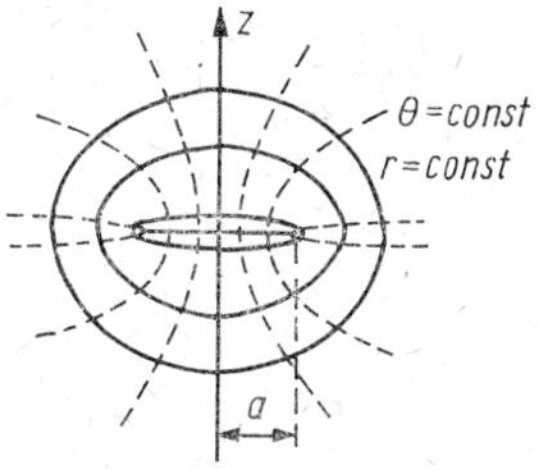

Fig. 31.24

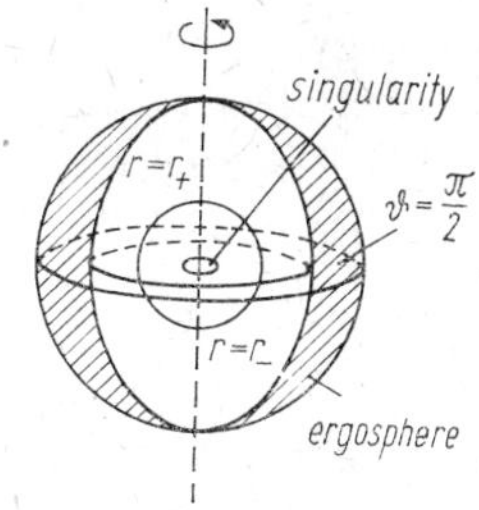

Fig. 31.25

The case $M > a$ (Fig. 31.25). In the hope of being able to describe the gravitational field of a rotating black hole, one will first ask for a reformulation of (1) that yields the Eddington metric for $a = 0$. The corresponding metric reads

$$\begin{aligned} ds^2 = &-\left(1 - \frac{2Mr}{\varrho^2}\right) dt^2 + \frac{4Mr}{\varrho^2}\, dt\,dr + \left(1 + \frac{2Mr}{\varrho^2}\right) dr^2 \\ &- 2a \sin^2 \vartheta \left(1 + \frac{2aM}{\varrho^2}\, r\right) dr\,d\varphi - 4\,\frac{a}{\varrho^2}\, Mr \sin^2\vartheta\, d\varphi\, dt \\ &+ \frac{1}{\varrho^2}\left[(r^2 + a^2)^2 - a^2 \Delta \sin^2 \vartheta\right] \sin^2\vartheta\, d\varphi^2 + \varrho^2 d\vartheta^2 , \end{aligned} \tag{5}$$

where ϱ and Δ have the same meaning as in (2), and $0 < r < \infty$, $0 < \vartheta < \pi$, whereas φ has a different meaning. For $a = 0$ one obtains (31.2.1/1). Letting $r \to \infty$ one approximately obtains the Minkowski metric again (slightly perturbed by φ), so that for large values of r this quantity can be regarded as the distance from the rotating body. As $r \to r_+$, distortions similar to those in the case $M < a$ are found to occur. It turns out that $B(t) = \{r \mid r < r_+\}$ is a stationary black hole. $r = r_-$ has no special meaning. In contrast to the Eddington solution (i.e., $a = 0$), for $a > 0$ one obtains a circular singularity (analogous to the case $M < a$) instead of a point singularity. An event at the point C will be found to appear in the circle ○ close to C after a short period (Fig. 31.26): a stationary behaviour is not possible, one cannot escape from $B(t)$. An observer A will indeed detect the rotation, but he can remain stationary. There is an interesting intermediate zone, the so-called *ergosphere*. It is bounded by $r = r_+$ on the inside and by $r = M + \sqrt{M^2 - a^2 \cos^2 \vartheta}$ on the outside. An observer B cannot remain stationary, but he can escape from this dangerous zone to the outside: violent rotation with the possibility to escape. According to Subsec. 31.1.5., $|\partial B(t)|$ is of interest. One obtains

$$|\partial B(t)| = 8\pi M(M + \sqrt{M^2 - a^2}) . \tag{6}$$

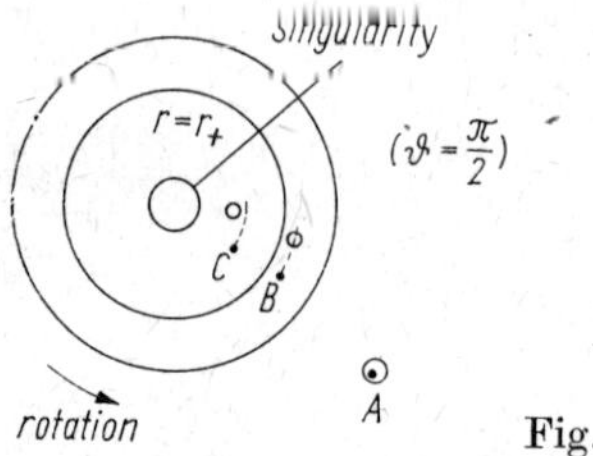

Fig. 31.26

Supply of energy. Reference [43], p. 906 and p. 908, tells the following story. An advanced civilization sends waste containers on a Penrose trajectory (see also Subsec. 31.2.5.) to a region near a rotating black hole (into the ergosphere). At a suitable point the waste is dumped into the black hole. The enormous rotation of the black hole causes the empty container to rush back to its starting point with a formidable speed. There its kinetic energy is utilized for, say, electric power generation. The final result is that not only the waste is converted into energy, but an additional amount of energy is tapped from the black hole.

31.2.5. Energy Balance of Black Holes

What happens when two black holes collide, or when a particle falls into a black hole?

Collision. Suppose that the collision of two neutral black holes (represented by Kerr metrics with the parameters M_1, a_1 and M_2, a_2) again produces a black hole (Kerr metric with the parameters M_3, a_3). This corresponds with Conjecture 31.1.5 as well as with Theorem 31.1.5. Then it follows from (31.1.5/1) and (31.2.4/6) that

$$M_3(M_3 + \sqrt{M_3^2 - a_3^2}) > M_1(M_1 + \sqrt{M_1^2 - a_1^2}) + M_2(M_2 + \sqrt{M_2^2 - a_2^2}) . \tag{1}$$

Here it is quite possible that $M_1 + M_2 - M_3 > 0$. According to Einstein's conceptions referred to in Subsec. 25.2.5., this mass defect is converted into energy. An extreme case is that $M_1 = M_2 = a_1 = a_2$ and $a_3 = 0$. Then it follows from (1) that $M_3 > M_1$. Hence it is possible that $M_1 + M_2 - M_3 \sim M_1$.

Particles. According to Penrose, a particle that falls into a black hole $B_1(t)$ leaves another, new black hole $B_2(t')$, where $|\partial B_2(t')| > |\partial B_1(t)|$. We again suppose that the two black holes are described by Kerr metrics with the parameters M_1,

a_1 and M_2, a_2. Then it follows from (31.2.4/6) that

$$M_2(M_2+\sqrt{M_2^2-a_2^2})=M_1(M_1+\sqrt{M_1^2-a_1^2}). \tag{2}$$

If the rotation is decelerated during this process, that is, if $a_1 > a_2$, then by (2) it is possible that $M_1 > M_2$. The mass defect $M_1 - M_2$ is converted into energy. This result forms the basis for the considerations on energy supply as described at the end of Subsec. 31.2.4.

> One thing I have learned in a long life: that all our science, measured against reality, is primitive and childlike – and yet it is the most precious thing we have.
> (A. Einstein)

31.3. Cosmology

31.3.1. Principles

From the very beginning, the general theory of relativity has claimed to be able to give information of the universe as a whole. Detached about local particularities (such as stars, galaxies, black holes etc.), one looks for global solutions of Einstein's equations,

$$R^{kl} - \frac{1}{2} R g^{kl} + \Lambda g^{kl} = \varkappa T^{kl} \tag{1}$$

making reasonable assumptions on the energy-momentum tensor T^{kl}. It is demanded that the cosmological principle is valid: the universe, when viewed from whichever of its points one may choose for observation, looks (globally) identical; there are no distinguished directions. Astronomical data furnish approximate support to this principle: galaxies, even clusters of galaxies, radio sources, and the so-called 3K-background radiation exhibit a largely isotropic (non-directional) distribution.

Data. There are only few data of interest for cosmological studies. Besides the 3K-background radiation, which shall be dealt with later on, these data include the following items.

1. The mean cosmic mass density ϱ. Its estimated value is $10^{-31} \frac{\text{g}}{\text{cm}^3} \leqq \varrho \leqq 10^{-29} \frac{\text{g}}{\text{cm}^3}$.
2. The frequencies ν of the spectral lines of the light of far galaxies (and of objects of radioastronomy) are shifted by $\Delta\nu$ to lower values (red shift). This phenomenon is described by Hubble's law, $\frac{\Delta\nu}{\nu} = -\text{H}r$, where H is Hubble's constant and r is the distance of the object from the earth (cf. Subsec. 31.2.3.). The present value is about $\text{H}^{-1} \sim 10^{28}$ cm $\sim 10^{10}$ light years (in [43], p. 732, the value of $\text{H}^{-1} = 18 \times 10^9$ l.y. is indicated).
3. For the age of the world, the following data are available. Investigations of terrestrial rocks on the basis of the half-times of radioactive elements yield a maximum age of 4.5×10^9 years. Star models yield age data of at most 3×10^{10} years.

Remark. Every model of the universe has to take account of these data (at least in terms of their order of magnitude). Conversely, when models of the universe are investigated which contain free parameters (and this will be the case), then these data can be used to determine the parameters in question.

Set-up for T^{kl}. We make the set-up (30.2.3/1) for T^{kl}, that is, $T^{kl}=(\mu+p)V^kV^l+$ $+pg^{kl}$, where, following the cosmological principle, we assume that $\mu=\mu(t)$ and $p=p(t)$ depend only on the time $t=x^4$, but not on the position (x^1, x^2, x^3). One then obtains the system (1)–(4) of Subsec. 30.2.3. For the streamlines V^k of Subsec. 30.2.3. we introduce 3 parameters: dilatation $V^k{}_{;k}$, vorticity $V_{[k;l]}$ and shear $\sigma_{kl}=V_{(k;l)}-\frac{1}{3}V^l{}_{;l}$.

31.3.2. The Robertson-Walker Metric

Theorem. *A non-vortical (irrotational) and shear-free model of the universe can be represented in suitable local co-ordinates by a Robertson-Walker metric* $\mathrm{d}s^2=$ $=S^2(t)\mathrm{d}\sigma^2-\mathrm{d}t^2$. *Here* $\mathrm{d}\sigma^2$ *is a positive definite three-dimensional line element whose associated curvature scalar is constant.*[1)]

Remark 1. What is desired is a solution of (31.3.1/1), where T^{kl} satisfies the set-up of Subsec. 31.3.1. and $V_{[k;l]}=\sigma_{kl}=0$.

Remark 2. One has $\mathrm{d}\sigma^2=\gamma_{\alpha\beta}\mathrm{d}x^\alpha\mathrm{d}x^\beta$, where the summation goes over α and β from 1 to 3. Here the fundamental tensor $\gamma_{\alpha\beta}$ is positive definite, i.e., its signature in the sense of Subsec. 29.5.1. is 3. If ε is the curvature scalar associated to $\gamma_{\alpha\beta}$ in the sense of Def. 29.5.7, then in the above theorem we can assume that $\varepsilon\equiv1$ or $\varepsilon\equiv0$ or $\varepsilon\equiv-1$. If this is not the case, then we modify $S(t)$ in the theorem by a positive constant factor.

31.3.3. The Dust Universe

We have to determine $S(t)$ of Theorem 31.3.2. That is, we ask for a cosmological model, i.e., a solution of (31.3.1/1), which is described by a Robertson-Walker metric and whose energy-momentum tensor T^{kl} satisfies the set-up of Subsec. 31.3.1. For a quantitative discussion of the empirical data concerning the energy density μ and the pressure p we refer to [43], pp. 711–713. The results presented there suggest the following set-up, which is called the dust universe: in (30.2.3/1), let $p(t)\equiv0$ and $\mu(t)\equiv\varrho(t)$, where $\varrho(t)$ is the mean cosmic mass density at the time t.[1)]

Theorem. *In the dust universe,*

$$\frac{4}{3}\pi S^3(t)\,\varrho(t)=M\,, \tag{1}$$

$$\dot{S}^2(t)=\frac{\varkappa}{4\pi}\frac{M}{S(t)}+\frac{\Lambda}{3}S^2(t)-\varepsilon\,. \tag{2}$$

Remark 1. (1) and (2) can be derived from (2)–(4) in Subsec. 30.2.3. and from the conventions made above. Here $\varkappa$ is the Einstein constant, Λ is the cosmological constant, and

---

1) See also [53, pp. 179/180].

$\varepsilon = -1, 0, 1$ in the sense of Remark 31.3.2/2. Further, $\dot{S}(t) = \frac{dS}{dt}$. The normalization of ε suggests that $d\sigma^2$ in Theorem 31.3.2 be regarded as dimensionless in the CGS system. Then $S(t)$ has the dimension of a length (like dt^2, in the sense of the convention that time = path). Hence $\frac{4}{3}\pi S^3(t)$ is the volume of a (Euclidean) ball. $S(t)$ is called the radius of the universe. (1) then means that the total mass M of the universe is independent of time (law of conservation of mass). M is an integration constant that can be fixed arbitrarily (from the mathematical point of view).

Remark 2. (2) is the famous Friedman equation. The case $\Lambda = 0$ is of special interest (last but not least for aesthetical reasons). Einstein, in his later years (after Hubble's discoveries) constantly pleaded that physical significance be only attributed to this case.

31.3.4. Hubble's Law

From now on we shall set $\Lambda = 0$ in (31.3.3/2). Our aim is as follows. From the present knowledge of the mean cosmic mass density ϱ and the Hubble constant H (cf. Subsec. 31.3.1.) we want to determine the still unknown parameters ε (with the possible values $-1, 0, 1$) and M of Theorem 31.3.3. Thereafter it will be possible to solve (31.3.3/2) (with $\Lambda = 0$). Then the age of the universe can be determined from the explicit formula for $S(t)$, where of course it is hoped that the theoretical value is in agreement with the empirical value indicated in Subsec. 31.3.1.

Hubble's constant. The light cones in a Robertson-Walker metric are determined from $ds^2 = 0$, i.e., $dt^2 = S^2(t)d\sigma^2$. From the isotropy of space, but also from the explicit line elements $d\sigma^2$ for each of the 3 cases $\varepsilon = -1, 0, 1$, it is observed that these light cones are translation-invariant in the σ-direction. If one considers two observers 1 and 2 (Fig. 31.27), then for infinitesimal conditions (on a cosmical scale this means distances like those between neighbouring galaxies) one obtains $a = S(t_1)d$ and $b = S(t_2)d$. The proper time of the two non-moving observers is t. If 1 emits light of the frequency ν_1, then 2 receives this light with the frequency ν_2, where $a\nu_1 = b\nu_2$ (= number of oscillations in the proper time intervals a and b, respectively). With $\Delta\nu = \nu_2 - \nu_1$ and $\nu = \nu_1$ this gives $\frac{\Delta\nu}{\nu} = \frac{S(t_1)}{S(t_2)} - 1$. For $t_1 \to t_2$ and $t = t_2$ one obtains

$$\frac{\Delta\nu}{\nu} = -\frac{\dot{S}(t)}{S(t)}\,r, \quad \text{and hence} \quad \mathrm{H}(t) = \frac{\dot{S}(t)}{S(t)}. \tag{1}$$

Here we have made use of $\Delta t = t_2 - t_1 = r$ (distance between the two observers).

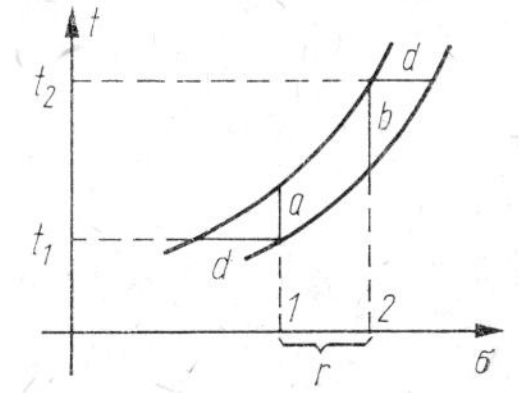

Fig. 31.27

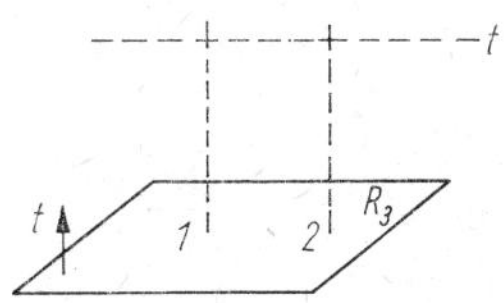

Fig. 31.28

Friedman's differential equation. From Theorem 31.3.3 (with $\Lambda=0$) it follows immediately that

$$\mathrm{H}^2(t)=\frac{\varkappa}{3}\varrho(t)-\frac{\varepsilon}{S^2(t)}. \tag{2}$$

This gives rise to a chain of conclusions that appears fantastic. If $\mathrm{H}(t)$ and $\varrho(t)$ are known at some time $t=t_0$ (our present time), then the quantities ε (that is, $\varepsilon=-1, 0, 1$) and $S(t_0)$ can be determined from (2). With $t=t_0$ in (31.3.3/1) one then calculates the total mass M. Thereafter (33.3.3/2) is solved for $\Lambda=0$. From the explicit solutions, it will then be possible (as we shall see) to determine t_0: the age of the universe.

Dilatation (Fig. 31.28). For the considerations to follow, the following remark will be useful. The distance between two observers in the Robertson-Walker metric is determined by $\mathrm{d}s^2=S^2(t)\mathrm{d}\sigma^2$. Consequently, with respect to the time independent standard distance σ we have $s=S(t)\sigma$. As $S(t)$ increases, s increases too. Thus the wavelength of the waves existing in the universe will increase in the course of time. This results in a red shift of the frequencies, which in its turn is described by (1).

31.3.5. Solutions of Friedman's Equation

Theorem. *The Friedman equation* (31.3.3/2) *with* $\Lambda=0$ *has the following solutions.*

$$S(\tau)=\frac{\mathrm{G}M}{2}\tau^2,\quad t=\frac{\mathrm{G}M}{6}\tau^3 \quad \text{for} \quad \varepsilon=0, \tag{1}$$

$$S(\tau)=\mathrm{G}M\left(\frac{\mathrm{e}^{\tau}+\mathrm{e}^{-\tau}}{2}-1\right),\quad t=\mathrm{G}M\left(\frac{\mathrm{e}^{\tau}-\mathrm{e}^{-\tau}}{2}-\tau\right) \quad \text{for} \quad \varepsilon=-1, \tag{2}$$

$$S(\tau)=\mathrm{G}M(1-\cos\tau),\quad t=\mathrm{G}M(\tau-\sin\tau) \quad \text{for} \quad \varepsilon=1. \tag{3}$$

Here G *denotes Newton's constant of gravitation.*

Remark 1. These are special solutions. The general solution is obtained if one replaces t by $-t$ and by $t+\text{const}$. In any case, $t=t(\tau)$ is a monotonic function. Hence $\tau=\tau(t)$ exists. Then the solution of (31.3.3/2) is $S(\tau(t))$.

Remark 2. In all cases one has $S(0)=0$. For $\tau\geqq 0$, the functions $S(\tau)$ in (1) and (2) are increasing functions of τ (and hence also of t). Hence the radius of the universe, $S(\tau(t))$, increases monotonically from 0 to ∞: the universe expands. (3) is the parametric representation of a cycloid; cf. Subsec. 11.2.2. One obtains a pulsating universe. Now it is also clear how the age of the universe, T, can be determined: if $\mathrm{H}(T)$ and $\varrho(T)$ are known, then ε and $S(T)$ are calculated from (31.3.4/2). This also determines T, where there are two possibilities in the case $\varepsilon=1$. Since Hubble's law corresponds to an expanding universe, for $\varepsilon=1$ one has to choose the smaller of the two possible T values (Fig. 31.29).

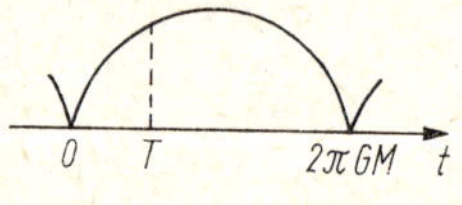

Fig. 31.29

Remark 3. 2500 years ago the greek philosopher Heraclides remarked: This world has been created neither by a god nor by a man. It was and will be a living fire in eternity, growing and ceasing in a temperate manner.

31.3.6. Friedman's Models

Critical mass density. If $\varepsilon = 0$, then $\ln S = c + \frac{2}{3} \ln t$, and hence, according to (31.3.4/1), $H(t) = \frac{2}{3t}$. With $H^{-1}(T) = 18 \times 10^9$ years one obtains $T = 12 \times 10^9$ years for the age of the universe. To our present knowledge this is quite a discussible value. If the associated mass density is calculated from (31.3.4/2), one obtains $\varrho_c = 6 \times 10^{-30} \frac{g}{cm^3}$. If $\varrho > \varrho_c$, then it follows from (31.3.4/2) that $\varepsilon = 1$ (pulsating universe). For $\varrho < \varrho_c$ one has $\varepsilon = -1$ (expanding universe). Finally, in all cases one can determine the radius of the universe, $S(T)$. The present experimental value indicated in Subsec. 31.3.1. for the mean cosmic mass density does not yet permit a final conclusion as to which of the three cases $\varepsilon = -1, 0, 1$ is the true one.

Line element $d\sigma^2$. The line element $d\sigma^2$ of Theorem 31.3.2 has the form

$$d\sigma^2 = dr^2 + f^2(r)\,(d\vartheta^2 + \sin^2\vartheta\, d\varphi^2)$$

where

$$f(r) = \sin r \quad \text{for} \quad \varepsilon = 1\,,$$
$$f(r) = r \quad \text{for} \quad \varepsilon = 0 \quad \text{and}$$
$$f(r) = \frac{1}{2}\,(e^r - e^{-r}) \quad \text{for} \quad \varepsilon = -1\,.$$

For $\varepsilon = 0$ this is the Euclidean line element in polar co-ordinate representation $0 < r < \infty$, $0 < \vartheta < \pi$, $0 \leqq \varphi < 2\pi$. For $\varepsilon = 1$,

$$d\sigma^2 = dr^2 + \sin^2 r\, d\vartheta^2 + \sin^2 r\, \sin^2 \vartheta\, d\varphi^2$$

with $0 < r < \pi$, $0 < \vartheta < \pi$, and $0 \leqq \varphi < 2\pi$, is the surface element of the unit sphere in the 4-dimensional Euclidean space. $S^2(t)\, d\sigma^2$ then is the surface element of a sphere of radius $S(t)$ in R_4. Thus the pulsating universe can be imagined as the surface of a pulsating sphere in R_4.

Our universe. The following data are thought to be realistic (with certain factors of uncertainty): radius of the universe 18×10^9 light years, age of the universe 11×10^9 years. Today the red shift discovered by Hubble has been widely accepted as an indication of an expanding universe.

31.3.7. The Big Bang

The cosmological models described above begin at the time 0, and at this time the radius of the universe is $S(0) = 0$. The universe springs into existence with a big bang: a huge fireball in an uproar of furious thermonuclear processes. The question is whether such an assumption can be supported by experiment today. Model calculations suggest that in this initial stage a relatively large quantity of hydrogen must have been converted into heavier elements, especially into helium. Further it can be assumed that a black body radiation developed, which obeys Planck's law, $u(\nu, T) = c\nu^3\,(e^{\frac{c'\nu}{T}} - 1)^{-1}$. Here ν is the frequency and

$u(\nu, T)$ is the energy density. c and c' are positive constants and T is a parameter, the radiation temperature. In fact a relative abundance of helium has been detected in the universe. What is much more amazing, however, is the prediction and the discovery of the 3K background radiation (K = kelvin scale of absolute temperature). In the nascent state the radiation temperature T must have been very high. According to the considerations on dilatation as described at the end of Subsec. 31.3.4., the expansion of the universe involves a large red shift, which is expressed by a decrease of radiation temperature in Planck's law. It has been predicted that the present value of the radiation temperature should be about 3 K. In 1965, Penzias and Wilson indeed detected this 3 K background radiation (the precise value is 2.7 K). In correspondence with the cosmological principle it is isotropic (independent of direction). This discovery was rewarded by the 1978 Nobel Prize for physics. It is probably the most impressive experimental support of the theory of the big bang. For descriptions of this theory, cf. [58, 70].

> We may indeed be very low in the cosmical intelligence scale, but it is a measure of our scientific advance during the last few centuries that we have come to realize how unimportant we really are.
> (P. Moore, [44, p. 223])

31.3.8. Birth of Life in the Universe

When we ask for the birth of life on the earth, beginning with the big bang, we can distinguish three stages.

1. Formation of the chemical elements.
2. Formation of those biochemical molecules which constitute the structural elements of life: polysaccharides (cellulose, starch), amino acids, chlorophyll etc.
3. The evolution of life from the most primitive forms to the flora and fauna of our time.

Today there is no one who seriously doubts that the first of the above stages is a cosmical matter, while the third stage is a terrestrial one. Modern astrophysics has detected all the stable chemical elements in the universe, although in very different quantities; cf. [28, p. 169]. Here it turns out that, besides the predominant elements hydrogen and helium, there are also carbon, nitrogen and oxygen in relative abundance. It can be imagined that the majority of the elements are formed within those processes which have been described in Subsecs. 31.2.2. and 31.2.3.: the life and the death of stars. As regards the second stage, there has been a remarkable reorientation of thinking in the past few years. The fantastic successes of radioastronomy and infrared astronomy of recent years have made it appear at least highly probable that the structural elements of life (or at least the component parts of these elements) are present in relative abundance in the universe. Books on astronomy published in the past few years give clear evidence of this: [44, Sec. 27], [20, pp. 85–87], [67, p. 165], and [80, p. 97, pp. 109–110]. The probably most impressive report on this problem, however, has been given in [28]. In what follows we shall make, in an outline form, at least some passing remarks on this subject, largely following the lines of the presentation given in [28].

Life on the earth. The oldest finds that contain traces of life on the earth have an age of 3.1×10^9 years. The age of the earth is 4.5×10^9 years. Thus it can be assumed that primitive forms of life developed on the earth very early.

Molecules in gas clouds. Since the late sixties, a multitude of inorganic and organic molecules have been detected in interstellar gas clouds. A long list is given in [28, pp. 180/181]. Water as well as a great number of carbon compounds, including alcohol, are listed there. Here the Orion nebula is a most productive object. On the other hand it is known that the Orion nebula is a region with extremely young stars. That means, formation of stars is observed in a nebula that contains a multitude of organic molecules. We also refer to [45, pp. 102–105]. It is assumed that about 4×10^9 to 5×10^9 years ago the sun and the earth were in a similar situation.

Dust. Today it is known that in the universe there exist dark nebulae of large extensions, which absorb the light emitted by stars. From this it is concluded that the size of the particles in question must be 10^{-4} cm: particles of metals, graphite and rocks. It is certainly one of the most remarkable successes of infrared astronomy that complex organic molecules have also been detected in these dust clouds: polysaccharides (such as cellulose) and structural elements for chlorophyll. Explanations of how it is possible that such complex molecules exist in the universe have been given in Secs. 9 and 10 of reference [28].

Birth of life. According to [28], life on the earth developed as follows. Shortly after the birth of the sun, it was surrounded by gas and dust, where (at greater distances from the sun), a fairly great number of complex organic molecules were present (analogously to the Orion nebula and the neighbouring dust nebulae). The earth, which is formed only a short time later, contains no complex molecules, because the latter were destroyed in the region near the sun. Subsequently, however, such molecules are carried along from greater distances in the interior of meteors which land on the earth. Under the terrestrial conditions, which are favourable at that time, finally primitive life develops from these molecules.

Comets and meteors. To support the above theory, the light of comets has been analyzed in the past few years. It could not be expected that complex organic molecules would be detected, because they are destroyed in the neighbourhood of the sun (provided they are not protected from the influences of sunlight). Organic decomposition products of such desirable molecules have, however, been detected; [28, p. 90]. The results of the investigation of meteorites are as follows. About 3 % of all meteorites belong to the class of "carbonaceous chondrites" (meteorites of grain structure, containing carbon). Their ages range between 4.5×10^9 and 4.7×10^9 years, and at least some of them do not belong to the solar system. In some of these meteorites, amino acids as well as cellulose-like molecules have been detected. These results support the above theory in a very impressive way.

Planetary systems. Since the biochemical structural elements of life are widespread in the universe, it can be assumed that life has evolved at other places too, provided that the conditions were suitable, maybe similar to terrestrial conditions. The theory of the formation of our planetary system, which is generally accepted today, suggests that planetary systems are abundantly present in the universe [28, Sec. 13]. This assertion is also supported by observations. So far a planetary system of another star has not yet been observed directly. But the existence of such a system should induce the star in question to perform a certain periodic trembling motion. And such tremblings have in fact been observed. Among our neighbouring stars it is already the nearest but two, Barnard's star at a distance of 5.88 light years, that should have at least two planets according to this interpretation [45, pp. 87–88]. In [28, p. 137], another five stars at distances between 8 and 27 light years are listed, which have planetary systems.

Habitable planets. According to [28], the estimated average distance between two planets in our galaxy that are host to higher civilizations at a given time is at least 200 light years. Mutual visits are banished into the realm of fantasy (Star War will not take place), whereas a certain chance, if a minute one, is attributed to the possibility of receiving signals of extraterrestrial civilizations.

Remark. F. Hoyle and N. C. Wickramasinghe continued their investigations from [28] in [29, 30]. We quote few sentences from [29]: "So it came about that the more we read and the more we probed many and diverse arguments, the more surely we were pressed to the strange conclusion that it was in the comets where we must seek for the early development of life . . . we find that the case for a cometary origin of life, as well as for a continuing infall onto the Earth of bacteria and viruses, is reasonably established . . .". This book is a painstaking report about indirect proofs under the assumption that the above hypothesis is true. However the authors have no direct proofs (at least at the moment when the book was written).

32. Geometry on Manifolds II (Forms)

32.1. Tensors and Differential Forms

32.1.1. The Vectors $\frac{\partial}{\partial x^k}$ and $\mathrm{d}x^k$. Tensor Products

We refer to the considerations described in Chap. 29. In Subsec. 29.2.2. we had introduced tensors (tensor fields, to be more precise) as local geometric objects with the transformation behaviour expressed by (29.2.2/1) ($w=0$). Now we seek to establish a more compact notation, dropping the representation in components and interpreting tensors as multilinear forms.

Basis vectors. Let M be an n-dimensional orientable C^∞-manifold in the sense of Def. 29.1.2, and let U be a local chart of an associated atlas. With respect to the Cartesian co-ordinates x^k in U, we form the vector ξ with the components $\xi_a=1$ for $a=l$ and $\xi_a=0$ for $a\neq l$. Here l is a given number, and a runs from 1 to n. (Following our previous convention, we speak of vectors and tensors instead of vector fields and tensor fields.) If, using the covariant transformation law (29.2.2/1), this vector is extended to all the local charts of the atlas, then a covariant vector is obtained, which shall be denoted by $\frac{\partial}{\partial x^l}$ from now on. The corresponding contravariant extension gives a contravariant vector, denoted by $\mathrm{d}x^l$. $\mathrm{d}x^l$ has nothing to do with an infinitesimal quantity. The notations $\frac{\partial}{\partial x^l}$ and $\mathrm{d}x^l$ are, however, very suggestive because they transform according to the usual rules (chain rule, total differential). $\frac{\partial}{\partial x^l}$ and $\mathrm{d}x^l$ are called basis vectors.

Basis tensors. $\frac{\partial}{\partial x^k}\otimes\ldots\otimes\frac{\partial}{\partial x^l}$ denotes the covariant extension, in the sense of (29.2.2/1), of the tensor ξ with $\xi_{a\ldots b}=1$ for $a=k$ to $b=l$, and $\xi_{a\ldots b}=0$ otherwise. Here k, ..., l are any fixed numbers between 1 and n. The contravariant extension is denoted by $\mathrm{d}x^k\otimes\ldots\otimes\mathrm{d}x^l$. Of course one may also form corresponding mixed-variant extensions, for instance $dx^k \otimes \ldots \otimes \frac{\partial}{\partial x^l}$. These tensors (more precisely: tensor fields) are called basis tensors.

Tensors (*new interpretation*). If $T_{k\ldots l}$ is a covariant tensor in the sense of Theorem 29.2.2, then

$$T = T_{k\ldots l} dx^k \otimes \ldots \otimes dx^l$$

is a scalar (Theorem 29.3.1/1(d)). In the sequel we shall call the scalar T, with the representation by basis tensors as given above, a covariant tensor (of the corresponding rank). This is a co-ordinate-free notation. Changing the co-ordinates from x^k to x'^k gives

$$T = T_{k\ldots l} dx^k \otimes \ldots \otimes dx^l = T'_{k\ldots l} dx'^k \otimes \ldots \otimes dx'^l .$$

If one inserts the transformations for dx^l, one automatically obtains the correct transformations (in the sense of Theorem 29.2.2). So it is possible to define tensors in this way too. Analogous interpretations shall be applied to contravariant and mixed-variant tensors, where one has to substitute, as applicable, $\frac{\partial}{\partial x^l}$ for dx^l.

Definition. (a) *If*

$$T = T_{a\ldots b} dx^a \otimes \ldots \otimes dx^b \quad \textit{and} \quad S = S_{a\ldots b} dx^a \otimes \ldots \otimes dx^b$$

are tensors of equal rank, and if λ and μ are real numbers, then

$$\lambda T + \mu S = (\lambda T_{a\ldots b} + \mu S_{a\ldots b})\, dx^a \otimes \ldots \otimes dx^b .$$

(b) (*Tensor product*). *If S and T are any two covariant tensors, then*

$$S \otimes T = S_{a\ldots b} T_{c\ldots d} dx^a \otimes \ldots \otimes dx^b \otimes dx^c \otimes \ldots \otimes dx^d .$$

Remark. This is the present variant of Theorem 29.3.1/1. In particular, $\lambda T + \mu S$ and $S \otimes T$ are again covariant tensors. It is clear how the definition for contravariant and mixed-variant tensors has to be modified.

32.1.2. The Alternating Product and the Exterior Product

If $\mathfrak{S}_k$ is the permutation group of Subsec. 28.6.2., then $(-1)^p$, for $p \in \mathfrak{S}_k$, has the meaning indicated there. In all of our considerations we again refer to an n-dimensional C^∞-manifold.

Definition. (a) (*Alternating product*). *If T is a covariant tensor of type* (0, k), *then*

$$\text{Alt } T = \frac{1}{k!} \sum_{p \in \mathfrak{S}_k} (-1)^p T_{p(a,\ldots,b)} dx^a \otimes \ldots \otimes dx^b . \tag{1}$$

(b) *A covariant tensor of type* (0, k) *is called a k-form if $T = \text{Alt } T$.*

(c) (*Exterior product, or wedge product*). *If T is a k-form, and if S is an l-form, then*

$$T \wedge S = \frac{(k+l)!}{k!\,l!} \text{Alt } (T \otimes S) .$$

Remark 1. (1) is the covariant version of (29.3.1/2). Alt T is again a tensor of Type (0, k), called the alternating part of T.

Remark 2. We make the convention that scalars shall also be called 0-forms; 1-forms are covariant vectors. If $S = \lambda$ is a 0-form, then we set $T \wedge S = S \wedge T = \lambda T$.

Lemma. (a) *T is a k-form if and only if $T_{p(a,\dots,b)} = (-1)^p\, T_{a\dots b}$ for all $p \in \mathfrak{S}_k$.*

(b) *If $k > n$, then there is no non-trivial k-form.*

(c) *If $0 \leqq k \leqq n$, then at every point of an n-dimensional C^∞-manifold there exist exactly $\binom{n}{k}$ linearly independent k-forms.*

Remark 3. If T is a k-form, then it follows from (a) that $T_{a\dots b\dots b\dots c} = 0$ (two equal indices, no summation). This proves (b). Further, in the case (c) it is noted that every k-form can, at every point of the manifold, be obtained as a linear combination of the $\binom{n}{k}$ special k-forms

$$\sum_{p \in \mathfrak{S}_k} (-1)^p\, \mathrm{d}x^{p(a)} \otimes \dots \otimes \mathrm{d}x^{p(b)} \quad \text{with} \quad 1 \leqq a < \dots < b \leqq n\,.$$

Here $p(a, \dots, b) = \big(p(a), \dots, p(b)\big)$.

Theorem. (a) *If T is a k-form, and if S is an l-form, then $T \wedge S$ is a $(k+l)$-form.*

(b) *Let $T_{(r)}$ be any k_r-forms. Then*

$$T_{(1)} \wedge (T_{(2)} \wedge T_{(3)}) = (T_{(1)} \wedge T_{(2)}) \wedge T_{(3)} = T_{(1)} \wedge T_{(2)} \wedge T_{(3)}$$

and (if $k_2 = k_3$)

$$T_{(1)} \wedge (T_{(2)} + T_{(3)}) = T_{(1)} \wedge T_{(2)} + T_{(1)} \wedge T_{(3)}\,.$$

Reduced representation. In particular one has $\mathrm{d}x^k \wedge \mathrm{d}x^l = -\mathrm{d}x^l \wedge \mathrm{d}x^k$. But this shows that every k-form can be written as

$$T = k! \sum_{1 \leqq a < \dots < b \leqq n} T_{a\dots b}\, \mathrm{d}x^a \wedge \dots \wedge \mathrm{d}x^b\,. \tag{2}$$

This is the reduced representation, because the summation is not, as prescribed by the sum convention, taken over the full range of the indices. The missing coefficients are calculated from $T_{p(a,\dots,b)} = (-1)^p\, T_{a\dots b}$. In particular, $\mathrm{d}x^a \wedge \dots \wedge \mathrm{d}x^b$ (k factors, with $1 \leqq a < \dots < b \leqq n$) is a (linearly independent) basis in the space of k-forms.

Remark 4. If one admits the summation in (2) to be taken over the full range of $a, \dots, b$ from 1 to n, then the factor $k!$ does not occur. Further when changing from the local co-ordinates x^k to the local co-ordinates x'^l one obtains

$$T = T_{a\dots b}\, \mathrm{d}x^a \wedge \dots \wedge \mathrm{d}x^b = T_{a\dots b} \left(\frac{\partial x^a}{\partial x'^c}\, \mathrm{d}x'^c\right) \wedge \dots \wedge \left(\frac{\partial x^b}{\partial x'^d}\, \mathrm{d}x'^d\right), \tag{3}$$

which is calculated in the usual, obvious way.

32.1.3. Exterior Derivative

Lemma. *If $T = T_{a\dots b}\, \mathrm{d}x^a \wedge \dots \wedge \mathrm{d}x^b$ is a k-form, then*

$$\mathrm{d}T = \frac{\partial T_{a\dots b}}{\partial x^c}\, \mathrm{d}x^c \wedge \mathrm{d}x^a \wedge \dots \wedge \mathrm{d}x^b$$

is a $(k+1)$-form (exterior derivative).

Remark 1. Recall $T_{a\dots b} = T_{a\dots b}(x)$. The lemma is a generalization of Theorem 29.3.2(a). We make the convention that $\mathrm{d}T = f_{,c}\, \mathrm{d}x^c$ if $T = f(x)$ is a 0-form (scalar).

Theorem. *If T is a k-form, and if S is an l-form, then*

$$\mathrm{d}(T+S)=\mathrm{d}T+\mathrm{d}S, \quad d^2T=\mathrm{d}(\mathrm{d}T)=0\,, \qquad (1)$$
$$\mathrm{d}(T\wedge S)=\mathrm{d}T\wedge S+(-1)^k T\wedge \mathrm{d}S$$

with $k=l$ in the first equality

Remark 2. It is of special importance that $\mathrm{d}^2T=0$. This property follows from the antisymmetry of the exterior product and from the commutativity of partial derivatives.

32.1.4. n-Forms

We consider an n-dimensional metric space as defined by Def. 29.5.6/1. The associated fundamental tensor shall now be written in the form $g=g_{kl}\mathrm{d}x^k\otimes \mathrm{d}x^l$. Further let $|g|=|\det g_{kl}|$. According to Theorem 29.3.1/2 one has $\sqrt{|g'(x')|}=$ $=\dfrac{\partial(x^p)}{\partial(x'^q)}\sqrt{|g(x)|}$, where the symbols have the meaning indicated previously. According to (32.1.2/2), an n-form T in its reduced representation has the form $T=n!\ T_{1\dots n}\mathrm{d}x^1\wedge\dots\wedge\mathrm{d}x^n$.

Theorem. $\mu=\sqrt{|g(x)|}\ \mathrm{d}x^1\wedge\dots\wedge\mathrm{d}x^n$ *is an n-form.*

Remark. What is meant is that, for co-ordinate transformations $x'^k=x'^k(x^l)$, the following relation holds:

$$\sqrt{|g(x)|}\ \mathrm{d}x^1\wedge\ \dots\ \wedge\mathrm{d}x^n=\sqrt{|g'(x')|}\ \mathrm{d}x'^1\wedge\ \dots\ \wedge\mathrm{d}x'^n\,.$$

This is easily verified on the basis of (32.1.2/3) and of the above transformation law for $\sqrt{|g(x)|}$.

32.1.5. Theorem of Poincaré

If A is a $(k-1)$-form in an n-dimensional orientable C^∞-manifold, then $\mathrm{d}T=0$ for the k-form $T=\mathrm{d}A$. We will state the converse to this proposition.

Theorem (*Poincaré*). *If T is a k-form with $\mathrm{d}T=0$, then locally (i.e., in a suitable neighbourhood of every point of the manifold) there exists a $(k-1)$-form A such that $T=\mathrm{d}A$.*

Remark 1. For a given T, $\mathrm{d}A=T$ is a system of linear partial differential equations of first order for the components of A. The theorem gives a positive answer to the question of whether the necessary integrability condition $\mathrm{d}T=0$ is also sufficient (at least locally). If in a local chart one has (in the reduced representation)

$$T=k!\sum_{1\leqq a<\dots<b\leqq n} T_{a\dots b}\mathrm{d}x^a\wedge\ \dots\ \wedge\mathrm{d}x^b\,,$$
$$A=(k-1)!\sum_{1\leqq c<\dots<d\leqq n} A_{c\dots d}\mathrm{d}x^c\wedge\ \dots\ \wedge\mathrm{d}x^d\,,$$

then, locally,

$$A_{c\dots d}=\frac{1}{k}\,x^l\int_0^1 T_{lc\dots d}(tx)\ t^{k-1}\mathrm{d}t$$

is a solution of $T=\mathrm{d}A$ (cf. [40], p. 218).

Remark 2. If A_0 is a special solution of $\mathrm{d}A=T$, then the general solution is given by $A=A_0+$ $+\mathrm{d}B$, where B stands for an arbitrary $(k-2)$-form.

32.2. Integral Calculus on Manifolds

32.2.1. Integrals of n-Forms

We shall now extend the investigations made in Subsec. 29.3.3. From now on we shall always assume that M be an n-dimensional metric space (in the sense of Def. 29.5.6/1) with the fundamental tensor $g = g_{kl} \mathrm{d}x^k \otimes \mathrm{d}x^l$. Given an n-form T in the reduced representation,

$$T = n!\, T_{1\ldots n} \mathrm{d}x^1 \wedge \ldots \wedge \mathrm{d}x^n \tag{1}$$

with a compact support supp T, where, for the time being, we assume that supp T lies in a suitable local chart U (Fig. 32.1). If $T_{1\ldots n}(x)$ is integrable in the Lebesgue sense, then we set

$$\int T = n! \int T_{1\ldots n}(x)\, \mathrm{d}x\,.$$

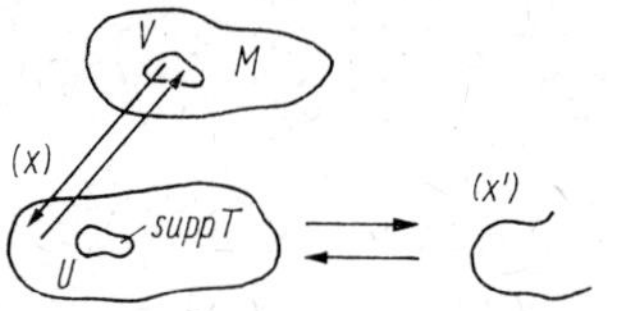

Fig. 32.1

It is easily seen that this definition is independent of the choice of the local co-ordinates. This follows from $T'_{1\ldots n}(x') = T_{1\ldots n}(x) \dfrac{\partial(x^l)}{\partial(x'^k)}$, where $x'^k = x'^k(x^l)$ is an admissible co-ordinate transformation.

Definition. (a) *If T is an integrable n-form on M with a compact support, and if $\{\varphi_k\}_{k=1}^{\infty}$ is a partition of unity in the sense of Lemma* 29.1.3, *then*

$$\int T = \sum_{k=1}^{\infty} \int T\varphi_k\,.$$

(b) *If f is an integrable function on M with a compact support, and if μ is the n-form of Theorem* 32.1.4, *then*

$$\int f\mu = \int T \quad \textit{where} \quad T = f(x)\mu\,.$$

Remark 1. T is said to be locally integrable if $T_{1\ldots n}(x)\,\varphi_k(x)$ is Lebesgue-integrable for each of the above functions φ_k in the associated local chart. Since T has a compact support, there is only a finite number of non-zero summands in $\sum_{k=1}^{\infty} \int T\varphi_k$. (a) is meaningful and does not depend on the particular choice of the partition of unity.

Remark 2. If $f(x)$ is an arbitrary locally integrable function on M, and if χ is the characteristic function of a compact set D in M, then we write $\int_D f\mu = \int f\chi\mu$.

32.2.2. The de Rham Operator

Theorem. *If $\xi = \xi^k \dfrac{\partial}{\partial x^k}$ is a contravariant vector, then*

$$*\xi = \sum_{r=1}^{n} (-1)^{r+1} \sqrt{|g(x)|}\, \xi^r(x)\, \mathrm{d}x^1 \wedge \ldots \wedge \mathrm{d}x^{r-1} \wedge \mathrm{d}x^{r+1} \wedge \ldots \wedge \mathrm{d}x^n \tag{1}$$

is an $(n-1)$*-form, and* $\mathrm{d}(*\xi)=(\operatorname{div}\xi)\mu$*, where* μ *is the n-form referred to in Theorem* 32.1.4.

Remark. This means that in changing the co-ordinates from x^k into x'^l in (1) one only needs to replace x by x' (as well as $g(x)$ by $g'(x')$ and $\xi^r(x)$ by $\xi'^r(x')$). According to Subsec. 29.3.2.,

$$\operatorname{div}\xi=\xi^k{}_{;k}=\frac{1}{\sqrt{|g|}}\frac{\partial}{\partial x^k}(\sqrt{|g|}\,\xi^k) \tag{2}$$

is a scalar. The last assertion of the theorem then follows from Lemma 32.1.3.

32.2.3. The Stokes Theorem

As usual, let M be an n-dimensional metric space (in the sense of Def. 29.5.6/1).

Definition. *If* $1\leq m<n$*, then the subset* D *of* M *is called an m-dimensional (smooth) submanifold of* M *if, by choosing a suitable atlas* $\{V_k;F_k\}$*, it is always possible to achieve that*

$$F_k(D\cap V_k)=\{x\mid x\in U_k,\ x^{m+1}=\ldots=x^n=0\}$$

in every chart $U_k=F_k(V_k)$ (Fig. 32.2).

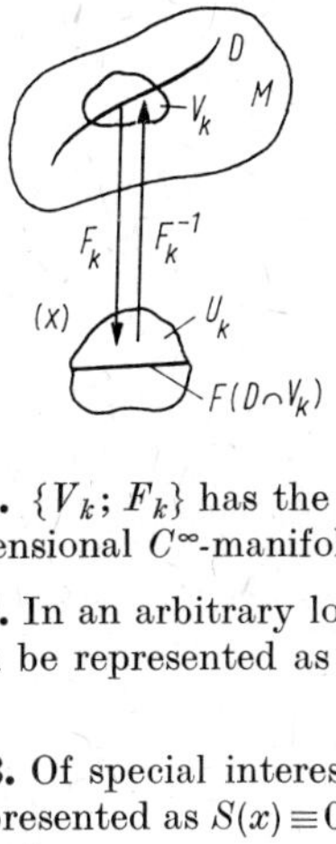

Fig. 32.2

Remark 1. $\{V_k;F_k\}$ has the meaning set out in Subsec. 29.1.2. It is easy to see that D is an m-dimensional C^∞-manifold in the sense of Def. 29.1.2.

Remark 2. In an arbitrary local chart, the image of D (as far as it is covered by this local chart) can be represented as $x^k=x^k(\lambda^r)$, $k=1,\ldots,n$, and $r=1,\ldots,m$ (parametric representation).

Remark 3. Of special interest are $(n-1)$-dimensional submanifolds (hypersurfaces), which can be represented as $S(x)\equiv 0$ or as $S(x)\equiv c$ in local charts, where $S(x)$ is a C^∞-function with grad $S(x)\neq 0$.

Lemma. *If* D *is an m-dimensional submanifold of* M *with the parametric representation* $x^k=x^k(\lambda^r)$ *in local charts, and if* $T=T_{a_1\ldots a_k}\mathrm{d}x^{a_1}\wedge\ldots\wedge\mathrm{d}x^{a_k}$ *is a k-form on* M*, then*

$$T_D=T_{a_1\ldots a_k}\left(\frac{\partial x^{a_1}}{\partial\lambda^{b_1}}\,\mathrm{d}\lambda^{b_1}\right)\wedge\ldots\wedge\left(\frac{\partial x^{a_k}}{\partial\lambda^{b_k}}\,\mathrm{d}\lambda^{b_k}\right) \tag{1}$$

$$=T_{a_1\ldots a_k}(x(\lambda))\,\frac{\partial x^{a_1}}{\partial\lambda^{b_1}}\cdots\frac{\partial x^{a_k}}{\partial\lambda^{b_k}}\,\mathrm{d}\lambda^{b_1}\wedge\ldots\wedge\mathrm{d}\lambda^{b_k}$$

is a k-form on D. Further,

$$(T+S)_D = T_D + S_D\,, \quad (T\wedge S)_D = T_D \wedge S_D\,, \quad (\mathrm{d}T)_D = \mathrm{d}(T_D)\,, \tag{2}$$

where S is an l-form (where k must be equal to l in the first formula of (2)).

Remark 4. The summation in (1) goes from 1 to n over a_s and from 1 to m over b_t. T_D is the restriction of T to D, the induced form.

Theorem. *Let Ω be an open set in M, such that $\bar{\Omega}$ (the closure of Ω) is compact. Further let the boundary $\partial\Omega = \bar{\Omega}\backslash\Omega$ be an $(n-1)$-dimensional hypersurface. If $\xi = \xi^k \dfrac{\partial}{\partial x^k}$ is a contravariant vector, then*

$$\int_{\partial\Omega} (*\xi)_{\partial\Omega} = \int_{\Omega} (\operatorname{div}\xi)\mu = \int_{\Omega} \mathrm{d}(*\xi) \tag{3}$$

Remark 5. $\partial\Omega$ is an $(n-1)$-dimensional compact orientable C^∞-manifold. Theorem 32.2.2 and the above lemma then show that the first integral in (3) can be formed by Def. 32.2.1. Further Theorem 32.2.2 shows that the last two integrals in (3) are meaningful and equal.

Remark 6. With the use of (32.2.2/2) and of local charts it is possible to reduce the theorem to Theorem 9.3.1/2. Here the direction of the normal ν in Theorem 9.3.1/2 must be chosen suitably (which corresponds to fixing an orientation on $\partial\Omega$; cf. [17], p. 26)[1])

32.2.4. Leray Forms

Given an open set Ω in the n-dimensional metric space M (Fig. 32.3). Further let $S(p)$ be a C^∞-function on Ω with grad $S \neq 0$ in Ω. Then

$$S_t = \{p \mid p\in\Omega,\ S(p) = t\}$$

is a family of $(n-1)$-dimensional C^∞-hypersurfaces (provided that S_t is not void).

Theorem 1. *If μ is the n-form referred to in Theorem 32.1.4, and if $\mathrm{d}S = S_{,k}\mathrm{d}x^k$, then in Ω there exists an $(n-1)$-form μ_S (Leray form) such that $\mu = \mathrm{d}S\wedge\mu_S$. Here the restriction $(\mu_S)_{S_t}$ of μ_S to S_t is unique.*

Remark 1. The geometric meaning of μ_S becomes clear if one compares $\mu = \mathrm{d}S\wedge\mu_S$ with the conventional infinitesimal classical analogue $\mathrm{d}x = \mathrm{d}\sigma\cdot\mathrm{d}\nu$. If, for instance, $|g(x)| = 1$, then μ corresponds to the n-dimensional volume element dx, $\mathrm{d}S$ corresponds to the vector $|\mathrm{grad}\, S|\mathrm{d}\nu$ (ν being the normal vector), and $(\mu_S)_{S_t}|\,\mathrm{grad}\, S|$ corresponds to the $(n-1)$-dimensional surface element $\mathrm{d}\sigma$ on the surface S_t (Fig. 32.4).

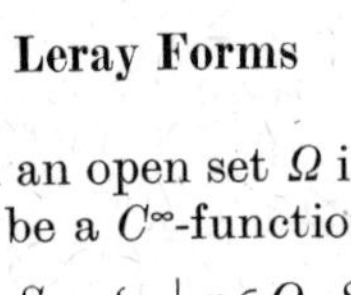

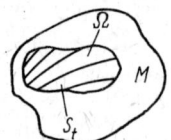

Fig. 32.3

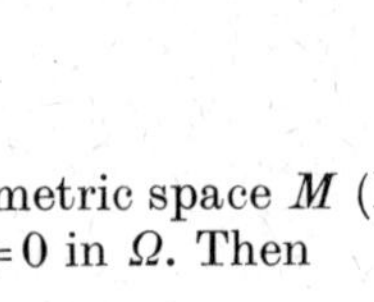

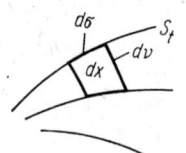

Fig. 32.4

Theorem 2. *If ψ is an integrable function with a compact support in Ω, then*

$$\int \psi\mu = \int \mathrm{d}t \int \psi\mu_S\,.$$

[1]) A recent but nevertheless elementary description of integration theory on manifolds (including Stokes' theorem) may be found in [60].

Further,

$$\frac{\mathrm{d}}{\mathrm{d}t}\int\limits_{S<t}\psi\mu=\int\limits_{S_t}\psi\mu_S \quad \textit{for almost all } t\,.$$

Remark 2. The theorem becomes clear if it is compared with the classical analogue in the sense of Remark 1. $S<t$ means the set $\{p \mid p\in\Omega,\, S(p)<t\}$. In the sense of Lemma 32.2.3, it would be more precise to write $(\mu_S)_{S_t}$ instead of μ_S in the above theorem. Since μ_S is an $(n-1)$-form, all integrals are defined. $\int \mathrm{d}t$ is an integral over R_1. Finally, in the deduction of the theorem it is seen that the orientation on S_t must be suitably chosen (otherwise one has to replace μ_S by $-\mu_S$).

Remark 3. Let $S(p)$ and $T(p)$ be any two C^∞-functions on Ω, with $\mathrm{d}S\neq 0$ and $\mathrm{d}T\neq 0$. Again let $S_t=\{p \mid S(p)=t\}$ and $T_u=\{p \mid T(p)=u\}$ (Fig. 32.5). One has

$$\mathrm{d}S\wedge\mathrm{d}T=S_{,k}T_{,l}\mathrm{d}x^k\wedge\mathrm{d}x^l=\sum_{1\leqq k<l\leqq n}(S_{,k}T_{,l}-T_{,k}S_{,l})\mathrm{d}x^k\wedge\mathrm{d}x^l\,.$$

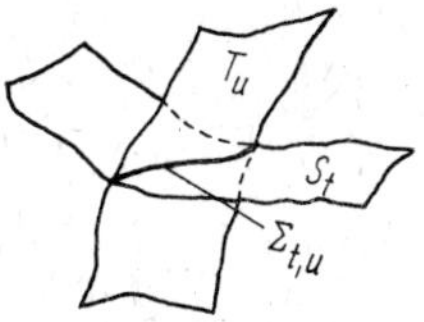

Fig. 32.5

Hence it follows that grad S is parallel to grad T if and only if $\mathrm{d}S\wedge\mathrm{d}T=0$. If one requires that $\mathrm{d}S\wedge\mathrm{d}T\neq 0$, then the surfaces S_t and T_u intersect at a positive angle of intersection (provided that t and u are chosen suitably). Let $\Sigma_{t,u}=S_t\cap T_u$. By iteration it follows from Theorem 1 and Theorem 2 that there exists an $(n-2)$-Leray form $\mu_{S,T}$ such that $\mu=\mathrm{d}S\wedge \mathrm{d}T\wedge\mu_{S,T}$. One has $\mu_{S,T}=-\mu_{T,S}$. Further, with the assumptions of Theorem 2 one has, for almost all t and u,

$$\int\limits_{\Sigma_{t,u}}\psi\mu_{S,T}=\frac{\partial}{\partial u}\frac{\partial}{\partial t}\int\limits_{\substack{S<t\\T<u}}\psi\mu\,.$$

32.3. Distributions on Manifolds

32.3.1. Scalar Distributions

In Chap. 22 we considered distributions defined on domains Ω of R_n. Now we are going to extend Def. 22.1.2/2 from $D'(\Omega)$ to C^∞-manifolds M. Here we are faced with difficulties which can be overcome in different ways. We shall follow the lines of the presentation given in [17]. We shall always assume that M is a metric space in the sense of Def. 29.5.6/1, with the fundamental tensor $g=g_{kl}\mathrm{d}x^k\otimes\otimes\,\mathrm{d}x^l$. As in Subsec. 32.1.4., let $|g|=|\det g_{kl}|$. Finally, we denote by $D(M)$ (by analogy with Subsec. 22.1.2.) the complex-valued variant of $C_0^\infty(M)$ as referred to in Subsec. 29.1.3., that means the class of all complex-valued functions with compact support that are differentiable on M up to arbitrary order.

Definition. *A complex-valued linear form T on $D(M)$ is called distribution on M if for every local chart U there exists a distribution $T_U\in D'(U)$ such that $T(\varphi)=$ $=T_U(\sqrt{|g(x)|}\,\varphi(x))$ for all $\varphi\in D(U)$. The space of the distributions on M is denoted by $D'(M)$.*

Remark 1. Here T is called a linear form if $T(\lambda_1\varphi_1+\lambda_2\varphi_2)=\lambda_1 T(\varphi_1)+\lambda_2 T(\varphi_2)$ for arbitrary complex numbers λ_1 and λ_2 and for $\varphi_1\in D(M)$ and $\varphi_2\in D(M)$. The local chart U is an open set in R_n. One has $U=F(V)$, where V is an open set in M and F is a homeomorphism (cf. Subsec. 29.1.2.; Fig. 32.6). So, in the above definition, it would be more precise to write $\varphi\in D(M)$ with $\operatorname{supp}\varphi\subset V$ and $(F\varphi)(x)$ instead of $\varphi(x)$. But there is no fear of confusion. Of course $D'(U)$ has the previous meaning of Subsec. 22.1.2. However, since $\sqrt{|g(x)|}\,\varphi(x)\in D(U)$, $T_U(\sqrt{|g(x)|}\,\varphi(x))$ is meaningful.

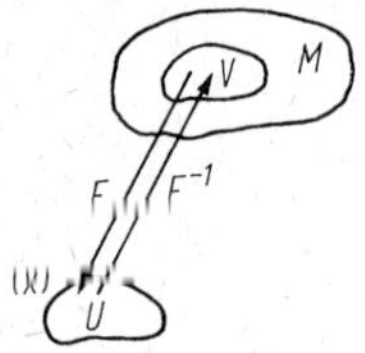

Fig. 32.6

Remark 2. Distributions are added and multiplied by complex-valued C^∞-functions in a way analogous to that set out in Def. 22.1.4/1(b, c). $D'(M)$ then becomes a linear space. The differentiation in the sense of Def. 22.1.4/1(a), however, is somewhat more complicated, leading to the definition of tensor distributions.

Regular distributions: By $L_1^{\text{loc}}(M)$ we denote the class of all complex-valued locally integrable functions on M, in the sense of Remark 32.2.1/1 (complex variant). If $f\in L_1^{\text{loc}}(M)$ and $\varphi\in D(M)$, then

$$T(\varphi)=\int f\varphi\mu \tag{1}$$

is meaningful according to Def. 32.2.1. One has $T\in D'(M)$. For, if $\varphi(x)\in D(U)$ in the sense of the above notation, then by Subsec. 32.2.1. one has

$$T(\varphi)=\int \sqrt{|g(x)|}\, f(x)\,\varphi(x)\,\mathrm{d}x\,,$$

where $f(x)$ is again the function transferred to U. So $T_U\in D'(U)$ equals $\sqrt{|g(x)|}\,f(x)$. This also gives us a motive for the above definition. Moreover it is seen that the correspondence between $f\in L_1^{\text{loc}}(M)$ and the associated distributions in the sense of (1) is one to one, provided that one identifies two functions $f(x)\in L_1^{\text{loc}}(M)$ and $g(x)\in L_1^{\text{loc}}(M)$ whose images $f(x)$ and $g(x)$ in local charts coincide a.e. (in the Lebesgue sense). In this sense (by analogy with Subsec. 22.1.3.) we write $L_1^{\text{loc}}(M)\subset D'(M)$, and $f\in L_1^{\text{loc}}(M)$ is called a regular distribution.

δ-distribution: If $p\in M$ and $\varphi\in D(M)$, then we set $\delta_p(\varphi)=\varphi(p)$. This is the analogue to the δ-distribution of Subsec. 22.1.3. One has $\delta_p\in D'(M)$. If U is a local chart with $Fp\in U$, then, for $\varphi\in D(U)$,

$$(F^{-1}\varphi)(p)=\delta_p(F^{-1}\varphi)=(\delta_p)_U\,(\sqrt{|g(x)|}\,\varphi(x)),\quad \text{and hence}$$

$$(\delta_p)_U(\varphi)=\frac{\varphi(Fp)}{\sqrt{|g(Fp)|}},$$

where for this time we have distinguished, for the sake of clearness, between $\varphi(x)\in D(U)$ and its image $F^{-1}\varphi\in D(M)$.

Compatibility (Fig. 32.7). If $V\cap V'\neq 0$ and $\varphi\in D(V\cap V')$, then let $\varphi(x)$ and $\varphi'(x')$ be the functions transferred to the local charts U and U'. If $T\in D'(M)$, then

$$T_U(\sqrt{|g(x)|}\,\varphi(x))=T_{U'}(\sqrt{|g'(x')|}\,\varphi'(x'))\,. \tag{2}$$

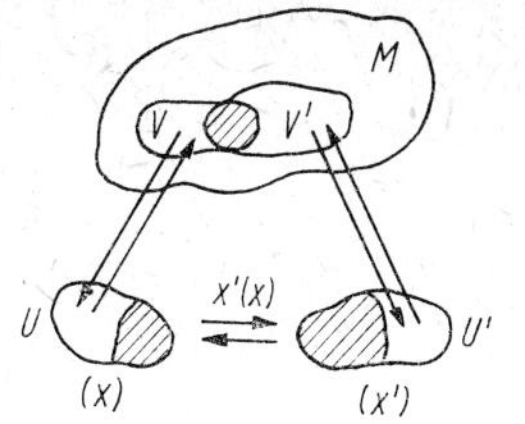

Fig. 32.7

Theorem. (**a**) *Let $\{V_k; F_k\}$ be an atlas of M with the local charts $U_k = F_k(V_k)$. If the distributions $T_{U_k} \in D'(U_k)$ satisfy the compatibility condition (2) for arbitrary pairs (U_k, U_l) instead of (U, U') then there exists a unique distribution $T \in D'(M)$ in the sense of the above definition (with U_k instead of U).*

(**b**) *A complex-valued linear form T on $D(M)$ is a distribution if and only if for every compact set Ω in M there exist a positive number C and a natural number N such that, for all $\varphi \in D(M)$ with* supp $\varphi \subset \Omega$,

$$|T(\varphi)| \leqq C \sup_{p \in \Omega} \sum_{|\alpha| \leqq N} |\varphi_{;\alpha_1 \ldots \alpha_n}(p)| . \tag{3}$$

Remark 3. Part (a) is a localization theorem in the sense of Theorem 22.1.4. So it will suffice to consider distributions locally, by analogy with the local geometric object considered in Subsec. 29.2.1. If $\{\varphi_k\}$ is a partition of unity associated to $\{V_k\}$, as referred to in Lemma 29.1.3, then it is possible to determine T by

$$T(\varphi) = \sum_{k=1}^{\infty} T_{U_k}(\sqrt{|g(x)|}\, \varphi_k(x)\, \varphi(x)), \quad \varphi \in D(M) .$$

Remark 4. (3) is best interpreted in local co-ordinates. It is the analogue to (22.1.5/1). As in that case, $\alpha = (\alpha_1, \ldots, \alpha_n)$ is a multiple index with $|\alpha| = \sum_{l=1}^{n} \alpha_l$. Further, $\varphi_{;\alpha_1 \ldots \alpha_n}$ is a tensor of type $(0, |\alpha|)$, where φ is subjected to a total of $|\alpha|$ covariant differentiations with respect to $\alpha_1, \ldots, \alpha_n$. The formula shows that distributions can also be differentiated, but the factor $\sqrt{|g(x|)}$ in the above definition suggests covariant differentiation.

32.3.2. Tensor Distributions

Definition. (**a**) *$D^{(r,s)}(M)$ is the class of all tensors φ of type (r, s) with compact support and complex-valued components that are differentiable up to arbitrary order.*

(**b**) *A complex-valued linear form T on $D^{(s,r)}(M)$ is called a tensor distribution of type (r, s) if for every local chart U there exists a system of distributions $T_U{}^{a_1 \ldots a_r}{}_{b_1 \ldots b_s} \in D'(U)$ such that*

$$T(\varphi) = T_U{}^{a_1 \ldots a_r}{}_{b_1 \ldots b_s} (\sqrt{|g(x)|}\, \varphi^{b_1 \ldots b_s}{}_{a_1 \ldots a_r}(x)) \tag{1}$$

for all $\varphi = \varphi^{a_1 \ldots a_s}{}_{b_1 \ldots b_r} \dfrac{\partial}{\partial x^{a_1}} \otimes \ldots \otimes \dfrac{\partial}{\partial x^{a_s}} \otimes \mathrm{d}x^{b_1} \otimes \ldots \otimes \mathrm{d}x^{b_r} \in D^{(s,r)}(M)$ *with compact support in U. The space of these tensor distributions is denoted by $D'^{(r,s)}(M)$.*

Remark 1. In part (a) we now admit (in contrast to previous definitions) that the components of the tensor

$$\varphi = \varphi^{a_1 \ldots a_r}{}_{b_1 \ldots b_s} \frac{\partial}{\partial x^{a_1}} \otimes \ldots \otimes \frac{\partial}{\partial x^{a_r}} \otimes \mathrm{d}x^{b_1} \otimes \ldots \otimes \mathrm{d}x^{b_s}$$

are complex-valued. By analogy with the notation in Subsec. 32.3.1., $\varphi^{a_1 \ldots a_r}{}_{b_1 \ldots b_s}(x)$ in (1) is the corresponding component in the local chart U. The requirement that these components be differentiable up to arbitrary order is meaningful, because M is a C^∞-manifold. The support of φ is again determined in the local charts as the closure of the set of all points x for which at least one of the components $\varphi^{a_1 \ldots a_r}{}_{b_1 \ldots b_s}(x)$ of φ is different from zero.

Remark 2. Part (b) of the definition generalizes Def. 32.3.1. Thus, in particular, $D'^{(0,0)}(M) = D'(M)$. Addition of tensor distributions and multiplication by complex numbers are defined in an obvious way. This makes $D'^{(r,s)}(M)$ a linear space.

Regular tensor distributions: If the tensor T of type (r, s) is locally integrable (i.e., if the components of T are locally integrable), then

$$T(\varphi) = \int T^{a_1 \ldots a_r}{}_{b_1 \ldots b_s} \varphi^{b_1 \ldots b_s}{}_{a_1 \ldots a_r} \mu \tag{2}$$

with $\varphi = \varphi^{b_1 \ldots b_s}{}_{a_1 \ldots a_r} \dfrac{\partial}{\partial x^{b_1}} \otimes \ldots \otimes \mathrm{d}x^{a_r} \in D^{(s,r)}(M)$, is a tensor distribution. If one again identifies locally integrable tensors whose components are equal a.e., then one obtains a one-to-one correspondence between these locally integrable tensors and the associated distributions in the sense of (2).

32.3.3. Covariant Derivative and Coderivative of Distributions

Definition 1 (*coderivative*). *For* $\varphi = \{\varphi^{a_0 a_1 \ldots a_s}{}_{b_1 \ldots b_r}\} \in D^{(s+1,r)}(M)$,

$$(\delta\varphi)^{a_1 \ldots a_s}{}_{b_1 \ldots b_r} = -\varphi^{a_0 a_1 \ldots a_s}{}_{b_1 \ldots b_r; a_0}$$

are the components of $\delta\varphi$.

Remark 1. It is immediately seen that $\delta\varphi \in D^{(s,r)}(M)$.

Lemma. *If* $T = \{T^{b_1 \ldots b_r}{}_{a_1 \ldots a_s}\} \in D^{(r,s)}(M)$, *and if* $\varphi = \{\varphi^{a_0 a_1 \ldots a_s}{}_{b_1 \ldots b_r}\} \in D^{(s+1,r)}(M)$, *then*

$$\int T^{b_1 \ldots b_r}{}_{a_1 \ldots a_s; a_0} \varphi^{a_0 a_1 \ldots a_s}{}_{b_1 \ldots b_r} \mu = \int T^{b_1 \ldots b_r}{}_{a_1 \ldots a_s} (\delta\varphi)^{a_1 \ldots a_s}{}_{b_1 \ldots b_r} \mu \,. \tag{1}$$

Remark 2. This formula is the starting point for the definition of derivatives of tensor distributions (especially also of scalar distributions).

Definition 2. (a) (*covariant derivative*). *Let* $T = \{T^{b_1 \ldots b_r}{}_{a_1 \ldots a_s}\} \in D'^{(r,s)}(M)$. *Then* $\nabla T = \{T^{b_1 \ldots b_r}{}_{a_1 \ldots a_s; a_0}\}$, *where*

$$(\nabla T)(\varphi) = T(\delta\varphi) \quad \textit{for} \quad \varphi \in D^{(s+1,r)}(M) \,.$$

(b) (*coderivative*). *If* $T = \{T^{a_0 a_1 \ldots a_s}{}_{b_1 \ldots b_r}\} \in D'^{(s+1,r)}(M)$, *then* $\delta T = \{(\delta T)^{a_1 \ldots a_s}{}_{b_1 \ldots b_r}\}$, *where* $(\delta T)(\varphi) = T(\nabla \varphi)$ *for* $\varphi \in D^{(r,s)}(M)$.

Remark 3. Here $\nabla\varphi = \{\varphi^{a_1 \ldots a_r}{}_{b_1 \ldots b_s; b_0}\}$ for $\varphi = \{\varphi^{a_1 \ldots a_r}{}_{b_1 \ldots b_s}\}$. The definitions are meaningful and agree with (1) for smooth distributions. In case (a) one has $\nabla T \in D'^{(r,s+1)}(M)$, and in case (b) one has $\delta T \in D'^{(s,r)}(M)$. Here s and r are integers, $r \geqq 0$ and $s \geqq 0$.

div and grad: If $T \in D'^{(1,0)}(M)$, then let $\operatorname{div} T = -\delta T$; so in particular one has $\operatorname{div} T \in D'(M)$.

If $T \in D'(M)$, then let grad $T = \nabla T$; so in particular one has grad $T \in D'^{(0,1)}(M)$. Hence

$$\operatorname{grad} T(\varphi) = -T(\operatorname{div} \varphi) \quad \text{for} \quad T \in D'(M), \quad \varphi \in D^{(1,0)}(M)\,, \tag{2}$$

$$\operatorname{div} T(\varphi) = -T(\operatorname{grad} \varphi) \quad \text{for} \quad T \in D'^{(1,0)}(M), \quad \varphi \in D(M)\,. \tag{3}$$

In particular, grad $\varphi = \{\varphi_{,k}\}$ for $\varphi \in D(M)$, and

$$\operatorname{div} \varphi = \varphi^k{}_{;k} = \frac{1}{\sqrt{|g(x)|}} \frac{\partial}{\partial x^k} \left(\sqrt{|g(x)|}\, \varphi^k(x)\right) \tag{4}$$

for $\varphi \in D^{(1,0)}(M)$. The validity of the last formula follows from Theorem 29.3.2(b), with the use of the normal co-ordinates of Lemma 29.5.6/2.

32.3.4. The Wave Operator

In this subsection, let M be an n-space-time, that is an n-dimensional metric space with the signature $n-2$ (Lorentz metric); cf. Subsec. 29.5.1. For $n=4$ one has the usual space-time.

Definition (*wave operator*). *If $T \in D'(M)$, then*

$$\Box T = \operatorname{div}\, (g^{kl}(\operatorname{grad}\, T)_l) = \operatorname{div}\, (g^{kl}(\nabla T)_l)\,. \tag{1}$$

Remark 1. The definition is meaningful. One has $g^{kl}(\nabla T)_l \in D'^{(1,0)}(M)$, where multiplications of tensor distributions by C^∞-functions are defined in the usual way. In particular, $\Box T \in D'(M)$ then is again a scalar distribution.

Remark 2. If T is a smooth function, for instance $T \in D(M)$, then it follows from (32.3.3/4) that

$$\Box T = \frac{1}{\sqrt{|g(x)|}} \frac{\partial}{\partial x^k} \left(\sqrt{|g(x)|}\, g^{kl}(x) \frac{\partial T}{\partial x^l}(x)\right). \tag{2}$$

In normal co-ordinates (Lemma 29.5.6/2) one then has

$$(\Box T)(x_0) = \sum_{r=1}^{n-1} \frac{\partial^2 T}{(\partial x_0^r)^2} - \frac{\partial^2 T}{(\partial x_0^n)^2}\,. \tag{3}$$

But this is the usual wave operator. Hence (2) is the invariant version of (3), and (1) is the extension of this version to distributions.

Remark 3. From (32.3.3/2) and (32.3.3/3) it follows that

$$\Box T(\varphi) = T(\Box \varphi) \quad \text{for} \quad T \in D'(M) \quad \text{and} \quad \varphi \in D(M)\,. \tag{4}$$

Hence $\Box$ is formally self-adjoint.

General wave operator. Let $a = \{a^l\}$ be a vector with C^∞-components, and let b be a C^∞-function, and $T \in D'(M)$. Then $PT \in D'(M)$, where

$$PT = \Box T + a^l(\nabla T)_l + bT\,. \tag{5}$$

The analogue to (4) now reads, for $T \in D'(M)$ and $\varphi \in D(M)$,

$$(PT)(\varphi) = T(P^*\varphi) \quad \text{with} \quad P^*\varphi = \Box \varphi - \operatorname{div}\,(a\varphi) + b\varphi\,. \tag{6}$$

32.3.5. Distributions of Type $f(S)$

Consider the same situation as in Subsec. 32.2.4. The open set Ω, the C^∞-function $S(p)$, the surfaces S_t, and the Leray form μ_S shall have the meaning indicated there.

Theorem. *If* $f \in D'(R_1)$, *and if*

$$f(S)(\varphi) = f(\int_{S_t} \varphi \mu_S) \quad \text{for} \quad \varphi \in D(\Omega) , \tag{1}$$

then $f(S) \in D'(\Omega)$, *and* $\nabla f(S) = f'(S) \nabla S$.

Remark 1. (1) is meaningful, because $\int_{S_t} \varphi \mu_S \in D(R_1)$. $f' \in D'(R_1)$ is the distribution derivative of f. Then $f'(S)$ has to be determined as shown for (1).

Remark 2. If $f \in L_1^{\text{loc}}(R_1)$, then it follows from Theorem 32.1.4/2 that

$$f(S)\,(\varphi) = \int f(t) \mathrm{d}t \int_{S_t} \varphi \mu_S = \int f(S(p))\, \varphi(p)\, \mu . \tag{2}$$

This justifies the set-up (1).

Remark 3. $f(S)$ depends continuously on f: $f_j \to f$ in $D'(R_1)$ (that is, $f_j(\psi) \to f(\psi)$ for all $\psi \in D(R_1)$) implies that $f_j(S) \to f(S)$ in $D'(\Omega)$ (that is, $f_j(S)(\varphi) \to f(S)(\varphi)$ for all $\varphi \in D(\Omega)$).

Remark 4. If S and T as well as $\varphi_{S,T}$ have the meaning indicated in Remark 32.2.4/3, and if $f \in D'(R_2)$, then it follows that $f(S, T) \in D'(\Omega)$, where

$$f(S, T)(\varphi) = f(\int_{\Sigma_{t,u}} \varphi \mu_{S,T}) \quad \text{for} \quad \varphi \in D(\Omega) .$$

33. The Wave Equation on Curved Space-Times

33.1. Characteristic Surfaces and Singularities

33.1.1. Characteristic Surfaces

In this chapter, an n-dimensional Lorentz-metric space, in the sense of Remark 29.5.6/2, will be called an n-space-time. If $n = 4$, we speak of a space-time, as in the chapters on General Relativity. In Sec. 33.1. we shall consider n-space-times, whereas later on we shall restrict ourselves to the physically interesting case $n = 4$, i.e., to space-times.

Definition. *Let M be an n-space-time.*

(a) *An m-dimensional submanifold of M is called space-like if every curve $x^k(s)$ that lies in it is space-like, that is if* $g^{kl}(x(s)) \dfrac{\mathrm{d}x^l}{\mathrm{d}s} \dfrac{\mathrm{d}x^k}{\mathrm{d}s} > 0$.

(b) *An $(n-1)$-dimensional surface $S(x) = 0$ is called characteristic surface if* $\mathrm{d}S \neq 0$ *and* $g^{kl}(x)\, S_{,k} S_{,l} = 0$ *on* $S(x) = 0$.

Remark 1. Except for the vertices of characteristic conoids (cf. Remark 2), we shall always assume that all the surfaces considered are smooth (differentiable up to arbitrary order). Then the definition is meaningful. Space-like surfaces have already been described in

Subsec. 31.1.3.; see also Def. 32.2.3, where the concept of a submanifold has been defined. In part (a) above, $1 \leq m \leq n-1$. Part (b) corresponds with Def. 29.5.3. $dS \neq 0$ means that at a point x where $S(x)=0$ not all the derivatives $S_{,k}(x)$ vanish simultaneously. $S(x)$ is always assumed to be an (at least locally defined) C^∞-function.

Remark 2. Characteristic conoids and rays shall have the same meaning as in Subsec. 29.5.3. $dS \neq 0$ is not satisfied at the origin of such a conoid (x_0 in Subsec. 29.5.3.).

Remark 3. If $S(x)=0$ is an $(n-1)$-dimensional surface such that $dS \neq 0$, and if $x^k(s)$ is a curve within this surface, i.e., if $S(x^k(s))=0$, then it follows that $S_{,k}\dfrac{dx^k}{ds}=0$. Hence grad $S = \{S_{,k}\}$ is orthogonal to $S(x)=0$ (in the Lorentz metric).

Theorem. (a) *If $S(x)=0$ is a characteristic surface, then the rays (null geodesics) lying in this surface are identical with the solution curves $x^k(s)$ of* $\dfrac{dx^k}{ds}=g^{kl}S_{,l}(x^r(s))$.

(b) *If $Pf=\Box f+a^l(\nabla f)_l+bf$ is the general wave operator of (32.3.4/5) for arbitrary C^∞-functions f, then an $(n-1)$-dimensional hypersurface (submanifold) $S(x)=0$ is a characteristic surface if and only if the restriction of Pf to $S(x)=0$ can be expressed by the tangential derivatives of the restrictions of $f(x)$ and $(\nabla f)(x)$ to $S(x)=0$.*

Remark 4. By Theorem 29.5.4, rays and null geodesics are identical. Now it is relatively easy to deduce part (a) from

$$g_{kl}\frac{dx^k}{ds}\frac{dx^l}{ds}=g_{kl}g^{ku}S_{,u}g^{lv}S_{,v}=g^{uv}S_{,u}S_{,v}=0\,.$$

Further it is seen that the contravariant version $g^{kl}S_{,l}$ of the orthogonal vector grad S (cf. Remark 3) is a tangential vector.

Remark 5. To explain part (b), we may assume that $S(x)\equiv x^1$. It then follows from (32.3.4/2) that $Pf=g^{11}\dfrac{\partial^2 f}{(\partial x^1)^2}+\ldots$, where the remainder $+\ldots$ includes terms of the form $f(x)$, $f_{,k}(x)$ and $f_{,kr}(x)$, where $k=1,\ldots,n$ and $r=2,\ldots,n$. Hence this remainder can be expressed by the tangential derivatives $\dfrac{\partial}{\partial x^2},\ldots,\dfrac{\partial}{\partial x^n}$ of f and ∇f. So the requirement that Pf be expressible in this way is identical with $g^{11}(x)\equiv 0$ for $S(x)=0$. But since $S(x)\equiv x^1$, the condition that $g^{11}(x)\equiv 0$ for $S(x)=0$ is equivalent with $S(x)=0$ being a characteristic surface. This proves (b).

Remark 6. Part (b) shows that the Cauchy problem for Pf in its usual formulation makes no sense for characteristic surfaces. $f(x)$ and $\nabla f(x)$ cannot be given arbitrarily on a characteristic surface $S(x)=0$ when solutions of $Pf(x)\equiv 0$ are desired in a neighbourhood of this surface. For it follows from $Pf\equiv 0$ and from part (b) that there is a dependence between $f(x)$ and $\nabla f(x)$ on $S(x)=0$. The typical initial value problem for the wave equation with data given on characteristic surfaces has only $f(x)$ given on these surfaces. Later on we shall preferably deal with the Cauchy problem with data given on space-like hypersurfaces. In this case the problem then has the usual formulation.

33.1.2. Initial Value Problems for Characteristic Surfaces and Null Fields

Theorem 1. *If σ is an $(n-2)$-dimensional space-like submanifold, then locally (i.e., in a suitable neighbourhood of an arbitrary point of σ) there exist exactly two different characteristic surfaces which contain σ* (Fig. 33.1).

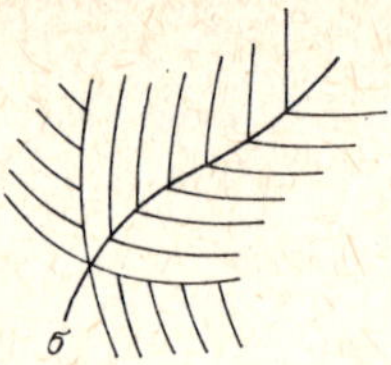

Fig. 33.1

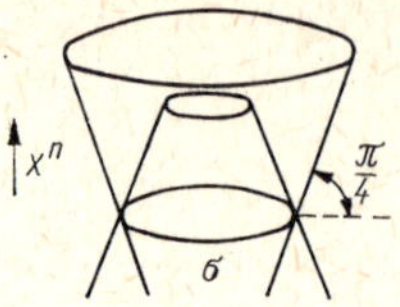

Fig. 33.2

Remark 1. If σ is a closed compact $(n-2)$-dimensional space-like submanifold, then the theorem holds globally, i.e., for a suitable neighbourhood of the total surface σ. Let $g_{kl}(x) \equiv 0$ for $k \neq l$, $g_{kk}(x) \equiv 1$ for $k = 1, \ldots, n-1$, and $g_{nn}(x) = -1$ (n-dimensional variant of the Minkowski space), and let σ be the surface of an $(n-1)$-dimensional ball in the plane $x^n = 0$. Then the two characteristic surfaces desired are the two circular cones that include an angle of $\frac{\pi}{4}$ with $x^n = 0$ and pass through σ (Fig. 33.2). The general case is the locally diffeomorphic distorted variant of this figure (Fig. 33.3).

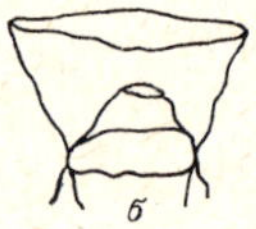

Fig. 33.3

Fig. 33.4

Nonlinear differential equations: Theorem 1 and Remark 1 are the local and the global version, respectively, of the solution of an initial value problem for a nonlinear partial differential equation of first order. Desired is a function $S(x)$ with $dS \neq 0$ which is differentiable up to arbitrary order in a (local or global) neighbourhood of σ, satisfies the equation

$$g^{kl}(x) \frac{\partial S}{\partial x^k} \frac{\partial S}{\partial x^l} = 0 \quad \text{for points } x \text{ with } S(x) = 0 , \tag{1}$$

and whose associated surface $S(x) = 0$ includes the submanifold σ (locally or globally). Other than in the linear theory, one obtains two solutions instead of a unique one (σ is space-like as in the above theorem).

Huygens' construction: As in geometrical optics, the two characteristic surfaces of Theorem 1 can be constructed as wave fronts. For every point $x \in \sigma$ one constructs the characteristic conoid K_x referred to in Theorem 29.5.3/1 (Fig. 33.4). The two characteristic surfaces of Theorem 1 are then (locally or globally) the enveloping surfaces of these conoids, that means $\partial(\bigcup_x K_x)$.

Null fields: The above problem (1) can be extended as follows. In an open set Ω of M, a C^∞-function $S(x)$ with $dS \neq 0$ is desired, such that

$$g^{kl}(x) \frac{\partial S}{\partial x^k} \frac{\partial S}{\partial x^l} = 0 \quad \text{for} \quad x \in \Omega \tag{2}$$

(null fields). Here initial data $S(x) = S_0(x)$ for $x \in \Lambda$ are given on an $(n-1)$-dimensional hypersurface Λ in Ω. $S_0(p)$ is a C^∞-function on Λ, and in addition it is required that $S_0(p) \equiv c = \text{const}$ yields an $(n-2)$-dimensional (smooth) submanifold of M for every admissible c (Fig. 33.5).

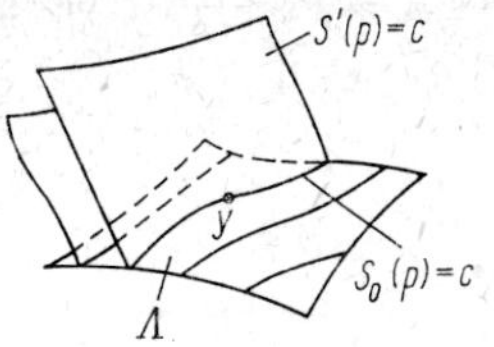

Fig. 33.5

Theorem 2. *Let $\{p \mid S_0(p) \equiv \mathrm{const}\}$ be $(n-2)$-dimensional submanifolds of M.*

(**a**) *If Λ is space-like, then in a suitable neighbourhood U of an arbitrary point $y \in \Lambda$ there exist exactly two null fields $S'(x)$ and $S''(x)$ such that $S'(x) - S''(x) = S_0(x)$ for $x \in U \cap \Lambda$.*

(**b**) *If Λ is a characteristic surface, then in a suitable neighbourhood U of an arbitrary point $y \in \Lambda$ there exists exactly one null field $S(x)$ such that $S(x) = S_0(x)$ for $x \in U \cap \Lambda$.*

Remark 2. The null fields can again be obtained by Huygens' construction.

33.1.3. Caustic

The characteristic surfaces and null fields of Theorems 1 and 2 of Subsec. 33.1.2. can be constructed as follows. (The constructions are started in the proofs of these theorems, and subsequently it is shown that the surfaces obtained in this way have the required properties.) Let σ again be an $(n-2)$-dimensional space-like submanifold (Fig. 33.6). If $y \in \sigma$, then let

$$g_{kl}(y)\, \xi^k t^l = 0$$

for all tangential vectors $t = \{t^l\}$ to σ at y. The set of all these orthogonal vectors $\xi = \{\xi^k\}$ span a 2-dimensional plane (in the space of the vectors). In this plane there are two null vectors, $\xi_{(1)}(y) = \{\xi^k{}_{(1)}(y)\}$ and $\xi_{(2)}(y) = \{\xi^k{}_{(2)}(y)\}$. Now one determines the two geodesics that pass through y and have the directions of these null vectors,

$$\frac{\mathrm{d}^2 x^k}{\mathrm{d}s^2} + \left\{ {k \atop lm} \right\} \frac{\mathrm{d}x^l}{\mathrm{d}s} \frac{\mathrm{d}x^k}{\mathrm{d}s} = 0, \quad x^k(0) = y, \quad \frac{\mathrm{d}x^k}{\mathrm{d}s}(0) = \xi^k , \tag{1}$$

where $\xi^k = \xi^k{}_{(1)}(y)$ or $\xi^k = \xi^k{}_{(2)}(y)$. By varying $y \in \sigma$ one then obtains two surfaces S_1 and S_2 which are spanned by the set of all these null geodesics. These are the desired characteristic surfaces of Theorem 33.1.2/1. If one of these two surfaces, say, $S = S_1$, is fixed, and if σ is parametrized by $x^k = x^k(\lambda^1, \ldots, \lambda^{n-2})$, $k = 1, \ldots, n$, then

$$p = x^k = x^k(s, \lambda^1, \ldots, \lambda^{n-2}) \quad \text{with} \quad k = 1, \ldots, n \tag{2}$$

is a parametrization of S. Here s is the arc length of the associated null geodesic in the sense of (1). At least locally this is a representation of the characteristic surface. In a neighbourhood of a point $y \in \sigma$, the $n-1$ vectors $\left\{\frac{\partial x^k}{\partial s}\right\}$, $\left\{\frac{\partial x^k}{\partial \lambda^1}\right\}$, ..., $\left\{\frac{\partial x^k}{\partial \lambda^{n-2}}\right\}$ are linearly independent. If the null geodesics (1) are elongated, then singularities may occur. One example of this is the vertex of a conoid in the Minkowskian space (Fig. 33.7), where σ is the surface of a 3-dimensional ball in the plane $x^4 = 0$.

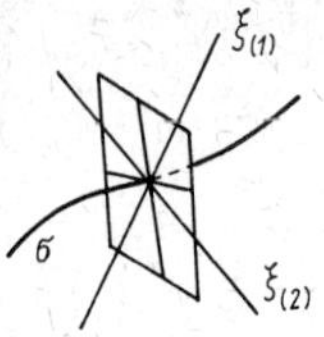

Fig. 33.6

Fig. 33.7

Definition. *The caustic (of the characteristic surface S) is the set of all points p for which (in local co-ordinates) the vectors* $\left\{\frac{\partial x^k}{\partial s}\right\}, \left\{\frac{\partial x^k}{\partial \lambda^1}\right\}, \ldots, \left\{\frac{\partial x^k}{\partial \lambda^{n-2}}\right\}$ *are linearly dependent.*

Remark 1. The above example of a cone in the Minkowskian space is not typical. Generally the caustic lying on S consists of several $(n-2)$-dimensional manifolds. The number of these $(n-2)$-dimensional manifolds generally amounts to $n-2$ (Fig. 33.8).

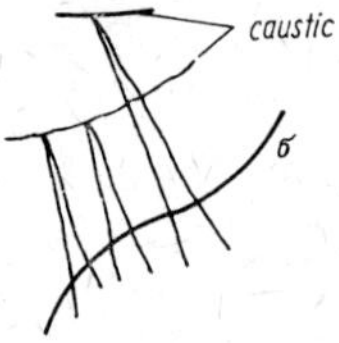

Fig. 33.8

Remark 2. If one deals with null fields in the sense of Theorem 33.1.2/2, then the caustic can be constructed on every characteristic surface $S'(x) \equiv c$. If c is varying then one obtains several $(n-1)$-dimensional surfaces in the general case.

Theorem. *A point $p = x^k(s, \lambda^1, \ldots, \lambda^{n-2})$ of the characteristic surface S belongs to the caustic if and only if*

$$\Delta(s, \lambda) = \det\left(g_{kl}\,\frac{\partial x^k}{\partial \lambda^\alpha}\,\frac{\partial x^l}{\partial \lambda^\beta}\right)^{n-2}_{\alpha,\beta=1}(s, \lambda) = 0\,.$$

Remark 3. For the calculation of $\Delta(s, \lambda)$ one has to use local charts. However, if $\Delta(s, \lambda) = 0$ holds in one local chart, then this is also true in all other admissible local charts.

33.1.4. The Caustic in the Minkowskian Space

If M is the (4-dimensional) Minkowskian space of the special relativity theory (see, for instance, (24.2.2/3)), then the rays coincide with the straight lines which include an angle of $\frac{\pi}{4}$ with the x^4-axis. It is our aim to determine the caustic of a 2-dimensional surface which lies in the plane $x^4 = 0$. For that purpose, some preliminary considerations are of advantage.

Evolute: The analogue to a caustic for a plane curve in the 2-dimensional Euclidean space is the evolute of this curve; cf. Subsec. 17.1.3.

Principal curvatures. Consider 2-dimensional smooth surfaces in the Euclidean 3-dimensional space. If $x = (x^1, x^2, x^3)$ is a point of such a surface, then let N be the associated normal vector. A normal section is a plane through x which contains N (Fig. 33.9). In this way a plane curve is cut out. Let $\varkappa$ be its curvature in the sense of Subsec. 17.1.3. Now suppose that $\varkappa$ be signed, as indicated in Fig. 33.10. If φ has the meaning indicated, then $\varkappa = \varkappa(\varphi)$ depends on φ. Points with $\varkappa(\varphi) \equiv \text{const}$ are called umbilical points. If x is not an umbilical point, then there exist exactly one maximum $\varkappa_1 = \varkappa(\varphi_1)$ and exactly one minimum $\varkappa_2 = \varkappa(\varphi_2)$, where $\varphi_2 = \varphi_1 + \frac{\pi}{2}$. These two values are called principal curvatures.

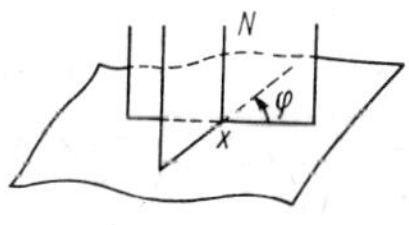

Fig. 33.9

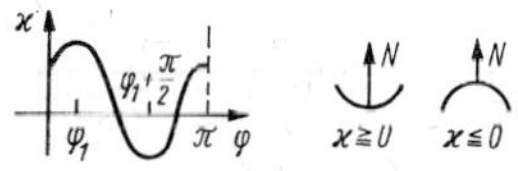

Fig. 33.10

Caustic. In the Minkowskian space, consider a 2-dimensional smooth surface which lies in the plane $x^4 = 0$ (Fig. 33.11). Then σ is space-like. Let σ be free of umbilical points, and let it be parametrized by $x^k = x^k(\lambda^1, \lambda^2)$, $k = 1, 2, 3$. Further let one of the two characteristic surfaces S be given by (33.1.3/2) with $n = 4$, where $s = 0$ corresponds to the surface σ. Let the principal curvatures of σ at the point (λ^1, λ^2) be $\varkappa_1(\lambda^1, \lambda^2)$ and $\varkappa_2(\lambda^1, \lambda^2)$. Finally, let $N = \{N^k(\lambda^1, \lambda^2)\}_{k=1}^3$ be the normal vector, normalized and fixed in its direction, in R_3 to σ at the point (λ^1, λ^2).

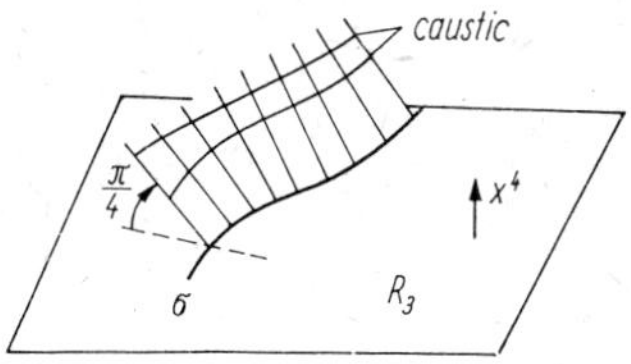

Fig. 33.11

Theorem. *With the above assumptions, the caustic on S consists of two surfaces given by*

$$y^k(\lambda^1, \lambda^2) = x^k(\lambda^1, \lambda^2) + \frac{N^k(\lambda^1, \lambda^2)}{\varkappa_r(\lambda^1, \lambda^2)} \quad \textit{for} \quad k = 1, 2, 3 \quad \textit{and} \quad y^4(\lambda^1, \lambda^2) = \frac{1}{\varkappa_r(\lambda^1, \lambda^2)}$$

where $r = 1, 2$.

Remark 1. This is the analogue to the construction of the evolute in Subsec. 17.1.3.

Remark 2. This construction can be extended to the n-dimensional Minkowskian space, $ds^2 = \sum_{k=1}^{n-1} (dx^k)^2 - (dx^n)^2$. The caustic then consists of $n-2$ surfaces.

33.1.5. Discontinuities of Solutions of the Wave Equation; Catastrophes

Let Ω be an open connected set in the n-space-time M. Further let Σ be an $(n-1)$-dimensional hypersurface which separates Ω into the open sets Ω_1 and Ω_2 and Σ (Fig. 33.12). Let $u(x)$ be twice continuously differentiable on both $\bar{\Omega}_1$ and $\bar{\Omega}_2$, and, for $p \in \Sigma$, let

$$[u(p)] = \lim_{\substack{q \to p \\ q \in \Omega_1}} u(q) - \lim_{\substack{q \to p \\ q \in \Omega_2}} u(q) \quad (\text{jump at } \Sigma)\,.$$

$[\nabla u(p)]$ is formed analogously. Let Pu be the general wave operator of (32.3.4/5).

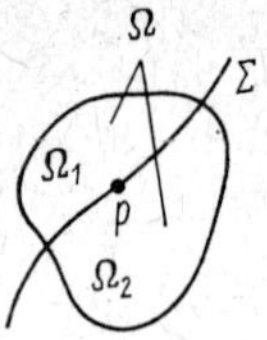

Fig. 33.12

Theorem 1. *If $u(x)$ fulfils the above assumptions, if $Pu=0$ in $D'(\Omega)$, and if, moreover, $|[u]|+|[\nabla u]|\neq 0$ on Σ, then Σ is a characteristic surface.*

Remark 1. Consequently, discontinuities of solutions of wave equations can only occur at characteristic surfaces. This cannot even be helped by resorting to the theory of distributions.

Classical variant: The classical variant of the above theorem reads as follows. Let $u(x)$ be a (classical) solution of $Pu=0$ which is twice continuously differentiable on Ω. Further let $u(x)$ be k times continuously differentiable on both $\bar{\Omega}_1$ and $\bar{\Omega}_2$, with $k\geq 3$ and $\sum_{|\alpha|=k}[D^\alpha u]\neq 0$ on Σ. Then Σ is a characteristic surface. Here $[D^\alpha u(p)]$ is defined analogously to $[u(p)]$.

Theorem 2. *Let the characteristic surface Σ in the domain Ω be parametrized according to (33.1.3/2) by $x^k=x^k(s,\lambda)$, with $\lambda=(\lambda^1,\ldots,\lambda^{n-2})$. If $Pu=0$ on Ω, and if s' and s'' are any two admissible parameters, then*

$$[u(x(s'',\lambda))]^2\,|\Delta(s'',\lambda)|^{\frac{1}{2}}=[u(x(s',\lambda))]^2\,|\Delta(s',\lambda)|^{\frac{1}{2}}\;e^{-\int_{s'}^{s''}a_l(x(s,\lambda))\frac{\partial x^l(s,\lambda)}{\partial s}ds}. \tag{1}$$

Remark 2. a_l is the covariant version of a^l. Further, $\Delta(s,\lambda)$ has the meaning indicated in Theorem 33.1.3. For $a^l=0$, which corresponds to the self-adjoint case $P=P^*$ of Subsec. 32.3.4., Eq. (1) assumes a particularly simple form.

Remark 3. For a fixed λ, $x^k=x^k(s,\lambda)$ is a null geodesic. Consequently, discontinuities of solutions $Pu=0$ of the wave equation are transported along null geodesics.

Catastrophes: Let $[u(x(s',\lambda))]\neq 0$. If s'' moves along the null geodesic $x^k(s,\lambda)$ towards a point of the caustic, then it follows that $\Delta(s'',\lambda)\to 0$ according to Theorem 33.1.3. From (1) one then finds that $[u(x(s'',\lambda))]\to\infty$. Hence, if u describes a physical process which exhibits a discontinuity along a characteristic surface, then a catastrophe will occur upon approach to the caustic.

33.2. Fundamental Solutions

33.2.1. The Problem

Fundamental solutions of partial differential equations were investigated in Sec. 23.1. within the theory of the distributions $D'(R_n)$ (Def. 23.1.1). Such solutions proved useful, especially in the treatment of initial value problems. Among the three types considered in that section, Δ, $\Delta-\frac{\partial}{\partial t}$ and $\Box$, we are now interested in the wave operator $\Box$. In our present terminology, Theorem 23.1.4. reads as follows. In the (4-dimensional) Minkowskian space with the line element

$$ds^2 = (dx^1)^2 + (dx^2)^2 + (dx^3)^2 - (dx^4)^2$$

$$G(\varphi) = -\frac{1}{4\pi} \int_0^\infty \frac{dx^4}{x^4} \int_{K_{x^4}} \varphi(x^1, x^2, x^3, x^4)\, ds_{x^4}\,, \tag{1}$$

where $\varphi \in D(R_4)$, is a solution of

$$\Box G = \frac{\partial^2 G}{(\partial x^1)^2} + \frac{\partial^2 G}{(\partial x^2)^2} + \frac{\partial^2 G}{(\partial x^3)^2} - \frac{\partial^2 G}{(\partial x^4)^2} = \delta\,. \tag{2}$$

Here K is the upper part of the light cone (characteristic cone) with origin 0, and K_{x^4} is the intersection with x^4, that means, the 3-dimensional sphere $\{(x^1, x^2, x^3) \mid (x^1)^2 + (x^2)^2 + (x^3)^2 = (x^4)^2\}$, with the surface element ds_{x^4} (Fig. 33.13). One has $\operatorname{supp} G = K$ (cf. Remark 23.1.4/1), and G is singular. Our further aim is to extend this result to the general wave operator P of (32.3.4/5) in an arbitrary space-time. Hence, in particular, from now on we shall always assume that $n = 4$, i.e., we shall consider space-times in the sense of the convention made at the beginning of Subsec. 33.1.1. The adjoint operator of (32.3.4/6) will be denoted by P^*.

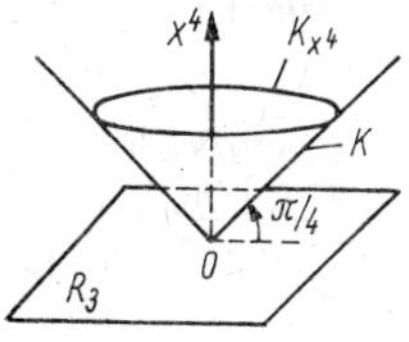

Fig. 33.13

Definition. *Let Ω be a connected open set in the space-time M. If $q \in \Omega$, then $G_q \in D'(\Omega)$ is called a fundamental solution of P in Ω if*

$$(PG_q)(\varphi) = G_q(P^*\varphi) = \varphi(q) \quad \text{for all} \quad \varphi \in D(\Omega)\,. \tag{3}$$

Remark 1. The formulation is meant to imply that for every point $q \in \Omega$ it is possible to find a corresponding distribution G_q. Referring to the δ-distribution to be considered in Subsec. 32.3.1., (3) then also means that $PG_q = \delta_q$ for all $q \in \Omega$. In particular, distributions on manifolds shall always have the meaning stated in Sec. 32.3. Of course the above definition can be immediately extended to n-space-times.

Remark 2. The above definition is the analogue to Def. 23.1.1 for space-times and general wave operators. Other than previously, we now consider a family $\{G_q\}_{q \in \Omega}$ of fundamental solutions rather than a single one. If M is the Minkowskian space, and if $P = \Box$, then one can obtain G_q from (1) by $G_q(\varphi) = G(\varphi(q + \cdot))$. In the general case, however, such an operation, $\varphi(x) \to \varphi(q + x)$ is not available. Other than in our previous investigations, now all of our considerations are local ones.

33.2.2. Causal Domains

By Theorem 29.5.5, for every point p of a given space-time M there exists a geodesically convex neighbourhood Ω. Now it is possible to introduce a time orientation in Ω by subdividing the light cones (characteristic conoids) in Ω continuously and consistently into future cones and past cones (cf. Subsec. 31.1.1.). To this end one starts the subdivision at a point $p_1 \in \Omega$ and continues it continuously along the unique geodesic from p_1 to p_2 (Fig. 33.14). Thus a time

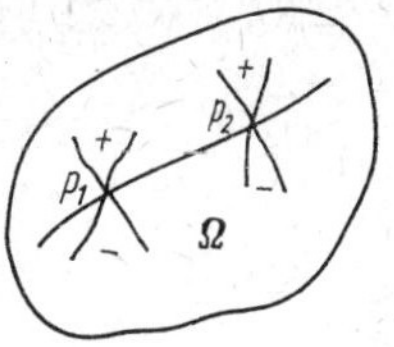

Fig. 33.14

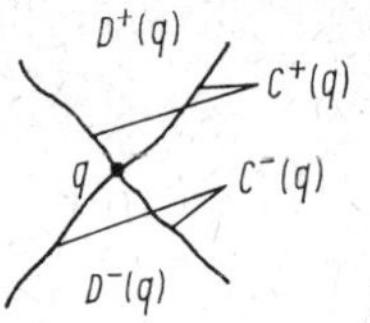

Fig. 33.15

orientation is always possible on a local scale, whereas globally it is an essential additional requirement.

Definition 1. *Let Ω be a geodesically convex domain of the space-time M, and let $q \in \Omega$. Then* (Fig. 33.15)
$D^+(q) = \{p \mid p \in \Omega$, *the geodesic from q to p being time-like and future-directed*$\}$,
$C^+(q) = \{p \mid p \in \Omega$, *the geodesic from q to p being null and future-directed*$\}$,
$J^+(q) = D^+(q) \cup C^+(q)$ (*future emission of q*).
$D^-(q)$, $C^-(q)$ *and* $J^-(q)$ *can be defined analogously.*

Remark 1. If $q \in \Omega$, then $D^+(q)$ is the interior of the future conoid with origin q, as far as it lies in Ω, and $C^+(q)$ is this future conoid, as far as it lies in Ω. One has $C^+(q) = \partial D^+(q)$ (in particular this implies that $q \in C^+(q)$).

Definition 2. *A connected open set Ω is called causal domain if it can be imbedded in a geodesically convex domain Ω_0, $\Omega \subset \Omega_0$, and if, for any two points $p_1 \in \Omega$ and $p_2 \in \Omega$, $J^+(p_1) \cap J^-(p_2)$ is either a compact subset of Ω or void.*

Remark 2. A causal domain for $p \in M$ can be constructed as follows. Let p_1 and p_2 be any points such that $p \in D^-(p_2)$ and $p \in D^+(p_1)$. For a suitable choice of p_1 and p_2, $D^-(p_2) \cap D^+(p_1)$ is then a causal domain containing p (Fig. 33.16).

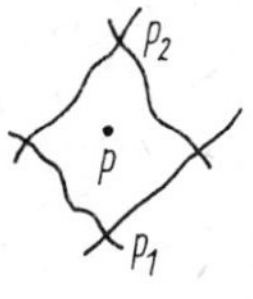

Fig. 33.16

33.2.3. The Distribution $\delta_{q^+}(\Gamma)$

Consider a causal domain Ω in a space-time M. If $q \in \Omega$ and $p \in D^+(q)$, then let s be the arc length (in the sense of Remark 29.5.4/4) of the unique time-like geodesic $x(s)$ that connects p with q (Fig. 33.17). If $q = x(s_1)$ and $p = x(s_2)$, then let $\Gamma(p, q) = |s_1 - s_2|$ be the geodesic distance of the two points p and q. Let $\Gamma(p, q) = 0$ for $p \in C^+(q)$, and let $\Gamma_\varepsilon = \{p \mid p \in D^+(q), \Gamma(p, q) = \varepsilon\}$ for any $\varepsilon > 0$.

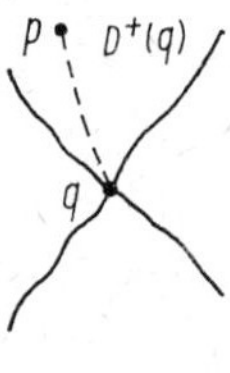

Fig. 33.17

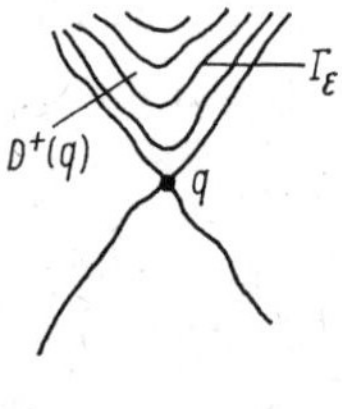

Fig. 33.18

(Since q is fixed, we do not note the dependence on q.) In the Minkowskian space one has $\Gamma_\varepsilon = \{x \mid (x^1)^2 + (x^2)^2 + (x^3)^2 - (x^4)^2 = \varepsilon^2\}$ for $q=0$. In the general case Γ_ε is the diffeomorphic image of such hyperboloids (Fig. 33.18).

Lemma. *If $S(p) = \Gamma(p, q)$ (where q is fixed, p variable), then $g^{kl} S_{,k} S_{,l} = 4S$ in $D^+(q)$.*

Remark 1. Hence, in particular, grad $S \neq 0$ in $D^+(q)$. Thus one has the situation described in Subsecs. 32.2.4. and 32.3.5. (the domain Ω considered there is now $D^+(q)$).

Definition. *If $\delta_\varepsilon \in D'(R_1)$ is the δ-distribution with respect to the point $\varepsilon > 0$, i.e., if $\delta_\varepsilon(\psi) = \psi(\varepsilon)$ for $\psi \in D(R_1)$, then $\delta_{q+}(\Gamma - \varepsilon) = \delta_\varepsilon(S)$ in the sense of Theorem 32.3.5.*

Remark 2. In Theorem 32.3.5. one has to substitute $D^+(q)$ for Ω. Hence one has $\delta_{q+}(\Gamma - \varepsilon) \in D'(D^+(q))$, and

$$\delta_{q+}(\Gamma - \varepsilon)(\varphi) = \int_{\Gamma_\varepsilon} \varphi(p) \mu_S \quad \text{for} \quad \varphi \in D(D^+(q)). \tag{1}$$

It is clear that (1) can be extended to $\varphi \in D(\Omega)$. Then $\delta_{q+}(\Gamma - \varepsilon) \in D'(\Omega)$.

Theorem. *$\delta_{q+}(\Gamma) = \lim_{\varepsilon \downarrow 0} \delta_{q+}(\Gamma - \varepsilon)$ converges in $D'(\Omega)$.*

Remark 3. What is meant is that $\delta_{q+}(\Gamma) \in D'(\Omega)$ and $\delta_{q+}(\Gamma - \varepsilon)(\varphi) \to \delta_{q+}(\Gamma)(\varphi)$ for $\varphi \in D(\Omega)$.

Remark 4. If M is the Minkowskian space and if $q=0$, then $\delta_{q+}(\Gamma)$ coincides with G of (33.2.1/1) except for a factor. Thus $\delta_{q+}(\Gamma)$ is the invariant (and local) version of G.

Remark 5. One may substitute $D^-(q)$ for $D^+(q)$. Then one obtains a corresponding distribution $\delta_{q-}(\Gamma)$.

33.2.4. Fundamental Solutions

Theorem. *If Ω is a causal domain in the space-time M, then there exists, in the sense of Def. 33.2.1, exactly one fundamental solution G_q of P in Ω, such that* supp $G_q \subset J^+(q)$ *for $q \in \Omega$. Denoting this fundamental solution by G_q^+, it can be represented as $G_q^+ = U\delta_{q+}(\Gamma) + V^+$. Here $\delta_{q+}(\Gamma)$ has the same meaning as in Theorem 33.2.3. Further, $U = U(p, q)$ is a function differentiable up to arbitrary order on $\Omega \times \Omega$, and $V^+ = V^+(p, q)$ is a function defined on $\Omega \times \Omega$, with* supp $V^+ \subset \Delta^+ = \{(p, q) \mid p \in \Omega, q \in \Omega, p \in J^+(q)\}$, *which is differentiable up to arbitrary order on Δ^+.*

Remark 1. It is possible to formulate a corresponding theorem for a fundamental solution G_q^-.

Remark 2. One has supp $\delta_{q+}(\Gamma) = C^+(q)$. The support of the singular part $U\delta_{q+}(\Gamma)$ of G_q^+ thus lies in $C^+(q)$, as in the case of the Minkowskian space, where G_q^+ in essence coincides with G of (33.2.1/1). In the general case this (Huyghenian) property is perturbed by the summand V^+.

Remark 3. Let M be a space of constant curvature, i.e., let $R_{abcd} = \frac{R}{12}(g_{ac}g_{bd} - g_{bc}g_{ad})$ in the sense of Remark 30.2.1/3. Further let $P = \Box + b$, i.e., $a^l = 0$ in (32.3.4/5). Then

$$G_q^+ = -\frac{1}{2\pi}\delta_{q+}(\Gamma) + \frac{R}{48\pi} H_{q+}(\Gamma), \quad \text{where} \quad H_{q+}(\Gamma) = \begin{cases} 1 & \text{for} \quad p \in J^+(q), \\ 0 & \text{for} \quad p \notin J^+(q). \end{cases}$$

$R=0$ is the Minkowskian space, that is, $G_q^+ = G$ of (33.2.1/1) for $q=0$.

33.3. Solutions of $Pu = f$, Cauchy Problems

33.3.1. Past-Compact Sets and Distributions

We identify the causal domain Ω of the space-time M by a local chart. This is not a restriction, since Ω is part of a geodesically convex domain, and such domains can always be represented by a single chart. G_q^+ shall have the meaning stated in Theorem 33.2.4.

Lemma 1. (**a**) *If* $\varphi \in D(\Omega)$, *then* $G_q^+(\varphi)$, *as a function of* q *on* Ω, *is differentiable up to arbitrary order, and* $\operatorname{supp} G_q^+(\varphi) \subset J^-(\operatorname{supp} \varphi)$.

(**b**) *If* K *and* K' *are any two compact subsets of* Ω, *and if* $N = 0, 1, 2, \ldots$, *then there exists a positive number* $c = c(K, K', N)$ *such that, for all* $\varphi \in D(\Omega)$ *with* $\operatorname{supp} \varphi \subset K$,

$$\sum_{|\alpha| \leq N} \sup_{q \in K'} |D_q^\alpha G_q^+(\varphi)| \leq c \sum_{|\alpha| \leq N} \sup_{p \in K \cap J^+(K')} |D_p^\alpha \varphi(p)| \,. \tag{1}$$

Remark 1. If K is a set in Ω, then $J^-(K) = \bigcup_{q \in K} J^-(q)$ and $J^+(K) = \bigcup_{q \in K} J^+(q)$ (Fig. 33.19). Further, D_q^α has the usual meaning, the differentiation being with respect to $q = (q^1, \ldots, q^n)$. The statement made on the support of $G_q^+(\varphi)$ in part (a) is an immediate consequence of the fact that $\operatorname{supp} G_q^+ \subset J^+(q)$. This also explains the restriction $p \in K \cap J^+(K')$ in (1).

Definition. (**a**) *A set* K *in* Ω *is called past-compact if* $J^-(p) \cap K$ *is either void or compact for all* $p \in \Omega$ (Fig. 33.20).

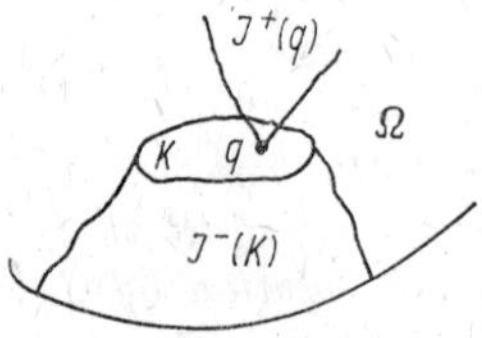

Fig. 33.19

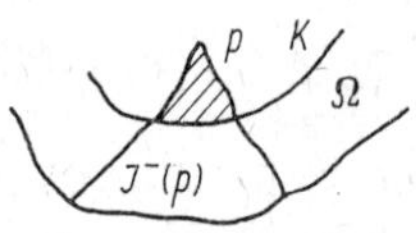

Fig. 33.20

(**b**) $D^+(\Omega)$ *is the class of all functions differentiable up to arbitrary order on* Ω *which have a past-compact support.*

(**c**) $D'^+(\Omega)$ *is the class of all distributions in* $D'(\Omega)$ *with a past-compact support.*

Remark 2. Every compact set is past-compact. It is, however, easy to find sets which are past-compact but not compact. $D'^+(\Omega)$ is not the space dual to $D^+(\Omega)$. Functions in $D^+(\Omega)$ need not have a compact support. Consider, for instance, the causal domain $\Omega = \{x \mid |x^4| < 2\}$ in the Minkowskian space, then $K = \{x \mid |x^4| \leq 1\}$ is a past-compact set, and it is easy to state C^∞-functions which have the support K.

Lemma 2. *If* K *is compact and* K' *is past-compact, then* $J^-(K) \cap K'$ *is compact.*

Remark 3. It can also be shown that $J^+(K')$ is past-compact if K' is past-compact.

Theorem. *If* $f \in D'^+(\Omega)$, *then* $f[G_q^+] \in D'(\Omega)$, *where* $f[G_q^+](\varphi) = f(G_q^+(\varphi))$ *for* $\varphi \in D(\Omega)$.

Remark 4. If $\varphi \in D(\Omega)$ with $\operatorname{supp} \varphi \subset K$, then, by Lemma 1(a), $\operatorname{supp} G_q^+(\varphi) \subset J^-(K)$. If K is compact, then it follows from Lemma 2 that $J^-(K) \cap \operatorname{supp} f$ is compact too (Fig. 33.21). The theorem is then deduced from (1), with $K' = J^-(K) \cap \operatorname{supp} f$, and from the known rules of the theory of distributions.

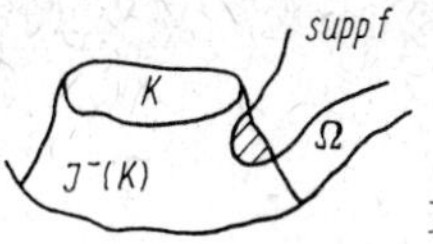

Fig. 33.21

33.3.2. An Existence and Uniqueness Theorem

Consider the general wave operator P of (32.3.4/5) on a causal domain Ω of the space-time M. If $p\in\Omega$, then $\Gamma(p, q)$ has the meaning indicated in Subsec. 33.2.3., where now we extend $\Gamma(p, q)$ to all points q with $q\neq p$. If $x^k(s)$ is the unique geodesic that connects p with q, where s is a geodesic parameter in the sense of Subsec. 29.4.5., with $p=x^k(0)$, then $\nabla\Gamma(p, q)=2\,|s|\,\dfrac{dx^k}{ds}$ (∇ is applied with respect to q). Thus in particular one has $\nabla\Gamma(p, q)\neq 0$ for $q\neq p$, and it is possible to apply Theorem 32.2.4/1 with $S(q)=\Gamma(p,q)$ (p fixed). Hence $\mu(q)=d_q\Gamma(p,q)\wedge\mu_\Gamma(q)$, where we have indicated the dependence on q for the sake of clearness (Fig. 33.22).

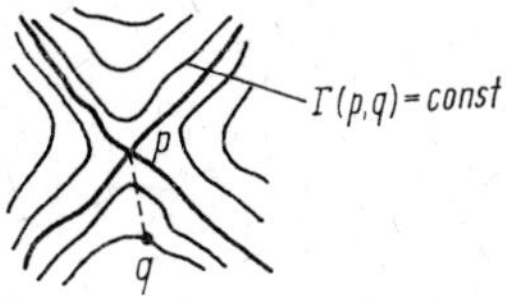

Fig. 33.22

Theorem. *If $f\in D'^+(\Omega)$, then $Pu=f$ has a unique solution in $D'^+(\Omega)$. This solution is $u=f[G_q^+]$, and one has* supp $u\subset J^+$(supp f). *If $f\in D^+(\Omega)$, then $u(p)$ belongs to $D^+(\Omega)$ and can be represented as*

$$u(p)=\frac{1}{2\pi}\int\limits_{C^-(q)} U(p, q)\, f(q)\, \mu_\Gamma(q)+\frac{1}{2\pi}\int V^+(p, q)\, f(q)\, \mu(q)\,. \tag{1}$$

Remark 1. $U(p, q)$ and $V^+(p, q)$ have the same meaning as in Theorem 33.2.4. Further one has to understand $\int\limits_{C^-(p)}\ldots$ as $\lim\limits_{\varepsilon\downarrow 0}\int\limits_{\substack{C^-(p)\\ |s|>\varepsilon}}\ldots$ In the Minkowskian space with the usual coordinate representation, μ_Γ coincides with $\dfrac{dx^1\wedge dx^2\wedge dx^3}{x^4}$ apart from a factor. In this case it is easily seen that the above limit exists. On a local scale, the general case is the diffeomorphic image of the situation found in the Minkowskian space. Hence it follows that the above integral exists in the general case, too.

Remark 2. The fact that $u=f[G_q^+]$ is a solution of $Pu=f$ follows from

$$(Pu)(\varphi)=u(P^*\varphi)=f(G_q^+(P^*\varphi))=f(\varphi)$$

(cf. (32.3.4/6), Theorem 33.3.1 and Def. 33.2.1). The statement made on the support of u follows from the construction of $f[G_q^+]$.

Remark 3. In the case of the Minkowskian space with the usual line element $ds^2=(dx^1)^2+(dx^2)^2+(dx^3)^2-(dx^4)^2$, one has $V^+\equiv 0$, $U\equiv 1$, and $\mu_\Gamma=c\,\dfrac{dx^1\wedge dx^2\wedge dx^3}{x^4}$, where c is a suitable constant. Then $u(p)$ of (1) is identical with the well-known retarded potentials (cf. 19.3.5/1) and the first term of (23.2.2/2)).

33.3.3. The Cauchy Problem: Existence and Uniqueness

The Cauchy problem (initial value problem) for the wave equation in the Minkowskian space (in our present terminology) was dealt with in Subsecs. 19.3.5. (classical solution) and 23.2.2. (solution by distributions). It is our aim to extend the previous considerations to arbitrary space-times M (that means, to 4-dimensional Lorentz-metric spaces) as well as to the wave operators P of (32.3.4/5). Here we confine ourselves to C^∞-functions, i.e., to classical solutions.

Cauchy data: Let Ω be a causal domain in a space-time M. We shall again assume that Ω is identical with its local chart. Let S be a 3-dimensional (smooth) hypersurface in Ω (Fig. 33.23). If u is a C^∞-function on S, and if t is a direction tangential to S, then $\frac{\partial u}{\partial t}$ is known if u is. Further, if $\frac{\partial u}{\partial v}$ is given, where v is not a tangential direction, then all first-order partial derivatives of u on S are known. We then say that Cauchy data are given on S.

Theorem. *Let Ω be a causal domain, and let S be a past-compact space-like 3-dimensional hypersurface in Ω, such that $\partial J^+(S)=S$ (Fig. 33.24). Further let f be a C^∞-function in Ω, and let C^∞-Cauchy data be given on S. Then the Cauchy problem*

$$Pu=f \quad \text{in} \quad J^+(S), \quad u \text{ being a } C^\infty\text{-function in } J^+(S)\,,$$
$$u\,|_S \text{ and } \nabla u|_S \quad \text{assuming given values on } S\,,$$

has one and only one solution.

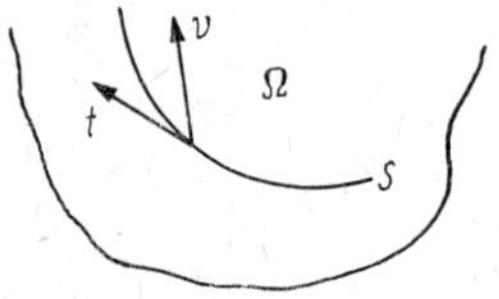

Fig. 33.23

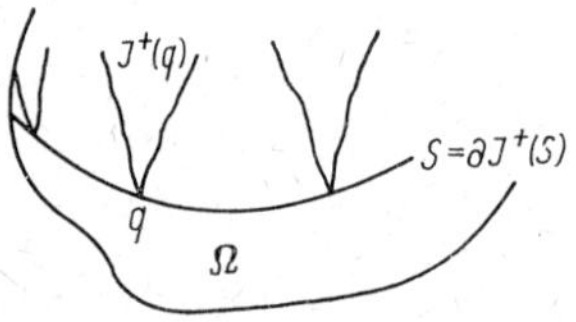

Fig. 33.24

Remark 1. The terms "space-like" and "past-compact" have been described previously (Def. 33.1.1 and Def. 33.3.1). As previously, $J^+(S)=\bigcup_{q\in S} J^+(q)$, and $\partial J^+(S)$ is the boundary of this set, where only points of the open set Ω come into question as boundary points (and not points of $\partial\Omega$ in the sense of the embedding of Ω in the R_n). The statement that u is a C^∞-function in $J^+(S)$ means that u is differentiable up to arbitrary order in the interior of $J^+(S)$ and that all of its derivatives can be continued continuously to the boundary of $J^+(S)$.

Remark 2. The assumptions made with respect to S are natural, especially the requirement that S be space-like. If $\Omega=M$ is the Minkowskian space, and if $P=\Box$ and $S=\{x \mid x^4=0\}$, then the above theorem in essence coincides with the existence and uniqueness proposition of Theorem 19.3.5/2. Our aim is to generalize the explicit construction of the solution as described in that latter theorem to our present case.

33.3.4. The Cauchy Problem: Representation

We make the same assumptions as in Subsec. 33.3.3.: Let Ω be a causal domain in the space-time M, which is identical with its local chart (Fig. 33.25). Let S again be a past-compact space-like 3-dimensional hypersurface in Ω, with

$\partial J^+(S)=S$. For $p\in J^+(S)$, $p\notin S$, we introduce the following notations:

$$D_p=J^-(p)\cap J^+(S),\quad C_p=C^-(p)\cap J^+(S)\,,$$
$$S_p=J^-(p)\cap S,\qquad \sigma_p=C^-(p)\cap S\,.$$

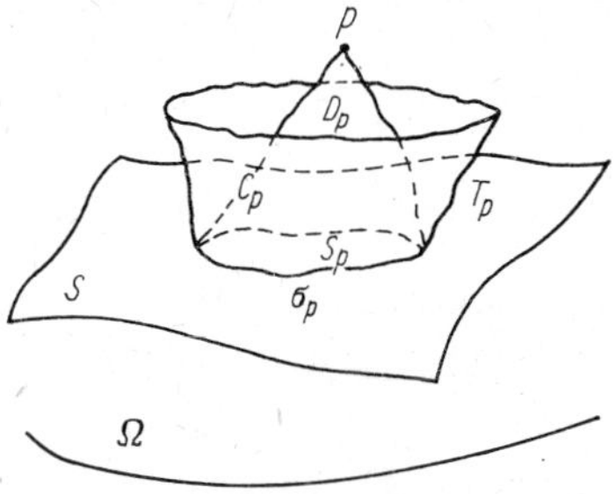

Fig. 33.25

Then $C_p\cup S_p=\partial D_p$ (four-dimensional), and $\sigma_p=\partial S_p$ (three-dimensional). σ_p is a 2-dimensional closed space-like compact submanifold. By Theorem 33.1.2/1 and Remark 33.1.2/1, there exist exactly two characteristic surfaces that contain σ_p; one of them is $C^-(p)$, the other is denoted by T_p. Let $\mu=\mu(q)$ be the invariant measure referred to in Theorem 32.1.4. On C_p, let $\mu_\Gamma=\mu_\Gamma(q)$ be the Leray form mentioned in Subsec. 33.3.2. Finally, let the 2-form $\mathrm{d}\sigma_p$ be the invariant measure on σ_p, where the metric in σ_p is induced by the metric in M. If η is a contravariant vector, then let $*\eta$ be the 3-form stated in Theorem 32.2.2. In the sense of Subsec. 32.2.3., let $(*\eta)_{S_p}$ be the restriction of $*\eta$ to S_p.

Dilatation factor: According to Subsec. 33.1.3, T_p can be generated by the null geodesics which pass through σ_p (Fig. 33.26). Let $\xi^l(q)$ be the contravariant tangent vector to these null geodesics, which is normalized by $\xi^l(q)\,(\nabla\Gamma(p,q))_l=-1$ (∇ is applied with respect to q; see also Subsec. 33.3.2.). $\Gamma(p,q)=\varepsilon$ cuts a 2-dimensional subset out of T_p. If ω is an open 2-dimensional subset of σ_p, then a set ω_ε is cut out of $T_p\cap\{\Gamma(p,q)=\varepsilon\}$ by the null geodesics which generate T_p and pass through ω. If $\mathrm{d}\sigma_p^\varepsilon$ is the invariant measure on $T_p\cap\{\Gamma(p,q)=\varepsilon\}$, then on σ_p there exists a function Θ (dilatation factor) such that

$$\lim_{\varepsilon\downarrow 0}\frac{1}{\varepsilon}\left(\int_{\omega_\varepsilon}\mathrm{d}\sigma_p^\varepsilon-\int_{\omega}\mathrm{d}\sigma_p\right)=-\int_{\omega}\Theta\,\mathrm{d}\sigma_p\,.$$

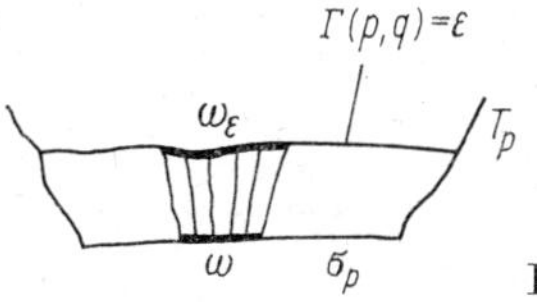

Fig. 33.26

All symbols used in the following theorem shall have the meaning indicated above. P is the general wave operator of (32.3.4/5). $U=U(p,q)$ and $V^+=V^+(p,q)$ have the same meaning as in Theorem 33.3.2 and are regarded as functions of q.

Theorem. *If u is differentiable on Ω up to arbitrary order, and if $p \in J^+(S)$, $p \notin S$, then $u(p)$ can be represented as*

$$u(p) = u^{(1)}(p) + u^{(2)}(p) + u^{(3)}(p) ,$$

$$u^{(1)}(p) = \frac{1}{2\pi} \int_{C_p} U P u \mu_\Gamma + \frac{1}{2\pi} \int_{D_p} V^+ P u \mu ,$$

$$u^{(2)}(p) = \frac{1}{2\pi} \int_{S_p} (*\eta)_{S_p}, \quad \text{where} \quad \eta = \{\eta^k\} ,$$

$$\eta^k = V^+ g^{kl} (\nabla u)_l - u g^{kl} (\nabla V^+)_l + a^k V^+ u ,$$

$$u^{(3)}(p) = \frac{1}{2\pi} \int_{\sigma_p} [2U\xi^k (\nabla u)_k + \Theta U u + a^l \xi_l U u + V^+ u] \, d\sigma_p .$$

Remark 1. If one substitutes f for Pu and the given Cauchy data for u and ∇u on S, then one obtains a representation of the unique solution u of the Cauchy problem of Theorem 33.3.3. Under the conditions of Theorem 33.3.2, with $f \in D^+(\Omega)$, one in essence obtains (33.3.2/1). The above theorem (being interpreted as a solution of the Cauchy problem) also generalizes Theorem 19.3.5/2. In that case $\Omega = M$ is the Minkowskian space, $P = \square$, $V^+ \equiv 0$, $U \equiv 1$, $a^l \equiv 0$, and $u^{(2)}(p) \equiv 0$.

Remark 2. For the determination of $u(p)$ one has to know Pu in D_p as well as the Cauchy data on S_p. If V^+ is not identically zero, then the knowledge of the Cauchy data on σ_p will not suffice (otherwise one has to do with the Huyghenian property mentioned in Subsec. 19.3.4.). Thus living in a curved space-time will not be as pleasant as in the Euclidean space. In the sense of the interpretation in Subsec. 19.3.4., a sound wave in a curved space-time will never die out completely.

33.4. Tensor Wave Equations

33.4.1. Definitions

So far we considered the wave operator P of (32.3.4/5) in $D'(M)$. In this section, too, M shall always be a space-time (i.e., a 4-dimensional Lorentz-metric space). Our previous investigations and many of the results can be transferred to tensor distributions in the sense of Def. 32.3.2 and to corresponding tensor wave operators. As tensor indices can be raised or lowered at choice, we shall confine ourselves to purely covariant and purely contravariant tensor distributions. The covariant derivative ∇ and the coderivative δ shall have the same meaning as in Def. 32.3.3/2.

Definition. *Let M be a space-time, and let $m = 0, 1, 2, \ldots$ For $T \in D'^{(0,m)}(M)$ we define $\square T = -\delta(g^{kl}(\nabla T)_l)$ (tensor wave operator).*

Remark 1. From $T \in D'^{(0,m)}(M)$ it follows that $\nabla T \in D'^{(0,m+1)}(M)$, $g^{kl}(\nabla T)_l \in D'^{(1,m)}(M)$, and $\square T = -\delta\, (g^{kl}(\nabla T)_l) \in D'^{(0,m)}(M)$. If $m = 0$, i.e., if $T \in D'(M) = D'^{(0,0)}(M)$, then $\square T$ is the wave operator of Def. 32.3.4. Naturally the above definition can be immediately extended to an arbitrary n-space-time.

Remark 2. If $T = \{T_{k_1 \ldots k_m}(x)\}$ is a tensor with components differentiable up to arbitrary order (we recall that we always speak of tensors rather than, more precisely, of tensor fields), then

$$(\square T)_{k_1 \ldots k_m} = (g^{kl} T_{k_1 \ldots k_m ; l})_{;k} = g^{kl} T_{k_1 \ldots k_m ; lk} \, .$$

Remark 3. The definition is also applicable if $T \in D'^{(r,s)}(M)$. If, for instance, $\varphi = \{\varphi^{k_1 \dots k_m}(x)\} \in D^{(m,0)}(M)$, then, by analogy with Remark 2,

$$(\Box\varphi)^{k_1 \dots k_m} = g^{kl} \varphi^{k_1 \dots k_m}{}_{;lk} . \tag{1}$$

Lemma. *Let $\{b^{r_1 \dots r_m}{}_{s_1 \dots s_m}\}$ be a tensor of type (m, m) with components differentiable up to arbitrary order. If $T = \{T_{r_1 \dots r_m}\} \in D'^{(0,m)}(M)$, and if $\varphi = \{\varphi^{s_1 \dots s_m}\} \in D^{(m,0)}(M)$, then*

$$(\Box T + \{b^{r_1 \dots r_m}{}_{s_1 \dots s_m} T_{r_1 \dots r_m}\})(\varphi) = T\,(\Box\varphi + \{b^{r_1 \dots r_m}{}_{s_1 \dots s_m} \varphi^{s_1 \dots s_m}\}) . \tag{2}$$

Remark 4. With the use of (1), the lemma can be derived from Def. 32.3.3/2. Here $\{b^{r_1 \dots r_m}{}_{s_1 \dots s_m} T_{r_1 \dots r_m}\} \in D'^{(0,m)}(M)$ is the tensor distribution whose components are $b^{r_1 \dots r_m}{}_{s_1 \dots s_m} T_{r_1 \dots r_m}$. By analogy, one has $\{b^{r_1 \dots r_m}{}_{s_1 \dots s_m} \varphi^{s_1 \dots s_m}\} \in D^{(m,0)}(M)$.

Remark 5. By analogy with our previous approach, for $T \in D'^{(0,m)}(M)$ we set

$$PT = \Box T + \{b^{r_1 \dots r_m}{}_{s_1 \dots s_m} T_{r_1 \dots r_m}\} . \tag{3}$$

The lemma then shows that for $\varphi \in D^{(m,0)}(M)$ the adjoint operator P^* can be calculated from

$$P^*\varphi = \Box\,\varphi + \{b^{r_1 \dots r_m}{}_{s_1 \dots s_m} \varphi^{s_1 \dots s_m}\} . \tag{4}$$

Then (2) reads $(PT)(\varphi) = T(P^*\varphi)$. Here the co- and contravariant components are interchanged when P is replaced by P^*.

Remark 6. By analogy with (32.3.4/5), it is possible to generalize PT of (3) by setting

$$PT = \Box\, T + \{a^{rr_1 \dots r_m}{}_{s_1 \dots s_m} (\nabla T)_{rr_1 \dots r_m}\} + \{b^{r_1 \dots r_m}{}_{s_1 \dots s_m} T_{r_1 \dots r_m}\}$$

for $T = \{T_{r_1 \dots r_m}\} \in D'^{(0,m)}(M)$. Here a and b are tensors with components differentiable up to arbitrary order. But (3) will suffice for treating Maxwell's equations in space-times. Therefore in the sequel we shall confine ourselves to this case.

33.4.2. Fundamental Solutions

The operators P and P^* shall have the same meaning as in Subsec. 33.4.1., formulas (3) and (4). By analogy with Def. 33.2.1 and Theorem 33.2.4, the problem is to find fundamental solutions.

Definition. *Let Ω be a connected open set in the space-time M. If $q \in \Omega$, then $G_q = \{G_{qs_1 \dots s_m}^{r_1 \dots r_m}\}$ is called a fundamental solution of P in Ω if $G_q^{r_1 \dots r_m} = \{G_{qs_1 \dots s_m}^{r_1 \dots r_m}\} \in D'^{(0,m)}(M)$ and*

$$(PG_q^{r_1 \dots s_m})(\varphi) = G_q^{r_1 \dots r_m}(P^*\varphi) = \varphi^{r_1 \dots r_m}(q)$$

for $\varphi \in D^{(m,0)}(\Omega)$ and for every fixed m-tuple of indices, $r_1, \dots, r_m$. Further, $G_{qs_1 \dots s_m}^{r_1 \dots r_m}$, with the m-tuple of indices $s_1, \dots, s_m$ being fixed, is supposed to transform as a contravariant tensor with respect to the variable q and the indices $r_1, \dots, r_m$.

Remark 1. For the scalar case $m = 0$ this coincides with Def. 33.2.1.

Bi-tensors: Let $\tau = \{\tau_{s_1 \dots s_m}^{r_1 \dots r_m}(p, q)\}$ be a system of functions on $\Omega \times \Omega$ (or on $M \times M$). Then τ is called a bi-tensor (more precisely, a bi-tensor field) if, for fixed

p and $r_1, \ldots, r_m$, $\tau^{r_1\ldots r_m}(p) = \{\tau^{r_1\ldots r_m}_{s_1\ldots s_m}(p, q)\}$ is a covariant tensor (with respect to q and $s_1, \ldots, s_m$) and if $\tau_{s_1,\ldots s_m}(q) = \{\tau^{r_1\ldots r_m}_{s_1\ldots s_m}(p, q)\}$, for fixed q and $s_1, \ldots, s_m$, is a contravariant tensor (with respect to p and $r_1, \ldots, r_m$). A corresponding definition is also possible in cases where the numbers of r- and s-indices are not equal, or when mixed-variant tensors occur instead of co- and contravariant ones. G_q is the distribution variant of a bi-tensor.

Theorem. *If Ω is a causal domain in the space-time M, then, in the sense of the above definition, there exists exactly one fundamental solution G_q of P in Ω, such that* $\operatorname{supp} G_q \subset J^+(q)$ *for $q \in \Omega$. Denoting this fundamental solution by G_q^+, one can represent it as*

$$G_q^{+\,r_1\ldots r_m}_{\;\;s_1\ldots s_m} = \tau^{r_1\ldots r_m}_{s_1\ldots s_m}(\cdot, q)\, \delta_{q+}(\Gamma) + V^{+\,r_1\ldots r_m}_{\;\;s_1\ldots s_m}(\cdot, q)\,. \tag{1}$$

Here $\delta_{q+}(\Gamma)$ has the same meaning as in Theorem 33.2.3. Further, $\{\tau^{r_1\ldots r_m}_{s_1\ldots s_m}(p, q)\}$ and $\{V^{+\,r_1\ldots r_m}_{\;\;s_1\ldots s_m}(p, q)\}$ are bi-tensors, and $\tau^{r_1\ldots r_m}_{s_1\ldots s_m}(p, q)$ is differentiable up to arbitrary order on $\Omega \times \Omega$. If Δ^+ has the same meaning as in Theorem 33.2.4, then $\operatorname{supp} V^{+\,r_1\ldots r_m}_{\;\;s_1\ldots s_m} \subset \Delta^+$, *and $V^{+\,r_1\ldots r_m}_{\;\;s_1\ldots s_m}$ is differentiable on Δ^+ up to arbitrary order.*

Remark 2. $\operatorname{supp} G_q \subset J^+(q)$ means that $\operatorname{supp} G_q^{r_1\ldots r_m} \subset J^+(q)$ (regarded as a covariant tensor distribution) holds for every m-tuple of indices $r_1, \ldots, r_m$. The theorem is the analogue to Theorem 33.2.4.

33.4.3. Solutions of $Pu = f$

Theorem 33.4.2 is the analogue to Theorem 33.2.4. This makes it possible to transfer the theory of solutions presented in Sec. 33.3. for scalar wave operators to the case of tensor wave operators. This applies to both the analogue to Theorem 33.3.2 and the Cauchy problems described in Subsecs. 33.3.3. and 33.3.4. Here we shall formulate the analogue to Theorem 33.3.2 and dispense with a treatment of Cauchy problems. We again consider a causal domain Ω in a space-time M. Further, P shall have the same meaning as in (33.4.1/3).

Theorem. *If $f \in D'^{(0,m)}(\Omega)$ has a past-compact support, then $Pu = f$ has exactly one solution $u = \{u_{s_1\ldots s_m}\} \in D'^{(0,m)}(\Omega)$ with past-compact support. This solution is given by*

$$u(\varphi) = f(G_q^+(\varphi)) = f(\{G_q^{+\,r_1\ldots r_m}(\varphi)\}) \quad \text{for} \quad \varphi \in D^{(m,0)}(\Omega) \tag{1}$$

and $\operatorname{supp} u \subset J^+(\operatorname{supp} f)$. *If, in addition, the components of f are differentiable on Ω up to arbitrary order, then the components of u in* (1) *are also differentiable on Ω up to arbitrary order, and*

$$u_{s_1\ldots s_m}(p) = \frac{1}{2\pi} \int_{C^-(p)} \tau^{r_1\ldots r_m}_{s_1\ldots s_m}(p, q)\, f_{r_1\ldots r_m}(q)\, \mu_\Gamma(q) \tag{2}$$
$$+ \frac{1}{2\pi} \int V^{+\,r_1\ldots r_m}_{\;\;s_1\ldots s_m}(p, q)\, f_{r_1\ldots r_m}(q)\, \mu(q)\,.$$

Remark. The theorem generalizes Theorem 33.3.2. Past-compact sets have been considered in Subsec. 33.3.1. So f belongs to the tensor variant of $D'^+(\Omega)$ as per Def. 33.3.1(c). Further, G_q^+ and the functions in (2) have the same meaning as in Theorem 33.4.2. To prove that

$u \in D'^{(0,m)}(\Omega)$, one has to verify that the correspondence $\varphi \in D^{(m,0)}(\Omega) \to f(G_q^+(\varphi))$ is a tensor distribution of $D'^{(0,m)}(\Omega)$. This is done in the same way as in Subsec. 33.3.1. Thereafter it is easily seen that u in (1) is a solution: for $\varphi \in D^{(m,0)}(\Omega)$ one has

$$(Pu)(\varphi) = u(P^*\varphi) = f(G_q^+(P^*\varphi)) = f(\varphi)\,,$$

cf. Lemma 33.4.1, Def. 33.4.2, and Theorem 33.4.2.

33.5. The Maxwell Equations

33.5.1. Definition

As usual, a 4-dimensional Lorentz-metric space will be called a space-time.

Definition. *Let M be a space-time, and let σ (current density) be a given vector whose components are differentiable on M up to arbitrary order. Desired a 2-form F (electromagnetic field) which satisfies*

$$\mathrm{d}F = 0 \quad \textit{and} \quad \delta F = \sigma \quad (\textit{Maxwell equations})\,, \tag{1}$$

and whose components are differentiable on M up to arbitrary order.

Remark 1. As a 2-form, $F = \{F_{kl}\}$ is a covariant tensor which, by Def. 32.1.2 and Lemma 32.1.2, satisfies the (necessary and sufficient) condition that $F_{kl} = -F_{lk}$. By Lemma 32.1.3 the first equation in (1) then means that the 3-form

$$\mathrm{d}F = F_{kl'm}\mathrm{d}x^m \wedge \mathrm{d}x^k \wedge \mathrm{d}x^l = 0\,. \tag{2}$$

In view of index shifting we do not distinguish between co- and contravariant components of a tensor. In the second equation in (1) one has to interpret F as a contravariant or mixed-variant tensor. Then by Def. 32.3.3/1 one obtains

$$F^{kl}{}_{;l} = -F^{lk}{}_{;l} = (\delta F)^k = \sigma^k\,. \tag{3}$$

Remark 2. In the above definition we have confined ourselves to vectors σ and forms F which are differentiable up to arbitrary order. In our further treatment we will follow this line. But it is also possible to extend the Maxwell equations (1) to tensor distributions σ and F.

Remark 3. Poincare's theorem in Subsec. 32.1.5. shows that locally $\mathrm{d}F = 0$ is equivalent with $F = \mathrm{d}A$, where A is a vector. Hence locally the Maxwell equations (1) are identical with

$$F_{kl} = A_{k'l} - A_{l'k}\,, \tag{4}$$

$$\frac{1}{\sqrt{|g|}}(\sqrt{|g|}\,F^{kl})_{'l} = F^{kl}{}_{;l} = \sigma^k\,. \tag{5}$$

This links our considerations with Subsec. 30.1.3. In the sense of Remark 30.1.3/3 one thinks of the Einstein-Maxwell field as being decoupled: the geometry of the space is determined locally by heavy masses and globally by the world models considered, and an electromagnetic field to be investigated does not influence this geometry.

33.5.2. Continuity Equation and Cauchy Data

Lemma. *The current density σ of the Maxwell equations* (33.5.1/1) *satisfies the continuity equation $\delta\sigma = 0$.*

Remark 1. The Lemma is derived from $F^{kl} = -F^{lk}$, (33.5.1/5) and Def. 32.3.3/1:

$$0 = (\sqrt{|g|}\,F^{kl})_{'lk} = (\sqrt{|g|}\,\sigma^k)_{'k} = \sqrt{|g|}\,\sigma^k{}_{;k} = -\sqrt{|g|}\,\delta\sigma\,.$$

Cauchy problem: The Maxwell equations in the Minkowskian space were dealt with in Sec. 25.3. We shall attempt to translate the problem setting considered there to the case of arbitrary space-times. As previously we shall take the example of the wave equation as a guide to our approach (cf. Subsecs. 33.3.3. and 33.3.4.). As in these subsections, let Ω be a causal domain in a space-time M, and let S be a 3-dimensional past-compact space-like hypersurface in Ω, with $\partial J^+(S) = S$ (Fig. 33.27). For $p \in J^+(S)$, $p \notin S$, let D_p and S_p have the same meaning as in Subsec. 33.3.4. The classical theory of Sec. 25.3. as well as physical reasons suggest the following question: if σ is given in Ω, and if F is given on S, will then the Maxwell equations (33.5.1/1) have one and only one solution F in $J^+(S)$ with initial data prescribed on S? It would further be desirable that at the point p this solution F depends only on σ in D_p and on F on S_p (the domains of dependence being analogous to those in Subsecs. 19.3.4. and 33.3.4.).

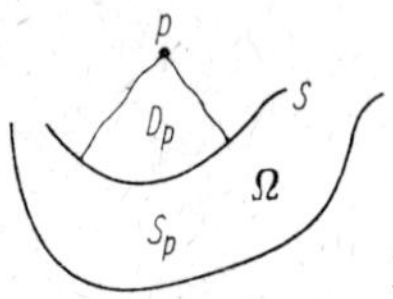

Fig. 33.27

Cauchy data: For the Cauchy problem to be posed reasonably, the following relations must be satisfied on Ω in addition to the continuity equation $\delta\sigma = 0$:

$$(\mathrm{d}F)(q) = 0 \quad \text{and} \quad (\delta F)(q) = \sigma(q) \quad \text{for} \quad q \in S\,. \tag{1}$$

For the discussion of these conditions we choose normal co-ordinates in the sense of Lemma 29.5.6/2 at the point $q = 0$. Further let S be part of the plane $\{x^4 = 0\}$ in a neighbourhood of 0. In view of (33.5.1/2) and (33.5.1/3), the relations (1) then reduce to

$$F_{[kl'm]}(0) = 0 \quad \text{and} \quad F^{kl}{}_{,l}(0) = \sigma^k(0)\,. \tag{2}$$

Since for the moment we assume that F is given only on S, only 2 of these 8 equations will result in additional conditions for the initial data, namely

$$F_{[12'3]}(0) = 0 \quad \text{and} \quad F^{4l}{}_{,l}(0) = \sigma^4(0) \quad \text{as well as} \quad F_{kl} = -F_{lk}\,. \tag{3}$$

The remaining 6 equations read

$$\begin{aligned} &F_{ab'4}(0) = -F_{4a'b}(0) - F_{b4'a}(0)\,,\\ &F^{a4}{}_{,4}(0) = \sigma^a(0) - F^{a1}{}_{,1}(0) - F^{a2}{}_{,2}(0) - F^{a3}{}_{,3}(0)\,, \end{aligned} \tag{4}$$

where a and b run from 1 to 3. If F^{kl} are known on S, then these equations make it possible to calculate the derivatives of F^{kl} with respect to x^4. As the derivatives of F^{kl} with respect to x^1, x^2, x^3 are also known, this means that all the partial first-order derivatives of F^{kl} on S are known. By iteration it then follows from (4) that all the partial derivatives of F^{kl} on S can be determined. In the sequel we shall speak of Cauchy data for the Maxwell equations if on S there is given an antisymmetric field $F^{kl}(q)$, that means, $F^{kl}(q) = -F^{lk}(q)$, which satisfies (1) (or (3) in the canonical form) and whose partial derivatives on S are all calculated from F^{kl} by the above method.

Remark 2. If one identifies F^{kl} with the electric field strength $\mathfrak{E}$ and the magnetic field strength $\mathfrak{B}$, as was done in Subsec. 24.3.3., and if $\sigma^4 = -\varrho$ (charge density), then (3) reads

$$(\operatorname{div} \mathfrak{B})(0) = 0 \quad \text{and} \quad (\operatorname{div} \mathfrak{E})(0) = \varrho(0) .$$

This corresponds with the conditions in Subsec. 25.3.2.

Definition (*Cauchy problem for Maxwell equations*). *Let Ω be a causal domain in the space-time M, and let S be a 3-dimensional past-compact space-like hypersurface in Ω, with $\partial J^+(S) = S$. Let Cauchy data differentiable up to arbitrary order, $F(q) = \{F^{kl}(q)\}$, be given on S for the Maxwell equations. Further let $\sigma = \{\sigma^k\}$, with $\delta\sigma = 0$, be given in $J^+(S)$, with components σ^k differentiable up to arbitrary order. Desired a 2-form F with components differentiable up to arbitrary order, which is a solution of*

$$\mathrm{d}F = 0 \quad and \quad \delta F = \sigma \quad in \quad J^+(S)$$

and assumes the given initial values on S.

Remark 3. This is the analogue to the Cauchy problem for the Maxwell equations in the Minkowskian space, as stated in Subsecs. 25.3.1. and 25.3.2. As was done there, a corresponding problem can also be formulated here on the basis of the theory of distributions.

33.5.3. Gauge Condition and Four-Potential

As has been stated in Remark 33.5.1/3, $\mathrm{d}F = 0$ is locally equivalent to $F = \mathrm{d}A$, where the 1-form A is called a four-potential. The Maxwell equations of (33.5.1/1) then reduce to $\delta\mathrm{d}A = \sigma$.

Cauchy data: Consider the same situation as in Def. 33.5.2. Let again Ω be a causal domain in the space-time M, and let S have the properties stated there. Let $F(q)$ be Cauchy data given for the Maxwell equations on S. Without loss of generality we assume that S be part of the plane $x^4 = 0$. Then the construction of A described in Remark 32.1.5 shows that, with the Cauchy data $F(q)$ being given on S, the components A_k of the four-potential A as well as all partial derivatives of these components on S are known, too. If A_k as well as all the partial derivatives of A_k on S are calculated from the Cauchy data $F(q)$ in this way, then we say that the Cauchy data for the four-potential A on S are given.

Gauge condition. According to Remark 32.1.5/2 it is possible to replace A, with $F = \mathrm{d}A$, by $A + \mathrm{d}B$, where B is a function. Now it is not very difficult to see that B can be so chosen that $(\delta A)(q) = -A^k{}_{;k}(q) = 0$ for $q \in S$ (gauge condition). From now on we shall always assume that A satisfies the gauge condition on S when the Cauchy data are given.

Lemma. *Written in component notation, the reduced Maxwell equation $\delta\mathrm{d}A = \sigma$ reads*

$$\sigma_k = (\Box A)_k + (\delta A)_{;k} + R^s{}_k A_s , \tag{1}$$

where $R^s{}_k$ is the Ricci tensor as per Def. 29.5.7.

Remark 1. The term $(\delta A)_{;k} = -A^l{}_{;lk}$ is annoying. Without it, Eq. (1), that is, $(\Box A)_k + R^s{}_k A_s$, would be a tensor wave operator of the form (33.4.1/3) applied to $A = \{A_k\}$. The question is whether the liberties one has in the choice of A can be utilized to neutralize the term $(\delta A)_{;k}$.

Definition (*Cauchy problem for the four-potential A*). *Let Ω be a causal domain in the space-time M, and let S be a 3-dimensional past-compact space-like hyper-*

surface in Ω, with $\partial J^+(S)=S$. In $J^+(S)$, let $\sigma=\{\sigma^k\}$ be given, with $\delta\sigma=0$ and with components differentiable up to arbitrary order. Further suppose that on S there are given Cauchy data differentiable up to arbitrary order for $A=\{A_k\}$, which can be determined from corresponding Cauchy data $F(q)$ for the Maxwell equations in the sense set out above. Suppose that the gauge condition $(\delta A)(p)=0$ be satisfied for $p\in S$. Desired are vectors A differentiable up to arbitrary order, with

$$\sigma_k=(\Box A)_k+R^s{}_kA_s\,,\quad \text{and}\quad \delta A=0\quad \text{in}\quad J^+(S)\,, \tag{2}$$

such that A_k and $A_{k;l}$ assume the given initial values on S.

Remark 2. (2) is the decoupling of (1). In the general case the system (2) is overdetermined. With arbitrary initial data being given on S, the first equation in (2), as a tensor wave equation, has a unique solution. The question is whether this solution also satisfies $\delta A=0$. It is clear that this can be the case only under special conditions.

Theorem. *The Cauchy problem for the four-potential A (in the sense of the above definition) has a unique solution $A=\{A_k\}$. If $p\in J^+(S)$, $p\notin S$, then $A_k(p)$ depends only on the values of σ_l in D_p and on the values of F^{lm} in S_p.*

Remark 3. D_p and S_p have the same meaning as in Subsecs. 33.3.4. and 33.5.2. As was already indicated in Remark 2, one first considers the Cauchy problem for the tensor wave equation

$$(\Box A)_k+R^s{}_kA_s=\sigma_k\,.$$

The tensor analogue to Theorem 33.3.3 shows that this problem has a unique solution. This solution A_k also depends on σ_l and F^{lm} in the form stated above. The proof that this solution automatically satisfies $\delta A=0$ in $J^+(S)$ is based on the gauge condition $(\delta A)(q)=0$ for $q\in S$ and $\delta\sigma=0$ in $J^+(S)$ and on the specific choice of the Cauchy data for A.

33.5.4. The Cauchy Problem for the Maxwell Equations

Theorem. *The Cauchy problem for the Maxwell equations in the sense of Def. 33.5.2 has a unique solution $F=\{F_{kl}\}$. If $p\in J^+(S)$, $p\in S$, then $F_{kl}(p)$ depends only on the values of σ_r in D_p and on the values of F_{rs} in S_p.*

Remark 1. D_p and S_p again have the same meaning as in Subsecs. 33.3.4. and 33.5.2. The theorem is reduced to Theorem 33.5.3 by setting $F=\mathrm{d}A$. Then $\mathrm{d}F=0$. The right-hand side of (33.5. 3/1) is identical with δF. From this, together with $\delta A=0$, it then follows that $\delta F=\sigma$. This proves the existence of the solution and the above-mentioned dependence of $F_{kl}(p)$ on the initial data. The uniqueness proposition is also reduced to Theorem 33.5.3.

Remark 2. The theorem is the analogue to Theorem 25.3.2/1. As in that case, it should also be possible here to state the solutions explicitly and to extend the considerations to distributions.

34. Singularity Theory

34.1. Local Mappings

34.1.1. Germs of Mappings, the Ideal $m(n)$

Let $C^\infty(R_n, R_l)$ denote the class of all mappings $y=f(x)$ of R_n into R_l which are differentiable up to arbitrary order. All the functions and mappings to be considered in this chapter are real-valued.

Definition 1. *Two mappings of $C^\infty(R_n, R_l)$ are said to be equivalent if they coincide in a suitable neighbourhood of the origin (in R_n). The equivalence classes with respect to this equivalence relation are called map-germs (or simply germs). The class of all germs is denoted by $\varepsilon(n, l)$.*

Remark 1. The above equivalence relation $f\sim g$ has the usual properties: 1. $f\sim f$, 2. $f\sim g$ implies $g\sim f$, and 3. $f\sim g$ and $g\sim h$ implies $f\sim h$. Hence the formation of equivalence classes is meaningful. From now on we shall no longer distinguish between equivalence classes and their representatives. Hence we speak of germs $f\in C^\infty(R_n, R_l)$ etc. (This convention is analogous to the notation $f\in L_p(R_n)$. In this case, too, f is identified with the equivalence class of those functions which coincide with f almost everywhere.) All statements to be made in this chapter are local ones (if not explicitly stated otherwise) and refer to germs.

Remark 2. Instead of considering functions differentiable up to arbitrary order, we might also consider continuous, differentiable, or analytic functions, and form the corresponding equivalence classes.

Definition 2. *Let* $\varepsilon(n)=\varepsilon(n, 1)$,

$$m^k(n)=\{f \mid f\in\varepsilon(n),\ (D^\alpha f)(0)=0 \quad \textit{for} \quad |\alpha|\leq k-1\}$$

for $k=1, 2, 3, \ldots$ *and* $m(n)=m^1(n)$.

Remark 3. In this definition we have already made use of the above convention to identify equivalenc classes with their representatives. Thus the elements of $\varepsilon(n)$ are function germs (or germs in the sense of the above terminology).

Theorem. *$\varepsilon(n)$ is a commutative ring with an identity element and a unique maximal ideal. This maximal ideal is $m(n)$, and is generated by $x_1, \ldots, x_n$.*

Remark 4. Addition and multiplication of germs are carried out point by point, being independent of the choice of the representatives. Since $f\cdot g=g\cdot f$, $\varepsilon(n)$ is a commutative ring. The identity element is $f(x)\equiv 1$. Further, $h(n)$ is called an ideal if $f\in h(n)$ and $g\in\varepsilon(n)$ implies $fg\in h(n)$, which is true for $h(n)=m(n)$. An ideal $h(n)$ is called maximal if $h(n)$ does not coincide with $\varepsilon(n)$ and if there is no ideal different from $\varepsilon(n)$ which is properly greater than $h(n)$. The theorem asserts that $h(n)=m(n)$ is the only maximal ideal in $\varepsilon(n)$. The assertion that $m(n)$ is generated by $x_1, \ldots, x_n$ means that $f\in m(n)$ can be represented as

$$f(x)=\sum_{r=1}^{n} x_r\alpha_r(x), \quad \text{where} \quad \alpha_r(x)\in\varepsilon(n)\,.$$

Remark 5. Correspondingly, it can be shown that $m^k(n)$ is an ideal in $\varepsilon(n)$ which is generated by the polynomials $x_1^{\alpha_1}\ldots x_n^{\alpha_n}$, where $\alpha=(\alpha_1, \ldots, \alpha_n)$ is a multiple index, $|\alpha|=\sum_{r=1}^{n}\alpha_r=k$.

34.1.2. Finitely Determined Germs

Definition. (a) *The germs* $f\in m(n)$ *and* $g\in m(n)$ *are called equivalent if there exists a germ* $y=\varphi(x)\in\varepsilon(n,n)$ *such that* $\varphi(0)=0$, $\dfrac{\partial(y_1,\dots,y_n)}{\partial(x_1,\dots,x_n)}(0)\neq 0$ *and* $f(x)=g(\varphi(x))$.

(b) $f\in m(n)$ *is called k-determined, with* $k=1,2,3,\dots$, *if every germ* $g\in m(n)$ *with* $(D^\alpha g)(0)=(D^\alpha f)(0)$ *for* $|\alpha|\leq k$ *is equivalent to* f.

(c) $f\in m(n)$ *is called finitely determined if there exists a number k such that f is k-determined.*

Remark 1. All these statements are to be understood locally, including the representation $f(x)=g(\varphi(x))$. As usual, $\dfrac{\partial(y_1,\dots,y_n)}{\partial(x_1,\dots,x_n)}$ is the Jacobian of $y=y(x)=\varphi(x)$ as defined in Subsec. 8.2.1. The requirements that $\varphi(0)=0$ and $\dfrac{\partial(y_1,\dots,y_n)}{\partial(x_1,\dots,x_n)}(0)\neq 0$ mean that $\varphi(x)$ is a mapping, one to one and differentiable up to arbitrary order, of a neighbourhood of $0\in R_n$ onto a neighbourhood of $0\in R_n$ (cf. Subsec. 8.2.2.). Such a mapping can also be interpreted as introducing new, curvilinear co-ordinates in a neighbourhood of $0\in R_n$. The co-ordinate surfaces are then $y_k(x)\equiv$ const. In this sense, f and g are said to be equivalent if f can be obtained from g by reparametrization.

Remark 2. If $g\in m(n)$ but $g\notin m^2(n)$, that means, if $\sum_{k=1}^{n}\left|\dfrac{\partial g}{\partial x_k}(0)\right|>0$, then g is equivalent to $f(x)=x_1$. From this normal form one can easily derive that every two germs g and h with $g\in m(n)$, $h\in m(n)$, $g\notin m^2(n)$, and $h\notin m^2(n)$ are equivalent. Hence it follows in particular that $g\in m(n)$ with $g\notin m^2(n)$ is 1-determined. Hence (b) and (c) are of interest mainly for singular germs $f\in m^2(n)$.

Remark 3 (examples). If $n=2$, then $f(x,y)=x^2$ is not finitely determined. This can be seen by comparing the zero manifolds of $f(x,y)=x^2$ and $f(x,y)=x^2-y^{2N}$. For $n=2$, $f(x,y)=x^2+y^3$ is an example of a 3-determined germ.

34.1.3. Criteria for Finitely Determined Germs

If $f\in\varepsilon(n)$, then let

$$\left\langle\frac{\partial f}{\partial x}\right\rangle=\left\{g\mid g\in\varepsilon(n),\ \exists\alpha_k(x)\in\varepsilon(n),\quad\text{where}\quad g(x)=\sum_{k=1}^{n}\alpha_k(x)\frac{\partial f}{\partial x_k}\right\}\tag{1}$$

denote the ideal generated by $\dfrac{\partial f}{\partial x_k}$ in $\varepsilon(n)$. We recall the previous convention that we do not distinguish between germs and their representatives. So the above representation for $g(x)$ needs only to hold locally in a neighbourhood of $0\in R_n$.

Remark 1. For the example $f(x_1,x_2)=x_1^2+x_2^3$ of Remark 34.1,2/3 one has $\left\langle\dfrac{\partial f}{\partial x}\right\rangle=\langle x_1,x_2^2\rangle$.

Theorem. $f\in m(n)$ *is finitely determined if and only if there exists a natural number k such that* $m^k(n)\subset\left\langle\dfrac{\partial f}{\partial x}\right\rangle$.

Remark 2. The theorem is not trivial. But the fact that such inclusions might play some role can be seen as follows. If $g\in m(n)$ is expanded in a Taylor series at 0, then the initial terms of the Taylor series for f and g are compared with each other on the basis of Def. 34.1.2(b), while the remainder of the Taylor series for g, which belongs to $m^k(n)$, is captured by f according to the theorem.

Remark 3. The theorem is derived from the following two partial assertions, which give a sharper formulation as well. (a) f is k-determined if $m^{k+1}(n) \subset m^2(n)\left\langle\frac{\partial f}{\partial x}\right\rangle$. (b) If f is k-determined, then $m^{k+1}(n) \subset m(n)\left\langle\frac{\partial f}{\partial x}\right\rangle$. Here $m(n)\left\langle\frac{\partial f}{\partial x}\right\rangle$ is the set of all products of elements of $m(n)$ and $\left\langle\frac{\partial f}{\partial x}\right\rangle$, by analogy with $m^2(n)\left\langle\frac{\partial f}{\partial x}\right\rangle$.

Definition. *If $f \in m^2(n)$, then* cod f *(codimension of f) is the minimum number of germs $g_l \in m(n)$ with $l = 1, \ldots, N$, such that every germ $g \in m(n)$ can be represented as*

$$g = \sum_{l=1}^{N} a_l g_l + \sum_{k=1}^{n} \alpha_k(x) \frac{\partial f}{\partial x_k}, \quad a_l \text{ real}, \quad \alpha_k(x) \in \varepsilon(n). \tag{2}$$

If $N = \operatorname{cod} f$, then $g_1, \ldots, g_N$ is called a basis of $m(n)$ with respect to $\left\langle\frac{\partial f}{\partial x}\right\rangle$.

Remark 4. Thus in the formal notation one has $\operatorname{cod} f = \dim_{R_1} m(n)/\left\langle\frac{\partial f}{\partial x}\right\rangle$, and $g_1, \ldots, g_N$ with $N = \operatorname{cod} f$ is a basis of $m(n)/\left\langle\frac{\partial f}{\partial x}\right\rangle$. Note that the a_l's in (2) are real numbers.

Remark 5. In the definition we implicitly assumed that there exists a natural number N with the above-mentioned properties. If $m(n) = \left\langle\frac{\partial f}{\partial x}\right\rangle$, then we set $\operatorname{cod} f = 0$. If there is no number $N = 0, 1, 2, \ldots$ in the sense of the definition, then we set $\operatorname{cod} f = \infty$.

Remark 6. For $n = 2$ one has $\operatorname{cod}(x^2y) = \infty$, that is $\dim_{R_1} m(2)/\langle xy, x^2\rangle = \infty$, because the germs $y, y^2, y^3, \ldots$ cannot be captured by finitely many adjunctions of suitable germs g_l. For $n = 2$ one has $\operatorname{cod}(x^3 + y^3) = 3$, that is $\dim_{R_1} m(2)/\langle x^2, y^2\rangle = 3$, where x, y, xy is a basis.

34.2. Stability

34.2.1. Definitions

Definition 1. (a) *Global diffeomorphism in R_n. A one-to-one C^∞-mapping $h(x)$ of R_n onto R_n is called a diffeomorphism.*

(b) *Local diffeomorphism in R_n. A C^∞-mapping $h(x)$ of R_n into R_n is called a local diffeomorphism at $x^0 \in R_n$ if a neighbourhood of x^0 is mapped one to one onto a neighbourhood $h(x^0) \in R_n$.*

(c) *Diffeomorphism on manifolds. A homeomorphic mapping h of an (n-dimensional orientable) C^∞-manifold M onto itself is called a diffeomorphism if $h(x)$ is a C^∞-mapping in every local chart.*

Remark 1. The term "mapping of R_n onto R_n" means that every point of the R_n is an image point (hence the range of such a mapping is the whole R_n). If this is not ensured, then we speak of a "mapping of R_n into R_n". By a C^∞-mapping we mean a mapping $h(x) = (h_1(x), \ldots, h_n(x))$ whose components are all differentiable up to arbitrary order. Part (b) means that $\varphi(x) = h(x + x^0) - h(x^0) \in \varepsilon(n, n)$ is a mapping with the properties stated in Def. 34.1.2. Manifolds M as well as homeomorphic and diffeomorphic mappings on manifolds have been defined in Subsec. 29.1.2.

Remark 2. If $f(x)=(f_1(x), \ldots, f_l(x))$ is a mapping of R_n into R_l, then from now on we shall set

$$|f(x)| = \sum_{k=1}^{l} |f_k(x)| \quad \text{and} \quad (D^\alpha f)(x) = (D^\alpha f_1(x), \ldots, D^\alpha f_l(x)) .$$

Definition 2. (a) *A C^∞-mapping $f(x)$ of R_n into R_l is said to be stable if there exist a positive number ε and a natural number k such that, for every C^∞-mapping $g(x)$ of R_n into R_l with*

$$\sup_{x\in R_n} \sum_{|\alpha|\leqq k} |D^\alpha f(x) - D^\alpha g(x)| < \varepsilon$$

there exist a diffeomorphism h_1 in R_n and a diffeomorphism h_2 in R_l for which the diagram

$$\begin{array}{ccc} R_n & \xrightarrow{f} & R_l \\ h_1\downarrow & & \downarrow h_2 \\ R_n & \xrightarrow[g]{} & R_l \end{array} \tag{1}$$

is commutative.

(b) *A C^∞-mapping $f(x)$ of R_n into R_l is said to be stable at the point $x^0\in R_n$ if there exists a positive number r_0 such that, for every ball $K_r=\{x \mid |x-x^0|<r\}$, where $0<r<r_0$, there exist a positive number $\varepsilon=\varepsilon(r)$ and a natural number $k=k(r)$ with the following property: for every C^∞-mapping $g(x)$ of R_n into R_l with*

$$\sup_{x\in K_r} \sum_{|\alpha|\leqq k} |D^\alpha f(x) - D^\alpha g(x)| < \varepsilon$$

there exist a local diffeomorphism h_1 at $x^0\in R_n$ and a local diffeomorphism h_2 at $f(x^0)\in R_l$, such that the diagram (1) *is locally commutative.*

Remark 3. The commutativity of (1) means that $g(h_1(x))=h_2(f(x))$ for $x\in R_n$, which is also written as $g\circ h_1=h_2\circ f$. $h_1{}^{-1}$ is also a diffeomorphism if h_1 is, so that (1) can also be written as $g=h_2\circ f\circ h_1{}^{-1}$. In other words, g is obtained from f by reparametrization (introduction of curvilinear co-ordinates) in R_n and R_l.

Remark 4. (b) is the local variant of (a). The family of balls K_r with $r\to 0$ is necessary because it is desirable to use only local properties of f at x^0 (or of comparison mappings g). The local variant of the diagram (1) is as follows. The local diffeomorphism h_1 maps a neighbourhood U of x^0, with $U\subset K_r$, onto $h_1(U)\subset K_r$. f and g map U and $h_1(U)$ into sets in R (Fig. 34.1). Finally, h_2 maps $f(U)$ diffeomorphically onto $h_2(f(U))$, and the above-mentioned relations hold true (Fig. 34.2). Thus g is obtained locally from f by reparametrization.

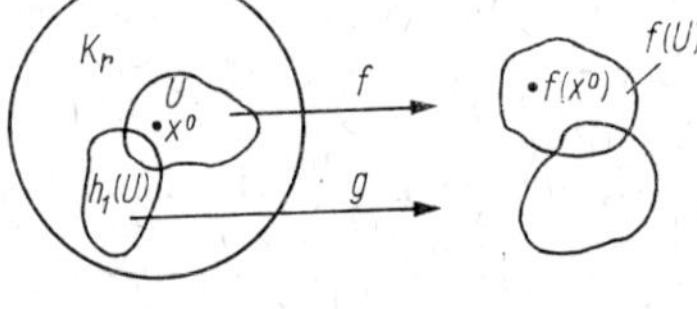

Fig. 34.1

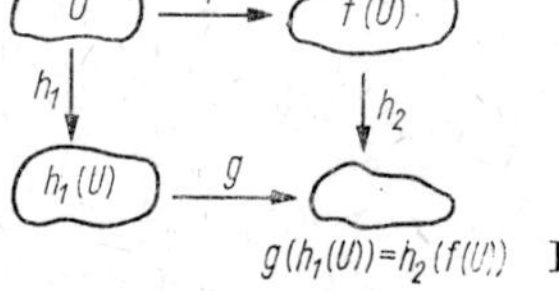

Fig. 34.2

Remark 5. The above concept of local stability suggests itself on the one hand, but it is also rather general on the other. We shall consider several specializations. One special case is Def. 34.1.2(a). There one has $l=1$, $h_1(x)=\varphi(x)$, with $\varphi(0)=0$, and h_2 is the identity.

Remark 6. The stability of C^∞-mappings between manifolds can be defined in an obvious way.

Remark 7. Regular mappings that are stable will be considered in the next subsection. Thus for the moment we confine ourselves to some singular mappings which will play an essential role later on. (The terms "regular" and "singular" will be explained later.) (a) $f(x)=x^2$ is a stable mapping of R_1 into R_1 (where $n=l=1$). (b) (*Whitney's cuspidal point of the second kind*).

$$u(x, y)=xy-x^3, \qquad v(x, y)=y\,, \tag{2}$$

is a stable mapping of R_2 into R_2 ($n=l=2$). (c) $f(x)=x^3$ is a mapping of R_1 into R_1 that is not stable ($n=l=1$). $f(x)$ is not even locally stable at $x^0=0$. From the assumption of local stability at 0 it would follow that there exist two local diffeomorphisms h_1 and h_2 such that $g\circ h_1=h_2\circ f$ for $g(x)=x^3+\varepsilon x$. The chain rule then shows that $(h_2\circ f)'$ has exactly one zero in a neighbourhood of 0, whereas $(g\circ h_1)'$ with $\varepsilon<0$ has two zeros. This is a contradiction.

34.2.2. Immersions and Submersions

If $f(x)=(f_1(x), \ldots, f_l(x))$ is a C^∞-mapping of R_n into R_l, then let

$$(Jf)(x)=\begin{pmatrix} \frac{\partial f_1}{\partial x_1} \cdots \frac{\partial f_1}{\partial x_n} \\ \vdots \qquad \vdots \\ \frac{\partial f_l}{\partial x_1} \cdots \frac{\partial f_l}{\partial x_n} \end{pmatrix}(x)$$

be the associated Jacobian matrix. If $n=l$ and $\det (Jf)(x^0)\neq 0$, then it follows from Theorem 8.2.2/1 that f is a local diffeomorphism at x^0. Hence it is at least plausible that the rank of Jf is of essential interest in the investigation of local mapping properties.

Definition. (a) *$x^0\in R_n$ is called a regular point of f if the rank of $(Jf)(x_0)$ is maximum, i.e., if rank $(Jf)(x^0)=\min(n, l)$.*

(b) *f is called an immersion if $n\leq l$ and f is regular at every point $x\in R_n$.*

(c) *f is called a submersion if $n\geq l$ and f is regular at every point $x\in R_n$.*

Remark 1. The definition is of a local nature. By means of local charts it can immediately be extended to mappings between manifolds in the sense of Def. 29.1.2.

Remark 2. A closed, smooth curve without double points can be immersed in very complex-structured sets in the plane (Fig. 34.3) (here $n=1$ and $l=2$ in the sense of the above definition and of Remark 1). On the other hand such a curve cannot be immersed in R_1. The global properties of immersions may be very complicated. The definition (as well as its extension to manifolds) covers regularity only on a local scale.

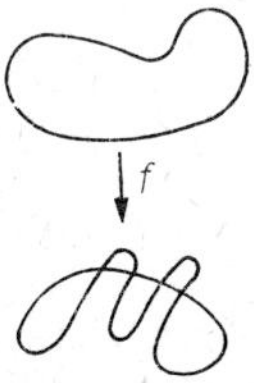

Fig. 34.3

Theorem 1. *If x^0 is a regular point of f in the sense of the above Definition, then f is stable at the point x^0 in the sense of Definition* 34.2.1/2(b).

Remark 3. For $n=l$ this is again easily deduced from Theorem 8.2.2/1.

Remark 4. The converse of the theorem is not true, as is shown by the examples given in Remark 34.2.1/7.

Remark 5. The theorem can be strengthened substantially. If x^0 is a regular point of f, and if $n \leqq l$, then there exists a local diffeomorphism g at $f(x^0) \in R_l$ such that, in a neighbourhood U of x^0,

$$(g \circ f)(x) = (x_1, \ldots, x_n, 0, \ldots, 0), \quad x \in U\,, \tag{1}$$

(local normal form for immersions). If x^0 is a regular point of f, and if $n \geqq l$, then there exists a local diffeomorphism h at $x^0 \in R_n$ such that, in a neighbourhood U of x^0,

$$(f \circ h)(x) = (x_{n-l+1}, \ldots, x_n) \quad \text{for} \quad x \in U \tag{2}$$

(local normal form for submersions). The theorem is a simple implication of (1) and (2).

The space $C^\infty(R_n, R_l)$: As in Subsec. 34.1.1., $C^\infty(R_n, R_l)$ consists of all mappings of R_n into R_l that are differentiable up to arbitrary order. Elements of $C^\infty(R_n, R_l)$ are added and multiplied by real numbers in a point-by-point manner. One then obtains a linear space. $C^\infty(R_n, R_l)$ becomes a Hausdorff space as defined in Subsec. 29.1.1. if the open sets are chosen to be arbitrary unions of

$$\{g \mid \sup_{x \in R_n} \sum_{|\alpha| \leqq k} |D^\alpha f(x) - D^\alpha g(x)| < \varepsilon\}\,. \tag{3}$$

Here $\varepsilon > 0$, $f \in C^\infty(R_n, R_l)$, $g \in C^\infty(R_n, R_l)$, and $k = 0, 1, 2, \ldots$ (cf. Remark 34.2.1/2). Such spaces are also called locally convex. If in the sequel we speak of sets dense in $C^\infty(R_n, R_l)$ then this will always be meant in the sense of the above topology. If N and L are any two manifolds of the dimensions n and l, respectively, then one can define $C^\infty(N, L)$ in a quite analogous way, and make it a Hausdorff space.

Theorem 2. *If $l \geqq 2n$, then the class of all immersions of R_n in R_l is dense in $C^\infty(R_n, R_l)$.*

Remark 6. Theorem 2 remains true if R_n and R_l are replaced by, respectively, any n- and l-dimensional manifold. In any case the immersions form an open set.

Remark 7. The restriction that $l \geqq 2n$ is annoying. Later on we shall deal with problems of denseness for two cases which are not included in the above theorem: (a) $l=1$, n arbitrary (Morse's theory), (b) $n=l=2$ (Whitney's theory). It will then turn out that in these cases sets dense in $C^\infty(R_n, R_l)$ can be constructed only if one also admits special singular mappings.

34.2.3. Global Theorems

The terms "stable" and "generic" are fundamental for the singularity theory. Local and global stability has been dealt with in detail in the preceding subsections. The term "generic" is not so clearly defined. We shall use it here in the following sense: desired are sets in $C^\infty(R_n, R_l)$ (or, more generally, in $C^\infty(N, L)$, where N is an n-dimensional, L an l-dimensional manifold in the sense of Def. 29.1.2) with two properties: (1) These sets shall be dense in $C^\infty(R_n, R_l)$ (or in $C^\infty(N, L)$). (2) The elements of these sets shall have as many good properties as possible. The meaning is clear: it is not desirable to struggle with pathological phenomena of isolated mappings if in an arbitrarily close neighbourhood (with respect to the topology of $C^\infty(R_n, R_l)$ or $C^\infty(N, L)$) there are mappings with much better properties. For the applications of the theory of singularities to problems of physics and biology, this point of view is reasonable. The two theorems stated in Subsec. 34.2.2. show that locally stable mappings are generic, provided that $l \geqq 2n$. It would be desirable to have analogous propositions on global stability and criteria for global stability. The following example shows that the situation on the global scale may be very complex (Fig. 34.4). Let $n=1$ and $l=2$ in the sense set out above. A closed C^∞-curve N without double points can be immersed by f in $L=R_2$ in the manner indicated. Slight modifications g of f (in the sense of $C^\infty(N, L)$) may lead to qualitatively new situations. The figures show that f, though being locally stable, is not globally stable, because double points, points of contact etc. are invariant under diffeomorphic mappings. This will raise the question for one-to-one immersions and whether they are generic and globally stable.

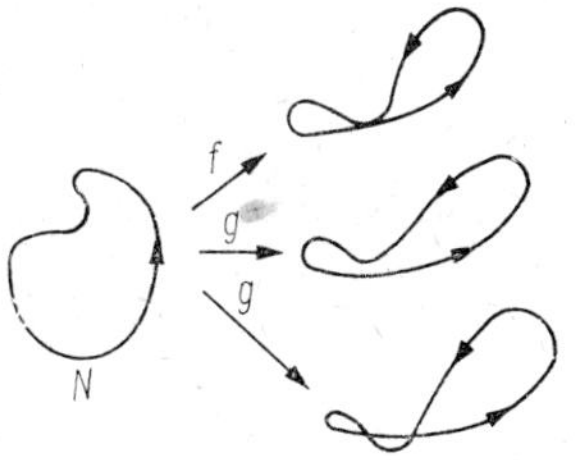

Fig. 34.4

Theorem. *Let N be an n-dimensional, and L an l-dimensional manifold. Further let $l \geqq n$.*

(a) *If N is compact, then every one-to-one immersion is globally stable.*

(b) *If N is compact, and if $l \geqq 2n+1$, then $f \in C^\infty(N, L)$ is globally stable if and only if f is a one-to-one immersion.*

(c) *If $l \geqq 2n+1$, then the class of all one-to-one immersions of N in L is dense in $C^\infty(N, L)$.*

Remark 1. We shall mainly be concerned with local properties. Global propositions like the above theorem will only play a minor role in the sequel.

Remark 2. In part (c) it is not required that N is compact. In particular it is thus possible that $N=R_n$ and $L=R_l$. However, if N is compact in addition, then the one-to-one immersions form an open set in $C^\infty(N, L)$.

34.3. Singularities and Morse Functions

34.3.1. Singularities

In Def. 34.2.2(a) we stated when $x^0 \in R_n$ is a regular point of $f \in C^\infty(R_n, R_l)$. f and Jf shall now have the same meaning as in that definition.

Definition 1. *$x^0 \in R_n$ is called a singular point of $f \in C^\infty(R_n, R_l)$ if the rank of $(Jf)(x^0)$ is not maximal, that is, if rank $(Jf)(x^0) < \min(n, l)$.*

Remark 1. The definition can be immediately transferred to $f \in C^\infty(N, L)$, where N is an n-dimensional, and L an l-dimensional C^∞-manifold.

Remark 2. If $f \in C^\infty(R_n, R_1)$ is a function (i.e., $l=1$ in the above definition), then $x^0 \in R_n$ is a singular point of $f(x)$ if and only if $\frac{\partial f}{\partial x_k}(x^0)=0$ for $k=1, \ldots, n$.

Remark 3. For the examples $f(x)=x^2$ and $f(x)=x^3$ given in Remark 34.2.1/7, $x=0$ is a singular point (here $n=l=1$). For the example (34.2.1/2) (where $n=l=2$) one has

$$(Jf)(x) = \begin{pmatrix} y-3x^2 & x \\ 0 & 1 \end{pmatrix}.$$

Hence a point $(x, y) \in R_2$ is singular if and only if it lies on the parabola $y=3x^2$.

Definition 2. *If $f \in C^\infty(R_n, R_1)$ (i.e., $l=1$ in the sense of Def. 1), then $x^0 \in R_n$ is called a non-degenerate critical point of f if x^0 is a singular point $\left(\text{i.e., if } \frac{\partial f}{\partial x_k}(x^0)=0 \text{ for } k=1, \ldots, n\right)$ and* $\det\left(\frac{\partial^2 f}{\partial x_k \partial x_l}\right)_{k,l=1}^n (x^0) \neq 0.$

Remark 4. $\left(\frac{\partial^2 f}{\partial x_k \partial x_l}\right)_{k,l=1}^n$ is also called Hessian matrix.

Remark 5. For $f(x)=x^2$, $x=0$ is a non-degenerate critical point. For $f(x)=x^3$, $x=0$ is a "degenerate" critical point (here $n=1$).

34.3.2. Morse Functions

Definition. *$f \in C^\infty(R_n, R_1)$ is called a Morse function if every point $x \in R_n$ is either a regular point or a non-degenerate critical point of f.*

Remark 1. $f(x)=x$ and $f(x)=x^2$ are Morse functions, whereas $f(x)=x^3$ (with $n=1$) is not.

Theorem. (**a**) *The Morse functions are dense in $C^\infty(R_n, R_1)$.*
(**b**) *A Morse function is stable at every point $x \in R_n$.*

Remark 2. Denseness in $C^\infty(R_n, R_1)$ is to be understood in the sense of the topology defined in Subsec. 34.2.2. Local stability has been defined in Def. 34.2.1/2(b). For regular points, part (b) is already known; cf. Theorem 34.2.2/1.

Remark 3. The property of being a Morse function is generic in $C^\infty(R_n, R_1)$ (in the sense of the terminology used in Subsec. 34.2.3.). Other than in Theorem 34.2.2/2 (mappings that consist of nothing but regular points are generic in the case $l \geqq 2n$), now one has to admit certain singular points in addition to regular ones in order to get functions with generic properties. It is clear that it will be attempted to keep these singularities as innocuous as possible. This leads to the concept of the non-degenerate critical point of Def. 34.3.1/2.

Remark 4 (normal forms). If $x^0 \in R_n$ is a regular point of $f \in C^\infty(R_n, R_1)$, then, by Remark 34.2.2/5, in a neighbourhood of x^0 it is possible to obtain the normal form $(f \circ h)(x)=x_n$, where h is a local diffeomorphism. If $x^0 \in R_n$ is a non-degenerate critical point of $f \in C^\infty(R_n, R_1)$, then in a neighbourhood of x^0 it is possible to obtain the normal form $(f \circ h)(x)=x_1^2+ \ldots +x_k^2-x_{k+1}^2-\ldots-x_n^2$, where k is a suitable number, $k=0, \ldots, n$, and h is a local diffeomorphism.

Remark 5 (*global version of the theorem*). $f \in C^\infty(R_n, R_1)$ is globally stable if and only if f is a

Morse function whose values at the singular points are pairwise different (this means that $f(x^1) \neq f(x^2)$ if x^1 and x^2 are singular points of f and $x^1 \neq x^2$). Since later on we shall mainly be interested in local effects, we have formulated this interesting proposition only as a remark.

34.4. Mappings in the Plane

34.4.1. Good and Excellent Mappings

We consider mappings $f \in C^\infty(R_2, R_2)$ of the plane into itself. Here we write $f = (u(x, y), v(x, y))$. Let

$$(Jf)(x, y) = \begin{pmatrix} \frac{\partial u}{\partial x} & \frac{\partial u}{\partial y} \\ \frac{\partial v}{\partial x} & \frac{\partial v}{\partial y} \end{pmatrix}, \quad \det (Jf)(x, y) = \frac{\partial u}{\partial x}\frac{\partial v}{\partial y} - \frac{\partial u}{\partial y}\frac{\partial v}{\partial x},$$

be the Jacobian matrix defined in Subsec. 34.2.2., and the Jacobian determinant, respectively.

Definition 1. *$f \in C^\infty(R_2, R_2)$ is called a good mapping if*

$$|\det (Jf)(x, y)| + \left| \frac{\partial}{\partial x} \det (Jf)(x, y) \right| + \left| \frac{\partial}{\partial y} \det (Jf)(x, y) \right| > 0$$

at every point $(x, y) \in R_2$.

Lemma. *If f is a good mapping, then the set of singular points consists of C^∞-curves in R_2 which do not intersect one another.*

Remark 1. By Def. 34.3.1/1, $(x, y) \in R_2$ is a singular point of $f \in C^\infty(R_2, R_2)$ if and only if $\det (Jf)(x, y) = 0$. Let $\gamma(t) = (\gamma_1(t), \gamma_2(t))$ be a smooth parametrization of a C^∞-curve of singular points of f.

Definition 2. (**a**) *Let f be a good mapping, and let $\gamma(t)$ be a curve of singular points. $p = \gamma(t_0)$ is called a fold point if*

$$\frac{\mathrm{d}}{\mathrm{d}t} f(\gamma_1(t_0), \gamma_2(t_0)) \neq 0 .$$

$p = \gamma(t_0)$ is called a cusp point if

$$\frac{\mathrm{d}}{\mathrm{d}t} f(\gamma_1(t_0), \gamma_2(t_0)) = 0 \quad \textit{and} \quad \frac{\mathrm{d}^2}{\mathrm{d}t^2} f(\gamma_1(t_0), \gamma_2(t_0)) \neq 0 .$$

(**b**) *A mapping $f \in C^\infty(R_2, R_2)$ is said to be excellent if f is a good mapping and its singular points are either fold points or cusp points.*

Remark 2. By 0 we always mean $0 \in R_2$. It is easily verified that the definition of the fold points and cusp points is independent of the manner in which the curve $\gamma(t)$ is parametrized.

Remark 3. Cusp points are always isolated points. If p is a cusp point on $\gamma(t)$, then the adjacent arcs of the curve $\gamma(t)$ in a neighbourhood of p consist of fold points alone (Fig. 34.5).

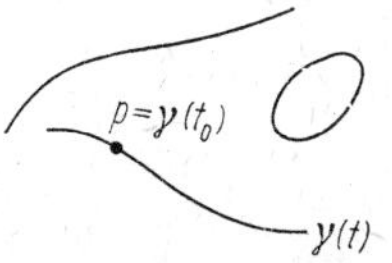

Fig. 34.5

34.4.2. Normal Forms of Fold Points and Cusp Points

Theorem. *If $f=(u(x, y), v(x, y))$ is an excellent mapping, then on the local scale it is possible to obtain the following normal forms by reparametrization:*
(a) $u=x$, $v=y$, *where* $(0, 0)$ *is a regular point.*
(b) $u=x^2$, $v=y$, *where* $(0, 0)$ *is a fold point.*
(c) $u=-xy+x^3$, $v=y$, *where* $(0, 0)$ *is a cusp point.*

Remark 1. Here reparametrization means the local variant of (34.2.1/1), where h_1 and h_2 are local diffeomorphisms of the x,y-plane and the u,v-plane, respectively. The point to be investigated is mapped onto $(0, 0)$.

Fold: The theorem suggests that the mappings in (b) and (c) should also be investigated globally. The mapping $f=(u, v)$ with $u=x^2$ and $v=y$ is called the fold. One has

$$(Jf)(x, y)=\begin{pmatrix} 2x & 0 \\ 0 & 1 \end{pmatrix}, \quad \det (Jf)(x, y)=2x, \quad \frac{\partial}{\partial x}\det (Jf)(x, y)=2.$$

f is an excellent mapping. The points $(x, y)\in R_2$ with $x\neq 0$ are regular. The points $(0, y)$ are singular, being fold points. f folds the x,y-plane, mapping it onto the half-plane $\{u\geqq 0, v\}$.

Cusp (*Whitney's cuspidal point of second kind*, cf. Remark 34.2.1/7): The mapping $f=(u, v)$ with $u=-xy+x^3$ and $v=y$ is called the cusp. One has

$$(Jf)(x, y)=\begin{pmatrix} -y+3x^2 & -x \\ 0 & 1 \end{pmatrix}, \quad \det (Jf)(x, y)=3x^2-y,$$

$$\frac{\partial}{\partial x}\det (Jf)(x, y)=6x, \quad \frac{\partial}{\partial y}\det (Jf)(x, y)=-1.$$

f is an excellent mapping. The singular points lie on the parabola $y=3x^2$, which can be parametrized by $\gamma(t)=(t, 3t^2)$ (Fig. 34.6). Then one has $f(\gamma(t))=(-2t^3, 3t^2)$ as well as

$$\frac{\mathrm{d}}{\mathrm{d}t}f(\gamma(t))=(-6t^2, 6t) \quad \text{and} \quad \frac{\mathrm{d}^2}{\mathrm{d}t^2}f(\gamma(t))=(-12t, 6).$$

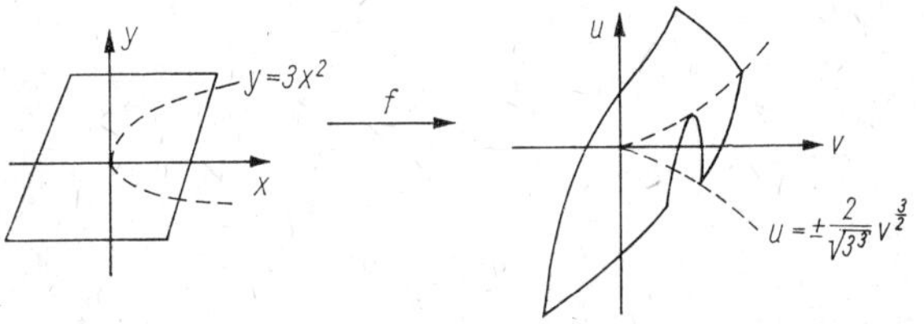

Fig. 34.6

Hence the points with $t\neq 0$ are fold points and the point with $t=0$ (that means $(0, 0)$) is an (isolated) cusp point. The mapping is folded along the parabola $y=3x^2$. The image of this parabola is the semicubical parabola (or Neil parabola, or cuspidal point of second kind) $u=\pm\frac{2}{\sqrt{3^3}}v^{\frac{3}{2}}$.

34.4.3. Whitney's Theory

Theorem. (a) *The mapping* $f \in C^\infty(R_2, R_2)$ *is stable at every point* $(x, y) \in R_2$ *if and only if* f *is excellent.*

(b) *The excellent mappings are dense in* $C^\infty(R_2, R_2)$.

Remark 1. The denseness in $C^\infty(R_2, R_2)$ is to be understood in the sense of the topology defined in Subsec. 34.2.2. Local stability was defined in Def. 34.2.1/2(b). Compare the above theorem with the considerations described in Subsec. 34.3.2. and with Theorem 34.2.2/2. Excellence is a generic property of mappings of the plane into itself.

Remark 2. (*global version of the theorem*). $f \in C^\infty(R_2, R_2)$ is globally stable if and only if f is excellent and has the following additional property: the images of the C^∞-curves consisting of fold points intersect at most pairwise but not at zero angles. Further, the set of images of the fold points and the set of images of the cusp points are disjoint.

34.5. Unfoldings

34.5.1. Definition

Besides the terms "stable" and "generic", unfolding is the third fundamental concept of singularity theory and catastrophe theory.

Definition. *If* $f \in m(n)$, *then* $F \in m(n+r)$ *is called unfolding of* f *if* $F(x, 0) = f(x)$ *for* $x \in R_n$ *and* $0 \in R_r$. r *is called codimension of the unfolding.*

Remark 1. $m(n)$ and $m(n+r)$ have the same meaning as in Def. 34.1.1/2. Hence it is also clear that all considerations are local, referring to a neighbourhood of $0 \in R_{n+r}$. Thus the germ $f(x) \in m(n)$ with $x \in R_n$ is extended to R_{n+r} by $F(x, u) \in m(n+r)$, with $x \in R_n$ and $u \in R_r$. $F(x, 0) = f(x)$ should also be understood in this sense. The values of r may be $r = 0, 1, 2, \ldots$ As the codimension r will play a major role later on, we also write (F, r) instead of F. There is no fear of confusion with $F(x, u)$.

Remark 2 (*examples*). (1) If $f(x) \in m(n)$, then

$$F(x, u) = f(x) \quad \text{for} \quad x \in R_n \quad \text{and} \quad u \in R_r$$

is the constant unfolding. (2) If $f(x) \in m(n)$, then

$$F(x, u) = f(x) + \sum_{k=1}^{r} b_k(x)\, u_k, \quad b_k \in \varepsilon(n), \quad u = (u_1, \ldots, u_r),$$

is an unfolding $F \in m(n+r)$ of f.

Remark 3. The meaning of this concept can be described as follows. In the theory of catastrophes, physical or biological processes that are discontinuous are described by unstable germs $f \in m(n)$: extremely small C^∞-perturbations effect substantial changes of the topological structure of the germs (the unperturbed and the perturbed germ can no longer be transformed into one another by reparametrization). One attempts to imbed this isolated catastrophe event in an r-parametric process which is described by a "structurally stable potential" $F(x, u)$. Catastrophes then occur at isolated points of the otherwise structurally stable process. If a catastrophe occurs at a certain position of the R_3 and at a certain time, then the process can be imbedded in a four-dimensional space-time, $r = 4$. All considerations to be made are local.

34.5.2. Associated and Equivalent Unfoldings

It is our aim to find minimal stable unfoldings of a given germ $f \in m(n)$. For that purpose it is necessary to define several new concepts. Two germs $f \in m(n)$ and $g \in m(n)$ will not be judged essentially different if they are equivalent in the sense of Def. 34.1.2(a).

Definition. *Let (F, r) be an unfolding of $f \in m(n)$, and let (G, s) be an unfolding of $g \in m(n)$.*

(**a**) *(G, s) is called associated to (F, r) if there exist germs*

$$\Phi \in \varepsilon(n+s, n+r) \quad \text{with} \quad \Phi(0)=0, \quad \psi \in \varepsilon(s, r) \quad \text{with} \quad \psi(0)=0,$$

and $\alpha \in m(s)$ which have the following two properties:
1. $\Phi(x, u) = (y(x, u), \psi(u))$; $x \in R_n$, $u \in R_s$, $y(x, u) \in R_n$, $\psi(u) \in R_r$;
2. $G(x, u) = F(y(x, u), \psi(u)) + \alpha(u)$.

(**b**) *(G, s) and (F, r) are called equivalent (or isomorphic) if (G, s) is associated to (F, r), $r=s$, and if $\psi(u)$ is a local diffeomorphism in R_r and $y(x, u)$, with u being fixed, is a local diffeomorphism in R_n.*

Remark 1. The requirement 1 means that $\Phi(x, u)$ is a fibre preserving mapping. The fibre R_n marked by "u" is transformed into the fibre R_n marked by "$\psi(u)$" (Fig. 34.7). For a fixed fibre u one then has $y(x, u) \in C^\infty(R_n, R_r)$. The requirement 2 means that G can be reduced to F. For $u=0$ one obtains

$$g(x) = G(x, 0) = F(y(x, 0), 0) + 0 = f(y(x)). \tag{1}$$

Hence, in particular, the germ g can be reduced to the germ f.

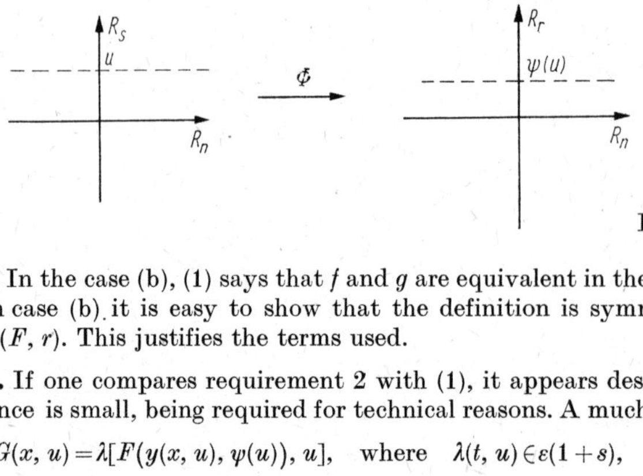

Fig. 34.7

Remark 2. In the case (b), (1) says that f and g are equivalent in the sense of Def. 34.1.2(a). Further, in case (b) it is easy to show that the definition is symmetrical with respect to (G, s) and (F, r). This justifies the terms used.

Remark 3. If one compares requirement 2 with (1), it appears desirable that $\alpha(u) \equiv 0$. But the difference is small, being required for technical reasons. A much more general version is

$$G(x, u) = \lambda[F(y(x, u), \psi(u)), u], \quad \text{where} \quad \lambda(t, u) \in \varepsilon(1+s), \quad \lambda(t, 0) = t,$$

which has, for instance, been considered in [65], p. 33.

34.5.3. Stable and Universal Unfoldings (Definition and Examples)

Definition (**a**) *An unfolding (F, r) of $f \in m(n)$ is called a stable unfolding of f if every unfolding (G, s) of f is associated to (F, r).*

(**b**) *An unfolding (F, r) of $f \in m(n)$ is called a universal unfolding of f if (F, r) is a stable unfolding of f and if r is minimal.*

Remark 1. Hence (F, r) is called a stable unfolding if every other unfolding (G, s) of f can be reduced to (F, r). In particular this means that small perturbations of (F, r), provided they are unfoldings of f, are reducible to (F, r).

Remark 2 (examples). 1 If $f \in m(n)$ is regular, i.e., if $f \notin m^2(n)$, then $(F, 0) = f$ is the universal unfolding of f (there is nothing to be unfolded). This follows from Remark 34.1.2/2 (here $\alpha(u) \equiv 0$ and $\psi(u) \equiv 0$ in Def. 34.5.2). Thus the concept of unfolding is only of interest for singular germs $f \in m^2(n)$. 2. For $f = x^2 \in m^2(1)$, the codimension of the universal unfolding is $r = 0$, just as in the first example. This can in essence be deduced from Theorem 34.3.2 and Remark 34.3.2/4, because x^2 is a Morse function. 3. According to Remark 34.2.1/7, $f(x) = x^3 \in m^3(1)$ is not stable. Hence it cannot be expected that $r = 0$ is the codimension of a universal unfolding. It turns out that $F(x, u) = x^3 - xu$ is a universal unfolding of $f(x) = x^3$ (thus $r = 1$ is the codimension of a universal unfolding). This is largely an implication of Theorem 34.4.3, because $F(x, u)$ corresponds to Whitney's cuspidal point of second kind, which is an excellent mapping in the plane.

34.5.4. Stable and Universal Unfoldings (Criteria)

If $F(x, u) \in m(n+r)$, where $x \in R_n$ and $u = (u_1, \ldots, u_r) \in R_r$, is an unfolding of $f(x) \in m(n)$, then we set

$$V_F = \left\{a_0 + \sum_{k=1}^{r} a_k \frac{\partial F}{\partial u_k}(x, 0),\ a_l \text{ real}\right\} \subset \varepsilon(n)\,.$$

Theorem 1. (**a**) *$f \in m^2(n)$ has a stable unfolding if and only if f is finitely determined.*
(**b**) *The unfolding (F, r) of $f \in m^2(n)$ is stable if and only if*

$$\varepsilon(n) = \left\langle \frac{\partial f}{\partial x} \right\rangle + V_F\,. \tag{1}$$

Remark 1. Finitely determined germs were considered in Subsec. 34.1.2. $\left\langle \frac{\partial f}{\partial x} \right\rangle$ was defined in Subsec. 34.1.3. Eq. (1) says that it must be attempted to extend $\left\langle \frac{\partial f}{\partial x} \right\rangle$ to $\varepsilon(n)$ by the adjunction of finitely many germs.

Theorem 2. *If $f \in m^2(n)$ is finitely determined, then* cod f, *as introduced in Def. 34.1.3, is the codimension of a universal unfolding of f.*

Remark 2. If $g_1, \ldots, g_r$, where $r = \operatorname{cod} f$, is a basis of $m(n)/\left\langle \frac{\partial f}{\partial x} \right\rangle$ in the sense of Def. 34.1.3, then

$$F(x, u) = f(x) + \sum_{k=1}^{r} u_k g_k(x), \quad \text{where} \quad x \in R_n \quad \text{and} \quad u = (u_1, \ldots, u_r) \in R_r$$

is a universal unfolding of the finitely determined germ $f \in m^2(n)$. This follows from Theorem 1.

Remark 3. In accordance with the examples given in Remark 34.5. 3/2 one has cod $x^2 = 0$ and cod $x^3 = 1$ (where $n = 1$).

34.5.5. Reduction of Unfoldings

Let $G(x, u)$ be an unfolding of $g(x) \in m(n)$, where $u \in R_r$. In the later applications, R_r will be the control space with u as a control parameter, and R_n will be the phase space with x as a phase or state parameter. $G(x, u)$ then is the potential which describes the process in question. We are interested to know the local relative minima of $G(x, u)$ with u being fixed (the corresponding x-values describe the possible states). As $u \to 0$, the system tends towards a catastrophe. Thus we look for x-values such that

$$\frac{\partial G}{\partial x_1}(x, u) = \dots = \frac{\partial G}{\partial x_n}(x, u) = 0, \quad u \in R_r \quad \text{fixed,} \tag{1}$$

where all considerations are confined to a neighbourhood of $0 \in R_{n+r}$. What is of real interest are the minima of $G(x, u)$ alone, so that any procedures which do not alter the topological structure of the set of all these minima are admissible. Let

$$h(x, y) = g(x) + \sum_{k=1}^{q} y_k^2 \in m(n+q)\,. \tag{2}$$

Then, if $x \in R_n$, $y \in R_q$, $u \in R_r$, and $v \in R_s$,

$$\hat{G}(x, y, u, v) = G(x, u) + \sum_{k=1}^{q} y_k^2 \tag{3}$$

is an unfolding of h, with the codimension $r+s$, which is the constant unfolding with respect to $v \in R_s$, in the sense of Remark 34.5.1/2. $(u, v) \in R_{r+s}$ are now the control parameters. The minima of $\hat{G}$ in (3) for fixed $(u, v) \in R_{r+s}$ are identical with the minima of G, provided that $y_1 = \dots = y_q = 0$.

Definition. *Let (G, r) be an unfolding of $g \in m(n)$, and let $(F, r+s)$ be an unfolding of $f \in m(n+q)$. Here $q \geqq 0$ and $s \geqq 0$.*
(a) *F reduces to G if F is equivalent to $\hat{G}$ in (3) (in the sense of an $(r+s)$-unfolding).*
(b) *G is called a proper reduction of F if F reduces to G and $q+s>0$.*
(c) *F is called irreducible if F has no proper reductions.*

Remark. The starting point lies in the germ f and its unfolding F. Desired are simple procedures which on the one hand do not alter the structure of the set of all minima, while reducing the number of phase parameters and the number of unfolding parameters on the other hand. This means that it is desirable to get rid of superfluous parameters which do not influence the qualitative picture of the set of all minima. It follows from Remark 34.5.2/2 that in case (a) the germs f and h in (2) are equivalent (in the sense of Def. 34.1.2 (a)). So one attempts to transform f into the form (2) by means of a local diffeomorphism, whereafter one can omit the uninteresting parameters $y_1, \dots, y_q$ and confine oneself to g. The second procedure consists in omitting those unfolding parameters in which the unfolding is constant.

34.5.6. Minima

Simple minimum: If $f \in m^2(n)$ is equivalent to $Q(x) = \sum_{k=1}^{n} x_k^2$ in the sense of Def. 34.1.2(a), then, by Theorem 34.1.1,

$$\left\langle \frac{\partial f}{\partial x} \right\rangle = \left\langle \frac{\partial Q}{\partial x} \right\rangle = \langle x_{,1} \dots, x_n \rangle = m(n)\,.$$

From Theorem 34.5.4/2 it then follows that f is its own universal unfolding (f is a Morse function, cf. Def. 34.3.2). If (F, r) is an unfolding of f, then it is easily seen that (F, r) is equivalent to the r-dimensional constant unfolding of f. In this case F is said to have a simple minimum at the point 0.

Definition. *Let (F, r) be an unfolding of $f \in m(n)$. Then F has a local minimum at $0 \in R_{n+r}$ if for every neighbourhood W of $0 \in R_{n+r}$ there exists a point $u_W \in R_r$ such that, for u_W being fixed, the function $F(x, u)$ has a local minimum in $W \cap \{u = u_W\}$.*

Remark. If (F, r) has a simple minimum (i.e., if $f \in m(n)$ is equivalent to $\sum_{k=1}^{n} x_k^2$), then F has a local minimum in the point 0. In this case one can always choose $u_W = 0$. But what we are interested in are just those cases in which $F(x, 0)$ has no minimum in the point $0 \in R_n$, whereas there are minima in an arbitrary x,u-neighbourhood of $0 \in R_{n+r}$. A typical example is

$$f(x) = x^3 \in m(1) \quad \text{and} \quad F(x, u) = x^3 + ux\ .$$

34.5.7. Thom's Theorem

Theorem. *Let $f \in m^2(n)$ be a finitely determined germ. Let (F, r) be a stable unfolding of f, with codimension $r \leq 4$, which has a local minimum at 0. Then F either has a simple minimum in the point 0 or F reduces to one of the following 7 irreducible (canonical) unfoldings G_k of the germs g_k*

Name	*Germ* g_k	*Unfolding* G_k	Codim. r
Fold	x^3	$x^3 + ux$	1
Cusp	x^4	$x^4 - ux^2 + vx$	2
Swallow tail	x^5	$x^5 + ux^3 + vx^2 + wx$	3
Hyperbolic umbilic	$x^3 + y^3$	$x^3 + y^3 + wxy - ux - vy$	3
Elliptic umbilic	$\frac{x^3}{3} - xy^2$	$\frac{x^3}{3} - xy^2 + w\,(x^2 + y^2) - ux - vy$	3
Butterfly	x^6	$x^6 + tx^4 + ux^3 + vx^2 + wx$	4
Parabolic umbilic	$x^2y + \frac{y^4}{4}$	$x^2y + \frac{y^4}{4} + tx^2 + wy^2 - ux - vy$	4

Remark 1. All the terms used in this theorem have been defined previously: finitely determined germs in Def. 34.1.2, stable unfoldings in Def. 34.5.3, the local minimum at 0 and the simple minimum in Subsec. 34.5.6., reduction of unfoldings in Subsec. 34.5.5. The three umbilics also answer to the following nicknames: *wave crest* (hyperbolic umbilic), *hair* (elliptic umbilic) and *mushroom* (parabolic umbilic). The poetic names will become clear when we shall draw some figures later on.

Remark 2. This is Thom's famous list of the seven elementary catastrophes. They form the foundation of catastrophe theory. The fact that G_k is a universal unfolding of g_k is rather easily verified on the basis of Subsec. 34.5.4. What is surprising with this list is its finiteness and its independence of the number n of the phase parameters. With the assumptions of the theorem, it turns out that only one or two of the n phase parameters are essential (in the sense of the reduction described in Subsec. 34.5.5.). This means that the plausible presumption that the list will grow longer and longer as n increases is not true.

Remark 3. The assumptions that (F, r) is stable and has a local minimum at 0 are natural. In the applications, (F, r) is a "structurally stable potential", and the stable states of the system are described by the minima of the potential in the sense of Subsec. 34.5.6. The restriction to $r \leq 4$ is reasonable in biological applications, because in such cases the unfolding parameters (control parameters) are identified with a space-time, while $x \in R_n$ describes, say, the biochemical state of a cell. But $r > 4$ is of interest, too, especially in physical applications. What is unclear for the moment is the role played by the assumption that f be finitely determined. As we shall see later on, this requirement is of generic nature, being useful for applications.

Remark 4. If one admits other codimensions r, then the following picture results:

Codim. r of the unfolding	1	2	3	4	5	6
Number of types	1	2	5	7	11	∞

Here the simple minimum has not been taken into account. The first cases can be taken from the above theorem.

Remark 5. The first step in the proof of the theorem is the so-called "splitting lemma", a refinement of the normal forms for Morse functions as per Remark 34.3.2/4: if 0 is a degenerate critical point of $f \in m^2(n)$, that is, a singular point with $\det \left(\frac{\partial^2 f}{\partial x_k \partial x_l}\right)_{k,l=1}^{n}(0) = 0$, then $f(x)$ is equivalent to $\sum_{k=1}^{r} \varepsilon_k x_k^2 + \psi(x_{r+1}, \ldots, x_n)$, where $r = 0, \ldots, n-1$, $\varepsilon_k = \pm 1$, $\psi \in m^3(n)$.

Remark 6. The presentation of this chapter follows the lines of [40]. We further refer to [7, 65]. Complete proofs of the above theorem have been given in [68, 78].

35. Catastrophes: Theory and Application[1)]

35.1. Principles and Models

35.1.1. General Principles and Fundamental Ideas

A detailed consideration of the possibilities of catastrophe theory can be found in the famous book of R. Thom [65]. The mathematical basis is the singularity theory presented in Chap. 34, especially Theorem 34.5.7. The aim of catastrophe theory is the mathematical description of discontinuous processes (especially in physics and biology), e.g., of the transition of a gas into the liquid

1) René Thom suggested that the combination of singularity theory and its applications should be called catastrophe theory.

phase, the buckling of an elastic rod, or the specialization of cells in a tissue. We shall order the heuristic considerations to follow by some striking keywords. Our aim is to make plausible that Theorem 34.5.7 is a useful instrument for modelling discontinuous natural processes.

Potential: Given a physical, chemical or biological system that depends on finitely many control parameters $u \in R_r$ (*control space*), e.g., a thermodynamic system (with temperature and pressure as control parameters) or a living cell (with space and time as control parameters). Let the state of the system be described by n state parameters $x \in R_n$ (*phase space*), e.g., the volume of a thermodynamic system of the biochemical state of a living cell. It is assumed that the system can be described by a potential $F = F(x, u)$, which assigns a real value $F(x, u)$ to every (admissible) $u \in R_r$ in the control space and to every (admissible) state $x \in R_n$ in the phase space. For a given $u \in R_r$, the possible stable states are determined by the relative minima of $F(x, u)$. This is a well-known procedure, which is analogous to the classical field theory. Naturally, the requirement that a potential exists involves a restriction, which can in part be relaxed if one uses the modern qualitative theory of dynamic systems; see for instance [8, 27]. If $u \in R_r$ is fixed, then a potential $F(x, u)$ generates a so-called gradient field,

$$\frac{\mathrm{d}x_k}{\mathrm{d}t} = \frac{\partial F}{\partial x_k}(u, x_1(t), \ldots, x_n(t)); \quad k = 1, \ldots, n .$$

The following considerations can at least partially be applied to potentials or gradient systems or even to more general dynamic systems,

$$\frac{\mathrm{d}x_k}{\mathrm{d}t} = X_k(u, x_1(t), \ldots, x_n(t)); \quad k = 1, \ldots, n ,$$

cf. [65]. But here we shall confine ourselves to potentials where recourse to dynamic systems is not necessary.

Structural stability: It is reasonable and plausible to require that small perturbations of the potential $F(x, u)$ of a real system do not change the topological character of $F(x, u)$. In particular, small perturbations of the potential are assumed to cause only small changes of the possible stable states of the system (that is, of the relative minima of $F(x, u)$ with $u \in R_r$ being fixed). This (somewhat imprecise) requirement is called structural stability. If a system is described by a potential that is not yet explicitly known, then one may try, for example, to determine $F(x, u)$ experimentally. In this case measuring errors will occur. Thus it is clear that very small changes of $F(x, u)$ must not have any substantial effect.

Local consideration. The above considerations are of local nature; $F(x, u)$ is investigated in the neighbourhood of a given point $(x_0, u_0) \in R_{n+r}$. Of course we can set $x_0 = 0 \in R_n$ and $u_0 = 0 \in R_r$ as well as $F(0, 0) = 0$ without loss of generality. The structural stability implies that $F \in C^\infty(R_{n+r})$ is a reasonable requirement. Hence $F \in m(n+r)$ is a germ in the sense of Def. 34.1.1/2. The control parameters $u \in R_r$ vary continuously (which is also a criterion of choice for these parameters), whereas the state parameters $x \in R_n$ may also vary discontinuously (jumping from one relative minimum to another). In the next subsection we shall describe the rules of this jumping in greater detail. We assume that $x = 0 \in R_n$ and $u = 0 \in R_r$ is an isolated point at which the system exhibits a degenerate behaviour, while there are relative minima of $F(x, u)$ in the above sense in every neighbourhood W of $0 \in R_{n+r}$. This suggests the following interpretation: $F(x, u) = (F, r)$ is that stable unfolding (of codimension r) of a germ $f \in m(n)$ which has a local minimum at $0 \in R_{n+r}$ in the sense of Def. 34.5.6. Thus the term "stable" has a precise mathematical meaning, meeting the sense of the above heuristic concept of "structural stability". This conception of stability fits the situation very well, because only those deformations are admitted which involve separate transformations of the control parameters on the one hand and the state parameters on the other.

Catastrophes: The above system realizes a process: every control parameter u is assigned a state $x(u)$. In the next subsection we shall further comment on this assignment. A regular point of the process is a parameter value u_0 in the control space, in the neighbourhood of which $x(u)$ varies continuously as u varies continuously. If u_0 is not a regular point, it is called a catastrophe point. For an unfolding of $f \in m(n)$ to have a local minimum at 0, it is necessary that $f \in m^2(n)$ (cf. Def. 34.1.1/2). The case where F, as an unfolding of $f \in m^2(n)$, has a simple minimum in the point 0 in the sense of Subsec. 34.5.6. is of no interest either. For in this case $u = 0$ is a regular point.

Elementary catastrophes: If we now assume in addition that $f \in m^2(n)$ is finitely determined and that $r \leqq 4$, then it is possible to apply Theorem 34.5.7 (the structure of the relative minima of $F(x, u)$, with u fixed, is not altered by the reduction in the sense of Theorem 34.5.7). In other words, the general catastrophe phenomena can be reduced to 7 basic types, provided that $f \in m^2(n)$ is finitely determined and $r \leqq 4$. What is annoying for the moment is the requirement that f be finitely determined. But we shall see later on in Subsec. 35.2.1. that this restriction is of generic character and can always be assumed to be fulfilled. The topological situation of catastrophe events is thus completely described by Thom's list, provided that $r \leqq 4$.

Global consideration: All the considerations made so far are of local nature, i.e. they refer to a neighbourhood of an (x, u) point. A global process has to be decomposed into local processes: in the neighbourhood of degenerate points in the x,u-space one has to find the associated "catastrophe potentials" (after being reduced, they are unfoldings in the sense of Theorem 34.5.7). Subsequently one has to combine the local catastrophes smoothly (and in a physically or biologically reasonable way). There are no recipes for such procedures.

Character recognition: Given a process $(x(u), u)$ of which we believe (or hope) that it can be described by a (still unknown) potential $F(x, u)$ in the sense set out above, and that $r \leqq 4$ with respect to the number of control parameters $u \in R_r$. Then one may try to determine all catastrophe points u experimentally, that is, all those points u for which $x(u)$ is possibly discontinuous. In the neighbourhood of a catastrophe point it can be expected that, in the course of a great number of repetitions of suitable experiments, all the stable states possible are in fact assumed. Now one can compare the data obtained with the 7 elementary catastrophes listed in Theorem 34.5.7 (where reductions, diffeomorphisms etc. must be eliminated by means of a computer). It can be hoped that in this way it is possible to calculate (guess) the catastrophe potential and to determine its type.

Formalization. If $(F, r)=F(x, u)$ is a stable unfolding of $f\in m^2(n)$, then we set

$$\Sigma_F=\left\{(x, u)\in R_{n+r}\,\middle|\,\frac{\partial F}{\partial x_1}(x, u)= \dots =\frac{\partial F}{\partial x_n}(x, u)=0\right\}, \tag{1}$$

$$\Delta_F=\left\{(x, u)\in \Sigma_F\,\middle|\,\det\left(\frac{\partial^2 F}{\partial x_k\partial x_l}\right)^n_{k,l=1}(x, u)=0\right\}, \tag{2}$$

$$D_F=\{u\in R_r\mid \exists x\in R_n \quad\text{with}\quad (x, u)\in \Delta_F\}\,. \tag{3}$$

The stable states of the system with the potential $F(x, u)$ form a subset of Σ_F, namely the minima contained in (1). If (x, u) is such a minimum, for which

$$\det\left(\frac{\partial^2 F}{\partial x_k\partial x_l}\right)^n_{k,l=1}(x, u)\neq 0\,,$$

then one has a simple minimum as defined in Subsec. 34.5.6. (cf. Remark 34.3.2/4). Then u is not a catastrophe point (at least as far as the delay rule to be considered in the next subsection is concerned). Here we have to choose our words with care, because the Maxwell convention to be stated in Subsec. 35.1.2. does not quite fit into this scheme. In any case it is clear, however, that the set D_F is of special interest in the investigation of possible catastrophe points.

Remark. All considerations are local. It is however reasonable to investigate the 7 elementary catastrophes listed in Theorem 34.5.7 on a global scale.

35.1.2. The Local Regime

As in Subsec. 35.1.1., we consider a (physical, chemical, or biological) system that has a potential $F(x, u)\in m(n+r)$ with the control parameters $u\in R_r$ and the state parameters $x\in R_n$. As was stated previously, a process realized by this system is described by $(x(u), u)$. For the investigation of the possible catastrophe phenomena we may confine ourselves to Thom's list given in Subsec. 34.5.7.

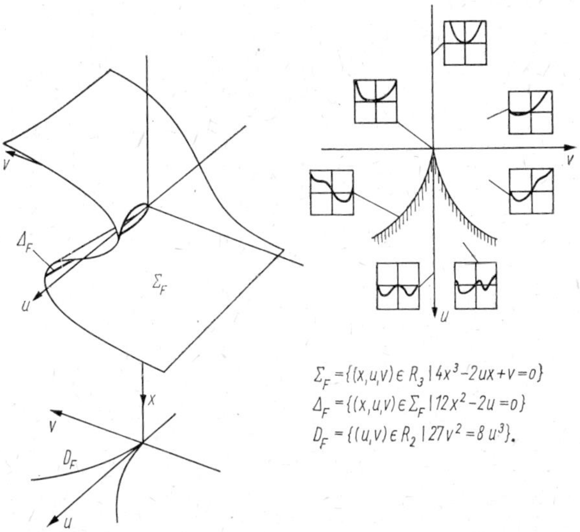
$\Sigma_F=\{(x,u,v)\in R_3\mid 4x^3-2ux+v=0\}$
$\Delta_F=\{(x,u,v)\in \Sigma_F\mid 12x^2-2u=0\}$
$D_F=\{(u,v)\in R_2\mid 27v^2=8u^3\}$.

Fig. 35.1

Example (*the cusp*). Let $f(x) = x^4 \in m^2(1)$, $F(x, u, v) = x^4 - ux^2 + vx$ (Fig. 35.1). Then Σ_F, Δ_F and D_F have the form indicated in the figure. In the left drawing, the u,v-axes are drawn twice, while the x-axis is the same in both co-ordinate systems. The right drawing shows curves of $F(x, u)$ for several parameter values of (u, v). (u, v) are control parameters, x is a phase parameter. The catastrophe in question is the second type of the elementary catastrophes of Theorem 34.5.7. Here D_F is a semicubical or Neil parabola. In the interior of the Neil parabola (shaded area), the function F, with fixed (u, v), has two minima for each parameter value, one of which disappears on D_F, and in the exterior of the Neil parabola there is only one minimum for each curve. The two minima in the interior of the Neil parabola are equal if and only if $v = 0$ (positive u-axis).

Local regime (Fig. 35.2): Now let $F(x, u)$, where $u \in R_r$ and $x \in R_n$, be an arbitrary potential. The process the system runs through in a neighbourhood of $x = 0$, $u = 0$ is described by $(x(u), u)$, where $x(u)$ is a relative minimum of $F(x, u)$ for u being fixed (local regime). The question is which minimum is assumed (in case there are more than one), and how the system develops in response to a variation of u. For this there are two rules:

Delay rule: For $u^0 \in R_r$, let the system be in the state $x(u^0)$ (relative minimum). *If u varies continuously* (in a neighbourhood of u^0), *then the system tends to change its state continuously, too.* It will remain as long as possible in a state $x(u)$ that can be changed continuously into $x(u^0)$. Catastrophes (jump-like changes of state) will occur if and only if in the variation of u the associated minimum $x(u)$ disappears. If, in the above example, $(u^0, v^0) \in R_2$ is a point of the interior shaded area, then catastrophes can occur on D_F (and only there). Whether or not a catastrophe will in fact occur when D_F is crossed depends on the particular minimum chosen. If D_f is crossed from the exterior, then a catastrophe will not occur.

Maxwell convention: *For each fixed $u \in R_r$, the system always realizes the state corresponding to the absolute minimum.* Thus catastrophes can occur only at those u-values where the absolute minimum is no longer uniquely determined. In the above example this is the positive u-axis.

System development (Fig. 35.3): If the control parameters are varied along the path $u(\tau)$, where τ is a curve parameter, then the system runs through the states $x = x(u(\tau))$. It is quite useful to fix a path in the u,v-plane of the above example and to observe how the system behaves. For the path indicated in the figure, catastrophes will occur at 1 and 2, provided that the system obeys the delay rule.

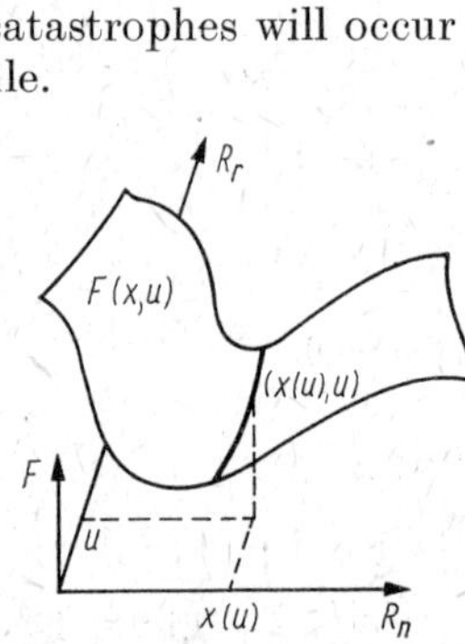

Fig. 35.2

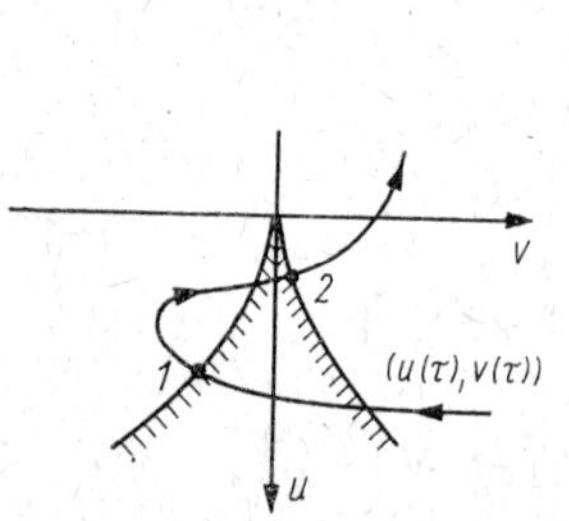

Fig. 35.3

Remark. At first glance, the Maxwell convention appears somewhat artificial. But it has a counterpart in Morse's theory. For each fixed $u \notin D_F$, $F(x, u) \in C^\infty(R_n, R_1)$ is, by Def. 34.3.2, a Morse function. But according to Remark 34.3.2/5 the global stability of $F(x, u)$ is lost if u is a point at which the absolute minimum is no longer uniquely determined. Thus the magnitude of the function values also plays a role in singularity theory.

35.1.3. Examples of Application

In this subsection we shall describe possible applications, whose detailed investigation will, however, be postponed.

Van der Waals equation: Given a real gas. The 3 characteristic parameters, the volume V, the pressure P, and the temperature T, fulfil the van der Waals equation of state for 1 mole,

$$\left(P+\frac{a}{V^2}\right)(V-b)=rT.$$

Here a, b and r are physical constants. One obtains the curve (1) in Fig. 35.4 for a fixed T below a certain critical temperature. As the pressure is increased at constant temperature, the volume decreases, and the gas changes into a liquid. The actual process follows curve (2), where the gaseous phase is described by section α, the liquid phase by section β. According to Maxewll's rule the straight line in (2) must be so drawn that the two shaded areas are equal. As we shall see later on, this corresponds to the Maxwell convention of Subsec. 35.1.2. The process described in (2) is reversible. When the experimental conditions are established with extreme care, it is possible to generate an irreversible thermodynamic process that corresponds to curve (3). In an analogous way the liquid phase can be changed into the gaseous one. By means of catastrophe theory it is possible to give a description satisfactory from both qualitative and quantitative aspects (cf. Subsec. 35.3.1.). The situation suggests regarding P and T as control parameters and V as a state parameter.

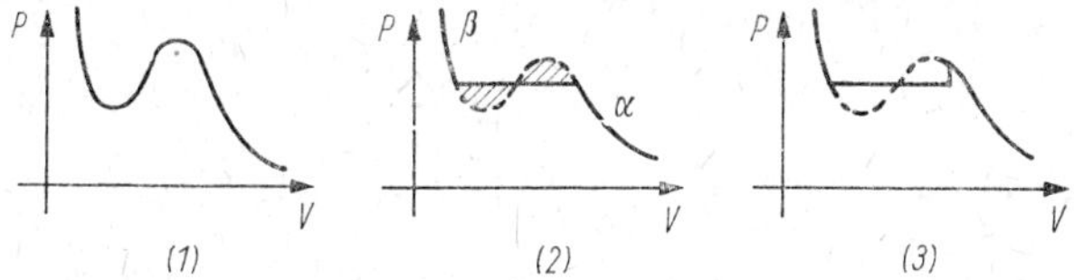

Fig. 35.4

Other physical examples. If one investigates a loaded elastic rod, then buckling and bulging will occur under critical load values. Such as well as similar phenomena of the theory of elasticity can be treated very well by means of catastrophe theory (cf. Subsec. 35.3.2.). The pressure load acting on the rod is the control parameter, the deviation from the normal position is the state parameter. In geometrical optics, the phenomenon of the caustic (cf. Subsecs. 33.1.3. and 33.1.4.) can be described by means of the catastrophe theory. We shall not, however, discuss this here. In Subsec. 35.3.3. we shall deal with the breaking of a water wave when it rolls on the shore.

Biological examples: R. Thom's book [65] makes propaganda for the application of catastrophe theory in biology. In the course of a living organism's evolution the cells specialize. If $u \in R_3$ is the position of a cell, and if t is the time, then $(u, t) \in R_4$ can be interpreted as control parameters. The state of the cell is characterized by biochemical parameters. The (u, t) points at which further specializations of the cell are decided are catastrophe points. By analogy with the electromagnetic field or the gravitational field, R. Thom says ([65], pp. 151–152), it is also possible to consider a "life field". Living organisms are then particles, or structurally stable singularities, of this field. All manifestations of life then result from the interaction of these particles and from their struggle with one another: eating and being eaten. The problem consists in the geometric description of this life field. In Subsec. 35.4.2. we shall discuss two examples: the gastrulation of amphibia and the voracity of phagocytes.

Geometry of forms: The discussion of the 7 elementary catastrophes leads to a certain stock of geometric formations (singularity surfaces etc.). If one considers a discontinuous natural process (e.g. geological dislocations, meteorological phenomena, medical catastrophes), then one may attempt to reduce the dynamics of the process to these basic geometric types. This is closely connected with the biological models. We refer to [65].

Applications: The catastrophe theory has been applied in medicine, biology, thermodynamics, the theory of elasticity, linguistics, meteorology, optics, geophysics, sociology and other fields. Apart from the two books mentioned above [40, 65] we refer to [50, 73, 77, 87] and [88, Sect. 5.5] in this connection. At present, however, it is only the physical applications that have been given a mathematically satisfactory treatment.

35.1.4. The Three Interpretations of Catastrophe Theory

In reference [21], J. Guckenheimer distinguishes between three interpretations of catastrophe theory.

1. (*V. I. Arnold*). Catastrophe theory is a singularity theory of surfaces and mappings, i.e., an intramathematical theory whose elements have been presented in Chap. 34. This does not exclude selected applications, especially to physical problems.
2. (*R. Thom*). Thom's point of view is more of philosophical nature. Processes in the living and lifeless nature shall be geometrized and represented as a succession of (elementary and general) catastrophes. Singularity theory appears as a mathematical tool, but is not the heart of the theory.[1])
3. (*E.C. Zeeman*). The geometric surfaces of singularity theory are used for concrete models in medicine, biology, sociology etc. These models are, on the one hand, speculative (as regards the hypotheses used), but on the other hand the conclusions drawn from them are concrete enough to allow of criticism (which is not always the case in Thom's approach). A typical example is the model describing the aggression of dogs (cf. Subsec. 35.4.3.). It is just these concrete models which have made catastrophe theory known beyond the circle of mathematicians and physicists. But they have also attracted much criticism; cf. [21].

[1]) "Poincaré ... once defined mathematics as the art of giving the same name to different things" (E.T. Bell, Men of Mathematics).

35.2. Elementary Catastrophes

35.2.1. The Generic Aspect

Let $F(x, u)$, where $x \in R_n$ and $u \in R_r$, be a structurally stable potential in the sense of Subsec. 35.1.1. We are interested in catastrophe points, where the rules stated in Subsec. 35.1.2. shall be taken into account. As stated previously, this leads to the following problem. $F(x, u) \in m(n+r)$ is the stable unfolding of a germ $f \in m^2(n)$, with a local minimum at $0 \in R_{n+r}$. If $r \leqq 4$, then it is possible to apply Theorem 34.5.7, provided that f is finitely determined. There remains the problem of whether or not this assumption (which is necessary according to Theorem 34.5.4/1) involves an excessive restriction of the set of admissible potentials.

Theorem. *Let $r = 1, 2, 3, 4$. In $C^\infty(R_{n+r}, R_1)$ there exists an open and dense set J_r whose elements $F(x, u) \in C^\infty(R_{n+r}, R_1)$, $x \in R_n$, $u \in R_r$, have the following properties:*
1. Σ_F, as defined in Subsec. 35.1.1., is an r-dimensional manifold.
2. If $(x^0, u^0) \in \Sigma_F$, then, in a neighbourhood of (x^0, u^0), $F(x, u)$ is a stable unfolding of the germ $F(x, u^0) \in C^\infty(R_n, R_1)$ (with respect to a neighbourhood of x^0 in R_n).

Remark 1. The point (x^0, u^0) now takes the role of the point $(0, 0)$ in Sec. 34.5. (cf. Def. 34.5.1). From Theorem 34.5.4/1 it follows that $f(x) = F(x + x^0, u^0) \in m^2(n)$ is finitely determined for $F \in J_r$ and $(x^0, u^0) \in \Sigma_F$. If there is a local minimum, then Thom's classification given in Theorem 34.5.7 can also be applied.

Remark 2. If a physical process is described by a potential $F_0(x, u) \in C^\infty(R_{n+r}, R_1)$, $r \leqq 4$, then there exists a potential $F \in J_r$ in an arbitrary neighbourhood of F_0. Hence, in the sense of structural stability it is reasonable to require directly that $F_0 \in J_r$.

Remark 3. Two points shall be emphasized once again: (a) Catastrophe phenomena are local in nature (even though the elementary types are investigated globally). (b) Thom's list is obtained only after a reduction. Hence for concrete problems one also has to take into account constant unfoldings as well as the addition of positive definite quadratic forms. For a detailed description of the corresponding mathematical operations, we refer to Appendix I in [40].

35.2.2. Pictures of Elementary Catastrophes

We are interested in a detailed geometric discussion of the unfoldings $F = G_k$ listed in Theorem 34.5.7 as well as of the corresponding formations Σ_F, Δ_F, and D_F mentioned in Subsec. 35.1.1. Especially for the delay rule of Subsec. 35.1.2., the possible catastrophe points are contained in D_F. The "cusp" has already been examined in this way in Subsec. 35.1.2. Here we shall confine ourselves to selected examples; the following pictures have been taken from [7] and [65].

The fold (Fig. 35.5):

$$f = x^3 \in m^2(1), \quad F(x, u) = x^3 + ux .$$
$$\Sigma_F = \{(x, u) \in R_2 \mid 3x^2 + u = 0\} ,$$
$$\Delta_F = \{(x, u) \in R_2 \mid x = u = 0\} ,$$
$$D_F = \{0\} .$$

The fold is a catastrophe not very interesting. For $u < 0$, $F(x, u)$ has exactly one minimum that disappears as $u \to 0$. Jumping from one minimum to another does not occur.

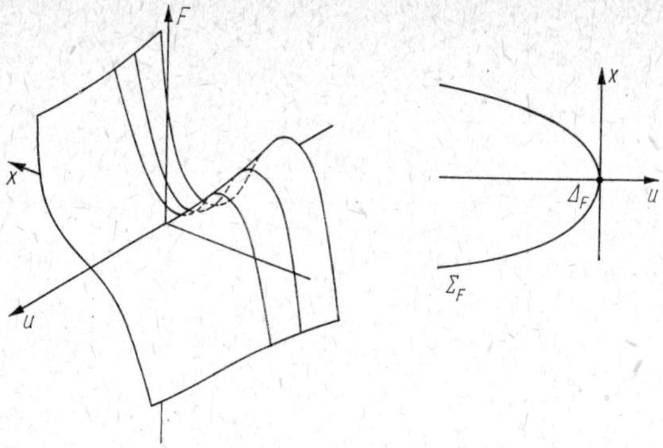

Fig. 35.5

The swallow-tail (Figs. 35.6 and 35.7):

$$f = x^5 \in m^2(1), \qquad F(x, u, v, w) = x^5 + ux^3 + vx^2 + wx ,$$
$$\Sigma_F = \{(x, u, v, w) \in R_4 \mid 5x^4 + 3ux^2 + 2vx + w = 0\} ,$$
$$\Delta_F = \{(x, u, v, w) \in \Sigma_F \mid 20x^3 + 6ux + 2v = 0\} ,$$
$$D_F = \{(u, v, w) \in R_3 \mid \exists x, \text{ where } 5x^4 + 3ux^2 + 2vx + w = 20x^3 + 6ux + 2v = 0\} .$$

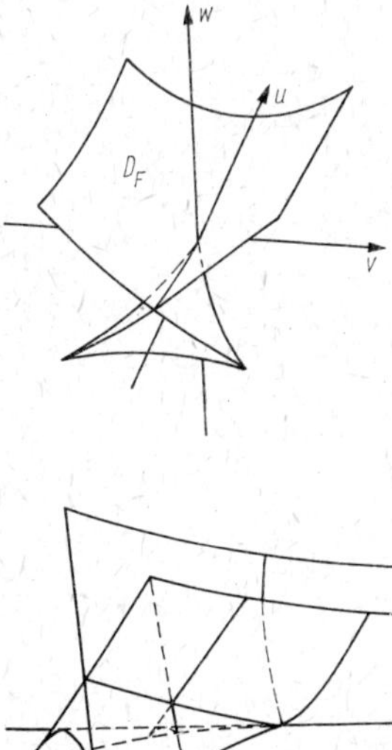

Fig. 35.6

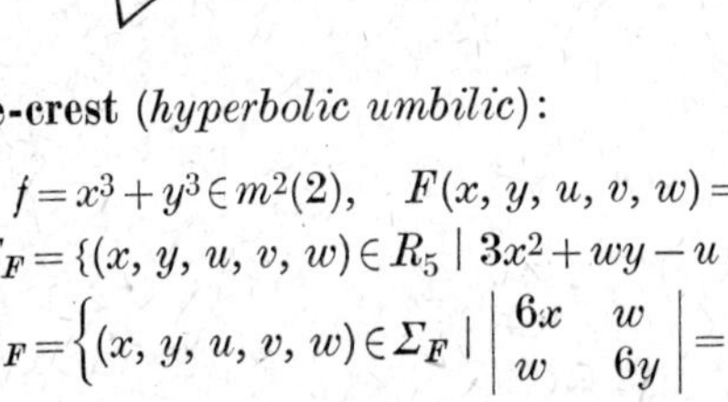

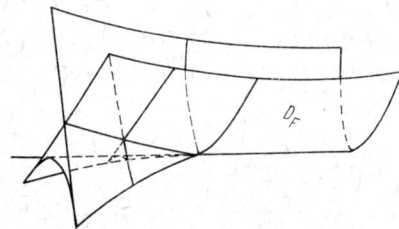

Fig. 35.7

The wave-crest (*hyperbolic umbilic*):

$$f = x^3 + y^3 \in m^2(2), \quad F(x, y, u, v, w) = x^3 + y^3 + wxy - ux - vy ,$$
$$\Sigma_F = \{(x, y, u, v, w) \in R_5 \mid 3x^2 + wy - u = 3y^2 + wx - v = 0\} ,$$
$$\Delta_F = \left\{(x, y, u, v, w) \in \Sigma_F \;\middle|\; \begin{vmatrix} 6x & w \\ w & 6y \end{vmatrix} = 0\right\} ,$$
$$D_F = \{(u, v, w) \in R_3 \mid \exists (x, y), \text{ where } u = 3x^2 + wy, v = 3y^2 + wx, w^2 = 36xy\} .$$

For a fixed w,

$$(x, y) \to (u, v) = (3x^2 + wy, 3y^2 + wx)$$

is a $C^\infty(R_2, R_2)$ mapping which transforms the square drawn in the x,y-plane into the corresponding figure in the u,v-plane (Fig. 35.8). This figure must not be understood to be three-dimensional, but only indicates the different coverings. $D_F \cap \{w = \text{const}\}$ is the image of the hyperbola $36xy = w^2$. For $w = 0$ this figure degenerates: the x,y-square is folded and mapped onto an 4-fold covered u,v-square: the folded handkerchief. Thus the u,v-figure drawn in the diagram is the generic image of the folded handkerchief, where the folding takes place along the hyperbola $36xy = w^2$.

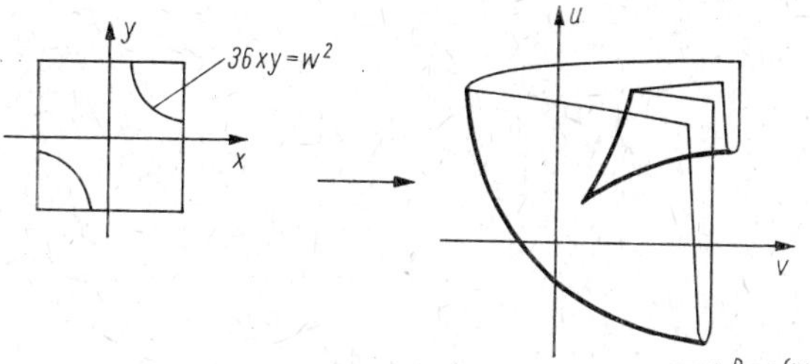

Fig. 35.8

The hair (*elliptic umbilic*) (Fig. 35.9):

$$f = \frac{x^3}{3} - xy^2 \in m^2(2), \quad F(x, y, u, v, w) = \frac{x^3}{3} - xy^2 + w(x^2 + y^2) - ux - vy\,,$$

$$D_F = \{(u, v, w) \in R_3 \mid \exists (x, y), \text{ where } u = x^2 - y^2 + 2wx,$$
$$v = -2xy + 2wy\,,\; x^2 + y^2 = w^2\}\,.$$

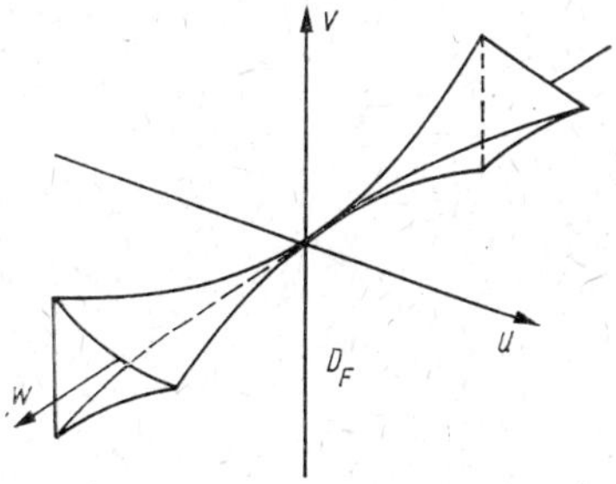

Fig. 35.9

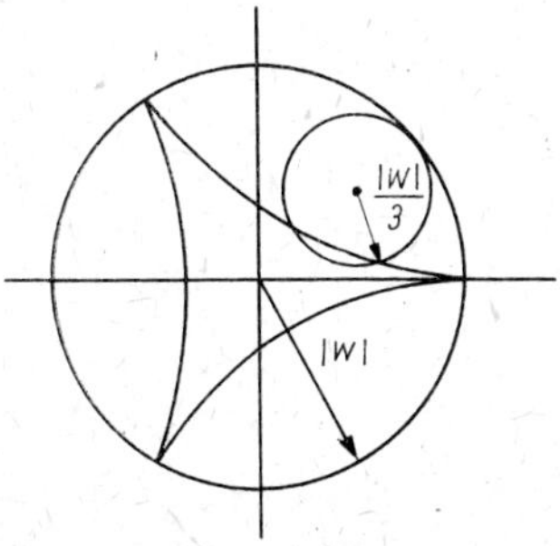

Fig. 35.10

D_F is a two-dimensional surface in the u,v,w-space. The intersection of D_F with $w=\text{const}$ is a hypocycloid which can be constructed as follows. A circle of radius $\frac{|w|}{3}$ rolls along the inside of a circle of radius $|w|$ (Fig. 35.10). If one fixes a point on the smaller circle, then the trace of this point obtained from the rolling circle is a hypocycloid.

The mushroom (*parabolic umbilic*) (Fig. 35.11):

$$f=x^2y+\frac{y^4}{4}\in m^2(2),\quad F(x,y,u,v,w,t)=x^2y+\frac{y^4}{4}+tx^2+wy^2-ux-vy\,,$$

$$\Sigma_F=\{(x,y,u,v,w,t)\in R_6 \mid 2x(y+t)-u=x^2+y^3+2wy-v=0\}\,,$$

$$\Delta_F=\left\{(x,y,u,v,w,t)\in\Sigma_F \,\Big|\, \begin{vmatrix} 2(y+t) & 2x \\ 2x & 3y^2+2w \end{vmatrix}=0\right\},$$

$$D_F=\{(u,v,w,t)\in R_4 \mid \exists(x,y),\text{ where } u=2x(y+t),\ v=x^2+y^3+2wy,$$
$$2x^2=(y+t)(3y^2+2w)\}\,.$$

The mushroom is the most complicated elementary catastrophe. D_F is a three-dimensional surface in R_4. Insertion of $x=x(y,t,w)$ into u and v gives $u=u(y,t,w)$ and $v=v(y,t,w)$. Then the intersections $D_F\cap\{w=\text{const},\ t=\text{const}\}$ are curves in the u,v-plane, with y as a curve parameter. If w and t are so varied that $w^2+t^2=c^2$ ($c>0$ small), then according to [65], pp. 86–88, one obtains a sequence of curves which correspond to the respective domains or curves in the w,t-plane (Fig. 35.12): if (w,t) belongs to domain 1, then the correspondence is to image 1, if (w,t) lies on curve 6, then the correspondence is to image 6, etc. The most

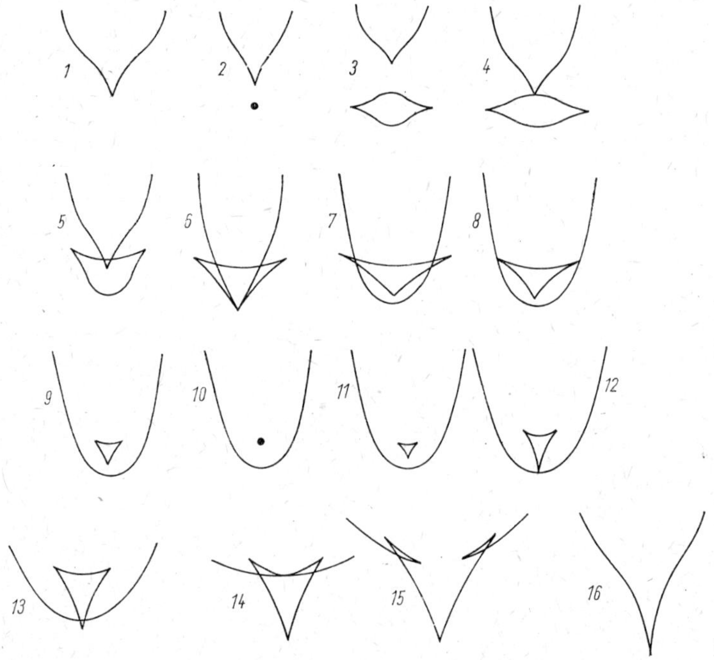

Fig. 35.11

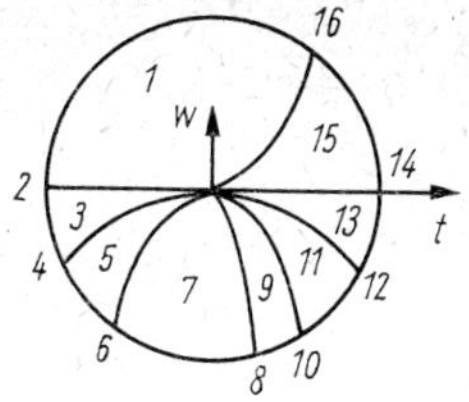

Fig. 35.12

interesting point is that a number of catastrophes considered previously reappear as special cases. 1 is a "cusp" (a diffeomorphically distorted Neil parabola); at 2 there appears a point which subsequently turns into the lip-type catastrophe and assumes a mushroom-like shape at 5 (this also accounts for the name). 6 corresponds to the wave-crest catastrophe (the generic image of the folded handkerchief). The adjacent curvilinear triangle is analogous to the hypocycloids of the hair catastrophe; the triangle disappears and immediately reappears thereafter. Details and formulas are given in [65], pp. 81–90.

Remark. The origin of the poetic names of the "swallow-tail", the "hair" and the "mushroom" becomes clear from a look at the figures. The figure of the butterfly catastrophe is similarly suggestive; cf. [7, 65]. Later on we shall return to the interpretation of "wave-crest".

35.3. Applications in Physics

35.3.1. The van der Waals Equation

The physical situation has been described in Subsec. 35.1.3. If, in the notation used there, P, T, and V are the pressure, temperature and volume, respectively, of a real gas, then the van der Waals equation of state for 1 mole is

$$\left(P+\frac{a}{V^2}\right)(V-b)=rT\,. \tag{1}$$

It provides a good description outside the range of transition from the liquid to the gaseous phase (or vice versa), that is, for the curve sections α and β in Fig. 35.4(2). We discuss this equation in two variants.

1st variant (*quantitative*). The critical point of (1) is

$$P_c=\frac{a}{27b^2}\,,\quad V_c=3b,\quad T_c=\frac{8a}{27br}\,.$$

Setting $p=\dfrac{P}{P_c}-1$, $t=\dfrac{T}{T_c}-1$, and $x=\dfrac{V_c}{V}-1$, one obtains

$$3x^3+(8t+p)x+(8t-2p)=0\,.$$

The substitution of $u=-\dfrac{2}{3}(8t+p)$, $v=\dfrac{4}{3}(8t-2p)$ gives

$$4x^3-2ux+v=0\,. \tag{2}$$

If $F = F(x, u, v) = x^4 - ux^2 + vx$, then F is the universal unfolding of $f = x^4 \in m^2(1)$ according to Theorem 34.5.7, i.e., the cusp. Then (2) agrees with Σ_F in Subsec. 35.1.1. Retransformation of F to P, V and T yields a function which provides a qualitative description of Gibbs' thermodynamic potential $G(V, P, T)$ (free enthalpy) in a neighbourhood of the critical point. The possible stable states of the system are described by (2), where the modification due to Maxwell's rule in the sense of Fig. 35.4(2) has to be taken into account. In terms of the Gibbs potential, Maxwell's rule reads as follows: for given control parameters P and T, the system realizes a state V which is given by an absolute minimum of $G(V, P, T)$ (with fixed P and T). In the above transformations, this property of $G(V, P, T)$ in a neighbourhood of (P_c, T_c, V_c) is transferred to $F(x, u, v)$ in a neighbourhood of $0 \in R_3$, with u, v as control parameters and x as a phase parameter. But this is the Maxwell convention for the second elementary catastrophe, the cusp. The terms "liquid" and "gas" in Fig. 35.13 indicate the liquid and the gaseous phase, respectively. The solid arrow indicates the usual catastrophe path with a sudden volume change that occurs when a point with $v = 0$ and $u > 0$ is reached. But the starting point and the end point of this process can also be reached on the non-catastrophe path indicated by the broken line. The term "gaquid" indicates the region mentioned previously in connection with Fig. 35.4(3), Subsec. 35.1.3., which can be reached under carefully chosen experimental conditions.

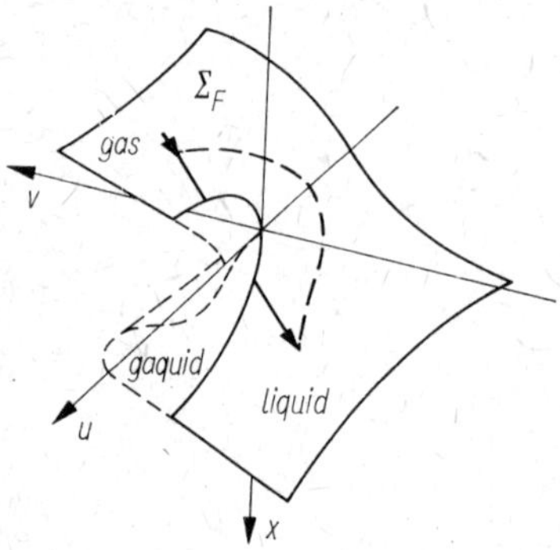

Fig. 35.13

2nd variant (*qualitative*). From the point of view of catastrophe theory, there is a very elegant, purely qualitative consideration which, to a certain degree of necessity, leads to the above picture without requiring knowledge of the van der Waals equation of state. This analysis is based on the following hypotheses. 1. The gas-liquid mixture has a critical point P_c, V_c, T_c (which has been established by experiment), and the behaviour of the system in the neighbourhood of this point is described by a structurally stable catastrophe potential $G(V, P, T)$. 2. It is expedient to regard the pressure P and the temperature T as continuous control parameters, and the volume V as a (possibly discontinuous) phase parameter. Now, by looking over Thom's list of Theorem 34.5.7, one finds that there is only one possibility to describe this situation, namely the second elementary catastrophe, the cusp. For the local regime one still has two possibilities, the Maxwell convention and the delay rule. The experimental data speak in favour of the Maxwell convention.

Conclusion. Catastrophe theory provides a description in the neighbourhood of the critical point (P_c, V_c, T_c); the van der Waals equation can be used for points far away from this critical point. The fact that these two entirely different considerations harmonize so wonderfully is another illustration of the well-known observation:

"Formulae are indeed wiser than men".

35.3.2. Eulerian Deformations

In this point we follow the description given by E. C. Zeeman [76].

Euler's bow: Suppose that two rods of equal length are connected by a joint (Fig. 35.14). Let the force β act upon the right- and the left-hand end of the bow in opposite directions. A possible buckling is counteracted by a spring attached to the joint. This enables the system to prevent buckling, provided that β remains smaller than a critical value β_c. For $\beta > \beta_c$ the jointed rod buckles up or down, not jerkily, but rather continuously as β increases continuously. The extent of buckling is described by the angle x. (We only consider stable equilibrium states, i.e., we exclude the case of unstable equilibrium, $x=0$ for $\beta > \beta_c$.) The joint of the buckled rod is acted upon by the force α. If α increases, then the system will snap into a lower, stable position when a critical value is reached.

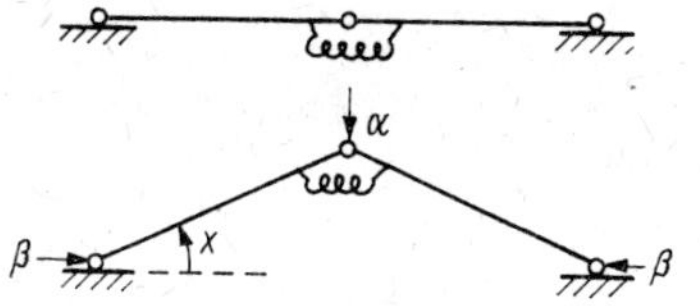

Fig. 35.14

Energy balance: An analysis shows that the system can be described by the potential

$$V(x, \alpha, \beta) = \frac{1}{2}\mu(2x)^2 + \alpha \sin x - 2\beta(1 - \cos x)$$

(total energy). Here $\mu > 0$ is a fixed spring constant. α and β are variable. For given α and β the system realizes a state of minimal energy; in particular one then has

$$0 = \frac{\partial V}{\partial x} = 4\mu x + \alpha \cos x - 2\beta \sin x\,.$$

This formula shows that $\beta = 2\mu$ is a specific value for $\alpha = 0$ and small values of x (which are considered here exclusively). As we shall see, one has $\beta_c = 2\mu$. With $\beta = 2\mu + b$ one obtains

$$V(x, \alpha, \beta) = \alpha x - bx^2 - \frac{\alpha}{6}x^3 + \frac{2\mu + b}{12}x^4 + O(x^5) \tag{1}$$

$$= \frac{\mu}{6}x^4 + \alpha x\left(1 - \frac{x^2}{6}\right) - bx^2\left(1 - \frac{x^2}{12}\right) + O(x^5)\,.$$

Of interest is the behaviour in a neighbourhood of $\alpha = b = 0$. We recall the cusp catastrophe and its universal unfolding, as stated in Theorem 34.5.7 and Subsec. 35.1.2.,

$$f(x) = x^4 \in m^2(1), \quad F(x, u, v) = x^4 - ux^2 + vx\,. \tag{2}$$

By means of diffeomorphic mappings, whose concrete form is of no interest here, it is now possible to transform (1) into the normal form (2), which means that $F(x, u, v) \sim V(x, \alpha, b+2\mu)$, with $\alpha \sim v$ and $b \sim u$ (where "$\sim$" means "approximately equal"). Hence (1) is a universal (and especially a structurally stable) unfolding of the germ $\frac{\mu}{6} x^4$ in the sense of Theorem 34.5.7. Thus, in particular, $V(x, \alpha, \beta)$ qualitatively exhibits the behaviour shown in Subsec. 35.1.2. The local regime is determined by the delay rule. A typical example is the path indicated in Fig. 35.15. One has $\alpha = 0$ and $\beta < 2\mu$ along section 1. Here V has exactly one minimum at $x = 0$. Along section 2 one has $\alpha = 0$ and $\beta > 2\mu$. V has two minima inside the (diffeomorphically distorted) Neil parabola. The corresponding (positive or negative) x-values indicate the continuous up- or down-buckling. Along section 3 the system is considered for increasing α, fixed β, and $x > 0$. According to the delay rule the system remains in a state with $x > 0$. This minimum disappears as the system approaches the point P (compare with the figures for the potential in Subsec. 35.1.2.). A catastrophe occurs, the system jumps into the remaining stable state with $x < 0$. Along section 4 the system is in a stable state with $x < 0$.

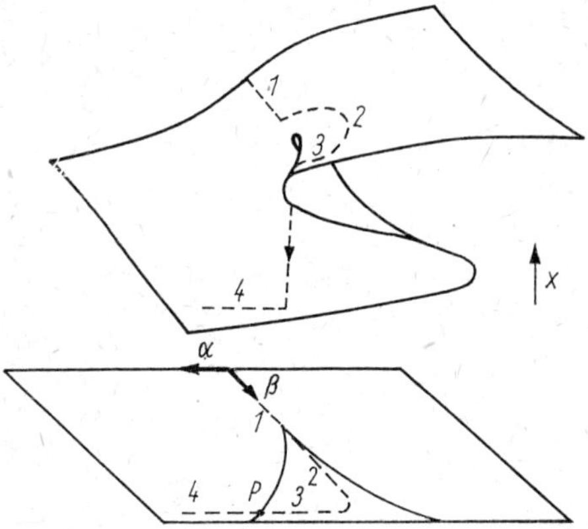

Fig. 35.15

Euler's buckling rod (Fig. 35.16): A homogeneous elastic rod with the modulus of elasticity μ and the length λ is acted upon by the forces β from both the right and the left end. According to Euler the rod withstands the load until the critical value $\beta = \mu \left(\frac{\pi}{\lambda}\right)^2$ is reached. Then the rod bows, assuming the shape $f(s) = x \sin \frac{\pi s}{\lambda} + O(x^3)$. Here x is the deflection amplitude (which is assumed to be small), s is the arc length, and $f(s)$ is the deflection from the normal position. If the bent rod is asymmetrically loaded by α, then initially one obtains a figure which corresponds to the sum of two sine curves with the amplitudes x and y (Fig. 35.17). Here $y = y(\alpha, \varepsilon)$. As α increases the system will approach a critical value at which the rod jumps into a lower, stable position. According to Zeeman, $y = y(\alpha, \varepsilon)$ is described by a dual cusp-type catastrophe (Fig. 35.18).

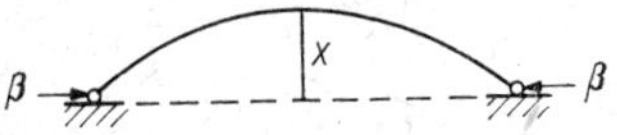

Fig. 35.16

"Dual" means that the stable position is represented by the y-values on the shaded intermediate surface. If α increases at constant ε, then a critical value is reached at Q. Then the rod jumps into a new stable position, which is not indicated in the figure.

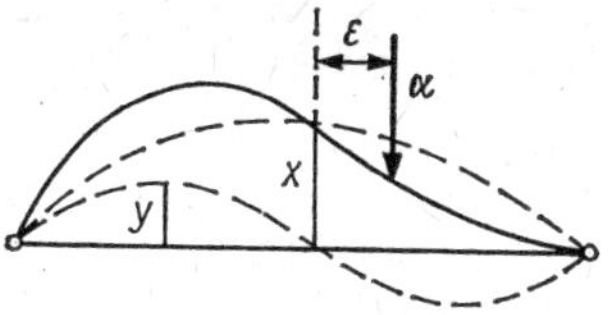

Fig. 35.17

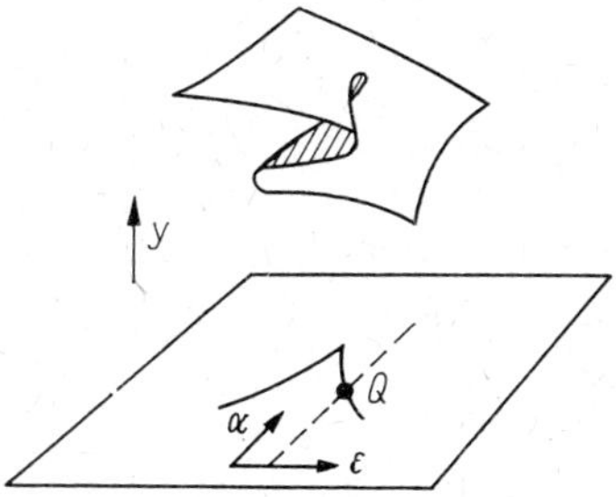

Fig. 35.18

35.3.3. Breaking of Water Waves

Here we shall follow the presentations of E. C. Zeeman [74] and Thom [65], pp. 78–79. A water wave that approaches the shore will break there. The geometry of this process can be described adequately by means of the wave-crest catastrophe and the associated unfolding according to Theorem 34.5.7 and Subsec. 35.2.2.,

$$f = x^3 + y^3 \in m^2(2), \quad F = x^3 + y^3 + wxy - ux - vy\ .$$

If one interprets w as a time parameter and u, v as space parameters, then according to Subsec. 35.2.2. one obtains the diagrams shown in Fig. 35.19 for

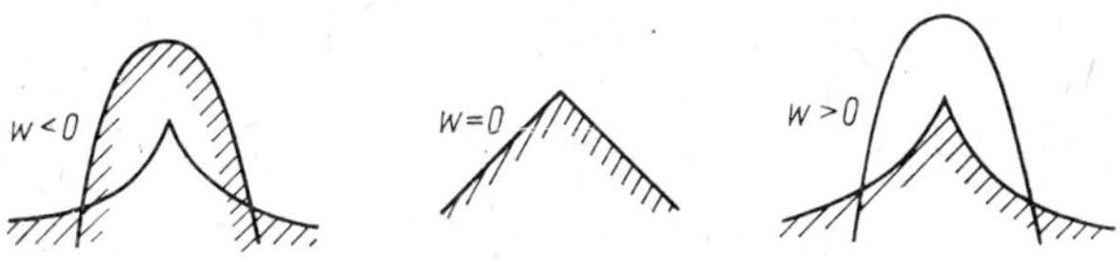

Fig. 35.19

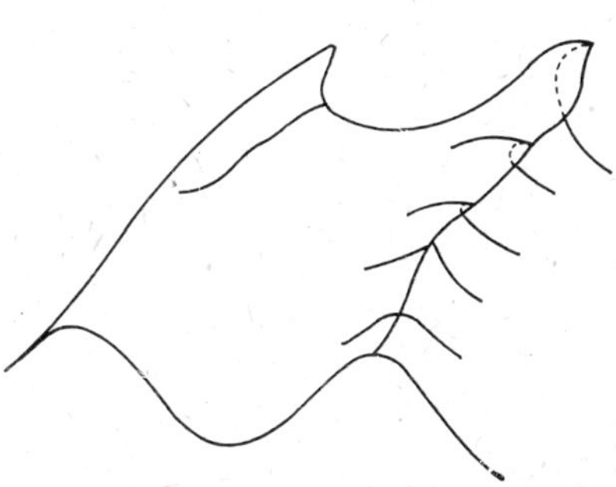

Fig. 35.20

$D_F \cap \{w = \text{const}\}$. For $w = 0$ the curves coincide, while for $w > 0$ they change parts. The change from a round (stable) shape to a cuspidal (unstable) one corresponds to the breaking of waves (Fig. 35.20). A generic distortion of this picture, combined with propositions of classical hydrodynamics (and a suitable identification of the occurring parameters with physical quantities) yields a qualitatively as well as quantitatively satisfactory picture (according to Zeeman): at the moment of breaking the wave is locally symmetric, the crest angle being 120°; after breaking the line section AB is a parabolic arc (Fig. 35.21).

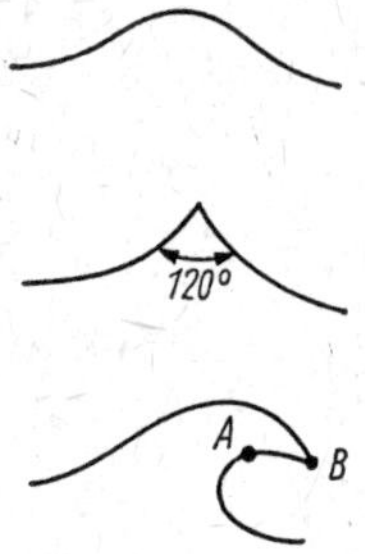

Fig. 35.21

35.3.4. Catastrophe Machines

In this subsection we shall base our considerations on the work of T. Poston [33]. Catastrophe machines are explicitly constructed physical (mostly mechanical) systems that are described by potentials corresponding to the unfoldings of the seven elementary catastrophes listed in Theorem 34.5.7. In particular these potentials depend on parameters (order of the catastrophe machine). A catastrophe machine of order zero corresponds to the simple minimum stated in Subsec. 34.5.6., a catastrophe machine of order one corresponds to the fold-type catastrophe, and so on. One may also generalize this concept by considering catastrophe machines which depend on infinitely many parameters (or functions). Examples of this are soap films in wire loops and their catastrophe behaviour when the wire loops are deformed. Here we shall confine ourselves to two simple cases.

Catastrophe machines of order zero (Fig. 35.22): Consider a strictly convex body B with the C^∞-surface $D = \partial B$. Then D is a C^∞-manifold. If $P \in D$, then let $f(P) \in C^\infty(D, R_1)$ be the potential energy of the body in the position in which P touches the ground. The body tends to assume a stable position by realizing a relative minimum of $f(P)$. Since according to Theorem 34.3.2 (with D instead of R_n) the Morse functions are dense in $C^\infty(D, R_1)$, from the generic point of view one may assume that $f(P)$ is a Morse function. The same Theorem 34.3.2, with D instead of R_n, also implies that every relative minimum is locally stable. After a slight deflection the body returns to its stable initial position in an oscillatory

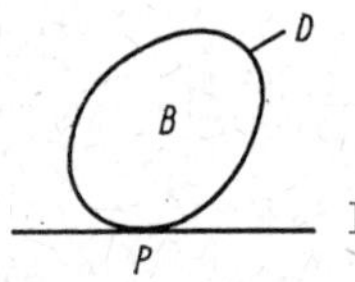

Fig. 35.22

motion. If a body has a symmetry axis, e.g. an egg, then it has unstable equilibrium positions (Fig. 35.23). Such a figure is not generic. This raises the following problem. If a (physical or other) problem exhibits natural symmetries, then, from the generic point of view, it appears expedient not to disturb these symmetries. The aim is a singularity theory in which symmetries are taken into account, a new version of Thom's list.

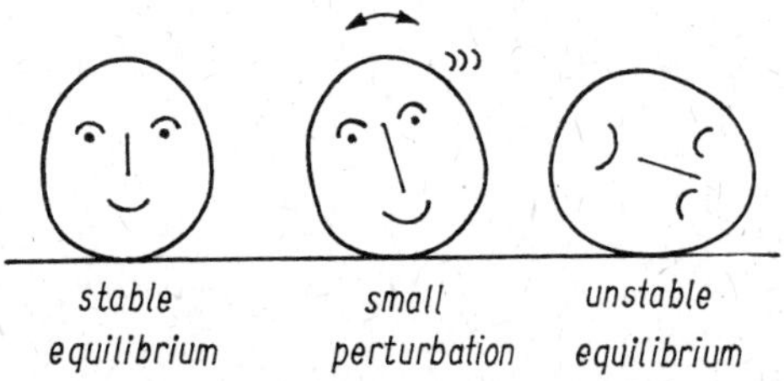

Fig. 35.23

Catastrophe machines of order one: Let $f = x^3 \in m^2(1)$ and $F = x^3 + ux$ be the fold catastrophe and its universal unfolding, respectively. It is our aim to find an example in the field of mechanics whose potential is essentially equal to F. Suppose that a weightless wheel (e.g. of thin sheet metal) of radius 1 is placed on an inclined plane with the slope angle α (Fig. 35.24). A heavy weight (e.g. a magnet) is attached to the wheel at a distance c, $0 < c < 1$, from the centre. If $c > \sin\alpha$, then there are two equilibrium states: a stable equilibrium with the weight arranged at the point A, and an unstable one, where the weight is at the point B. For $c = \sin\alpha$ the system has a singular point, and we set $c = -u + \sin\alpha$. If $H(\varphi, u)$ is the quantity indicated in Fig. 35.25, then the system is in equilibrium if and only if $H(\varphi, u) = 0$. Of interest is the behaviour in the neighbourhood of the critical

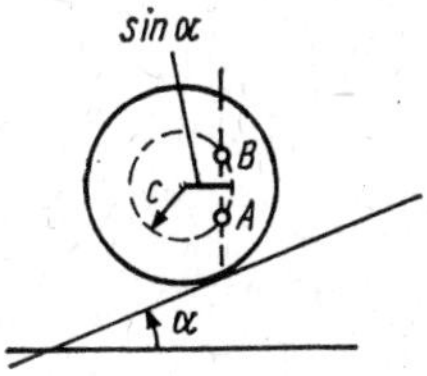

Fig. 35.24

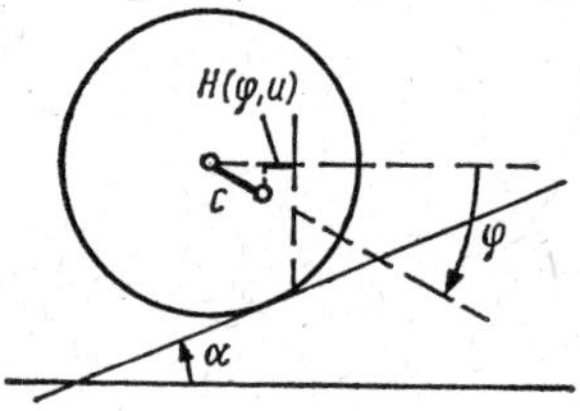

Fig. 35.25

point $u = \varphi = 0$ (Fig. 35.26). One obtains

$$H(\varphi, u) = \sin\alpha - c\cos\varphi = \sin\alpha\,(1 - \cos\varphi) + u\cos\varphi\,,$$

$$F(\varphi, u) = \sin\alpha\,(\varphi - \sin\varphi) + u\sin\varphi = \frac{\sin\alpha}{6}\,\varphi^3 + u\varphi + \dots\,,$$

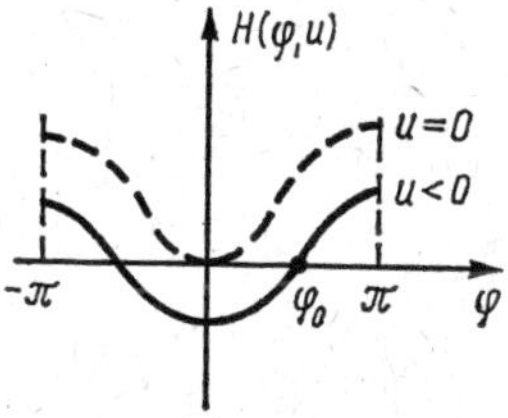

Fig. 35.26

where F is a potential with $\frac{\partial F}{\partial \varphi} = H$. But this is just the unfolding of the fold catastrophe. For $u < 0$ there are two φ-values such that $\frac{\partial F}{\partial \varphi} = H(\varphi, u) = 0$. But a minimum is only obtained for φ_0, because $\frac{\partial^2 F}{\partial \varphi^2}(\varphi_0, u) = \frac{\partial H}{\partial \varphi}(\varphi_0, u) > 0$.

35.4. Other Applications

35.4.1. Taylor Series and Cells

Thom also regards the catastrophe theory as a theory of analogies. From this point of view, any two catastrophe events are analogous if they can be described by one and the same elementary catastrophe. But these analogies go much farther, being in part of a very formal nature, as will be shown by the following example (cf. [65], pp. 30/31). Let $f(x, y) \in C^\infty(R_2, R_1)$, with $f(0, 0) = 0$, where f can be formally expanded in a Taylor series,

$$f(x, y) \sim \sum_{k,l=0}^{\infty} a_{k,l} x^k y^l, \quad a_{0,0} = 0 .$$

As regards the topological nature of $f(x, y) \in m(2)$ (normal forms that can be obtained by local diffeomorphisms), the following can be said: if $|a_{1,0}| + |a_{0,1}| > 0$, then, according to Remark 34.3.2/4, the topological structure of $f(x, y)$ is determined by $a_{1,0}x + a_{0,1}y$. The normal form is $f(x, y) = x$. For $a_{1,0} = a_{0,1} = 0$ the topological structure is unknown for the present. If

$$a_{2,0}a_{0,2} - a_{1,1}^2 \neq 0 ,$$

then $f(x, y)$ is (locally) a Morse function in the sense of Def. 34.3.2. The topological structure is then determined by $a_{2,0}x^2 + 2a_{1,1}xy + a_{0,2}y^2$, and according to Remark 34.3.2/4 the normal forms are $f(x, y) = \pm x^2 \pm y^2$. On the other hand, if $a_{2,0}a_{0,2} - a_{1,1}^2 = 0$, then the topological structure is unclear for the present, and the problem must be further investigated. One obtains the following picture: the topological structure of the Taylor series with $|a_{1,0}| + |a_{0,1}| > 0$ is determined by the linear terms alone, while higher terms have no influence. On the next level of expansion, the topological structure of the Taylor series is determined by $a_{1,0} = a_{0,1} = 0$ and $a_{2,0}a_{0,2} - a_{1,1}^2 \neq 0$, and so on. The analogy to cell evolution is as follows. At a first stage of evolution the function of certain cells is fixed, so that subsequent biochemical perturbations cannot affect the function of these cells. Other cells designed for higher tasks have their functions not yet fixed. At a second stage (which largely corresponds with the Morse functions in the above considerations) the functions of other cells get fixed and immune to subsequent biochemical perturbations. Other cells designed for even higher tasks have their functions not yet fixed, and so on. The analogy is clear, though being of a very formal nature.

35.4.2. Applications in Biology

It is Thom's idea to draw conclusions about the dynamics of the process from the geometry of the catastrophe event (to guess the topological form of the potential). In particular he applies this method to biological processes. Isolated elementary catastrophes will not suffice for such a program. One has to consider collisions of elementary catastrophes, accumulation points of elementary catastrophes, elementary catastrophes occurring along lines and surfaces, etc. This leads to highly speculative investigations in evolution biology: specialization of cells, formation of bones etc. Without going into detail, we shall give two examples, which are discussed in [65], pp. 169–171 and pp. 187–189.

Gastrulation of amphibia: The cells of the embryo sac of amphibia flow into the interior of the embryonic sac at a certain point. Part of the sac is introverted, forming the blastopore (Fig. 35.27). According to Thom the geometry of this process can be described by a swallow-tail catastrophe.

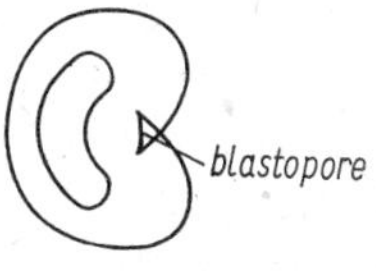

Fig. 35.27

Phagocytes: Phagocytes take up foreign matter most easily. The geometry of this process can be described by the mushroom catastrophe. The mushroom catastrophe (cf. Subsec. 35.2.2.) allows the structurally stable transition from the wave-crest catastrophe (hyperbolic umbilic) to the hair catastrophe (elliptic umbilic). Following Thom's presentation ([65], p. 187) (Fig. 35.28), we consider a situation which results from a generic deformation of the hair catastrophe (a). As the prey approaches, this elliptic state (under tension) turns into a hyperbolic one (b). Regime 1 is replaced by the simpler, more relaxed regime 2. The outer lips of 1 move outward and close in order to catch something from outside. The prey is entrapped in a cavity, and digested (c). Finally the phagocyte returns to the elliptic state (d).

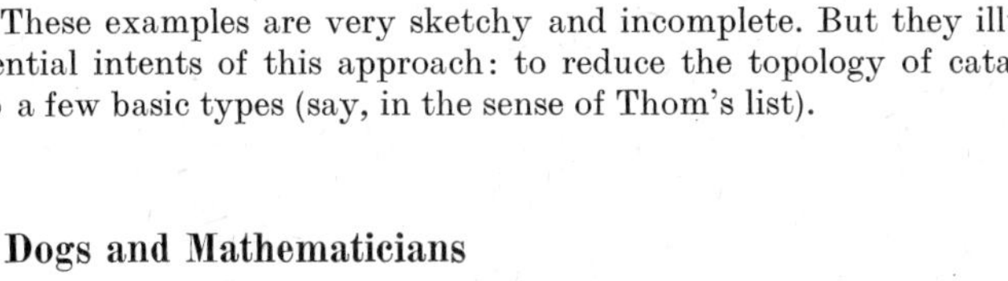

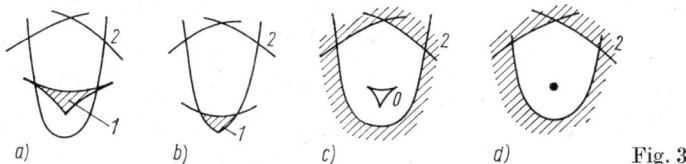

Fig. 35.28

Remark. These examples are very sketchy and incomplete. But they illustrate one of the most essential intents of this approach: to reduce the topology of catastrophe events in biology to a few basic types (say, in the sense of Thom's list).

35.4.3. Dogs and Mathematicians

The preceding section gave an impression (if a very rough, incomplete, and sketchy one) of the intent of R. Thom. The applications presented by E. C. Zeeman are frequently much more concrete, being designed for the geometrization of isolated factual situations that depend on a few parameters (which are sometimes only qualitatively detectable). We shall give two examples: a (bestially) earnest one, and another (of human nature) that is meant not quite so earnestly.

Aggression behaviour of dogs (Fig. 35.29): The cusp-type catastrophe serves as the model surface. Fear and rage are the continuous control parameters. The behaviour of the dog is the (possibly discontinuous) phase parameter. The picture of Zeeman's dog model has been taken from [21].

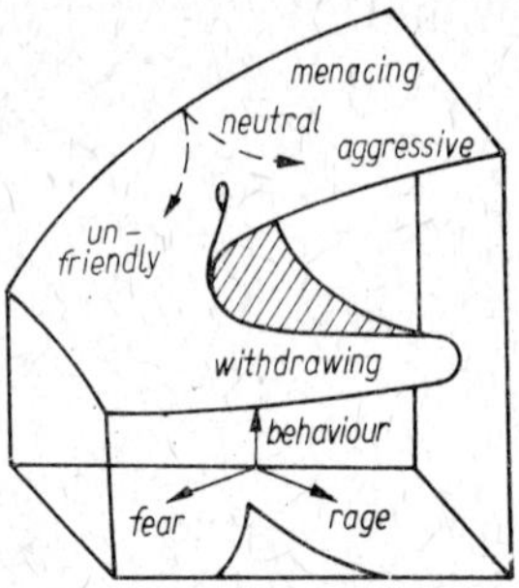

Fig. 35.29

Mathematicians (Fig. 35.30): Corresponding to this dog model, there is an analogue (meant not quite seriously) that concerns the mental powers of a mathematician; it is likewise due to E. C. Zeeman [75]. The model is also based on the cusp-type catastrophe. The control parameters are the mathematical and the speculative content (of, say, a publication). The phase parameter (which possibly varies discontinuously) will answer the question of whether that author is a common-or-garden mathematician, a genius, or a fool. Common-or-garden mathematicians don't speculate, but a speculating mathematician may very easily turn from a genius to a fool. E. C. Zeeman writes: "Thom does not belong to the common-or-garden mathematicians, as is shown by his enterprising speculative excursions. However closely he sails to the edge, anyhow he always manages to stay on the upper surface."

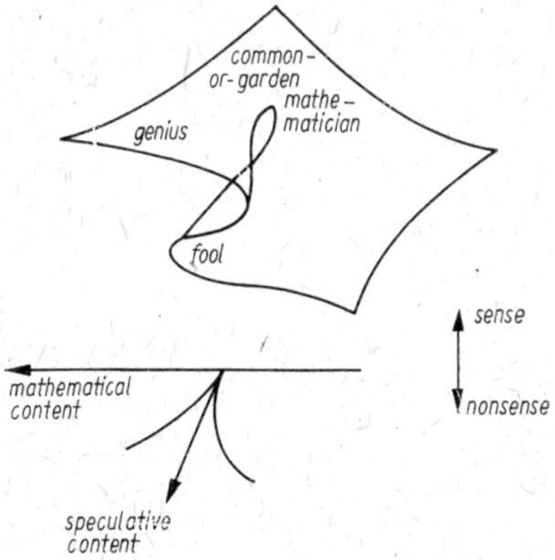

Fig. 35.30

Appendix: On the Relation between Geometry and Reality during Time's Change[1])

1. The Emancipation of Mathematics

The history of modern mathematics begins with *Pythagoras* (about 540 B.C.). Up to that time mathematics, but especially geometry, had been something like a natural science. There were empirically found "laws", verifiable by experiment at any time, e.g., geometric theorems for the right-angled triangle etc. Pythagoras was the first to point out that mathematical theorems are based on assumptions (axioms), from which they have to be derived by purely logical deduction. He introduced the abstract proof into mathematics: one of the greatest achievements in the history of mathematics, which has, however, become almost a matter of course today. But Pythagoras deserves mention for another reason, too. He and his disciples (the Pythagoreans) believed that the universe is "governed" by the natural numbers 1, 2, 3, ... Using simple procedures (admitted by the Pythagoreans) one can derive the negative integers, and hence the rational numbers. The second great discovery at the same time meant the end of this dream. In our present language it reads: $\sqrt{2}$ is irrational. This proposition is fundamental for the understanding of what modern mathematics is. For up to that time all geometric theorems (no matter whether they had been found empirically or logically deduced in the sense of Pythagoras) had been verifiable by experiment (like laws of natural sciences). The question whether a given straight line segment is the rational or irrational multiple of another given line segment cannot be decided by experiment. This means that the above fact (that $\sqrt{2}$ is irrational) is an intramathematical theorem that bears no relation to reality (provided that reality is assigned the attribute of experimental verifiability). Mathematics had emancipated itself: abstract proofs and the existence of intramathematical theorems gave that stamp to it which makes it stand out among the large family of all sciences in a unique way up to this day.

2. Euclid's Elements

Euclid's Elements date from about 325 B.C.; they represent an attempt to axiomatically formulate the geometry of the plane (as far as it concerns points, straight lines, and circles). Following the lines of Pythagoras, some few assumptions (axioms) are placed at the top, all other theorems being derived from them by purely logical deduction. (The fact that, from our present point of view, this attempt exhibits gaps does not make any difference to its supreme importance.) The theorems in question in essence concern straight lines, triangles and circles in the plane, and have an immediate intuitively geometric meaning. The empirical origin (at least of the great majority) of the geometric theorems is not denied.

---

[1]) Paper read within the lecture on "History of Mathematics" on the occasion of the celebration of the 100th anniversary of A. Einstein's birth, on March 14th, 1979.

The intent is, rather, to systematically order the arsenal of available theorems of the geometry of the plane and to reduce them to a few axioms. Here geometry becomes an idealized and mathematized reality.

3. Analytical Geometry

On June 8th, 1637, *Descartes* published the work which laid the foundations of analytical geometry. The plane, the three-dimensional Euclidean space (and its n-dimensional generalization) were equipped with Cartesian co-ordinates (which has likewise become a matter of course meanwhile). The intuitive, synthetically constructive methods of the Greeks were replaced, or at least decisively complemented, by analytical methods. This made it possible to describe geometric figures analytically and to reduce geometric problems to analytical-algebraic ones.

4. Newton's Mechanics

Together with *Archimedes* and *Gauss*, *Newton* is counted among the greatest mathematicians that ever lived. His masterpiece, "*Philosophiae Naturalis Principia Mathematica*", was published in 1687, counting among the greatest scientific books that have ever been written, [46]. It contains, among other things, the fundamentals of classical point mechanics, especially an analytical description of planetary motion. The three-dimensional Euclidean space and a "*uniformly flowing*" time are regarded as being given. They are, so to speak, divine institutions which enable Man to make physical investigations. Equipped with the tools of analytical geometry, Newton developed his mechanics from a few fundamental principles. The impression upon his contemporaries and immediate successors must have been enormous. And even today it may well be a good test for a young student of mathematics or physics whether or not he becomes enthusiastic at the grandiose, self-contained system of classical point mechanics (for it will not do without enthusiasm in mathematics). In 1942, on the occasion of the three-hundredth anniversary of Newton's birth, Einstein devoted the following verses to his great predecessor.

> "*Look to the Heavens, and learn from them*
> *How one should really honour the Master*
> *The stars in their courses extol Newton's laws–*
> *In silence eternal.*"

The reality of the three-dimensional Euclidean space as a frame in which physics was going on soon became undisputed. (Objections raised by Leibniz were soon forgotten. After all, there were no reasons for doubts, for the latter would have had to be based on experimentally verifiable facts, at least from the point of view of natural science. And there were no such facts at that time.) Geometry, especially analytical geometry, was the set of tools to be used for a mathematical description of this reality.

5. The Parallel Axiom and its Implications

As was told in Section 2, Euclid has axiomatized the geometry of the plane. His axioms are altogether very simple, with the exception of the parallel axiom. It reads as follows. *If g is a straight line* (of infinite extension to both sides), *and if P is a point that does not lie on g, then there exists one and only one straight line g′* (of infinite extension to both sides) *which goes through P and does not intersect g* (Fig. A1). Such a straight line is called a parallel. Over 2000 years mathemati-

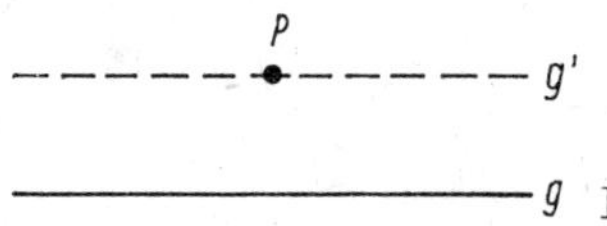

Fig. A1

cians tried hard to deduce this complicated axiom from the other, simpler axioms of Euclid's system, but without any success. The situation about 1800 was as follows: Newton had been very cautious in his expressions concerning space and time. But his successors dogmatized the three-dimensional Euclidean space. It was in this space, and only in it, that natural processes could take place. From the philosophical side, Kant declared the three-dimensional Euclidean space logically necessary. In this situation, the two alternatives to the above parallel axiom, namely

1. *There is no parallel g′*

and

2. *There are more than one parallels g′*

sound like heresy. *Gauss* worked intensively at the parallel axiom, and about 1816 he (as we know today) arrived at the conception that geometries other than the Euclidean are logically possible, too. To put it more precisely: if one replaces the parallel axiom in Euclid's system by either of the two alternatives indicated above, then one obtains a consistent mathematical theory, a non-Euclidean geometry. Gauss did not publish anything in order to save himself malicious criticism. Thus the possibility of non-Euclidean geometries was discovered independently of him (and somewhat later) by *J. Bolyai* and *Lobačevskij*. The first paper on this subject was published by Lobačevskij in the "Kasan Herald" in 1829/30. After the ban had been broken, numerous geometries organized on purely intramathematical lines were developed in the second half of the past century. Geometry no longer reflected reality; geometry had emancipated itself and separated from physics. It had become an intramathematical subdiscipline (which does not exclude physical applications). On the other hand it is most remarkable that Gauss identified straight lines with rays of light, and that he wondered whether reality is in fact described by the Euclidean geometry. Einstein was to be the first to resume this point.

6. Differential Geometries

Differential geometries are of special interest for the further considerations. If one considers a two-dimensional surface in the three-dimensional Euclidean space, e.g., the surface of a sphere, then one may ask for the intrinsic geometric conditions of this surface. Generally these conditions will differ from the familiar ones in the two-dimensional plane. The world of hypothetical two-dimensional beings on a spherical surface differs from the Euclidean world of corresponding two-dimensional beings in the plane. These two-dimensional beings must build their geometry on the basis of their (two-dimensional) experience. We, being gifted with a third dimension, look at these two-dimensional worlds "from above". We are the gods of these beings. Our knowledge that this two-dimensional world is a surface in the three-dimensional Euclidean space is of no use for these beings, because they have no sense for a third dimension. The idea of "intrinsic geometries" goes back to Gauss. He largely discussed the intuitive (or, to put it otherwise, the real) case of two-dimensional surfaces in the three-dimensional Euclidean space. Here local differential properties of functions play a decisive role, which also accounts for the name differential geometry. Decisive generalizations are due to *Riemann*, a disciple of Gauss. Differential geometries were investigated in arbitrary n-dimensional spaces. The formulations of the problems (and, as it will be expected, the answers too) were purely intramathematical. Geometry and reality (in the form of physics) had separated definitely, striking out for themselves. At least this was the way things looked like at the end of the 19th century.

God is cunning, but He is not malicious.
(A. Einstein)

7. How the Gods Were Robbed of the Time

About the turn of the century, leading physicists held the view that their discipline was on the verge of completion (as far as the theoretical foundations were concerned). The few problems still open were hoped to be soluble within the available theories. The mathematics underlying these theories was relatively simple (as measured by the level of mathematics of that time, or compared with the present inventory of mathematical tools of theoretical physics) and had little concern with the problems the majority of the mathematicians were facing at that time. One of the still open problems of physics resulted from the different space-time structures of Newton's mechanics and Maxwell's electrodynamics. Newton's conception of space and time (which had proved its worth so brilliantly in the course of centuries) was considered unimpeachable. But now it led to difficulties. In 1905, *Einstein* replaced Newton's conception of space and time by a new one that led to the special theory of relativity. The ban was broken, and a new conception of space and time gained ground. According to Newton, time was absolute, and there was nothing that had any influence on it. Einstein degraded time. It no longer existed a priori, but rather was determined by a periodic natural process, i.e., by a physical procedure. Two observers merely stipulate that their "*proper times*" shall be be determined by one and the same physical procedure. Subsequently they can communicate about their proper times, comparing them experimentally. In this case the positions of these two observers in space and their speeds relative to one another will then play a decisive role. If these two observers convert their space and time co-ordinates into one another, then space and time become intermixed (Lorentz transformations). This gave rise to the following comment of Minkowski in 1909:

> *"From this hour on, space as such and time as such shall recede to the shadows and only a kind of union of the two retain significance."* [1])

Time was, so to speak, stolen from the gods and surrendered to the physicists. As a consolation, the gods were allowed to keep hold of the three-dimensional Euclidean space till 1915. In other words, in the special theory of relativity physics likewise takes place in the three-dimensional Euclidean space, which is unaffected by physical processes. The formulae of the special theory of relativity are relatively simple. It was the interpretation of these formulae which (together with the quantum hypothesis) rang in a new era of physics.

8. The General Theory of Relativity[2])

In 1915, one of the most glorious and ingenious theories ever devised by human imagination was completed: *Einstein's* general theory of relativity. It is a self-contained, complicated, highly elegant mathematical-physical theory. Its language is that of the Riemannian geometry (in a generalized sense), which had once been believed to have nothing to do with reality. In this theory, Newton's conception of an eternal, really existing three-dimensional Euclidean space finally had to be abandoned, too (so the gods are unemployed now). The structure of space is influenced by physical processes, e.g., by heavy masses. If one considers, for example, the hypothetical two-dimensional beings of Section 6., then one can imagine that there are many two-dimensional worlds of such kind, with different geometries. A general theory of relativity for these two-dimensional beings would make, say, the following statement: the surface in which these two-dimensional beings are living is not given for ever and a day, but will change in the course of time, being influenced by (two-dimensional) physical processes (heavy masses, electromagnetic waves). Our real world is three-dimensional. Since in General Relativity space and time are coupled, one has to add the time as the fourth dimension. Thus physical events take place in a four-dimensional space-time whose structure is determined by a (generalized) Riemannian geometry which, in its turn, is influenced by physical processes.

9. Outlook

Geometry as a language of physics has proven extremely fruitful. It is just in our days that the geometrization of physical processes approaches another culmination. As one would expect, this is once again connected with an increase in complexity of the inventory of mathematical tools used. One of the latest developments is the catastrophe theory of the French mathematician *R. Thom*. His book "*Stabilité structurelle et morphogénèse*" (1972) has created a sensation. Again it is geometric theories that play a decisive role. The following aphorism is due to Thom:

---

[1]) Wilhelm Busch, in his "Tobias Knopp", gave the following (unconscious) comment on the subject of proper time: "One-two-three, in hasty race, time is running, we keep pace." (Well, I have to apologize to serious readers.)

[2]) „In Einstein's conception space is no longer the stage on which the drama of physics is performed: it is one of the performers“ (Sir E. Whittaker, Space and Spirit).

> *"Geometry is magic that works . . . Isn't all magic geometry in the same measure as it is successful?"*

Is this the Pythagoreans' mysticism of numbers in the guise of geometry? Couldn't this phrase have been written by Einstein too?

10. Epilogue

Prometheus robbed the gods of fire
and taught man how to use it.
Einstein robbed the gods of space-time
and taught man how to use it.

References

[1] *Abraham, R.; Marsden, J. E.:* Foundations of Mechanics. London, Amsterdam, Don Mills, Sydney, Tokyo: Benjamin 1978.

[2] *Achieser, N. I.; Glasman, I. M.:* Theorie der linearen Operatoren im Hilbert-Raum. 5. Aufl. Berlin: Akademie-Verlag 1968.

[3] *Anderson, J.L.:* Principles of Relativity Physics. New York, London: Academic Press 1967.

[4] *Arnol'd, V. I.:* Gewöhnliche Differentialgleichungen. Berlin: VEB Deutscher Verlag der Wissenschaften 1979.

[5] *Arnol'd, V. I.:* Complementing chapters of the theory of ordinary differential equations. Moskva: Nauka 1978. (Russian). (English translation: Geometrical methods in the theory of ordinary differential equations. New York, Heidelberg, Berlin: Springer-Verlag 1983).

[6] *Behnke, H.; Sommer, F.:* Theorie der analytischen Funktionen einer komplexen Veränderlichen. Berlin, Göttingen, Heidelberg: Springer-Verlag 1962.

[7] *Bröcker, T.; Lander, L.:* Differentiable Germs and Catastrophes. London Math. Soc. Lect. Notes Ser. 17, Cambridge Univ. Press 1975.

[8] *Chillingworth, D. R. J.:* Differential Topology with a View to Applications. London, San Francisco, Melbourne: Pitman Publ. 1977.

[9] *Choquet-Bruhat, Y.; DeWitt-Morette, C.; Dillard-Bleick, M.:* Analysis, Manifolds and Physics. Amsterdam, New York, Oxford: North-Holland Publ. Company 1977.

[10] *Christie, A.:* The Clocks. Fontana Books 1966.

[11] *Dixon, W. G.:* Special Relativity. Cambridge: Cambridge Univ. Press 1978.

[12] *Dodson, C. T. J.; Poston, T.:* Tensor Geometry (The Geometric Viewpoint and its Uses). London, San Francisco, Melbourne: Pitman 1979.

[13] *Dukas, H.; Hofmann, B.:* Albert Einstein, the Human Side. Princeton: University Press 1979.

[14] *Dunford, N.; Schwartz, J. T.:* Linear Operators, I, II. New York: Interscience Publ. 1958, 1963.

[15] *Duschek, A.:* Vorlesungen über höhere Mathematik, I, II. Wien: Springer-Verlag 1960, 1963.

[16] *Fichtenholz, G. M.:* Differential- und Integralrechnung I–III. 9., 6., 8. Aufl. Berlin: VEB Deutscher Verlag der Wissenschaften 1975–1977.

[17] *Friedlander, F. G.:* The Wave Equation on a Curved Space-Time. Cambridge, London, New York, Melbourne: Cambridge Univ. Press 1975.

[18] *Friedman, A.:* Partial Differential Equations of Parabolic Type. Englewood Cliffs: Prentice-Hall 1964.

[19] *Gelfand, I. M.; Schilow, G. E.:* Verallgemeinerte Funktionen, I, II. 2. Aufl. Berlin: VEB Deutscher Verlag der Wissenschaften 1967, 1969.

[20] *Golden, F.:* Quasars, Pulsars, and Black Holes. Pocket Books. New York: Gulf and Western Corporation 1977.

[21] *Guckenheimer, J.:* The catastrophe controversy. The Mathematical Intelligencer 1, **1** (1978), 15–20.

[22] *Günther, P.; Beyer, K.; Gottwald, S.; Wünsch, V.:* Grundkurs Analysis, I–IV. Leipzig: BSB B. G. Teubner Verlagsgesellschaft 1972–1974.

[23] *Halmos, P. R.:* Measure Theory. New York, Cincinnati, Toronto, London, Melbourne: Van Nostrand Reinhold Comp. 1973.

[24] *Hawking, S. W.; Ellis, G. F. R.:* The Large Scale Structure of Space-Time. Cambridge: Cambridge Univ. Press 1973.
[25] *Hellwig, G.:* Partial Differential Equations, 2nd ed. Stuttgart: B. G. Teubner 1977.
[26] *Herlt, E.; Salié, N.:* Spezielle Relativitätstheorie. Berlin: Akademie-Verlag 1978.
[27] *Hirsch, M. W.; Smale, S.:* Differential Equations, Dynamical Systems, and Linear Algebra. New York: Academic Press 1974.
[28] *Hoyle, F.; Wickramasinghe, N. C.:* Life Cloud. London: Sphere Books Ltd. 1979.
[29] *Hoyle, F.; Wickramasinghe, N. C.:* Diseases from Space. London, Toronto, Melbourne: J. M. Dent & Sons Ltd. 1979.
[30] *Hoyle, F.; Wickramasinghe, N. C.:* Evolution from Space. London, Toronto, Melbourne: J. M. Dent & Sons Ltd. 1981.
[31] *Hund, F.:* Einführung in die theoretische Physik. I. Mechanik. Leipzig: Bibliographisches Institut 1945.
[32] *Kamke, E.:* Differentialgleichungen, Teil 1. 6. Aufl. Leipzig: Akademische Verlagsgesellschaft Geest & Portig 1969.
[33] *Kobayashi, S.; Nomizu, K.:* Foundations of Differential Geometry, I, II. New York, London: Interscience Publ. 1963, 1969.
[34] *Kolmogorov, A. N.; Fomin, S. V.:* Reelle Funktionen und Funktionalanalysis. Berlin: VEB Deutscher Verlag der Wissenschaften 1975.
[35] *Kramer, D.; Stephani, H.; MacCallum, M.; Herlt, E.:* Exact Solutions of Einstein's Field Equations. Berlin: VEB Deutscher Verlag der Wissenschaften 1980.
[36] *Kufner, A.; Kadlec, J.:* Fourier Series. Prague: Academia 1971.
[37] *Landau, L. D.; Lifschitz, E. M.:* Theoretische Physik kurzgefaßt. I, II. Berlin: Akademie-Verlag 1973, 1975.
[38] *Leopold, H.-G.:* Die Cauchyaufgabe für die Maxwellschen Gleichungen in Distributionen. Math. Nachr. **78** (1977), 231–247.
[39] *Ljusternik, L. A.; Sobolev, W. I.:* Elemente der Funktionalanalysis. 4. Aufl. Berlin: Akademie-Verlag 1968.
[40] *Lu, Y.-C.:* Singularity Theory and an Introduction to Catastrophe Theory. New York, Heidelberg, Berlin: Springer-Verlag 1976.
[41] *v. Mangoldt, H.; Knopp, K.:* Einführung in die höhere Mathematik, I–III, 16. bzw. 15. Aufl. Leipzig: S. Hirzel Verlag 1979, 1978, 1981.
[42] *Markuševič, A. I.:* Theory of Analytic Functions, II. Moskva: Nauka 1967. (Russian)
[43] *Misner, C. W.; Thorne, K. S.; Wheeler, J. A.:* Gravitation. San Francisco: W. H. Freeman and Comp. 1973.
[44] *Moore, P.:* The Story of Astronomy. 5. ed. London: MacDonald and Jane's Publ. Ltd. 1977.
[45] *Murdin, P.; Allen, D.:* Catalogue of the Universe. Cambridge, London, Melbourne: Cambridge Univ. Press 1979.
[46] *Newton, Sir I.:* Principia, I (The Motion of Bodies). Berkeley: University of California Press 1934.
[47] *Petrowski, I. G.:* Vorlesungen über die Theorie der gewöhnlichen Differentialgleichungen. Leipzig: B. G. Teubner Verlagsgesellschaft 1954.
[48] *Petrowski, I. G.:* Vorlesungen über partielle Differentialgleichungen. Leipzig: B. G. Teubner Verlagsgesellschaft 1955.
[49] *Poston, T.:* Various catastrophe machines. In "Structural Stability, the Theory of Catastrophes, and Applications in Sciences". Lecture Notes in Math. 525. New York, Heidelberg, Berlin: Springer-Verlag 1976.
[50] *Poston, T.; Stewart, I.:* Catastrophe Theory and its Applications. London, San Francisco, Melbourne: Pitman Publ. Ltd. 1978.
[51] *Priwalow, I. I.:* Einführung in die Funktionentheorie. I–III. B. G. Teubner Verlagsgesellschaft 1958/59.
[52] *Riesz, F.; Sz.-Nagy, B.:* Vorlesungen über Funktionalanalysis. 2. Aufl. Berlin: VEB Deutscher Verlag der Wissenschaften 1968.

[53] *Sachs, R. K.; Wu, H.:* General Relativity for Mathematicians. New York, Heidelberg, Berlin: Springer. Verlag 1977.
[54] *Schechter, M.:* Operator Methods in Quantum Mechanics. New York, Oxford: North Holland 1981.
[55] *Schmutzer, E.:* Relativistische Physik. Leipzig: B. G. Teubner Verlagsgesellschaft 1968.
[56] *Sciama, D. W.:* Modern Cosmology. Cambridge: Cambridge Univ. Press 1973.
[57] *Segrè, E.:* From X-Rays to Quarks. San Francisco: W. H. Freeman and Comp. 1980.
[58] *Silk, J.:* The Big Bang. San Francisco: W. H. Freeman and Comp. 1980.
[59] *Smirnow, W. I.:* Lehrgang der höheren Mathematik. I–IV. 12., 13., 8., 6., Aufl. Berlin: VEB Deutscher Verlag der Wissenschaften 1973–1977.
[60] *Spivak, M.:* Calculus on Manifolds (A Modern Approach to Classical Theorems of Advanced Calculus). New York: W. A. Benjamin, Inc. 1965.
[61] *Stephani, H.:* Allgemeine Relativitätstheorie. Berlin: VEB Deutscher Verlag der Wissenschaften 1977.
[62] *Sulanke, R.; Wintgen, P.:* Differentialgeometrie und Faserbündel. Berlin: VEB Deutscher Verlag der Wissenschaften 1972.
[63] *Sullivan, W.:* Black Holes. New York: Warner Book 1980.
[64] *Thirring, W.:* Lehrbuch der mathematischen Physik. II. Wien, New York: Springer-Verlag 1978.
[65] *Thom, R.:* Structural Stability and Morphogenesis. London, Amsterdam, Don Mills, Sydney, Tokyo: Benjamin 1976.
[66] *Triebel, H.:* Höhere Analysis. Berlin: VEB Deutscher Verlag der Wissenschaften 1972.
[67] *Verschuur, G. L.:* The Invisible Universe. New York, Heidelberg, Berlin: Springer-Verlag 1974.
[68] *Wassermann, G.:* Stability of Unfolding. Lecture Notes in Math. 393. New York, Heidelberg, Berlin: Springer-Verlag 1974.
[69] *Weidmann, J.:* Lineare Operatoren in Hilberträumen. Stuttgart: B. G. Teubner 1976.
[70] *Weinberg, S.:* The First Three Minutes. New York: Bantam Books 1979.
[71] *Weinberg, S.:* Gravitation and Cosmology. New York: Wiley 1972.
[72] *Wladimirow, W. S.:* Gleichungen der mathematischen Physik. Berlin: VEB Deutscher Verlag der Wissenschaften 1972.
[73] *Woodcock, A.; Davis, M.:* Catastrophe Theory. Harmondsworth: Penguin Books Ltd. 1980.
[74] *Zeeman, E. C.:* Breaking of waves. Symposium on Differential Equations and Dynamical Systems. Lecture Notes in Math. 206. New York, Heidelberg, Berlin: Springer-Verlag 1971, 2–6.
[75] *Zeeman, E. C.:* Catastrophe theory: a reply to Thom. Manifold 15, Univ. Warwick 1974.
[76] *Zeeman, E. C.:* Euler buckling. In "Structural Stability, the Theory of Catastrophes, Applications in Sciences". Lecture Notes in Math. 525. New York, Heidelberg, Berlin: Springer-Verlag 1976.
[77] *Zeeman, E. C.:* Catastrophe Theory (Selected Papers). London: Addison-Wesley Publ. Comp. 1977.
[78] *Zeeman, E. C.; Trotman, D. J. A.:* The classification of elementary catastrophes of codimension $\leqq 5$. In "Structural Stability, the Theory of Catastrophes, and Applications in Sciences". Lecture Notes in Math. 525. New York, Heidelberg, Berlin: Springer-Verlag 1976, 263–327.
[79] *Zygmund, A.:* Trigonometric Series, I, II. Cambridge: Cambridge Univ. Press 1977.
[80] Encyclopedia of Astronomy and Space (Editor *I. Ridpath*). London: Macmillan Ltd. 1976.
[81] General Relativity. An Einstein Centenary Survey (Editors: *S. W. Hawking*, and *W. Israel*). Cambridge, New York: Cambridge Univ. Press 1979.
[82] Einstein Symposium, Berlin 1979 (Editors: *H. Nelkowski, A. Hermann, H. Poser, R. Schrader, R. Seiler*). Lecture Notes in Physics 100. Berlin, Heidelberg, New York: Springer-Verlag 1979.

[83] Einstein, A Centenary Volume (Editor *A. P. French*). Cambridge (U.S.A.): Harvard Univ. Press 1979.
[84] *Albert Einstein,* Akademie-Vorträge. Berlin: Akademie-Verlag 1978.
[85] Die Hilbertschen Probleme (Editor: *P. S. Alexandrov*). Leipzig: Akademische Verlagsgesellschaft Geest & Portig K.-G. 1971.
[86] Science & The Universe. The Mitchell Beazley Joy of Knowledge Library. London 1977.
[87] *Arnol'd, V. I.:* Catastrophe Theory. Berlin, Heidelberg, New York, Tokyo: Springer-Verlag 1984.
[88] *Haken, H.:* Synergetik. 2. Aufl. Berlin, Heidelberg, New York, Tokyo: Springer-Verlag 1983.

Hints for the Use of the References

Chaps. 1–9: 15, 16, 22, 41, 59
Chap. 10 (also 4): 4, 5, 22IV, 32, 47
Chaps. 11, 12: 31, 37
Chaps. 13, 14: 23, 34
Chap. 15: 6, 22IV, 51
Chap. 16: 42, 51II
Chap. 17: 51I, 66
Chap. 18: 36, 79
Chap. 19: 25, 48
Chaps. 20, 21, 26 (also 6 and 17.3): 2, 14, 39, 66, 69
Chaps. 22, 23: 19, 66, 72
Chaps. 24, 25: 3, 11, 26, 37I
Chaps. 27, 28: 54, 66
Chap. 29: 1, 3, 9, 12, 33, 60, 61
Chaps. 30, 31: 3, 24, 35, 43, 53, 55, 61, 71, 81, 82, 83
Chap. 32: 1, 9, 12, 17, 33, 60
Chap. 33: 17
Chaps. 34, 35: 7, 40, 50, 65, 73, 78, 87.

Index

Symbol Index
